PLANET EARTH 2011 – GLOBAL WARMING CHALLENGES AND OPPORTUNITIES FOR POLICY AND PRACTICE

Edited by **Elias G. Carayannis**

Planet Earth 2011 – Global Warming Challenges and Opportunities for Policy and Practice
Edited by Elias G. Carayannis

Published by InTech

Janeza Trdine 9, 51000 Rijeka, Croatia

Publishing Process Manager Davor Vidic

Technical Editor Teodora Smiljanic

Cover Designer InTech Design Team

Additional hard copies can be obtained from orders@intechopen.com

Planet Earth 2011 – Global Warming Challenges and Opportunities
for Policy and Practice, Edited by Elias G. Carayannis
p. cm.
ISBN 978-953-307-733-8

For EU product safety concerns:
IN TECH d.o.o., Prolaz Marije Krucifikse Kozulić 3, 51000 Rijeka, Croatia,
info@intechopen.com or visit our website at intechopen.com.

Meet the editor

Dr. Elias G. Carayannis is Full Professor of Science, Technology, Innovation and Entrepreneurship, as well as co-Founder and co-Director of the Global and Entrepreneurial Finance Research Institute (GEFRI) and Director of Research on Science, Technology, Innovation and Entrepreneurship, European Union Research Center, (EURC), at the School of Business of the George Washington University in Washington, DC. Dr. Carayannis' teaching and research activities focus on the areas of strategic Government-University-Industry R&D partnerships, technology road-mapping, technology transfer and commercialization, international science and technology policy, technological entrepreneurship and regional economic development.

Contents

Preface XI

Chapter 1 **Climate Change in the Mediterranean over the Last Five Hundred Years 1**
Dario Camuffo, Chiara Bertolin, Antonio della Valle, Claudio Cocheo, Nazzareno Diodato, Silvia Enzi, Mirca Sghedoni, Mariano Barriendos, Roberto Rodriguez, Fernando Dominguez-Castro, Emmanuel Garnier, Maria Joao Alcoforado and Maria Fatima Nunes

Chapter 2 **Recent Global Warming Induced Climate Changes 27**
P. C. Werner, F.-W. Gerstengarbe, H. Österle and M. Wodinski

Chapter 3 **Environmental Stability for Convective Precipitation Under Global Warming 57**
Tetsuya Takemi, Syohei Nomura and Yuichiro Oku

Chapter 4 **Effects of Global Warming on Climate Conditions in the Japanese Alps Region 73**
Keisuke Suzuki

Chapter 5 **Melting of Major Glaciers in Himalayas: Role of Desert Dust and Anthropogenic Aerosols 89**
Anup K. Prasad, Hesham M. Elaskary, Ghassem R. Asrar, Menas Kafatos and Ashok Jaswal

Chapter 6 **Accounting the Carbon Storage in Disturbed and Non-Disturbed Tropical Andean Ecosystems 123**
Enrique Peña, Orlando Zúñiga and Julián Peña

Chapter 7 **Modeling the Monsoons in a Changing Climate 141**
H. Annamalai

Chapter 8 **Climatic Effect of the Greenhouse Gases Clusterization 157**
Alexander Y. Galashev

Chapter 9 **California Coastal - Cooling a Reverse Reaction from Global Warming General Circulation and Mesoscale Effects 193**
Bereket Lebassi-Habtezion, Jorge González and Robert Bornstein

Chapter 10 **Causes and Consequences of the Late 1960s Great Salinity Anomaly 213**
Mihai Dima and Gerrit Lohmann

Chapter 11 **Tropical Dry Forests in the Global Picture: The Challenge of Remote Sensing-Based Change Detection in Tropical Dry Environments 231**
Betsabé de la Barreda-Bautista, Alejandra A. López-Caloca, Stephane Couturier and José Luis Silván-Cárdenas

Chapter 12 **Cirrus Clouds and Climate Engineering: New Findings on Ice Nucleation and Theoretical Basis 257**
David L. Mitchell, Subhashree Mishra and R. Paul Lawson

Chapter 13 **Destruction of Fluorinated Greenhouse Gases by Using Nonthermal Plasma Process 289**
Young Sun Mok

Chapter 14 **The Choice of Biofuels to Mitigate Greenhouse Gas Emissions 311**
Rogério Cezar de Cerqueira Leite, Manoel Regis Lima Verde Leal and Luís Augusto Barbosa Cortez

Chapter 15 **Contribution of the Atmospheric Chlorine Reactions to the Degradation of Greenhouse Gases: CFCs Substitutes 329**
Iván Bravo, Yolanda Díaz-de-Mera, Alfonso Aranda, Elena Moreno and Ernesto Martínez

Chapter 16 **The Importance of Advective Fluxes to Gas Transport Across the Earth-Atmosphere Interface: The Role of Thermal Convection 353**
Uri Nachshon, Noam Weisbrod, Maria I. Dragila and Yonatan Ganot

Chapter 17 **Reforming CO_2 into Fuel Using a TiO_2 Photocatalyst Membrane Reactor 373**
Akira Nishimura and Eric Hu

Chapter 18 **Power Generation Using Nonconventional Renewable Geothermal & Alternative Clean Energy Technologies 399**
Basel I. Ismail

Chapter 19 **Four Steps to the Hydrogen Car 425**
Paul McGuiness and Urša Pirnat

Chapter 20 **Using Micro Cogeneration Technologies to Enhance the Sustainable Built Environment 447**
Esam Elsarrag and Yousef Alhorr

Chapter 21 **Nuclear Methodology for Non-Destructive Multi-Elemental Analysis of Large Volumes of Soil 467**
Lucian Wielopolski

Chapter 22 **The Role of Plant Breeding and Biotechnology in Meeting the Challenge of Global Warming 493**
Nigel G. Halford

Chapter 23 **Modeling Renewable Energy Consumption for a Greener Global Economy 507**
Perry Sadorsky

Chapter 24 **Influence of Global Warming on the RC Structures and Durability Monitoring in Civil Engineering 525**
Guofu Qiao, Tiejun Liu, Guodong Sun, Yi Hong, Baoguo Han, Huigang Xiao, Zhichun Zhang and Jinping Ou

Chapter 25 **Tuvalu Visualization Project - Net Art on Digital Globe: Telling the Realities of Remote Places 557**
Makiko Suzuki Harada, Hidenori Watanave and Shuuichi Endou

Chapter 26 **The Role of Methane Emissions on Ancient and Present Climatic Changes 573**
Guido Barone, Elena Chianese and Angelo Riccio

Chapter 27 **Enteric Methane Emission from Pigs 605**
Henry Jørgensen, Peter K. Theil and Knud Erik Bach Knudsen

Chapter 28 **Human Security and Global Warming: Challenges Before Christianity and Islam in Nigeria 623**
Dokun Oyeshola

Preface

The "Planet Earth 2011 - Global Warming Challenges and Opportunities for Policy and Practice" project is a collection of a truly trans-disciplinary perspectives on global warming causes and effects as well as the insights and implications for policy, polity and practice.

Following the December 2009 failure to coordinate among the major nations of the world in Denmark, there is clear and present set of challenges and opportunities that Europe in particular is trying to address in the face of short-termism and narrow-minded behaviors of other governments and countries who try to forget that we are all on the same planet whether we like it or not and as a result we are all exposed to the follies of individual peoples, governments or industries.

The need to develop and / or the love of free markets as well as other socio-economic and socio-political agendas do not in any way justify self-destructive, narrow-minded obstructionism based on "tainted science" - there is a clear and present need for as accurate a truth as we can make it whether this is painful, scary or costly to address as it will only become more so over time.

This effort will hopefully contribute towards a better understanding and more focused and effective dialog involving all stakeholders from government and civil society as well as NGOs, industry and academia.

Elias G. Carayannis
George Washington University, School of Business
USA

1

Climate Change in the Mediterranean over the Last Five Hundred Years

Dario Camuffo et al.*
National Research Council of Italy (CNR)
Institute of Atmospheric Sciences and Climate (ISAC), Padua
Italy

1. Introduction

Global Warming (GW) is expected to affect the Mediterranean area with three major challenges, i.e. increase in temperature, decrease in precipitation and sea level rise that will likely submerge the coastal areas, including Venice. Aim of this Chapter is to discuss the expected changes under the light of long-term observations. Documentary proxies and instrumental readings in Portugal, Spain, France and Italy have been recovered and analysed. These observations cover the last five centuries from the Little Ice Age (LIA) to the present-day GW.

This Chapter is based on documentary proxies and instrumental series collected over the Mediterranean area, i.e. Portugal, Spain, France and Italy (Fig.1) within the EU funded ADVICE, IMPROVE, MILLENNIUM, and Climate for Culture projects. A huge effort was made to seek for written sources and original logs with early and less recent instrumental readings. The next steps were to recover, correct, adjust to modern standards, homogenize and analyse the earliest data and most of the longest European series. The detailed study of the history of the series (e.g. instrument type, calibration, observational methodology, sampling time, exposure, location) and the recovery of any related metadata were fundamental to apply and perform the due corrections to the series. The methodology was presented in previous papers (Camuffo and Jones, 2002, Camuffo et al. 2010a)

*Chiara Bertolin[1], Antonio della Valle[1,2], Claudio Cocheo[3], Nazzareno Diodato[4], Silvia Enzi[5], Mirca Sghedoni[5], Mariano Barriendos[6], Roberto Rodriguez[6], Fernando Dominguez-Castro[7], Emmanuel Garnier[8], Maria Joao Alcoforado[9] and Maria Fatima Nunes[10]
[1]*National Research Council of Italy (CNR), Institute of Atmospheric Sciences and Climate (ISAC),Padua,Italy*
[2]*Now: Department of Astronomy, University of Bologna, Italy*
[3]*Salvatore Maugeri Foundation IRCCS, Padua, Italy*
[4]*Monte Pino Met Research Observatory, Benevento, Italy*
[5]*Kleiò, Padua, Italy*
[6]*Department of Modern History, University of Barcelona, Spain*
[7]*Department of Physics, University of Extremadura, Badajoz, Spain*
[8]*Laboratory of Climate and Environmental Sciences, UMR CEA-CNRS, Saclay*
Research Centre of Quantitative History, University of Caen, France
[9]*Centre of Geographic Studies, University of Lisbon, Portugal*
[10]*History and Philosophy of Science Research Unit, University of Évora, Portugal*

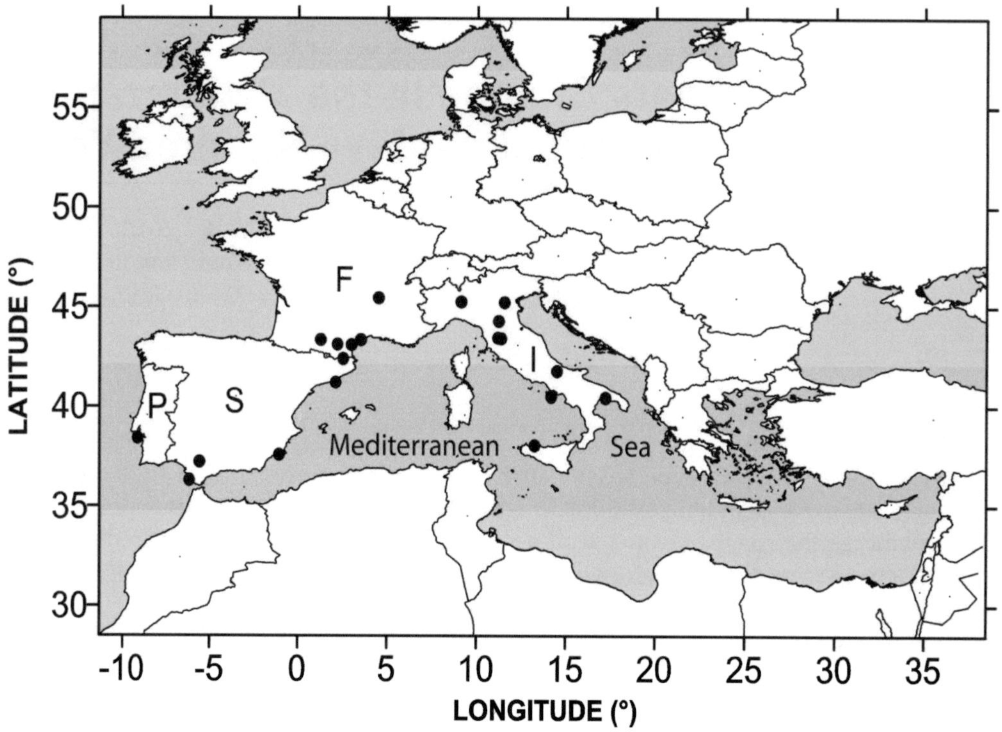

Fig. 1. The Mediterranean Basin with the indicated locations (black circles) where documentary proxies and/or instrumental observations have been retrieved for use in this Chapter, divided by countries, i.e.: Portugal (P), Spain (S), France (F) and Italy (I). The stations are: Lisbon (P), Cadiz (S), Seville (S), Murcia (S), Barcelona (S), Perpignan (F), Narbonne (F), Carcassonne (F), Toulouse (F), Montpellier (F), Lyon (F), Milan (I), Padua (I), Bologna (I), Vallombrosa (I), Florence (I), Benevento (I), Naples (I), Locorotondo (I) and Palermo (I).

2. Weather in the Mediterranean

The Northern part of the Mediterranean Basin is under the influence of the European continental climate, the shield of extended mountain chains and the penetration of external air masses through the mountain gates. The rest of the Basin is strongly conditioned by the difference in temperature between the air and the seawaters and the expected increase in temperature could reduce the annual precipitation, increasing the aridity in the warmest areas.

In the warm season (May-October) the Azores High is well developed and prevents external air from entering the Mediterranean. The local air is warmer than the seawater and blows over the Mediterranean with reduced heat and moisture exchanges, forming thermal layering, atmospheric stability and clear sky. In the warm season, precipitation is possible only on the Northern side, for the passage of Atlantic disturbances or the formation of convective clouds and afternoon thunderstorms especially in the mountain areas.

In the mid seasons, when the Azores High weakens, Atlantic disturbances penetrate and diagonally cross the Mediterranean, bringing cold air and stormy weather. The southern part of the Basin remains untouched, with no precipitation.
In the cold season, when continental Europe is cold, a high-pressure ridge forms over Europe, joining the Russian with the Azores High, and tends to block the passage of fronts and disturbances coming from West. Polar or arctic air blows from North or from East. During their motion towards lower latitudes, the cold air masses increase in temperature and depart more and more from saturation, so that no precipitation is possible on the Northern coast. However, when cold air blows over the Mediterranean Sea, the air gains heat and moisture from the warm water, forming atmospheric mixing, large clouds and precipitation on the downwind regions. Warm water feeds clouds and enhances depressions and storms, especially in the occasion of cold air inflows from North or East. The mountain chains surrounding the Mediterranean shield most of the basin but leave some open gates through which external air may enter with violence, determining some local strong winds e.g. Mistral, Libeccio, Bora, Etesians, Vardar and Sirocco. However, the worst stormy weather is associated with the penetration of low pressures from North or West. When a northern trough enters the Mediterranean in the cold season, it is fed by the warm water and generates violent storms on the Eastern coast of the Spain, Southern France and Italy. Atlantic depressions enter the Mediterranean crossing diagonally France and penetrating through the gate between the Pyrenees and the Alps, or crossing the Iberian Peninsula with a westerly circulation. Westerly depressions bring heavy precipitation over Portugal and Spain with decreasing intensity while advancing. When they enter the Mediterranean, they are fed by the warm seawater, increase in strength and cause heavy precipitation on the Eastern coast of Spain, Southern France and Northern-Central Italy. More details about the particularities of the Mediterranean climate can be found for instance in UK Meteorological Office (1962); Reiter (1975); Wallén (1970; 1977), Jeftic et al. (1992), Bolle (2003), Xoplaki et al. (2003; 2004); Fletcher et al. (2005), Lionello et al. (2006) Camuffo et al. (2010a,b), Glaser et al. (2010), Luterbacher et al. (2010).

3. Documentary proxies

3.1 Documentary sources in the Mediterranean area

In the period before the instrumental observations, the climate in the Mediterranean area is known after written sources that fully cover the Medieval Optimum and the LIA. This area is rich in documents concerning descriptions of exceptional or regular weather events and natural hazards, e.g. intense rain and rivers in flood; aridity, famine and rivers in low; severe cold killing people, animals and trees; freezing over large water bodies; regular weather and abundant yield. These documents may be manuscript or printed press and can be classified according their character and purposes, as follows.

- Narrative sources: generic descriptions of events, such as chronicles, annals, diaries, correspondence, poems and compilations of remarkable events written for historical purposes, the pleasure of informing or disseminating news.
- Ecclesiastic sources: registers noting liturgical services and rogation ceremonies commissioned by the local community or authority in the case of adverse conditions. The most relevant topics were: to beg for rain (pro pluvia) in view of the yield, or to

stop precipitation (pro serenitate) especially in the case of rivers in flood, or in the occasion of famine, plague, locust invasions or any other challenge.
- Administrative sources: official documents (e.g. diplomatic letters, municipal registers, inspection reports) written by public officers to describe some local catastrophe happened, its impact on the society and the landscape and the remedy actions to undertake, e.g. repair and maintenance, temporary reduction of taxes.

In principle, the most accurate sources are the Administrative ones, being written with the purpose of being absolutely precise and objective; then the Ecclesiastic ones reflecting the duration and severity of the extreme meteorological events, because the clergy followed a rather standardized style combining some liturgical formats with challenge severity level, being more solemn and complex with increasing hazard severity and risk. Abundant literature exists on the above subjects (Camuffo and Enzi, 1992a,b, 1994, 1996, 2010a; Enzi and Camuffo, 1995; Martín-Vide & Barriendos, 1995; Brazdil, 1999; Barriendos, 1997, 2005; Pfister et al., 1999; Alcoforado et al., 2000; Piervitali & Colacino, 2001; Barriendos & Llasat, 2003; Chuine et al., 2004; Brazdil et al., 2005, 2010; Luterbacher et al., 2006; Dominguez-Castro et al., 2008; Rodrigo & Barriendos, 2008). Written documents describe events with emphasis proportional to the impact that the event had on the society, the landscape or something else that the observer considered highly relevant to his advice. A number of severe events, confirmed by a number of dramatic consequences, were caused by some relatively short-term peaks, or drops, in temperature, precipitation or other meteorological variables. For instance, severe floods were generated by rain persisting over a limited number of days. The greatest winters in the history were renamed for some dramatic and spectacular effects (e.g. large water bodies unusually frozen over and supporting people) but were caused by two, three, maximum four weeks of very intense cold. However, from the comparison between documentary sources and instrumental observations, we see that such peaks and drops appear evident on the daily series, but tend to disappear when increasing the averaging period, i.e. they are almost damped on monthly averages. Therefore, some extreme events that are well known after written sources, when compared with instrumental data, may appear fully or partially justified depending on the temporal window used for the statistical analysis of the instrumental observations. Documentary sources are useful to establish the occurrence, and the frequency, of extreme events, the short-term variability, the persistence of some dry or wet, cold or hot periods particularly relevant to the agriculture, the landscape or the society. However, they are unable to provide long-term trends.

3.2 From written documents to indexes, and from indexes to proxy data

It is obvious that the written documents are extremely relevant because they qualitatively inform us about the weather and the climate in the period before the instrumental observations. Their value is even greater if they can be transformed from literary items into numbers to be used instead of instrumental readings. If such a case, the above written sources are transformed into "proxy data" that constitute an indirect way of assessing what historical temperature or precipitation might have been. Literally, "proxy" is a person having power of attorney, i.e. authorized to act for, or to represent another person on a single occasion. In the case some items are objectively related to temperature or precipitation, they can be considered "proxies", i.e. a valid replacement of real instrumental observations if the latter are missing. If we can express proxies in numerical terms, and if we can in some way control and validate the numbers that we will obtain after having applied some transformation, we will obtain sound "proxy data" that will provide the needed information about past climate.

Written sources are transformed into proxy data following a careful series of transformations. First the sentence is interpreted, analysed and classified in terms of exceptionality looking at the description of the event and on the ground of objective facts, e.g. the effects it has produced. The classification is made attributing levels from +3 to a really extreme and well-documented event (e.g. extremely high temperature or abundant precipitation), as we can expect to occur not more than two or three times per century, to -3 to the symmetrical, but negative case (e.g. extreme cold or aridity). Levels ±2 and ±1 are used for intermediate levels at decreasing severity and of course level 0 is "normality". The various severity levels are inspired to the departure of readings from the average, expressed in terms of standard deviation (SD), e.g. level 0 lies between ± 0.5 SD; levels ±1 lie externally to level 0 but are topped by +1 SD; and similarly with levels +2 topped by +2 SD; the levels ±3 being external to ±2 SD. In this way the episodes taken from written sources are indexed into severity levels, but they are not yet expressed in quantitative terms of degrees of temperature (°C) or precipitation amount (mm, or % in comparison with the precipitation occurred in a selected reference period). The transformation of an index into numerical values of temperature or precipitation, i.e. into a proxy data, is made with the help of a common period in which we have both such indexes and instrumental observations. In such a case it is immediate to find a correspondence between the two and know the transformation.

Proxies are useful in the absence of instrumental observations and their transformation is made in an objective way. However, they have some weak points, as follows:

- In some periods the documentation is scarce and we cannot be sure that the totality of the events, or what part of them, has been represented. In these periods it is impossible to know if the 0 level is "normality" or missing value.
- It is obvious that all the data are expressed in relative terms, the absolute reference level being missed. The transformed data are similar to the so called "anomaly", i.e. the difference between the selected level and the 1961-1990 reference period for temperature, and the ratio between these two values for precipitation, but reference is made with the personal experience of the writer and not with a standard period.
- The severity is based on the witness of people living at that time, and reflects what was considered regular or extraordinary at that time. This means that the zero level, i.e. the "normality" remains flat: we can evaluate high-frequency fluctuations, but we miss long-term trends or cycles.
- Calibration and validation are based on a relatively recent period when simultaneous instrumental observations were available. However, going back in the time, we automatically extend the results to earlier periods in which the perception of man was different and conditioned by the culture of his time. This means that the calibration becomes uncertain when we extend it back over the previous centuries.

However, despite the above limits, proxy data constitute a very valuable source of information for the climate of the past centuries. The data analysis has been performed as already discussed in Camuffo et al. (2010a).

3.3 Results from the documentary proxies

This research was able to recover a satisfactory amount of proxies, sufficient to provide a reasonable documentation for all the four seasons in Northern-Central Italy followed by Southern France for temperature, and in Spain, Southern France and Northern-Central Italy for precipitation. The period after 1700 is better documented by instrumental records, but it has been useful for the calibration and validation of the proxies.

The temperature in Southern France is shown in Fig.2 and in Northern-Central Italy in Fig.3. In general, temperature is best documented in winter, followed by spring, for many complaints concerning cold severity, snow and frost. The less documented season is autumn, because in this season temperature is not a critical factor. In the period from 1500 to 1700, winter and spring seem having been characterized by cold episodes more frequently than today both in Southern France and in Northern-Central Italy. Hot summers have been more frequent than fresh summers, at least in Italy.

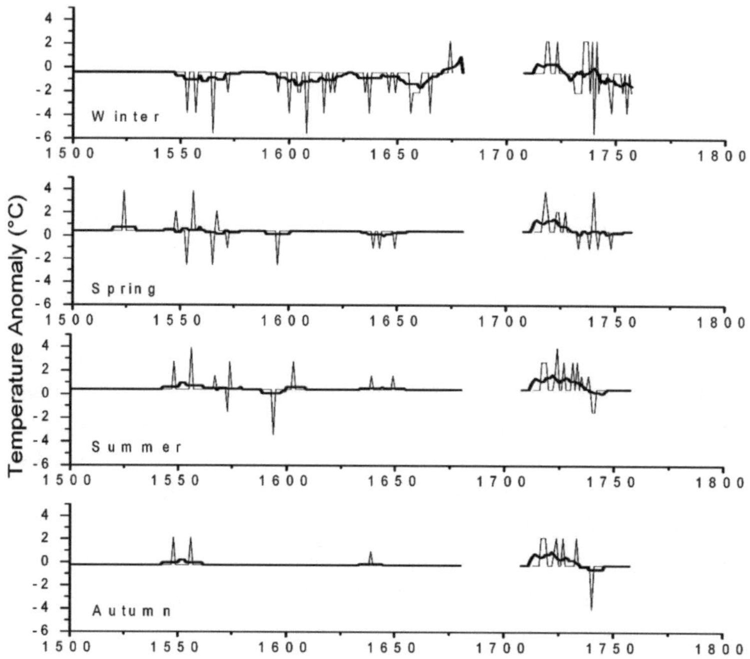

Fig. 2. Seasonal temperature anomaly (°C) from documentary proxy data in Southern France. The baseline has been set to be correspondent to the average of the whole instrumental period. Thin lines refer to proxies, thick lines to 11-year running averages. Seasons in the plots are related to DJF (Winter), MAM (Spring), JJA (Summer) and SON (Autumn).

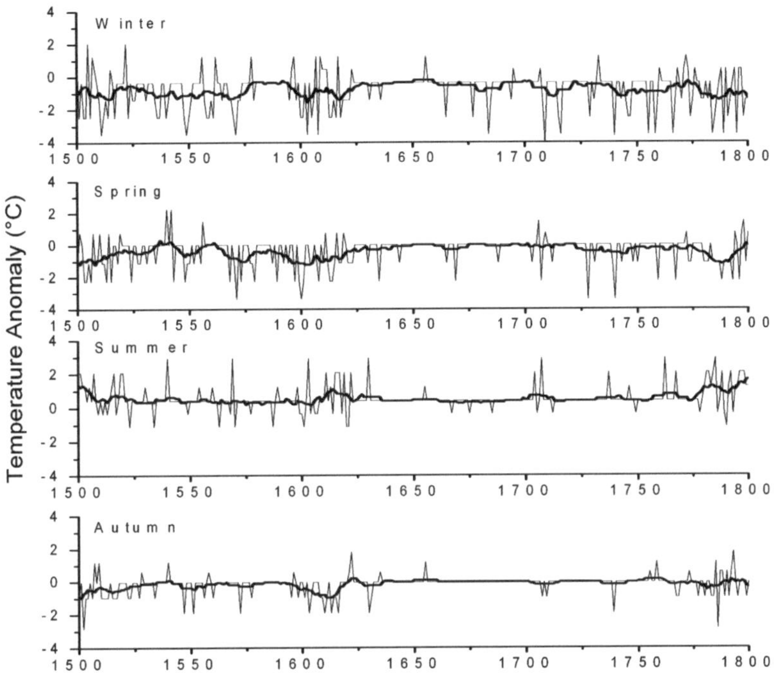

Fig. 3. Seasonal temperature anomaly (°C) from documentary proxy data in Northern-Central Italy. Symbols as in Figure 2.

The precipitation too is presented in terms of anomaly, but expressed as normalized ratio, i.e. the ratio between the observed values and the related ones in the 1961-1990 reference period, i.e. 0 means no precipitation, 1 the same precipitation as in the reference period, 2 is twice the precipitation observed in the reference, etc. This style follows IPCC 2007 (Le Treut et al., 2007).

The information concerning precipitation is more abundant in Spain (Fig.4) because this area is dryer than Southern France and Northern-Central Italy, with the consequence that people made frequent rogations and ceremonies to implore rain for crops. When rain occurs, it is intense or long lasting, and rivers risk being in flood. The risk of spot floods easily brings the local inhabitants to pray to stop rain. Documents are almost equally distributed over the four seasons, and in all of them ultra-decadal swings between dryer and wetter are visible. In Southern France (Fig.5) the precipitation is a less critical factor and for this reason proxies are less frequent. Rainfall is better documented in spring and summer because in these seasons rainfall is necessary for the growth and maturation of the agricultural products. In the second half of the 1500s and in the first half of the 1700s when the information density is higher, we can recognize two oscillations. In particular the 1550-1570 period rainfall was particularly abundant in summer and autumn.

In Northern-Central Italy (Fig.6) the information is quite regularly distributed along the whole period and among all the seasons. The most severe impacts of rain and dryness were from 1500 to 1620, with particular relevance in the two decades 1600-1620. Moderate swings of the 11-yr running average are visible.

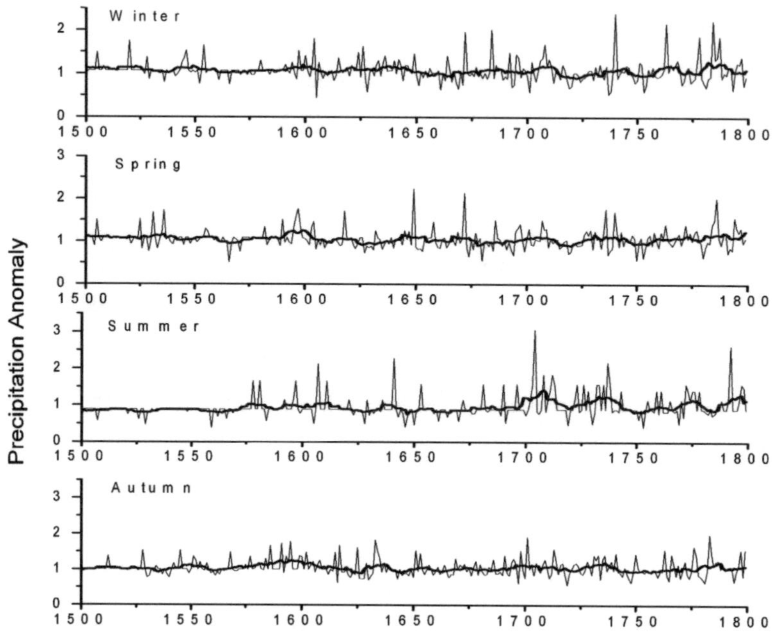

Fig. 4. Seasonal precipitation anomaly from documentary proxy data in Spain. Symbols as in Figure 2.

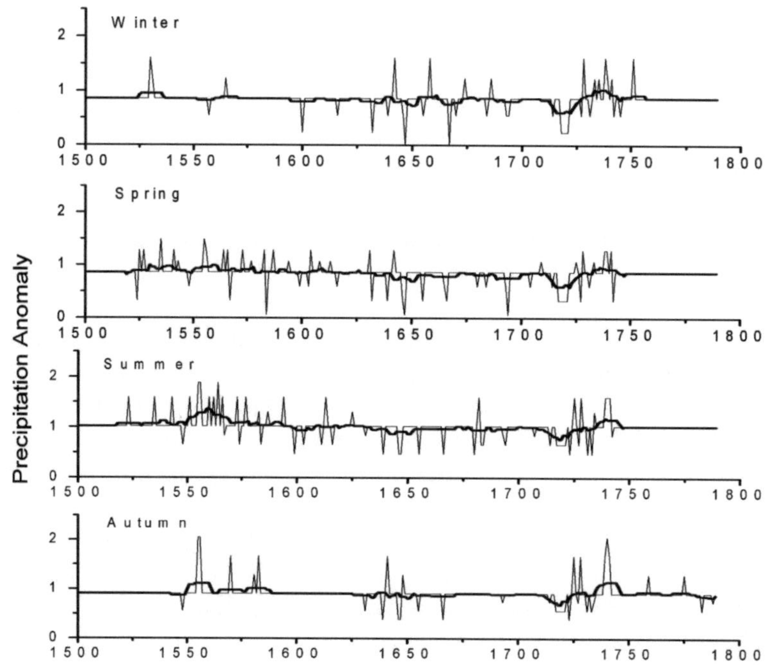

Fig. 5. Seasonal precipitation anomaly from documentary proxy data in Southern France. Symbols as in Figure 2.

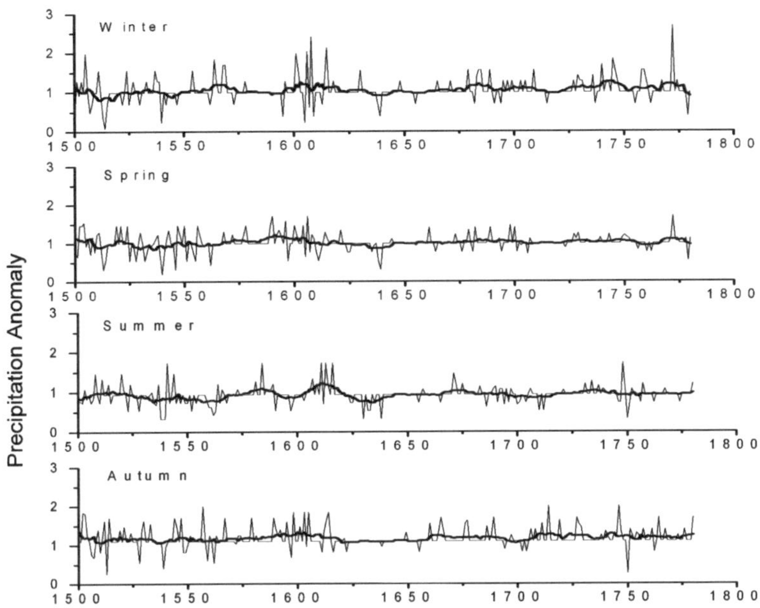

Fig. 6. Seasonal precipitation anomaly from documentary proxy data in Northern-Central Italy. Symbols as in Figure 2.

4. Instrumental series

The last period of the LIA and the present-day GW are well documented by the longest instrumental series that cover three centuries.

4.1 The earliest series of instrumental observations

The spirit-in-glass thermometer, the barometer, the rain gauge and some hygrometers were invented in Florence, Italy in the first half of the 17th century, where science flourished with Galileo, his pupils and the Grand Duke of Tuscany, Ferdinand 2nd De' Medici (Magalotti, 1666, Targioni Tozzetti, 1780) who was a supporter of the scientific research. He founded the first "modern" scientific Academy, i.e. the "*Accademia del Cimento*" (i.e. Academy of Experiment, flourished 1657 - 1667) with the first aim of investigating the Nature with particular reference to Mathematics, Physics, Astronomy, Meteorology, Biology and Medicine with instrumental observations and the second aim of expressing these disciplines in numbers and formulae.

In this context the Grand Duke organized the first meteorological network, called *Rete Medicea* (i.e. Medici Network), active for the 1654-1670 period with regular temperature readings taken 6 to 8 times a day in a number of stations. Unbroken series were Florence and Vallombrosa on the mountain slope near to Florence. All stations received from the Grand Duke thermometers with the same calibration and an operational protocol for the exposure in order to obtain strictly comparable readings. The thermometers used in the Medici Network are known as the Little Florentine Thermometer, entirely made of glass except for the spirit, and resistant to any kind of weather. The readings taken by this Network are the earliest regular observations in the world, and are of excellent quality (Camuffo, 2002; Camuffo et al., 2010°, Camuffo and Bertolin, 2011). The activity of the Medici Network was stopped by the Inquisition that hardly accepted the idea that somebody wanted to know the Nature from sources and modalities other than the Holy Bible.

Next century, James Jurin coordinated the next international network, with climate and health purposes, on behalf of the *Royal Society*, London. Unfortunately, Jurin (1723) suggested indoor readings for two reasons: people lived in unhealthy houses and most of the early thermometers, made with a glass capillary fixed to a wooden tablet with a iron wire were unable to resist to outdoors because the wooden shrinkage and swelling varied the iron wire tension, breaking the glass tube. Fortunately, some Italian observers made parallel observations indoors and outdoors following the local tradition. The Network was active from 1724 to 1735.

Some fifty years later, Vicq d' Azyr and Father Louis Cotte coordinated the next international network, also with climate and health purposes, on behalf of the *Royal Society of Medicine*, Paris. The Society published the meteorological observations from 1777 to 1786.

In the same period, the German Prince Karl Theodor von Pfalz launched another international Network, i.e. the *Societas Meteorologica Palatina*, Mannheim, with secretary Jacob Hammer (1783) who published readings in the "*Ephemerides Societatis Meteorologicae Palatinae*" from 1781 to 1792. This Network was of excellent quality with a precise sampling (i.e. three readings a day) and exposition protocol, and distribution of instruments to the observers who needed them. The thermometers were weatherproof, so that outdoor measurements were possible without problems.

The above meteorological networks had the primary merit of raising the public interest to regular daily observations, made with standardised instruments. All observers operated in the same way, with readings performed at the same sampling times and all instruments had similar exposure.

In the next century the meteorology was better developed, with several national and international initiatives, so that the number of available data was largely increased. The first most important milestone of the nineteen century was the creation of the national weather services around 1860, with the production of many high-quality series of regular instrumental observations made with standardized features. The second milestone was the foundation in 1873 of the *International Meteorological Committee* that in 1950 was transformed into the *World Meteorological Organization*.

4.2 Results from the instrumental observations of temperature

In this Chapter the temperature is presented in terms of anomaly, expressed as the difference between the observed values and the related values in the 1961-1990 reference period, following the IPCC 2007 style (Le Treut et al., 2007).

If we compare the instrumental records with the above documentary proxies, the first remarkable difference with the documentary proxies is the high time resolution and the high density of readings. All the plots show high-frequency fluctuations and ultradecadal swings, i.e. 12.7, 26.5, 34.4 and 57.3 yr over the whole set of data, that are characteristic of the Mediterranean climate (Camuffo et al., 2010a,b).

The temperature in Spain is reported in Fig.7 and refers only to the long time series of Cadiz-S. Fernando. The plots show that autumn and winter months experienced a colder period that reached the minimum around 1850. Spring and summer months are linearly interpolated with no trends but with some persistent swings alternating colder and warmer periods, with a marked warming in the most recent decades. However, the period of the GW, i.e. after 1980 is evident in the cold season and spring, more exactly from November to June, but it is not dissimilar from other previous warm periods, and it is not the warmest one in the Spanish series.

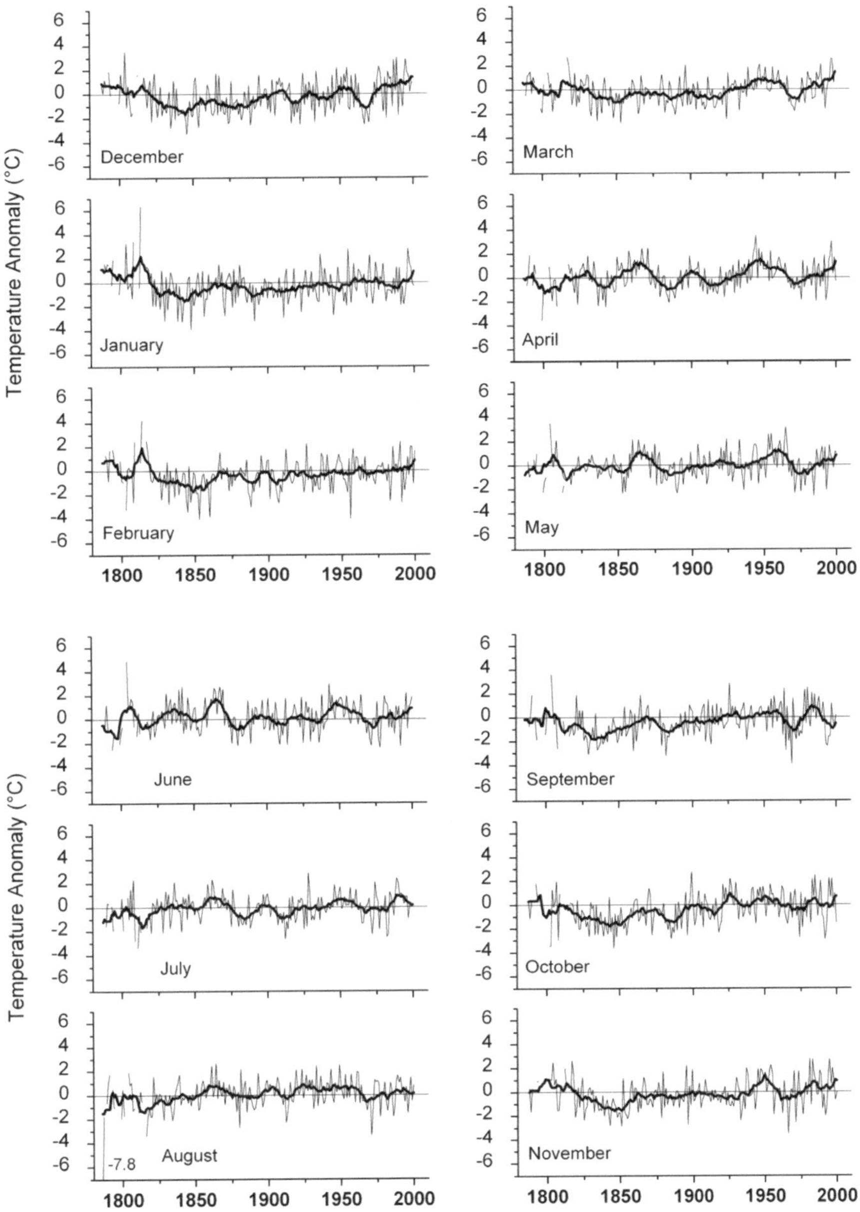

Fig. 7. Monthly temperature anomaly (°C) from instrumental data in Spain. The zero level corresponds to the average of the 1961-1990 reference period. Thin black lines refer to instrumental observations, thick black lines to 11-year running averages.

In Southern France (Fig.8), December and January undergone a change in temperature around 1900, passing from colder to milder winters. On the other hand, from the beginning of the series to 1800, the warm season from May to September was warmer than the 1961-1990 reference period, reaching a maximum slightly after 1750 with temperature levels higher than today. All the months are characterized by absence of long period trends but include several warm-cold swings.

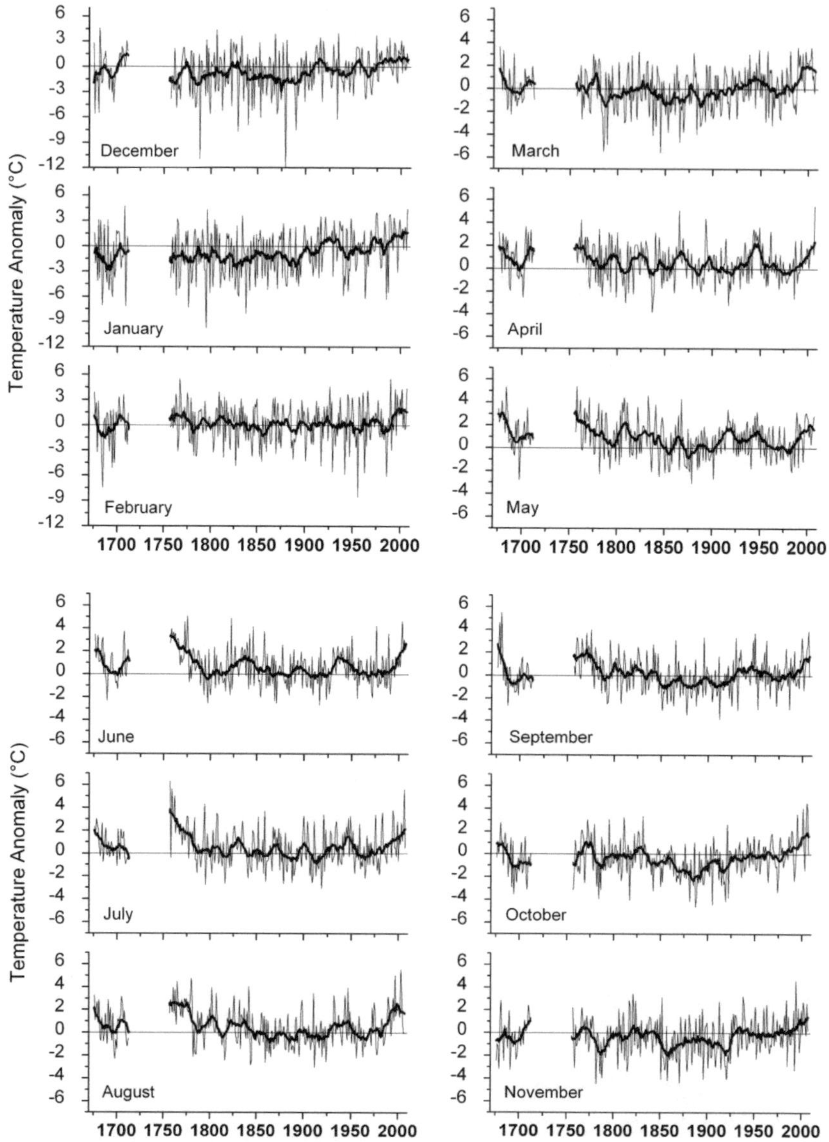

Fig. 8. Monthly temperature anomaly (°C) from instrumental data in Southern France. Zero level, thin and thick lines as in Fig.7.

In Northern-Central Italy (Fig.9) January, February and March show warming trends from 1739 to nowadays. No trends in other months. Warm-cold swings are evident in all months. In most months, and especially in May, June and August, the temperature increased after 1980 following the GW. However, around 1725-1730 the temperature was similar, or even higher than today in a number of months, i.e. May, September, October, November and December. The great winter in December 1738 and January 1739 is clearly visible; as opposed, the great winter 1929 with a deeper extreme in temperature the first half of February 1929 is not visible in the monthly averages being masked by the contribution of the second half of the month that was milder.

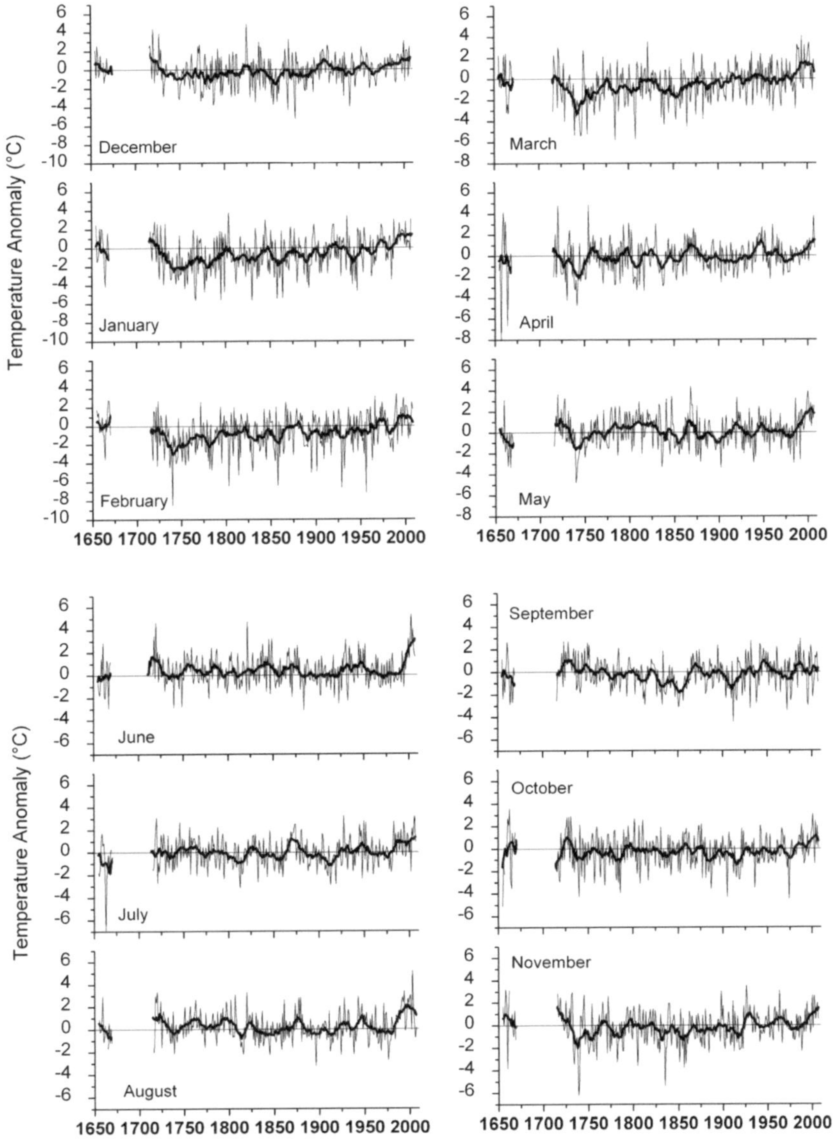

Fig. 9. Monthly temperature anomaly (°C) from instrumental data in Northern-Central Italy. Zero level, thin and thick lines as in Fig.7.

Southern Italy is located in the middle of the Mediterranean and its climate is strongly influenced by the difference between air and seawater temperature, differently from the Northern-Central part of the Italian Peninsula. The monthly temperature of Southern Italy is reported in Fig.10. The earliest period before 1800 is not homogeneous with the rest of the series and seems to overestimate the readings for some instrumental bias or relocation problem, and should be disregarded. The series seem more or less stationary till 1920, followed by a positive trend, i.e. increasing temperature. However, the most recent period, i.e. 1990-today, is not the warmest one in the series, maybe for the effect of change of phase of decadal swings or other chaotic factors.

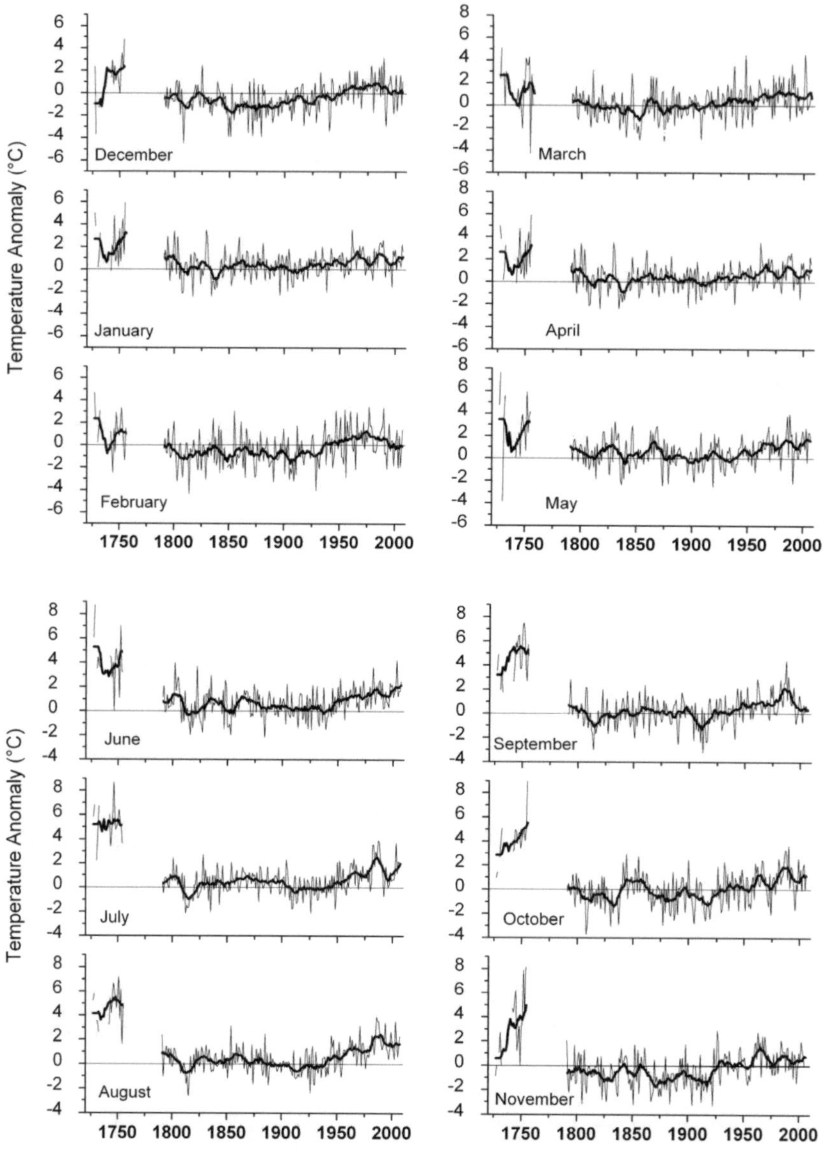

Fig. 10. Monthly temperature anomaly (°C) from instrumental data in Southern Italy. Zero level, thin and thick lines as in Fig.7.

4.3 Results from the instrumental observations of precipitation

The precipitation is presented in terms of anomaly as normalized ratio following the IPCC 2007 style (Le Treut et al., 2007). This style is particularly convenient in the case of arid regions, e.g. Spain and Southern Italy. The precipitation in the Mediterranean is characterized by large high-frequency variability, with irregular rainy-dry swings and 93 yr periodicity that has already been observed over the whole set of data (Camuffo et al., 2010a,b).

The precipitation in Portugal (Fig.11) is characterized by huge variability, with several extreme episodes from no rain at all to extremely intense rain episodes, in particular July and August exceeding in a number of episodes the average of the 1961-1990 reference period by an order of magnitude. The high frequency fluctuations are superimposed to a decadal swing. In the crucial last three decades of GW, i.e. after 1980, the situation is apparently contradictory, i.e. March, September and October seem to be characterised by positive trend, i.e. rainy; as opposed, February, May and July by negative trend, i.e. aridity, the other months having no or unclear trends.

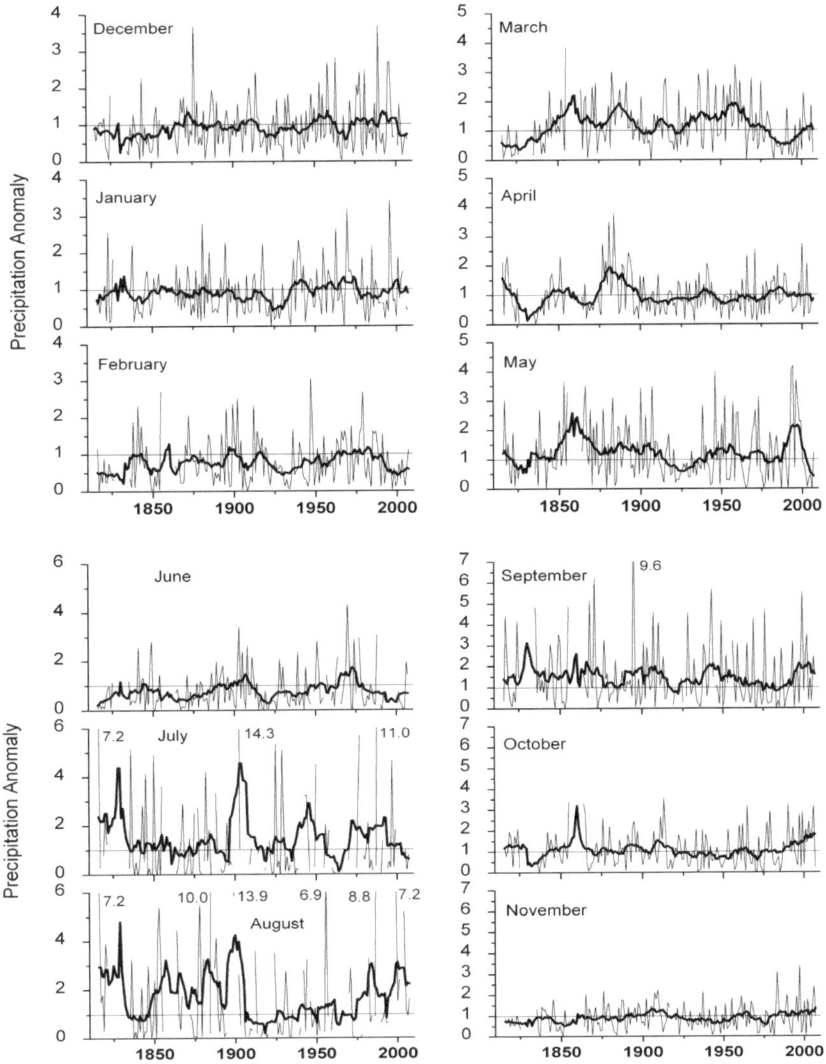

Fig. 11. Monthly precipitation anomaly from instrumental data in Portugal. Non dimensional units, the anomaly being expressed as a ratio with the 1961-1990 reference period. The 1-line level corresponds to the 1961-1990 average. Thin black lines refer to instrumental observations, thick black lines to 11-year running averages. Numbers on the side of peaks indicate the top level, external to the scale.

The precipitation (Fig.12) in all the Spanish stations shown in Fig.1 largely differs from month to month, including the reaction to the last three decades of GW, i.e. after 1980. In particular, February, March, April and July seem to go toward dryness; as opposed January, September and December toward wetness, the other months being unchanged or unclear. Some extreme episodes appear especially in June, July, August and September.

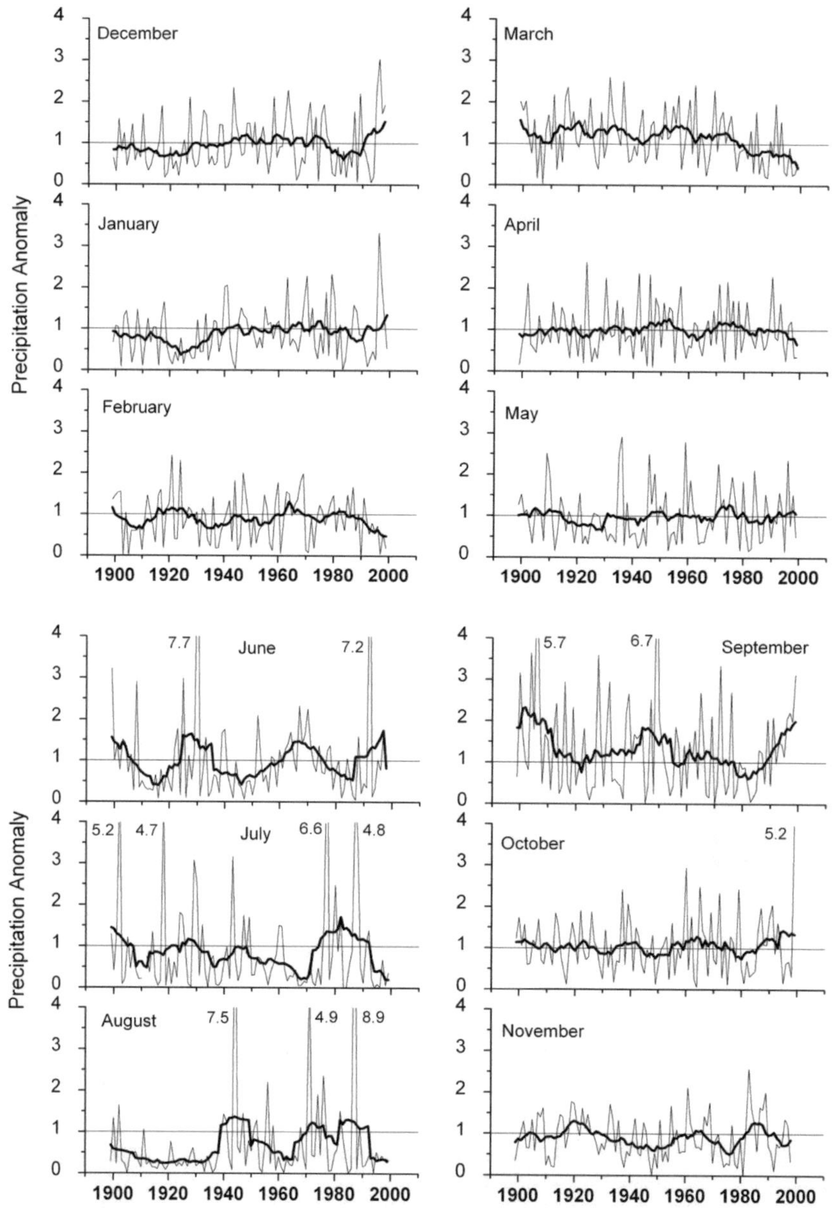

Fig. 12. Monthly precipitation anomaly from instrumental data in Spain. Units and symbols as in Fig.11.

The precipitation in Southern France (Fig.13) is more abundant in autumn, winter and spring, and less frequent in summer. All the monthly series are characterized by high frequency fluctuations and wet-dry swings for the whole period. December, January and February had less precipitation from the beginning of the instrumental records to 1900 and a wetter regime after 1900. The recent decades in the GW period are not characterized by any specific trends.

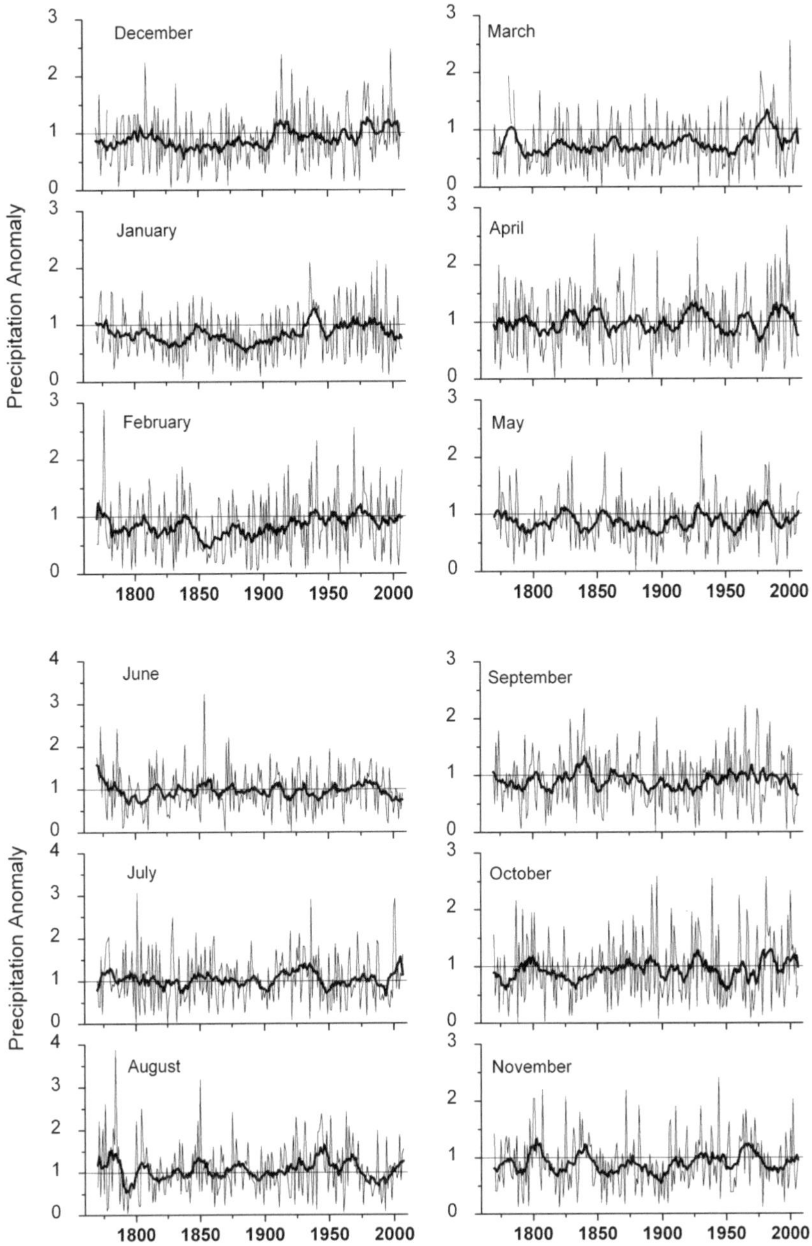

Fig. 13. Monthly precipitation anomaly from instrumental data in Southern France. Units and symbols as in Fig.11.

The rainy season in Northern Central Italy (Fig.14) is late spring (May-June) and autumn (October-November). In these months the series are almost regular from the statistical point of view, with high frequency fluctuations and some dry-wet decadal swings but no major changes. In general, peaks of abundant rain and episodes of aridity appear repeatedly, over the whole series. The last three decades of GW, i.e. after 1980, do not present any specific character.

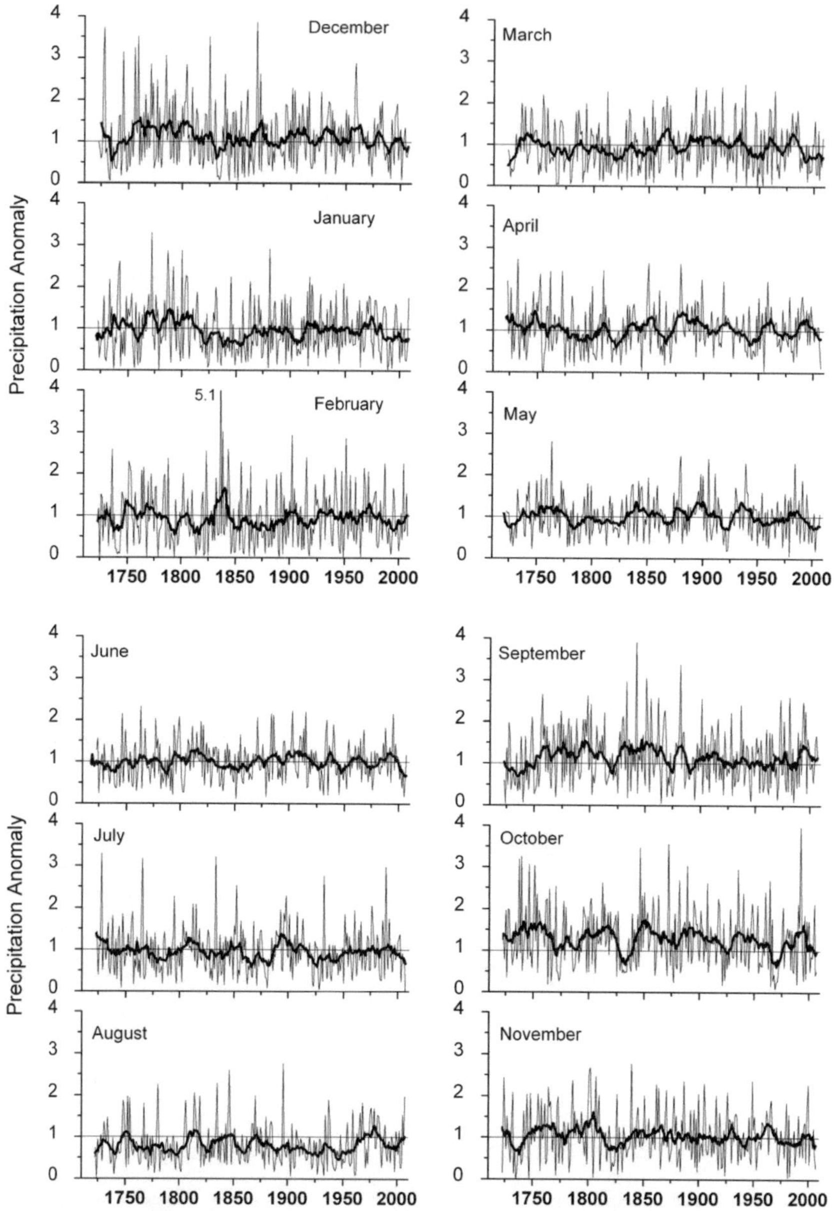

Fig. 14. Monthly precipitation anomaly from instrumental data in Northern-Central Italy. Units and symbols as in Fig.11.

The rainy season in Southern Italy (Fig.15) is the cold season, from October to April. The series are characterized by chaotically distributed short-term fluctuations and longer-term variability, with unclear or contradictory response to GW, especially after 1980. In this recent period, some months, e.g. June, September and December seem to be characterized by a positive trend, October and November by a negative one, the other months having no specific trends.

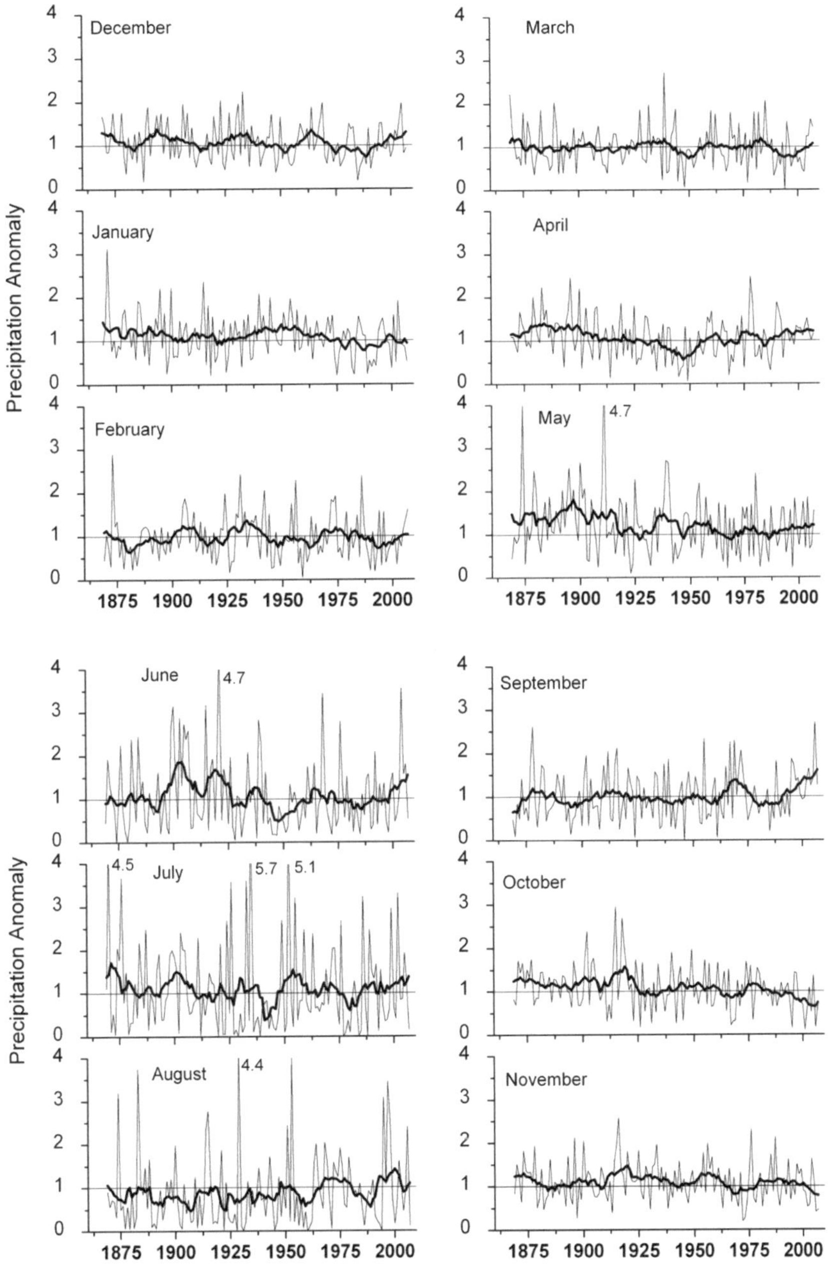

Fig. 15. Monthly precipitation anomaly from instrumental data in Southern Italy. Units and symbols as in Fig.11.

5. Biological proxies and the sea level rise

At Venice, regular tide gauge records are available since 1872 and show that the city is sinking at impressive rate, i.e. 2.4 mm yr^{-1} 1872-2009 average rate. Sinking is due to the combined effect of local land subsidence and sea level rise. The land subsidence is mainly due to tectonic drift, at constant rate, with some minor departures due to other factors, e.g. ground compaction, underground water extraction. The other key variable is the sea level rise for thermal expansion of the ocean water on the global scale (i.e. eustatic sea rise) and is expected to worsen for GW. The tide gauge measures the relative sea level (RSL), i.e. the sea level measured in relation with a frame fixed to the soil, that is a mobile reference.

The period before instrumental records is known after the fortunate combination of two available factors: a biological marker constituted by green algae, and their precise reproduction in some particular paintings. The methodology is simple. Buildings facing the canals have a green belt of algae living on walls at levels periodically reached by tides, and the belt front corresponds to the average of the high tides of the year. When the sea level changes, the green belt follows it by the same amount. The front of the algae on the Venice buildings was accurately reproduced by Paolo Caliari, nicknamed Veronese (1528-1588), and especially by the painters of Vedutas Giovan Antonio Canal, nicknamed Canaletto (1697-1768) and his nephew Bernardo Bellotto (1720-1780). All of them used a Camera Obscura as a tool to obtain precise reproductions of the views and were extremely accurate in reporting the algae levels as it has been verified with historical and statistical tests (Camuffo and Sturaro, 2003, 2004; Camuffo et al., 2005). After having determined the level of the algae as it was in the paintings, and how it is today, it is possible to obtain the RSL rise occurred in the meantime after the difference between the two levels.

The first painting useful to our aims is by Veronese and is dated 1571. It accurately reproduces the Coccina Palace, facing the Grand Canal, with people standing on the staircase that has five steps clear from algae. Nowadays the algae infest all the steps of this staircase. The height of each step is 18 cm, so that the algae belt has displaced by 5×18 cm = 90 cm. We should, however, correct this finding for two factors that have caused an additional rise of the algae belt. The first factor is that modern motorboats generate waves higher than the rowboats in use at the XVIII century. The difference in wave height was assessed after wave gauge monitoring in the Grand Canal during normal business days and in the occasion of the *'Regatta'*, i.e. the historical happening repeated every year, in which accurate replicas of rowboats and characters from the XVIII century rowing on the Grand Canal as in the times of the Venice Republic, ended 1797. The result was 5 cm additional wall wetting for motorboat waves. The second factor is the dynamic increase of the tidal wave for the excavation of deep and wide canals in the Lagoon and was evaluated to cause 3 cm additional wetting. If we subtract to the finding after the Veronese painting the 8 cm for the above corrections, we get RSL = 82±9 cm where we have attributed the uncertainty of one step, i.e. ±9 cm (Camuffo, 2010). The result from twelve Canaletto and Belotto paintings with the algae belt clearly visible was that the RSL rose by 69±11 cm. After having subtracted 8 cm for the above corrections, the bulk submersion of Venice, i.e. the RSL estimated from the Canaletto and Bellotto paintings is 61±11 cm.

If we combine in the same graph the 1872-2010 tide gauge record in Venice with the biological proxies related to the tide level in 1571, i.e. Veronese, and around 1750, i.e. Canaletto and Bellotto, we can assess the past sea level at Venice back to 1571 as summarized in Fig.16.

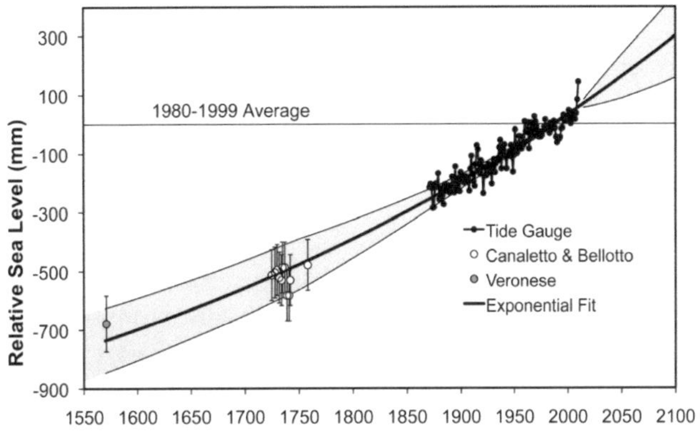

Fig. 16. Relative sea level (RSL) in Venice. Small black dots: tide gauge readings; white circles: proxies from Canaletto and Bellotto paintings; grey circle: proxy from Veronese painting. Thick black line: exponential interpolation with equation RSL = 71.196 exp (0.0016 X) where X is the year AD; thin grey lines: 95% confidence interval.

In the period from Veronese to Canaletto the average yearly trend is 1.2 mm yr^{-1}; from Canaletto to nowadays the average yearly trend is 1.9 mm yr^{-1} (Camuffo and Sturaro, 2003, 2004; Camuffo et al., 2005). The tide gauge and the displacement of the algae front show that the relative sea level has increased at exponential rate (i.e. acceleration for increasing thermal expansion) over the last five centuries. This is a real problem for the city.
A dramatic consequence of SLR is that exceptionally high tides, i.e. surges generated by the Sirocco wind when a low pressure lies in the western Mediterranean, become more and more frequent, with seawaters flooding Venice. Flooding tides, locally called "*acqua alta*" invade the lowest part of the city when the sea water exceeds 110 cm above the average tide level in 1897 established as a reference. The long series of flooding surges from documentary sources for the 792–1867 period (Camuffo, 1993; Enzi and Camuffo, 1995), combined with the tide- gauge records for the 1872–2010 period, show that the frequency of the flooding tides too is exponentially increasing (Fig.17).

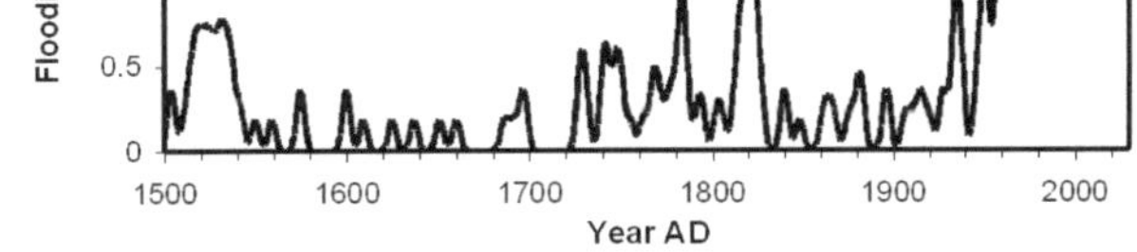

Fig. 17. Occurrence of the flooding tides (i.e. sea level 110 cm higher than the average in 1897) at Venice over the last 500 years. The graph reports the number of independent floods, i.e. one flooding tide per surge disregarding the short-term interruptions and repetitions for the astronomical tide modulation.

6. Conclusions

This Chapter has presented the trends in temperature, precipitation and sea level rise over the last five centuries, i.e. from the Medieval Optimum to the present-day GW.
The air temperature series do not provide clear evidence for marked increasing trends over the long period (e.g. one or more centuries), except for the recent GW over the last 30-40 years (Le Treut et al., 2007). Short term fluctuations and decadal or pluridecadal cycles are not a novelty for the Mediterranean. All the series show that both the temperature and the precipitation are characterized by repeated swings. Individual swings may differ in amplitude and duration, depending on the historical period. In previous papers it has been demonstrated that the periodicity of temperature is 12.7, 26.5, 34.4 and 57.3 years (Camuffo et al., 2010a,b). In conclusion, it is probable that the Mediterranean temperature will not increase like a hockey stick as expected for the GW over the Northern Hemisphere but will continue to swing maybe in association with the GW trends.
The precipitation series too are characterized by huge variability, not by specific decreasing trends. Decadal dry-wet swings are chaotically distributed and disappear for the change of phase when common periodicities are searched in the whole set of data, except for the 90 year period (Camuffo et al., 2010a,b). Air temperature and precipitation evolve in an apparently independent way, changing in phase and having correlation coefficient variable over time. This means that the two phenomena are not fully controlled by the same forcing factors. The crucial factor is that air temperature and precipitation are related between each other through the action of the sea, and the mechanism will depend on the thickness of the thermocline and the resulting Sea Surface Temperature (SST). The Mediterranean water constitutes a huge reservoir of heat, and the weather conditions depend on the seawater temperature profile i.e. the thermocline. In the case summer and early autumn have been relatively calm, the summer has accumulated heat near the surface, forming a thin, warm layer and the higher SST will start earlier and will favour the rainy season. On the other hand, if the summer had a violent storm with high waves that mixed the surface waters and deepened the thermocline redistributing in depth the heat, the SST will be lower and the winter rains will be less. In conclusion, any feedback governed by a number of interacting factors will necessarily be complex because of the chaotic variability in correlation and phase, making difficult any forecast.
Swings and chaos make obscure any tendency that may derive from the GW. Today temperature and precipitation have opposing trends, i.e. hot and dry. However, the observations suggest that no any combination is stable and persistent for a long time. Today we see hot and dry, but in the past we also had hot and wet, cold and wet and cold and dry periods. The observations suggest that the local climate system is not characterized by any stable equilibrium and in the future it will continue to swing and change temperature and precipitation coupling.
Although the temperature in the Mediterranean will increase less than in the Northern Hemisphere, nevertheless the thermal expansion of oceanic waters on the global scale will govern the water fluxes through the Gibraltar Strait, increasing the level of the Mediterranean Sea and flooding the coastal areas. The problem will be especially relevant in the areas affected by land subsidence because of the synergistic effect of both factors. The most famous example is the city of Venice, where the local sea level, observed over the last 500 years, is exponentially increasing and the expectation by 2100 will likely be 30 cm higher than the 1980-1999 average level. In the future, the most dramatic impact will affect

historical buildings. The problem is that the Venice buildings originally had an impermeable basement and a protruding protection against splashing waves. At present, the protective elements sunk below the average high tide level, with the consequence that walls are impregnated with seawater, with internal migration of NaCl and crystallisation cycles which loose rendering, destroy mortars, bricks and stones. Briefly, this is a tremendous, powerful mechanism that is destroying at slow rate all buildings reached by seawaters.

7. Acknowledgements

Most of this his work was supported by the EU, projects MILLENNIUM (Contract 017008-2) and Climate for Culture (Grant 226973). Previous EU contracts mentioned in the text are ADVICE (ENV4-CT95-0129) and IMPROVE (ENV4-CT97-0511). The French contribution was within the framework of climate history OPHELIE and RENASEC. The following instrumental data are due to the courtesy of: Bologna series (1813-2007): Dr. T. Nanni, Dr. M. Brunetti, CNR-ISAC, Bologna; Milan series (1763-2007): Prof. M. Maugeri, University of Milan; Florence (1889-2007): Prof. GP. Maracchi and Dr. A. Crisci, CNR-IBIMET, Florence; Vallombrosa (1872-2007): Dr. M. Sulli, Istituto Sperimentale per la Selvicoltura, Florence; Locorotondo (1884-2007): Prof. L. Ruggiero, University of Lecce; Palermo (1791-2007): Prof G. Micela, Prof. V. Iuliano and Dr. D. Randazzo, Astronomical Observatory "G.S.Vaiana", Palermo. For the archival research, data recovery and any other kind of help, special thanks are due to: Historical Archive of the Astronomical Observatory (INAF), Padova: Dr V. Zanini and Dr L. Pigatto; University Library, Padova: Dr L. Prosdocimi; Library of the Astronomy Department, Bologna University: Dr R. Stasi; National Central Library, Florence (BNCF): Dr. P. Pirolo and Dr S. Pelle; Museo Galileo - Institute and Museum of the History of Science (IMHS), Florence: Prof. P. Galluzzi, Dr. A. Lenzi, Dr. G. Strano, Dr. A. Saviori; Library of the Vallombrosa Abbey: Father P.D. Spotorno; Ximenian Observatory, Florence: Prof. E. Borchi and Prof. R. Macii. Italian Corps of Foresters (GFS), Office for the Biodiversity, Vallombrosa: Mrs L. Persia; Naples: Capodimonte Astronomical Observatory (INAF).

8. References

Alcoforado, M.J., Nunes, M.F., & Garcia, J.C. (2000). Temperature and precipitation reconstruction in southern Portugal during the late Maunder Minimum (AD 1675-1715). *The Holocene*, Vol. 10, No. 3, (April 2000), pp. (333-340)

Barriendos, M. (1997). Climatic variations in the Iberian Peninsula during the late Maunder Minimum (AD 1675-1715): an analysis of data from rogation ceremonies. *The Holocene*, Vol. 7, No. 1, (March 1997), pp. (105-111)

Barriendos, M., & Llasat, M.C. (2003). The Case of the `Maldá' Anomaly in the Western Mediterranean Basin (AD 1760–1800): An Example of a Strong Climatic Variability. *Climatic Change*, Vol. 61, No. 1-2, (November 2003), pp. (191-216)

Barriendos, M. (2005). Climate and Culture in Spain. Religious Responses to extreme Climatic Events in the Hispanic Kingdoms (16th-19th c). In: *Kulturelle Konsequenzen der "Kleinen Eiszeit"*, Behringer W, Lehmann H, Pfister C (ed.), Vandenhoeck & Ruprecht, Göttingen

Bolle, H.J. (2003). *Mediterranean Climate*, Springer-Verlag, Berlin

Brazdil, R., Glaser, R., Pfister, C., Dobrovolny, P., Antoine, J.M., Barriendos, M., Camuffo, D., Deutsch, M., Enzi, S., Guidoboni, E., Kotyza, O., & Rodrigo, F.S. (1999). Flood Events of Selected European Rivers in the Sixteenth-Century. Special Issue: Climatic Variability in Sixteenth-Century Europe and Its Social Dimensions. *Climatic Change*, Vol. 43,No. 1, (September 1999), pp. (239-285)

Brázdil, R., Pfister, C., Wanner, H., von Storch, H., & Luterbacher, J. (2005). Historical climatology in Europe - The State of the Art. Climatic Change, Vol. 70, No. 3, (June 2005), pp. (363-430)

Brázdil, R., Dobrovolný, P., Luterbacher, J., Moberg, A., Pfister, C., Wheeler, D., & Zorita, E. (2010). European climate of the past 500 years: new challenges for historical climatology. *Climatic Change*, Vol. 101, No. 1, (May 2010), pp. (7-40)

Camuffo, D. (1993). Analysis of the Sea Surges at Venice from A.D. 782 to 1990. *Theoretical and Applied Climatology*, Vol. 47, pp. (1-14)

Camuffo, D. (2002). History of the long series of the air temperature in Padova (1725-today). *Climatic Change*, Vol. 53, No. 1-3, (April 2002), pp. (7-76)

Camuffo, D. (2010). Le niveau de la mer à Venise d'après l'œuvre picturale de Véronèse, Canaletto et Bellotto. Révue d'Histoire Moderne et Contemporaine, Vol. 57, No. 3, pp. (92-110)

Camuffo, D., & Bertolin, C. (2011). The earliest temperature observations in the world: the Medici Network (1654–1670). *Climatic Change*, DOI 10.1007/s10584-011-0142-5

Camuffo, D., & Enzi, S. (1992a). Critical Analysis of Archive Sources for Historical Climatology of Northern Italy, In: *European Climate Reconstructed from Historical Documents: Methods and Results*. Frenzel B., Paleoclimate Research, Special Issue 7, Fischer Verlag, Stuttgart

Camuffo, D., & Enzi, S. (1992b). Reconstructing the Climate of Northern Italy from Archive Sources, In: *Climate since 1500 A.D.*. Bradley RS, Jones PD (eds.) Routledge, London

Camuffo, D., & Enzi, S. (1994). The Climate of Italy from 1675 to 1715, In: *Climatic Trends and Anomalies in Europe 1675-1715*. Frenzel B., Paleoclimate Research, Special Issue 8, Fischer Verlag, Stuttgart

Camuffo, D., & Enzi, S. (1996). The Analysis of two Bi-millenary Series: Tiber and Po River Floods, In: *Climatic Variations and Forcing Mechanisms of the last 2000 years*, Jones P.D., Bradley R.S., Jouzel J., NATO ASI Series, Series I: Global Environmental Change, Vol. 41, Springer Verlag, Stuttgart

Camuffo, D., & Jones, P. (2002). *Improved Understanding of Past Climatic Variability from Early Daily European Instrumental Sources*. Kluwer Academic Publishers, Dordrecht, Boston, London

Camuffo, D., & Sturaro, G. (2003). Sixty-cm submersion of Venice discovered thanks to Canaletto' s paintings. *Climatic Change*, Vol. 58, No. 3, (June 2003), pp. (333-343)

Camuffo, D., & Sturaro, G. (2004). Use of proxy-documentary and instrumental data to assess the risk factors leading to sea flooding in Venice. *Global and Planetary Change*, Vol. 40, pp (93-103)

Camuffo, D., Pagan, E., & Sturaro, G. (2005). The extraction of Venetian sea-level change from paintings by Canaletto and Bellotto, in: *Flooding and Environmental Challenges for Venice: State of Knowledge*, Fletcher C.A. & Spencer T., pp. (129-140), Cambridge University Press, Cambridge

Camuffo, D., Bertolin, C., Barriendos, M., Dominguez-Castro, F., Cocheo, C., Enzi, S., Sghedoni, M., della Valle, A., Garnier, E., Alcoforado, M.J., Xoplaki, E.,

Luterbacher, J., Diodato, N., Maugeri, M., Nunes, M.F., & Rodriguez, R. (2010a). 500-year temperature reconstruction in the Mediterranean Basin by means of documentary data and instrumental observations. *Climatic Change*, Vol. 101, No. 1-2, (July 2010), pp. (169-199)

Camuffo, D., Bertolin, C., Diodato, N., Barriendos, M., Dominguez-Castro, F., Cocheo, C., della Valle, A., Garnier, E., & Alcoforado, M.J. (2010b). The Western Mediterranean Climate: how will it respond to Global Warming? *Climatic Change*, Vol. 100, No. 1, (May 2010), pp. (137-142)

Chine, I., Yiou, P., Viovy, N., Seguin, B., Daux, V., & Le Roy Ladurine, E, (2004). Grape ripening as a past climate indicator. *Nature* ,Vol. 432, No. 7015, (November 2004), pp. (289-290)

Dominguez-Castro, F., Santisteban, J.I., Barriendos, M., & Mediavilla, R. (2008). Seasonal reconstruction of drought episodes for central Spain from rogation ceremonies re.corded at Toledo Cathedral from 1506 to 1900: a methodological approach. *Global and Planetary Changes*, Vol. 63, pp (230–242)

Enzi, S., & Camuffo, D. (1995). Documentary sources of the sea surges in Venice from A.D. 787 to A.D. 1867. *Natural Hazards*, Vol. 12, No. 3, (November 1995), pp. (225-287)

Fletcher, C.A., & Spencer, T. (2005). *Flooding and Environmental Challenges for Venice and its Lagoon: State of Knowledge*. Cambridge University Press, Cambridge

Glaser, R., Riemann, D., Schönbein, J., Barriendos, M., Brazdil, R., Bertolin, C., Camuffo, D., Deutsch, M., Dobrovolny, P., Engelen van, A., Enzi, S., Halićhová, M., Koenig, S., Kotyza, O., Limanowka, D., Macková, J. & M. Sghedoni. 2010. The variability of European floods since AD 1500. *Climatic Change*, Vol. 101, No. 1-2, (July 2010), pp. (235-256).

Jeftic, L., Milliman, D.J., & Sestini, G. (1992). *Climatic Change and the Mediterranean*. Edward Arnold, London

Jurin, J. (1723) . Invitatio ad observationes Meteorologicas communi consilio instituendas a Jacobo Jurin M.D. Soc. Reg. Secr. et Colleg. Med. Lond. Socio. *Philosophical Transactions*, Vol. 379, pp. (422-427)

Le Treut, H., Somerville, R., Cubasch, U., Ding, Y., Mauritzen, C., Mokssit, A., Peterson, T., & Prather, M. (2007). Historical Overview of Climate Change. In: *Climate Change 2007: The Physical Science Basis. Contribution of Working Group I to the Fourth Assessment* Report of the Intergovernmental Panel on Climate Change, Solomon S., Manning M., Chen Z., Marquis M., Averyt K.B., Tignor M., & Miller H.L., Cambridge University Press, Cambridge, United Kingdom and New York, NY, USA

Lionello, P., Malanotte-Rizzoli, P., & Boscolo, R. (2006). *The Mediterranean Climate: an overview of the main characteristics and issues*. Elsevier, Amsterdam

Luterbacher, J., Xoplaki, E., Casty, C., Wanner, H., Pauling, A., Kuettel, M., Rutishauser, T., Broennimann, S., Fischer, E., Fleitmann, D., Gonzalez-Rouco, J.F., Garcìa-Herrera, R., Barriendos, M., Rodrigo, F.S., Gonzalez-Hidalgo, J.C., Saz, M.A., Gimeno, L., Ribera, P., Brunet, M., Paeth, H., Rimbu, N., Felis, T., Jacobeit, J., Duenkeloh, A., Zorita, E., Guiot, J., Turkes, M., Alcoforado, M.J., Trigo, R., Wheeler, D., Tett, S.F.B., Mann, M.E., Touchan, R., Shindell, D.T., Silenzi, S., Montagna, P., Camuffo, D., Mariotti, A., Nanni, T., Brunetti, M., Maugeri, M., Zerefos, C., De Zolt, S., Lionello, P., Nunes, M.F., Rath, V., Beltrami, H., Garnier, E., & Le Roy Ladurie, E. (2006). Mediterranean climate variability over the last centuries: A review, In: *The*

Mediterranean Climate: an overview of the main characteristics and issues, Lionello P., Malanotte-Rizzoli P., Boscolo R., Elsevier, Amsterdam, the Netherlands

Luterbacher, J., Koenig, S.J., Franke, J., van der Schrier, G., Della-Marta, P.M., Jacobeit, J., Küttel, M., Gonzalez-Rouco, F.J., Zorita, E., Xoplaki, E., Stössel, M., Rutishauser, T., Wanner, H., Pfister, C., Brázdil, R., Dobrovolný, P., Camuffo, D., Bertolin, C., Moberg, A., Leijonhufvud, L., Söderberg, J., Allan, R., Wilson, R., Wheeler, D., Barriendos, M., Glaser, R., Riemann, D., Nordli, O., Limanówka, D., van Engelen, A., & Zerefos, C.S. (2010). Circulation dynamics and its influence on European and Mediterranean January-April climate over the past half millennium: Results and insights from instrumental data, documentary evidence and coupled climate models *Climatic Change*, Vol. 101, No. 1-2, (July 2010), pp. (201-234)

Magalotti, L. (1666). *Saggi di naturali esperienze fatte nell'Accademia del Cimento*, Giuseppe Cocchini, Florence

Martín Vide, J., & Barriendos, M. (1995). The use of rogation ceremony records in climatic reconstruction: a case study from Catalonia (Spain). *Climatic Change*, Vol. 30, No. 2, (June 1995), pp. (201-221)

Pfister, C., Brazdil, R., Glaser, R., Barriendos, M., Camuffo, D., Deutsch, M., Dobrovolny, P., Enzi, S., Guidoboni, E., Kotyza, O., Militzer, S., Racz, L., & Rodrigo, F.S. (1999). Documentary Evidence on Climate in the Sixteenth-Century Europe. *Climatic Change*, Vol. 43, No. 1, (September 1999), pp. (55-110)

Piervitali, E., & Colacino, M. (2001). Evidence of drought in Western Sicily during the period 1565-1915 from Liturgical Offices. *Climatic Change*, Vol. 49, No. 1-2, (April 2001), pp. 225-238

Reiter, E.R. (1975). *Weather Phenomena of the Mediterranean Basin*. EPRFNPS, Montrey, California

Rodrigo, F.S., & Barriendos, M. (2008) . Reconstruction of seasonal and annual rainfall variability in the Iberian Peninsula (16th – 20th Centuries) from documentary data. *Global and Planetary Changes* ,Vol 63, pp (243–257) doi:10.1016/j.gloplacha.2007.09.004

Targioni Tozzetti, G. (1780). *Notizie degli aggrandimenti delle Scienze Fisiche accaduti in Toscana nel corso di anni LX del secolo XVII*. Tomo I. Bouchard, Florence

UK Meteorological Office (1962). *Weather in the Mediterranean*. Her Majesty Stationery Office, London

Xoplaki, E., Gonzalez-Rouco, F.J., Luterbacher, J., & Wanner, H. (2003). Mediterranean Summer air temperature variability and its connection to the large-scale atmospheric circulation and SSTs. Climate Dynamics, Vol. 20, No. 7-8, (May 2003), pp. (723-739)

Xoplaki, E., Gonzalez-Rouco, J.F., Luterbacher, J., & Wanner, H. (2004), Wet season Mediterranean precipitation variability: influence of large-scale dynamics and trends. Climate Dynamics, Vol. 23, No. 1, (August 2004), pp. (63-78)

Wallén, C.C. (1970). Climates of Northern and Western Europe, Elsevier, Amsterdam

Wallén, CC. (1977) . Climates of Central and Southern Europe. Elsevier, Amsterdam

2

Recent Global Warming Induced Climate Changes

P. C. Werner, F.-W. Gerstengarbe, H. Österle and M. Wodinski
Potsdam Institute for Climate Impact Research
Germany

1. Introduction

Climate should be regarded as a complex system, thus requiring an investigation of its complex changes. For this purpose, it is obvious to define climate types that are determined by a number of factors to be able to afterwards investigate the way they are changing. The most popular procedure so far consists in defining threshold values - as a rule for temperature and precipitation for the seasons - to determine those areas as a climate type that are within these threshold values. A popular example of this procedure is the frequently used Köppen climate classification (Köppen, 1931; Fraedrich et al., 2001). The drawback of this methodology consists in the fact that the threshold values can be absolutely plausible but need to be given subjectively in any case. The first question is: Can be a methodology developed here for the identification of climate changes in an objective way? First, all regions with a structural behavior as similar as possible are combined in one climate type and the pertaining threshold values of the meteorological parameters used are determined afterwards. Regions or areas with the same structure can be determined by means of a pattern recognition method. An extended non-hierarchic cluster analysis will be used for this purpose in the investigation. This cluster analysis provides a number of statistically separated patterns (climate types) for the climate classification on earth. So the characteristics of each climate type can be described. The second question that needs to be answered now can be formulated as follows: Are there any changes in the location and dimension of individual climate types due to the global climate changes presently observed? In order to answer this question, the shifting of the climate types during the observed period must be calculated. After calculation, an analysis of the results follows, first for the global scale as a whole and second more detailed for each continent and selected regions. So the chapter gives an overview on the actual climate changes for different scales.
Section 2 describes the basis of the data and its validation. The methodology can be found in section 3 followed by the analysis of the actual climate changes for the whole earth and each continent in section 4. Finally some conclusions are made in section 5.

2. Data

2.1 Introduction

The first step to prove climate change is to provide a data base, i.e., to choose adequate parameters for describing climate types. These parameters must be capable to indicate annual and seasonal differences. As regarding temperature, the following parameters were chosen:

- annual mean value
- 3 months mean, calculated from the 3 highest monthly values
- 3 months mean, calculated from the 3 lowest monthly values
- spread between highest and lowest monthly value

For precipitation, the parameters are defined as follows:

- annual sum
- 3 months sum, calculated from the 3 highest monthly sums
- 3 months sum, calculated from the 3 lowest monthly sums

On the basis of the so chosen parameters, one can produce an equivalent climate classification. At the same time, the demand for a cluster analysis developed under conditions of parameters highly independent from each other is fulfilled.

To provide these parameters, an appropriate data set for temperature and precipitation based on monthly values had to be produced. On the one hand, we used a temperature data set (TS3.0, 1901–2005) compiled by the Climate Research Unit (CRU) of the University of East Anglia (Mitchell & Jones, 2005), and, on the other hand, we employed a data set of the Global Precipitation Climatology Centre (GPCC, 1901–2009) for precipitation (Rudolf et al., 2010). In both cases, the values of the data sets were available at a grid of 0.5° x 0.5°.

Investigations of the CRU temperature data set and the GPCC precipitation data set revealed that, with regard to the temporal structure, there were a number of grid points with inhomogeneities (leaps) that could not be neglected. Hence, both data sets were statistically homogenized and regularly actualized at PIK until 2009.

2.2 Methods to identify inhomogeneities

We will just briefly go into the method how to check inhomogeneity in data sets. There are lots of publications on this. An elaborate survey can be found, e.g., in Peterson et al. (1998). For the investigation presented, the t-test or Student's t-test (Taubenheim, 1969; Österle et al., 2003) was applied. For this test, both data sets were "seasonally cleaned" in a first step, i.e., the long-time mean value (here: 1961–1990) was derived from the particular value of the single month.

By the t-test one checks if the mean value from two random samples come from one and the same basic entity, i.e., they derive from each other only within a defined probability value. The floating t-test was applied here: two random samples of the length 5 years lying side by side in the particular time series were defined, e.g., 1/1901–12/1905 and 1/1906–12/1910, and the t-value was identified. After this, the random samples were shifted by a month (2/1901–1/1906 and 2/1906–1/1911), and the t-value was identified once more, and so on. In the end, one receives a time series of t-values. If the local maxima resp. minima of these t-values exceed the defined significance level of 0.1 %, this points to one or more leaps within the time series. These leaps separate quasi-homogenous periods from each other. Starting with the last homogenous period within the time series, the homogenous partial periods which lie before were adapted by a bias correction.

2.3 Temperature

As remarked above, CRU contains temperature data for the period 1901–2009. The particular outliers were identified by simple $\pm 5 \cdot$ Sigma test and the monthly values in which the deviations from the long-time monthly values exceeded 10K were equalized to the mean value.

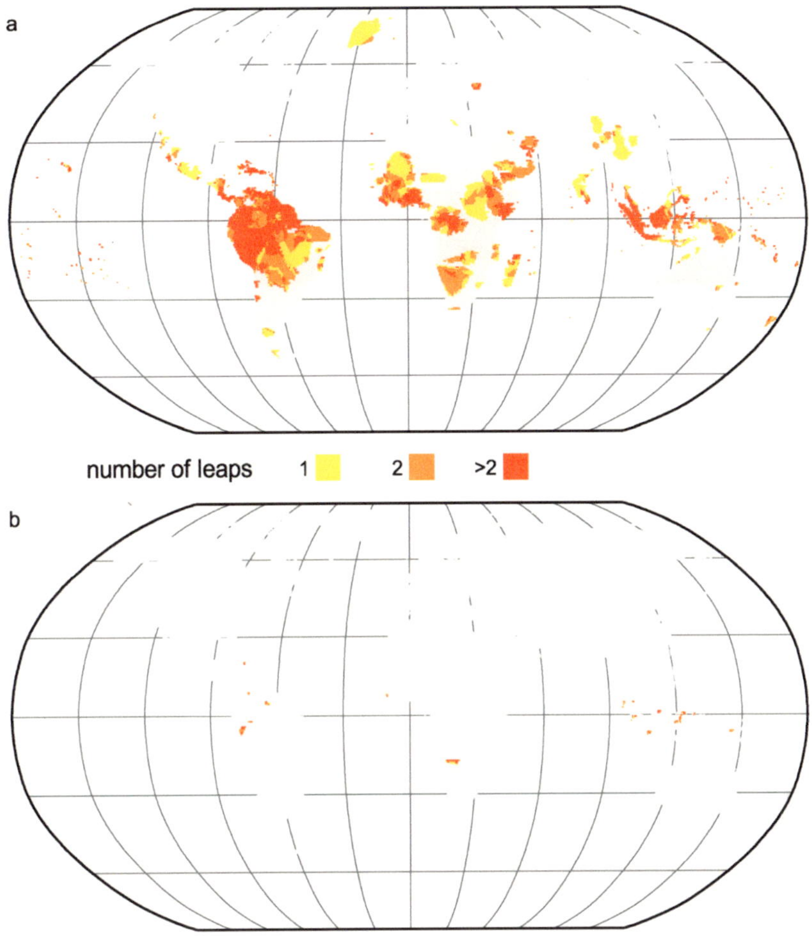

Fig. 1. (a) Regions with inhomogeneities in CRU temperature time-series 1901–2009, localized by the t-test, (b) homogenized data set of temperatures

An investigation of data concerning its temporal behavior revealed that there are a number of grid points with inhomogeneities (leaps) that cannot be neglected. In the TS3.0 version, approximately 19 % of the grid rows were found with inhomogeneities (t-test with a critical t-value = 6.0). The affected area is indicated in Figure 1a. The areas with a relatively bad observation grid also show the most inhomogeneities, such as parts of South America, Southeast Asia, and Africa. This means that, in most cases, inhomogenous grid points do not occur insulated.

The data set was homogenized with the method as explained before.

The homogenization resulted in a significant improvement of the spatial structure as a whole (Figure 1b).

4.500 stations, 3.200 out of which on a daily-value basis, delivered the initial dates for the temperature updates from 2006 on. The rest was derived from the journal Monthly Climate Reports. The data from these 4.500 stations were interpolated to the grid and adapted to the TS3.0 data set.

2.4 Precipitation

The precipitation data used for this investigation come from the data set "Full Data Reanalyse (V.5)" with a grid of 0.5° x 0.5° developed at GPCC in Offenbach. The data set contains data for all months within the period 1901 to 2009 basing on quality-controlled data of the available number of stations with an irregular temporal coverage during the analysis period (max. number of approx. 45.000 stations in 1986/87). This product is optimized for best spatial coverage and usage in water balance assessments but it contains gridded data sets (Rudolf et al., 2010; Rudolf & Schneider, 2005).

Akin to temperature, the analysis of the precipitation data sets showed some inhomogeneities (Figure 2a). As already described, these were homogenized as well so that a significant reduction of inhomogeneities was achieved (Figure 2b).

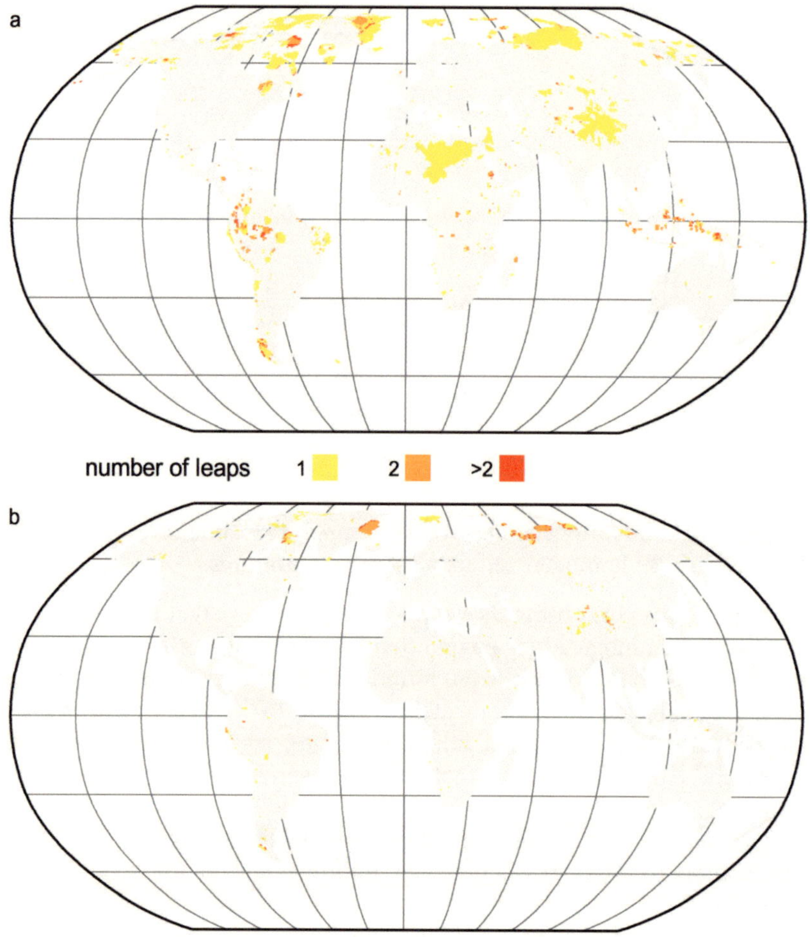

Fig. 2. (a) Data set GPCC V 5.0; grid points with inhomogenous time series found by floating t-test, (b) homogenized data set GPCC V 5.0

3. Methodology

3.1 Introduction

The main idea of the cluster analysis is to relate to each other an existing number M of elements e_i which are each described by N parameters p, i.e.:

$$\mathrm{e_i} = f(p_{i1}, ..., p_{iN}) \tag{1}$$

Two main techniques are possible:
Using hierarchical methods, different sequences of groups on different levels are constructed. The result is a hierarchy of clusters in a "tree structure". The disadvantage of this technique lies in the fact that an exchange of elements is impossible if the "tree structure" is built up. This disadvantage restricts the application.
With the non-hierarchical methods, the elements e_i are simultaneously partitioned into a given number of clusters K: by displacing the elements between the clusters in case of a given quality criterion, a given initial partition is built up step by step, and developed into steadily improving groupings until reaching the optimum. For more details, see Steinhausen & Langer (1977). The starting point for concerning the description of the following method is the non-hierarchical minimum-distance method according to Forgy (1965). The starting condition when applying the above method is to have the elements e_i equally distributed over a number K of given clusters (initial partition). In the case of M given elements and K clusters each cluster receives $L = M/K$ elements as follows:

$$\begin{matrix} e_1, ..., & e_L & \ni & c_1 \\ e_{L+1}, ..., & e_{2L} & \ni & c_2 \\ \vdots & & & \vdots \\ e_{(k-1)L+1}, ..., & e_{kL} & \ni & c_k \end{matrix} \tag{2}$$

(The number of clusters K must be defined empirically; the number of elements depends on the data series and the problem which has to be investigated.)
A so-called group centroid $\overline{e}_k$ is then calculated for each k of the K clusters (cluster mean value under consideration of those existing parameters that have to be normalized accordingly in the case of different scalings):

$$\overline{e}_k = \frac{1}{L} \sum_{i=(k-1)L+1}^{kL} e_i. \tag{3}$$

By applying the Euclidean distance, the following objective function $a(g)$ for each grouping step g can be defined:

$$a(g) = \sum_{k=1}^{K} \sum_{i \in k} \left| e_i - \overline{e}_k \right|^2. \tag{4}$$

By considering the Euclidean distance, each grouping step can be seen as a displacement of the element e_i into that cluster which contains the respective nearest centroid. The objective function can thus be minimized:

$$a(g)\ \forall g \rightarrow \min. \tag{5}$$

This procedure is repeated until a local minimum of the objective function is reached. The objective function reaches a local minimum if two successive grouping steps show the same result; the iteration is in this case discontinued, i.e., the optimum classification with respect to the given number of clusters has been reached.
An important disadvantage of this method is that one does not know whether an absolute or just a secondary minimum of the objective function has been obtained (Fovell & Fovell, 1993). That is why the quality of separation is unknown, as is the objective number of clusters. The following procedure shows a solution of this problem.

3.2 Definition of quality criterion to separate clusters

The quality criterion represents the statistical security of the cluster separation. The basic idea to define this criterion can be described as follows (Gerstengarbe & Werner, 1997):
After having reached the local minimum, each cluster is equipped with a generally varying number of elements. Each element is defined by N parameters, i.e., it is located in a N-dimensional parameter space. As each cluster consists of a certain number of elements, they each represent a scatterplot of elements in the above space. If the clustering leads to a local secondary minimum, overlaps occur between the scatterplots of single clusters. The principle of this method is presented in Figure 3, which depicts the projection of two parameters within the N-dimensional space.

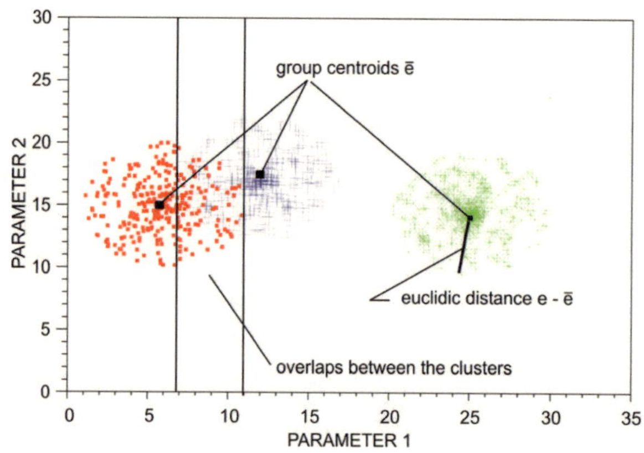

Fig. 3. Principle scheme of the description of the clustering quality (red/blue - overlapped clusters, green - full separated cluster)

The number of overlaps O of the two clusters a and b of N parameters can accordingly be defined as follows:

$$O_{a,b} = \sum_{i_a=1}^{L_a} \sum_{i_b=1}^{L_b} \sum_{j=1}^{N} o_{i_a,i_b,j} \qquad \begin{matrix} a = 1, \ldots, k \\ b = 2, \ldots, k \end{matrix} \tag{6}$$

with

$$o_{i_a,i_j,j} = \begin{cases} 1 & p_{i_b,j} \geq p_{i_a,j} \\ 0 & p_{i_b,j} < p_{i_a,j} \end{cases} \tag{7}$$

under the additional condition

$$\bar{e}_1 > \bar{e}_2 > \ldots > \bar{e}_k \tag{8}$$

If $O_{a,b} = 0$, than the clusters a and b are completely separated from each other. The maximum possible number of overlaps is

$$O_{a,b}^{\max} = NL_a L_b \tag{9}$$

This number is reached if both clusters cover the same region within the N-dimensional space.

Thus by applying the equations (6) to (9) the quality of the separation of clusters can be determined statistically by the following steps:

- Calculation of the mean number of the maximum possible overlaps $\bar{O}^{\max}$ as well as the mean actual number of overlaps $\bar{O}$ over all combinations of cluster pairs.
- Subsequently, a test is carried out to see whether $\bar{O}$ and $\bar{O}^{\max}$ originate from the same basic population. Assuming that there is a normal distribution, Student's t-test can be used. (Because of the necessary normalization of the parameters, a normal distribution is generally realized.) The null hypothesis implies that both mean values originate from the same population. The clusters can be separated only when the null hypothesis is rejected. Otherwise, the procedure is as follows:
- The ratio $v_{a,b}$ of the actual to the maximum possible number of overlaps is determined for each cluster pair:

$$v_{a,b} = \frac{O_{a,b}}{O_{a,b}^{\max}} \tag{10}$$

- The mean value $\bar{v}$ over all $v_{a,b}$ is calculated. It is the empirical estimate of the actual occurrence probability of overlaps.
- In the case that not all mean values $\bar{v}$ are identical, the second bullet point implies that there is - according to the chosen level of significance - a statistically significant separation of those clusters for which $v_{a,b} < \bar{v}$.
- The quality of separation in the case $v_{a,b} > \bar{v}$ still needs to be determined. The point is hence to clarify whether a certain value of the number of the actual overlaps $O_{a,b}$ is compatible with the mean value of all numbers of the actual overlaps $\bar{O}$ or not. If one interprets the overlaps as empirical occurrence frequencies, a statistical comparison between both is possible. This can be done for instance by the χ^2-test (e.g. Taubenheim, 1969) which can be written as follows:

$$\chi^2 = \frac{(O_{a,b} - \bar{O})^2 \cdot (2\,O_{a,b}^{\max} - 1)}{(O_{a,b} + \bar{O}) \cdot (2\,O_{a,b}^{\max} - O_{a,b} - \bar{O})} \tag{11}$$

with the degree of freedom $d_f = 1$.

The result of the test can be interpreted in the following way: If the calculated χ^2-value is greater than a given threshold of significance, the frequency of overlaps exceeding the mean value $\bar{O}$ differs significantly from the χ^2-value. The separation between the clusters is hence statistically not significant, in contrast to the other case where a statistically reliable separation exists.

3.3 Determination of an optimum number of clusters

The optimum number of clusters is defined as that number which realizes the best separation between all clusters. The method presented above allows the optimum number of clusters for the non-hierarchical clustering to be determined in the best possible way. The following procedure is required to this end:

- If a clustering with a given initial number of clusters does not lead to a separation, then the initial number of clusters is varied until at least a single statistically reliable separation between one cluster and the rest exists.
- If the first bullet point. is fulfilled, the elements of the separated clusters are noted as being a final partial result.
- The initial series is reduced by the separated cluster elements.
- This algorithm is repeated using the method presented above until all clusters are statistically reliably separated.
- The optimum number of clusters results from the amount of clusters separated per algorithm step.

Nevertheless, some problems applying this method remain:

- a correct prevision of the initial partition (Is valid also for all other cluster analysis methods).
- an estimation of the optimal initial number of clusters
- a reduction of the error appearing in connection with the delimitation of the level of significance for cluster separation.

In the following, solutions to these problems are proposed and discussed. Chapter 3.4 contains the theoretical basis of the improvements. In Chapter 3.5 the theoretical mechanisms are discussed by a one parameter oscillation.

3.4 Theoretical basis of improvements

3.4.1 The initial partition

For each statistical investigation, the elements of the sample must be independently and identically distributed. This principle is also valid for cluster analysis. If neglected the following course of events may appear (s. also Gerstengarbe et al., 1999):

In the first step of the clustering, the elements of the sample are regularly distributed in the initial number of clusters. In this case, the sequence of the distribution depends on the position of each sample element. That is, in each cluster of the initial partition there is a number of elements which are sorted one following the other in the sample. Thus these elements are not necessarily independent which means that the structure of the sample may create "pre-grouping". As a consequence, a greater number of dependent elements must exist within a sample. Then, a secondary minimum of the target function can be reached already after only a few iterations devoid of an optimal grouping. This defect can be avoided in a simple way by a random ranking of the elements of the sample so that the persistence of the series tends to 0.

3.4.2 The initial number of clusters

Given is a sample with a limited number n of elements and their regular distribution in the initial number of clusters. This means that a too large or too small initial number of clusters leads to a situation in which some clusters can be separated significantly before the optimum distribution of the elements was been reached. If, for example, the initial number of clusters is too small, the number of elements within a single cluster is relatively large. As a result possible internal structures of a separated cluster cannot be considered. In the other case, artificial structures can be occurred. To estimate the optimum initial number of clusters the following procedure can be carried out:

The starting point for the calculation of the initial cluster number is the target function (eq. 4). The target function is constructed in such a way that the partition for which the function reaches a minimum defines the most favorable grouping of the clusters. If, for a varying number of clusters (from 2 to m), the value of the target function is calculated, a series is obtained whose values can be included for the estimation of the optimum initial number of clusters. As each value of the target function is equivalent to a specific number of clusters, the initial number of clusters can be defined as that inflection point within the series (of target function values) where a trend disappears. From this point on significant changes within the series do not exist. This idea can be solved practically with the following steps:

- Calculation of the differences between neighboring values of the target function series and creation of a differential series with $m1 = m - 1$ values,
- Using the Pettitt-test (Pettitt, 1979) to estimate the beginning of a trend (inflection point) within the differential series.

The Pettitt-test can be derived from the U-test (Mann & Whitney, 1947), based on the ranks of the series. The inflection point is defined as that point for which the absolute value of X_k reaches a maximum with

$$X_k = 2 \cdot R_k - k \cdot (m1 + 1) \tag{12}$$

where

$$R_k = \sum_{i=1}^{k} r_i \tag{13}$$

k is the position within the series, $m1$ is the number of values of the differential series, and r_i is the rank of the ith target function value. Continuously increasing the initial number of clusters, the Pettitt-test finally defines that position within the series of the target function values where the series is divided into a part with significant changes of the target function values and another one without changes.

3.4.3 The error margin

In general, the χ^2-test is connected with an error probability of 1 % or 5 %. That is, in spite of a statistically significant separation of two clusters, a small number of overlaps can occur so that some clusters may contain "strange" elements. In a statistical sense, this case is without any consequence. In some cases, however, such outliers can have a negative influence on the clustering.

This problem can be circumvented as follows using the definition of an outlier as a value deviating significantly from the basic sample: After a significant separation of all clusters has been achieved, the distance between each element within the cluster and the group centroid is calculated. These distances within each cluster are defined as a basic sample and utilized for identifying outliers. Here we suggest the square of the Euclidean distance as the measure for the estimation of the outliers. For each element of a cluster, we calculate the sum of Euclidean distances between the single parameters and the group centroid. This leads to a sample of these sums for each cluster. Using the Thompson-rule (Müller et al., 1973) we can estimate the outliers of the clusters. The test value is defined as:

$$t_i = \frac{x_i - \overline{x}}{s^*} \quad (i = 1, ..., n) \tag{14}$$

where $\overline{x}$ is the arithmetical mean of the sample and s^* the standard deviation of the sample. Outliers are all values $x_i (i = 1, ..., n)$ for which $|t_i| > z_{m,\alpha}$ is valid, with $m = n - 2$ ($z_{m,\alpha}$ = critical value). In this sense the Thompson rule is a two-sided test to examine the hypothesis H_0: "The sample has no outliers for a chosen level of significance α". If outliers exist, the Euclidean distance makes it possible to test a better assignment of the outlier to another cluster.

3.5 Validation of the method

As an example for a one parameter oscillation, a simple sine-oscillation is selected and described by 200 values. Its regular course is replaced by 10-value steps in form of stairs. In case of clustering of the new curve the boundaries between the clusters have to be identical with those between the steps of the curve. The partition of the clusters must be symmetric in two respects: First, the positive part of the oscillation must be symmetric as well as the negative. Second, the positive region must mirror the negative region symmetrically. Three variants of clustering are investigated on the basis of the discussed procedures:

1. the defined initial number of clusters is set to $k_0 = 8$; the initial partition consists of random ranked values
2. the optimal initial number of clusters is counted; the values of the initial partition are ordered from 1-200 in the same course like the sine oscillation
3. the optimal initial number of clusters is calculated; the initial partition consists of random ranked values.

Figure 4a shows the result of variant 1. One can see that the boundaries of the clusters are coincided with the spots. Additionally, the symmetry is fulfilled within the positive part as well as within the negative part. The positive and negative parts are asymmetric with respect to each other. If we define cluster 4 as "neutral", 3 clusters remain in the positive part, 4 in the negative one, while cluster 1 contains 5 steps and cluster 8 as the pendant only 3.

The results of version 2. are presented in Figures 4b and 5. Figure 5 gives an overview of the course of the target function values with respect to the number of clusters. Also included is the result of the Pettitt-test with an optimal initial cluster number of 5. This number agrees coincidentally with the optimal separated number of clusters (Figure 4b). The symmetry in the positive and negative parts is fulfilled, but a "neutral" cluster does not exist. Thus an asymmetry exists between both parts: the positive one includes 3 clusters (1–3), while the negative one only 2 (4–5). This is why the ranges of the clusters are different.

For variant 3. we start with the same initial number of clusters as calculated for 2. The number of statistically separated clusters is also 5. In this case all conditions of symmetry are fulfilled (Figure 4c).

This example shows that a correct solution exists for the clustering, if data of the initial partition are ranked randomly and the optimal initial number of clusters is used.

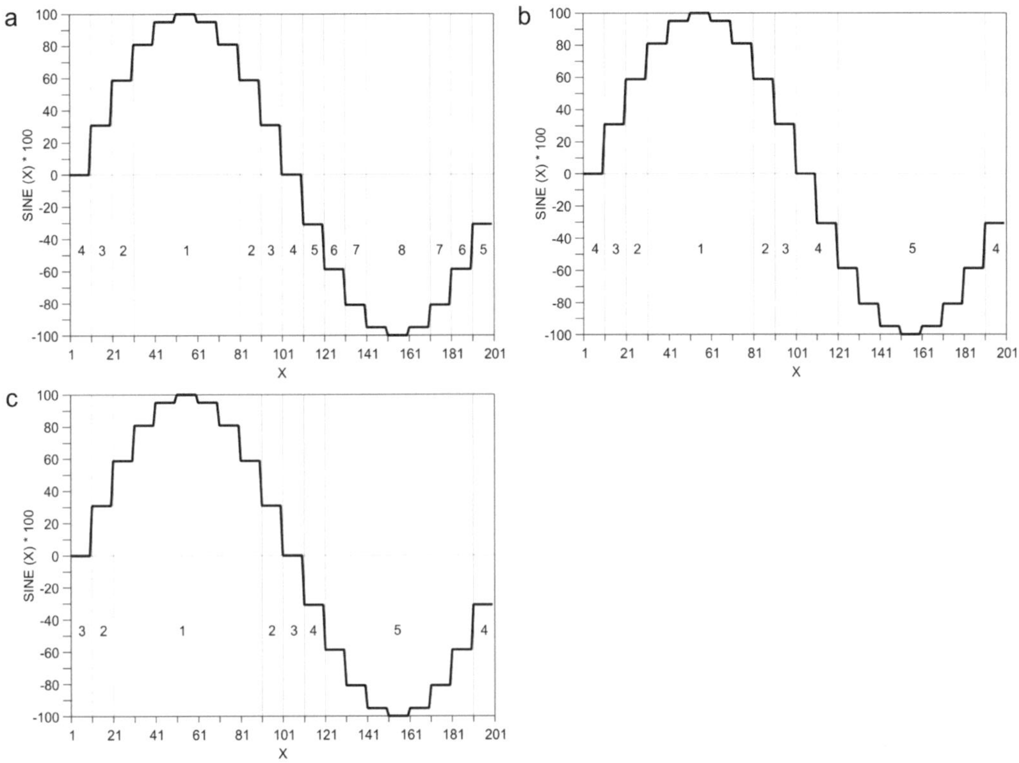

Fig. 4. Theoretical test calculation; random ranked values of the initial partition (a): a defined initial number of clusters $k_0 = 8$; (b): optimal initial number of clusters; the values of the initial partition are ordered from 1-200 in the same course like the sine oscillation, (c): optimal initial number of clusters

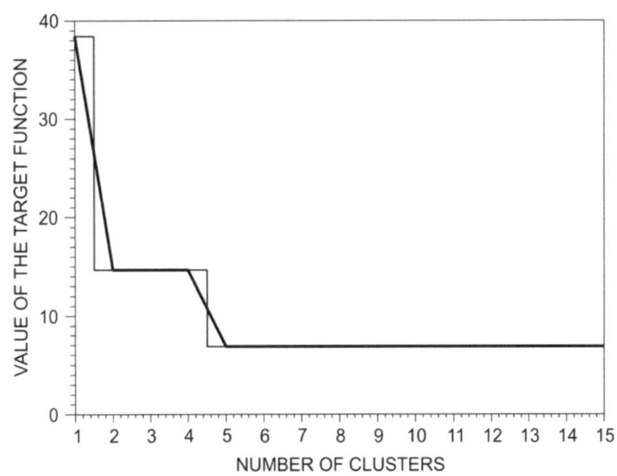

Fig. 5. Result of the Pettitt-test for the estimation of the initial number of clusters (sine oscillation)

The presented results show that the suggested procedure is the first which allows the quality of the separation of clusters to be calculated in a statistically well-founded way; it replaces the often adverse effects of a given number of clusters when employing the non-hierarchical cluster analysis by the application of the optimum number of clusters guaranteeing a statistically reliable separation of all clusters from each other. Additionally for all cluster analysis methods the following conclusions can also be drawn:

- Each method has to guarantee the statistically significant separation of the clusters.
- The ranking of the data within the initial partition must be random.
- A computer program for a cluster analysis has to be built up in such a way that the access to the elements is random.
- For the optimum cluster separation the initial number of clusters is of great importance. It can be calculated using the target function values.
- It is recommendable that existing outliers sort into the cluster with the smallest distance between the outlier parameters and the respective group centroid.

Considering these aspects yields a cluster analysis method which fulfills all the demands of an optimum multivariate climate classification.

4. Analysis

4.1 Climate type characterization

4.1.1 Global

If one calculates a global climate classification for the reference period 1901–1994 with the method as described in 3., and the parameters as provided in 2., this operation produces 30 climate types altogether. These climate types have to be classified in the following (Gerstengarbe & Werner, 2008). For this, the marginal thresholds for every climate type and every parameter will be derived from the data of the respective area. Since it is common in climatology to determine a verbal characterization for the regions in addition to the exact indication of threshold values, the particular parameters are separated into categories - from very cold to very hot for temperature, from very dry to very humid for precipitation, and from very low to very high for the temperature amplitude. For better illustration, the climate types will afterwards be numbered according to the ascending-ordered values of the parameter annual mean temperature.

The thresholds and the verbal characteristics of the single climate types as well as the synopsis of climate types are captured in a table.

In a second step, the climate types for partial periods of 15 years are calculated. This length of partial periods was chosen in order to guarantee for a stable estimation of the employed parameters. To capture the spatial alteration of climate types for the time elapsed, the 15-year window was shifted floating over the whole period 1901–2009 at shifting-time steps of one year. Finally, the current climate changes were determined by calculating the differences between the last shifting period (1995–2009) and the reference period (1901–1994).

If one would cluster each of the partial periods according to the method described, this would result in different group centroids, in addition to different cluster numbers. This will make it nearly impossible to compare a partial period to the whole period. One can elude this by scaling the values of the whole period and allocating the elements of the particular partial period to the group centroids of the reference period, instead of clustering the partial periods. For this, the minimum of the Euclidean distance (equation 4) has to be determined. The results are presented in Chapter 4.2.1.

4.1.2 Continental

In referring to the continents, the global results were augmented, and, thus, the following regions were evaluated: Europe, Asia, Africa, North America, South America, Australia/Oceania. For each of these regions, the areas of the several climate types were determined in a first step, based on the global results. Analogically to the global analysis, the change in area over the last 15-year period was determined and evaluated per climate type in relation to the reference period. So a detailed survey of climate changes per continent could be established. The results are discussed in Chapter 4.2.2. (Preliminary investigations were carried out by Werner et al., 2002; Gerstengarbe & Werner, 2003, 2004.)

4.2 Results

4.2.1 Global

Table 1 contains the thresholds and their verbal characterization for the single climate types. Table 2 comprises the synopsis of climate types. Figure 6a depicts the global distribution of climate types in the reference period. Figure 6b displays the areas where climate types have changed between the last 15-year period and the reference period. Table 3 lists the areas covered by the climate types during the reference period. The dimensions of these areas differ largely, they vary from 556,000 km² (type 23 - monsoon-influenced luffs) to 15,082,000 km² (type 27 - deserts). Moreover, this table contains the changes in area which emerged between the reference period and the last period 1995-2009. Type 24 with 900,400 km², type 27 with 829,500 km², and type 30 with 659,300 km² encountered the biggest gains. These are tropical and subtropical steppe and desert areas.

parameter	very cold VC	cold C	cool c	moderate M	warm W	hot H
annual mean temperature T	< − 10.0	< − 5.0	< 5.0	< 12.0	<20.0	≥ 20.0
Tmax	< 0.0	< 8.0	< 12.0	< 21.0	< 28.0	≥ 28.0
Tmin	< − 25.0	< − 10.0	< 0.0	< 12.0	< 20.0	≥ 20.0
	very small vs	small s	mean m	large l	very large vl	extra large el
amplitude T	< 5.0	< 10.0	< 21.0	< 30.0	< 40.0	≥ 40.0
	very dry VD	dry D	moderately dry MD	moderately humid MH	humid H	very humid VH
annual precipitation sum N	< 200.0	< 400.0	<600.0	< 1000.0	< 2000.0	≥ 2000.0
Nmax	< 30.0	< 60.0	< 100.0	<300.0	<500.0	≥ 500.0
Nmin	< 6.0	< 16.0	< 25.0	< 40.0	< 100.0	≥ 100.0
characteristics	1	2	3	4	5	6

Table 1. Thresholds for the climate type characterization

type	Parameter						
	Tmean	Tmax	Tmin	Ampl	Nyear	Nmax	Nmin
1	VC	VC	VC	l	D	VD	D
2	VC	C	VC	vl	VD	VD	VD
3	VC	c	VC	vl	D	D	D
4	VC	C	VC	vl	D	D	D
5	C	C	C	l	D	D	D
6	C	M	VC	el	D	MD	D
7	c	M	C	vl	MD	MD	MD
8	c	C	C	l	MD	MD	MD
9	c	M	C	vl	MH	MD	MD
10	c	M	C	vl	D	D	D
11	c	M	c	m	H	MH	H
12	M	M	c	m	VH	MH	VH
13	M	M	c	l	MH	MD	MH
14	M	M	M	s	D	D	D
15	M	M	c	l	D	D	D
16	M	M	M	m	MH	MD	H
17	M	W	c	vl	VD	VD	VD
18	W	M	M	s	MH	MH	D
19	W	W	M	l	MH	MH	MD
20	W	W	M	m	H	MH	H
21	W	W	M	m	D	D	D
22	H	W	H	vs	H	MH	H
23	H	H	W	m	VD	VD	VD
24	H	W	H	s	VH	VH	MD
25	H	W	H	s	H	MH	D
26	H	W	H	vs	VH	H	VH
27	H	W	H	s	MH	MH	D
28	H	W	H	vs	VH	H	MH
29	H	W	H	vs	VH	MH	VH
30	H	H	H	s	MD	MD	VD

Table 2. Characterization of the climate types

With 569,000 km², type 13 from the temperate zone gains remarkably, too. According to area, type 22 (subtropics) suffers the biggest loss with 978,200 km². According to percentage, the cold types 2 and 3 with minuses of 27.7 % resp. 15.8 % encountered the biggest losses. This is a significant indicator for global warming. Another sign of the impact of warming appears when one considers the shift directions of climate types between the reference period and the last period, with the annual mean temperature as the point of reference. Only 5,479,360 km² (3.81 % of the whole land area) have changed to colder types but 14,815,086 km² (10.31 %) have changed to warmer ones. So, during a relatively short time, climate types have changed in 14.12 % of the area. Only the continents of Africa and South America (with no or a just little share in cold climate types) have a lesser tendency towards a change to warmer types (see Chapter 4.2.2). This points out to the fact that the rise in temperature is stronger in higher latitudes than in the lower ones.

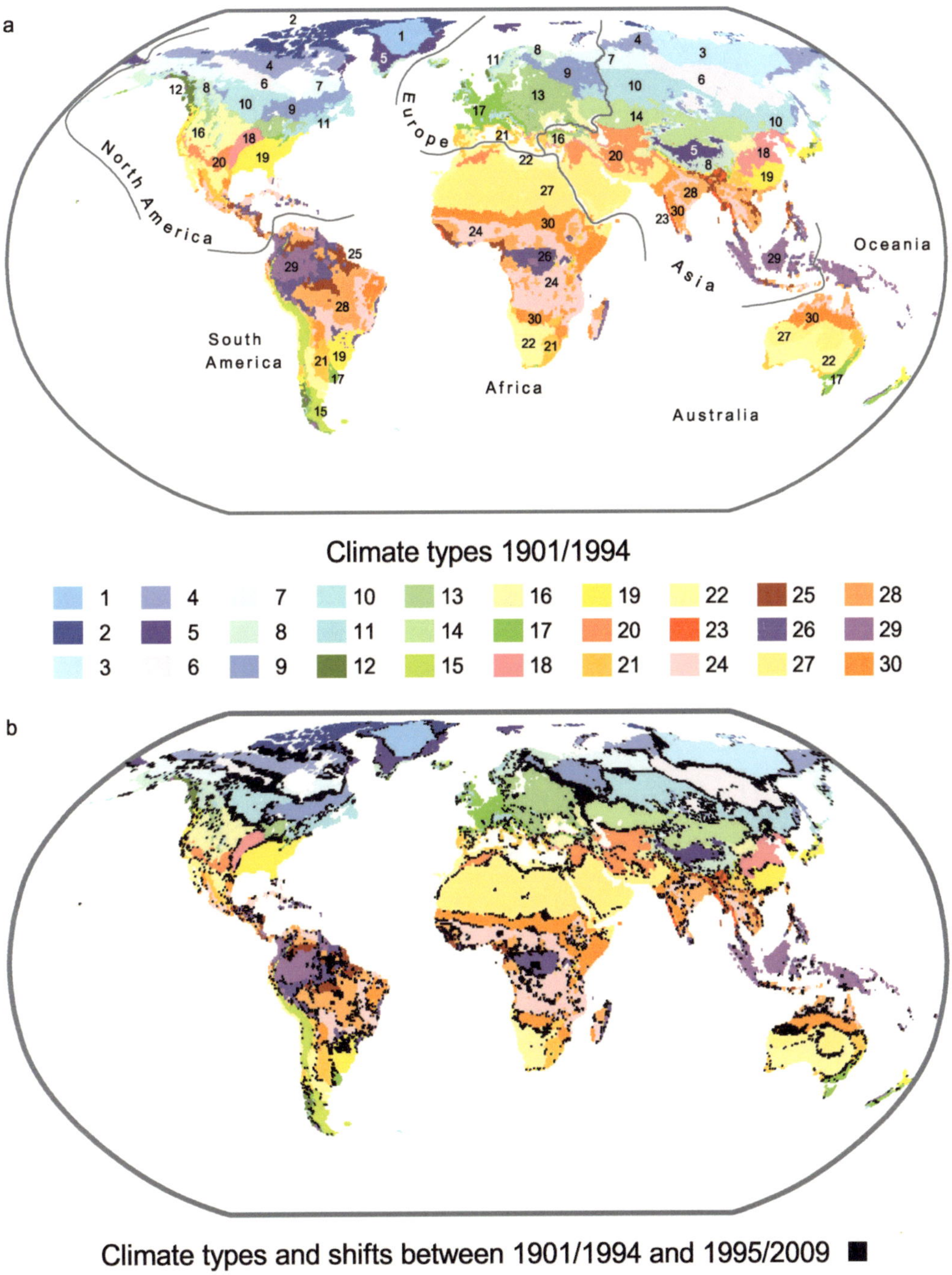

Fig. 6. (a) Global climate type distribution, (b) climate type shifts 1995/2009 vs. 1901/1994

climate type	area * 1901/1994 [10^3 km²]	difference *1995/2009–1901/1994 [10^3 km²]	change * [%]
1	1090	−34.9	-3.2
2	2028	−561.8	-27.7
3	3503	−553.5	−15.8
4	4141	−335.4	-8.1
5	2473	-9.9	−0.4
6	4706	−296.6	-6.3
7	4643	−120.7	−2.6
8	3540	−155.8	-4.4
9	3886	124.4	3.2
10	7161	43.0	0.6
11	2503	−12.5	−0.5
12	799	10.4	1.3
13	4516	569.0	12.6
14	4694	93.9	2.0
15	1867	67.2	3.6
16	4435	323.8	7.3
17	2856	−194.2	−6.8
18	2691	−88.8	−3.3
19	5594	−139.8	−2.5
20	3780	192.8	5.1
21	4908	−88.3	−1.8
22	8084	−978.2	-12.1
23	556	−30.0	−5.4
24	13049	900.4	6.9
25	3461	83.1	2.4
26	5137	−354.4	−6.9
27	15082	829.5	5.5
28	8303	107.9	1.3
29	5568	−44.5	−0.8
30	8675	659.3	7.6

* rounded values

Table 3. Changes of climate type areas 1995/2009 - 1901/1994

The correlation between the time series of global warming and the area of the cold type 3 for 15-year estimation intervals (Figure 7a) is −0.83 and statistically firm with a probability value of 1 %. The hot type 27 (Figure 7b) develops with a correlation of 0.72 that is nearly so high. Further close interrelations emerged for the types 2, 4, 5, 6, 14, 16, 17 23, 27, and 28. All firm correlations are collected in Table 4.

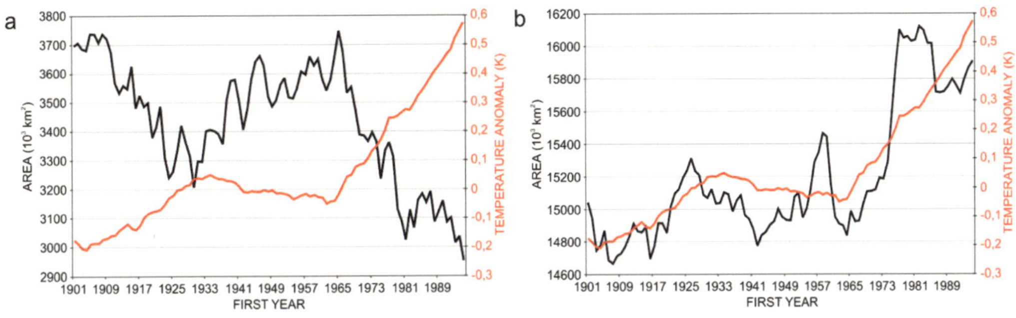

Fig. 7. Area of climate types (black) and anomaly of the global mean temperature 1901–2009 (shifting window 15 years), (a) type 3, (b) type 27

Tendencies did not come out so clear when examining the change to dryer or more humid types. With reference to the annual sum of precipitation, 6.11 % of the area changed to dryer types, and 8.01 % to more humid ones. The gain in area among more humid types, though rather little, is, however, an indicator that, together with warming, precipitation increases globally - with an aggravation in extremes, though, since deserts have grown simultaneously.
(Remark to the shifts cold-warm, dry-humid: The conclusions drawn are only tendencies since one cannot reconstruct in every particular case which of the seven parameters is crucial for the change.)

type	2	3	4	5	6	8	10	13
correlation	−0.43	−0.83	−0.65	0.52	−0.65	−0.37	0.34	0.34
probability value (%)	1	1	1	1	1	5	5	5
type	14	16	17	19	21	23	27	28
correlation	0.46	0.52	−0.49	0.52	−0.37	−0.49	0.72	−0.49
probability value (%)	1	1	1	1	5	1	1	1

Table 4. Significant correlations between the global mean temperature and the area extension of the climate (estimation interval 15 years)

4.2.2 Continental

In the following, the impacts of global warming on the climate types of the several continents will be examined. Therefore, the continents were separated against each other as shown in Figure 8. The analyses were done as described above.

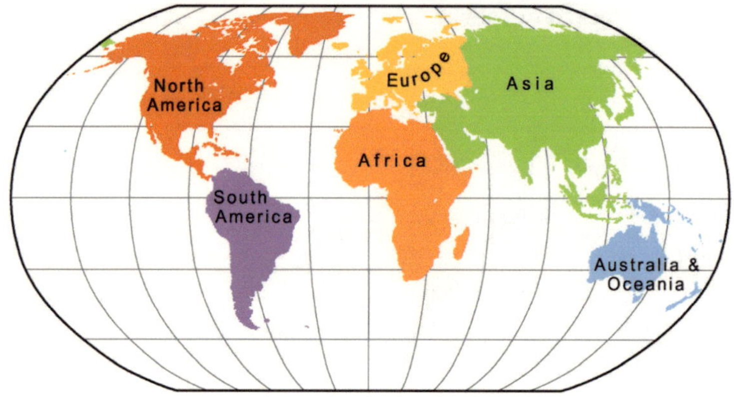

Fig. 8. Continent areas

4.2.2.1 Europe

The distribution of climate types and their shifts in the period 1995/2009 - 1901/1994 is depicted in Figure 9 and all numerical values are noted in Table 5.

type	area (10^3 km²) 1901–1994	area (%) 1901–1994	area (10^3 km²) 1995–2009	change (10^3 km²)	change (%) *	change (%) **
2	22.7	0.20	16.1	−6.6	−0.06	−29.07
4	67.4	0.59	73.8	6.4	+0.06	+9.50
5	140.1	1.23	135.4	−4.7	−0.04	−3.35
7	392.8	3.46	279.9	−112.9	−0.99	−28.74
8	808.6	7.12	670.1	−138.5	−1.22	−17.13
9	1629.4	14.35	1411.7	−217.7	−1.92	−13.36
10	337.2	2.97	61.2	−276.0	−2.43	−81.85
11	655.1	5.77	614.7	−40.4	−0.36	−6.17
12	99.1	0.87	120.9	21.8	+0.19	+22.00
13	3325.0	29.28	3901.3	576.3	+5.08	+17.33
14	618.7	5.45	531.5	−87.2	−0.77	−14.09
15	15.3	0.13	16.6	1.3	+0.01	+8.50
16	650.3	5.73	625.0	−25.3	−0.22	−3.89
17	1623.0	14.29	1696.2	73.2	+0.64	+4.51
18	56.3	0.50	117.8	61.5	+0.54	+109.24
19	105.9	0.93	156.0	50.1	+0.44	+47.31
20	69.8	0.06	47.5	−22.3	−0.20	−31.95
21	597.5	5.26	599.9	2.4	+0.02	+0.40
22	204.3	1.80	279.7	75.4	+0.66	+36.91

*change 1995/2009 – 1901/1994 in relation to the whole area of the continent
**change 1995/2009 – 1901/1994 in relation to the area of the type 1901/1994

Table 5. Climate type extension and change, Europe

The extension of the continental-moderate climate type 13 towards the east can clearly be seen. This type enlarges its whole area about 576,300 km² which makes a gain of 17.33 % resp. an area which is 25,000 km² larger than France. On the other hand, the climate types 7 to 10 ranging from cold to moderately cold suffer from area losses which sum up to approximately 745,000 km². The largest alteration in percentage with 109.24 % can be found in the warm-moderate climate type 18. However, this is "only" an extension of 61,500 km² since the area of this climate type was "only" 56,300 km² during the reference period.

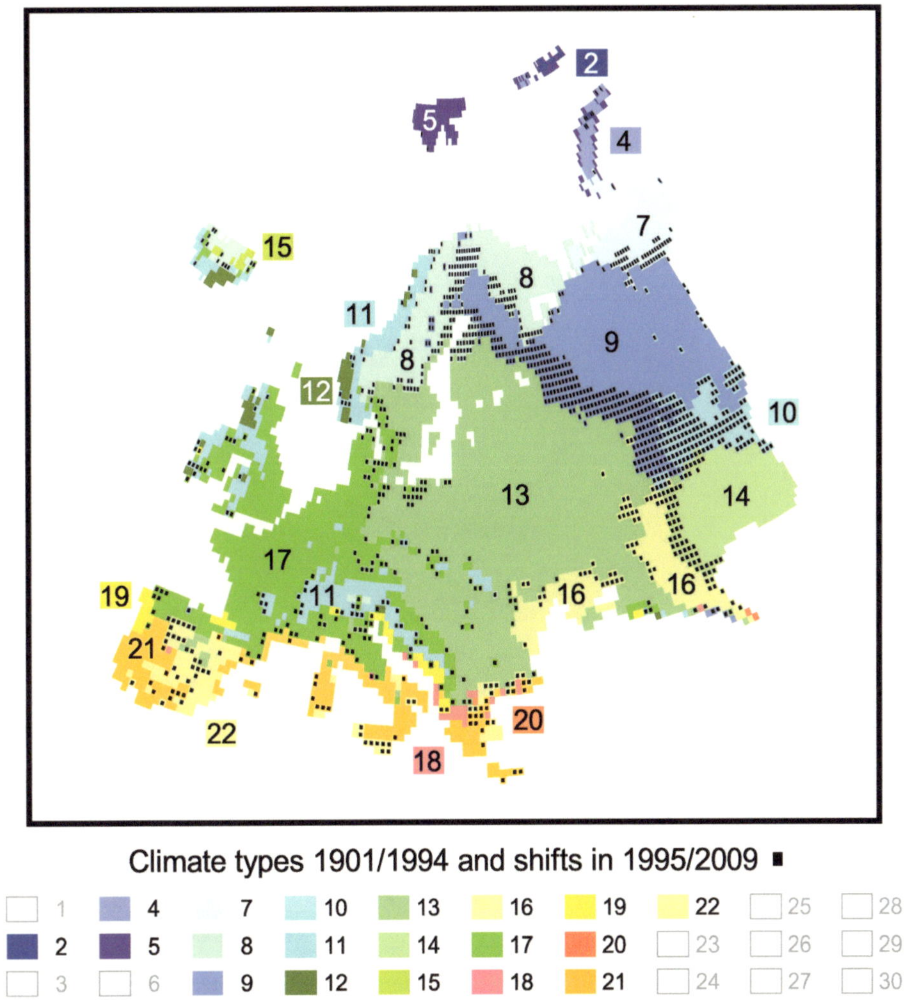

Fig. 9. Climate type distribution and shifts (black squares), Europe, 1995/2009 – 1901/1994

4.2.2.2 Asia

On the Asian continent, all climate types apart from type 1 are affected by area shifts. The widest changes can again be found in the more northern climate types: type 3 with a loss of 551,700 km², and type 4 with a loss of 274,900 km². But also climate type 18 (warm-moderate) with a minus of 187.100 km², the warm-dry type 22 with a minus of 153,900 km², and type 26 (hot-humid) with a minus of 160,700 km² are affected by considerable area losses. Type 27 (hot-very dry; desert) shows the biggest gain with 395,500 km², followed by type 9 (moderate-cold, moderately humid) with 290,700 km² and type 14 (from moderately warm to cool and very dry) with 271,100 km². The relatively greatest changes can be found in type 15 (moderate-cool, dry) with 109.32 %. Nevertheless, the default value of the area is very low in this case. All data and changes can be found in Table 6 and Figure 10.

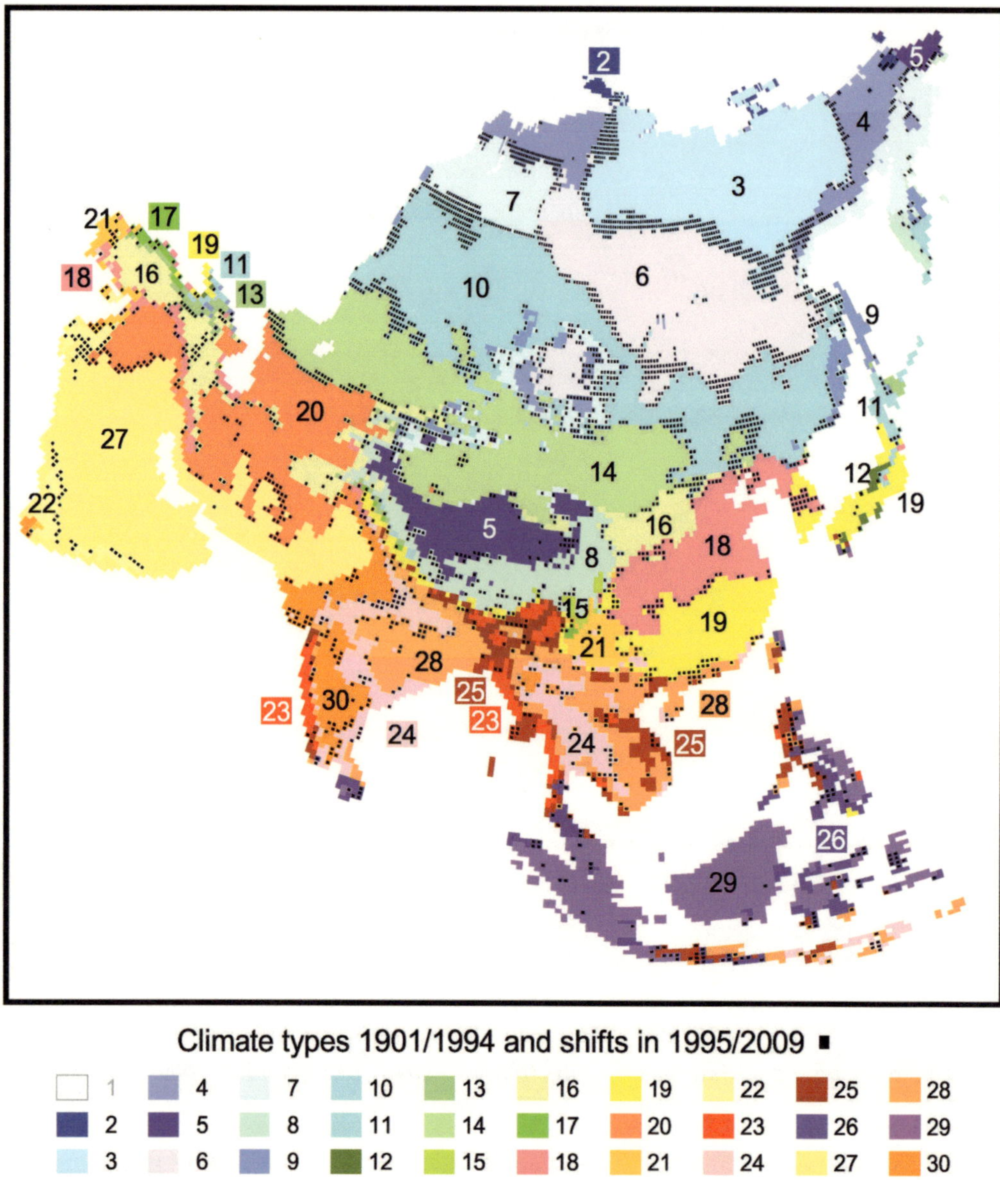

Fig. 10. Climate type distribution and shifts (black squares), Asia, 1995/2009 – 1901/1994

type	area (10^3 km^2) 1901–1994	area (%) 1901–1994	area (10^3 km^2) 1995–2009	change (10^3 km^2)	change (%) *	change (%) **
2	135.0	0.29	115.2	−19.8	−0.04	−14.67
3	3503.0	7.56	2951.3	−551.7	−1.19	−15.75
4	1685.6	3.64	1410.7	−274.9	−0.59	−16.31
5	1382.9	2.98	1311.7	−71.2	−0.15	−5.15
6	3775.3	8.15	3817.0	41.7	+0.09	+1.10
7	2200.1	4.75	2082.0	−118.1	−0.25	−5.37
8	1409.9	3.04	1395.3	−14.6	−0.03	−1.04
9	489.0	1.06	779.7	290.7	+0.62	+59.45
10	5078.7	10.96	5051.8	−26.9	−0.06	−0.53
11	313.3	0.68	278.3	−35.0	−0.08	−11.17
12	77.6	0.17	77.1	−0.5	negligible	−0.64
13	231.0	0.50	254.1	23.1	+0.05	+10.00
14	3967.6	8.56	4238.7	271.1	+0.58	+6.83
15	59.0	0.13	123.5	64.5	+0.14	+109.32
16	1617.7	3.49	1753.2	135.5	+0.29	+8.38
17	97.0	0.21	72.7	−24.3	−0.05	−25.05
18	1857.5	4.01	1670.4	−187.1	−0.40	−10.07
19	1755.0	3.79	1757.0	2.0	negligible	+0.11
20	2722.6	5.88	2863.1	140.5	+0.30	+5.16
21	466.8	1.01	458.5	−8.3	−0.02	−1.78
22	434.1	0.94	280.2	−153.9	−0.33	−35.45
23	500.8	1.08	494.9	−5.9	−0.01	−1.18
24	1567.9	3.38	1698.6	130.7	+0.28	+8.34
25	1105.0	2.38	1127.4	22.4	+0.05	+2.03
26	776.2	1.68	615.5	−160.7	−0.35	−20.70
27	4025.6	8.69	4421.1	395.5	+0.85	+9.82
28	2138.2	4.61	2193.0	54.8	+0.12	+2.56
29	1975.1	4.26	2110.0	134.9	+0.29	+6.83
30	986.6	2.13	932.3	−63.3	−0.12	−5.50

*change 1995/2009 – 1901/1994 in relation to the whole area of the continent
**change 1995/2009 – 1901/1994 in relation to the area of the type 1901/1994

Table 6. Climate type extension and change, Asia

4.2.2.3 Africa

The climate types in Africa are characterized by two significant features. On the one hand, an extreme reduction of the Savanna has to be observed (climate type 22 with a minus of 662,800 km^2 = 21.21 % of the whole area). On the other hand, the desert enlarges to the same extent (climate type 27 with a plus of 689,500 km^2). This trend towards a shift from "humid" to "dry" can be noticed with the other climate types which are characteristic of Africa. So the predominantly humid climate types 21, 25, 26, 28, and 29 loose an area of 671,400 km^2 altogether. It is remarkable here that the loss in area in climate type 29 is 77.66 %. In contrast to this development, the more dry types 24 and 30 show an increase in area of 698,400 km^2.

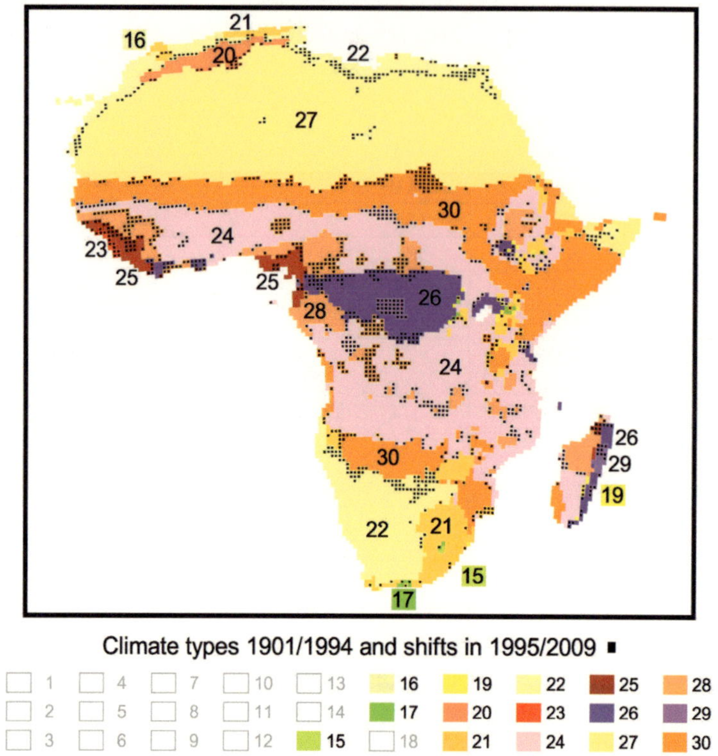

Fig. 11. Climate type distribution and shifts (black squares), Africa, 1995/2009 – 1901/1994

type	area (10^3 km^2) 1901-1994	area (%) 1901-1994	area (10^3 km^2) 1995-2009	change (10^3 km^2)	change (%) *	change (%) **
15	11.0	0.04	8.0	−3.0	−0.01	−27.27
16	15.5	0.05	7.9	−7.6	−0.02	−40.03
17	45.4	0.15	40.2	−5.2	−0.02	−11.45
19	86.6	0.28	75.0	−11.6	−0.04	−13.39
20	372.7	1.20	389.2	16.5	+0.05	+4.43
21	1388.7	4.49	1245.0	−143.7	−0.46	−10.35
22	3124.4	10.11	2461.6	−662.8	−2.14	−21.21
23	48.8	0.16	6.1	−42.7	−0.14	−87.50
24	7661.3	24.78	8001.0	339.7	+1.10	+4.43
25	483.2	1.56	363.9	−119.3	−0.38	−24.69
26	1651.5	5.34	1483.0	−168.5	−0.54	−10.20
27	8751.4	28.31	9440.9	689.5	+2.23	+7.88
28	2001.1	6.47	1895.0	−106.1	−0.34	−5.30
29	172.3	0.56	38.5	−133.8	−0.43	−77.66
30	5098.3	16.49	5457.0	358.7	+1.16	+7.04

*change 1995/2009 – 1901/1994 in relation to the whole area of the continent
**change 1995/2009 – 1901/1994 in relation to the area of the type 1901/1994

Table 7. Climate type extension and change, Africa

Since these two climate types already covered large areas in Africa between 1901 and 1994, the relative increase is "only" 4.43 % resp. 7.04 %. Figure 11 and Table 7 give an overview of the shifts. In summary, one can state that the climatic situation in Africa has remarkably changed for the worse during the period 1995/2009.

4.2.2.4 Australia and Oceania

If one considers the climate development in Australia and Oceania, the changes are little in Papua New Guinea and New Zealand. Nevertheless, the situation in Australia is different. Here we find distinct losses of area within the types 17 (−109,400 km^2), 22 (−151,500 km^2), and 27 (−310,500 km^2). There are gains for the types 24 (+165,200 km^2) and 30 (+272,300 km^2). Type 27 (desert) looses area to the types 22 and 30, and type 30, in turn, to type 24. The great relative changes in the climate types 15 (+206.15 %), 23 (320.69 %), and 25 (242.86 %) are of only little influence since their particular shares in the whole area are only very little during the period 1901/1994. Overall one can say that there is a tendency towards a change from the very hot and dry climate types in Australia to the less dry types. All changes are summarized in Figure 12 and Table 8.

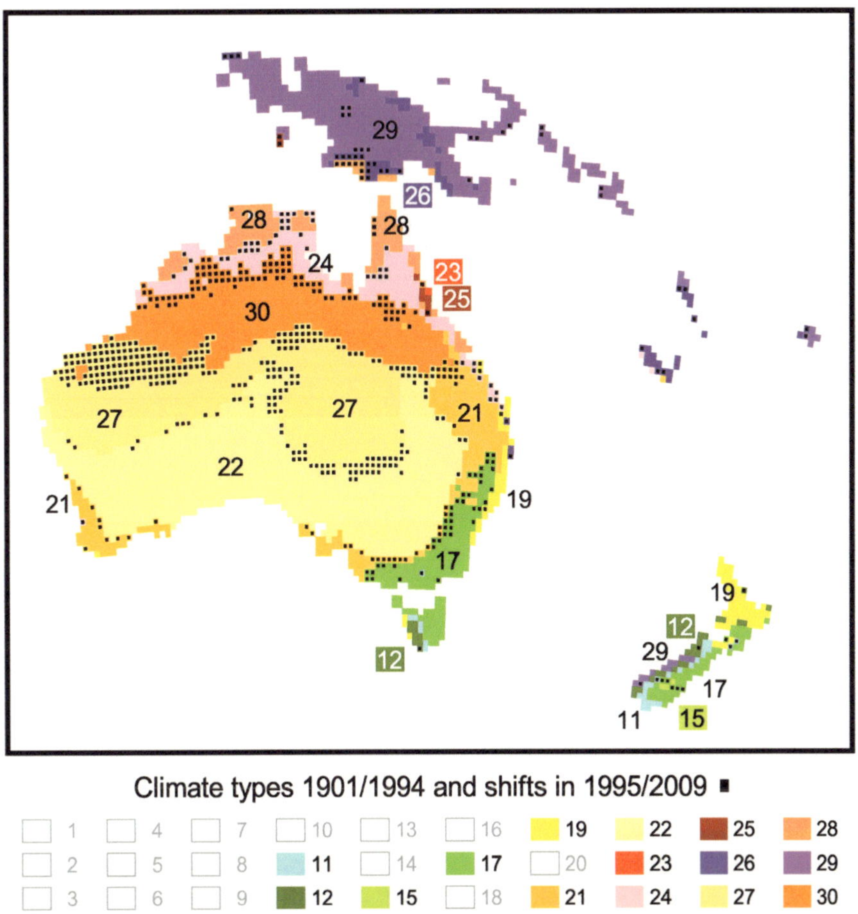

Fig. 12. Climate type distribution and shifts (black squares), Australia and Oceania, 1995/2009 – 1901/1994

type	area (10^3 km^2) 1901–1994	area (%) 1901–1994	area (10^3 km^2) 1995–2009	change (10^3 km^2)	change (%) *	change (%) **
11	66.9	0.66	59.9	−7.0	−0.07	−10.46
12	73.2	0.72	68.3	−4.9	−0.05	−6.69
15	6.5	0.06	19.9	13.4	+0.13	+206.15
17	564.0	5.59	454.6	−109.4	−1.08	−19.40
19	216.5	2.15	225.8	9.3	+0.09	+4.30
21	665.4	6.60	702.8	37.4	+0.37	+5.62
22	2676.5	26.54	2525.0	−151.5	−1.50	−5.66
23	2.9	0.03	12.2	9.3	+0.09	+320.69
24	571.2	5.66	736.4	165.2	+1.64	+28.92
25	23.8	0.24	81.6	57.8	+0.57	+242.86
26	216.1	2.14	180.9	−35.2	−0.35	−16.29
27	2078.1	20.61	1767.6	−310.5	−3.08	−14.94
28	387.0	3.84	464.6	77.6	+0.77	+20.05
29	1161.9	11.52	1138.1	−23.8	−0.24	−2.05
30	1374.7	13.63	1647.0	272.3	+2.70	+19.81

*change 1995/2009 – 1901/1994 in relation to the whole area of the continent
**change 1995/2009 – 1901/1994 in relation to the area of the type 1901/1994

Table 8. Climate type extension and change, Australia and Oceania

4.2.2.5 North America, Central America and Greenland

Except for climate type 3, all other types can be found in this region. The two cold or very cold and, in addition, dry climate types 2 (−535,700 km^2) and 6 (−340,000 km^2) show significant losses in area, with climate type 2 loosing substantial shares to type 4. The latter, nevertheless, increases only about 69,400 km^2 which is due to the fact that, in turn, type 6 shifts to the north. Concurrently, the neighboring climate type 10 spreads to the same direction (+345,800 km^2) so that type 6 shrinks at the already mentioned value, which brings about a loss of area of 36.51 %. This development continues into the Midwest of the USA which is basically represented by climate type 16 (moderately warm, dry) and expands to the north for 222,600 km^2. Climate type 19 (warm, humid) behaves in a similar way, and it also spreads to the north for 157,200 km^2. Concerning the changes in Central America, only type 28 is of importance with a plus of 117,600 km^2. This equates to an increase in area of 30.88 %. Figure 13 and Table 9 summarize all changes.

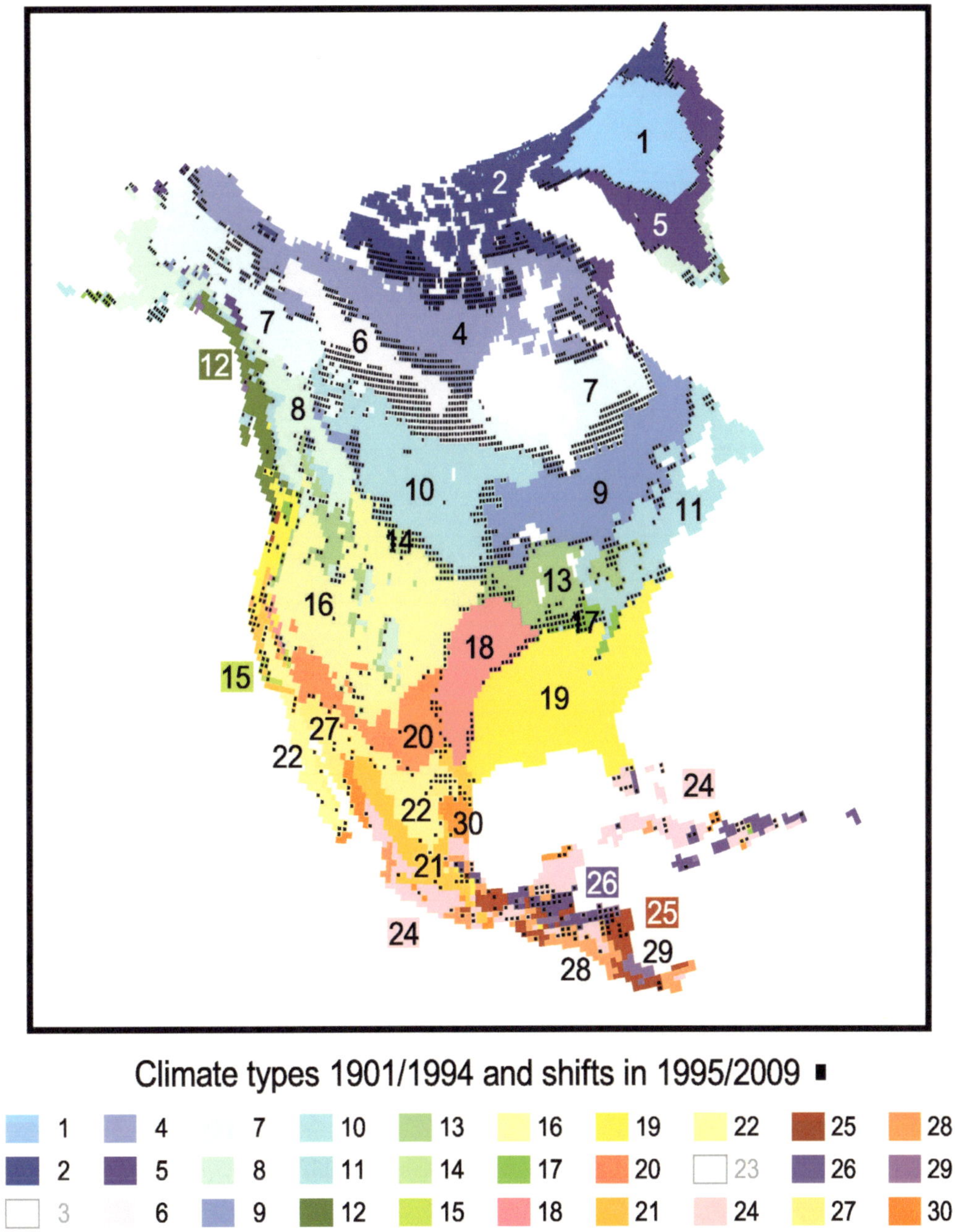

Fig. 13. Climate type distribution and shifts (black squares), North America, 1995/2009 – 1901/1994

type	area (10^3 km^2) 1901–1994	area (%) 1901–1994	area (10^3 km^2) 1995–2009	change (10^3 km^2)	change (%) *	change (%) **
1	1089.9	4.11	1055.4	−34.5	−0.13	−3.16
2	1870.3	7.05	1334.6	−535.7	−2.02	−28.64
4	2388.2	9.01	2321.5	−66.7	−0.25	−2.79
5	947.2	3.57	1016.4	69.2	+0.26	+7.30
6	931.3	3.51	591.3	−340.0	−1.28	−36.51
7	2050.0	7.73	2158.6	108.6	+0.41	+5.30
8	1311.7	4.95	1306.7	−4.1	−0.02	−0.38
9	1768.1	6.67	1818.7	50.6	+0.19	+2.86
10	1744.9	6.58	2090.7	345.8	+1.30	+19.82
11	1398.3	5.27	1471.0	72.7	+0.27	+5.20
12	321.3	1.21	352.6	31.3	+0.12	+9.74
13	960.6	3.62	930.9	−29.7	−0.11	−3.09
14	107.9	0.41	16.8	−91.1	−0.34	−84.43
15	37.6	0.14	20.9	−16.7	−0.06	−44.41
16	2147.1	8.10	2369.7	222.6	+0.84	+10.37
17	153.1	0.58	125.6	−27.5	−0.10	−17.96
18	777.0	2.93	814.8	37.8	+0.14	+4.86
19	2091.8	7.89	2249.0	157.2	+0.59	+7.52
20	677.4	2.56	674.1	−3.3	−0.01	−0.49
21	628.9	2.37	596.4	−32.5	−0.12	−5.17
22	777.2	2.93	719.1	−58.1	−0.22	−7.48
23	0.0	0.00	2.9	2.9	+0.01	new
24	853.5	3.22	818.3	−35.2	−0.13	−4.12
25	305.2	1.15	318.0	12.8	+0.05	+4.19
26	313.5	1.18	269.6	−43.9	−0.16	−14.00
27	226.6	0.85	279.9	53.3	+0.20	+23.52
28	308.8	1.16	426.4	117.6	+0.44	+30.88
29	95.3	0.36	77.7	−17.6	−0.07	−18.47
30	230.3	0.87	285.4	55.1	+0.21	+23.92

*change 1995/2009 – 1901/1994 in relation to the whole area of the continent
**change 1995/2009 – 1901/1994 in relation to the area of the type 1901/1994

Table 9. Climate type extension and change, North America

type	area (10^3 km^2) 1901-1994	area (%) 1901-1994	area (10^3 km^2) 1995-2009	change (10^3 km^2)	change (%) *	change (%) **
5	2.6	0.01	0.0	−2.6	−0.01	not to apply
8	9.8	0.05	10.2	0.4	negligible	+4.08
11	69.8	0.38	67.4	−2.4	−0.01	−3.44
12	228.2	1.23	191.0	−37.2	−0.20	−16.30
15	1737.2	9.38	1745.6	8.4	+0.04	+0.48
16	4.8	0.03	5.0	0.2	negligible	+4.17
17	373.8	2.02	271.4	−102.4	−0.55	−27.39
19	1338.1	7.22	990.1	−348.0	−1.88	−26.01
21	1160.7	6.26	1216.4	55.7	+0.30	+4.80
22	867.2	4.68	842.1	−25.1	−0.14	−2.89
23	3.1	0.02	9.2	6.1	+0.03	+196.77
24	2395.4	12.93	2692.5	297.1	+1.60	+12.40
25	1543.8	8.33	1653.7	109.9	+0.59	+7.12
26	2179.8	11.76	2233.3	53.5	+0.29	+2.45
28	3467.8	18.71	3429.3	−38.5	−0.21	−1.11
29	2162.9	11.67	2159.4	−3.5	−0.02	−0.16
30	985.1	5.32	1013.3	28.2	+0.15	+2.86

*change 1995/2009 – 1901/1994 in relation to the whole area of the continent
**change 1995/2009 – 1901/1994 in relation to the area of the type 1901/1994

Table 10. Climate type extension and change, South America

4.2.2.6 South America

In South America, there are two climate types with relatively big area losses; on the one hand, climate type 17 (moderately warm, moderately humid with a minus of 102,400 km^2), on the other hand, climate type 19 (warm, humid, with a minus of 348,000 km^2).

In essence, type 17 looses area to the types 19 and 21. The losses of climate type 19 are caused by an extension of the neighboring types 24 and 26 into this region. Climate types 24 and 25 are the types with the biggest gain in area (type 24 with a plus of 297,100 km^2, and type 25 with a plus of 109,900 km^2). Overall, the observed shifts on the South American continent are moderate, in comparison to the changes on other continents. Figure 14 and Table 10 give a detailed survey of the changes.

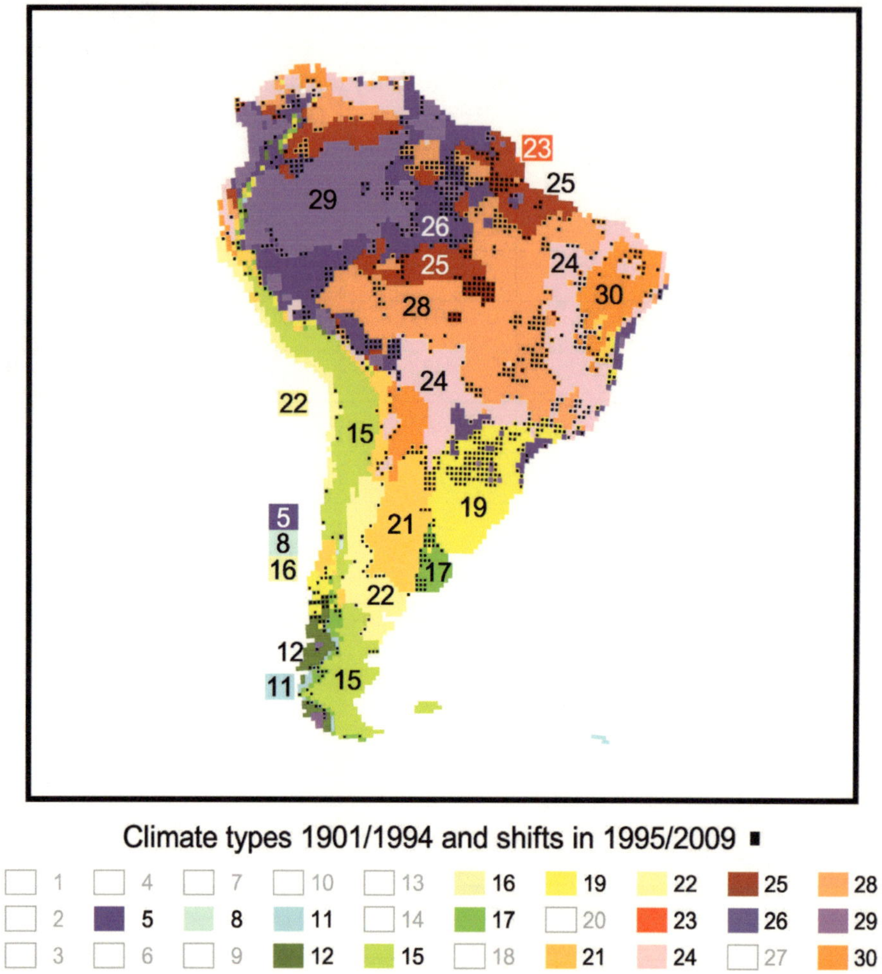

Fig. 14. Climate type distribution and shifts (black squares), South America, 1995/2009 – 1901/1994

5. Conclusions

It could be shown that the new developed cluster analysis algorithm leads to objective classifications. On this basis it was possible to define a new climate type classification of the earth. Since the beginning of the last century and lasting until today, an increase of the global annual mean temperature of approximately 0.8 K has been observed. This change can be classified as moderate in comparison to the temperature deviations which occurred in the history of the earth. However, this moderate rise in temperature has already remarkable impacts on climate as a complex factor. Because of this fact and the use of the new developed methodology an investigation was carried out to estimate the shifts of single climate types during the observation period.

In principle, one could expect that the "cooler" climate types would give off area to the "warmer" types, in case of an increase in temperature. This could be proved. It is remarkable that the ice and tundra climates have disproportionately lost area when compared to other climate types (Europe - type 7; Asia - types 3, 4, 7; North America - types 2, 6). At the same time, the dry climates in Asia and Africa have significantly expanded (Asia - types 14, 16, 20, 27; Africa - type 27). However, there are considerable gains in the more humid climate types (Europe - type 13; Asia - types 9, 24, 29; Africa - type 24; Australia/Oceania - type 24; North and Central America - types 19, 28; South America - types 24, 25). So, by trend, the development expected by climatologists (e.g., Flohn, 1988) that humid regions will get still more humid and dry ones will still get dryer proves true. It is also remarkable that the percentaged change in area of a climate type is greatest in Europe, in relation to the whole area of the particular continent (type 13 +5.08 %, type 10 −2.43 %). For Asia, all values of change lie below one percent due to the dimension of the continent. In Africa, the maximum gains and losses keep the balance with +2.23 (type 27) und −2.14 % (type 22). For Australia/Oceania, the decrease in type 27 of 3.08 % and the increase in type 30 of 2.70 % are of especial importance. There are only 4 climate types on the whole American continent with changes of above one percent (type 6 = −1.28 %, type 10 = +1.30 %, type 19 = −1.88 %, type 24 = +1.60 %). If one considers the percentaged changes in relation to the whole area of the climate type, the margins of deviation are particularly wide. Their values lie between 0.11 % (Asia, type 19) and 320.69 % (Australia/Oceania, type 23). Altogether, one can state that the shifts in climate regions which have taken place in the investigation period during the last 15 years are a factor that can no longer be neglected for both humankind and environment. This holds true in particular because of the high velocity of the changes observed, and also because of the dimension of these changes in relation to the comparatively short period.

6. References

Flohn, H. (1988). *Das Problem der Klimaänderungen in Vergangenheit und Zukunft*, Wiss. Buchges., Darmstadt

Forgy, E. W. (1965). Cluster Analysis of Multivariate Data: Efficiency versus Interpretability of Classifications (abstract). *Biometrics*, Vol. 21, pp. 768–769

Fovell, R. G., Fovell, M. C. (1993). Climate Zones of the Conterminous United States Defined Using Cluster Analysis. *Journal of Climate*, Vol. 6, pp. 2103–2135

Fraedrich, K., Gerstengarbe, F.-W., & Werner, P. C. (2001). Climate Shifts During the Last Century. *Climatic Change*, Vol. 50, pp. 405–417

Gerstengarbe, F.-W., Werner, P. C. (1997). A Method to Estimate the Statistical Confidence of Cluster Separation. *Theoretical and Applied Climatology*, Vol. 57, No. 1-2, pp. 103–110

Gerstengarbe, F.-W., Werner, P. C., & Fraedrich, K. (1999). Applying Non-Hierarchical Cluster Analysis Algorithms to Climate Classification: Some Problems and their Solution. *Theoretical and Applied Climatology*, Vol. 64, pp. 143–150

Gerstengarbe, F.-W., Werner, P. C. (2003). Klimaänderungen zwischen 1901 und 2000. In: *Nationalatlas Bundesrepublik Deutschland*, Spektrum Akademischer Verlag, Heidelberg, Berlin

Gerstengarbe, F.-W., Werner, P. C. (2004). Klimaentwicklung im südlichen Afrika im 20. Jahrhundert. *Geographische Rundschau*, Vol. 56, No. 1, pp. 18–24

Gerstengarbe, F.-W., Werner, P. C. (2008). Climate development in the last century - Global and regional. *International Journal of Medical Microbiology*, Vol. 298, No. S1, pp. 5–11

Köppen, W. (1931). *Grundriss der Klimakunde*, Walter de Gruyter & Co., Berlin

Mann, H. B., Whitney, D. R. (1947). On a test of whether one of two random variables is stochastically larger than other. *Annals of Math. Statistics*, Vol. 18, No. 1, pp. 50–60

Mitchell, T. D., Jones, P. D. (2005). An improved method of constructing a database of monthly climate observations and associated high-resolution grids. *International Journal Climatology*, Vol. 25, pp. 693–712

Müller, P. H., Neumann, P., & Storm, R. (1973). *Tafeln der mathematischen Statistik*, VEB Fachbuchverlag, Leipzig

Österle, H., Gerstengarbe, F.-W., & Werner, P. C. (2003). Homogenisierung und Aktualisierung des Klimadatensatzes der Climate Research Unit der Universität of East Anglia, Norwich. *Terra Nostra*, Vol. 6, pp. 326–329

Peterson T. C., Easterling, D. R., Karl, T. R., Groisman, P., Nicholls, N., Plummer, N., Torok, S., Auer, I., Boehm, R., Gullet, D., Vincent, L., Heino, R., Tuomenvirta, H., Mestre, O., Szentimrey, T., Salingen, J., Førland, E. J., Hanssen-Bauer, I., Alexandersson, H., Jones, P., & Parker, D. (1998). Homogeneity adjustments of in situ atmospheric climate data: a review. *International Journal of Climatology*, Vol. 18, pp. 1493–1517

Pettitt, A. N. (1979). A Non-parametric Approach to the Change-point Problem. *Applied Statistics*, Vol. 28, pp. 126–135

Rudolf, B., Schneider, U. (2005). Calculation of Gridded Precipitation Data for the Global Land-Surface using in-situ Gauge Observations. *Proceedings of the 2nd Workshop of the International Precipitation Working Group IPWG*, Monterey, October 2004

Rudolf, B., Becker, A., Schneider, U., Meyer-Christoffer, A., & Ziese, M. (2010). The new "GPCC Full Data Reanalysis Version 5" providing high-quality gridded monthly precipitation data for the global land-surface is public available since December 2010, In: *GPCC Status Report December 2010*, Available from: <gpcc.dwd.de>

Steinhausen, D., Langer, K. (1977). *Clusteranalyse – Einführung in Methoden und Verfahren der automatischen Klassifikation*, Walter de Gruyter, Berlin

Taubenheim, J. (1969). *Statistische Auswertung geophysikalischer und meteorologischer Daten*, Akad. Verlagsges, Geest & Portig, Leipzig

Werner, P. C., Gerstengarbe, F.-W., & Österle, H. (2002). Klimatypänderungen in Deutschland im 20. Jahrhundert. In: *Klimastatusbericht 2001*, DWD, Offenbach a. M., pp. 185–194

3

Environmental Stability for Convective Precipitation Under Global Warming

Tetsuya Takemi, Syohei Nomura and Yuichiro Oku
Disaster Prevention Research Institute
Kyoto University
Japan

1. Introduction

Severe convective weather induces various types of local-scale meteorological disasters, e.g., torrential rain, flush flooding, tornadoes, and damaging winds, which have significant impacts on our societal environments. Such convective weather is caused mostly by atmospheric deep convection that accompanies cumulonimbus clouds. With the explosive development and extension of densely populated, metropolitan areas worldwide, vulnerability of such areas to extreme convective weather is rapidly increasing. Therefore, accurate short-term forecasting and reliable long-term projections of extreme convective weather are critically important for the prevention and mitigation of induced disasters.

As far as precipitation is concerned, cumulonimbus clouds primarily spawn extreme precipitation events. In a future climate under global warming, it is anticipated that convective weather including extreme precipitation is more prevalent and more intensified in various regions of the world. Previous studies indicated that extreme events producing heavy rainfall are projected to increase in a global warming climate (Kimoto et al., 2005; Kamiguchi et al., 2006). In the Fourth Assessment Report (AR4) from the Intergovernmental Panel on Climate Change (IPCC) (2007), it is summarized that under future global warming the intensity of precipitation events is projected to increase particularly in tropical and high latitude areas. Furthermore, the IPCC report says that in the warming climate the precipitation extremes are more enhanced than the precipitation means in most tropical and mid-/high-latitude areas. However, such future projections on precipitation characteristics are based on the numerical experiments by general circulation models (GCMs) that have horizontal resolutions of much coarser than the scales of individual cumulonimbus clouds, which means that the convective precipitation is not explicitly resolved but only represented by a parameterized way. Characteristics of extreme precipitation events due to deep convection under global warming, therefore, require further investigations in order to gain deeper understandings of convective precipitation under global warming.

Even if we are not concerned with the long-term changes of convective precipitation but the short-term forecasts, we are still far from accurately predicting the times and locations of the occurrence of convective precipitation. Such difficulty in predicting the behaviour of cumulonimbus clouds not only in climatic timescales but also in daily-weather timescales is due to the chaotic nature and randomness of convection development. In this sense, quantitative evaluations of convective precipitation become harder if the area of interest is smaller. On the other hand, we need regional-scale evaluations of convective precipitation not only in weather forecasting but also in climate projections. At this time we are facing critical challenges in dealing with the quantitative evaluations of convective precipitation at regional scales not only in short timescales but also in long-range climate timescales. In contrast, the horizontal scales of the environments for the development of deep convective clouds are in general an order of magnitude larger than the scales of individual deep clouds. The environmental conditions for convective weather have been widely investigated in the literature (Bluestein, 1993; Emanuel, 1994; Cotton et al., 2011) and are relatively tractable owing to their longer timescales and larger spatial-scales than convective clouds themselves. Therefore, in this study we deal with the environmental conditions for deep convection rather than consider deep convection itself.

Since deep convection develops under unstably stratified conditions, the vertical stability of the stratified atmosphere plays a critical role in determining the development of deep convection. Therefore, in order to understand how and to what degree cumulonimbus clouds develop, it is important to obtain the information on the stability of the atmosphere that is the environment for such deep convection. The environmental conditions for the occurrence of cumulonimbus clouds have been investigated by a large number of studies in the atmospheric science community, which can be found in standard meteorological textbooks (e.g., Bluestein, 1993; Emanuel, 1994; Cotton et al., 2011), and it is widely understood that the environmental conditions strongly control the development, structure, and organization of deep convective clouds and hence the characteristics of associating precipitation (see the review of Houze, 2004 and the references therein). As mentioned above, the future changes of the characteristics of convective precipitation, therefore, should be investigated not only in terms of the precipitation characteristics themselves but also in terms of the environmental stability conditions for deep convective clouds.

Of course, the activity of deep convective clouds is not only regulated by the vertical environmental stability but also by larger-scale weather disturbances such as tropical and extra-tropical cyclones, fronts, and planetary waves. Generally speaking, the control of large-scale weather disturbances in determining the development and organization of deep convective clouds is considered to be much stronger than that without such large-scale effects. Therefore, the precipitation characteristics under global warming have been in most cases investigated in the context of synoptic-scale and/or planetary-scale controls. For example in Japan, the features of monsoon-related rainfall (i.e., Baiu-period rainfall) in future warming climates have been extensively investigated (e.g., Yoshizaki et al. 2005; Kusunoki et al. 2006; Ninomiya 2009; Kusunoki et al. 2011). In other words, the precipitation characteristics significantly depend on the representations of such background large-scale disturbances for which legitimate validation for their representations in climate models is not well founded. Considering these uncertainties, it should not be ruled out to investigate the environmental stability conditions for deep convection without the presence of major large-scale disturbances. It is clear that there are also cases in which deep convective clouds develop under synoptically undisturbed conditions. In these cases, the intensity and organization of deep convective clouds depend on the degree of a convectively unstable state. Del Genio et al. (2007) considered how the intensity of deep convection changes under global warming by examining updraft velocity in response to temperature lapse rate estimated from a simple buoyant energy formulation. Their theoretical approach is based on the idea that the intensity of deep convection depends on the vertical stability that would be determined by synoptic-scale meteorological settings with both disturbed and undisturbed natures. Although external forcings such as synoptic-scale disturbances will play a critical role in determining the intensity of deep convection, the consideration based on a simple theoretical/analytical/experimental framework like the one done in Del Genio et al. (2007) will give theoretical foundations and basic understandings on the changes in deep convection and associating precipitation in a future warmer climate. Furthermore, the changes of vertical stability in the tropical atmosphere were also focused by Santer et al. (2008) in order to understand the modelled temperature trends in GCMs. In this sense, we consider that synoptically undisturbed conditions are a good test bed for discussing convective precipitation under global warming, since we are able to regard the effects of external forcings to be minimal.

In this study, we investigate the environmental stability for precipitating deep convection in synoptically undisturbed conditions under the influence of global warming. Synoptically undisturbed conditions are specifically focused on in order to distinguish the influences of environmental stability on deep convection from large-scale effects.

This chapter is organized as follows. Firstly, the rationale of the present analyses is described in Section 2. We will choose the Tokyo metropolitan area over a flat plain as our analysis region. A wide flat area is favorable for the present study, since the effects of complex topography will not be a major factor. In addition, there are a number of previous studies that investigated the environmental conditions for summertime thunderstorms in that area (Yonetani 1975; Taguchi et al. 2002; Kawano et al. 2004; Nomura and Takemi 2011, hereafter NT11). Next, the changes of the environmental stability in the analysis region are examined from the past observed records. The temporal variability of the environmental conditions in the Tokyo area over the last 35 years are described with the use of surface and upper-air meteorological observations, which will be described in Section 3. In Section 4, we extend our considerations on convective precipitation under future climate conditions. We examine the projected changes of the environmental stability under future global warming by using the outputs of high-resolution global climate simulations at a 20-km resolution. In Section 5, thoughts from these analyses are given as concluding remarks.

2. Environmental stability and deep convection

The intensity of convectively induced precipitation is directly controlled by the structure, organization, and intensity of a single and/or a system of deep cumulonimbus clouds and thus can be related to and diagnosed by the conditions of the convective environments. Vertical wind shear determines the structure and organization of convective systems and hence their intensity, while temperature and moisture conditions characterize static stability that would control the development and maintenance of convective systems.

In our previous works on the sensitivity of mesoscale convective systems (MCSs) to environmental shear, moisture, and temperature profiles, Takemi (2006) examined the effects of tropospheric moisture profile under a single temperature environment. Takemi (2007a), extending the study of Takemi (2006), compared the intensity of MCSs simulated in two contrasting temperature environments characteristic of the Tropics and the midlatitudes under a comparable CAPE condition and found that the midlatitude system is significantly stronger than the tropical one. Takemi (2007a) concluded that the static stability can be regarded as a key parameter that describes the MCS intensity. Takemi (2007b) further investigated numerically the effects of static stability on the MCS intensity by systematically changing temperature lapse rate with CAPE being unchanged. He showed that a colder environment (which is more unstable) is favorable for generating stronger cold pool, which will highly control the scale and strength of convective updrafts and hence the organization and intensity of MCSs. Furthermore, Takemi (2010) concluded that temperature lapse rate in the convectively unstable troposphere is useful in comparing the characteristics of precipitation produced by MCSs that occur in various climate regions of the world. It should be pointed out that the experimental setup employed in those numerical studies was a highly idealized one which assumes horizontally homogeneous stratifications as basic environmental states. Accordingly, the MCS behaviors in such horizontally homogeneous states are regarded as representing deep convection under synoptically undisturbed conditions.

Recent study of NT11 has investigated the environmental stability for afternoon convective rainfalls during summertime under synoptically undisturbed environments over a plain region that includes the Tokyo metropolitan area in Japan. The conventional surface observation data as well as weather charts were used to extract the hot, sunny days under synoptically undisturbed conditions, and the gridded mesoscale analysis data (which have a horizontal resolution of 10 or 5 km) that cover the Tokyo area were used to examine the difference of the characteristics of environmental stability between no-rain, rain, and strong-rain events in the afternoon by calculating stability indices and parameters. The stability indices are useful diagnostic parameters in identifying locations and/or times with high convective potential and therefore are commonly used in the meteorological community. The stability indices and parameters examined in NT11 are convective available potential energy (CAPE), convective inhibition (CIN), lifting condensation level (LCL), level of free convection (LFC), level of neutral buoyancy (LNB), Showalter stability index (SSI), lifted index (LI), K-index (KI), total-totals index (TT), temperature lapse rate from 950 hPa to 500 hPa (TLR), and precipitable water (PW), most of which have been widely used in the literature. Definitions and meanings of the stability indices can be found in Bluestein (1993). In the study of NT11, statistical analysis by t-test statistic was conducted to determine the significance of the distinguishing features of the stability parameters among the no-rain, rain, and strong-rain events. Among the parameters, K-index (George, 1960) indicated the highest significance level. The analyses on the difference of temperature and humidity at each height among the events indicated that the temperatures and moistures at low to middle levels clearly distinguish the stability conditions for the afternoon rain events.

In this way, the environmental thermodynamic stability controls the development, intensity, and organization of deep convective clouds, especially under synoptically undisturbed conditions. Moreover, some studies emphasized the importance of environmental conditions for extreme events in simulated future climates under global warming. For the cases of tropical cyclones, Sugi et al. (2002) noted that the decrease in the occurrence of tropical cyclones in their simulated future climates is due to more stable environments. For the cases of convective precipitation events, on the other hand, Kanada et al. (2010) found that static stability as well as moisture content is more increased in the future climate simulated by a non-hydrostatic regional climate model. In addition, they pointed out that the increased CAPE would result in stronger updrafts and hence intense precipitation. The results of these studies strongly suggest that the environmental stability is closely related to the intensity and behavior of extreme weather.

Based on the above studies, the changes of the environmental stability from the past and in the future will be investigated in the following sections. The Tokyo metropolitan area is chosen as the analysis region, since there are a number of previous studies that investigated the stability conditions for deep convection or thunderstorms over the region (e.g., Yonetani, 1975). The Tokyo metropolitan area is located in the Kanto Plain, the largest plain area in Japan. The reason why a mesoscale region is chosen is because stability conditions widely change even over Japan depending on locations and seasons (Chuda and Niino, 2005). Thus it is convenient to focus on a mesoscale region for the analysis on convective environments.

3. Variation of environmental stability from the past observations

This section provides the results of the analyses on the environmental stability with the use of past records. Conventional meteorological observations during 35 years are examined to demonstrate the variability of the environmental stability under synoptically undisturbed conditions.

3.1 Data and analysis procedure

The environmental stability over the Tokyo metropolitan area or the Kanto Plain is examined here. There is a dense network of surface meteorological observations at horizontal spacings of about 20 km in Japan, called the Automated Meteorological Data Acquisition System (AMeDAS) in addition to the upper-air observations by radiosondes. The data by these conventional observations are used in the present section.

The AMeDAS stations over the Kanto Plain around Tokyo (i.e., within the area of latitude 35.2-36.2 degrees north and longitude 139.00-142.25 degrees east) with their elevations of below 100 m above the mean sea level are selected. On the other hand, the upper-air data used for the calculations of the stability indices and parameters are those obtained at the Tateno station (36.057 degrees north and 140.125 degrees east) located at about 50 km distance northeast of Tokyo. The location is within the Kanto Plain and far from the mountains; therefore, even at one upper-air station, the observations are considered to represent the characteristics over the whole analysis region at least in a climatological sense. The period of these surface and upper-air data used is during the years of 1976-2010, and the July and August data are chosen in order to focus on typical summertime undisturbed conditions.

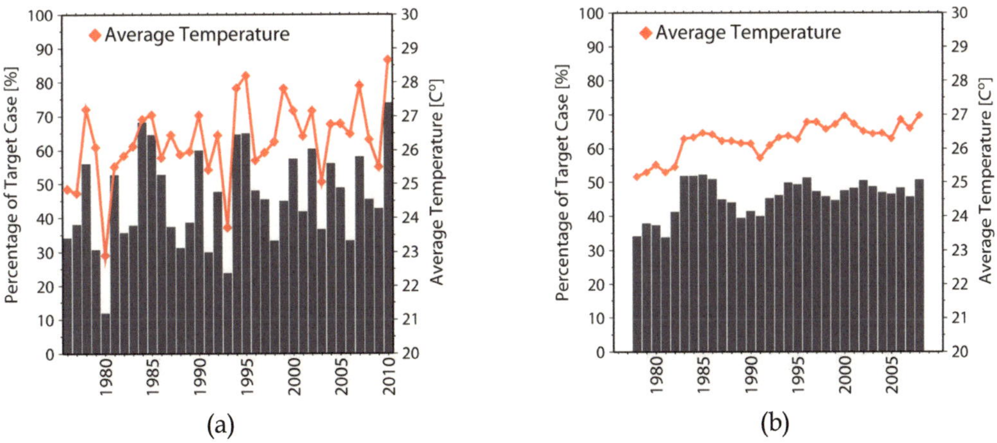

Fig. 1. Time series of (a) the percentage number of the undisturbed days among the total number of days in July and August during 1976-2010 and the annual-mean temperature averaged over the chosen days and (b) the 5-year running means for those shown in (a).

The AMeDAS data as well as surface weather charts issued by Japan Meteorological Agency (JMA) are examined to extract synoptically undisturbed conditions during July and August for the period of 1976-2010. The details of the procedure to extract the undisturbed conditions are given in NT11, but for readers' convenience the procedure is briefly described here. At first, days with the daily maximum temperature exceeding 30 degrees Celsius at least one AMeDAS station in the analysis region are chosen. Secondly, among the chosen days, the days without precipitation during the morning hours at any AMeDAS stations are selected. Thirdly, the days influenced by synoptic-scale disturbances (tropical cyclones, stationary fronts, and extra-tropical cyclones) are excluded by looking at surface weather maps at 0900 Japan Standard Time (hereafter referred to as JST) on those days. After these screenings, we are able to extract hot and sunny days (at least in the morning hours) which are considered to match synoptically undisturbed conditions. Totally 699 days are chosen from the 35-year dataset.
Actually, the number of the selected days widely varies year by year. Figure 1 shows the variation of the number (in percentage) of the undisturbed days among the total number of days in July and August during the period of 1976 and 2010 and their annual-mean temperature for the chosen days. Figure 1a clearly indicates a large variability of these properties throughout the time period. There seems to be a close correlation between the percentage number of the undisturbed days and the average temperature. This suggests that the average temperature increases with the increase in the number of the undisturbed days. Actually, the running means shown in Fig. 1b clearly demonstrate that the average temperature increases with the increase in the number of the undisturbed cases.
Although there is a large variability in the number of the undisturbed days during the analysis period, it was indicated that there is an increasing trend of the surface air temperature irrespective of the variability of the undisturbed occurrence (not shown). These results suggest that the warming trend appears also in the undisturbed situations. Based on this situation, the variability of the environmental stability is examined from the upper-air observations.
For reference, the frequency distributions of maximum hourly precipitation on the chosen days during the period of 2002-2010 (which was investigated in NT11) and the period of 1976-2010 are exhibited in Figure 2. It is clearly seen that intense precipitation events do occur even under synoptically undisturbed conditions.

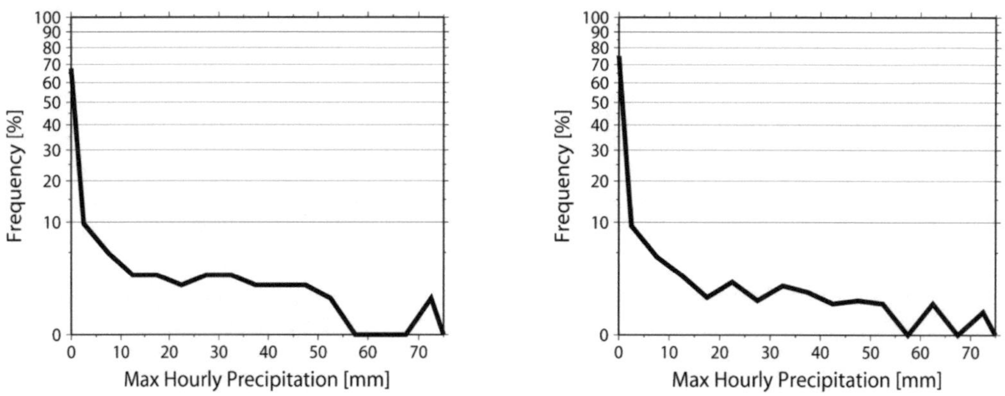

(a) (b)

Fig. 2. The frequency distribution of maximum hourly precipitation for (a) the period of 2002 and 2010, which was examined in NT11, and (b) the period of 1976 and 2010.

3.2 Results

The stability indices and parameters mentioned in Section 2, .i.e.: CAPE; CIN; LCL; LFC; LNB; SSI; LI; KI; TT; TLR; and PW, are calculated from the observation data that meet the synoptically undisturbed criteria. Figure 3 shows the time series of the annual means of these stability parameters. In order to demonstrate the long-term trend, 5-year running

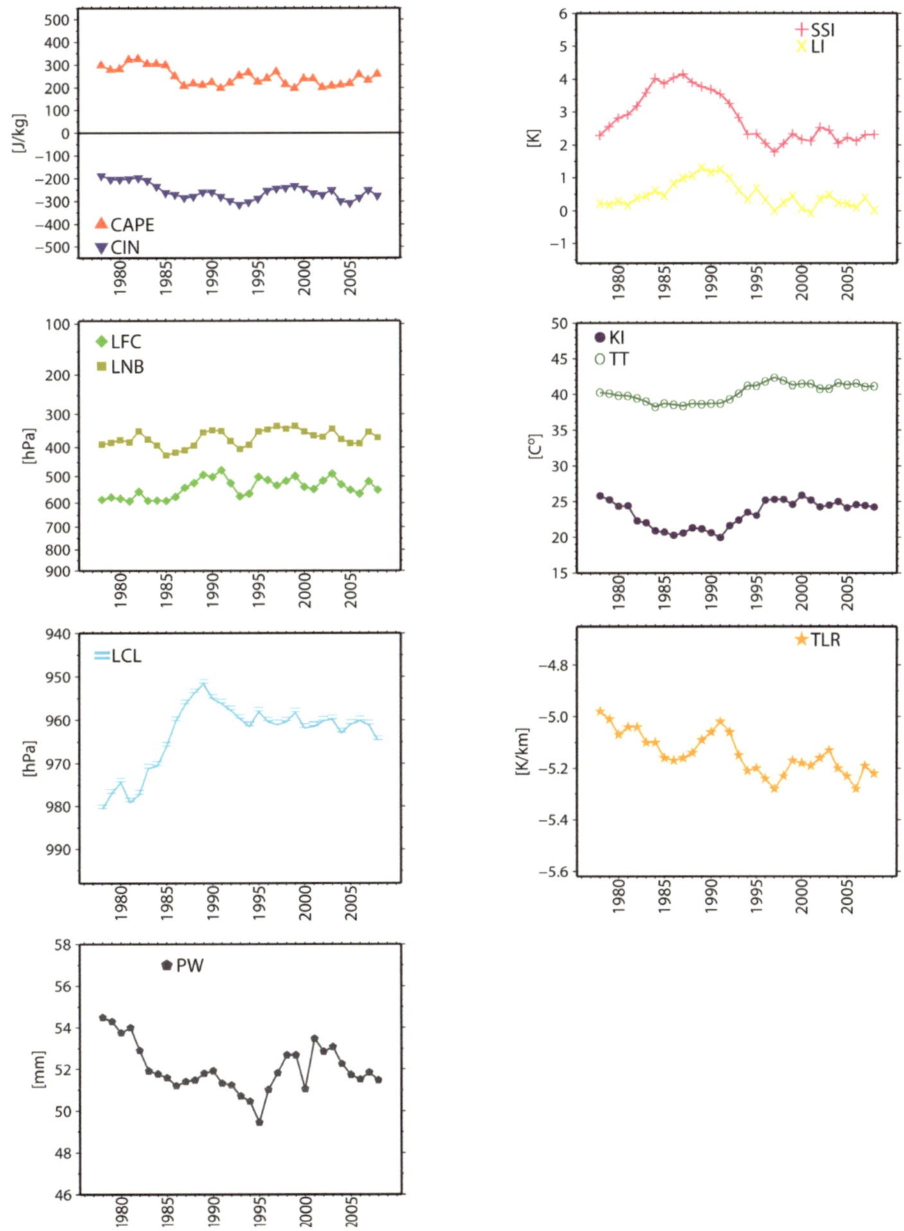

Fig. 3. Time series of the environmental indices and parameters from 1976 to 2010. The data are running means for 5-year period.

means were taken for the annual mean values. Noticeable features are seen for CAPE, CIN, TLR, and PW. CAPE and PW appear to be smaller with time, while the absolute values of CIN and TLR are seen to become larger with time. Smaller CAPE and larger CIN suggest that the atmosphere becomes more stabilized in terms of buoyant energy perspective, while larger TLR indicates that the atmosphere becomes more statically unstable. These points seem to be contradicting, but the reason why this happens is understood by the change in PW. The decrease in PW, indicating that the atmospheric moisture becomes smaller, suggests that it is getting drier. The lower moisture content is considered to be more dominant in reducing the buoyant energy than the contribution from the increase in the temperature lapse rate. It is also found that the LFC and LCL become larger with time, again indicating that the atmosphere is more stabilized.

In order to examine the statistical significance of the changes in these stability parameters, we use the t-test statistic for the differences between the averages during the first 10 years (1976-1985) and during the recent 10 years (2001-2010). The period of the first 10 years is referred to as Past, while the recent 10-year period as Present. The number of the chosen days is 192 for the Past period and is 217 for the Present period.

Test static, T, is defined as:

$$T = \left(x_A - x_B\right)\left(\frac{\sigma_A^2}{n_A} + \frac{\sigma_B^2}{n_B}\right)^{-\frac{1}{2}}, \quad (1)$$

where x_A and x_B are the means for category A and B, respectively, σ_A and σ_B the standard deviations, n_A and n_B denote the number of cases in each category. In this test, if T is larger than 1.96, a significant difference between the categories is statistically indicated. Otherwise, there is no significant difference between each category. A larger value of T means that the difference between each category is more significant.

Figure 4 exhibits the T-values obtained from the t-test statistical analysis as their properties as well as the bar plot. More unstable states in the Present (Past) period than in the Past (Present) are indicated as larger T-values in the Present (Past). It is shown that the

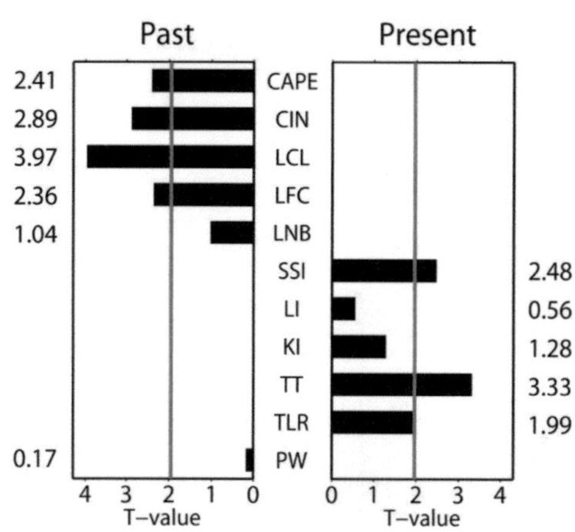

Fig. 4. T values from the t-test statistic for the difference between the averages during 1976-1985 (referred to as Past) and during 2001-2010 (referred to as Present). If the average during the Past indicates a more unstable situation than that during the Present, the bar is plotted in the Past panel. The significance level (1.96) is indicated by a red line.

parameters that exceed the significance level (1.96) for the present data are CAPE, CIN, LCL, LFC, SSI, TT, and TLR. This seems to be consistent with the trend shown in Figure 3. In terms of CAPE, CIN, LCL, and LFC, the atmosphere becomes more stabilized in the Present period than in the Past. In contrast, the temperature lapse rate becomes more unstable in the Present. Actually, the change in the moisture content (i.e., PW) is not statistically significant. The statistically significant trend indicated in Figure 3 is shown to be for CAPE, CIN, LCL, LFC, and TLR.

To see how the vertical profiles affect the significance of the temperature and moisture differences between the Past and the Present periods, we calculate the test statistic T for those differences at each height level. Figure 5 indicates the vertical distributions of the T values for the differences in temperature and relative humidity between the Past and the Present periods. The significant differences are seen in the middle layers of 800-600 hPa and at the 500-hPa level for temperature and in the lowest layer of below 900 hPa for relative humidity. In the Present period the 800-600 hPa layer becomes warmer while the 500-hPa level becomes colder. This leads to the increase in temperature lapse rate with time, as indicated in Figures 3 and 4. On the other hands, the reduction of relative humidity is only significant in the lowest levels, indicating that no significant difference is found for the PW change.

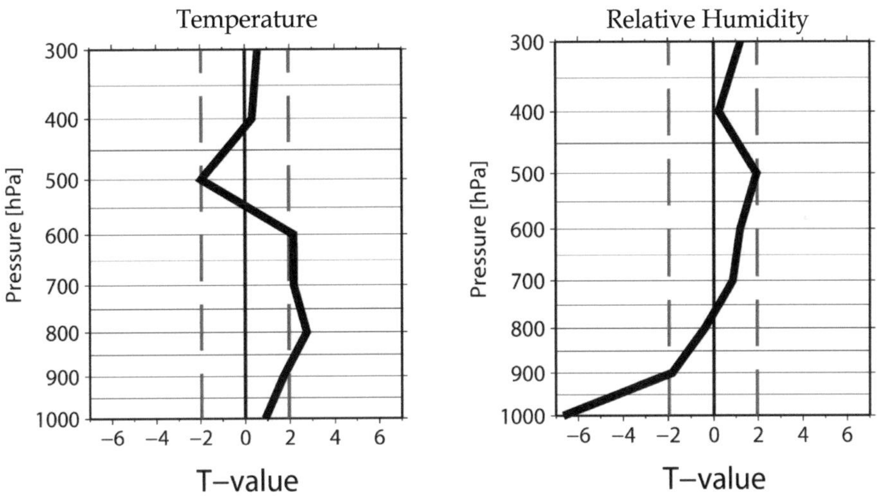

Fig. 5. The vertical profiles of the T values for the temperature and relative humidity differences between the Past and the Present periods. The red long-dashed lines indicate the significance level. The positive (negative) T value indicates that temperature and relative humidity are larger in the Present (Past) period than in the Past (Present).

4. Projected changes in environmental stability

In this section, we examine the future changes in the environmental stability under the undisturbed conditions by using the simulations outputs from a global circulation model (GCM) for the present and future climates.

4.1 Data and analysis procedure

The data for the present analysis are the gridded outputs at the 20-km horizontal resolution from an atmospheric global circulation model (AGCM) with the resolution of T959L60 (which means triangular truncation of 959 and 60 vertical levels) for present climate and future warming climates. The AGCM was developed by Meteorological Research Institute (MRI) of JMA under one of the projects in the Innovative Program of Climate Change Projection for the 21st Century (the KAKUSHIN program) (Kitoh et al., 2009) and is called as MRI-AGCM3.2S. The original version of this GCM was developed by Mizuta et al. (2006) and has been updated for the KAKUSHIN program (Murakami and Wang, 2010; Kusunoki et al., 2011).

The simulations with MRI-AGCM3.2S were performed for the climate periods of 30 years to represent the present climate (corresponding to the period of 1979-2008), the near-future climate (corresponding to the period of 2015-2044), and the future climate (corresponding to the period of 2074-2104). The near-future and future climate simulations assume a global warming scenario, the A1B IPCC emission scenario (IPCC, 2007). The preliminary experiments (Kitoh et al., 2009) indicated that the simulated present climate was almost the same as that in Mizuta et al. (2006) who validated their GCM simulations with the observed data. Therefore, the present-climate simulation with MRI-AGCM3.2S is considered to be successful in representing the existing climate states. In this study, we use the data periods of 1980-2004 for the present climate, 2020-2044 for the near-future climate, and 2075-2099 for the future climate. The time intervals of the outputs are 1 hour for the surface variables and 6 hours for the upper-air variables.

In order to extract undisturbed conditions from the climate simulation data in which no weather charts and no surface observations are available (although of course there are surface-level outputs from the GCM simulations), we need to employ an objective approach. For this purpose, we examined the meteorological characteristics representative of the synoptically undisturbed conditions from the radiosonde observations and found that the vertical shears of the horizontal winds as well as the wind speeds at upper levels are quite small. This character is clearly distinct from the wind conditions for severe storms and squall lines (Bluestein and Jain, 1985). Therefore, to objectively choose the undisturbed conditions, we apply criteria on wind speed and shear from the GCM outputs. By taking the spatial averages for the wind speeds over the analysis area (i.e., the Tokyo area, part of the Kanto Plain), we examine the following criteria: 1) to use the data only in August to avoid monsoonal stationary fronts; 2) to exclude days with the 500-hPa wind speed of equal to or greater than 10 m s^{-1} and the wind shear of less than 8 m s^{-1} in the layer of 500 and 975 hPa; and 3) to choose days having rainfall of less than 1 mm during the morning period at all the grid points in the analysis area.

To validate this objective approach, we examined the 35-year upper-air observation data and compared the mean vertical profiles of temperature and moisture averaged for the undisturbed conditions independently extracted by the NT11 observation-based approach and by the present objective approach. The number of days chosen by the NT11 approach was 699 (as mentioned in Section 3), while that by the objective approach was 344. Three hundred cases out of the 344 cases extracted by this objective approach satisfy also the NT11 conditions. The mean profiles for the objectively chosen cases were found to be very similar to those for the chosen cases as extracted in Section 3. In this way, the present objective approach is useful in identifying the undisturbed conditions from the GCM outputs.

By applying the objective criteria for the simulation data during the present-climate 25-year period, 182 days were extracted. The vertical profiles of temperature and moisture averaged for the extracted undisturbed cases were compared with the radiosonde observations for the undisturbed days extracted by the same objective approach. From the comparison, it was indicated that the vertical distributions of temperature and moisture obtained from the GCM simulation well capture the existing climate revealed by the upper-air observations. We have also compared the representations of the stability indices and parameters in the GCM simulation with those calculated by the observations and found that the GCM reproduces the general characteristics of the stability conditions. Based on these comparisons, the GCM outputs were demonstrated to be useful in investigating the characteristics of the environmental stability under the undisturbed conditions. In the following, we examine the projected changes in the environmental stability revealed by the GCM climate simulations.

The same criteria were also applied for the simulated near-future and future climates: 160 days were chosen for the near-future climate; and 128 days were chosen for the future climate. In the following, the changes in the stability parameters from the present climate to the future are examined.

4.2 Results

From the analyses on the upper-air observations described in Section 3, it was indicated that the changes in CAPE, CIN, and TLR over the last 35 years demonstrate a statistically significant trend. Thus, the characteristics of the representations of these three parameters in each simulated period are shown here.

Figure 6 compares the frequency distributions of the stability parameters, i.e., CAPE, CIN, and TLR, during the 25-year simulation periods of the present climate, the near-future climate, and the future climate. The higher values of CAPE become more frequent in the future climate, while at the same time the higher values of CIN become more frequent in the future. This result indicates that the potential instability for convection increases but that the energy required to raise low-level air parcels above LFC also increases. In terms of temperature lapse rate, the atmosphere becomes more stable in the future climate than in the present. We have also examined the frequency distributions of PW and found that the atmospheric moisture content increases.

The statistical significance for the differences in the means of the stability parameters between the present and the future climates is examined by t-test static. The T value indicating the significance level for this analysis is 1.96. The parameters that exceed the significance level for the differences between the present and the future climates are TLR (T=3.53), PW (T=7.11), CAPE (T=4.27), CIN (T=2.73), LNB (T=2.23), and LI (T=2.14): CAPE, CIN and PW become larger with the future period; TLR and LI become smaller; and LNB becomes higher.

This statistical analysis as well as the result shown in Figure 6 indicates that the atmosphere becomes more unstable in terms of CAPE, PW, and LNB with the future while becomes more stable in terms of temperature lapse rate and CIN. These results seem to be contradicting with each other, although some interpretations can be made. From the lapse rate change, the activity of deep convection is generally inhibited in the future climate. However, once deep convection develops after penetrating the level of free convection, the intensity of convection will be more enhanced with the consumption of CAPE.

The present analyses on the environmental stability under the synoptically undisturbed conditions from the GCM outputs showed that with the future period the temperature lapse rate in the lower troposphere becomes smaller while precipitable water vapor and CAPE become larger. By examining the statistical significance on the differences of temperature and moisture between the present and the future climates, it was indicated that the increase in precipitable water vapor was due to the increase in water vapor mixing ratio at all the vertical levels while the decrease in temperature lapse rate was due to the larger increase in temperature at higher levels. In addition, it was found that LNB becomes higher with the simulated period. On the other hand, LFC did not vary among the climate periods, although the statistical significance was not identified.

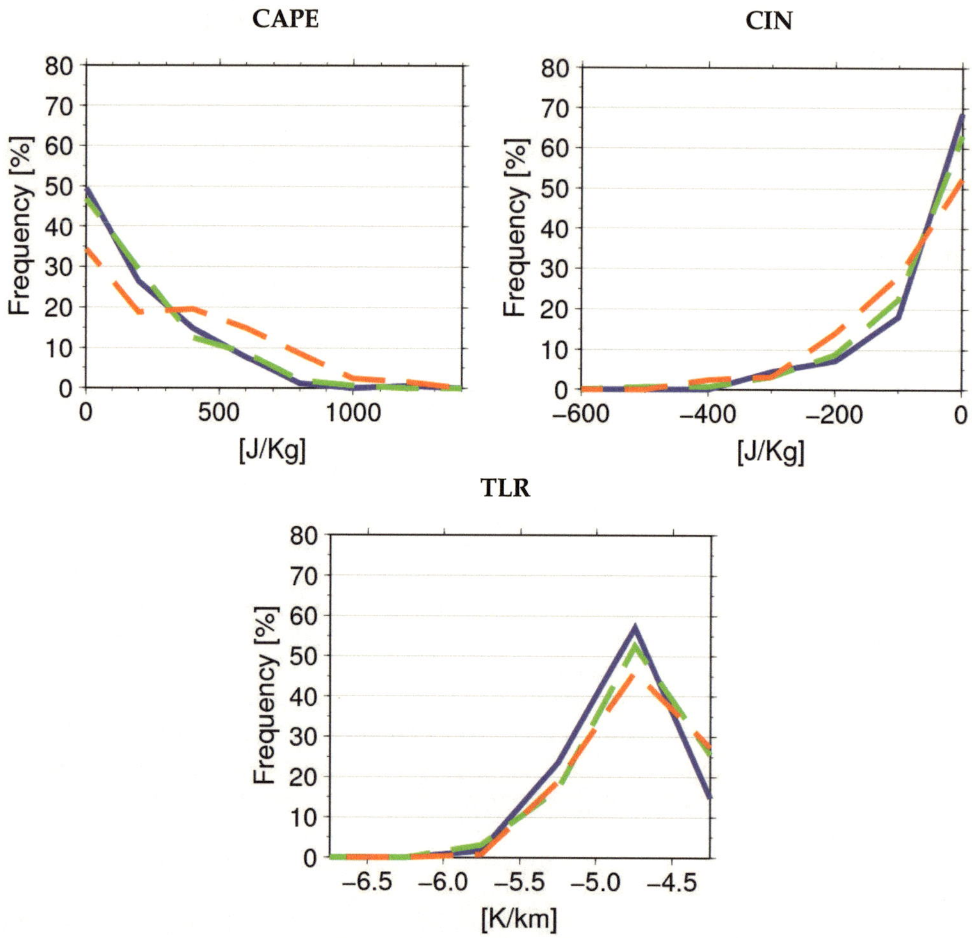

Fig. 6. The frequency distributions of CAPE (upper), CIN (middle), and TLR (lower) averaged over the analysis area during the 25-year periods of the present climate (blue line), the near future climate (green line), and the future climate (red line).

The unchanged height of LFC and the elevated height of LNB suggest that the integrated vertical path in calculating CAPE becomes deeper even with the decreased lapse rate. Considering that CAPE is the integrated buoyancy between LFC and LNB and that buoyancy will not be larger in a lower lapse rate than in a higher lapse rate, the increase in CAPE is considered to be primarily due to the increase in LNB. At the same time, the mean buoyancy between LFC and LNB (like the one defined as normalized CAPE in Blanchard 1998) will not significantly change among the climate periods. The maintained buoyancy with the future period is favorable for developing strong updrafts, since it cancels negative effects of enhanced entrainment in environments with lower lapse rates (Takemi 2007b). In this way, it is considered that the seemingly contrasting features of temperature lapse rate and CAPE in future climates enhance the activity of convection and precipitation.

The precipitation characteristics that will result from the projected changes in the environmental stability are discussed here. Takemi (2007b; 2010) investigated the precipitation characteristics within the MCSs by changing temperature lapse rate and moisture content as well as shear profile in an idealized simulation setup where the base-state atmosphere is horizontally uniform without any external forcing. This idealized setup is regarded as synoptically undisturbed, since only convective forcing is initial thermal and the resultant self-organizing effects. The weak shear condition (i.e., 5 m s^{-1} difference only in the lowest 2.5 km) examined in Takemi (2007b; 2010) corresponds to the synoptically undisturbed conditions examined in this study. Their results indicated that with the same CAPE environment convection intensity (represented as updraft strength) and mean precipitation decrease as temperature lapse rate decreases. In the environments projected in the future periods, temperature lapse rate decreases but CAPE increases. Consequently, convection intensity and mean precipitation does not decrease, according to the results of Takemi (2010). Precipitation amount will be maintained or even intensified in the future climates.

One of the interesting points in Takemi (2010) was that the maximum precipitation intensity increases with the decrease in temperature lapse rate while with CAPE being unchanged. On the other hand, the present analyses of the GCM data indicated that both static stability and CAPE increases. Applying the result of Takemi (2010) to the projected changes in the environmental stability, the precipitation intensity in short time-scales is suggested to increase.

5. Concluding remarks

From the analyses in the GCM climate simulation data, it was indicated that in the future climates temperature lapse rate decreased in the lower troposphere while water vapor mixing ratio increased throughout the deep troposphere. The changes in the temperature and moisture profiles resulted in the increase in both precipitable water vapor and CAPE, which were evaluated as statistically significant. These projected changes will be enhanced with the future period.

However, these projected changes are totally opposite from the analyses on the past 35-year observations. This observational evidence strongly suggests that the GCM simulations need to be investigated with sufficient cautions.

The environmental stability, especially in undisturbed cases, strongly controls the structure, development, and organization of deep convective clouds and associating precipitation. Long-term variability of the general characteristics of precipitation for cases in Japan has been investigated from statistical viewpoints by many studies such as Yonetani (1982), Iwashima and Yamamoto (1993), Fujibe (1998), Sato and Takahashi (2000), Kanae et al. (2004), Fujibe et al. (2005; 2006; 2009), Kamiguchi et al. (2010), and Iwasaki (2010). There have been arguments on the reason for the change in the frequency and intensity of precipitation: some studies attributed to global climate change (i.e., warming); and others implied the effects of urban heat island. However, there have been few studies on the changes in background conditions for such precipitation events in changing climates, that is, environmental characteristics for the development of rain-producing thunderstorms. Studies which investigate the connection between environmental stability and convective precipitation under global warming require sound observational evidence and/or implication.

6. Acknowledgments

We would like to thank Drs. Sachie Kanada and Masuo Nakano for their comments on the analyses of the GCM simulation data. This work was conducted partly under the framework of the "Projection of the Changes in Future Weather Extremes Using Super-High-Resolution Atmospheric Models" supported by the KAKUSHIN program of the Ministry of Education, Culture, Sports, Science, and Technology (MEXT) of Japan. This work was also supported by a Scientific Research grant from Japan Society for Promotion of Sciences (JSPS). The GCM computations were conducted on the Earth Simulator.

7. References

Blanchard, D. O., 1998: Assessing the vertical distribution of convective available potential energy. *Weather and Forecasting*, vol. 13, pp. 870-877, ISSN 0882-8156

Bluestein, H. B. (1993). *Synoptic-Dynamic Meteorology in Midlatitudes, Vol. II*, Oxford University Press, ISBN 0-19-506268-2, New York, USA

Bluestein, H. B. & Jain, M. H. (1985). Formation of mesoscale lines of precipitation: Severe squall lines in Oklahoma during the spring. *Journal of the Atmospheric Sciences*, Vol. 42, pp. 1711-1732, ISSN 0022-4928

Chuda, T. & Niino, H. (2005). Climatology of environmental parameters for mesoscale convections in Japan. *Journal of the Meteorological Society of Japan*, Vol. 83, pp. 391-408, ISSN 0026-1165

Cotton, W. R.; Bryan, G. B. & Van Den Heever, S. C. (2011). *Storm and Cloud Dynamics*, Academic Press, ISBN 978-0-12-0885428, San Diego, USA

Del Genio, A. D.; Yao, M.-S. & Jonas, J. (2007). Will moist convection be stronger in a warmer climate? *Geophysical Research Letters*, Vol. 34, L16703, ISSN 0094-8276

Emanuel, K. A. (1994). *Atmospheric Convection*, Oxford University Press, ISBN 0-19-506630-8, New York USA

Fujibe, F. (1998). Spatial anomalies and long-term changes of precipitation in Tokyo. *Tenki*, Vol. 45, pp. 7-18, ISSN 0546-0921

Fujibe, F.; Yamazaki, N.; Katsuyama, K. & Kobayashi, K. (2005). The increasing trend of intense precipitation in Japan based on four-hourly data for a hundred years. *Scientific Online Letters on the Atmosphere*, Vol. 1, pp. 41–44, Online ISSN 1349-6476

Fujibe, F.; Yamazaki, N.; Katsuyama, M. & Kobayashi, K. (2006). Long-term changes in the diurnal precipitation cycles in Japan for 106 years (1898-2003). *Journal of the Meteorological Society of Japan*, Vol. 84, pp. 311-317, ISSN 0026-1165

Fujibe, F.; Togawa, H. & Sakata, M. (2009). Long-term change and spatial anomaly of warm season afternoon precipitation in Tokyo. *Scientific Online Letters on the Atmosphere*, Vol. 5, pp. 17–20, Online ISSN 1349-6476

George, J. J. (1960). *Weather Forecasting for Aeronautics*, Academic Press

Houze, R. A., Jr. (2004). Mesoscale convective systems, Reviews of Geophysics, Vol. 42, RG4003, doi:10.1029/2004RG000150, ISSN 8755-1209

IPCC (2007). *Climate Change 2007: The Physical Science Basis. Contribution of Working Group I to the Fourth Assessment Report of the Intergovernmental Panel on Climate Change*, Cambridge University Press, ISBN 978-0-521-70596-7, Cambridge, United Kingdom

Iwasaki, H. (2011). Recent positive trend in heavy rainfall in eastern Japan and its relation with variations in atmospheric moisture. *International Journal of Climatology*, doi: 10.1002/joc.2269, in press.

Iwashima, T. & Yamamoto, R. (1993). A statistical analysis of the extreme events: long-term trend of heavy daily precipitation. *Journal of the Meteorological Society of Japan*, Vol. 71, pp. 637-640, ISSN 0026-1165

Kamiguchi, K.; Kitoh, A.; Uchiyama, T.; Mizuta, R. & Noda, A. (2006). Changes in precipitation-based extremes indices due to global warming projected by a global 20-km-mesh atmospheric model. *Scientific Online Letters on the Atmosphere*, Vol. 2, pp. 64–67, Online ISSN 1349-6476

Kamiguchi, K.; Arakawa, O.; Kitoh, A.; Yatagai, A.; Hamada, A. & Yasutomi, N. (2010). Development of APHRO_JP, the first Japanese high-resolution daily precipitation product for more than 100 years. *Hydrologic Research Letters*, Vol. 4, pp. 60–64, Online ISSN 1882-3416

Kanada, S.; Nakano, M. & Kato, T. (2010). Changes in mean atmospheric structures around Japan during July due to global warming in regional climate experiments using a cloud-system resolving model. *Hydrologic Research Letters*, Vol. 4, pp. 11–14, Online ISSN 1882-3416

Kanae, S.; Oki, T. & Kashida, A. (2004). Changes in hourly heavy precipitation at Tokyo from 1890 to 1999. *Journal of the Meteorological Society of Japan*, Vol. 82, pp. 241-247, ISSN 0026-1165

Kawano, K.; Hirokawa, Y. & Ohno, H. (2004). Diagnosis of air-mass thunderstorm days using radiosonde data: The summer Kanto area under the Pacific subtropical anticyclone. *Tenki*, Vol. 51, pp. 17-30, ISSN 0546-0921

Kimoto, M.; Yasutomi, N.; Yokoyama, C. & Emori, S. (2005). Projected changes in precipitation characteristics around Japan under the global warming. *Scientific Online Letters on the Atmosphere*, Vol. 1, pp. 85–88, Online ISSN 1349-6476

Kitoh, A.; Ose, T.; Kurihara, K.; Kusunoki, S.; Sugi, M. & KAKUSHIN Team-3 Modeling Group (2009). Projection of changes in future weather extremes using super-high-resolution global and regional atmospheric models in the KAKUSHIN program: Results of preliminary experiments. *Hydrologic Research Letters*, Vol. 3, pp. 49–53, Online ISSN 1882-3416

Kusunoki, S.; Yoshimura, J.; Yoshimura, H.; Noda, A.; Oouchi, K. & Mizuta, R. (2006). Change of Baiu rain band in global warming projection by an atmospheric general circulation model with 20-km grid size. *Journal of the Meteorological Society of Japan*, Vol. 84, pp. 581-611, ISSN 0026-1165

Kusunoki, S.; Mizuta, R. & Matsueda, M. (2011). Future changes in the East Asian rain band projected by global atmospheric models with 20-km and 60-km grid size. *Climate Dynamics*, in press.

Mizuta, R.; Oouchi, K.; Yoshimura, H.; Noda, A.; Katayama, K.; Yukimoto, S.; Hosaka, M.; Kusunoki, S.; Kawai, H. & Nakagawa, M. (2006). 20-km-mesh global climate simulations using JMA-GSM Model -mean climate states-. *Journal of the Meteorological Society of Japan*, Vol. 84, pp. 165-185, ISSN 0026-1165

Murakami, H. & Wang, B. (2010). Future change of north Atlantic tropical cyclone tracks: Projection by a 20-km-mesh global atmospheric model. *Journal of Climate*, Vol. 23, pp. 2699-2721, ISSN 0894-8755

Ninomiya, K. (2009). Characteristics of precipitation in the Meiyu-Baiu season in the CMIP3 20th century climate simulations. *Journal of the Meteorological Society of Japan*, Vol. 87, pp. 829-843, ISSN 0026-1165

Nomura, S. & Takemi, T. (2011). Environmental stability for afternoon rain events in the Kanto Plain in summer. *Scientific Online Letters on the Atmosphere*, Vol. 7, pp. 9–12, Online ISSN 1349-6476

Santer, B. D.; Thorne, P. W.; Haimberger, L.; Taylor, K. E.; Wigley, T. M. L.; Lanzante, J. R.; Solomon, S.; Free, M.; Gleckler, P. J.; Jones, P. D.; Karl, T. R.; Klein, S. A.; Mears, C.; Nychka, D.; Schmidt, G. A.; Sherwood, S. C. & Wentz, F. J. (2008). Consistency of modelled and observed temperature trends in the tropical troposphere. *International Journal of Climatology*, Vol. 28, pp. 1703-1722, ISSN 1097-0088

Sato, N. & Takahashi, M. (2000). Long-term changes in the properties of summer precipitation in the Tokyo area. *Tenki*, Vol. 47, pp. 643-648, ISSN 0546-0921

Sugi, M.; Noda, A. & Sato, N. (2002). Influence of the global warming on tropical cyclone climatology: An experiment with the JMA global model. *Journal of the Meteorological Society of Japan*, Vol. 80, pp. 249-272, ISSN 0026-1165

Taguchi, A.; Okuyama, A. and Ogura, Y. (2002). The thunderstorm activity observed by SAFIR and its relation to the atmospheric environment over the Kanto area in the summer. Part II: Thunderstorm prediction by stability indices. *Tenki*, Vol. 49, pp. 649-659, ISSN 0546-0921

Takemi, T. (2010). Dependence of the precipitation intensity in mesoscale convective systems to temperature lapse rate. *Atmospheric Research*, Vol. 96, pp. 273-285, ISSN 0169-8095

Takemi, T. (2007a). A sensitivity of squall line intensity to environmental static stability under various shear and moisture conditions. *Atmospheric Research*, Vol. 84, pp. 374-389, ISSN 0169-8095

Takemi, T. (2007b). Environmental stability control of the intensity of squall lines under low-level shear conditions. *Journal of Geophysical Research*, Vol. 112, D24110, doi:10.1029/2007JD008793, ISSN 0148-0227

Takemi, T. (2006). Impacts of moisture profile on the evolution and organization of midlatitude squall lines under various shear conditions. *Atmospheric Research*, Vol. 82, pp. 37-54, ISSN 0169-8095

Yonetani, T. (1982). Increase in number of days with heavy precipitation in Tokyo urban area. *Journal of Applied Meteorology*, Vol. 21, pp. 1466-1471, ISSN 1558-8424

Yonetani, T. (1975). Characteristics of atmospheric vertical structure on days with thunderstorms in the northern Kanto Plain. *Journal of the Meteorological Society of Japan*, Vol. 53, pp. 139-148, ISSN 0026-1165

Yoshizaki, M.; Muroi, C.; Kanada, S.; Wakazuki, Y.; Yasunaga, K.; Hashimoto, A.; Kato, T.; Kurihara, K.; Noda, A. & Kusunoki, S. (2005). Changes of Baiu (Mei-yu) frontal activity in the global warming climate simulated by a nonhydrostatic regional model. *Scientific Online Letters on the Atmosphere*, Vol. 1, pp. 25–28, Online ISSN 1349-6476

4

Effects of Global Warming on Climate Conditions in the Japanese Alps Region

Keisuke Suzuki
Faculty of Science and Institute of Mountain Science
Shinshu University
Japan

1. Introduction

The region of Japan that lies along the Sea of Japan is known to experience some of the heaviest snowfall in the world. In this region, precipitation brought by snowfall is more important as a water resource than rainfall. During winter, a thermal anticyclone is formed over the Siberian continent by strong radiative cooling. The Tibetan Himalayas towering to the south block the anticyclone southward advance, the advance is enhanced during the winter at low solar elevation angle. When the warm Tsushima Current pours into the Sea of Japan heading north, the cold, dry air mass blowing out of the Siberian anticyclone toward the east becomes unstable, absorbing heat and moisture vapour from the lower layer. This produces cumulus convections in sequence, which land on the northwest coast of the Japanese islands. In cumulus clouds, high moisture rates and cold temperature accelerate the formation of snow particles. In addition, these clouds hit the high mountains that run along the center of the Japanese Archipelago and are forced to climb upward, bringing a much larger amount of snowfall. Therefore, the mountain range experiences exceptionally heavy snowfall that is extreme even by world standard, and in spring, the melting snow becomes a valuable water resource for the region. Snow plays the role of a natural white dam by accumulating in watersheds during winter.

Recent studies have reported that the amount of snowfall in Japan will decrease as a result of global warming (Inoue and Yokoyama, 2003). However, these studies used data observed at low altitudes. The question arises whether the same theory can be applied to high-altitude mountain areas. In a temperate snow-covered area where rainfall is observed in winter or where the temperature is above 0 °C and the snow melts, the temperature often reaches the threshold point, which is the boundary line between snow and rain. At this point, a slight temperature increase or decrease can change snow into rain or, conversely, increase the snow. This means that global warming seems to cause a decrease in snowfall even if the amount of precipitation stays the same, because the snowfall particles melt, transforming into rainfall during precipitation. However, it is possible that even with a temperature rise of several degrees 100 years from now, the amount of snowfall could increase in mountain areas at high altitude where we predict that the temperature will not increase up to the threshold temperature between snow and rain. Inoue and Yokoyama (2003) stated that the amount of snowfall in Hokkaido, at the northern latitude of 41.5° or above, will not change much even after 100 years of global warming. However, in a chart presented in their report, we can see that snowfall is decreasing in coastal areas and increasing on the mountains in Hokkaido. A report published by the Ministry of Education, Culture, Sports, Science and Technology also states that "the amount of snowfall will decrease except for Hokkaido. This is because areas south of Tohoku will get more rain than snow with global warming; on the other hand, in Hokkaido, the cold atmosphere will be maintained even with global warming, and the increase in moisture vapour in the air caused by global warming will increase snowfall." However, it is reasonable to consider that at high altitude regions in Honshu (the main island of Japan), it is "cold enough for snowfall to prevail even with global warming"; therefore, the amount of snowfall will increase as in Hokkaido.

The results of a study on cumulative snowfall variation in recent years using observation data from the Japan Meteorological Agency indicate that the increasing tendency is recognized only at Suttsu, Hokkaido, and that in the decreasing sites, the winter mean temperature is more than 3 ºC. The winter mean temperature in these sites is close to the threshold temperature of snow/rain, which determines whether the precipitating particles will from snow or rain. The rise in temperature in recent years has resulted in a decrease in the accumulated snow depth (Ishii and Suzuki, 2011).

Observations of the amount of snow have not been carried out in high-altitude mountains in Japan where the temperature is colder than the threshold temperature of snow/rain even with the temperature rise observed in recent years. Therefore, we cannot discuss the effect of global warming on the change in the amount of snow in the mountainous region of Japan based on observation data. Therefore, in this study, we discuss the relationship between temperature and the amount of snow using observation data for the Japanese Alps region and present the results of some meteorological observations we carried out at high-altitude sites in the Japanese Alps region.

2. Characteristics of the Japanese Alps region

The Japanese Alps is a collective name for the Hida Mountains, Kiso Mountains, and Akaishi Mountains, spreading approximately 200 km north to south and 100 km east to west. In a wider sense, the Japanese Alps region includes the Ryohaku Mountains in the west and Mt. Fuji, Mt. Yatsugatake, and the Mikuni Mountains in the east, bringing the geographical span of the region to more than 200 km in the east west direction. The altitude of the Japanese Alps summit is around 3,000 m; however, the climate conditions vary widely because it lies in the center of Honshu between the Japan Sea coast and the Pacific coast, with the northern area being known to experience one of the heaviest snowfalls in the world. The vegetation is a mix of circumpolar, northern, and continental elements. It is the habitat of many treasured alpine butterflies and is the southern-most habitat of the Ptarmigan on the planet. Because there are no glaciers in the Japanese Alps, the ecological system constituents such as alpine butterflies or the Ptarmigan are considered to be the remaining survivors from the last glacial period. However, the ecological system is quite sensitive to global-scale environmental changes. Therefore, the mountain ecological system exists under strict and critical environmental conditions (air temperature or hydrological factors). Additionally, global warming has a stronger influence in a smaller space because the temperature decreases more at higher altitudes. Next, we will show that the temperature change is dramatically larger with a change in elevation than with a horizontal change. Figure 1 shows the relationship between the latitude of the meteorological bureau office and the annual mean air temperature. Naturally, the further north the obserbation site is, the bigger the decrease in the annual mean air temperature. We can recognize a good correlation between the latitude and annual mean air temperature, except for the high-altitude Mt. Fuji and the observation sites marked on Fig. 1. According to this relation, for a

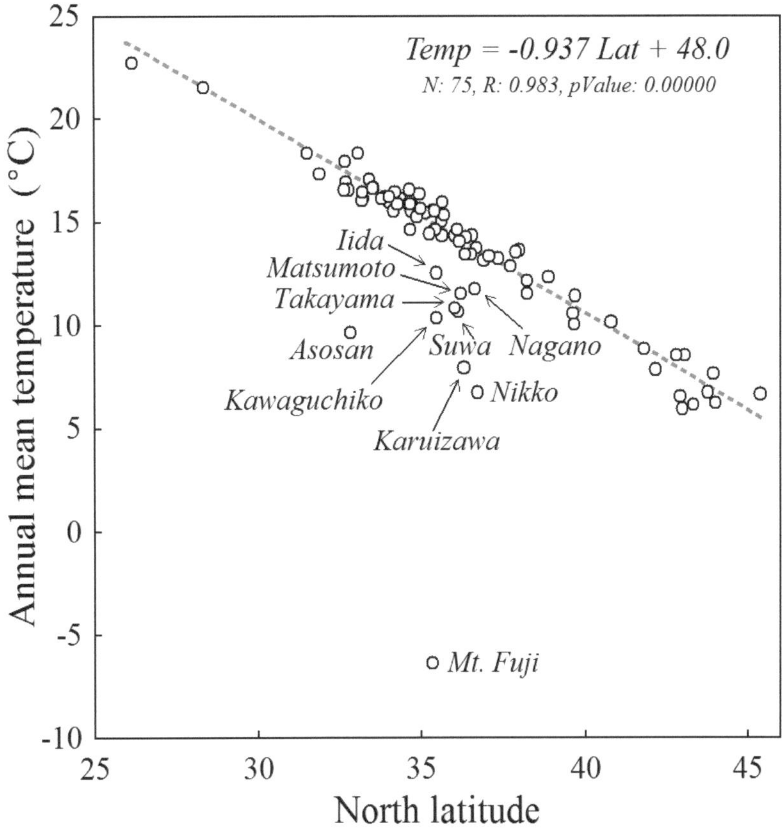

Fig. 1. Relationship between the latitude of the meteorological bureau office and the annual mean air temperature

change of 1 °C in the annual mean air temperature, one must move 118 km north or south. However, if the lapse rate of the air temperature is 0.65 °C/100 m, then a 154 m changes in elevation will bring about a change of 1 °C in air temperature. Thus, the air temperature change due to altitude variation is almost 800 times more drastic than that due to horizontal variation. The distribution of vegetation is perspectively determined by air temperature, so it is difficult to realize the horizontal vegetation change; however, in mountain areas where the air temperature changes rapidly with elevation, the vegetation is very sensitive to global warming. Furthermore, because of the warming, vegetation that has adapted to the high altitude and colder environment could lose its habitat. Changes in vegetation will change the distribution of insects and consequently affect larger animals. Thus, global-scale environmental change will impact the mountain area directly.

In Fig. 1, data that are out of regression line have been named and their locations (circles) and elevations are shown on the map in Fig. 2. The altitude of the observation site on Mt. Fuji is 3,775 m, and it is the highest observation site of the Japan Meteorological Agency. The second highest one, which records temperature, humidity, wind, precipitation, and snow depth, is Nikko, whose altitude is 1,292m. Ten sites, including Nagano with an altitude of 418 m, are located at high altitudes, and the effect of high altitude is coupled with that of temperature change; hence we can see that they are out of the regression line in Fig. 1. All of the sites except for Asosan and Nikko are located in the Japanese Alps region.

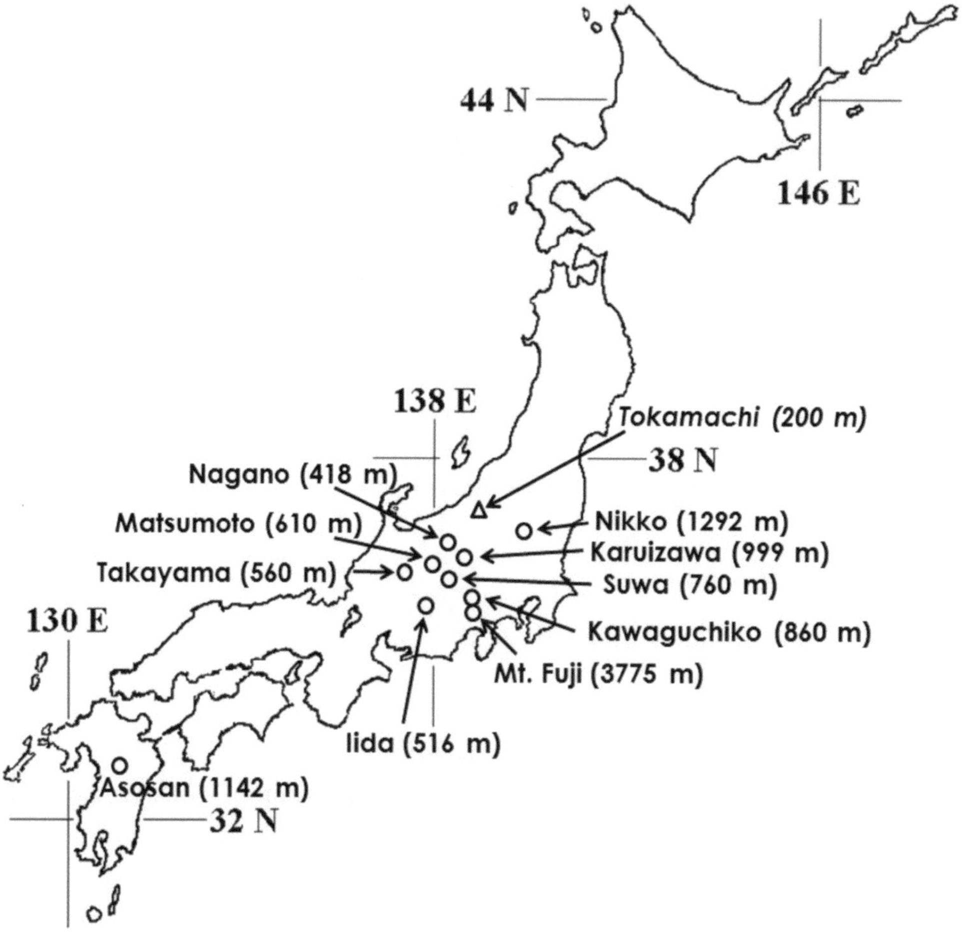

Fig. 2. Location map of the meteorological sites and their elevation

3. Variation in temperature and snow accumulation in the Japanes Alps region

As mentioned above, the high-altitude locations are selected from meteorological observation sites; we will discuss the variation tendencies of temperature and snow accumulation in resent years at these sites. It is well known that global warming trends are notably recognized in minimum temperature rather than mean temperature or maximum temperature, so the annual minimum temperature is discussed for the warming index. Though Nikko and Asosan are not located in the Japanese Alps region, ten sites including these two sites are discussed.

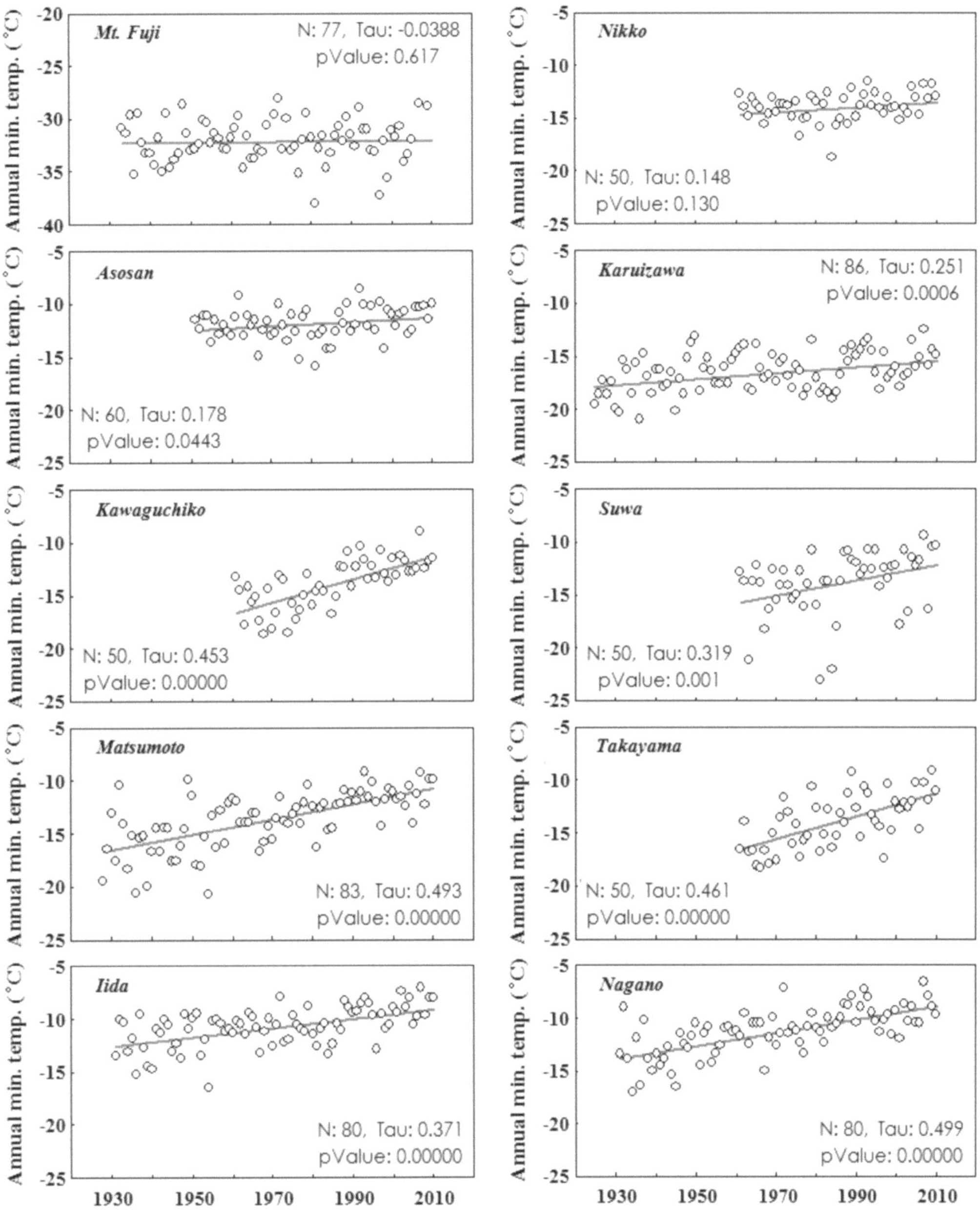

Fig. 3. The varifation in the annual minimum temperature and the results of the Mann-Kendall test at each site

The variation in the annual minimum temperature at each site is shown in Fig. 3, using data from the Japan Meteorological Agency. We used observations up to 2010 for all sites, with the longest dataset going back 86 years to 1925 at Karuizawa. The shortest observation period is 50 years at Nikko, Kawaguchiko, Suwa, and Takayama. In Fig. 3, the results of the

Mann-Kendall test are also shown to demonstrate the statistical significance of the annual minimum temperature variation over long periods. At Mt. Fuji, tau = -0.0388, and shows a slightly decreasing trend; however, P = 0.617 and the change in the annual minimum temperature for 77 years is statistically insignificant. At Nikko, tau = 0.148, which shows a

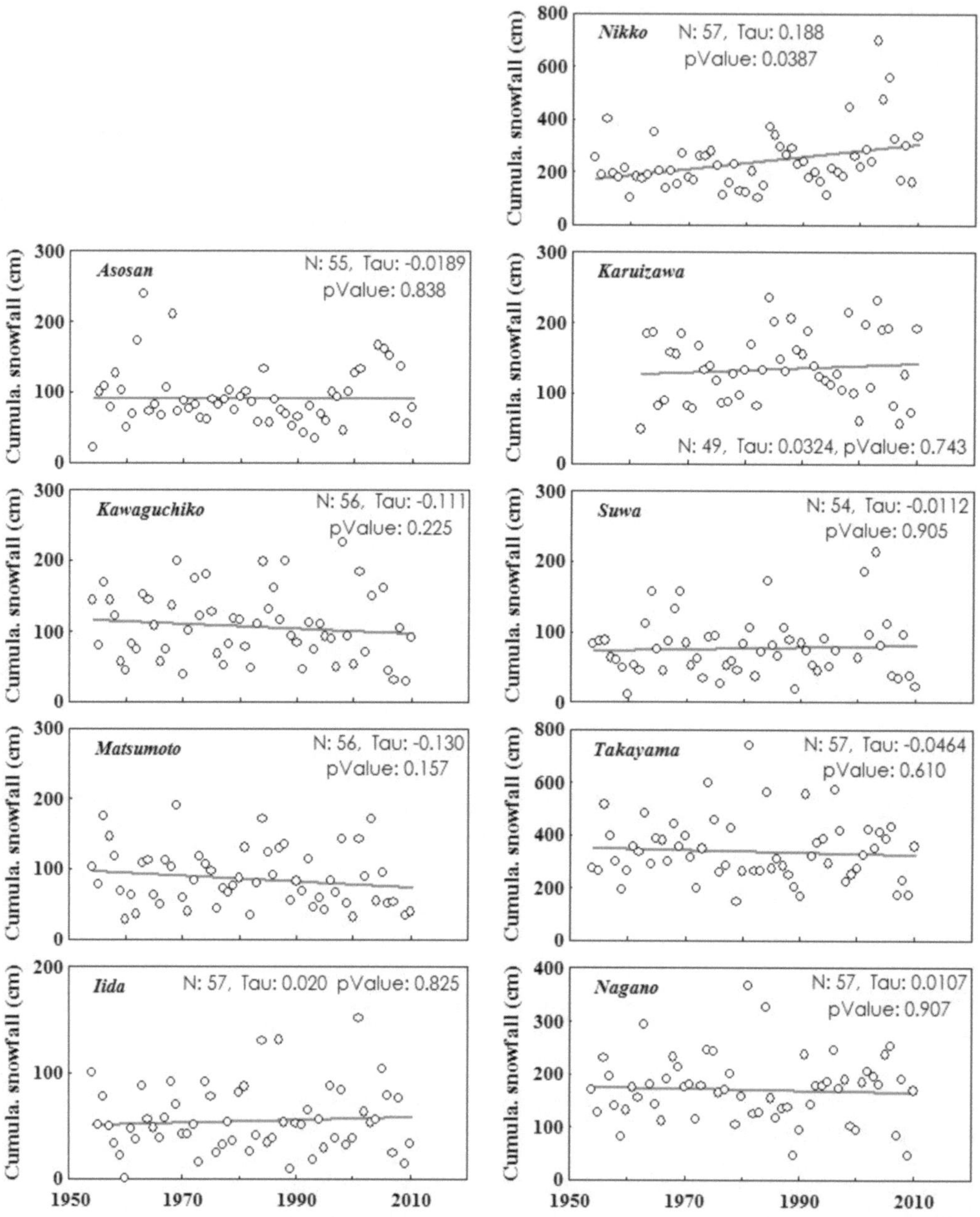

Fig. 4. The variation in the annual cumulative snowfall and the results of the Mann-Kendall test at each site

weak increasing trend of annual minimum temperature; however, P = 0.130 and it does not show a statistically significant trend in 50 years of annual minimum temperature. At Asosan, tau = 0.178 and P = 0.0443, indicating a statistically increasing trend of annual minimum temperature in the 5 % significant level. In seven sites, from Karuizawa to Nagano, an increasing trend of annual minimum temperature is recognized at the 1 % significant level, which is statistically significant. As can be seen, at a location with extreme high altitude such as Mt. Fuji, the annual minimum temperature over the last several decades does not increase or decrease. At the second highest altitude, Nikko, the changing trend is insignificant; however, at altitudes lower than Asosan, the increasing trend of annual minimum temperature during the last several decades is statistically significant. Nonetheless, it should be noted that the observation sites at Mt. Fuji, Nikko, and Asosan are set up in places with either no population or very small population; in contrast, the observation sites from Karuizawa to Nagano are set up in urban environments. Thus, the influence of urban climate is added to the notable increasing trend of annual minimum temperature in Fig. 3.

The variation in the annual cumulative snowfall is shown in Fig. 4 for each observation sites using the data from the Japan Metrological Agency. Winter generally means from November to April, hence, cumulative snowfall here indicates the cumulative value of daily snowfall from November (sometimes October) to next April (sometimes May). The annual cumulative snowfall in 2010 is calculated from the daily value between November 2009 and April 2010. There is no snowfall observation at Mt. Fuji, so it is not shown in Fig. 4. In each site, the aggregation period is almost the same, 54-57 years. The results of the Mann-Kendall test are also shown in Fig. 4 to consider the statistical significance of the variation in the annual cumulative snowfall over a long period. According to these results, at Nikko, tau = 0.188, P = 0.0387, and the increasing trend of annual cumulative snowfall is statistically significant at the 5 % level. At the other eight sites, it is shown that the recent annual cumulative snowfall does not show any increasing or decreasing trend. At Nikko, the annual minimum temperature shows an insignificant, slightly increasing trend; however, the annual cumulative snowfall has a significant increasing trend. At the other eight sites, the annual minimum temperature shows a significant increasing trend; however, the annual cumulative snowfall has no increasing or decreasing trend. Thus, an increase in the annual minimum temperature does not necessarily lead to a decrease in the annual cumulative snowfall.

Figure 5 shows the variation in the annual maximum snow depth in each site using the data from the Japan Metrological Agency. The snow depth and temperature at Karuizawa have been observed for 86 years, since 1925, however, in the other sites the snow depth had been observed only for around 50 years. At Mt. Fuji, observations of snow depth ended in 2004. The results of the Mann-Kendall test are also shown in Fig. 5 for the statistical significance of the long period data of variations in the maximum snow depth. Accordingly, in Mt. Fuji, tau = 0.239, P = 0.032, and the increasing trend of the annual maximum snow depth is statistically significant at the 5 % level. At the other nine sites, the absolute value of tau is small and the P value is large, which means the annual maximum snow depth is not showing a statistically significant change in recent decades. At Nikko, despite the significant increasing trend of annual cumulative snowfall, for the annual maximum snow depth, tau = -0.121 and P = 0.219, which does not show a statistically significant change.

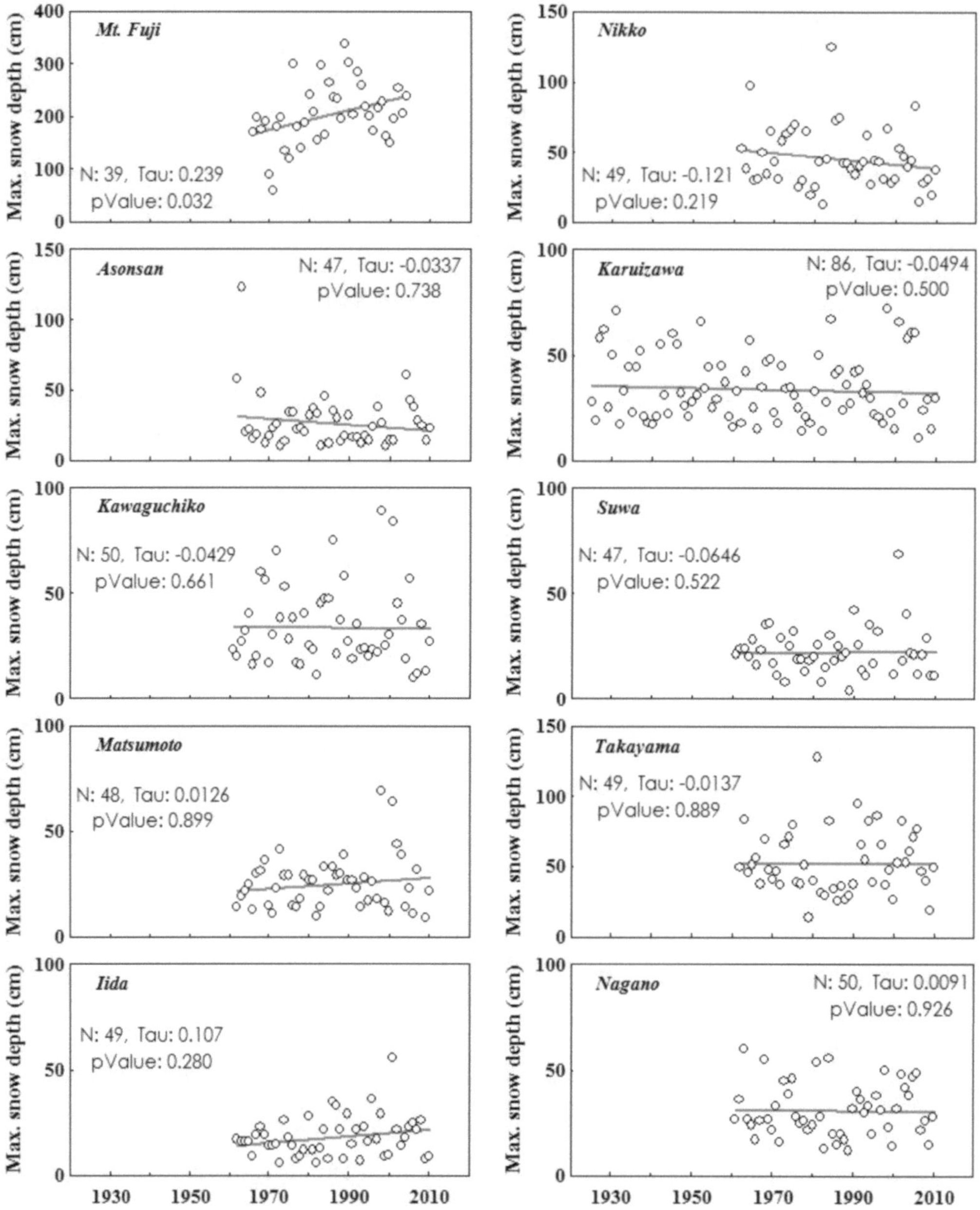

Fig. 5. The variation in the annual maximum snow depth and the results of the Mann-Kendall test at each site

4. Relationship between winter mean air temperature and winter precipitation in the Japanese Alps region

The IPCC 4th report also states that future warming will increase precipitation in mid-high latitudes in winter. To examine this issue with regards to the Japanese Alps region, the

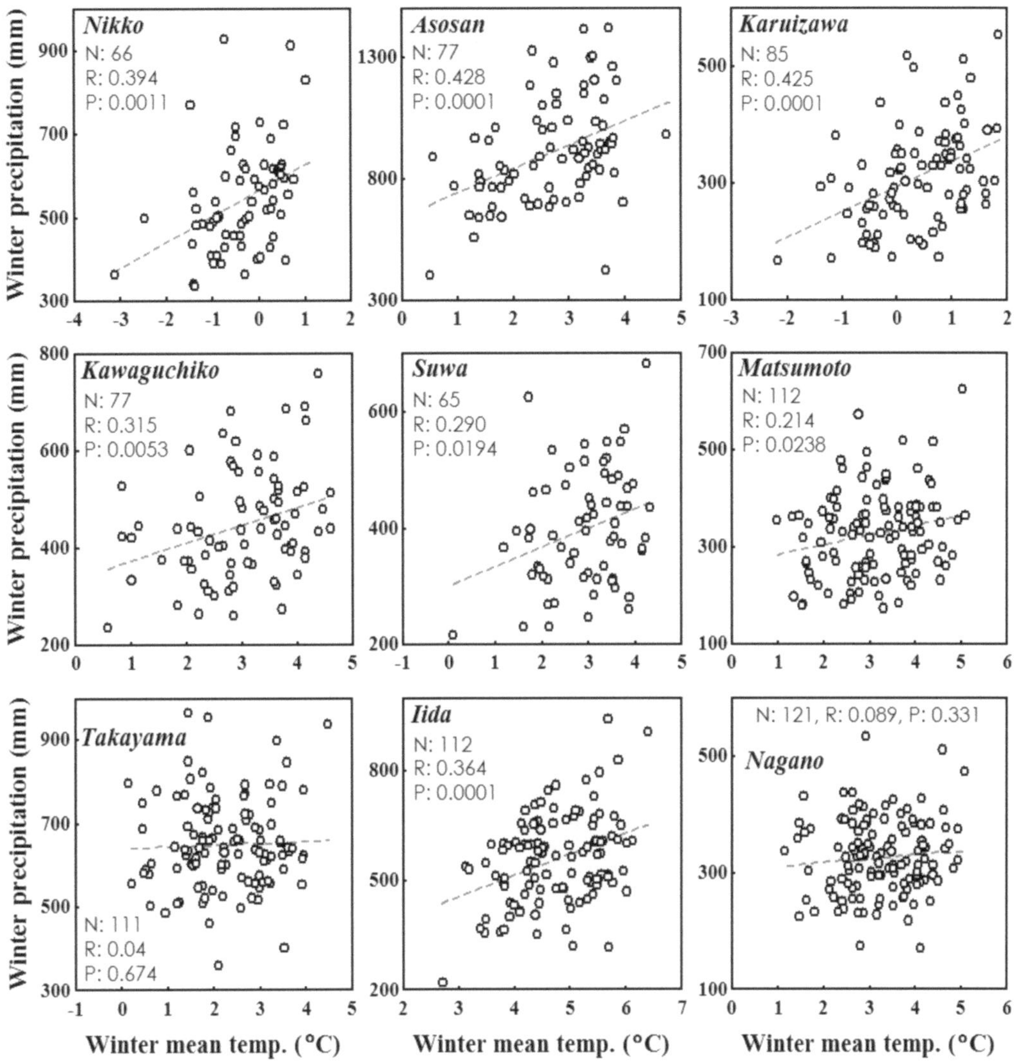

Fig. 6. Relationship between the winter mean air temperature and the winter precipitation at the observation sites

interannual variability of the winter mean air temperature and winter precipitation are discussed. The Japan Meteorological Agency has carried out precipitation observations in all the sites except Mt. Fuji, and these were used to calculate the precipitation in winter (from November to April) which is shown along with the winter mean air temperature in Fig. 6.

The monthly mean air temperature and monthly precipitation have been observed for a long period by the Japan Meteorological Agency. Both data-sets were obtained spanning 121 years from 1890 at Nagano, 112 years (from 1988) at Matsumoto and Iida, and for 111 years (since 1900) at Takayama. Even the shortest record at Suwa goes back 65 years. At all of the nine sites, winter precipitation shows an increasing trend with an increase in the winter mean air temperature. The mutual relations are statistically significant at a 1 % level at

Nikko, Asosan, Karuizawa, Kawaguchiko, and Iida. At Suwa and Matsumoto, the mutual relation is statistically significant at a 5 % level. At Takayama and Nagano, the relation is not statistically significant. Iida is exceptional, in favourable correlation of winter mean air temperature and winter precipitation, we note that the statistical significance of the correlation is reduced as the observations move from high altitude to lower altitudes.

At observational sites except for Mt. Fuji and Nikko, we recognize an increasing trend of winter mean air temperature in recent years along with an increase in winter precipitation. We note, however, that the annual cumulative snowfall or the annual maximum snow depth has not increased. This is because in the area of the Japanese Alps that is below 1,000 m, the changes in air temperature in winter are around 0 °C, a temperature that is close to the precipitation temperature boundary, which determines whether particles formed in the cloud will fall as snow or rain. That is, if the air temperature increases, winter precipitation would increase; however, this affects rainfall but does not increase snowfall.

However, a relatively large part of the Japanese Alps region is within the altitude range of 1,000–3,000 m and the snowfall there serves as a vital water resource. Therefore, the effect of global warming on snow accumulation in mountain areas is a crucial issue. At places with high altitude in the Japanese Alps, it is rare for the air temperature to rise to around 0 °C at the current levels of temperature increase. An increase in winter precipitation should lead directly to an increase in snow accumulation in the high altitude area. Unfortunately, currently there are no observational records to base our discussion on. We must note that lack of meteorological observation data from high altitudes is a very unfavorable situation when attempting to evaluate the effect of global warming on the ecological system and water resources in mountainous regions.

5. Variation in snow depth at Tokamachi

As mentioned above, observations of snow depth by the Japan Meteorological Agency have been recorded for over 86 years since 1925, even in Karuizawa. At the Tokamachi Experimental Station of the Forestry and Forest Products Research Institute (represented by a triangle in Fig. 2), snow depth has been continuously observed since 1917 (Takeuchi, *et al.*, 2008).

The record of daily snow depth is shown in Fig. 7. Figure 7 also shows the annual maximum snow depth for each winter, we can see the total amount of snow in each winter from the graphical form. The annual maximum snow depth in the winter of 1944–45 was 425 cm and is the highest recorded at Tokamachi. Snow began falling at the beginning of November 1944 and fell continuously until the middle of May 1945. We can see in the graph that the snow is also extremely deep. The next highest record of annual maximum snow depth was 381 cm in the winter of 1937–38, 380 cm in the winter of 1933–34. In Tokamachi, years with exceptionally heavy snowfall are often observed in the 1930s and 40s. After 1949, there is an increase in lighter snowfall years; however, in the 1980s extraordinarily heavy snow years are more common such as 377 cm in the winter of 1980–81 and 367 cm in the winter of 1983–84. After 1987, the lighter snowfall trend returned. The annual maximum snow depth was only 81 cm in the winter of 1988–89. Then, the trend of heavy snow years resumed, with 305 cm in the winter of 2004–05 and 313 cm in the winter of 2005–06.

As mentioned above, the interannual variation in snowfall at Tokamachi has wide fluctuations; some years have an annual maximum snow depth of 425 cm while on other

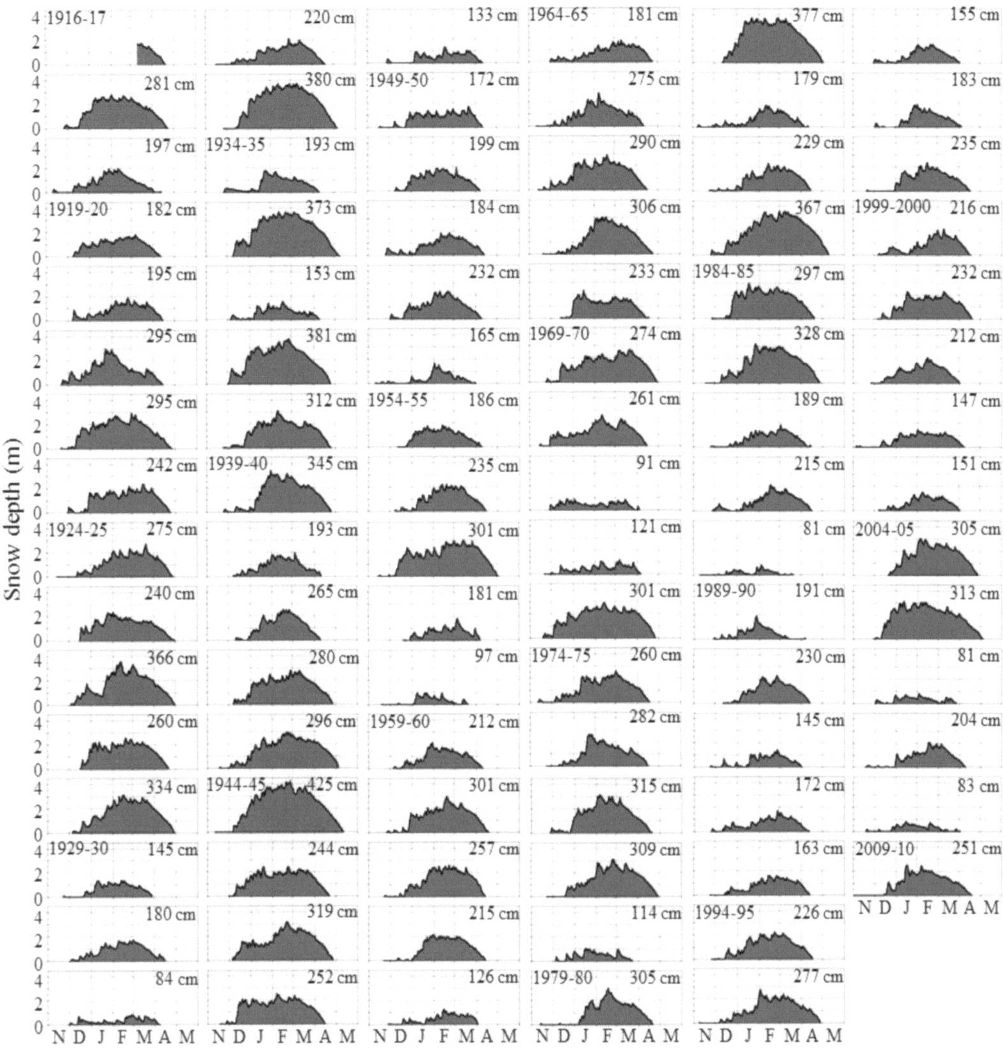

Fig. 7. The record of the daily snow depth in Tokamachi

years it is less than 90 cm. This is because the altitude of the observation site at Tokamachi is 200 m and the winter temperature is relatively high. This area has rainfall even in winter. That is because in a warm winter year, a slight increase in the air temperature creates suitable conditions for precipitation of rather than snow, so the accumulated snow depth does not get deeper.

Because the winter air temperature may affect snow accumulation at Tokamachi; we will discuss the relation between them. Figure 8 shows the relationship between the winter mean air temperature (from November to April) and the annual maximum snow depth from 1918 to 2010 at Tokamachi. As Fig. 8 shows, the statistics for the data spanning 93 years at Tokamachi gives R = 0.749 and P = 0.0000, which statistically shows quite significant.

It is clear that in the winter, when the winter mean air temperature is low, the annual maximum snow depth is greater and when the winter mean air temperature is high, the

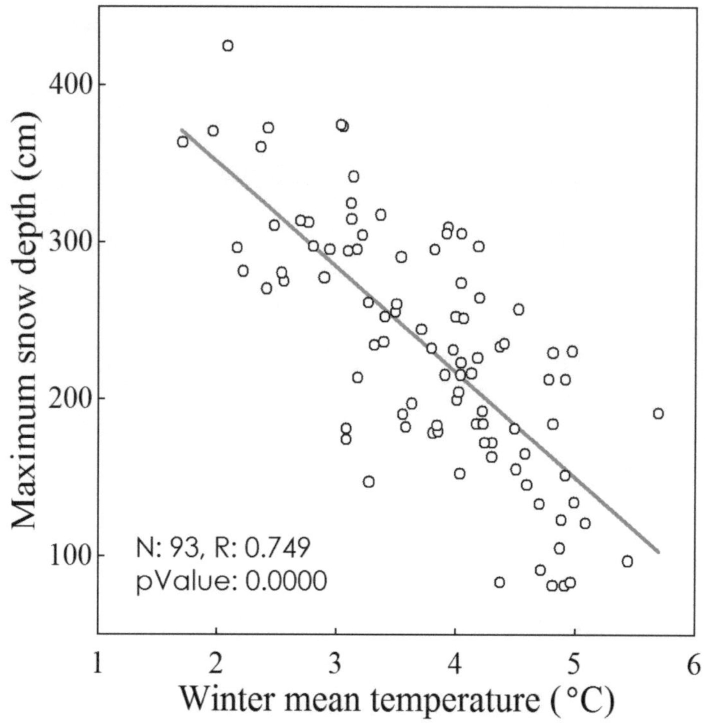

Fig. 8. The variation in the winter mean air temperature (from November to April) and the annual maximum snow depth from 1918 to 2010 in Tokamachi

annual maximum snow depth becomes shallower. We can conclude that snow accumulation is defined by winter air temperature at Tokamachi.

We now consider the interannual variation in the winter mean air temperature and the annual maximum snow depth. Figure 9 shows the variation in the winter mean air temperature and the annual maximum snow depth from 1918 (November 1917–April 1918) to 2010. In Fig. 9, the straight lines that show each variation trend are also displayed, and the result of the Mann-Kendall test which shows the statistics of the long-term variability. Considering the winter mean air temperature for 93 years, tau = 0.262 and P = 0.0002, with a statistically significant increasing trend. The variation in the annual maximum snow depth for 93 years is tau = -0.120 and P = 0.089, with a weak decreasing trend, though statistically significant level was not confirmed. However, there is possibly a gap before and after a certain year in both the winter mean air temperature and the annual maximum snow depth. As mentioned before, years with extraordinary heavy snowfall occurred frequently until 1948, however afterwards, years with lighter snowfall increased. By evaluating the Mann-Kendall test for 1918–1948 and 1949–2010 separately, we find that the annual maximum snow depth in the former period is on a weak increasing trend, tau = 0.166, P = 0.190 while in the latter period no variation in trend can be identified, with tau = 0.0005 and P = 0.995. As for the winter mean air temperature, tau = -0.046 and P = 0.717 in the former period, and tau = -0.009 and P = 0.916 in the latter period with no obvious trend in the variation. Climate is known to change with some gap, but not to shift unilaterally. The winter air temperature and snow accumulation at Tokamachi is characterized by a clear gap.

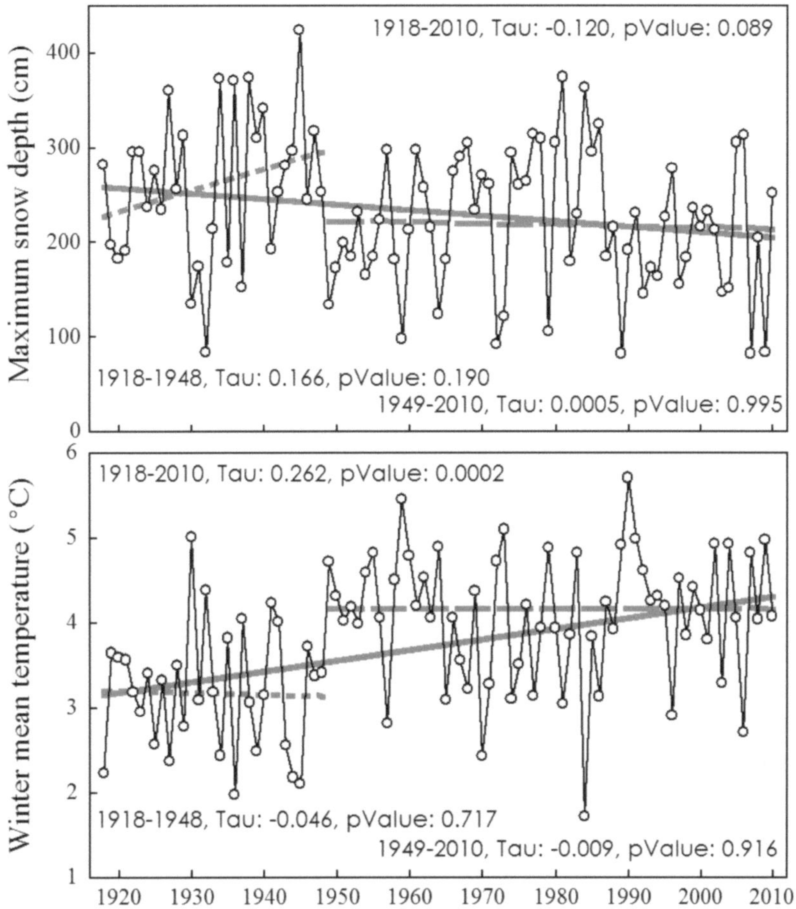

Fig. 9. The variation in the winter mean air temperature and the annual maximum snow depth from 1918 to 2010 in Tokamachi

As mentioned previously, the northwest coast of Honshu is one of the snowiest regions in the world. However, at lower altitudes where the winter temperature is relatively high, the interannual variation in snow accumulation is wide and the variation in snow accumulation is not necessarily a simple upward or downward trend.

6. Development of meteorological observation stations at high altitudes in the Japanese Alps region

It is said that mountainous areas are quite sensitive to global-scale environmental change such as warming. The ecological system in mountainous areas is strongly affected not only by air temperature, wind, or solar radiation, but also by snow cover. Snow accumulation in mountainous areas plays the role of a white dam, which is an essential water resource. Therefore, the effect of global warming on these meteorological elements is a critical issue. However, Mt. Fuji Weather Station, which used to be a symbol of meteorological observation in mountainous areas, has been unmanned since August 2004. Since then, only temperature, humidity, atmospheric pressure, and hours of daylight (summertime only) are monitored. Of the other observation sites of the Japan Meteorological Agency, Nobeyama, at 1350 m elevation, is the highest. On evaluating the effect of a global-scale warming event on environmental change at a regional scale of the Japanese Alps at a high altitude of 3,000 m, we must note that the lack of meteorological observation data for high altitudes is a very unfavorable situation for any effort to evaluate the effect of warming against the ecological system and water resources in mountainous ranges.

The Institute of Mountain Science, Shinshu University has developed a network of meteorological observations in the Japanese Alps region. The ten stations shown in Fig. 10 have already started recording observations. The highest observation site is Yari at 3125 m.

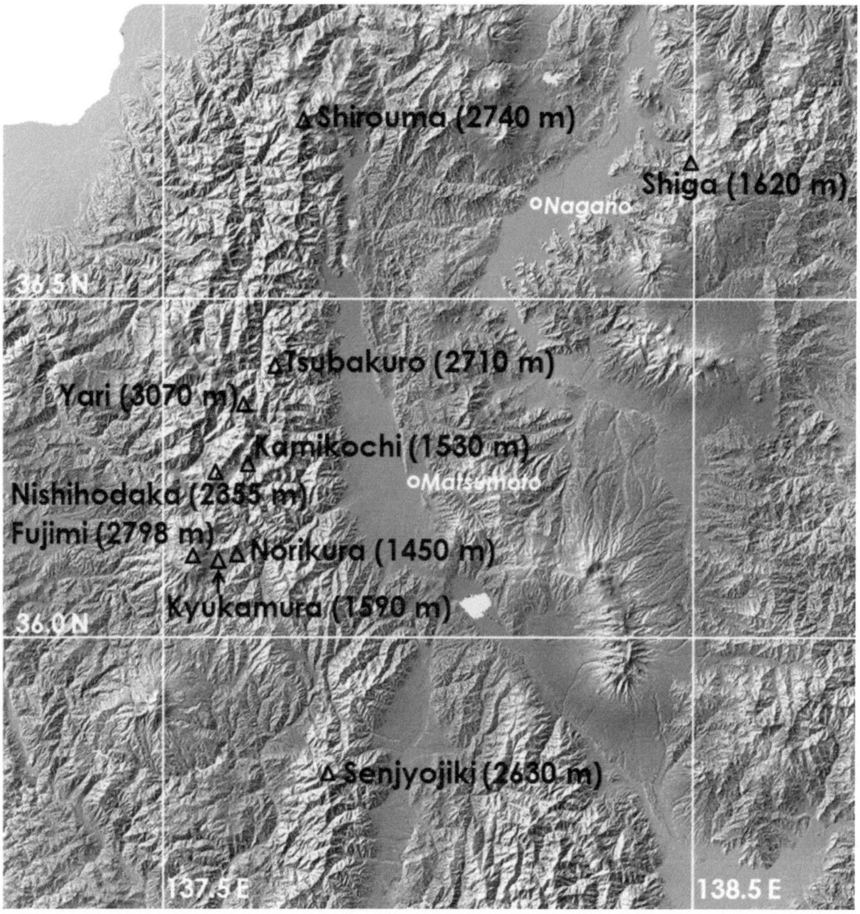

Fig. 10. Location map of the meteorological station developed by the Institute of Mountain Science, Shinshu University

Observation data from these sites are sent to a computer at the Institute via a data communication mobile telephone network or a phone line year around. These meteorological observation data are available on the institute's website in quasi-real time. Table 1 lists the location and observational components at each observation site.

Observation site	Height	North lat.	East lon.	Start at	Obserbation component						
	(m)	(degree)	(degree)	(y/m/d)	Temp.	Humid.	Wind	Press.	Solar rad.	Precip.	Snow dep.
Yari	3070	36.340	137.646	2008/10/22	O	O	O	O	O	△	×
Fujimi	2798	36.121	137.558	2003/10/15	O	O	O	O	O	△	×
Shirouma	2740	36.752	137.753	2010/09/26	O	O	O	O	×	△	×
Tsubakuro	2710	36.399	137.715	2008/11/14	O	O	O	O	O	△	×
Senjyojiki	2630	35.778	137.814	2006/11/08	O	O	O	O	O	O	×
Nishihodaka	2355	36.265	137.617	2008/10/15	O	O	O	O	O	△	O
Shiga	1620	36.711	138.495	2007/11/21	O	O	O	O	O	O	O
Kyukamura	1590	36.114	137.613	2002/11/18	O	O	O	O	O	△	O
Kamikochi	1530	36.253	137.669	2008/09/08	O	O	O	O	O	O	O
Norikura	1450	36.122	137.630	2006/03/18	O	O	O	O	O	O	×

Table 1. Location and observational components at each observation site

Figure 11 shows the daily mean air temperature at the Fujimi observation site as an example of an observation result. At the beginning of the observation, some data were missing in winter due to battery condition, *etc.*; however, since spring 2005, continuously observation data have been available. So far, the maximum value of the daily mean air temperature at Fujimi is 15.0 °C and the minimum value is -24.1 °C. Considering the air temperature inter-annual variation trend in winter and summer, there is no significant increase or decrease during this period.

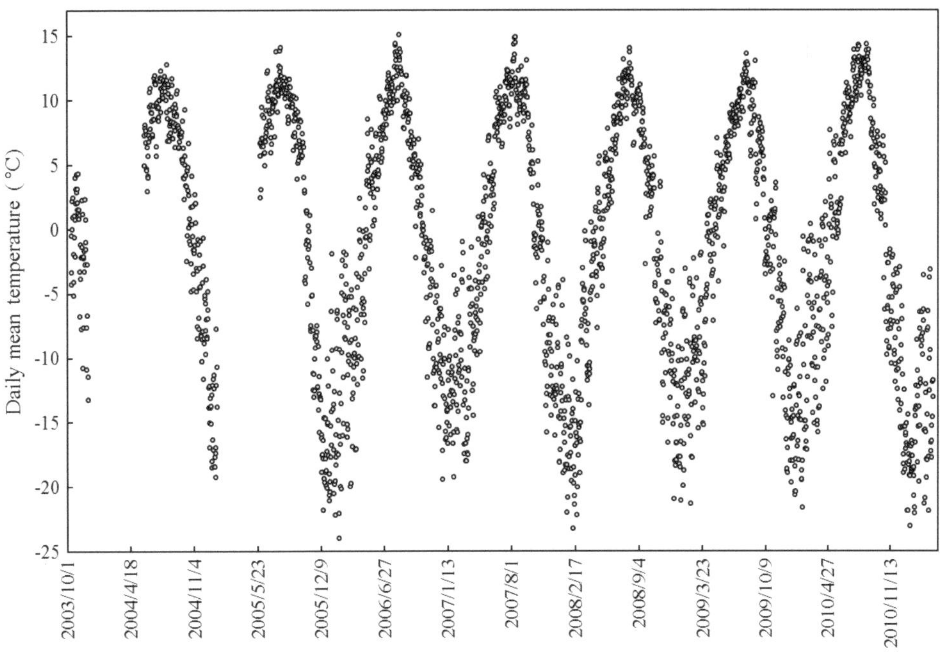

Fig. 11. The variation of the daily mean air temperature at the Fujimi observation site

It is difficult to observe snow depth in a mountain region with extreme heavy snowfall. Moreover, it is almost impossible to observe snowfall as precipitation in a mountain location with no commercial power source. Therefore, a method for estimating winter solid precipitation in mountain regions, using a snow chemical technique is being developed (Suzuki *et al.*, 2011).

7. Conclusion

Under the present circumstances, the role of the mountain regions as a system for purification of water and air that is essential for human beings' existence is not duly recognized and is undervalued. However, there is no doubt that we need to understand the response of mountain regions to global-scale environmental change in the near future. To achieve that, we need to discuss the matter based on proper observation data. We therefore plan to continue our meteorological observations at high–altitude mountain regions, overcoming various issues that arise under the extreme conditions of low temperature and gale-force winds (including powerful blizzards in winter). The steady recording of observations does not get immediate results such as publishing articles, *etc.* so it trends to be overlooked, however, we must not forget that there is no reliable future prospect for understanding this topic without obtaining the above mentioned data on site.

8. References

Inoue, S. & Yokoyama, K. (2003). Estimates of snowfall depth, maximum snow depth, and snow pack conditions in Japan by using five global warming predicted data, *Journal of Agricultural Meteorology*, Vol. 59, pp. 227-236, ISSN 0021-8588

Ishii, H. & Suzuki, K. (2011). Regional Characteristics of Variation of Snowfall in Japan, *Journal of Japanese Association of Hydrological Sciences*, Vol. 41, ISSN 1342-9612

Suzuki, K., Ikeda, A., Kaneko, Y., Suzuki, D. & Maki, T. (2011). Estimation of winter precipitation at a mountainous site by snow chemical analysis, *Journal of the Japanese Society of Snow and Ice*, Vol. 73, ISSN 0373-1006

Takeuchi, U., Niwano, S., Murakami, S., Yamanoi, K., Endo, Y. & Kominami, Y. (2008). Meteorological statistics during the past 90 years from 1918 to 2007 at Tohkamachi in Niigata Prefecture, Japan, *Bulletin of the Forestry and Forest Products Research* Institute, Vol. 7, pp. 187-244

5

Melting of Major Glaciers in Himalayas: Role of Desert Dust and Anthropogenic Aerosols

Anup K. Prasad[1,2], Hesham M. Elaskary[1,2], Ghassem R. Asrar[1],
Menas Kafatos[1,2] and Ashok Jaswal[3]
[1]*School of Earth and Environmental Sciences, Schmid College of Science*
Chapman University, Orange
[2]*Center of Excellence in Earth Observing, Chapman University, Orange*
[3]*India Meteorological Department, Pune*
[1,2]*USA*
[3]*India*

1. Introduction

The Himalayan and Tibet Glaciers, that are among the largest bodies of ice and fresh water resource outside of the polar ice caps, face a significant threat of accelerated meltdown in coming decades due to climate variability and change (Hasnain et al., 2002; Lau et al., 2010; Shrestha et al., 2011; Scherler et al., 2011). The rate of retreat of these glaciers and changes in their terminus (frontal dynamics) is highly variable across the Himalayan range (Raina, 2010; Scherler et al., 2011). These large freshwater sources are critical to human activities for food production, human consumption and a whole host of other applications, especially over the Indo-Gangetic (IG) plains. They are also situated in a geo-politically sensitive area surrounded by China, India, Pakistan, Nepal and Bhutan where more than a billion people depend on them. The major rivers of the Asian continent such as the Ganga (also known as Ganges), Brahmaputra, Indus, Yamuna, Sutluj etc., originate and pass through these regions (Kulkarni et al., 2010; Kehrwald et al., 2008; Bookhagen & Burbank, 2010; Immerzeel et al., 2009, 2010) and they have greater importance due to their multi-use downstream: hydro power, agriculture, aquaculture, flood control, and as a freshwater resource. Recent studies over the Himalayan Glaciers using ground-based and space-based observations, and computer models indicate a long-term trend of climate variability and change that may accelerate melting of the Himalayan Glaciers (Lau et al., 2010; Prasad et al., 2009). Other studies also suggest a decreasing trend in snowfall, which has historically served as a main source of precipitation for maintaining the glaciers and fresh water resources in this region. Short-term studies of terminus and mass balance of the Himalayan Glaciers, based on in situ observations, show an accelerated rate of melting (Berthier et al., 2007; Das et al, 2010; Raina, 2010). However, several studies report the rate of melting and the corresponding change in temperature is found to vary across the entire Himalayan range (Naz et al., 2011a; Raina, 2010; UNEP, 2008). Observations from space-based lasers altimeters (such as GLASS/ICESat), show glacial thickening in certain areas such as the Karakoram (Naz et al., 2011b, personal communication) but increased melting in its surrounding regions. The recent increase in the lake water level over the Tibetan Plateau using ICESat altimetry data (2003-2009) emphasize the effect of global warming on the glaciers (Zhang et al., 2011).

The major research objectives (and organization of sections of the chapter) of the present study are:

1. Brief overview of the source and influence of desert dust and anthropogenic aerosols on the atmosphere, regional temperature change, evidences from ice-core studies, and retreat pattern of Himalayan and Tibet glaciers. The results from recent studies on these aspects are briefly discussed in section 2.
2. Study of the aerial extent of major Himalayan and Tibetan Plateau snow cover and glaciers. We have used space-based sensors to study the several indicators of glacier dynamics such as glacier front, glacier lakes (melt-water lakes), and indicators of changes in the temperature of glaciers etc. We have used a combination of Landsat (since 1970s) and Advanced Spaceborne Thermal Emission and Reflection Radiometer (ASTER) observations (2000-2010). The findings from the current study are discussed in section 3.
3. Source and transport of desert dust aerosols and anthropogenic pollutants over the Indian sub-continent. The inter-annual variability in the aerosol loading over the Indian sub-continent and the surrounding regions, such as Arabian Sea, have been studied using Moderate Resolution Imaging Spectroradiometer (MODIS) observations during the last decade. The role played by increasing anthropogenic activity, especially thermal power plants, industries, and vehicular emissions (since 1980) in changing the regional atmospheric (tropospheric) chemistry have been discussed. The influence of dense networks of coal-fired thermal power plants along the pathways of transport of desert dust has been discussed using observations from OMI Aura tropospheric NO_2. The results from MODIS AOD (2000-2008) and OMI NO_2 observations are discussed in section 4. The increasing awareness about the impact of dust storms due to daily observations from multiple environmental satellites and dissemination of news by media (print and television) since the last decade is also discussed briefly in section 4.
4. Whether the major dust storms are capable of reaching up to high-altitude Himalayas (~4-6 km above the mean sea level - msl). What is the height of vertical mix-up of aerosols during the dust storms over the IG plains, Himalayan and Tibet regions and its vertical range from the start of pre-monsoon season (April), through May month, and prior to the arrival of monsoon rainfall (late June or early July). Further, the variation of vertical mix-up across the western, central, and eastern regions is not well known.

To answers these questions, several case studies showing evidence of dust storms that can reach up to high altitude Himalayas (~4-6 km above msl) have been presented using MODIS Terra, and Aqua observations during the pre-monsoon season (please see section 5.1). The MODIS Terra and Aqua (deep-blue and deep-target aerosol optical depth - AOD) observations showing high AOD over Himalayan cryosphere regions is illustrated in section 5.1. The significance of band 3 (blue band) in aerosol retrieval over land is also discussed in this section.

The results from several Cloud-Aerosol Lidar and Infrared Pathfinder Satellite Observations (CALIPSO) observations (night-time) that show the vertical structure of major dust storms across the IG plains, Himalayas and Tibet Plateau is shown in this article (please see section 5.2). The selected case studies of dust storms from CALIPSO observations for year 2010 not only cover the western, central, and eastern IG plains and Himalayas but also the entire pre-monsoon season (April to June) (section 5.2). The impact of dust storms on the western, central and eastern Himalayas and IG plains has been discussed using multiple parameters specifically developed for the detection of aerosols and cloud from CALIPSO observations.

5. Impact of the entrainment of anthropogenic aerosols with the desert dust, during the long-range transport that leads to the complex climate forcing scenarios is discussed in section 6.

2. Aerosols, temperature change, and retreat of glaciers

2.1 Mid-latitude westerlies and dust over the Himalayas and IG plains

The dust storms originating from Africa (Sahara desert), Arabia, Middle East, Afghanistan, and the Thar desert regions are transported by wind over to the IG plains and Himalayas (also called as Himalaya) due to the dominant westerly wind during the pre-monsoon season (Kayetha et al., 2007; Prasad et al., 2006b; Prasad and Singh, 2007a; Singh et al., 2004). The western and central part of Himalayas and IG plains are affected more than the eastern part as they are closer to the source of dust storms (Prasad and Singh, 2007a, 2007b). The glacier retreat pattern is different over the western and eastern Himalayas and Tibet plateau (He et al., 2003). The observed differences in the warming trends and glacial retreat of the western and eastern Himalayan range is very important in view of this contrast in aerosol loading over the region. A combination of the desert dust and aerosols resulting from anthropogenic sources (i.e. industrial, automobiles and crop, wood and animal waste burning) further complicates the interactions between the atmosphere and dynamics of glaciers in Himalayas (Ramnathan et al. 2005).

2.2 Ice-cores: Evidence of dust and anthropogenic aerosols

The influence of dust storms and anthropogenic activities on glacier dynamic has also been documented in the analysis of ice-cores from Tibet and Himalayan region (Duan, et al., 2007; Kang et al., 2000; Lee et al., 2008; Xu et al., 2007). The paleoclimatic records indicate that the South Asian monsoon and the mid-latitude westerlies affect the Himalayan glacier dynamics (Benn and Owen, 1998). The analysis of ice cores for multiple trace elements from the northeastern slope of Mt. Everest (central Himalayas) indicate high crustal enrichment factors that are attributed to the transport of anthropogenic aerosols to the high altitude region (Lee et al., 2008). The ice-cores from central Himalayas (Dasuopu glacier, 1988-1997) show seasonal variation of oxygen isotope and major ion concentrations (Ca^{2+}, Mg^{2+}, NH_4^+, SO_4^{2-} and NO_3^-) that coincide with the pre-monsoon dust storm period (Kang et al., 2000). The increasing effect of anthropogenic aerosols due to increased burning of fossil fuel in the last decade is reflected in some Himalayan ice-core analysis. For instance, the sulfate record (1000-1997) from ice cores of Dasuopu glacier show increased anthropogenic influence since the 1930s. Moreover, the doubling of sulfate concentration since 1970 coincides with the increased anthropogenic activities such as growth of coal-fired power plants in and around the IG plains. Thus, the transport of anthropogenic aerosols together with dust storm is evident based on the analysis of ice-cores in the Himalayan range (Duan et al., 2007).

2.3 Temperature change over Himalayan-Tibet glaciers

Recent studies show an increasing temperature trend over the Himalayas, Tibetan Plateau and IG plains (Shrestha et al., 1999; Liu & Chen, 2000; Rikiishi & Nakasato, 2006; Arora et al., 2005; Prasad et al., 2009). The Nepal Himalayas show a warming trend ranging between 0.068 to 0.128 °C (1971-1994) (Shrestha et al., 1999). The Tibetan Plateau also show a warming trend of 0.16°C/decade for the annual mean during period 1955-1996 (Liu & Chen,

2000). A significant increase in the maximum temperature over the Kashmir region (+0.04 to 0.05 °C/year) and the minimum temperature over the Jammu region (+0.03 to +0.08 °C/year) have been observed for period 1976-2007 (Jaswal & Rao, 2010). The northwestern Himalayan region shows significant rise in the air temperature by 1.6 °C during the last century based on data from three stations (Shimla in Himachal Pradesh, Srinagar and Leh in Jammu and Kashmir) (Bhutiyani et al., 2007). The inequality of tropospheric warming trend between the western and eastern IG plains and Himalayas has been observed from Microwave Sounding Unit (MSU) (Prasad et al., 2009). The western Himalayas show annual mean Temperature Middle Troposphere (TMT) warming of 0.48 °K (0.016±0.005 °K/year) during 1979-2008 compared to 0.33 °K (0.011±0.005 °K/year) over the eastern Himalayas. Similarly, the annual mean TMT trend (0.018±0.005 °K/year) is higher over the western IG plains compared to the eastern IG plains (0.013±0.004 °K/year). Over the IG plains, the warming trend is prominent and statistically significant (>0.030 °K/year) during the months of December, February, March and May. The warming trend is also generally positive during period December-May with higher values over the western side compared to the eastern side. In general, the warming trend over the northern India (the IG plains) during the winter months has been found to be positive and significant (maximum temperature trend at +0.29 °C/decade and minimum temperature trend at +0.38 °C/decade during February) compared to the southern parts of India (Jaswal, 2010). The ground stations based winter-time warming trend over the north-western Himalayas (Bhutiyani et al., 2007) is also corroborated by the MSU tropospheric temperature trends (Prasad et al., 2009) implying that the warming trend is also significant at the elevated levels of the atmosphere. Recently, Kulkarni et al. (2010) have found significant changes in the snow cover over western and central Himalayan river basins especially during the winter season. The warming trend is usually 2-3 times higher over the individual months (December-May). The maxima of the mean monthly warming TMT trend is around 0.48 and 0.51 °K/decade (or 0.048±0.026 and 0.051±0.024 °K/year) over the Himalayas and IG plains, respectively (Prasad et al., 2009). In contrast to the western side (Jammu and Kashmir region), the temperature trends have been found to be trendless over the stations of northeastern India (Jhajharia & Singh, 2010).

2.4 Retreat of glaciers

The Himalayan glaciers are among of the fastest receding glaciers of the world. The increasing rate of retreat of glaciers, that varies with the region, has a potentially detrimental effect on the available freshwater (river) resources especially in India, Pakistan, Bangladesh, and China (Kulkarni, 2007; Kulkarni & Bahuguna, 2002; Kulkarni et al., 2005; Krishna, 2005; Winiger et al., 2005; Rees & Collins, 2006; Kehrwald, et al., 2008). Recent studies document the alarming retreat of Parbati Glacier and Chenab, Parbati and Baspa basins since 1962 (Kulkarni & Bahuguna, 2002; Kulkarni & Alex, 2003; Kulkarini et al., 2005). The Rongbuk Glacier (Mount Everest, central Himalayas) shows a retreat of 170-270 m in 30 years (1966-1997) that is equivalent to a retreat speed of 5.5-8.7 m a^{-1} (Qin et al., 2000). Based on geological record, the Siachen Glacier in the western Himalayas, a 74 km long valley glacier which is also the largest glacier in the Karakoram and second largest glacier known outside the polar and sub-polar region, has receded by approximately 76 km compared to the past inter-glacial period (Raina & Sangewar, 2007; Upadhyay, 2009). In the last century, the Siachen glacier has receded during the period 1929 to 1958 compared to 1862 data, but it does not show any retreat since 1958 (Raina & Sangewar, 2007). The rate of retreat of several glaciers over the Himalayan and Tibetan Plateau has been updated and discussed by several recent studies (Prasad et al., 2009; Raina, 2010; Scherler et al., 2011). The degradation of a glacier is accompanied by the debris cover around the glacier termini and the formation of lakes (Ageta et al., 2000). An 8 % loss of glacier area has been observed between 1963-1993 over Bhutan (eastern Himalaya) (Karma et al., 2003).

3. Glaciers and snow cover regions: Landsat and ASTER

The monitoring of Himalayan glaciers started in India (1st phase: early 2000 - 1950 by the Geological Survey of India, GSI) with approximately 20 glaciers from Jammu and Kashmir in the west to Sikkim in the east. This was followed by a more systematic 2nd phase from the year 1957 (International Geophysical Year) to 1970. The third phase started with the Induction of the International Hydrological Programme (1970) that covered a number of major glaciers: Siachen, Mamostang, Kumdan, Machoi in J&K, Barashigri, Sonapanii, Guglu in Himachal Pradesh, Gangotri, Arwa, Poting, Milam, Pindari, Shankalpa, Kalganga, Bamlas, Safed, Bhilmagwar, Pachu, Burphu in Uttarakhand and Zemu in Sikkim (Raina, 2010). All the glaciers under observation by Indian agencies (particularly GSI) show negative mass balance during 1970-2000. The decline in glacial mass is found to be highest in the western Himalayas (Jammu and Kashmir) compared to the eastern Himalayas (Sikkim), with a declining trend from west to east. However, some glaciers, such as Siachen in the western Himalayas, do not show appreciable change since 1970 (Raina, 2010; Raina & Sangewar, 2007).

Himalayan snow cover and glaciers are a major source of water for major rivers of Asia, such as Indus, Ganga (or Ganges), Brahmaputra, Sutluj, Yamuna etc. Among them, the major rivers of Asia, such as Indus, Ganga and Brahmaputra, originate at the border region of India, Nepal and China (Prasad et al., 2009; Prasad & Singh, 2007b) (Figure 1, 2a). The point of origin of some of these rivers is shown in Figure 1. The surface reflectance (true color) images of a part of this region, based on the Landsat observations (since 1972) and ASTER (since 2000) show a conspicuous decline in the snow cover and the formation of numerous new lakes especially at the terminus of receding glaciers (Figure 2a). The Landsat images from 1972, 1989, 2000 and 2006 and ASTER images of 2000 show a gradual decline in the snow cover compared to the earliest image from 1972 (Figure 1b). The ASTER image obtained in 2008 show, however, an increased snow cover compared to year 2000 due to an increased snowfall over the region during 2008 which is corroborated with the maximum snow cover extent product derived from MODIS Terra at 500m resolution. The Landsat and ASTER images over the western Himalayas (Figure 2b) and central Himalayas (Figure 2c) also show significant change in the snow cover since 1972. The monitoring of the Bara Shigri shows the glacier frontal retreat as well as vertical shrinkage since 1950 (Raina, 2010). The photographs over the western Himalayas (nearby Bara Shigri Glacier region, Himachal Pradesh, Figure 2b) show contamination of glaciers and their discoloring (yellowish layer over the white snow) (Figure 2d). The nature of material as seen over the peak Himalayan region is not known and needs further investigation. Figure 2c show the glaciers situated in the Tibet region that is ~85 km north of Kathmandu, Nepal (near the Nepal and Tibet border). These glaciers south of Lake Paiku Tso (Tibet) show the retreat pattern as well as the formation of numerous new glacial lakes (at their terminus) due to the melting of snow and glaciers. The change in color (light blue to deep blue) in the satellite images indicates the melting of snow (ice phase) to water phase that usually appears as deep blue to black in true color images. The eastern Himalayas, border of Arunachal Pradesh (India) and Tibet (China) also show a decline in the snow cover as seen in the Landsat images taken during year 1988 and 2001 (Figure 2e). The major glaciers of the eastern Himalayas are showing signs of melting and formation of new lakes in the region.

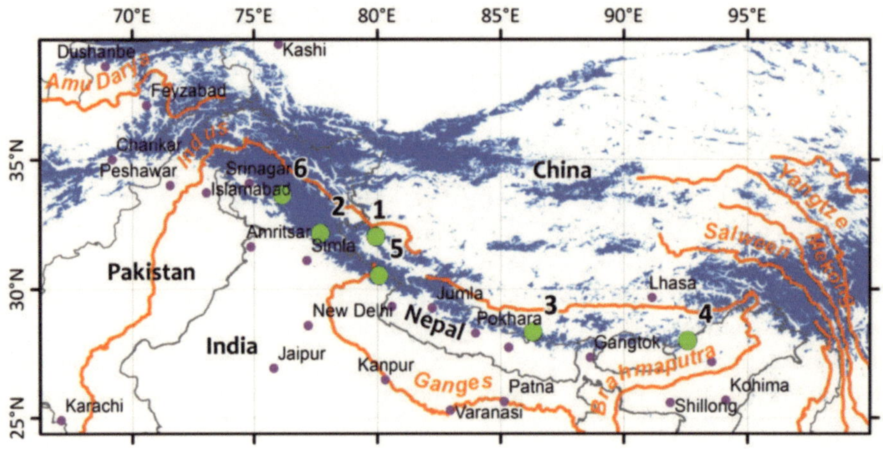

Fig. 1. The point of origin and pathways of flow of major rivers of Asia (marked as orange lines) are shown against the background of maximum snow cover extent from MODIS Terra (marked as blue color). The violet dots represent major cities in the region and the green dot (or circle) represents the location of major glaciers that are covered in Figures 2a-f.

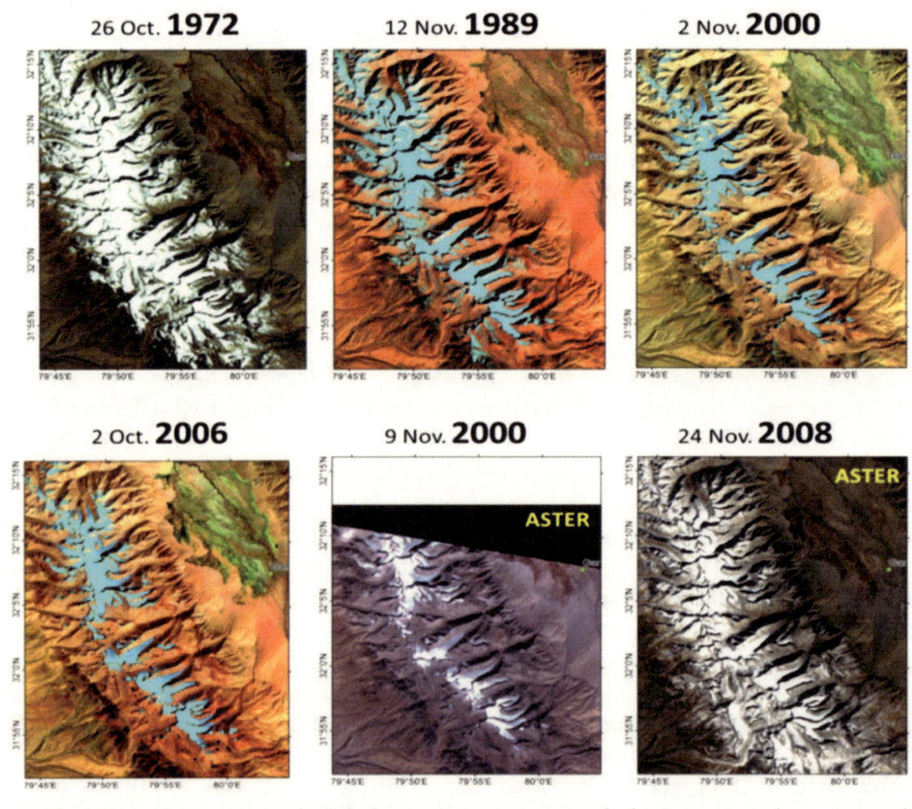

Fig. 2a. The true color images of Himalayan snow cover and glaciers, near the source of origin of three major rivers of Asia (Indus, Ganga and Brahmaputra), from the Landsat series (1972, 1989, 2000, 2006) and ASTER (2000, 2008). (*Location 1, in Figure 1*).

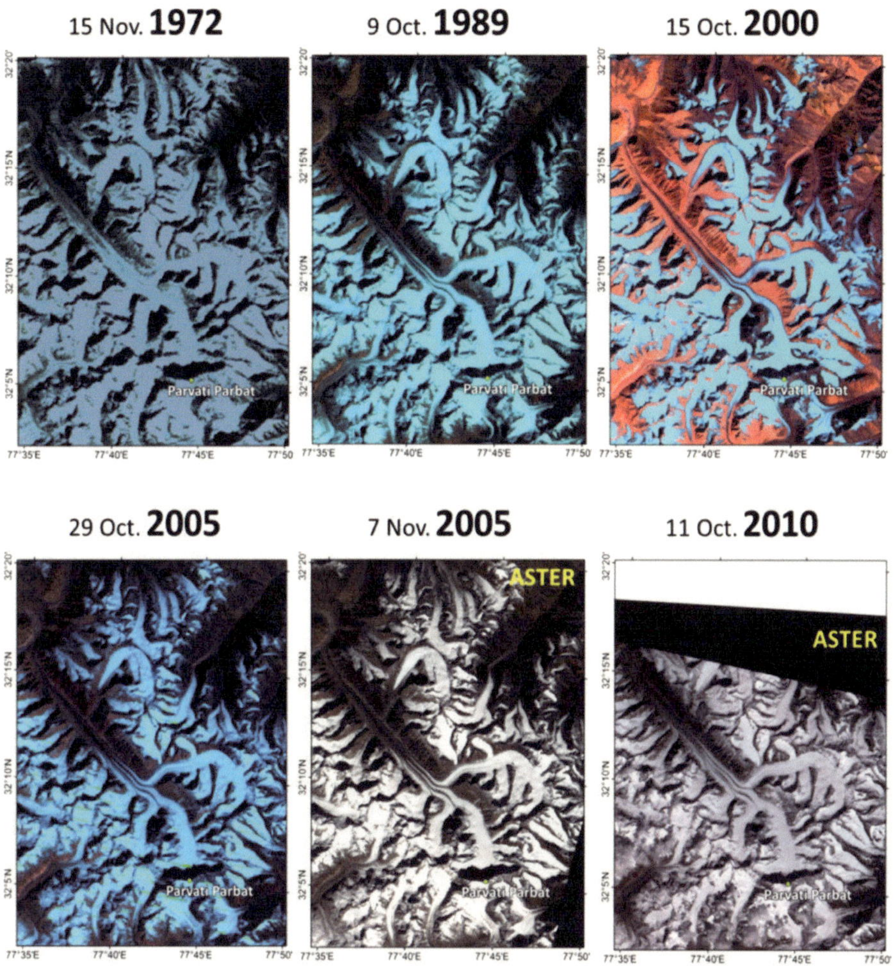

Fig. 2b. The true color images of western Himalayan glaciers (Bara Shigri Glacier, Spiti Valley, Himachal Pradesh, near Parvati Parbat) and surrounding regions from the Landsat series (1972, 1989, 2000, 2005) and ASTER (2005, 2010). (*Location 2, in Figure 1*).

The Goriganga river originates from the Milam Glacier at an altitude of ~3600 m and joins the Kali (Sharda) river downstream (after approximately 90 km) which finally merges into the Ganga river. The Landsat images from 1976, 1990, and 1999 depict changes in the snow cover and glacier over this region (Figure 2f). The Milam glacier is located NNE of Nanda Devi (7816 m, second highest mountain in India), a part of Garhwal Himalayas in the Uttarakhand State, India. Nanda Devi, means Bliss-Giving Goddess, is also regarded as the patron-goddess of the Uttarakhand Himalayas. The snow cover changes in numerous tributaries such as Goriganga and Kali River are important as they eventually feed into the main river Ganga. Moreover, small hydroelectric projects such as the Karmoli Lumti Tulli Hydroelectric project (55 MW) are proposed on the river Goriganga. The glaciers in the western Himalayas (near Doda, Jammu and Kashmir) show a decline in the snow cover along with the retreat of glaciers (Figure 2g). A significant trend in the warming of atmosphere (near-ground air temperature) has been observed over Jammu and Kashmir for period 1976-2007 (Jaswal & Rao, 2010).

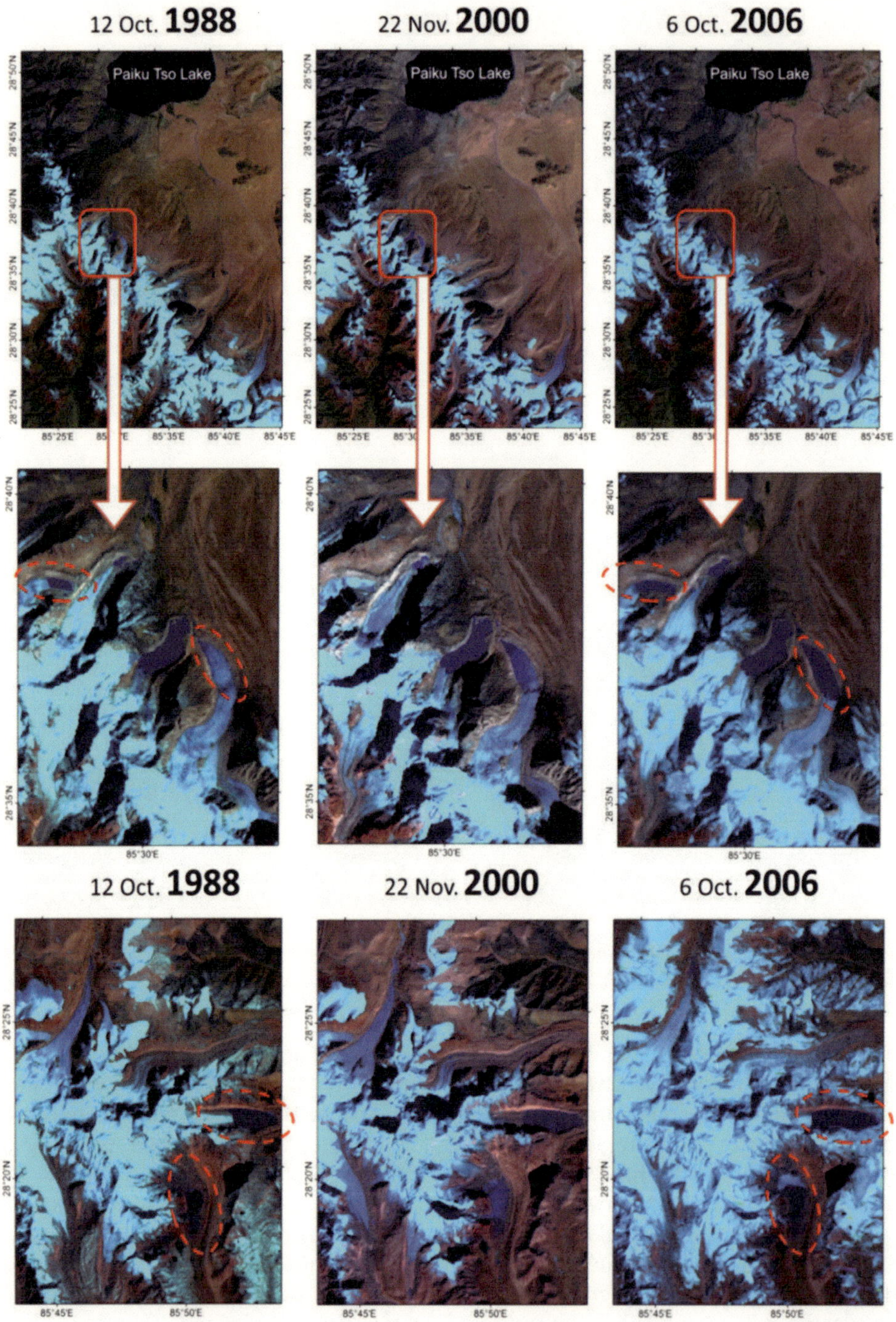

Fig. 2c. The true color images of central Himalayan glaciers (near Shisha Pangma which is the 14th highest mountain in the world - 8013m) from the Landsat series (1988, 2000, 2006) show the melting and formation of new lakes in the region during 2000 and 2006 compared with 1988. (*Location 3, in Figure 1*).

Fig. 2d. The photographs over the western Himalayan region (approximately around the Parvati Parbat, Bara Shigri Glacier region, Himachal Pradesh, Figure 2b) show contaminated glaciers with a yellowish layer over the white snow and glacier. Photo Courtesy: Alexander Naumov. Date of the photographs: 28 Aug. 2010.

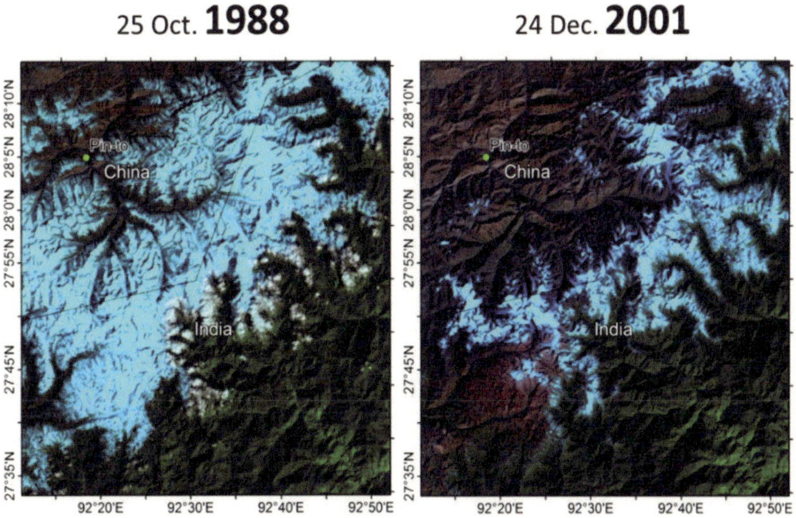

Fig. 2e. The Landsat images during year 1988 and 2001 show a sharp decline in the snow cover and enhanced melting of glaciers in the eastern Himalayas near the state of Arunachal Pradesh (India) and Tibet (China) border. (*Location 4, in Figure 1*).

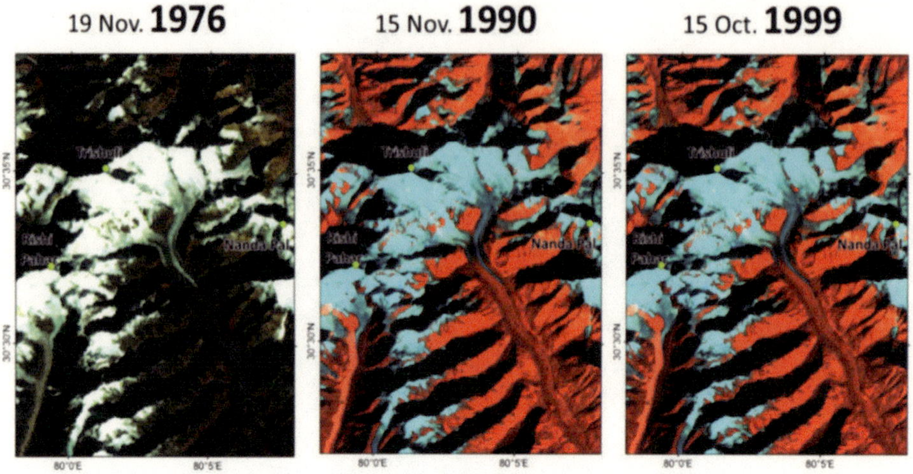

Fig. 2f. The Landsat images for year 1976, 1990 and 1999 over the Milam Glacier (Goriganga basin) in the central Himalayas (Uttarakhand State). (*Location 5, in Figure 1*).

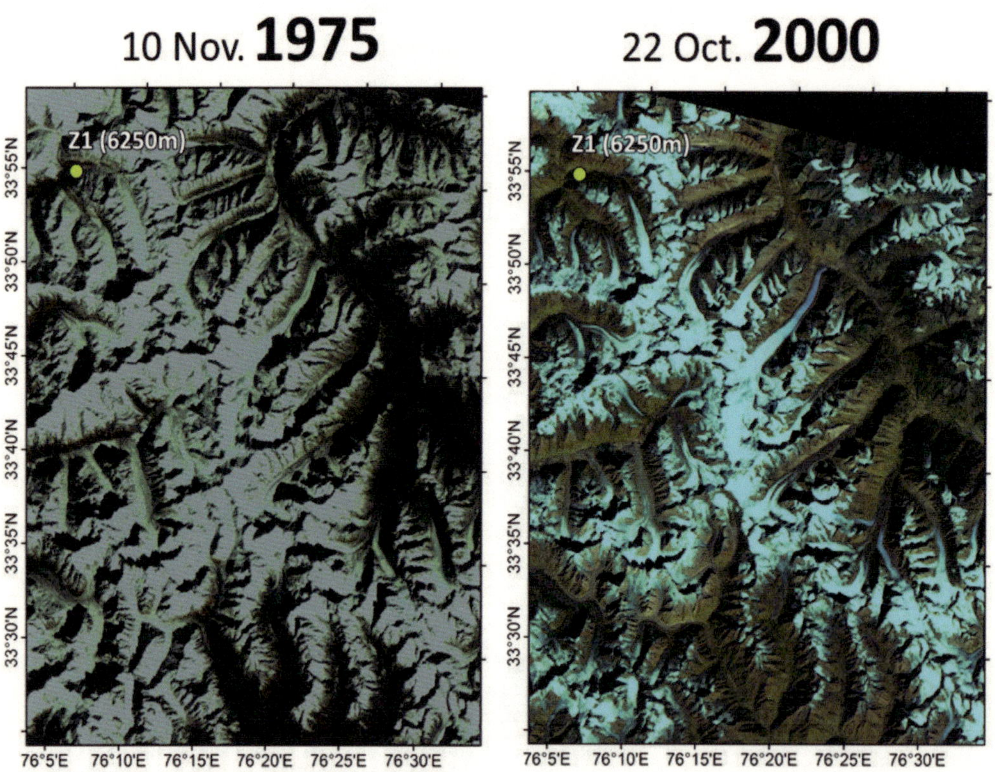

Fig. 2g. A comparison of the Landsat images from year 1975 and 2000 show a decline in the snow cover and enhanced melting of the western Himalayan glacier in between Doda (6573 m) and Z1 peak (6250 m) in the State of Jammu and Kashmir (India). (*Location 6, in Figure 1*).

4. Journey of dust storm over the Indian sub-continent

The first retrieval of aerosol loading over the land using POLDER (POLarization and Directionality of the Earth's Reflectances) data showed a dense concentration of aerosols over the Indo-Gangetic (IG) plains. A number of planned space-based sensors (MODIS Terra, MODIS Aqua, AIRS, PARASOL, ENVISAT, CALIPSO) and ground observations (AErosol RObotic NETwork - AERONET, Multi-wavelength radiometer - MWR, Microtops) have led to a better understanding of the amount, type and changes in aerosols over the Indian sub-continent and also the discovery of major aerosol "hot spots" over other regions of the globe. The desert dust, mixed with black carbon, is found to be one of the major sources of aerosols that affect the regional climate and precipitation patterns worldwide (Gautam et al., 2009; Lau & Kim, 2006; Lau et al., 2006, 2008, 2010; Ramanathan et al., 2005; UNEP, 2009).

4.1 Aerosol loading (dust and pollution) over the Indian sub-continent

The Intergovernmental Panel on Climate Change (IPCC) report of 2001 (IPCC, 2001) outlined a low level of understanding on the aerosols and their potential impact on the regional vegetation and climate. Since 2000, the availability of daily global data on the aerosols and associated variables from MODIS Terra (since 2000) and Aqua (since 2002) have greatly improved understanding of the aerosols, major sources, intra- and inter-annual variability over the Indian sub-continent (Bhattacharjee, 2007; Prasad & Singh, 2007a, 2007c; Prasad et al. 2006a, 2006b). Further, the ground-based observations from stations such as Delhi (Microtops) and Kanpur (AERONET) since 2001 have been used extensively to understand the physical and optical characteristics of aerosols over the western and central IG plains respectively (Prasad & Singh, 2007a; Prasad et al., 2007; Srivastava et al., 2011).

4.1.1 Pre-monsoon season: Dominance of desert dust

The vast agricultural land, plains of Indus and Ganga river, suffer from high concentration of desert dust and anthropogenic aerosols. Figure 3 show very high aerosol loading (appear as orange to red color, AOD > 0.5) over the IG plains during the summer and winter seasons (mean of 2000-2008). This belt of high AOD is observed to be running parallel to the Himalayan Mountain range (Figure 3a). The retrieval of AOD, as shown in figures 3a,b is based on the dark-target (DT) algorithm which is more sensitive to the ocean or lake water and dense vegetation covered regions over the land. The seasonal distribution of aerosols shows a striking contrast between the summer and winter aerosol loadings over the IG plains. The winter season, which is largely devoid of the transported desert dust, show mostly fine anthropogenic aerosols while the summer season is dominated by the coarse desert dust aerosols that get mixed with the local anthropogenic aerosols (Prasad et al., 2006a, 2006b). The long range transport of desert dust from the western sources gradually raises the aerosols loading over the IG plains during the period April to June. The AERONET data show that the mean AOD rises from 0.4-0.5 to 0.6-0.7 (>0.8-0.9 over the western side) from April to June months respectively due to the influence of these episodic dust storms (Prasad & Singh, 2007a). These pre-monsoon dust storms have been found to reach up to high altitude over Himalayas (Prasad & Singh, 2007b). The pre-monsoon dust storms are thereby affecting the regional atmospheric temperature and snow cover and hence glaciers dynamics over the Himalayas and Tibetan Plateau, especially the western and central regions (Lau et al., 2010; Prasad et al., 2009).

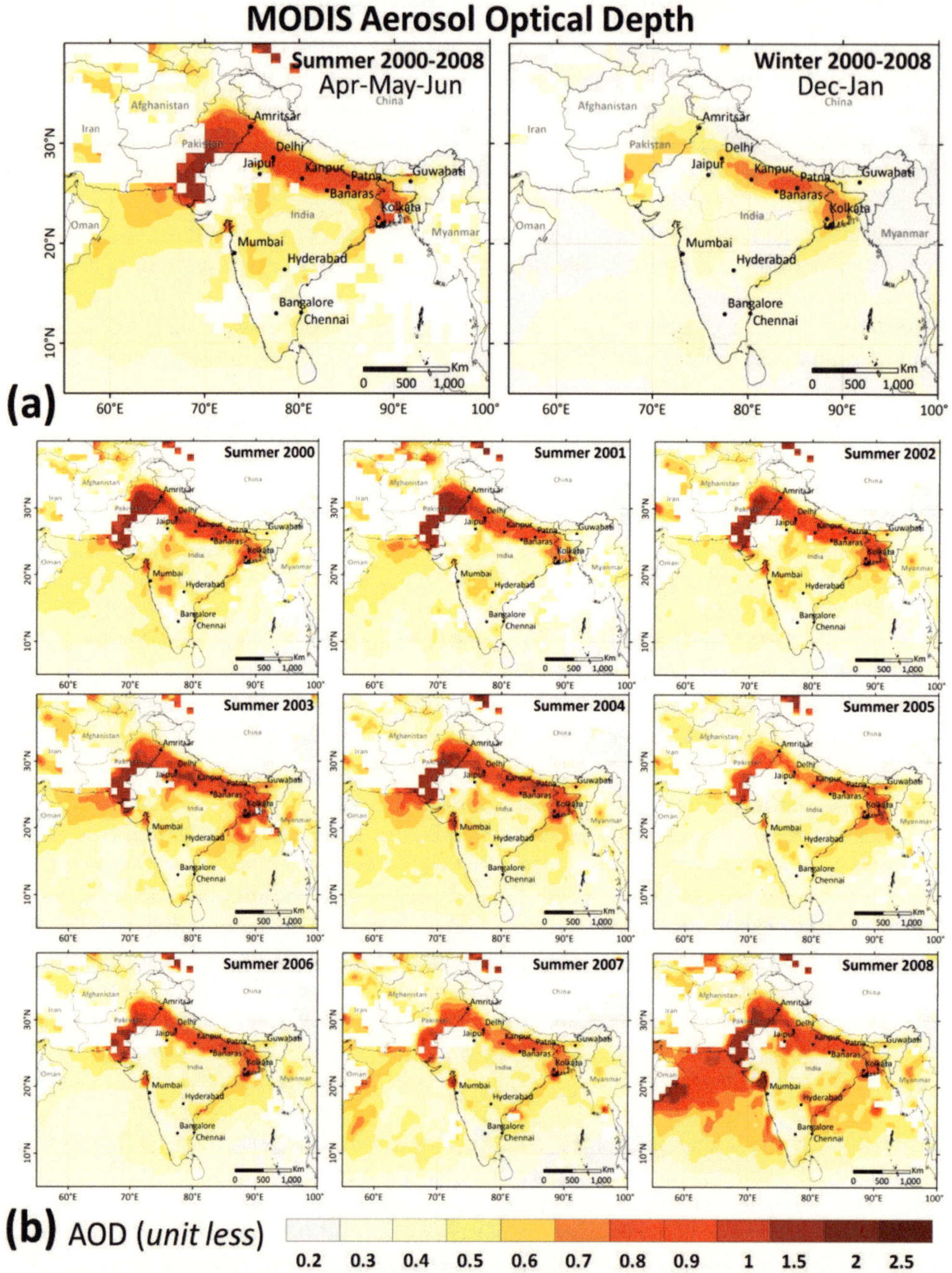

Fig. 3. (a) Mean MODIS Terra derived dark-target (DT) AOD (summer and winter season) over the Indian sub-continent and surrounding regions (Arabian Sea) during 2000-2008. (b) Inter-annual variability (2000 to 2008) of aerosol loading over the Arabian Sea and IG plains during the pre-monsoon season (April to June).

Figure 3b shows strong inter-annual variation of aerosol loading over the Indian sub-continent and its surroundings. For instance, the year 2008 shows the highest aerosol loading over the Arabian Sea which is situated between the major dust sources and sink regions. This anomaly indicates that the emission as well as the transport of dust varies greatly every year. Thus the effect of desert dust on the regional climate of the IG plains varies year-to-year with the variation of influx of transported aerosols.

Figure 4 show a simplified map of the location of major sources of dust over the IG plains. The satellite aerosol data as well as the dust transport model (HYbrid Single-Particle Lagrangian Integrated Trajectory - HYSPLIT) leads to the identification of these major sources. The HYSPLIT back-trajectory shows that the dust storm reaches over India in 2-5 days depending upon the distance of the source of desert dust (Prasad & Singh, 2007a). The dust storm from the arid regions of Afghanistan, Pakistan, Thar Desert (India) usually passes over land (land route) while the dust storm from Iran, other parts of Middle East and Sahara usually passes over the Arabian Sea (sea route) before reaching the IG plains (Figure 4). Thus the moisture associated with these dust storms varies with their route (Prasad & Singh, 2007a).

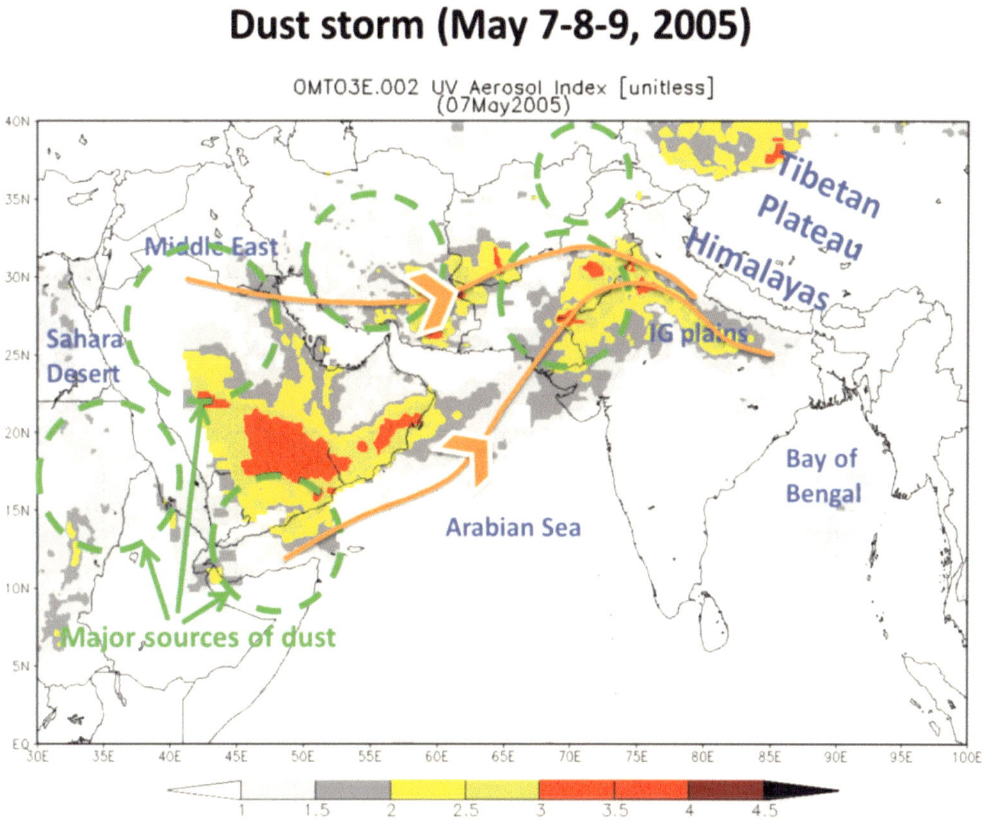

Fig. 4. The location of major sources of dust storms over the Indian sub-continent is marked as green broken circles (Prasad and Singh, 2007a). The two major pathways of dust storms, the sea route passing through the Arabian Sea and the land route, are marked by the orange line. The background show Aura derived UV aerosol index during a dust storm event (May 7, 2005) that illustrates the transport of dust aerosols through the Arabian Sea before reaching the IG plains.

The valley-like topography of the IG plains bounded by high altitude Himalayas (~ 4-6 km) in the north and Vindhyan mountain range (~700m) and Chotanagpur plateau in the south and Rajmahal hills in the southeast causes channelization of the dust storm as it moves from the west (Delhi and Jaipur) to the east (Kolkata and Dhaka) (see Figure 5a). This figure shows the three-dimensional topographical set-up of the study region (the Himalayas and IG plains). The west to east journey of the desert dust in the IG plains, capped by towering Himalayas in the north, not only affects the aerosol loading in the entire IG plains but also the Himalayas because a portion of the desert dust (mixed with local pollutants) also reaches the high altitude region over Himalayas. Because the dust storm is much stronger over the western and central IG plains as compared with the eastern zone, the direct and indirect effect of dust related loading is more prominent and frequent in the western and central Himalayas. However, some of the episodes of dust have been found to be reaching up to the Eastern Himalayas (East Nepal and Sikkim Himalayas). The evidence for transport of dust up to Himalayas is discussed in section 5.

4.1.2 Winter season: Dominance of anthropogenic emissions

The winter aerosol loading, mostly anthropogenic, is also high in the entire IG plains (valley region) (Figures 3a, 5a). During the winter season, fine aerosols emitted by the thermal power plants, industries, vehicles, coal-fired brick kilns, and bio-fuel burning stoves dominate the total aerosol loading over the IG plains (Di Girolamo et al., 2004; Prasad et al., 2006a; Ramanathan & Ramana, 2005). Although the dust storms are rare during this period, the satellite observations and overall assessment of aerosols indicate a dominant anthropogenic component exists over the IG plains. The bio-mass and bio-fuel burning and vehicular emission are traditionally considered to be one of the highest contributors to the winter aerosols (Di Girolamo et al., 2004; Garg et al., 2001; Gadi et al., 2003; Ramanathan & Ramana, 2005). Recent studies on the emissions from the major thermal power plants over India show that the coal fired power plants and similar industries, such as smelters, are one of the major contributors of gaseous (tropospheric NO_2) and particulate (black carbon, fly ash from coal) pollution (Ghude et al., 2008; Prasad 2007; Prasad et al., 2011) (Figure 5b). The NO_2 emissions (Ozone Monitoring Instrument, OMI Aura) from these coal based power plants situated in low populated regions such as Agori in the central India are found to be much higher than the largest city such as Delhi (with ~25 million human and ~5.6 million vehicular population) (Prasad et al., 2011). The presence of fine anthropogenic aerosols and gaseous pollutants such as ozone, NOx and SOx, very high humidity (>90%), near zero or calm wind, and near-ground (10-100 m) boundary layer leads to a severe problem of dense fog and high ozone during December and January (Di Girolamo et al., 2004; Prasad et al., 2006a; Ramanathan & Ramana, 2005). The development of dense fog with increasing intensity, duration, and frequency over the last two decades (1990-2010) is a relatively recent phenomena compared to the beginning of the Industrial (power utility) revolution in India (1970-1980) because most of the power plants were setup in the region since 1980 (Prasad et al., 2011) along with the gradual increase in the vehicular population. The disruption of air-traffic, ground-traffic and railways due to the dense fog during the winter season has increased the awareness to this problem among the public and media in recent years.

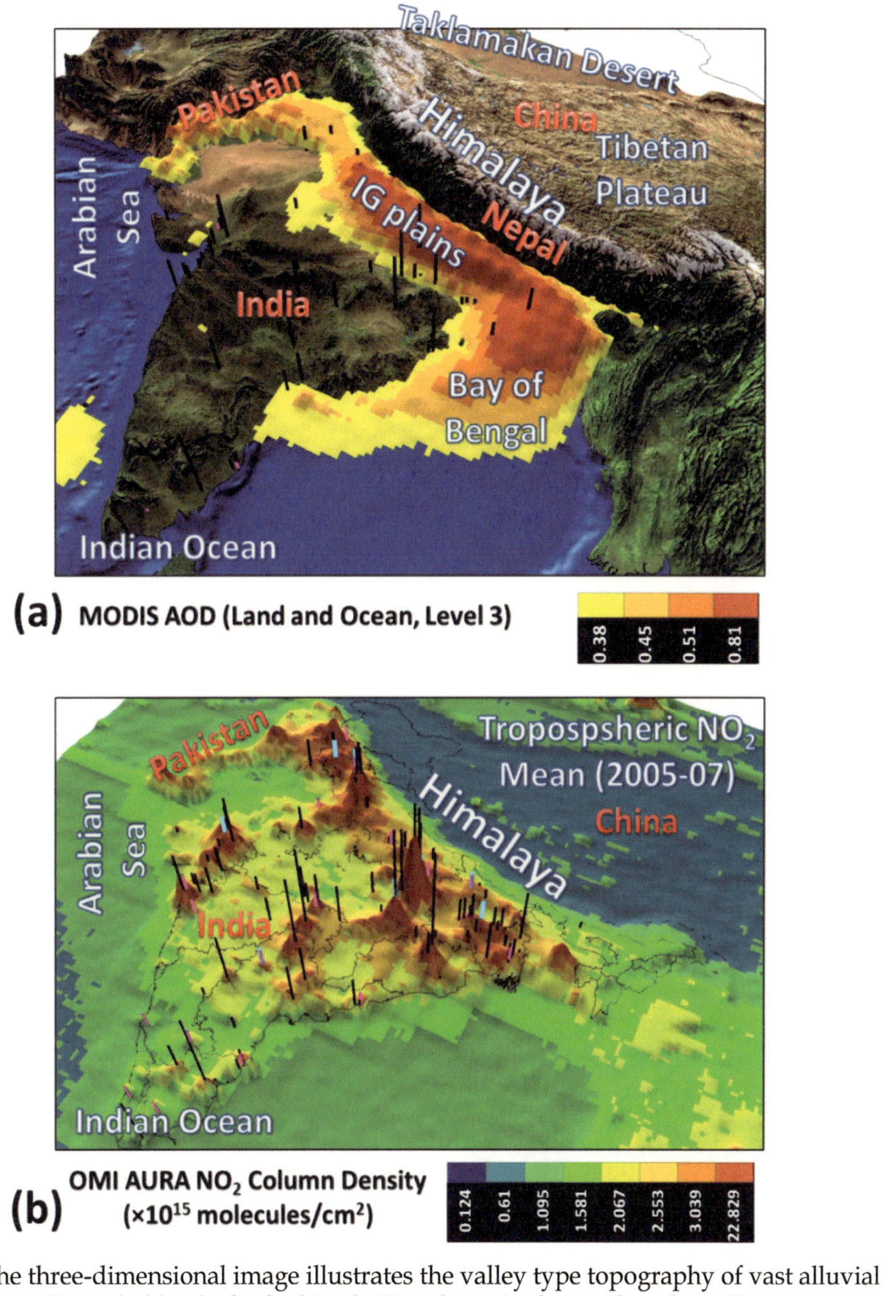

Fig. 5. (a) The three-dimensional image illustrates the valley type topography of vast alluvial IG plains that is bounded by the high altitude Himalayas in the north and Vindhyan mountain range in the south. The mean aerosol loading over the Indian sub-continent during the winter season (December and January, 2004-2008) illustrates high AOD over the entire IG plains. (b) The mean annual tropospheric NO_2 over the Indian sub-continent as measured by OMI Aura. The black lines or bar represent major (>100 MW) thermal power installations in the region where the length of bar is directly proportional to its capacity. The seasonal distribution of NO_2 is available at Prasad et al. (2011).

The recent advances made by the atmospheric aerosols science community through the space based multi-sensor monitoring, dust modeling and ground dust observations had led to an increased awareness among the public and media (newspapers, magazines, radio, and television). Figure 6 shows the arrival of a dust storm over the Delhi region on May 13, 2008 (using Total Ozone Monitoring Satellite - TOMS aerosol index) and its after-effect on the Delhi metropolitan area on May 13-14, 2008. Similar reports appeared in the print and television media from several cities west of Delhi (Lucknow and Kanpur, located in central IG plains) on the subsequent day. The reports of loss of life, property, respiratory problems, and air and ground traffic disruptions etc, in various media sources show the unpreparedness of the agencies and organizations affected or responsible for responding to such events. Though the awareness about the after-effects of dust storms are increasing in the region, an integrated approach (a decision support system) to tap the potential of forecasting dust storms using near real-time satellite observations, such as the MODIS direct, broadcast together with the dust transport models and ground based observations to warn the public and industrial sectors affected by these events is still lacking.

Fig. 6. The left panel shows arrival of the dust storm over Delhi region on May 13, 2008 using TOMS aerosol index (AI). The high values of TOMS AI (2.5-5) illustrate the aerosol loading associated with the dust storm. The right panel shows the after-effects of the dust storm over Delhi that was reported in several leading national newspapers (such as Hindustan Times, New Delhi edition on May 14, 2008).

5. Journey of dust to high altitudes over Himalayas and Tibetan Plateau

The episodes of dust storms are very frequent during the pre-monsoon season. The Figure 3 emphasizes the seasonal and inter-annual variation and distribution of aerosols over the land and oceans. Figure 3 also shows some notable blank regions in Pakistan, Iran, Afghanistan, Oman and Sahara. This is due to the fact that the traditional Dark Target (DT) AOD retrieval algorithms do not capture the aerosol loading over very bright arid or semi-arid desert surfaces that have sparse or negligible vegetation cover.
To overcome this limitation, incorporating the blue spectral band (Deep-Blue, DB algorithm) enhances the accuracy and frequency of the retrieval of AOD over such land surfaces as they are very sensitive to the dust (Hsu et al., 2006; Ginoux et al., 2010). For instance, a comparison of a true color image (composite of red, green, and blue - RGB: Band 1, 4, and 3), reflectance from the blue band (MODIS Band 3, 479 nm), MODIS Terra AOD (only DB), MODIS Terra AOD (both DT and DB) and AOD-like scale for values derived from the blue band (level 2, collection 5.1) is shown for a dust storm that affected the IG plains on June 7, 2003 (Figure 7). The area covered by the dust storm which is visible as a very bright surface in the true color image appears as a prominent region with very high reflectance in the band 3. The MODIS Terra AOD, based on the DB approach, retrieves the dust loading over bright surfaces, such as Thar desert, while a combination of both DT and DB approach gives better results (Figure 7). However, there is a scope for an improvement in the AOD retrieval procedure as both DB and DT misses the central median line of dust storm that appears as blank region bounded by very high AOD values. The scaling of values like AOD from the blue band (no other correction is applied) shows the continuity of dust storm and its perimeter. The band 3 by itself, presents scope for improvement in the aerosol retrieval along with other bands such as band 11 and 12 (used for cloud detection) that helps in distinguishing the dust aerosols from surroundings.
Therefore, to identify and quantify the aerosol loading due to the major dust storms, we have utilized a combination of DT and DB algorithm to retrieve the AOD over land and ocean using the MODIS sensor on the Terra and Aqua. In this section, we present dust storm cases that highlight the journey of dust up to high altitudes over Himalayas using the MODIS (column AOD) and CALIPSO (vertical profile) observations.

5.1 Dust over Himalayas: Evidences from MODIS Terra and Aqua

A day to day analysis of aerosol data from MODIS during May 28 - June 15 show a number of dust storm events that affected the IG plains and Himalayas during the year 2003. The ground-based AERONET station at Kanpur also confirms these dust storms (Prasad & Singh, 2007a). Figure 8 shows a three day composite (maxima) of aerosol loading (Level 2, version 5.1, 10km grid) using both DT and DB algorithm to maximize the coverage over land and ocean. The composite of May 28-30 show very high AOD over the land and Arabian Sea. Similarly, a composite of June 6-8 also show very high AOD over the land and Arabian Sea. The path and source of the dust storm is emphasized by very high AOD over the western region that shows continuity up to the IG plains. During this period, the ground observed AOD (approximately every 15 minutes) from the Kanpur AERONET station (Prasad & Singh, 2007a) also showed arrival of a number of dust storms over the Kanpur city (central IG plains).

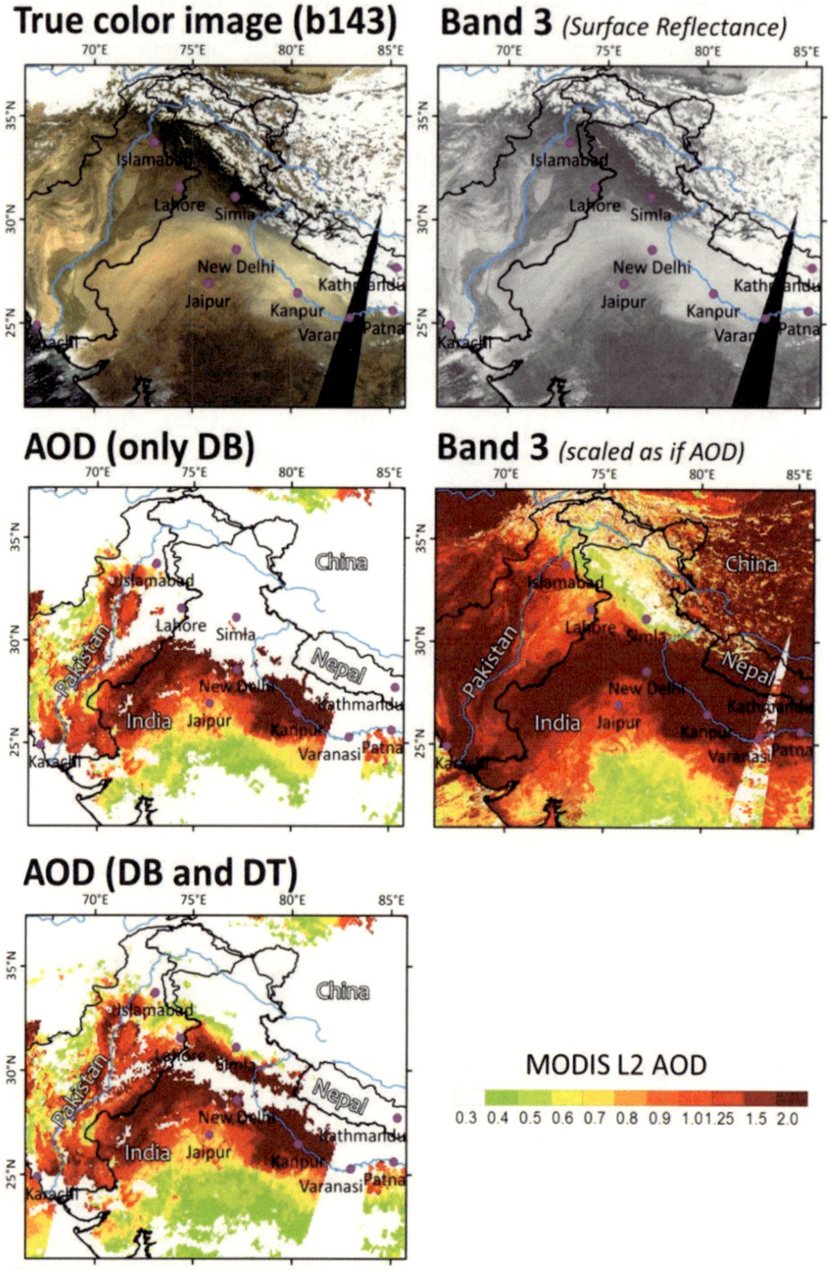

Fig. 7. The true color image (band combination: RGB: 1,4, and 3) based on the MODIS Terra observations during a dust storm event on June 7, 2003. The dust affected region is prominent in the band 3 (blue band: 479 nm). The AOD (only DB) and AOD (combination of DB and DT) from level 2 Terra (collection 5.1) show quantitative aerosol loading due to the dust storm (marked as green to deep red color). The MODIS AOD like scaling using only MODIS band 3 (blue band) without any other filtering for cloud etc., illustrates its usefulness in deriving AOD over bright (arid) and low vegetative surfaces.

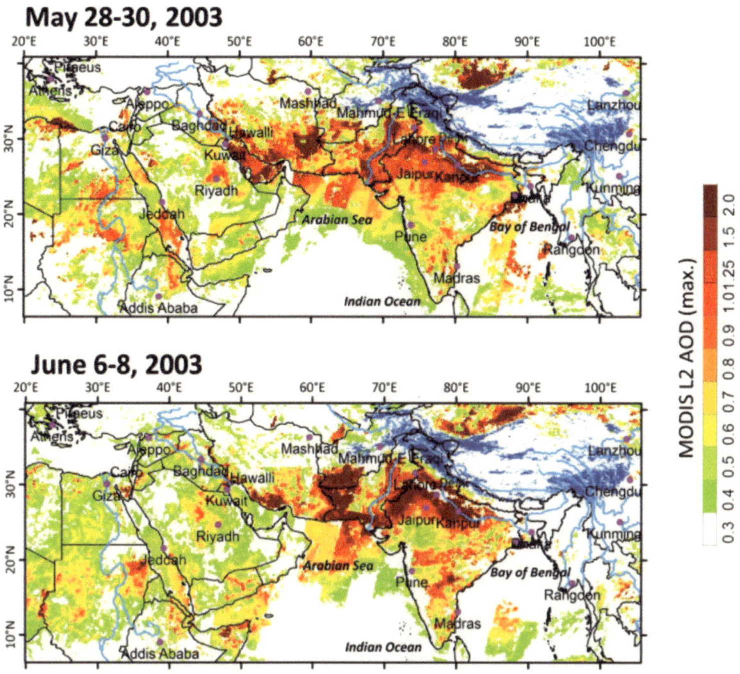

Fig. 8. Three day maxima of aerosol loading derived from MODIS Terra level 2 AOD (10 km grid, collection 5.1) during May 28-30 (top panel) and June 6-8 (bottom panel) illustrates source and path of transport of dust storms through the land and sea route as marked in Figure 4. The high aerosol loading over the IG plains (appear as deep red color) is primarily attributed to the desert dust brought by such dust storms.

5.1.1 Dust over the major rivers of Asia

A closer look at the daily aerosol retrieval (at 10 km grid resolution) using both DB and DT algorithm clearly shows a very high AOD over the high-altitude snow covered and glacier regions of Himalayas (Figure 9). The data used is a daily level 2 AOD, collection 5.1 from MODIS Terra and Aqua (Figure 9). The white color features in figure 9 represents the high altitude snow cover and Himalayan glaciers that are situated at an average height of ~4-6 km above the msl. Further, the region shown in these figures is interesting and valuable as three major rivers of Asia (Indus, Ganga - Brahmaputra) originate from here and depend on the snow cover and glacier melt for base river flow. Numerous other rivers such as Yamuna (passing through Delhi), Sutluj, and tributaries of river Ganga also originate from here. Very high AOD over the Himalayas is clearly visible during these dust storm events (May 31 to June 2, 2003 and June 12-13, 2003) (Figures 9a,b). Both Terra and Aqua (pre-noon and after-noon overpass) data show high AOD, visible as green-yellow-orange-red, over the Himalayan region. The MODIS AOD images also show that the dust storm also affects the Nepal Himalayas (both western and eastern Nepal) as it moves from the west to the east. The level-2 AOD is highly useful for such studies as it captures the dust storm and its reach in greater detail compared to the level 3 AOD at 100 km grid that has been found to be grossly missing the event of dust over Himalayas primarily due to its coarser resolution, incomplete retrieval over all pixels, and the binning method used to produce the level 3 data. The vertical profile of dust from CALIPSO is not available for these events as it was launched during the year 2006.

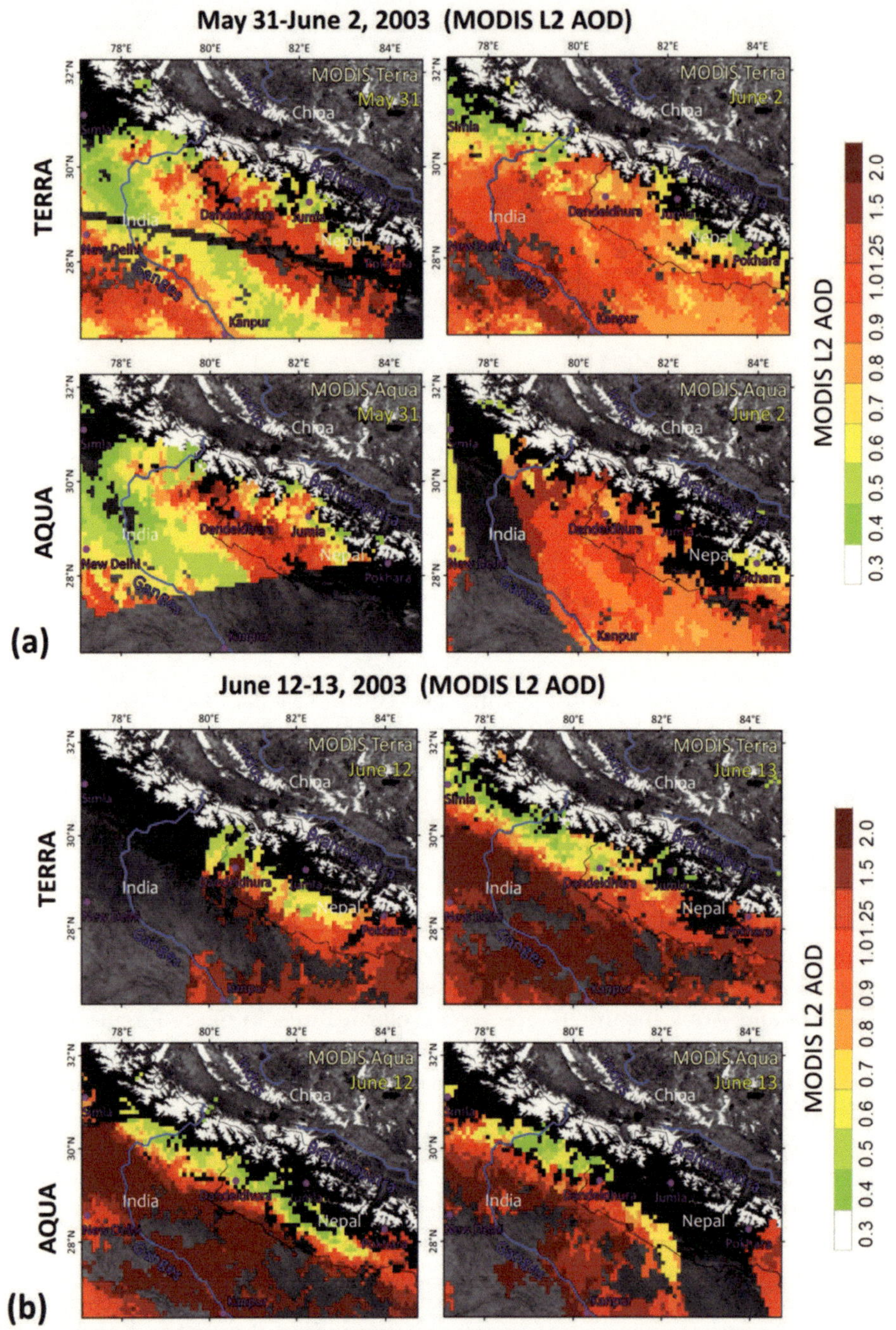

Fig. 9. The MODIS Terra and Aqua level 2 column AOD (collection 5.1) images during a dust storm episode on (a) May 31 and June 2, 2003 and (b) June 12-13, 2003, illustrates that the dust storm reaches up to the high-attitude Himalayan snow covered and glacier regions. The white color objects along the border of India, Nepal and China show the Himalayan snow cover and glaciers near the origin of three major rivers of Asia (Indus, Ganga, and Brahmaputra).

5.2 Dust over Himalayas: Evidences from CALIPSO vertical profiles

CALIPSO uses an active laser beam that provides unique information about the vertical structure of aerosols and clouds in the atmosphere. A number of parameters from CALIPSO such as the total attenuated backscatter at 532 nm, 1064 nm, their attenuated color ratio or c-ratio (1064/532nm), perpendicular backscatter at 532nm, and depolarization ratio or d-ratio provides critical information and identification of the dust (or clouds) and also its vertical structure. The aerosol related parameters from other sensors onboard Terra and Aqua (MODIS), CNES/Myriade (PARASOL), AURA, Earth Probe -EP (TOMS) etc., gives information only about the total column of atmosphere. To obtain evidence of dust storms reaching up to the high altitude Himalayas and Tibetan Plateau (~4-6 km), a vertical profile of the atmosphere when the major dust storms hits the region is vital. Further, such vertical profiles of the dust storms are needed over different regions (western, central and eastern) to assess the impact of dust. In this section, we present some evidence of dust reaching Himalayas in all three aforementioned regions during the entire pre-monsoon period (April, May and June) when the dust storm activity is highest over the IG plains. All the CALIPSO vertical profiles (0-20 km or 0-8 km) presented here depict the time of overpass (in UTC), and latitude-longitude (location) in the X-axis (bottom and top of x-axis respectively). The surface elevation along the path of CALIPSO is marked as a thick black line in Figures 10-15. The inset shows the path of CALIPSO overpass over the Globe (black line) and the study region (pink line) (Figures 10-15).

5.2.1 Dust and anthropogenic aerosols over Central Himalayas and IG plains

The vertical profile of the atmosphere (0-20 km), as measured by CALIPSO, over passes through the central IG plains (State of Uttar Pradesh), Himalayas (State of Uttarakhand - formerly Uttaranchal, and western Nepal), Tibetan Plateau and Taklamakan desert is shown in Figure 10. The vertical profile from 41°-16° N (during 20:34 to 20:41 UTC or 2:04 to 2:11 am local India Standard Time - IST) show the vertical structure of a major dust storm passing through the central IG plains on April 22, 2010. Very high concentration of dust is observed from night-time CALIPSO backscatter (532nm) over the IG plains that are reaching up to the 4-5 km height near Himalayas and its foothill region. Dust is also observed over the Taklamakan desert. Other parameters from CALIPSO, such as perpendicular attenuated backscatter (532nm), total attenuated backscatter at 1064nm, c-ratio, and d-ratio, also support the presence of a major dust storm (Figure 12a). The cloud, with very high backscatter, at ~20° N latitude, blocks the backscatter from the atmosphere (appear as deep blue) below it. The perpendicular attenuated backscatter (532nm) show the scattering of laser beams by the dust particles which appear as colored dots against a blue background in Figure 12a.

5.2.2 Dust and anthropogenic aerosols over Western Himalayas and IG plains

During year 2010, one of the major dust storms is found to be reaching up to ~7 km over the western IG plains and western Himalayas on May 27 (Figure 11). The ceiling of the dust storm over the states of Punjab and Rajasthan (part of the western IG plains) is between 5-7 km. The height of dust (~7 km) is extending higher than the range of height of the Kashmir and Himachal Pradesh Himalayas (~2-5.5 km) as seen from the CALIPSO profile along with the surface elevation profile (Figure 11). This night-time profile was obtained during 21:05 to 21:11 UTC (or 2:35 to 2:41 am local time or IST).

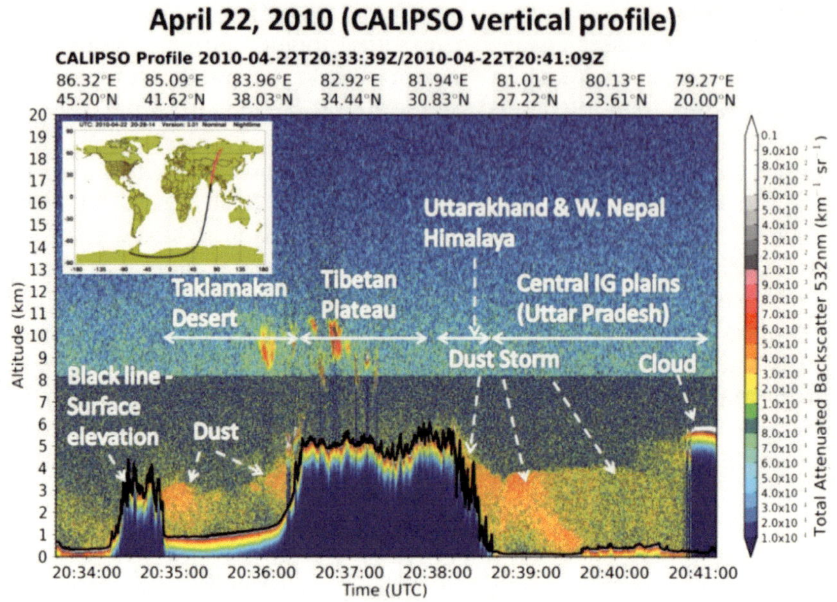

Fig. 10. The atmospheric profile (total attenuated backscatter at 532 nm, 0-20 km) of a dust storm event as measured by the night-time CALIPSO overpass over the central IG plain, Himalayas and Tibetan during April 22, 2010 (at 2:04 to 2:11 am IST).

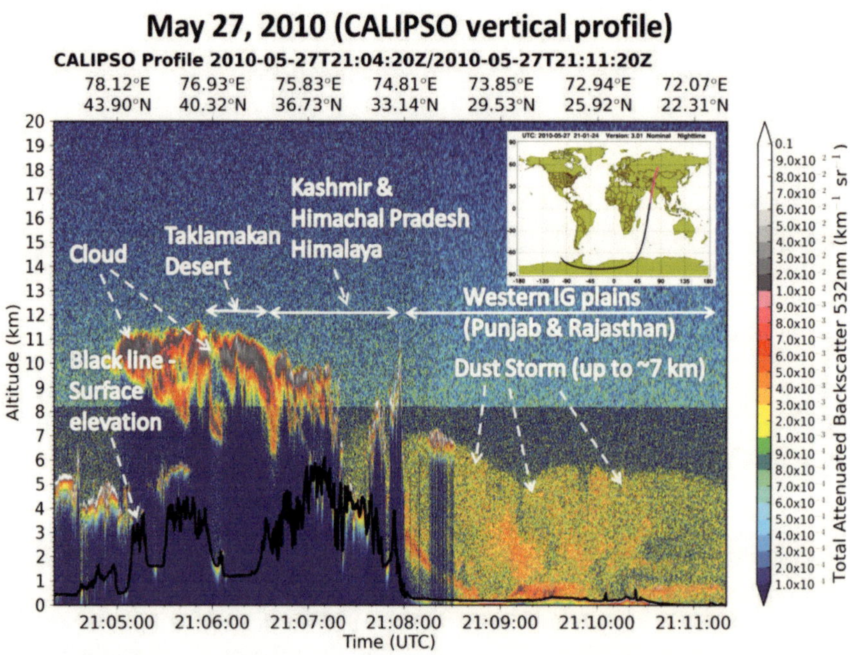

Fig. 11. The atmospheric profile (total attenuated backscatter at 532 nm, 0-20 km) of a dust storm event as measured by the night-time CALIPSO overpass over the western IG plain and western Himalayas during May 27, 2010 (at 2:35 to 2:41 am IST).

No dust is observed over the Taklamakan desert (near ground) during this overpass compared to the previous example (April 22, 2010, Figure 10). However, a thick layer of very high scattering particles (or cloud) is observed over the Himalayas and Taklamakan desert between altitudes 6-12 km that needs further investigation as the dust may get mixed with clouds that change the characteristics of aerosols and clouds (Figure 11). Other parameters from CALIPSO, shown in Figure 12b, mark the presence of various dense layers of dust between 26°-30° N (at 0-1 km height and from 1-8 km) that shows the state of vertical mix up of the dust in the atmosphere. Further, the dense layer of dust between 0-1 km show a gradual increase in the height to 0-4 km as it moves northwards between 30°-33.15° N.

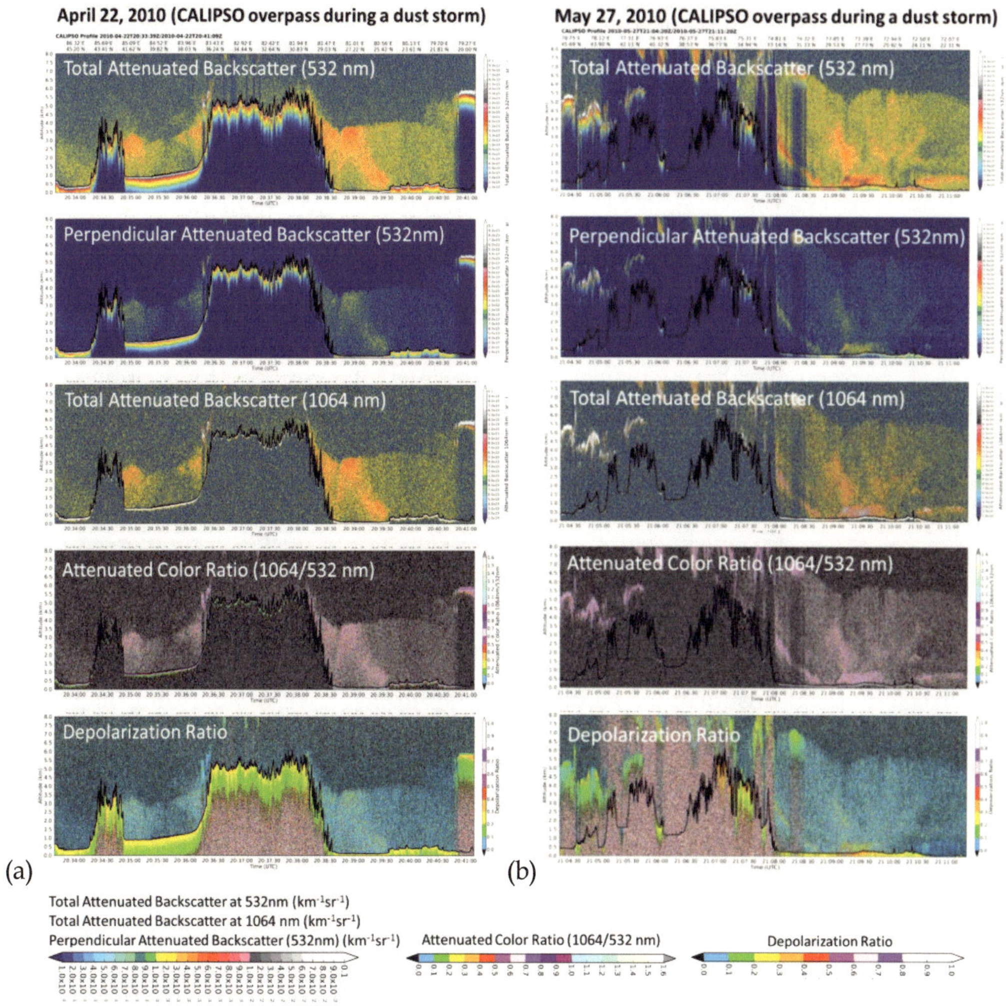

Fig. 12. The atmospheric profiles (0-8 km, y-axis) of a dust storm event using multiple parameters as measured by the night-time CALIPSO overpass over (a) the central IG plain and Himalayas during April 22, 2010 (at 2:04 to 2:11 am IST), and (b) the western IG plain and Himalayas during May 27, 2010 (at 2:35 to 2:41 am IST). The x-axis (top and bottom) of Figures 12(a) and 12(b) is same as that of Figures 10 and 11 respectively.

5.2.3 Dust and anthropogenic aerosols over Eastern Himalayas and IG plains

The night-time profile obtained from CALIPSO on the next day (May 28, 2010) shows the vertical structure of the same dust storm as it passes over the eastern IG plains (State of Bihar and Jharkhand) (Figure 13). The upper limit of dust storm in the eastern side of the IG plains is observed to be ~5 to 5.5 km while a dense layer of dust is observed at 2-3 km between 20°-26° N latitude. The vertical profile was obtained during the night-time over pass of CALIPSO (20:10 to 20:17 UTC or 1:40 to 1:47 am local time or IST). The dense layers of the dust storm are clearly visible in other parameters obtained from CALIPSO (Figure 15a). The CALIPSO profiles corroborate the transport of aerosols over the eastern Nepal Himalayas and its foothill region, as obtained from MODIS Terra and Aqua column AOD. High column AOD (at 10 km horizontal grid) from Terra and Aqua is observed over the snow and glacier cover regions during the episodes of major dust storms (May-June 2003) in the central and eastern Himalayas (Figure 3b, 3c). However, the dust over the eastern Himalayas (eastern Nepal and Sikkim) is relatively less prominent than the western Himalayas as the average height of the Himalayas is more in the eastern side and the dust storm gets weakened in strength (dust load, wind speed) and height as it moves to the eastern side which is approximately 1000-1500 km eastwards from the western end.

5.3 Dust storms during June (prior to the arrival of monsoon system)

The dust storms are also common during June which is followed by intense rains due to the arrival of monsoon over the IG plains. Figure 14 shows the vertical profile of one of the major dust storms hitting the western IG plains during the late June (June 28, 2010). The ceiling of the dust storm over the western IG plains is ~6 km. Thick dense layer of dust is observed between 0-2 km at 25°N latitude that gradually increases in height to 2-4 km at 32°N latitude. The density of dust changes with the height and latitude which is clearly visible in all the parameters shown in Figure 15b. Presence of dust is also observed in the atmosphere over the Taklamakan Desert (Figures 14, 15a).

6. The contribution by anthropogenic aerosols

The arrival of desert dust over the IG plains, either through the sea route (Arabian Sea) or land route (Rajasthan, Pakistan), leads to a mixing of coarse desert dust with the finer anthropogenic emissions (such as black carbon) from the major point sources. Figure 5b shows the presence of large networks of power plants around Gujarat (near Arabian Sea), north Pakistan, and the western IG plains (Punjab, Delhi). The passage of dust storms through these pockets of major point sources, before reaching the Himalayan range, increases the complexity of the physical (optical) and chemical nature of otherwise dust dominant aerosols.

Desert dust brought by the dust storms is rich in minerals such as quartz, feldspar, mica, alumino-silicates, calcite, carbonates, iron oxides etc. The anthropogenic emissions, emitted by burning of fossil fuel (coal and petroleum), biomass and biofuel burning, are mainly comprised of black carbon – BC (elemental and organic carbon), sulfate, and nitrate aerosols. The BC is widely known as a major climate forcing agent as it strongly absorbs the solar radiation which also varies with the nature, type, and source of black carbon aerosols (Lau et al., 2010; Yasunari et al., 2010).

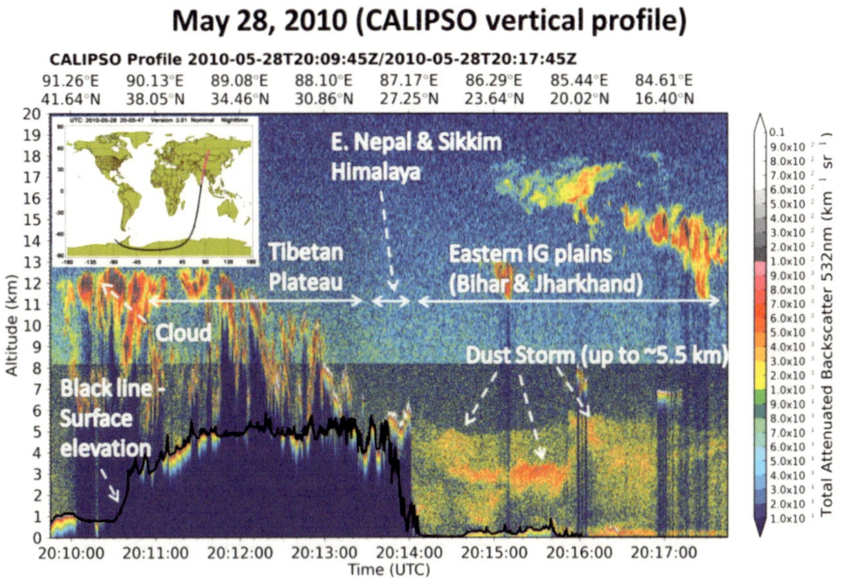

Fig. 13. The atmospheric profile (0-20 km) of a dust storm event as measured by the night-time CALIPSO overpass over the eastern IG plain, Himalayas and Tibetan Plateau during May 28, 2010 (at 1:40 to 1:47 am IST). The total attenuated backscatter (at 532 nm) shows the vertical structure of pre-dominantly dust aerosols up to ~5.5 km.

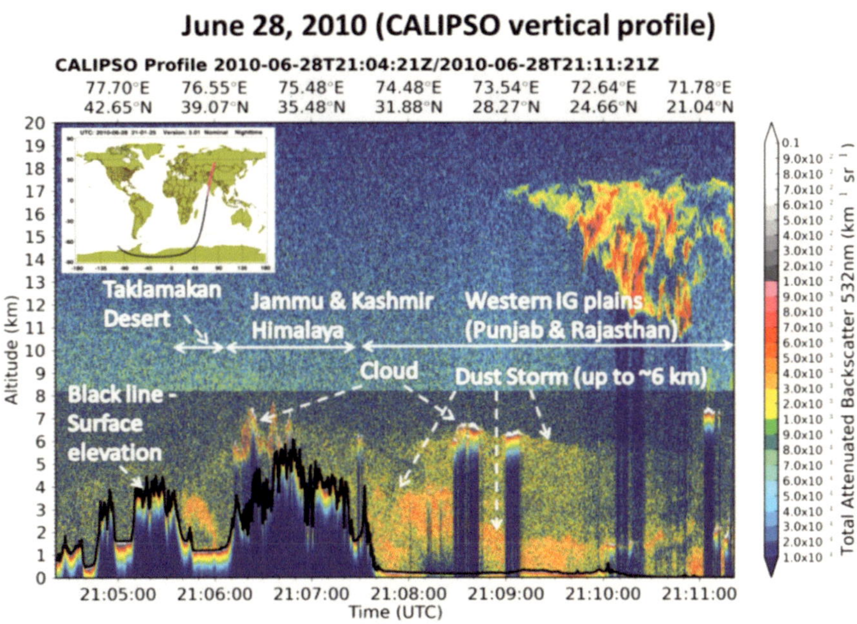

Fig. 14. The atmospheric profile (0-20 km) of a dust storm event as measured by the night-time CALIPSO overpass over the western IG plain, and Himalayas during June 28, 2010 (at 2:35 to 2:41 am IST). The total attenuated backscatter (at 532 nm) shows the vertical structure of pre-dominantly dust aerosols up to ~6 km.

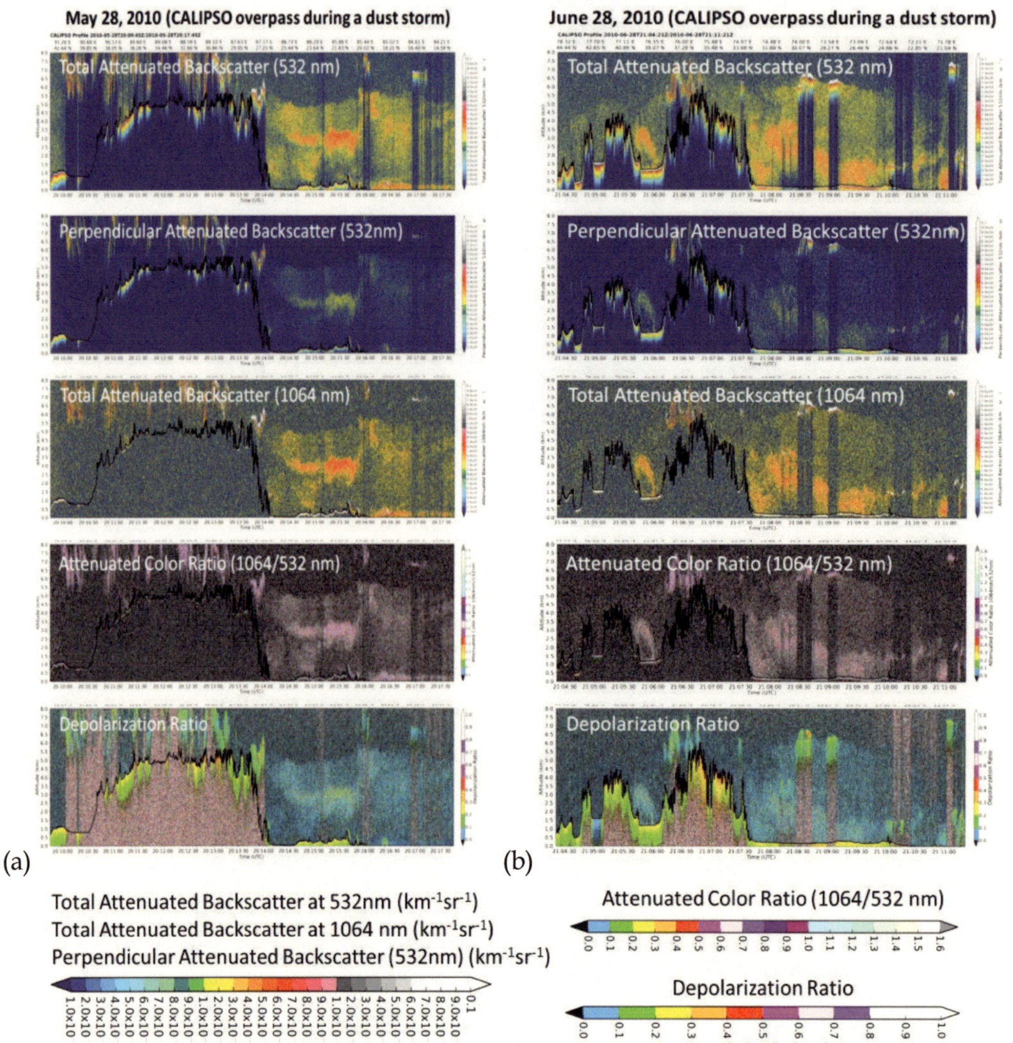

Fig. 15. The atmospheric profiles (0-8 km, y-axis) of a dust storm event using the multiple parameters as measured by the night-time CALIPSO overpass over (a) the eastern IG plain and Himalayas during May 28, 2010 (at 1:40 to 1:47 am IST), and (b) the western IG plain and Himalayas during June 28, 2010 (at 2:35 to 2:41 am IST). The x-axis (top and bottom) of Figures 15(a) and 15(b) is same as that of Figures 13 and 14 respectively.

The entrainment of anthropogenic aerosols with the desert dust, during the long-range transport, leads to the complex climate forcing scenarios because the radiative impact of the mixed aerosols depends on the chemical composition and optical characteristics of the fraction of individual components and also on the type of mixed state - internal (one or more of smaller aerosol particles are embedded in larger host particles), external (various aerosol species exists independently), and semi-external (aerosol aggregate - physical contact between two or more aerosol particles) mixed state. Recent studies involving the microscopic examination of the desert dust over polluted environments (anthropogenic) have found coating of desert dust with black carbon and other compounds of nitrate and sulfate that significantly alters the optical and radiative properties (such as single scattering albedo, extinction efficiency) of dust aerosols (Bauer et al., 2007; Buseck & Posfai, 1999; Chylek et al., 1995; Huang et al., 2010; Li & Shao, 2009; Yasunari et al., 2010). The desert dust also causes reduction in the albedo of contaminated snow over the Himalayas (Negi & Kokhanovsky, 2011).

Surface roughness of the snow increases due to the deposition of dust over the smooth snow surface. Over the years, the cycle of snow melting, dust deposition and snowfall leads to the formation of alternate layers of dust and snow. This may lead to increased melting as the presence of a very thin dust layer changes the friction within a snow pack. Thus, the snow pack in the desert dust affected region is more prone to melting than a dust-free region.

The vertical mixing (elevated layer, up to ~7 km) of dust with anthropogenic emissions have an impact on the radiation budget and precipitation over the IG plain and the high altitude Himalayan mountain range (Lau et al., 2006, 2008; Lau & Kim, 2006; Prasad & Singh, 2007; Prasad et al, 2009; Ramanathan et al., 2005). The deposition of desert dust and other pollutants as well as the changes in atmospheric temperatures (lower and middle tropospheric temperature) over the region due to the presence of mixed aerosols negatively impacts the snow cover and glaciers of Himalayas and Tibetan Plateau, particularly the western (Kashmir and Himachal Pradesh) and central (Uttarakhand, and west Nepal) Himalayas which are closer to the source and experience more frequent dust storms compared to the eastern Himalayas (eastern Nepal, Sikkim, Bhutan, and Arunachal Pradesh) (Das et al., 2010; Prasad & Singh, 2007b; Prasad et al., 2009; Yasunari et al., 2010).

7. Summary and conclusions

The Himalayan and Tibet Glaciers, source and origin of major rivers of Asia, are showing a variable rate of change of the snow cover and glacier retreat during the last several decades (since 1970) due to increasing effects of the climate variability and change, and ever increasing anthropogenic aerosols. The true color images from the earliest available satellite records from Landsat series (since 1972) and ASTER (since 2000) show substantial changes in the snow cover and glacial termini, with some spatial variability, over different regions of the Himalayas. The formation of numerous lakes, especially at the terminus of numerous glaciers, is a common feature compared to the 1970-1980 images. The ASTER images (2000-2010) also show inter-annual changes in the snow cover besides decadal changes since 1972.

The presence of dense networks of coal-fired power plants over the Indian sub-continent (emitting black carbon and various other aerosols) along the pathways of transport of desert dust leads to the mixing of dust and anthropogenic aerosols before reaching the Himalayan-Tibet region. As the anthropogenic activities, such as burning of fossil fuel, have tremendously increased during 1980-2010 over the IG plains, their adverse impact on the snow-cover and glaciers are discernible in space-based observations over the Himalayas, especially the western and central Himalayas. These observations are corroborated with independent measurement of concentrations of anthropogenic particles (aerosols) in the ice-cores that show an increase over the recent decades as compared with preceding decades.

The daily aerosol observations from MODIS Terra and Aqua show long-range transport of desert dust to high altitude Himalayas. This is also supported by air-mass transport models such as HYSPLIT. Models and satellite observations indicate that the arid and desert regions of Sahara, Middle East, Iran, Afghanistan, Pakistan and Thar-desert are the major sources of desert dust. The major pathways of transport of dust (land and sea route) show large inter-annual variability in the dust concentration. For instance, the aerosol loading during year 2008 was abnormally high compared to years 2000-2007 over the land and sea-route (Arabian Sea) and over the dust sink region (IG plains).

The long term aerosol observations from MODIS Terra and Aqua and the vertical profile of pre-monsoon (April-June) dust storms from CALIPSO show direct evidence of the transport of mixed dust and anthropogenic aerosols over the high altitude Himalayas (approximately 4-6 km above the msl). A combination of DT and DB AOD from MODIS Terra and Aqua satellites provides better results as compared with previous approaches to aerosol retrieval methods using the same observations (i.e. DT AOD method only). The current study shows that the elevated aerosol layer is visible up to ~7 km above the msl during the dust storm episodes. The vertical mixing of dust and pollutants also changes the radiation budget of the troposphere. The desert dust mixed with anthropogenic aerosols (black carbon) affects the atmospheric conditions (enhanced heating, change in the temperature gradient, and the monsoon circulation pattern), the reduction in snow albedo, and the roughness of layered snow deposits, leading to an increased melting of the snow pack. The melting of the cryosphere regions in the Himalayas is likely to accelerate due to growing anthropogenic aerosols from emissions in the IG plains. This, in turn, may result in the warming of the troposphere during December-May as indicated by atmospheric temperature trends derived from the Microwave Sounding Unit (MSU) during 1979-2008 (Prasad et al., 2009). Such changes would have profound socio-economic implications over the IG plains in the future.

8. Acknowledgements

We are thankful to the Land Processes Distributed Active Archive Center (LP DAAC), LAADS WeB for providing data from various NASA Earth Observing System (EOS) missions (Terra, Aqua, landsat and ASTER) (https://lpdaac.usgs.gov). ASTER GDEM is a product of METI and NASA. These data are distributed by the Land Processes Distributed Active Archive Center (LP DAAC), located at the U.S. Geological Survey (USGS) Earth Resources Observation and Science (EROS) Center (lpdaac.usgs.gov). We are also thankful to MODIS Rapid Response System (http://rapidfire.sci.gsfc.nasa.gov/). Special thanks to Alexander Naumov for providing the photographs over the western Himalayan Glaciers. We acknowledge the financial support from Institute for Global Change (IGC) and Science Systems and Applications (SSAI # 2804-09-024).

9. References

Ageta, Y. & Fujita, K. (1996). Characteristics of mass balance of summer-accumuation type glaciers in the Himalayas and Tibetan Plateau, *Zeitschrift für Gletscherkunde und Glazialgeologie*, 32 (2), pp. 61-65.

Arora, M.; Goel, N. K. & Singh, P. (2005). Evaluation of temperature trends over India, *Hydrological Sciences Journal-Journal Des Sciences Hydrologiques*, 50(1), pp. 81-93.

Bauer, S. E.; Mishchenko, M. I.; Lacis, A. A.; Zhang, S.; Perlwitz, J. & Metzger, S. M. (2007). Do sulfate and nitrate coatings on mineral dust have important effects on radiative properties and climate modeling?, *Journal of Geophysical Research-Atmospheres*, 112(D6).

Berthier, E.; Arnaud, Y.; Kumar, R.; Ahmad, S.; Wagnon, P. & Chevallier, P. (2007). Remote sensing estimates of glacier mass balances in the Himachal Pradesh (Western Himalaya, India), *Remote Sensing of Environment*, 108(3), 327-338.

Bhattacharjee, P. S.; Prasad, A. K.; Kafatos, M. & Singh, R. P. (2007). Influence of a dust storm on carbon monoxide and water vapor over the Indo-Gangetic Plains, *Journal of Geophysical Research-Atmospheres*, 112(D18).

Bhutiyani, M. R.; Kale, V. S. & Pawar, N. J. (2007). Long-term trends in maximum, minimum and mean annual air temperatures across the Northwestern Himalaya during the twentieth century, *Climatic Change*, 85(1-2), 159-177.

Bookhagen, B. & Burbank, D. W. (2010). Toward a complete Himalayan hydrological budget: Spatiotemporal distribution of snowmelt and rainfall and their impact on river discharge, *Journal Of Geophysical Research-Earth Surface*, 115.

Buseck, P. R. & Posfai, M. (1999). Airborne minerals and related aerosol particles: Effects on climate and the environment, *Proceedings of the National Academy of Sciences of the United States of America*, 96(7), 3372-3379.

Chylek, P.; Videen, G.; Ngo, D.; Pinnick, R. G. & Klett, J. D. (1995). Effect of black carbon on the optical-properties and climate forcing of sulfate aerosols, *Journal of Geophysical Research-Atmospheres*, 100(D8), 16325-16332.

Das, S. K.; Dobhal, D. P. & Juyal, N. (2010). Variability of aerosol optical depth and recent recessional trend in Dokriani Glacier, Bhagirathi Valley, Garhwal Himalaya, *Current Science*, 99(12), 1816-1821.

Di Girolamo, L.; Bond, T. C.; Bramer, D.; Diner, D. J.; Fettinger, F.; Kahn, R. A.; Martonchik, J. V.; Ramana, M. V.; Ramanathan, V. & Rasch, P. J. (2004). Analysis of Multi-angle Imaging SpectroRadiometer (MISR) aerosol optical depths over greater India during winter 2001-2004, *Geophysical Research Letters*, 31(23).

Duan, K.; Thompson, L. G.; Yao, T.; Davis, M. E. & Mosley-Thompson, E. (2007). A 1000 year history of atmospheric sulfate concentrations in southern Asia as recorded by a Himalayan ice core, *Geophysical Research Letters*, 34(1).

Gadi, R.; Kulshrestha, U. C.; Sarkar, A. K.; Garg, S. C. & Parashar, D. C. (2003). Emissions of SO2 and NOx from biofuels in India, *Tellus Series B-Chemical And Physical Meteorology*, 55(3), pp. 787-795.

Garg, A.; Shukla, P. R.; Bhattacharya, S. & Dadhwal, V. K. (2001). Sub-region (district) and sector level SO2 and NOx emissions for India: assessment of inventories and mitigation flexibility, *Atmospheric Environment*, 35(4), pp. 703-713.

Gautam, R.; Hsu, N. C.; Lau, K. M.; Tsay, S. C. & Kafatos, M. (2009). Enhanced pre-monsoon warming over the Himalayan-Gangetic region from 1979 to 2007, *Geophysical Research Letters*, 36.

Ghude, S. D.; Fadnavis, S.; Beig, G.; Polade, S. D. & van der A, R. J. (2008). Detection of surface emission hot spots, trends, and seasonal cycle from satellite-retrieved NO(2) over India, *Journal Of Geophysical Research-Atmospheres,* 113(D20).

Ginoux, P.; Garbuzov, D. & Hsu, N. C. (2010). Identification of anthropogenic and natural dust sources using Moderate Resolution Imaging Spectroradiometer (MODIS) Deep Blue level 2 data, *Journal of Geophysical Research-Atmospheres,* 115.

Goloub, P.; Deuze, J. L.; Herman, M.; Marchand, A.; Tanre, D.; Chiapello, I.; Roger, B. & Singh, R. P. (). Aerosol remote sensing over land from the spaceborne polarimeter POLDER, *in* W. L. Smith & Y. M. Timofeyev, ed., IRS 2000: Current Problems in Atmospheric Radiation, A Deepak Publishing, pp. 113-116.

Hasnain, S. J. (2002). Himalayan glaciers meltdown: impact on south Asian rivers, *in* H. A. J. VanLanen & S. Demuth, ed., FRIEND 2002-Regional Hydrology: Bridging The Gap Between Research And Practice, Int Assoc Hydrological Sciences, pp. 417-423.

He, Y.Q., Z.L. Zhang, W.H. Theakstone, T. Chen, T.D. Yao, H.X. Pang (2003). Changing features of the climate and glaciers in China's monsoonal temperate glacier region, *Journal of Geophysical Research – Atmospheres,* 108(D17), article number 4530.

Hsu, N. C.; Tsay, S.-C.; King, M. D. & Herman, J. R. (2006). Deep blue retrievals of Asian aerosol properties during ACE-Asia, *IEEE Transactions on Geoscience and Remote Sensing,* 44(11, Part 1), 3180-3195.

Huang, K.; Zhuang, G.; Li, J.; Wang, Q.; Sun, Y.; Lin, Y. & Fu, J. S. (2010). Mixing of Asian dust with pollution aerosol and the transformation of aerosol components during the dust storm over China in spring 2007, *Journal of Geophysical Research-Atmospheres* 115.

Immerzeel, W. W.; Droogers, P.; de Jong, S. M. & Bierkens, M. F. P. (2009). Large-scale monitoring of snow cover and runoff simulation in Himalayan river basins using remote sensing, *Remote Sensing Of Environment,* 113(1), 40-49.

Immerzeel, W. W.; van Beek, L. P. H. & Bierkens, M. F. P. (2010). Climate Change Will Affect the Asian Water Towers, *Science,* 328(5984), 1382-1385.

IPCC (2001). Climate Change 2001: Impacts, Adaptation and Vulnerability - Contribution of Working Group II to the Third Assessment Report of Intergovernmental Panel on Climate Change, Cambridge University Press, UK. http://www.grida.no/climate/ipcc_tar/wg2/index.htm

Jaswal, A. K. & Rao, G. S. P. (2010). Recent trends in meteorological parameters over Jammu and Kashmir, *Mausam,* 61(3), pp. 369-382.

Jaswal, A. K. (2010). Recent winter warming over India - spatial and temporal characteristics of monthly maximum and minimum temperature trends for January to March, *Mausam,* 61(2), pp. 163-174.

Jhajharia, D. & Singh, V. P. (2010). Trends in temperature, diurnal temperature range and sunshine duration in Northeast India, *International Journal of Climatology,* DOI: 10.1002/joc.2164.

Kang, S. C.; Wake, C. P.; Qin, D. H.; Mayewski, P. A. & Yao, T. D. (2000). Monsoon and dust signals recorded in Dasuopu glacier, Tibetan Plateau, *Journal of Glaciology,* 46(153), 222-226.

Karma, Ageta, Y.; Naito, N.; Iwata, S. & Yabuki, H. (2003). Glacier distribution in the Himalayas and glacier shrinkage from 1963 to 1993 in the Bhutan Himalayas, 29-40.

Kayetha, V. K.; Senthilkumar, J.; Prasad, A. K.; Cervone, G. & Singh, R. P. (2007). Effect of dust storm on ocean color and snow parameters, *Photonirvachak-Journal of the Indian Society of Remote Sensing,* 35(1), 1-9.

Kehrwald, N. M.; Thompson, L. G.; Tandong, Y.; Mosley-Thompson, E.; Schotterer, U.; Alfimov, V.; Beer, J.; Eikenberg, J. & Davis, M. E. (2008). Mass loss on Himalayan glacier endangers water resources, *Geophysical Research Letters,* 35(22).

Krishna, A. P. (2005). Snow and glacier cover assessment in the high mountains of Sikkim Himalaya, *HYDROLOGICAL PROCESSES* 19(12), pp. 2375-2383.

Kulkarni, A. & Alex, S. (2003). Estimation of recent glacial variations in Baspa basin using remote sensing techniques, Photonirvachak-Journal of the Indian Society of Remote Sensing, 31, pp. 81-90.

Kulkarni, A. (2007). Effect of Global Warming on the Himalayan Cryosphere, *Jalvigyan Sameeksha,* 22, 93-108.

Kulkarni, A. V. & Bahuguna, I. M. (2002). Glacial retreat in the Baspa basin, Himalaya, monitored with satellite stereo data, *Journal of Glaciology,* 48(160), pp. 171-172.

Kulkarni, A. V.; Rathore, B. P.; Mahajan, S. & Mathur, P. (2005). Alarming retreat of Parbati Glacier, Beas basin, Himachal Pradesh, *Current Science,* 88(11), pp. 1844-1850.

Kulkarni, A. V.; Rathore, B. P.; Singh, S. K. & Ajai (2010). Distribution of seasonal snow cover in central and western Himalaya, *Annals of Glaciology,* 51(54), 123-128.

Lau, K. M. & Kim, K. M. (2006). Observational relationships between aerosol and Asian monsoon rainfall, and circulation, *Geophysical Research Letters,* 33(21).

Lau, K. M.; Kim, M. K. & Kim, K. M. (2006). Asian summer monsoon anomalies induced by aerosol direct forcing: the role of the Tibetan Plateau, *Climate Dynamics,* 26(7-8), pp. 855-864.

Lau, K. M.; Ramanathan, V.; Wu, G. X.; Li, Z.; Tsay, S. C.; Hsu, C.; Sikka, R.; Holben, B.; Lu, D.; Tartari, G.; Chin, M.; Koudelova, R.; Chen, H.; Ma, Y.; Huang, J.; Taniguchi, K. & Zhang, R. (2008). The Joint Aerosol-Monsoon Experiment - A new challenge for monsoon climate research, *Bulletin of the American Meteorological Society,* 89(3), 369-+.

Lau, W. K. M.; Kim, M.-K.; Kim, K.-M. & Lee, W.-S. (2010). Enhanced surface warming and accelerated snow melt in the Himalayas and Tibetan Plateau induced by absorbing aerosols, *Environmental Research Letters,* 5(2).

Lee, K.; Do Hur, S.; Hou, S.; Hong, S.; Qin, X.; Ren, J.; Liu, Y.; Rosman, K. J. R.; Barbante, C. & Boutron, C. F. (2008). Atmospheric pollution for trace elements in the remote high-altitude atmosphere in central Asia as recorded in snow from Mt. Qomolangma (Everest) of the Himalayas, *Science of the Total Environment,* 404(1), pp. 171-181.

Li, W. J. & Shao, L. Y. (2009). Observation of nitrate coatings on atmospheric mineral dust particles, *Atmospheric Chemistry and Physics,* 9(6), pp. 1863-1871.

Liu, X. D. & Chen, B. D. (2000). Climatic warming in the Tibetan Plateau during recent decades, *International Journal of Climatology,* 20(14), pp. 1729-1742.

Naz, S. B.; Bowling, L.C. & Crawford, M. M. (2011a). Spatial and temporal glacier changes in the Karakoram Himalaya derived from Landsat satellite data. Journal of Glaciology, (*Submitted*, December, 2010).

Naz, S. B.; Bowling, L.C. & Crawford, M. M. (2011b). Quantification of glacier changes using ICESat elevation data and the SRTM digital elevation model in the Upper Indus Basin, *(under preparation)*.

Negi, H. S. & Kokhanovsky, A. (2011). Retrieval of snow albedo and grain size using reflectance measurements in Himalayan basin, *Cryosphere*, 5(1), 203-217.

Prasad, A. K. & Singh, R. P. (2007a). Changes in aerosol parameters during major dust storm events (2001-2005) over the Indo-Gangetic Plains using AERONET and MODIS data, *Journal of Geophysical Research-Atmospheres*, 112(D9).

Prasad, A. K. & Singh, R. P. (2007b). Changes in Himalayan Snow and Glacier Cover Between 1972 and 2000, *Eos Trans. AGU*, 88(33), 326.

Prasad, A. K. & Singh, R. P. (2007c). Comparison of MISR-MODIS aerosol optical depth over the Indo-Gangetic basin during the winter and summer seasons (2000-2005), *Remote Sensing of Environment*, 107(1-2), pp. 109-119.

Prasad, A. K. (2007). Multi-sensor appraisal of aerosols, vegetation and monsoon dynamics over Indian subcontinent. *Ph.D. Thesis*, Indian Institute of Technology Kanpur, pp. 1-238 *(Unpublished)*.

Prasad, A. K.; Singh, R. P. & Kafatos, M. (2006a), Influence of coal based thermal power plants on aerosol optical properties in the Indo-Gangetic basin, *Geophysical Research Letters*, 33(5).

Prasad, A. K.; Singh, R. P. & Singh, A. (2006b), Seasonal climatology of aerosol optical depth over the Indian subcontinent: trend and departures in recent years, *International Journal of Remote Sensing*, 27(12), pp. 2323-2329.

Prasad, A. K.; Singh, S.; Chauhan, S. S.; Srivastava, M. K.; Singh, R. P. & Singh, R. (2007). Aerosol radiative forcing over the Indo-Gangetic plains during major dust storms, *Atmospheric Environment*, 41(29), pp. 6289-6301.

Prasad, A. K.; Yang, K. H. S.; El-Askary, H. M. & Kafatos, M. (2009). Melting of major Glaciers in the western Himalayas: evidence of climatic changes from long term MSU derived tropospheric temperature trend (1979-2008), *Annales Geophysicae*, 27(12), pp. 4505-4519.

Prasad, A.; Singh, R. & Kafatos, M. (2011). Influence of Coal Based Thermal Power Plants on the Spatial-Temporal Variability of Tropospheric NO2 Column over India, *Environmental Monitoring and Assessment*, DOI: 10.1007/s10661-011-2087-6 *(in press)*.

Qin, D. H.; Mayewski, P. A.; Kang, S. C.; Ren, J. W.; Hou, S. G.; Yao, T. D.; Yang, Q. Z.; Jin, Z. F. & Mi, D. S.Steffen, K., ed., (2000). Evidence for recent climate change from ice cores in the central Himalaya, *Annals of Glaciology*, 31, pp. 153-158.

Raina, V. (2010). Himalayan Glaciers A State-of-Art Review of Glacial Studies, Glacial Retreat and Climate Change, Geological Survey of India. Retrieved January 10, 2010. Available at http://gbpihed.gov.in/MoEF%20Dissussion%20Paper%20on %20Himalayan%20Glaciers.pdf

Raina, V. K. & Sangewar, C. (2007). Siachen glacier of Karakoram Mountains, Ladakh its secular retreat, *Journal of the Geological Society of India*, 70(1), 11-16.

Ramanathan, V. & Ramana, M. V. (2005). Persistent, widespread, and strongly absorbing haze over the Himalayan foothills and the Indo-Gangetic Plains, *Pure and Applied Geophysics,* 162(8-9), pp. 1609-1626.

Ramanathan, V.; Chung, C.; Kim, D.; Bettge, T.; Buja, L.; Kiehl, J. T.; Washington, W. M.; Fu, Q.; Sikka, D. R. & Wild, M. (2005). Atmospheric brown clouds: Impacts on South Asian climate and hydrological cycle, *Proceedings of the National Academy of Sciences of the United States of America,* 102(15), pp. 5326-5333.

Rees, H. G. & Collins, D. N. (2006). Regional differences in response of flow in glacier-fed Himalayan rivers to climatic warming, *Hydrological Processes,* 20(10), pp. 2157-2169.

Rikiishi, K. & Nakasato, H. (2006). Height dependence of the tendency for reduction in seasonal snow cover in the Himalaya and the Tibetan Plateau region, 1966-2001, *Annals of Glaciology, 43,* pp. 369-377.

Scherler, D.; Bookhagen, B. & Strecker, M. R. (2011). Spatially variable response of Himalayan glaciers to climate change affected by debris cover, *Nature Geoscience,* 4(3), pp. 156-159.

Shrestha, A. B. & Aryal, R. (2011). Climate change in Nepal and its impact on Himalayan glaciers, *Regional Environmental Change,* 11, S65-S77.

Shrestha, A. B.; Wake, C. P.; Mayewski, P. A. & Dibb, J. E. (1999). Maximum temperature trends in the Himalaya and its vicinity: An analysis based on temperature records from Nepal for the period 1971-94, *Journal of Climate,* 12(9), pp. 2775-2786.

Singh, R. P.; Dey, S.; Tripathi, S. N.; Tare, V. & Holben, B. (2004). Variability of aerosol parameters over Kanpur, northern India, *Journal of Geophysical Research-Atmospheres,* 109(D23).

Srivastava, A. K.; Tiwari, S.; Devara, P. C. S.; Bisht, D. S.; Srivastava, M. K.; Tripathi, S. N.; Goloub, P. & Holben, B. N. (2011). Pre-monsoon aerosol characteristics over the Indo-Gangetic Basin: implications to climatic impact, *Annales Geophysicae,* 29, pp. 789-804.

UNEP (2008). U.N. Environmental Program and World Glacier Monitoring Service, Global Glacier Change: Facts and Figures UNEP Publ., http://www.grid.unep.ch/glaciers/ (2008).

UNEP (2009). Recent trends in melting glaciers, tropospheric temperatures over the Himalayas and summer monsoon rainfall over India. Available at http://www.unep.org/dewa/Portals/67/pdf/Himalayas.pdf

Upadhyay, R. (2009). The melting of the Siachen glacier, *Current Science,* 96(5), pp. 646-648.

Winiger, M.; Gumpert, M. & Yamout, H. (2005). Karakorum-Hindukush-western Himalaya: assessing high-altitude water resources, *Hydrological Processes,* 19(12), pp. 2329-2338.

Xu, J.; Hou, S.; Qin, D.; Kang, S.; Ren, J. & Ming, J. (2007). Dust storm activity over the Tibetan Plateau recorded by a shallow ice core from the north slope of Mt. Qomolangma (Everest), Tibet-Himal region, *Geophysical Research Letters,* 34(17).

Yasunari, T. J.; Bonasoni, P.; Laj, P.; Fujita, K.; Vuillermoz, E.; Marinoni, A.; Cristofanelli, P.; Duchi, R.; Tartari, G. & Lau, K. M. (2010). Estimated impact of black carbon deposition during pre-monsoon season from Nepal Climate Observatory - Pyramid data and snow albedo changes over Himalayan glaciers, *Atmospheric Chemistry and Physics,* 10(14), pp. 6603-6615.

Zhang, G.; Hongjie, X,; Kang, S.; Yi, D. & Stephen, F. A. (2011). Monitoring lake level changes on the Tibetan Plateau using ICESat altimetry data (2003-2009), *Remote Sensing of Environment* 115, pp. 1733-1742.

6

Accounting the Carbon Storage in Disturbed and Non-Disturbed Tropical Andean Ecosystems

Enrique Peña, Orlando Zúñiga and Julián Peña
Universidad del Valle
Colombia

1. Introduction

Global temperature has increased by about 0.7°C over the last century, a figure that is considered disproportionally large (Zhang et al., 1996). It is projected that climate changes will have profound biological effects, including the changes in species distributions as well as in carbon storage in forest ecosystems. The observed increase in emission of greenhouse gases, with attendant effects on global warming, have raised interests in identifying sources and sinks of carbon in the environment. A significant number of papers have demonstrated the history of the carbon dioxide content in the atmosphere over the last 4 billion years ant its relationship with the climatic history of the Earth, but it is less documented the effect of global warming on the carbon storage patterns in high Andean ecosystems. Recently, it has demonstrated that organic matter decomposition could increase under warmer climates, which may cause carbon flux and energy flow changes in alpine ecosystems (Kato et al., 2006). Simultaneously, Zhang et al., 1996 registered findings that are derived from a short-term responses to simulated environmental warming, focusing on aboveground biomass of three dominated life forms and community compositional attributes.

Carbon storage in the Andean region involves different ecosystems such as the tropical montane cloud forests, the high-altitude wetlands, and the paramos ecosystem (Earle, et al., 2003, Peña et al., 2009). The total ecosystem carbon stock on these high mountains ecosystems is large and they are playing recently an important role in the global carbon balance. Although they cover only about 3% of the land area, they store about 30% of the global carbon storage of terrestrial ecosystems (IPCC, 2007). Most of the cover natural areas in the Andean regions are represented by paramos, high mountains, and wetlands (Peña et al., 2009).

The paramo is a unique ecosystem of High Mountain between 3000 and 4800 meters above the sea level. The vegetation is composed of shrubs and grasses, and the thermal condition is continuously cool to cold, and relatively dry because of rain shadow effects from the surrounding mountains (Van Der Hammen, 1997). In the Andes, the paramo ecosystem ranges from Merida, Venezuela throughout Colombia, Ecuador down to Huancabamba (Perú). The low temperatures and the soil capacity to retain water lead this ecosystem to have a relatively low mineralization and nutrient recycling rates (Brady, N. & Weil, R., 2002; Lal, 2004). These features are related with the ability of paramos to sequester atmospheric

CO_2 between 466 a 239 t ha^{-1} at 1 m deep soils (Hofstede et al., 2003) and 20 t ha^{-1} in vegetation (Hofstede, 1999).

The tropical montane cloud forests range between 2.800 and 3.200 meters above sea level (IDEAM, 2002). The carbon deposit in those areas is higher at higher levels at relatively lower temperatures and lower humidity. Carbon storage in those systems have ranged between 155 to 231 t ha^{-1} in soils a 1 meter deep and 167 to 249 t ha^{-1} in vegetation (Amezquita et al., 2006).

The high-altitude wetlands are related to the formation of water sources characteristic of the region of the Andean paramos (> 3000 m altitude), which are manifested in the form of ponds, swamps, lakes and springs that emerge from underground (Van Der Hammen & Hooghiemstra, 2003). High-altitude wetlands cover only approximately 3% of the total land area (Maltby & Immirzi, 1993), but their importance in the carbon cycle has been recognized because they can store approximately 30% of the global terrestrial carbon, equivalent to 455 Pg C (Gorham, 1991; Blodau, 2002) (1 Pg C = 1 Gt C = 10^{15} g of carbon). This percentage of carbon is sequestered primarily via the process of transforming the organic matter in plant biomass (Blodau et al., 2004, Peña et al., 2009).), reaching total levels of 0,5-0,7 t of carbon ha^{-1} (Heathwaite, 1993).

In most Andean countries, human population is concentrated in the surrounding areas of the tropical montane cloud forests. Consequently, there is an increasingly use for intensive cattle grazing, cultivation, and pine planting on these natural areas (Van Der Hammen 1995; Castaño et al., 2002; Verweij et al., 2003). Particularly, removal of vegetation cover and intensified land use on the Colombian Andean montane cloud forests have been causing erosion and pollution of the surface water. Recently, investigations are still being undertaken to quantify the impact of human activities on the carbon flux on impacted and non-impacted high mountain forests (Peña et al., 2009). Monitoring the carbon content in high mountain ecosystems has become of great global importance, given the potential role that these ecosystems can play as sinks or sources of greenhouse gas emissions (CO_2). In Colombia, the high mountains ecosystems are related to the formation of the Andean Paramos (> 3000 m altitude) which are manifested in the form of typical vegetation (MAVDT et al., 2001). Simulation scenarios for Colombian have shown that for 2050, the mean annual temperature will rise up between 1 and 2 °C, and the mean annual rainfall around ± 15%. Those scenarios will bring up a decreasing percentage in paramos vegetation and snow mountains in about 15 and 78 % respectively (IDEAM, 2002, UNIVALLE & IDEAM, 2008).

The study was developed in two natural protected areas from the Colombian National Parks Systems, Chingaza (PNN), and Los Nevados (PNN) national parks. In disturbed and non-disturbed forests of both areas, the carbon storage on soil, biomass and wetland land-water interaction zones were evaluated. The study also quantified the degree to which human impacts and global warming influence the carbon sink/sources on those compartments at each ecosystem.

2. The study area

Carbon storage in three typical high mountain ecosystems was monitored; the paramo vegetation, the tropical montane cloud forests, and high altitude wetlands in disturbed and non-disturbed natural areas (Fig. 1 and Fig. 2). Tab. 1 summarized the geographical distribution of each representative ecosystem.

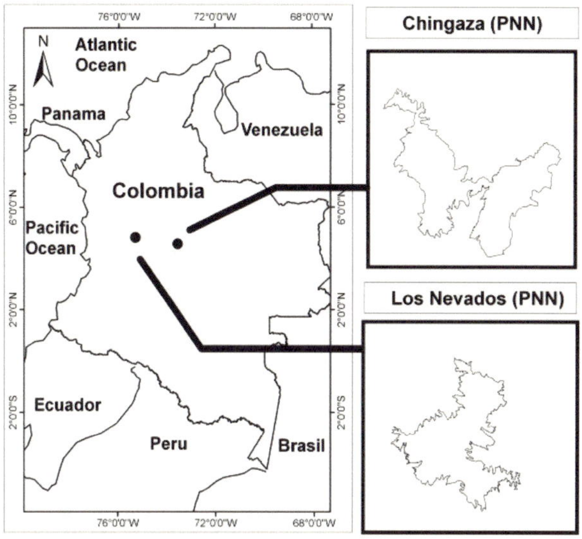

Fig. 1. Maps of studied National Parks, Chingaza (PNN) and Los Nevados (PNN)

(a) tropical montane cloud forest

(b) Paramo vegetation

(c) High altitude wetlands

Fig. 2. High mountain ecosystems

National Park	Ecosystem Type	Geographycal Location	Altitude (meters above sea level)
Chingaza (PNN)	Non-disturbed Paramo	73° 48' 3,4" W 4° 40' 58,8" N	3611
	Disturbed Paramo	73° 49' 32,6" W 4° 39' 51,7" N	2291
	Non-disturbed Forest	73° 50 '52" W 4° 39' 54" N	3041
	Disturbed Forest	73° 51' 28,4" W 4° 40' 6,4" N	2650
	Wetland	73° 48' 36" W 4° 40' 21,7 " N	3200
Los Nevados (PNN)	Non-disturbed Paramo 1	75° 22 ' 11,3" W 4° 50' 38,8" N	4250
	Non-disturbed Páramo 2	75° 23' 12,2" W 4° 50' 22,6" N	4173
	Disturbed Paramo	75° 23' 39" W 4° 51' 9,5" N	3900
	Non-disturbed Forest	75° 25' 30,9" W 4° 53' 16,8" N	3194
	Disturbed forest	75° 25' 43" W 4° 53' 14,8" N	3152
	Wetland 1	75° 21' 33,5" W 4° 49' 54,2" N	4456
	Wetland 2	75° 22' 19" W 4° 50' 57" N	4080

Table 1. Geographical location of studied ecosystems

2.1 Carbon storage in soils

Two soil cores on a core region of each ecosystem were taken along a transect, parallel to the vegetation of non-disturbed and disturbed forest areas. Each core was 1 m x 1 m x 1 m. In each core site a detailed description of soil features was done and samples for organic matter content, microbiological analysis, bulk density, and soil structural characteristics were performed (UNIVALLE & IDEAM, 2008). Total Organic Carbon (COS) content (t ha^{-1}) was estimated by using equation 1 (Rosenzweig & Hillel, 2000):

$$COS = A * Fc * p * Da \quad (1)$$

where,

A: sampling area (ha); Da: soil apparent Density (t m^{-3})

Fc: carbón fraction (%); p: sampling horizon soil (m)

Soil features of each study site are different among them. Chingaza (PNN) has typically soils with low bulk density (-0,35 kg L^{-1}, oven-dried basis) and a very fine structure that adsorbs large amounts of water and higher carbon content and humus (Fig. 3a). In contrast, soils from Los Nevados (PNN) National Park are characterized by having high amount of volcanic ash with relatively low carbon content, typically from areas of high volcanic activity (Fig. 3b).

COS content at one meter deep in Paramo soils of Chingaza (PNN), and in Los Nevados (PNN) is shown in Fig. 4. These data are consistent in those found in similar areas of high mountains systems, with values ranged between 239 and 479 t ha^{-1}, respectively (Hofstede et al., 2003).

(a) Chingaza (PNN) (b) Los Nevados (PNN)

Fig. 3. Cross section of soil conditions in a) Chingaza (PNN) and b) Los Nevados (PNN)

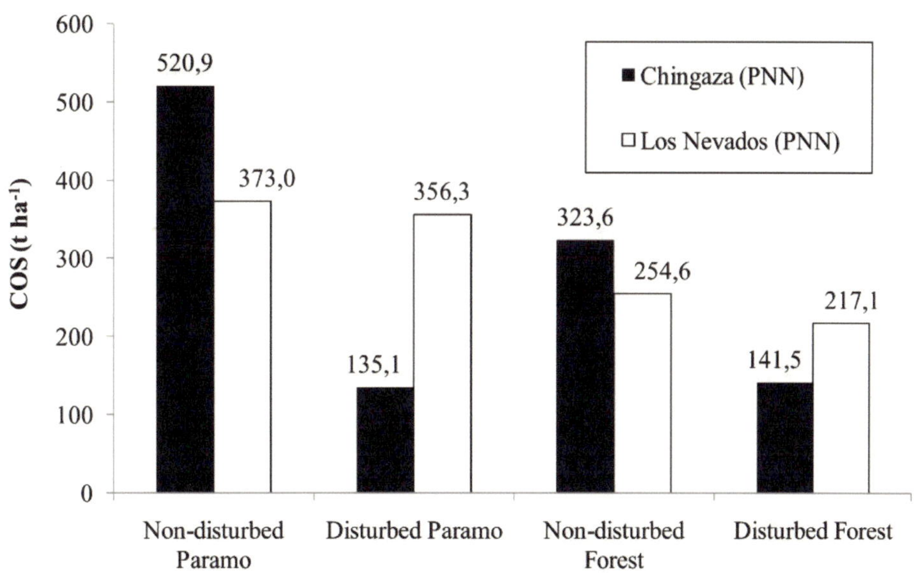

Fig. 4. Comparative analysis of total organic content (COS) in soils of disturbed and non-disturbed areas in Chingaza (PNN) and Los Nevados (PNN).

Carbon storage in non-disturbed sites are higher compared those with values from disturbed sites. This pattern was similar for both protected areas. However, COS values turned out higher in non-disturbed cloud forests from Chingaza (PNN). Hofstede (1999) found similar values of COS in tropical forests around 50 t ha^{-1}. Batjes (1999) registered storage values around 80 to 100 t ha^{-1}, which were significantly lower than those found in this study. In Colombia, mean values between 52 t ha^{-1} and 75 t ha^{-1} of COS in rain forests was reported by Ibrahim et al., (2007). Other studies reported mean values of 102 t ha^{-1} (Escobar & Toriatti Dematté, 1991). Carvajal et al., (2009), registered mean values of 107 t ha-1, and Amezquita et al., (2006) around 181 t ha-1. Other reports such as, Kanninen (2003), Brown and Lugo (1992) and Brown et al. (1969) in the Peruvian Andes registered carbon storage in tropical montane cloud forest ranged between 190 to 230 t ha-1, which are close to those ones found in those similar natural areas. In Guatemala, mean values of COS were around 130,4 t ha-1 (Arreaga, 2002). In Costa Rica, around 82,2 t ha-1 (Powers & Schlesinger, 2002, López-Ulloa et al. 2005). In Venezuela around 125 t ha-1 (Delaney et al., 1997). In Brazil, values around 162 t ha-1 (IPCC, 2000). In Hawai, values around 62,5 t ha-1 (Schuur et al., 2001).

Globally, over 69% of the total C pool in forest ecosystems is stored in soil, especially the soil organic carbon (SOC) fraction, which higher amounts are related with relatively higher organic matter decomposition rates in those areas favored by the humidity and temperature conditions Glenday (2006). These soil horizons are generally considered to be older, more stable, and slower to turn over than labile C at the forest floor. In the African tropics, Glenday (2006) found values of COS up to 100 t ha^{-1} in Kenyan ecosystem forests. In subtropical forests, COS values have been reported to be significantly lower, below 70 t ha^{-1} (Dupouey et al., 1999, IPCC, 2000, Montagnini & Jordan 2002, Moretto et al., 2005).

Results showed significant higher COS levels in non-disturbed forests of both natural areas than in disturbed forest areas. Similar trends were reported by Carúa et al. (2008) and Hernández & Triana (2009) in Los Nevados (PNN) National Park. This pattern was consistent with increased organic matter content in the paramos and the tropical montane cloud forests floor (Lal, 2004, Diaz, 2008). The relationship is also consistent with soil particles size below 0,125 mm, typically found in the cloud forest floor. These conditions may favor microbiological processes which in turn increases carbon sinks in the soil ecosystem compartment (Lal, 2004). Andean montane cloud forest have suffered recurrent anthropogenic fire during the last decades (Diaz, 2008), and those affected forest areas, such as the paramo vegetation of Chingaza (PNN) National Park, have show carbon depletion, containing 25% less total carbon than the least disturbed sites (Laegaard, 1992; Vargas 2000; Vargas et al., 2002).

2.2 Carbon storage in forest biomass

Nine sampling quadrant plots of 500 m^2 or 0,05 ha (20 m x 25 m) with 20 sub-quadrant of 5 m x 5 m were taken along a transect, parallel to the vegetation of non-disturbed and disturbed forest areas. In each sampling plot a detailed description of above forest biomass and vegetation structure was performed. The botanical survey was taken during a one year period and permanent plots were established as follows: in the paramos ecosystem five permanent plots in the non-affected area and one permanent plot in the affected areas. Four permanent plots was placed up in non-disturbed cloud forest areas and one plot in disturbed forest areas, according to the protocol proposed by study of UNIVALLE & IDEAM (2008).

On each plot information of total height, stem height, stem diameter at breast height, and wood density from all individuals were performed. Specimens of each plant species were deposited in the Herbarium of Universidad del Valle (CUVC). The model proposed by

Brown et al. (1989) was applied for biomass estimates, as a non-destructive measurement of tree biomass, using equation 2 (Dauber et al., 2000):

$$BAA = e\left(-2{,}4090 + 0{,}9522\left(d^2 * h * D\right)\right) \tag{2}$$

Where:
BAA = above-ground biomass (kg)
e = base of natural logarithm (2,718271)
d = diameter at breast height (cm)
h = total tree height (cm)
D = basic wood density (kg cm^{-3})
Non-destructive methods were used to evaluate ground biomass of thicker roots (> 5 mm thick), using equation 3, which also was used to calibrate tree size, based on samples of trees in secondary and primary forests (Sierra et al., 2003).

$$BSA = -\exp\left(-4{,}394 + 2{,}693\ln(d)\right) \tag{3}$$

Where:
BSA = ground biomass of tree > 5 mm (kg)
d = diameter at breast height (cm)
Due to the absence of equations for estimating grassland biomass of paramo vegetation or bush type, which it is not available in the literature, it was necessary to establish equations 4 and 5 developed to estimate aboveground and belowground biomass in the paramo ecosystem. Biomass was measured directly in the four most common species of shrub-type vegetation. Samples were taken outside permanent plots. The regression equation was modeled with the total plant height in centimeters and therefore it was the input unit for the equation.

$$BAP = -2074{,}48 + 24{,}49 * h \tag{4}$$

$$BSP = -920{,}51 + 10{,}83 * h \tag{5}$$

Where:
BAP = above-ground biomass shrub heath (kg)
BSP = ground biomass of shrub heath (kg)
h = total plant height (cm)
Above and underground biomass of grassland-type vegetation from paramo was directly measured along a transect of 100 m long using sampling plots of 1 m^2 every 10 m.
The dominated forest type of vegetation in Chingaza (PNN) were represented by the genera *Clusia*, *Weinmannia* and species of Melastomataceae, Meliaceae and Rubiaceae. The dominant species of the forest vegetation in Los Nevados (PNN) were represented mainly by the genera *Saurauia*, *Oreopanax*, *Cyathea* and species of Melastomataceae.
Results showed that the carbon storage in affected forested areas represented only between 10 and 20% of the carbon stored in undisturbed forests for both, Chingaza (PNN) and Los Nevados (PNN) national parks (Fig. 5). The greater pressure was noted in Chingaza (PNN), compared it with Los Nevados (PNN), mainly due to the surrounding areas of Chingaza (PNN) have been used for agricultur between 3000 and 3500 meters, as well as the recent introduction of improved pastures for intensive cattle activities (Vargas & Pedraza, 2004). Similar trends were found by Lapeyre et al. (2004) in Peru, where carbon sequestration in disturbed tropical forests were significantly lower than carbon accumulation in pristine areas.

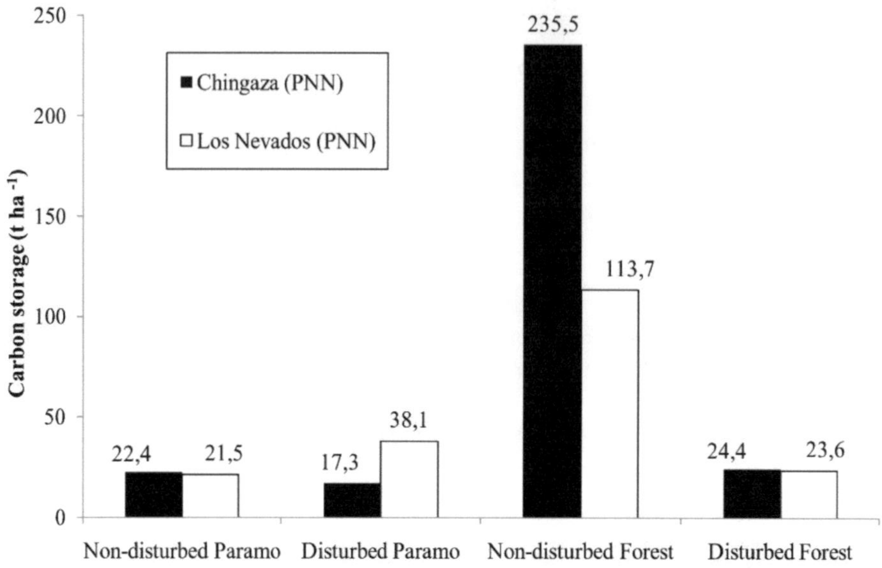

Fig. 5. Carbon storage in paramo and montane cloud forest areas in Chingaza (PNN) and Los Nevados (PNN), Colombia

Grassland and shrubes were the most common vegetation type in paramo ecosystems of Chingaza (PNN). The were represented by shrubs such as *Hypericum goyanesii* (Cuatrc.), *Espeletia argentea* (Bonpl.), *Espeletia grandiflora* (Bonpl.), *Bejaria resinosa* (Mutis), *Berberis glauca* DC. and *Aragoa sp.* In Los Nevados (PNN) the most common species were *Baccharis latifolia* (Ruiz and Pav.) Pers, *Escallonia myrtoidea* (Bertero), *Diplostephium sp. Gynoxis sp.* and *Espeletia spp.* Pastures type of vegetation was the common botanical ecological unit in disturbed areas. It was mainly represented by genera such as *Agrostis, Calamagrostis* (Poaceae), *Cortaderia sp.* (Poaceae), and the african grassland species, *Pennisetum clandestinum* (Hochst), which is more abundant in Los Nevados (PNN). This grass forage is very appreciated by farmers for their high production of biomass due its impact on milk production. It also has a fairly aggressive rhizomatous growth that makes it easy to spread and very difficult to eradicate. The results of carbon storage by vegetation in disturbed and non-disturbed areas showed similar trends with those observed in other tropical montane forest areas Hofstede (1999) where mean values ranged around 250 t ha^{-1} and paramo vegetation ranged around 20 t ha^{-1} (Fig. 6).

The relatively higher carbon storage of paramo vegetation in non-disturbed forest areas of Chingaza (PNN) is probably explained by more stable humid conditions that favors the successful establishment and growth of forest biomass, compared those with a more drier environment in Los Nevados (PNN). This is particularly important when comparing carbon storage and type of vegetation in both natural areas. Forest biomass in permanent plots of Los Nevados (PNN) showed smaller woody plant species and lower values of above-ground biomass. The lower carbon storage is also related with a high intensive cattle activity which has brought changes in the natural successional processes of grassland vegetation in this area. The relatively lower biomass production due to the type of vegetation has showed a strong relationship with carbon sequestration in paramo natural areas, which was observed by (Guhl, 1982, Vargas & Pedraza, 2004, Lotero et al., 2006).

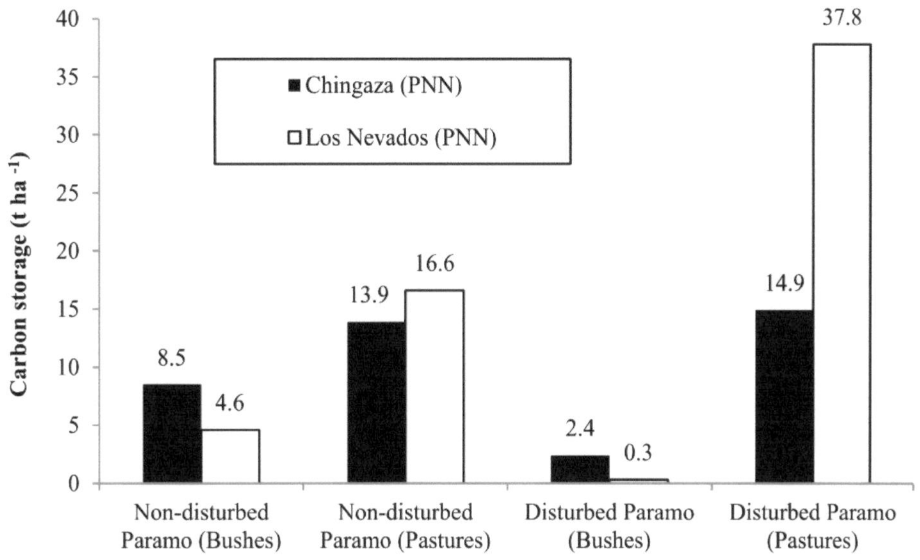

Fig. 6. Carbon storage in disturbed and non-disturbed forest areas of Chingaza (PNN) and Los Nevados (PNN) National parks, Colombia.

2.3 Carbon storage in high-altitude wetlands

In each of the wetlands studied, three samples were collected of water, were collected in 1-liter jars in order to conduct physicochemical tests of their quality (Peña et al., 2009). Measurements of dissolved (DOC) and total organic carbon (TOC) were done using the equipment TOC-5050 (Shimadzu). Curves for measuring DOC and TOC concentrations were calculated for each wetland, based on the relation between area and carbon concentration in accordance with the methodology proposed by Wetzel & Likens (2002). For the analyses of hardness and alkalinity, lab analyses were done using the EPA protocol; and the results were expressed as ppm (mg L^{-1}) CaCO3. The Winkler method (Wetzel & Likens, 2002) was used to measure the dissolved oxygen (DO). Batimetric data were recorded in each system based on points selected in the transects laid out over the total area of each wetland (Peña et al., 2009).

In the selected transects, 1-m^2 quadrants were placed to determine the organic carbon of the plant biomass. In total 18 quadrants were established for the wetlands studied (Peña et al., 2009). The biomass was collected using a machete for the tall vegetation and manually for the ground-level and submerged vegetation. The samples were placed in sacks in order to transport them and were then weighed fresh, using an industrial platform-type scale (Bosche IPS-C). The plant material was then dried in ovens at an average temperature of 40-45°C for approximately two weeks until it was totally dry. The dry weight was measured with an electronic scale (Nobelsound NS-SM 788). The dry weight values of the plant biomass were then multiplied by a factor of 0,5 to obtain the amount of carbon present. This factor is based on the principle that the plant matter of any ecosystem contains 50% carbon in its biomass once the water has been removed (Vallejo et al., 2005).

Leaf material of the plant species in the selected quadrants was submitted to P and N analyses using the procedure stipulated by ICONTEC under Standard Specification 5167 (norm for N and P analyses) (Peña et al., 2009). The data obtained for dissolved (DOC) and total (TOC) organic carbon were significantly lower in Los Nevados (PNN) wetlands than in the Chingaza (PNN) wetlands: 1,2 mg L^{-1} vs. 2,8 mg L^{-1}, respectively (Tab. 2).
The highest DOC and TOC concentrations (4,2 mg L^{-1} for both) were found in Los Nevados (PNN) lagoon (Tab. 2). Similarly within the wetlands the DOC and TOC concentrations were greater at the entrance sampling point than at the exit point of the water.

Study Site	Sampling Points	COT (mg L^{-1})	COD (mg L^{-1})	Dureza CaCO3 (mg L^{-1})	Acidez $CaCO_3$ (mg L^{-1})	OD (mg L^{-1})	pH	Temp. (°C)
Los Nevados (PNN) (Lagoon)	P1	4,4	4,4	0,044	---	1,1	6,8	5,3
	P2	4,2	4,2	0,031	0,02	1,1	6,7	5,6
	P3	4,0	4,0	0,034	0,02	1,1	6,7	5,8
	P4	---	---	---	---	1,1	6,6	5,5
	P5	---	---	---	---	1,1	6,8	5,5
Los Nevados (PNN) (Wetlands)	Entry	1,303	1,303	0,053	0,02	6,46	6,8	4,3
	Exit	1,178	1,178	0,052	0,03	5,79	6,9	4,6
Chingaza (PNN) (Wetlands)	Entry	3,2	3,2	0,03	0,02	---	---	---
	Exit	2,3	2,3	0,014	0,01	---	---	---

Table 2. Effect of different wetlands and lagoon sampling points on water quality parameters.

The DOC, which is found in all ecosystems, is an important component in the global carbon cycle in aquatic flows (Giesler et al., 2007). The processes of mineralization of the DOC have received special attention due to the effect of carbon dioxide (CO_2), as a gas related to the greenhouse effect and its role in global warming (Suhett et al., 2007). The reservoirs and concentration of carbon is associated with the transformation of the organic matter, particularly in the case of water, either by exogenous processes (material from runoff) or by endogenous processes derived from the transformation of the biological matter existing in the water column (Wetzel, 2000). The results obtained with the dissolved (DOC) and total (TOC) organic carbon in the water column in the wetlands studied (Los Nevados and Chingaza) had relatively low values (< 5mg L^{-1}) in comparison with other similar ecosystems, where values from 20 up to 60 mg L^{-1} have been recorded (Blodau, 2002; Giesler et al., 2007; Suhett et al., 2007). These differences could be based on the factors that determine the concentrations of DOC and TOC, where the temperature regulates the transformation of DOM, either by decomposition of plant litter/humus or by bacterial necromass). This latter factor has been considered to be the principal process that contributes to the concentrations of DOC and TOC in the wetlands (Giesler et al., 2007).
Consequently the low temperatures in the study sites are the factors that regulate the DOC and TOC concentrations and also explain the low contents of these values in the ecosystems

studied (Moore & Dalva, 2001). This can be evidenced taking into account the average temperature in the water column from all the study sites such as the case of Los Nevados (PNN) wetlands, which had the lowest concentrations of DOC and TOC (1,1-1,3 mg L^{-1}). The highest temperatures were recorded in Los Nevados (PNN) lagoon (5,3-5,8°C), which also had the highest concentrations of DOC and TOC (4,0-4,4 mg L^{-1}).

The concentrations of carbon in plant biomass in Los Nevados (PNN) wetlands (15 t ha^{-1}) were higher than for the Chingaza (PNN) wetlands (7 t ha^{-1}). Similarly, differences in the carbon concentrations can be seen with respect to the transects marked for both wetlands, where they were highest at the entrance of the water, followed by Transect 2 and lastly by the transect farthest from the entrance point of the water. For Los Nevados (PNN) lagoon the concentrations of carbon in plant biomass were higher in the emerging plants (0,925 t ha^{-1}) than in those that were submerged (0,15 t ha^{-1}).

Global changes have strong effects on terrestrial ecosystems but with significant regional differences. For instance, fluctuations in temperature have had significant effects on alpine tundra ecosystem, which produces the important changes in the global energy balance and carbon budget (Kato et al., 2006). Salas & Infante, 2006 demonstrated that regional vegetation biomes participate in the balance and sequestration of carbon in specific areas through different processes based on (a) the area of their biomass and (b) the decomposing material or necromass (Schroeder & Winjum, 1995). In both wetlands there was a gradient in the distribution of the vegetation with a dominance of mountain bamboo (*Chusquea* sp.) observed in the water entrance zone (Transect 1), where the highest average concentration of carbon in the biomass was obtained (16 t ha^{-1}); vs. the exit zone (Transect 3) with a dominance of herbaceous vegetation, where the lowest average of carbon in the biomass (2 t ha^{-1}) was obtained. In comparison with the study sites, the levels of carbon in Los Nevados (PNN) wetlands (15 t ha^{-1}) were higher than the concentration of carbon in the Chingaza (PNN) wetlands (7 t ha^{-1}). Consequently, an extrapolation of the data would make it possible to predict that within the wetlands under study, the plant biomass reaches sequestration levels of 7-15 t of carbon ha^{-1}, being much greater than that described by Heathwaite (1993). Levels of 0,5-0,7 t of carbon sequestered ha^{-1} in similar ecosystems evidence the great importance of the high Andean wetlands with respect to the sequestration and storage of carbon and as buffers of the global warming effect.

The concentrations of nitrogen (N) reached values higher than 0,5 t ha^{-1}, whereas for phosphorus (P) the highest value did not surpass 0,2 t ha^{-1} (Fig. 7). Similarly, significant differences were found between the N and P concentrations in relation to the study sites, being highest in Los Nevados (PNN) wetlands, followed by the Chingaza (PNN) wetlands and Los Nevados (PNN) lagoon.

The N and P values in the plant biomass for the sites studied had ranges similar to those of other ecosystems (Blodau, 2002; Blodau et al,. 2004), averaging 0,32 t ha^{-1} (N) and 0,16 t ha^{-1} (P). The higher N and P concentrations in the wetlands vs the lagoon are partly due to the fact that the wetlands have greater plant biomass density as they are not totally flooded by the water column; whereas in the lagoon the sheet of water covers the whole area, which hinders dense growth of the vegetation, except for a few emerging and submerged plants.

In general the plant biomass of the wetlands under study can concentrate an average of up to 3290 t ha^{-1} (N) and 1250 t ha^{-1} (P) every year vs only 170 t ha^{-1} (N) and 60 t ha^{-1} (P) in the lagoon, evidencing concentrations for these types of ecosystems. In relation to forest ecosystems, these concentrations are much higher for both N and P (Bragazza et al., 2006).

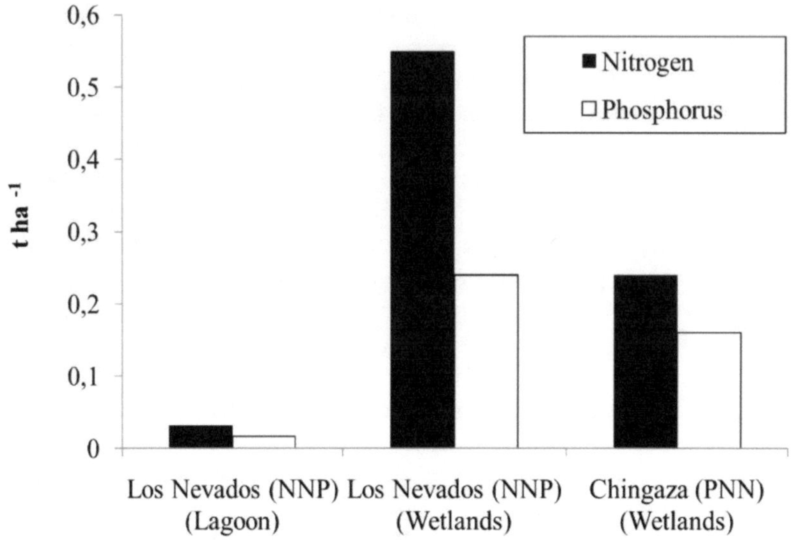

Fig. 7. Relation between the concentration of nitrogen (N) and phosphorus (P) (t ha^{-1}) and the study sites (p=0,013, CI = 95%).

The ecological consequences of global climate changes in any ecosystem environmental conditions will be manifested in a shift of trace gas fluxes, in plant and soil mineral nutrition, and in leaf carbon isotope discrimination (Welker et al., 2005). Changes in low biomass productivity and in levels of N and P in the studied ecosystems are consistent with those changes of ecological properties reported in different world ecosystems. However, it is necessary to validate the dynamics of the carbon flux in these types of high altitude systems, as principal reservoirs of carbon in high Andean zones.

3. Conclusions

Carbon storage and changes of forest cover due to anthropogenic impacts were sensitive and vulnerable to global change, since forest biomass and plant growth depended on local climate conditions. Based on this hypothesis, the comparative analysis of carbon storage of disturbed and non-disturbed forest areas in tropical Andean ecosystems showed a relationship between carbon storage content and the degree of physical disturbance, with a consistent depletion in approximately 10 to 20 % of carbon sequestration in soils, forest biomass and land-water interaction of wetland zones in each studied ecosystems. The net carbon storage (COS) in non-disturbed forests of Chingaza (PNN) was calculated in about 1102,4 t ha^{-1}, and was significantly higher than the mean values observed in forest areas of Los Nevados (PNN), estimated in 762,8 t ha^{-1}.

The carbon sink dynamics in each of the compartments of the studied ecosystems is also related to three predominant climatic variables: humidity, temperature and soil moisture availability. Those variables are dependent of global warming changes at both, regional and global climatic scales. Thus, forest areas with drier climatic conditions common in Los

Nevados (PNN) paramo, registered lower mean values of carbon storage. The dominant plant species on each forestry plots indicated that woody vegetation type favors carbon accumulation in system soils. Therefore, the progressive deforestation and the increased anthropogenic activities in those high mountain ecosystems suggest a strong effect on the carbon fluxes.

The observed data in the wetlands of the water quality parameters s and the dynamics of the plant biomass reflected the significance of both components in the carbon cycle in both systems, especially the wetlands area covered by vegetation and the decomposing material (necromass) accumulated in the soil in the form of plant litter and roots. The total organic carbon in the systems concentrated in a range between 3290 t ha^{-1} (N) and 1250 t ha^{-1} (P) every year vs only 170 t ha^{-1} (N) and 60 t ha^{-1} (P) in the water column of the limnetic zone in the wetland, evidencing spatial differences in carbon concentrations for these types of ecosystems. Consequently, results revealed that these systems participate in the balance and sequestration of carbon in the Colombian Andes.

4. Future directions

The results showed here can be expected to make significant progress to estimate carbon balance on Andean tropical ecosystems and its relationship with global warming. For studies based on COS content, there needs to be an expansion of data points in these poorly monitored regions, including tropical montane cloud forest, the paramos ecosystems and wetlands systems at a local level including micrometeorological studies at each ecosystem. Long-term tropical forest plots offer great potential for direct monitoring of aboveground C stocks. There will be a substantial expansion of available data when all historical long-term forest plot records are compiled. However, these historical data sets will have several statistical and methodological problems. In order to construct future scenarios of global warming effects on carbon flux of Andean forests, it is important to assess regional and global climatic patterns of those ecosystems, specifically to measure changes in atmospheric CO_2 concentrations.

A more rigorous approach should be performed to evaluate the effect of anthropogenic activities on COS content, including mass balance analysis of carbon dynamics. The final point of convergence will be when field studies, laboratory studies and physiological models converge on a consistent picture of the C balance of each studied ecosystem. Continuing studies of accounting the net carbon balance in those systems will bring a more consistent picture of the net carbon balance of tropical forests, and of other biomes, will emerge over the next years.

5. Acknowledgements

This research was sponsored by the Japanese-Colombian Cooperation Agency JICA through a grant given to the Instituto de Estudios Ambientales IDEAM. A Research Agreement between IDEAM and research groups of the Sciences School from Universidad del Valle: Applied Plant Biology and High mountain ecosystem research groups were able to develop this study within the project "National Pilot Program to evaluate the impacts of global change on Tropical Andean ecosystems, Caribbean areas and human health in Colombia".

6. References

Amezquita, M. C., Putten, B. van, Ibrahim, M., Ramirez, B.L., Giraldo, H. & Gomez, M.E. (2006). Recovery of Degraded Pasture Areas and C-sequestration in Ecosystems of Tropical America. *WSEAS Transactions on Environment and Development* 2(8): 1085-1091.

Arreaga, WE. (2002). *Almacenamiento de carbono en bosques con manejo forestal sostenible en la Reserva de Biosfera Maya, Petén, Guatemala.* Tesis Mag. Sc. Turrialba, CR, CATIE. 73 p.

Batjes, N.H. (1999). *Management options for reducing CO2- concentrations in the atmosphere by increasing carbon sequestration in the soil.* ISRIC. Wageningen, The Netherlands. 114 pp.

Blodau, C. (2002): Carbon cycling in peatlands: A review of processes and controls. *Environmental Reviews* 10: 111-134

Blodau, C., N. Basiliko & T. R. Moore. (2004). Carbon Turnover in Peatland Mesocosms Exposed to Different Water Table Levels. *Biogeochemistry* Vol. 67 (3): 331 - 351.

Brady, N. & Weil, R. (2002). *The nature and properties of soils.* 13 ed. Prentice-Hall, New Jersey.

Bragazza, L., C. Siffi, P. Iacumin & R. Gerdol. (2006). Mass loss and nutrient release during litter decay in peatland: the role of microbial adaptability to litter chemistry. *Soil Biology & Biochemistry* Vol. 39: 257 - 267.

Brown, S., A.J Gillespie & A.E. Lugo. (1989). Biomass estimation methods for tropical forests with applications to forest inventory data. *Forest Science.* 35(4).

Carvajal, A., Feijoo A., Quintero H. & Rondón M. (2009). Carbono orgánico del suelo en diferentes usos del terreno de paisajes andinos Colombianos. *Rev. Cienc. Suelo Nutr.* 9(3): 222-235.

Carúa, J., Proaño, M., Suarez, D. & Podwojevski, P. (2008). Determinación de retención de agua en los suelos de los páramos: estudio de caso en la subcuenca del Río San Pedro, Cantón Mejía, Pichincha, Ecuador. pp. 27-39. En: *Forestación. Serie Páramo 26. GTP/Abya Yala.* Quito.

Castaño, C., R. Carrillo & F. Salazar. (2002). Sistema de Información Ambiental de Colombia Tomo III. *Perfil del estado de los recursos naturales y del medio ambiente en Colombia 2001.* IDEAM. Ministerio del Medio Ambiente.

Dauber, E.; Terán, J. & Guzmán, R. (2000). Estimaciones de biomasa y carbono en bosques naturales de Bolivia. *Revista Forestal Iberoamerica.* 1(1): 1-10.

Delaney, M., Brown, S., Lugo, A. E., Torres, A. & Bello, N. (1997). The distribution of organic carbon in major components of forest located in five zones of Venezuela. *Journal of Tropical Ecology* 13:697-708.

Diaz, E., (2008). *Distribución del contenido de carbono orgánico en agregados de diferentes tamaños, procedentes de varios sistemas de uso y altitudes en suelos de la cuenca del Rio Cauca, Colombia.* Tesis de maestría. Universidad Nacional de Colombia.

Dupouey, J.L., Siguand, G., Bateau, V., Thimonier, A., Dhole, J.F., Nepveu, G., Bergé, L. Augusto, L., Belkacem, S. & Nys, C. (1999). Stocks et flux de carbone dans les forêts françaises. *C.R. Acad. Agric. Fr.* 85 (6): 293-310.

Earle, L.R., Warner B.G. & Aravena R. (2003). Rapid development of an unusual peat accumulating ecosystem in the Chilean Altiplano. *Quaternary Research,* 59, 2–11.

Escobar, C.J. & Toriatti Dematté, J.L. (1991). Distribución de la materia orgánica y del carbono-13 en un Ultisol del piedemonte amazónico. *Pasturas Tropicales* 13 (2), 27-30.

Giesler, R., M. Högberg, B. Strobel, A. Richter, A. Nordgren & P. Högberg. (2007). Production of dissolved organic carbon and low-molecular weight organic acids in soil solution driven by recent tree photosynthate. *Biogeochem* 84 (1): 1-12.

Glenday, J. (2006). Carbon storage and emissions offset potential in an East African tropical rainforest. *Forest Ecology and Management* 235, 72–83.

Gorham, E. (1991). Northern peatlands: Role in the global carbon cycle and probable response to climate warming. *Ecol. Appl* 1: 182–193.

Guhl, (1982). *Los páramos circundantes de la Sabana de Bogotá.* Jardín Botánico José Celestino Mutis: Bogotá.

Heathwaite A. L. (1993). Disappearing Peat-Regenerating Peat? The Impact of Climate Change on British Peatlands. *The Geographical Journal* Vol. 159 (2): 203 - 208.

Hernández, F. & Triana, A. (2009). Efecto de actividades agropecuarias en la capacidad de infiltración de los suelos del Páramo del Sumapaz. *Revista EIDENAR* (8), 29-38.

Hofstede, R. (1999). El páramo como espacio para la fijación de carbono atmosférico. En: Medina, G y P. Mena (Eds). 1999. *El páramo como espacio para la fijación de carbono atmosférico. Serie Páramo 1. GTP/Abya Yala.* Quito. Ecuador.

Hofstede, R.G.M. & P. Mena. 2000. Los beneficios escondidos del páramo: Servicios ecológicos e impacto humano. En: Recharte, J. J. Torres & G. Medina (eds.). *II Conferencia electrónica sobre Usos Sostenibles y Conservación del Ecosistema Páramo en los Andes: Los Páramos Como Fuente de Agua, Mitos, Realidades, Retos y Acciones.* CONDESAN, Lima. (URL: http://www.condesan.org/e-foros/bishkek/Bishkek%20B1-Caso(R.Hofstede-P.Mena).htm)

Hofstede, R., P. Segarra, & P. Mena, V. (Eds.). (2003). *Los Páramos del Mundo. Proyecto Atlas Mundial de los Páramos.* Global Peatland Initiative/NC-IUCN/EcoCiencia. Quito.

Houghton, J. T., Y. Ding, D. J. Griggs, M. Noguer, P. J. van der Linden, X. Dai, K. Maskell & C. A. Jonhnson. (2001). *The scientific basis: IPCC Intergovernmental Panel on Climate Change. Contribution of Working Group 1 to the IPCC Third Assessment.* Cambridge University Press, Cambridge, UK. 944 p.

IDEAM, (2002). *Páramos y ecosistemas alto andinos de Colombia en Condición HotSpot & Global Climatic Tensor.* IDEAM (Instituto de Meteorología, Hidrología y Estudios Ambientales), Bogotá, Colombia.

Ibrahim, M., Chacón, M., Cuartas, C., Naranjo, J., Ponce, G., Vega, P., Casasola, F. & Rojas, J. (2007). Almacenamiento de carbono en el suelo y la biomasa arbórea en sistemas de usos de la tierra en paisajes ganaderos de Colombia, Costa Rica y Nicaragua. *Agroforestería en las Americas* 45: 27-36.

IPCC, (2000). *Land use, land-use change, and forestry special report.* Cambridge University Press 377 pp.

IPCC, (2007): Cambio climático 2007: *Informe de síntesis. Contribución de los Grupos de trabajo I, II y III al Cuarto Informe de evaluación del Grupo Intergubernamental de Expertos sobre el Cambio Climático* [Equipo de redacción principal: Pachauri, R.K. y Reisinger, A. (directores de la publicación)]. IPCC, Ginebra, Suiza, 104 p.

Kanninen, M. (2003). Secuestro de Carbono en bosques, su papel en el ciclo global. En: M.D. Sánchez y M. Rosales (eds.). *Agroforestería para la Producción Animal en América Latina - II - Memorias de la Segunda Conferencia Electrónica.* Organización de las Naciones Unidas para la Agricultura y la Alimentación (FAO). Roma.

Kato, T., Y. Tang, S. Hirota, M., Du, M., Li, X. Zhao. (2006). Temperature and biomass influences on interannual changes in CO2 exchange in an alpine meadow on the Qinghai-Tibetan Plateau. *Global Change Biology* 12: 1285-1298.

Laegaard, S. (1992). Influence of fire in the grass páramo vegetation of Ecuador. Pp. 151-170. En: Balslev H. y J. L. Luteyn (Eds.). *Páramo: An Andean ecosystem under human influence.* Academic Press, San Diego.

Lal, R. (2004). Soil carbon sequestration to mitigate climate change. *Geoderma* 123, 1-22.

Lapeyre T., Alegre J. & Arévalo L. (2004). Determinación de las reservas de carbono de biomasa aérea, en diferentes sistemas de uso de la tierra en San Martín, Perú. *Ecología Aplicada* 3(1,2):35-44.

López, M., Veldkamp, E. & De Koning, G.H.J. (2005). Soil carbon stabilization in converted tropical pastures and forests depends on soil type. *Soil Science Society of America Journal* 69 (4), 1110-1117.

Lotero J., Dossman M., Castillo C. & Giraldo M. (2006). *Plan básico de manejo 2006- 2010 Parque Nacional Natural Los Nevados.* Documento preparado por Parques Nacionales Naturales de Colombia Dirección Territorial Noroccidente, Colombia, Medellín.

Maltby E., & P. Immirzi. (1993). Carbon dynamics in peatlands and other wetland soils, regional and global perspectives. *Chemosphere* 27: 999–1023.

MAVDT, IDEAM & PNUD, (2001). *Primera Comunicación Nacional ante la Convención Marco de las Naciones Unidas sobre el Cambio Climático.* Ministerio de Ambiente, Vivienda y Desarrollo Territorial (MAVDT), Instituto de Hidrología, Meteorología y Estudios Ambientales (IDEAM), Programa de las Naciones Unidas para el Desarrollo (PNUD). Bogotá, Colombia.

Moore T. R., & Dalva M. (2001). Some controls on the release of dissolved organic carbon by plant tissues and soils. *Soil Sci.* 166: 38- 47.

Moreno, F. & W. Lara. (2003). Variación del carbono del suelo en bosques primarios intervenidos y secundarios. En: Orrego, S. A., J. I. Del Valle y F. H. Moreno (eds). *Medición de la captura de carbono en ecosistemas forestales de Colombia: Contribuciones para la mitigación del cambio climático.* Universidad Nacional de Colombia, sede Medellín, Departamento de Ciencias Forestales y Centro Andino para la Economía en el Medio Ambiente, Bogotá, DC. 314 p.

Moretto, A., Lázzari, A. & O. Fernández. (2005). *Calidad y cantidad de nutrientes de la hojarasca y su posterior mineralización en bosques primarios y bajo manejo con distintos sistemas de regeneración.* Módulo Lenga. Subproyecto 5. PIARFON BAP p. 1-17.

Montagnini F. & C. F. Jordán. (2002). Reciclaje de nutrientes. En: Guariguata, M., Y. Kattan (eds). *Ecología y conservación de bosques neotropicales.* San José: Libro Universitario Regional (LUR).

Powers, J.S. & Schlesinger, W.H. (2002). Relationships among soil carbon distributions and biophysical factors at nested spatial scales in rain forests of northeastern Costa Rica. *Geoderma* 109, 165-190.9

Peña, J. E., Sandoval, H., Zúñiga, O. & Torres, A., (2009). Estimates of carbon reservoirs in high-altitude wetlands in the Colombian Andes. *Journal of Agriculture and rural development in the tropics and subtropics* 110: 115-126.

Rosenzweig, C. & D. Hillel. (2000). Soils and global climate change: Challenges and opportunities. *Soil Science* 165: 47-56.

Salas R.J & A. Infante. (2006). Producciòn Primaria Neta aérea en algunos ecosistemas y estimaciones de biomasa en plantaciones forestales. *Rev, For. Lat.* (40): 47 - 70.

Schuur, E., Chadwick, O. & Matson, P. (2001). Carbon cycling and soil carbon storage in mesic to wet Hawaiian montane forests. *Ecology* 82 (11): 3182–3196.

Schroeder P. E. & J. K. Winjum (1995). Assessing Brazil's carbon budget: II. Biotic fluxes and net carbon balance. *Forest Ecology and Management.* 75(1-3): 87-99

Sierra, C.A., J.I. del Valle & S.A. Orrego. (2003). Ecuaciones de biomasa de raíces en bosques primarios intervenidos y secundarios. PP. 169-189. En: S.A. Orrego, J.I. del Valle y F.H. Moreno (eds.). *Medición de la captura de carbono en ecosistemas forestales tropicales de Colombia: Contribuciones para la mitigación del cambio climático.* Universidad Nacional de Colombia sede Medellín, Centro Andino para la Economía en el Medio Ambiente, Bogotá. 314 p.

Suhett A.L., A.A. Megali, E. Prast, A. Esteves, F. Farjalla & V. Fortes. (2007). Seasonal changes of dissolved organic carbon photo-oxidation rates in a tropical humic lagoon: The role of rainfall as a major regulator. *Can. J. Fish. Aquat. Sci.* 64 (9): 1266-1272.

UNIVALLE & IDEAM, (2008). *Validación del protocolo para la caracterización de los ciclos de agua y carbono en ecosistemas de alta montaña y diseño, instalación y puesta en operación de la red de monitoreo para determinar los impactos del cambio climático en dichos ciclos.* 456p.; Universidad del Valle (UNIVALLE), Instituto de Meteorología, Hidrología y Estudios Ambientales (IDEAM). Cali, Colombia.

Vallejo M.I., A.C. Londoño, R. López, G. Galeano, E. Álvarez & W. Devia. (2005). *Establecimiento de parcelas permanentes en bosques de Colombia.* Instituto de Investigación de Recursos Biológicos Alexander von Humboldt. Bogotá D. C., Colombia. 310 p. (Serie: Métodos para Estudios Ecológicos a Largo Plazo; No. 1);.

Van Der Hammen T. (1997). Ecosistema terrestre: Páramo. En Chaves, M. E. Y N. Arango (eds.), *Informe nacional sobre el estado de la biodiversidad.* Instituto de Investigación de Recursos Biológicos Alexander von Humboldt. Instituto Humboldt, PNUMA, Minambiente. Pag 9 - 37.

Van Der Hammen, T. (1995). La dinámica del medio ambiente en la alta montaña colombiana: historia, cambio global y biodiversidad. En: Lozano, J.A., J.D. Pabón. (Eds.). *Memorias del Seminario Taller sobre alta montaña Colombiana.* Academia Colombiana de Ciencias Exactas, Físicas y Naturales. Colección Memorias No 3. Santafé de Bogotá, 13-15 de octubre de 1993. pp. 11-15.

Van Der Hammen T. & H. Hooghiemstra. (2003). Interglacial–glacial Fuquene-3 pollen record from Colombia: an Eemian to Holocene climate record. *Global and Planetary Change* 36 (3): 181-199.

Vargas, O. (2000). *Sucesión regeneración del páramo después de quemas.* Tesis de Maestría. Departamento de Biología. Universidad Nacional de Colombia. Bogotá.

Vargas, O., J. Premauer & M. Zalamea. (2002). *Impacto de fuego y ganadería sobre la vegetación de páramo.* Memorias del Congreso Mundial de Páramos. Paipa. Tomo II pp 819-841 Ministerio del Medio Ambiente, CAR, IDEAM, Conservación Internacional.

Vargas, O. & Pedraza, P., (2004). *Parque Nacional Natural Chingaza.* Primera edición. Gente Nueva Editorial. Bogotá D. C., Colombia. 197 p.

Verweij, P.A., K. Kok & P.E. Budde. (2003). Aspectos de la transformación del páramo por el hombre. En: Van der Hammen, T., A.G. Dos Santos (eds.). *Estudios de ecosistemas tropoandinos Volumen 5.* La Cordillera Central Colombiana. Transecto Parque Los Nevados. Berlin-Stuttgart.

Watson, R. T., I. R. Noble, B. Bolin, N. H. Ravindranath, D. J. Verardo & D. J. Dokken. (2000). En: *IPCC Special Report, Intergovernmental Panel on Climate Change.* Cambridge University Press, Cambridge, United Kingdom and New York, 377 p.

Welker, J. M., J.T.,Fahnestock, P. Sullivan, R.A. Chimner, (2005). Leaf mineral nutrition of arctic plants in response to long-term warming and deeper snow in N. Alaska. *Oikos* 109: 167-177.

Wetzel R. G. & G. E. (2002). *Likens; Limnological Analyses.* 3a ed. Springer-Verlag, NY. 429 p.

Zhang, Y., D.A. Zhou, G.S., Liu, C.Y., J. Zhang. (1996). Model expectation of impacts of global climate change on biomes of the Tibetan Plateau. In: Omasa, K., Global Warming 250 Kai, K., Taoda, H., Uchijima, Z. & Yoshino, M. (eds.) *Climate change and plants in East Asia* pp. 25-38. Springer-Verlag, Tokyo.

7

Modeling the Monsoons in a Changing Climate

H. Annamalai
IPRC/SOEST, University of Hawaii
USA

1. Introduction

Over half of the World's population resides in regions influenced by the Asian Summer Monsoon (ASM). The seasonal mean (June through September) rainfall associated with the ASM is the primary source of water for the agrarian society. To a great extent, the livelihood and socio-economic issues of the people are influenced by the vagaries of the monsoon rainfall. Given their anticipated population rise, the countries experiencing the monsoon will likely face increased stress in issues such as drinking water supply, among other factors. It is therefore important to know how the ASM and its variability would respond in a warmer climate.

Climate change refers to various facets of climate, but the effect receiving the widest attention is referred to as global warming. Today, the principal concern is to understand the physical processes that determine the mean temperature of the Earth's surface, and interpret how this may change in response to changes in atmospheric composition. Since the temperature and atmospheric moisture content are inherently related the obvious question is how the global hydrological cycle will respond to changes in temperature. Observations over the last 1000 years indicate that carbon dioxide (CO_2) amount has increased by 25% since the industrial revolution, and continues to increase at a rate of 0.5 per annum. In other words, by releasing considerable amount of CO_2 into the atmosphere, humans have applied a stronger forcing with the potential for irreversible effects. To address and assess the impact of this additional forcing, scientists around the world turn to mathematical models of the Earth's climate system. These climate models are constructed based on the physical principles and executed on the most powerful computers. Various hypotheses are tested to understand the present climatic condition. A classic example is that the observed increase in global mean surface temperature since 1950s is reproducible *only* if the greenhouse gases are included in the model simulations (IPCC 2001). This surface warming is projected to increase into the 21st century (IPCC 2007). These models also suggest that under increased CO_2 forcing, global hydrological cycle will intensify (Held and Soden 2006). However, regional precipitation projections from these models have large uncertainties.

Recognizing the strengths and weaknesses of the models, appropriate tools are employed to interpret the future projections. This is the "need of the hour" as policy and decision-makers seek to mitigate and adapt to the undesirable impact of climate change caused by human intervention. Over half of the World's population lives in regions influenced by the Asian monsoon (Webster et al. 1998). The seasonal mean rainfall, both summer (June through September) and winter (October through December), associated with the monsoons influence the people lives, particularly the small-scale farmers. For a variety of reasons, it is desirable to know whether the time-mean monsoon precipitation will increase or decrease as global temperature continues to warm, and understand how the increase in time-mean atmospheric water vapor would influence the severity of droughts and floods, and other monsoon characteristics such as extreme rainfall events, number of rainy days, and extended breaks.

The remaining of the chapter is organized as follows. In Section 2, our current understanding of the monsoon as a coupled ocean-atmosphere-land system is presented followed by the ability of the climate models in simulating its current mean monsoon precipitation and the future time-mean monsoon rainfall projections. In Section 3, monsoon variability is presented starting with the ENSO-monsoon association, the projected floods and droughts, monsoon onset and subseasonal variations and finally their future projections. In Section 4, the uncertainties in model projections and policy matters are mentioned. In Section 5, the way forward in monsoon modeling is presented.

2. Mean monsoon in CMIP3 models and its future projection

Of all the regional monsoon systems around the globe, the ASM is the most energetic and vigorous monsoon system dominating the global circulation during boreal summer. The seasonal cycle in solar forcing during spring-early summer (March through early June) results in the greater warming of the land in the Northern Hemisphere than the adjoining oceans. This is primarily related to the lower heat capacity of the land. The north-south thermal gradient due to land-sea thermal contrast induces a large-scale pressure gradient. Consequently, in the lower troposphere a strong cross-equatorial flow sets-up in the tropical Indian Ocean (Fig. 1c). This flow is instrumental in transporting the required moisture to feed the monsoon rainfall (Fig. 1a). The warm sea surface temperature (SST) in the northern Indian Ocean and tropical western Pacific (Fig. 1b) and the barrier layer beneath it maintain the deep monsoon convection (Shenoi et al. 2006). In retrospect, the in-situ heavy rainfall and river run-off maintain the barrier layer over the northern Bay of Bengal, and the cycle continues. Briefly, the coupled processes that determine the mean evolution of the ASM are varied and interactive.

Our current understanding is that the monsoon is an integral component of the climate system and is manifested due to ocean-land-atmosphere interaction. Therefore, our trust in projections of how the monsoon will respond to Earth's warming depends on the ability of the atmosphere-ocean general circulation models (AOGCMs) or alternately referred to as coupled models, in capturing the present-day monsoon climate and its complex space-time variations. The availability of the simulations with numerous global AOGCMs conducted recently for the third coupled model inter-comparison project (CMIP3) allows for analysis of the various facets of the ASM, namely, the mean monsoon precipitation, historical and projected variations in El Niño Southern Oscillation (ENSO)-monsoon coupling, monsoon onset, subseasonal variations and response of these aspects to global warming.

The simulation of regional monsoon precipitation climatology has proven to be rather difficult and therefore provides a severe test of the climate models. Modeling groups around the world have participated in many international programs to assess and examine the models' skill in simulating the mean monsoon in the atmosphere only models forced with observed SST (e.g., Sperber and Palmer 1996). Recognizing the fact that coupled processes are inherent for the monsoon, the output of 18 models used to simulate the climate of the

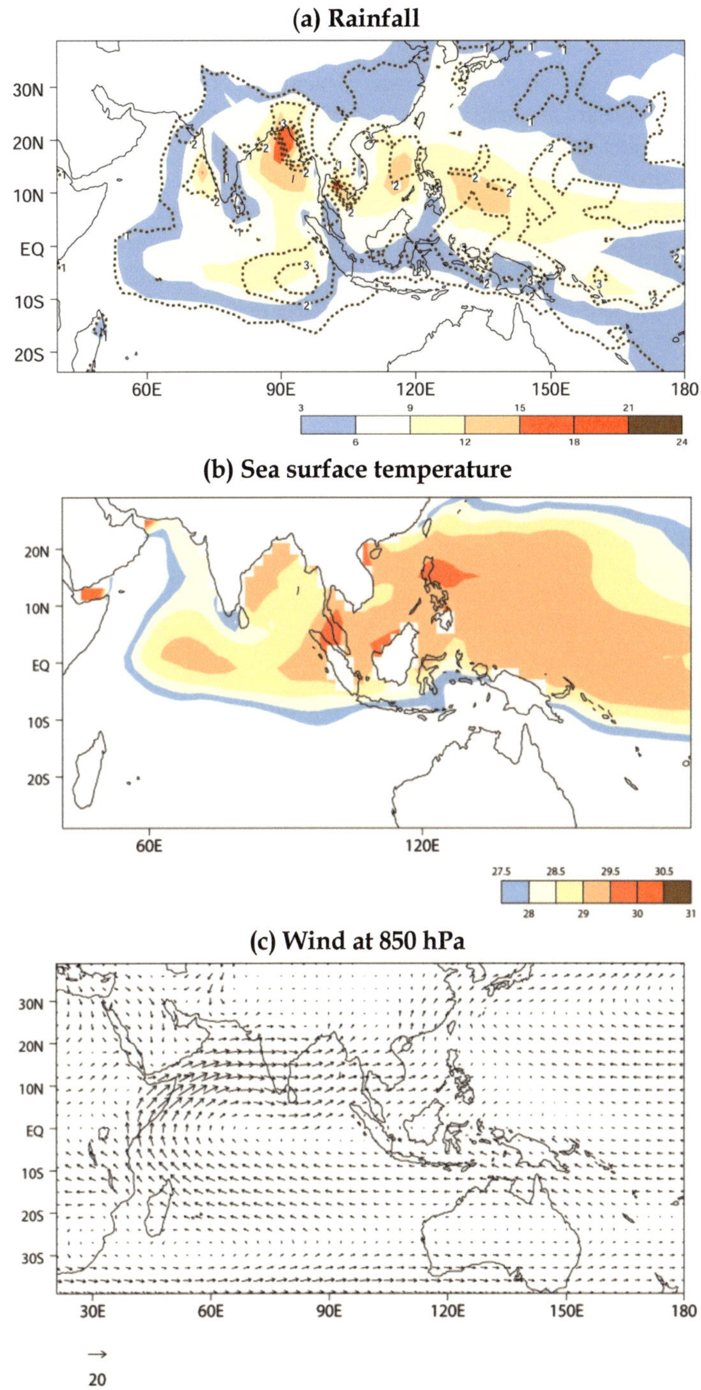

Fig. 1. Observed June-September averaged climatology over the Asian summer monsoon region: (a) precipitation (mm/day, shaded); (b) SST (oC), and (c) 850hPa wind (m/s). In (a) contours represent interannual standard deviations, and in (c) reference wind vector is also shown.

20th Century (20c3m) are examined. The 20c3m simulations attempt to replicate the overall climate variations during the period ~1850-present by imposing each modeling groups best estimates of natural (e.g., solar irradiance, volcanic aerosols) and anthropogenic (e.g., greenhouse gases, sulfate aerosols, ozone) climate forcing during this period. For each model, June–September averaged rainfall climatology of the last 30-years from the 20c3m simulations was constructed and compared with observations.

For climate scientists, one of the challenging issues is to identify and employ metrics in choosing models so that uncertainty in future monsoon projections can be reduced. While models are generally successful in capturing the large-scale circulation features, simulating the details in regional rainfall patterns and intensities are still in their infancy (Annamalai et al. 2007; Turner et al. 2007). Yet, modest progress has been made in recent decades. The observations (Fig. 1a) show that intense rainfall (> 6 mm/day) occurs over three regions in Asia, which represent (1) the Indian Summer Monsoon (ISM, 70°E–100°E, 10°N–25°N,) (2) the Western North Pacific Monsoon (WNPM, 110°E–150°E, 10°N–20°N), and (3) the eastern equatorial Indian Ocean (EEIO, 10°S–0, 90°E–110°E). Because these centers do not respond in unison to any internal or external forcing (Annamalai and Sperber 2005; Annamalai et al. 2007; Annamalai 2010), and influence each other at all time scales, a realistic representation of these three regional centers is needed if the models are to adequately project the future time-mean changes in rainfall as well as reproduce the variability of the monsoon. One metric based on the pattern correlations between simulated and observed rainfall estimates and their root mean square differences (RMSDs) for the chosen period indicated that only six models are capable of capturing the mean monsoon precipitation (Annamalai et al. 2007). The multi-model mean rainfall climatology constructed from averaging the climatology of these six models is shown in Fig. 2a. The agreement with observations in particular the representation of three regional rainfall centers is encouraging but diversity between model and observation still exists in the strength of rainfall over these regional centers. Therefore, in these models a large diversity in the projected intensity of monsoon rainfall is to be expected introducing uncertainties in future projections.

We are confident that due to anthropogenic forcing global mean surface temperature and associated water vapor will increase but large uncertainties exist in quantifying how various elements of the monsoon will respond to these changes. As mentioned above, due its low heat capacity land surface warming will be more pronounced over the land than the neighboring oceans, introducing a larger land-sea thermal contrast over the monsoon region. It is a well-accepted fact that sufficient land-sea thermal contrast is a necessary condition for the monsoon circulation to develop (Li and Yanai, 1996). Thus, consistent with the notion of rich-get-richer mechanism (Held and Soden 2006; Chou et al. 2009) one could expect monsoon rainfall to increase due to enhanced transport of moisture to South Asia from the tropical Indian Ocean (Kitoh et al. 1997; Meehl and Arblaster 2003; Stowasser et al. 2009; May 2010). Based on these arguments, most climate models do suggest for an increase in time-mean monsoon rainfall over tropical west Pacific and south Asia (Turner and Annamalai 2011). The projected rainfall changes over the tropical Indo-Pacific region from two CMIP3 models that both demonstrated realistic simulation of mean monsoon precipitation and ENSO-monsoon association (Annamalai et al. 2007; Turner et al. 2007) is shown in Figs. 2b-c. While both models indicate for an increase in rainfall over south Asia, the large diversity in the rainfall amplitude over the tropical Indo-Pacific region suggests the uncertainty in future quantitative projections. Despite both models experience similar

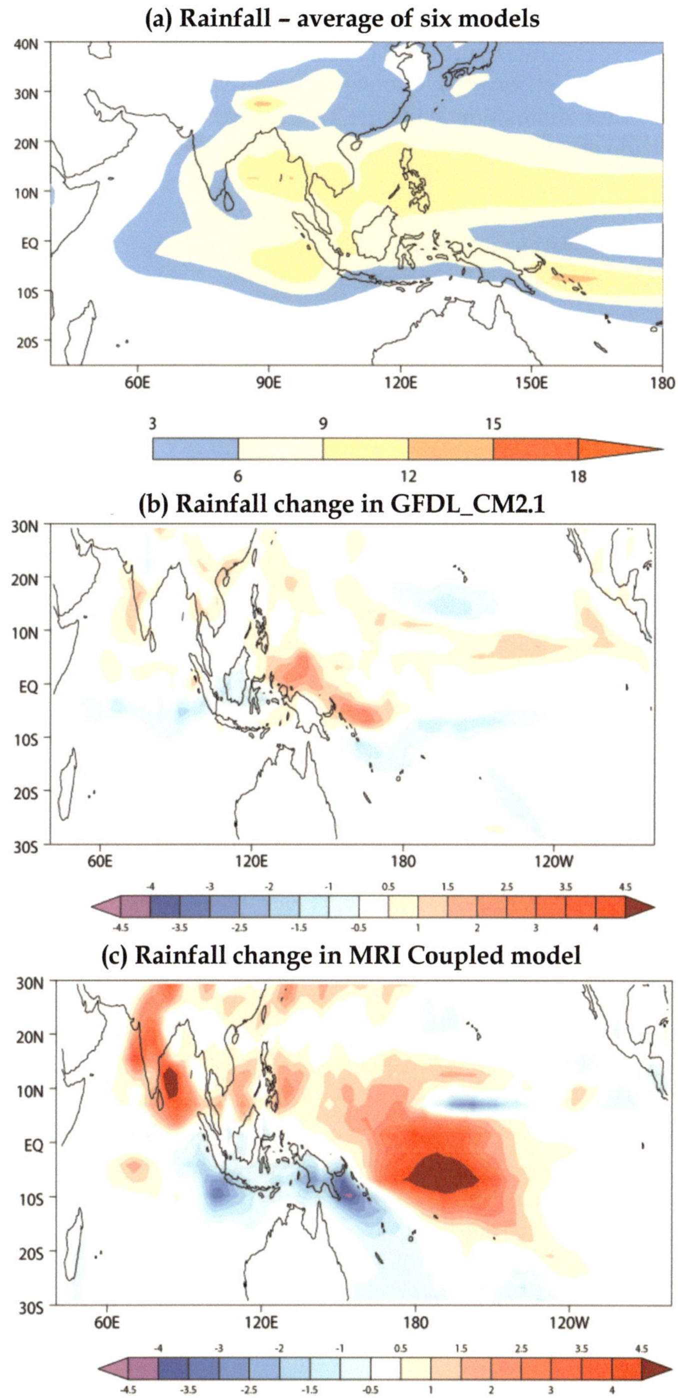

Fig. 2. (a) Mutli-model mean June-September precipitation (*mm/day*) climatology from 203m simulations; projected time-mean change in rainfall in doubled CO_2 simulations (1pccto2x minus 20c3m simulations), (b) GFDL_CM2.1 coupled model and (c) MRI coupled model

anthropogenic forcing in the historical simulations, significant rainfall differences imply the challenge in proper representation of important physical processes and their feedbacks in climate models. While moist convective processes and their interaction with circulation, role of mountains and their interaction with the flow, oceanic processes such as salinity and barrier layer over the warm pool of the Indo-Pacific are few elements that are thought to influence the monsoon (as mentioned above) but realistic representation of all these interactive processes in a model is a demanding task for climate scientists.

Due to the presence of multiple regional heat sources and their interaction (Stowasser et al. 2009), the simulated monsoon response to anthropogenic forcing is not that simple where rainfall increases despite a weakened monsoon circulation (Kitoh et al. 1997; Udea et al. 2006). Stowasser et al. (2009) argued that the reduced rainfall over the equatorial Indian Ocean rainfall (Fig. 2b) forces a low-level anomalous circulation that opposes the climatological cross-equatorial flow (Fig. 1b) but the reduced circulation is richer in moisture resulting in an increase in rainfall. They highlighted that changes in time-mean circulation, through local air-sea interaction, are instrumental in anchoring SST warming over the Arabian Sea and subsequent increase in evaporation. It is suggested that to understand the monsoon response to anthropogenic forcing a clear understanding and quantification of dynamic and thermodynamic feedbacks are necessary. While most models capture the thermodynamic feedbacks (rich get richer mechanism), realistic representation of dynamic feedbacks in climate models will require realistic simulation of regional heat sources.

3. Monsoon variability

In this section, we start with year-to-year variations in the seasonal mean monsoon rainfall that accounts for floods and droughts (Section 3.1), followed by monsoon onset (Section 3.2), and end with subseasonal variations (Section 3.3). In all subsections, we will briefly review the models' ability in simulating these aspects in the current climate and discuss their future projections.

3.1 Interannual variability

One of the major components of the ASM is the Indian Summer Monsoon whose strength is often represented by the all-India Rainfall (AIR) index. From historical record, the observed AIR departures from seasonal mean are shown in Fig. 3a. While droughts (years of rainfall below 1 standard deviation) and floods (years of rainfall above 1 standard deviation) are clearly noticeable, there is a tendency for droughts (e.g., 1871, 1877, 1963; 1972) and floods (e.g., 1988), to occur during ENSO years. A similar rainfall time series from a coupled model simulation for the 20th century (Fig. 3b) also suggests that occurrence of droughts and floods are more common during ENSO years. It needs to be mentioned here that since the seminal work of Walker (1924) and Walker and Bliss (1932) the contemporaneous SST anomalies in the central-eastern Pacific associated with ENSO have been the predominant forcing of the AIR variability supporting the hypothesis that slowly varying boundary conditions exert a significant impact on monsoon variability (Charney and Shukla, 1981). In particular, both in observations and in a coupled model Pillai and Annamalai (2011) noted that all severe weak and strong monsoons are associated with the developing phase of ENSO. While a weakening in the ENSO-monsoon association in recent decades (Shukla 1995) was

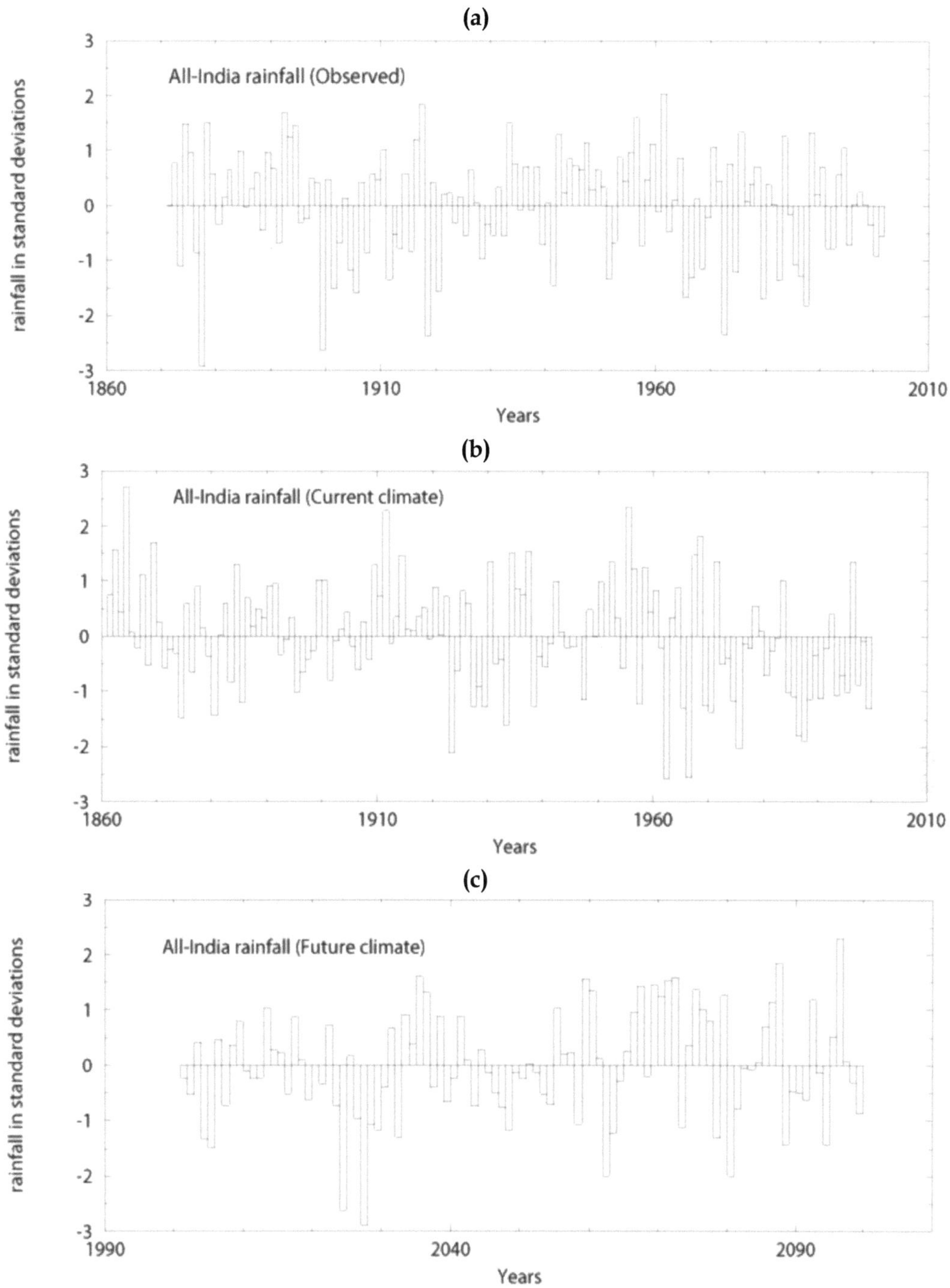

Fig. 3. Interannual variations in monsoon rainfall over India: (a) observations; (b) GFDL_CM2.1 simulated current climate, and (c) same as (b) but for future climate.

attributed to global warming (Krishnakumar et al. 1999), one emerging result from coupled modeling studies is that the ENSO-monsoon association remains intact in a warmer world (Annamalai et al. 2007; Turner et al. 2007). Given the large diversity in the simulation of ENSO characteristics in CMIP3 models (Achcutha Rao and Sperber 2006; Joseph and Nigam 2006), uncertainty exists in the projection of monsoon interannual variability. However, while the magnitude of interannual standard deviation of monsoon rainfall depends on ENSO variance, the frequency of droughts and floods may not be influenced as shown next. Fig. 3c shows monsoon interannual variability as simulated by a coupled model when subjected to doubled CO_2 forcing. While droughts and floods do occur in a warmer world, there is no clear suggestion for them to either increase or decrease. This result is consistent with that of Turner and Annamalai (2011) suggesting that the monsoon is a very robust system. As in observations (Fig. 3a), both in the current and future climates, there is a tendency for decadal clustering of monsoon floods and droughts (Figs. 3b-c), a feature that is perhaps unrelated to global warming (Annamalai et al. 2007).

3.2 Onset

One of the spectacular aspects of the monsoon is its onset. Usually, the monsoon rains and associated moisture laden winds first hit the plains of Indo-China and gradually move westwards. Over India, the monsoon rains usually arrive at the southern state of Kerala around 01 June with an interannual variations of about 7 days. The monsoon onset is closely monitored because it offers remedy from pre-monsoon heat waves, sets the pace of agricultural practices such as soil moisture increase and sowing period, and drinking water needs. Under anthropogenic forcing, if one expects stronger land-sea thermal contrast and abundant moisture then a natural question is: Will the monsoon onset occur earlier and with greater intensity? If so, will the rainfall amount during the onset spell lead to changes in agricultural practices? For instance, if the onset is accompanied with heavy rainfall leading to soil erosion and run-off, what measures need to be taken to avoid loss in farming?

Annamalai et al. (2007) analyzed the CMIP3 models to check if the onset occurs earlier in a warming world. Their findings suggest that models that realistically simulate current monsoon precipitation climate, project an enhanced land-sea thermal contrast between the Indian land-mass and the southern Indian Ocean in the range of ~1.5 - 3°C. They noted that the MRI coupled model with strongest gradient suggests onset 2-3 pentads earlier and a 15% increase in rainfall strength (Fig. 2c) while the model with moderate gradient suggests a delayed onset. This suggests that factors other than land-sea thermal contrast plays important roles and thus introduces uncertainty in our projections of change to the monsoon onset. One such candidate is that feedback between dynamics and moist convection anchors the onset process, and therefore how moist process are represented in models will perhaps determine the onset dates.

3.3 Intraseasonal variability and extreme events

Rainfall during the summer monsoon is not steady, consisting of a sequence of active and break periods, one cycle of which lasts 40–60 days. During the ASM season there are two preferred locations of convection over the Indian longitudes, over the continent and over the equatorial Indian Ocean (Fig. 1a). The transition of convection from its oceanic to continental regime often involves intraseasonal northward propagation over India and the Bay of Bengal (Yasunari 1979). Similar north-northwestward migration of convection is also

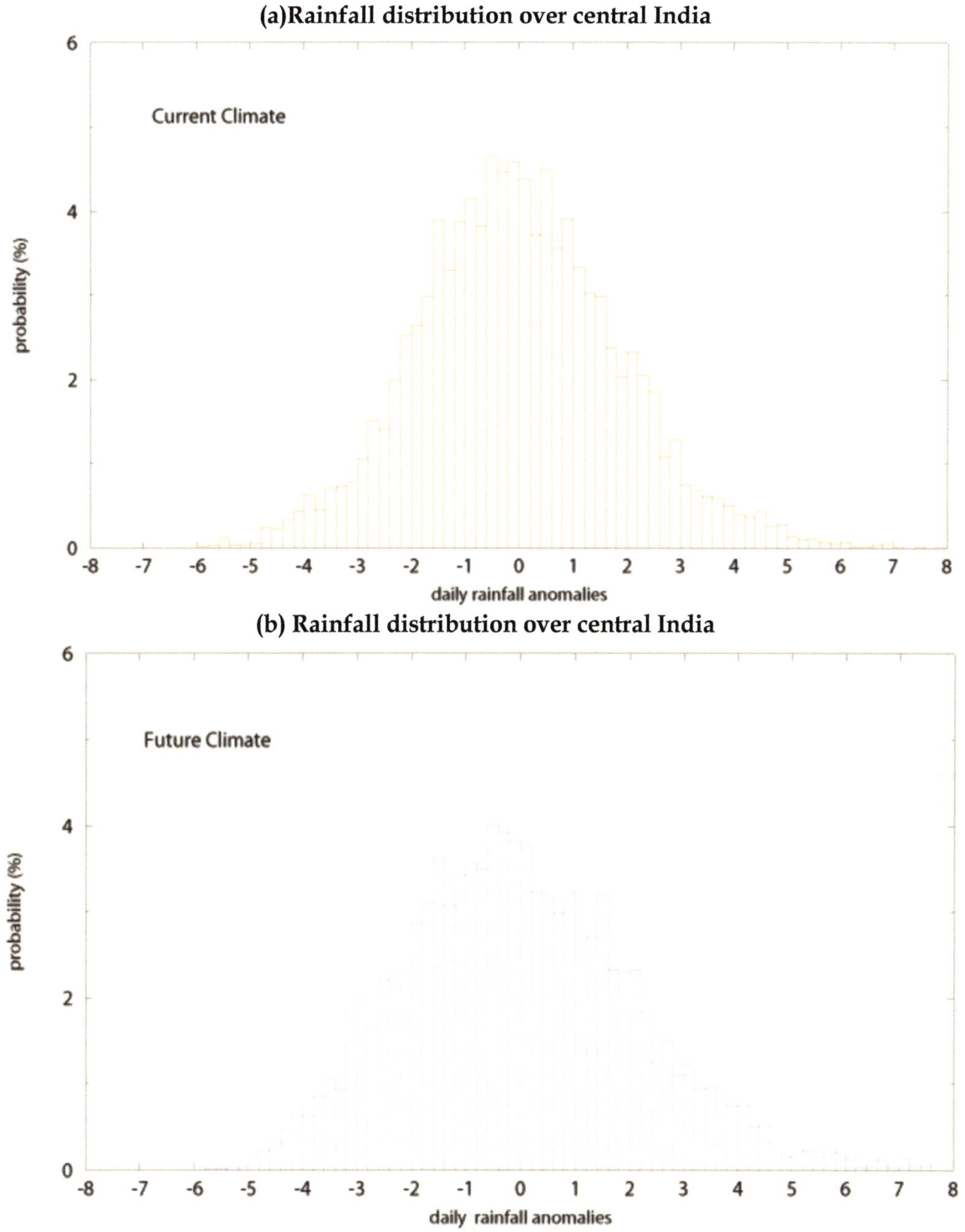

Fig. 4. Probability of occurrences of active and break events over central India as simulated by GFDL_CM2.1 coupled model: (a) current climate (1961-2000), and (b) future climate (2061-2100)

observed over the western Pacific (Annamalai and Slingo 2001). These northward components co-exist with the eastward propagating Madden-Julian oscillation (MJO), and they encompass the boreal summer intraseasonal variability (BSISV). For climate models to project the future behavior of extended monsoon breaks (breaks or lull in rainfall lasting for 7 days or more) and extreme rainfall events (daily rainfall exceeding 120.0 mm/day), it is a pre-requisite that they capture the space-time evolution of BSISV, as well as monsoon synoptic systems (discussed below). Sperber and Annamalai (2008) examined the ability of CMIP3 models in capturing BSISV and noted that almost all of them (with various degrees of fidelity) simulated the equatorial component over the Indian Ocean with appreciable intensity of convection, a feature absent in models forced by observed SST (Waliser et al. 2003). It is argued that realistic simulation of equatorial component is a pre-requisite for BSISV simulation. Of the models, the GFDL_CM2.1 shows reasonable skill in capturing the poleward propagation of rainfall over the Indian longitudes, while all others have limitations (Sperber and Annamalai 2008). Turner and Slingo (2009) noted that both active and break events will become more intense (wetter and drier respectively relative to the seasonal cycle) consistent with general intensification of tropical rainfall. A simple histogram of daily rainfall anomalies averaged over central India for current (Fig. 4a) and future climate (Fig. 4b) by the GFDL_CM2.1 coupled model does suggest that probability of less intense breaks (anomalous rainfall less than 2-3 mm/day) is expected to decrease in a warmer world but heavy rainfall events (anomalous rainfall greater than 6-8 mm/day) increase only marginally. While the news that extreme rainfall events do not increase dramatically is encouraging, the limitations are that coarse model resolutions employed in climate models may not be adequate enough to capture the genesis and intensification of monsoon depressions (as discussed below) and also that intensity of mean monsoon rainfall over central India is not represented realistically in climate models.

At sub-seasonal time scales, synoptic systems such as monsoon depressions form over the quasi-stationary seasonal monsoon trough. During the peak monsoon season (July-August), a majority of them form over the warm waters of northern Bay of Bengal, and move in a west-northwesterly direction, and in most places over central India, almost all extreme rainfall events there are attested to depressions (Sikka 2006). Therefore, to know the future changes in the number extreme rainfall events, it is important to examine climate models' projections in the depressions' strength. Stowasser et al. (2009) examined synoptic systems in ECMWF Reanalysis (ERA-40) products and output from selected CMIP3 models that showed skill in simulating the mean monsoon precipitation (Fig. 2a). The monsoon synoptic systems thus identified were then analyzed as a function of maximum wind speed and as the number of 12-hour time-steps each system was present in the domain. The frequency distributions of maximum storm speed for the *20th century* GFDL model runs (blue bars) over the larger domain (Figure 5a) and over only the Bay of Bengal (Figure 5b) are asymmetric. The distributions peak around 12 m/s after which they fall off rapidly. The model simulates very few intense storms (wind speed > 15m/s). In summary, of the AOGCMs analyzed, the CM2.1 coupled model captures, in relative terms, the mean monsoon precipitation and the spectrum of its variability realistically. But the coarse resolution global model does not project any increase in the depressions strength. Is it due to the limitation in model resolutions that the genesis and intensification of synoptic systems are not well represented? To address this, a high-resolution regional climate model study was undertaken. Using CM2.1 lateral boundary conditions, the current and future climate (i.e., $4xCO_2$) states were simulated at 100 km resolution by a regional model developed at the International Pacific Research Center (IPRC-RegCM). The simulated storms from this regional model indicate that those with intense magnitude (blue bars in Fig. 5 bottom panel) become much stronger (red bars in Fig. 5 bottom panel). These results would then suggest for the possible occurrence of more extreme rainfall events over central India. However, this result is from one model and therefore needs to be confirmed with other regional model experiments.

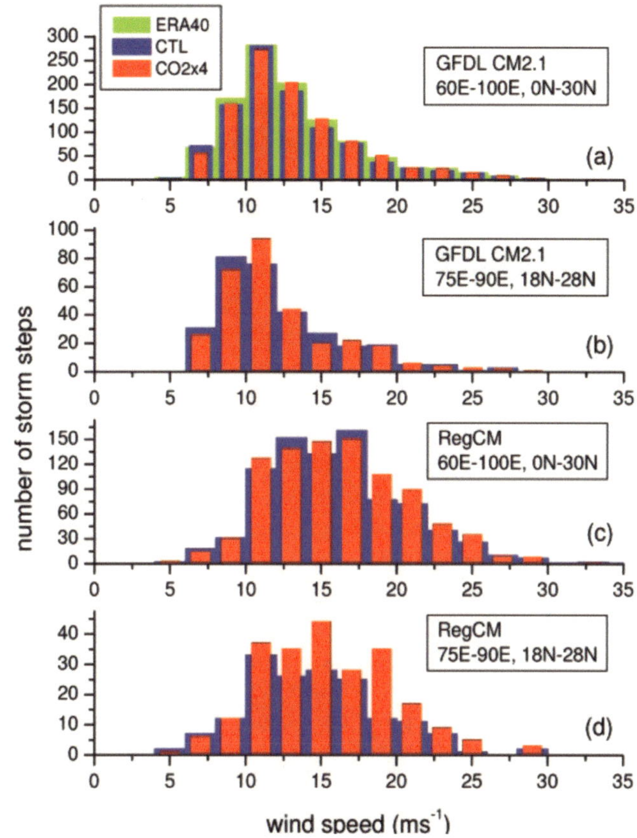

Fig. 5. Frequency distribution as a function of maximum wind speed of each synoptic system estimated for the entire northern Indian Ocean (a and c), and only over Bay of Bengal (b and d): Panels (a) and (b) results from CM2.1, and panels (c) and (d) results from IPRC_RegCM. The blue bars indicate the control runs (CTL) and the red bar the 4xCO2 (four times increase in CO2 forcing) integrations, respectively. The frequency is given in numbers of 12h time-steps a system was present in the domain (adopted from Stowasser et al. 2009).

4. Sources of uncertainties

Climate scientists depend on models that incorporate sophisticated understandings of the coupled behavior of the climate system. The fact that these models reproduce the time evolution of global mean temperature over the last century provides credibility. However, whether at global or regional scales uncertainty in the model precipitation projections still exist. Here, we summarize the most relevant sources of uncertainties that hamper our confidence in climate model projections.

The leading factor is perhaps cloud feedback. Briefly, the changes in atmospheric temperature lead to increase in water vapor content that subsequently results in changes in cloud properties. It is well known that water vapor is the most dominant variable in influencing green house effect (more effective than CO_2). Therefore, how the changes in cloud properties influence the Earth's energy balance and how that feedback loop operates is poorly understood and hence difficult for climate models to handle. Second, apart from increases in

CO_2 concentrations, increases in black carbon aerosols (or soot) in recent decades in the Northern Hemisphere and in particular over the ASM region pose another source of uncertainty. Depending on their concentration and vertical distribution, the aerosols either absorbing or scattering influence the radiation budget in a complex way. For example, both scattering and absorbing aerosols reduce the solar insolation reaching the surface. Over the ASM region, reduced solar insolation tends to reduce the land-sea thermal contrast, and thereby weaken the monsoon circulation. On the other hand, absorbing aerosols warm the lower atmosphere and can induce a stronger monsoon circulation. Undoubtedly aerosols add a source of uncertainty in model monsoon projections (Menon 2004). In summary, we are unsure of how much warming due to greenhouse gases has been cancelled by cooling due to aerosols, and to what extent the future projections are influenced by systematic errors in climate models. Thus, the magnitude of global warming in the next century is very uncertain, and thus the response of the monsoon and its variability across all time scales are bound to have large uncertainties. Above all, monsoon itself is a very complex phenomenon to model.

While uncertainty glooms for future monsoon projections, it is important to recognize that human practices may add to its future behavior. While it may not be important for humans to influence a single monsoon depression but can and do have a long-term influence on the climate that causes the depression. For instance, land-use changes in agricultural practices can potentially influence the surface energy balance and can have a direct impact on the regional atmospheric circulation and rainfall. Since the industrial revolution, hard observed evidences are available for increased burning of fossil fuels and subsequent increase (about 30%) in the concentration of CO_2 in the atmosphere. The booming economy in many countries means that more demand for better quality of life that include improved transportation, industrial expansion and so on - all elements capable of enhancing the CO_2 levels. While human civilization does influence the global and regional climates, it is rather difficult to predict the future human practices and thus emission scenarios causing another source of uncertainty.

It is a challenge for climate researchers to address the uncertainties and provide accurate projections to the policy makers. The uncertainty has contributed to disagreement between groups that want immediate action and reduce GHGs emissions and develop adaptation measures, and those who want to wait and watch. While the debate on uncertainty prevails, policy makers need to focus on adaptation measures. The confidence on scientific evidence on the magnitude of long term changes in climate is expected to increase in the next two decades but the ongoing changes in the frequency and intensity of certain weather events such as tropical cyclones have already raised the concern to develop adaptation measures One way the climate scientists adopt to reduce uncertainty is to produce climate projections from many different climate models, and provide a range of possible outcomes. Based on them, identifying a set of policies for practical solutions will have far reaching implications.

5. Summary

It is now recognized that any future predictions or projections of long-term changes in climate can only by made by complex climate models that depend on super computers. While the last 2 decades have witnessed a phenomenal advancement in computing facilities and access of climate model data (Meehl et al. 2007) predicting or projecting the regional features that are directly relevant to humans and policy makers need ultra supercomputers that are not currently available. The world modeling summit held in May 2008 argued that "climate models will, as in the past, play an important, and perhaps central role in guiding the trillion dollar decisions that the peoples, governments and industries of the world will be making to cope with the consequences of changing climate" (Shukla et al. 2009). The summit called for "Governments should work together to build the supercomputers needed for future predictions that can capture the detail required to inform policy".

It is fair to say that in climate science there is no bigger problem than modeling the monsoons. Even after the availability of super computers, there are many other factors that will limit our ability to model the monsoons. While the importance of large-scale forcing such as orography and boundary forcing such as SST are fairly understood (Hoskins and Rodwell 1995; Charney and Shukla 1981), how the various moist processes interact among them, as well as the specific role of oceanic processes responsible in shaping the mean monsoon precipitation and its spectrum of variability are just beginning to unfold. The slow progress in modeling the monsoon is perhaps due to lack of quality and sustained observations over the monsoon region including that over the tropical Indian Ocean. It is important to have 3-dimensional observational network to monitor the horizontal and vertical extent of moist processes, careful monitoring of evaporative processes over the tropical Indian Ocean, and check the changes for SST threshold for the occurrence of deep convection are some of the observational program to be given priority. At the same time, improvements in model physical parameterizations are continually getting improved. Both new observational information with model improvement needs to go hand-in-hand for making progress in monsoon modeling. Another stumbling factor is that the monsoon itself is a complex phenomenon and poses a monumental challenge for climate scientists.

While we are confident that global hydrological will intensify in a warmer atmosphere, any future projections of regional rainfall changes have large uncertainties. This is because, our current understanding of the various physical processes and their feedback in the climate system is limited to make any significant progress in the prediction of future climate change at regional spatial scales. One of the major uncertainties is due to cloud feedback. Briefly, the changes in atmospheric temperature lead to increase in water vapor content and that subsequently increases changes in cloud properties. It is well known that water vapor is the most dominant variable in influencing green house effect (more effective than CO_2). Therefore, how the changes in cloud properties influence the Earth's energy balance and how that feedback loop operates is poorly understood and hence difficult for climate models to handle. Efforts are underway to bridge this gap, and it is expected that with the availability of ultra supercomputers and very high resolution (e.g. cloud resolving) models the future changes in regional climate can be predicted with an uncertainty envelope that will be useful for policy making.

6. Acknowledgement

H. Annamalai acknowledges the support of the Office of Science (BER) U.S. Department of Energy, Grant DE-FG02-07ER6445 and the institutional grants (JAMSTEC, NOAA and NASA) of the IPRC. Partial support from Climarice project funded by the Royal Norwegian Embassy, New Delhi, India is also acknowledged.

7. References

AchutaRao, K. & Sperber, K. R., 2006: ENSO simulation in coupled ocean-atmosphere models: are the current models better? *Climate Dynamics* 27, 1-15, doi:10.1007/s00382-006-0119-7.

Annamalai, H., K. Hamilton, and K.R. Sperber, 2007: South Asian summer monsoon and its relationship with ENSO in the IPCC-AR4 simulations. *J. Climate,* 20, 1071-1092.

Annamalai, H., and K. R. Sperber, 2005: Regional heat sources and the active and break phases of boreal summer intraseasonal (30-50 day) variability. *J. Atmos. Sci.,* 62, 2726-2748.

Annamalai, H., and J.M. Slingo, 2001: Active/break cycles: Diagnosis of the intraseasonal variability over the Asian summer monsoon region. *Clim. Dyn.,* 18, 85-102.

Annamalai, H., 2010: Moist dynamical linkage between the equatorial Indian Ocean and south Asian monsoon trough. *J. Atmos. Sci.,* 67, 589-610.

Annamalai, H., J. Hafner, K.P. Sooraj, and P. Pillai, 2011: Global warming shifts the monsoon circulation, drying south Asia. (submitted)

Charney, J. G., and J. Shukla, 1981: Predictability of monsoons, Monsoon Dynamics, Cambridge University Press, New York, Pp 99-109.

Chou, C., Neelin, J. D., Chen, C. A. & Tu, J. Y. Evaluating the "Rich-Get-Richer" Mechanism in Tropical Precipitation Change under Global Warming. *J. Climate* 22, 1982-2005, doi:10.1175/2008jcli2471.1

Held, I.M., and B. Soden 2006: Robust Responses of the Hydrological Cycle to Global Warming, *J. Climate,* 19, 5686-5699.

Hoskins, B.J., and M. Rodwell, 1995: A model of the Asian summer monsoo. Part 1: The global scale. *J. Atmos. Sci.,* 52, 1329-1340.

IPCC 2001: Intergovernmental Panel on Climate Change third assessment report

IPCC 2007: Intergovernmental Panel on Climate Change fourth assessment report

Joseph, R. & Nigam, S., 2006: ENSO evolution and teleconnections in IPCC's twentieth-century climate simulations: Realistic representation? *Journal of Climate* 19, 4360-4377.

Kitoh, A., Yukimoto, S., Noda, A. & Motoi, T., 1997: Simulated changes in the Asian summer monsoon at times of increased atmospheric CO2. *Journal of the Meteorological Society of Japan* 75, 1019-1031.

Krishnakumar K., and Coauthors., 1999: On the weakening of the ENSO-monsoon relationship. *Science,* 284, 2156-2159.

Li, C. F. & Yanai, M., 1996: The onset and interannual variability of the Asian summer monsoon in relation to land sea thermal contrast. *Journal of Climate* 9, 358-375.

May, W., 2010 The sensitivity of the Indian summer monsoon to a global warming of 2°C with respect to pre-industrial times. *Climate Dynamics,* 1-26, doi:10.1007/s00382-010-0942-8.

Meehl, G. A. & Arblaster, J. M., 2003: Mechanisms for projected future changes in south Asian monsoon precipitation. *Climate Dynamics* 21, 659-675, doi:10.1007/s00382-003-0343-3.

Meehl, G. A. *et al.* 2007: The WCRP CMIP3 multimodel dataset - A new era in climate change research. *Bulletin of the American Meteorological Society* 88, 1383-+, doi:10.1175/bams-88-9-1383.

Menon, S., 2004: Current uncertainties inassessing aerosol effects on climate. *Annual Review of Environment and Resources* 29, 1-30.

Pillai, P.A., and H. Annamalai, 2011: Moist dynamics of severe monsoons over south Asia: Role of the tropical SST. *J. Atmos. Sci.,* (revised)

Shenoi, S. S. C., D. Shankar and S.R. Shetye, 2002. Differences in heat budgets of the near-surface
Arabian Sea and Bay of Bengal: Implications for the summer monsoon. *Journal of Geophysical Research-Oceans* 107, doi:10.1029/2000jc000679 .
Shukla, J and Coauthors 2009: Revolution in climate prediction is both necessary and possible. *Bull. Amer. Met. Soc.*, 16-19
Shukla, J., 1995: Predictability of the tropical atmosphere, the tropical oceans and TOGA: Proceedings of the International conference on the tropical ocean global atmosphere (TOGA) Programme, Vol. 2., WCRP-91, 725-730 pp. World Climate Research Programme, Geneva, Switzerland, 1995.
Sikka, D.R., 2006: A study on the monsoon low pressure systems over the Indian region and their relationship with drought and excess monsoon seasonal rainfall. COLA Report No. 217, Available at: COLA, 4041, Powder Mill Road, Calverton, MD, USA.
Sooraj, K.P., H. Annamalai, A. Kumar, and H. Wang, 2011: A comprehensive assessment of CFS seasonal forecast over the tropics. *Weather and Forecasting* (revised)
Sperber, K.R.,and T.N. Palmer, 1996: Interannual tropical rainfall variability in general circulation model simulations associated with the Atmospheric Model Intercomparison Project. *J. Climate,* 9, 2727-2750.
Sperber, K.R., and H. Annamalai, 2008: Coupled model simulations of boreal summer intraseasonal (30-50 day) variability, Part I: Systematic errors and caution on use of metrics. *Clim. Dyn.* 31, 345–372.
Stowasser, M., H. Annamalai, and J. Hafner, 2009: Response of the south Asian summer monsoon to global warming: Mean and synoptic systems. *J. Climate,* 22 , 1014–1036
Turner, A. G., Inness, P. M. & Slingo, J. M., 2005: The role of the basic state in the ENSO-monsoon relationship and implications for predictability. *Quarterly Journal of the Royal Meteorological Society* 131, 781-804, doi:10.1256/qj.04.70.
Turner, A. G., Inness, P. A. & Slingo, J. M., 2007: The effect of doubled CO2 and model basic state biases on the monsoon-ENSO system. I: Mean response and interannual variability. *Quarterly Journal of the Royal Meteorological Society* 133, 1143-1157, doi:10.1002/qj.82.
Turner, A. G. & Slingo, J. M. , 2009: Subseasonal extremes of precipitation and active-break cycles of the Indian summer monsoon in a climate-change scenario. *Quarterly Journal of the Royal Meteorological Society* 135, 549-567, doi:10.1002/qj.401.
Turner, A.G., and H. Annamalai, 2011: Climate change and the Asian summer monsoon (submitted)
Ueda, H., Iwai, A., Kuwako, K. & Hori, M. E. , 2006: Impact of anthropogenic forcing on the Asian summer monsoon as simulated by eight GCMs. *Geophysical Research Letters* 33, doi:10.1029/2005gl025336.
Waliser, D.E., and Coauthors 2003: AGCM simulations of intraseasonal variability associated with the Asian summer monsoon, *Clim. Dyn.,* 21, 423-446.
Walker, G. T. , 1924: Correlations in seasonal variations of weather, IX- a further study of world weather. *World Weather II, Memoirs of the India Meteorological Department,* 275-332.
Walker, G. T., and E. W. Bliss, 1932: World Weather V. *Mem. Roy. Meteor. Soc.,* 4, No. 36, 53-84.

Webster, P. J. *et al.* 1998: Monsoons: Processes, predictability, and the prospects for prediction. *Journal of Geophysical Research-Oceans* 103, 14451-14510.

Yasunari, T., 1979: Cloudiness fluctuations associated with the Northern Hemisphere summer monsoon. *J. Met. Soc. Japan,* 57, 227-242.

8

Climatic Effect of the Greenhouse Gases Clusterization

Alexander Y. Galashev
Institute of Industrial Ecology of the Ural Branch of the Russian Academy of Sciences
Russia

1. Introduction

The Earth's atmosphere is a complex dynamic system, which protects the biosphere. One of the significant factors impacting the Earth's radiation balance is the greenhouse effect. The ability of the atmosphere to capture and recycle energy emitted by the Earth's surface is the defining characteristic of the greenhouse effect. Greenhouse gases (GHG) dissipate heat in the atmosphere. The greenhouse effect is observed only when radiant heat stays in the troposphere, instead of leaving into Space. Layers lying above the troposphere, such as stratosphere and thermosphere, serve as an obstacle for heat loss into Space. At low temperatures, characteristic of stratosphere and the lower thermosphere, water clusters, capable of absorbing molecules of other gases are formed. Many chemical compounds present in Earth's atmosphere behave as greenhouse gases.

The fact that the Earth's climate is altered by a change in atmospheric composition is well known. The analyses of various paleorecords, such as ice cores, have shown that atmospheric composition and climate have been correlated over the past 100,000 years. Water Vapor is the most abundant greenhouse gas in the atmosphere. However, changes in its concentration is also considered to be a result of climate feedbacks related to the warming of the atmosphere rather than a direct result of industrialization. The feedback loop in which water is involved is critically important to projecting future climate change, but as yet is still fairly poorly measured and understood.

Carbon dioxide is a greenhouse gas, and the increased concentration of carbon dioxide in the atmosphere must influence earth's radiation balance. The rising concentration of atmospheric CO_2 in the last century is not consistent with supply from anthropogenic sources. Such anthropogenic sources account for less than 5% of the present atmosphere, compared to the major input/output from natural sources (~95%). Hence, anthropogenic CO_2 is too small to be a significant or relevant factor in the global warming process, particularly when comparing with the far more potent greenhouse gas water vapor. The rising atmospheric CO_2 is the outcome of rising temperature rather than vice versa. The time taken for atmospheric gases to adjust to changes in sources or sinks is known as the atmospheric lifetime of a gas. The atmospheric lifetime of carbon dioxide is estimated as 5-200 years (Essenhigh, 2009). An individual molecule of CO_2 has a short residence time in the atmosphere. However, in most cases when a molecule of CO_2 leaves the atmosphere it is simply swapping places with one in the ocean. The CO_2 molecules cannot accumulate in the atmosphere, because of Henry's Law, which says that most of the atmospheric CO_2 must be exchanged with the ocean water. CO_2 is essentially chemically inert in the atmosphere and is only removed by biological uptake and by dissolving into the ocean. Dissolution of CO_2 into the oceans is fast but the transfer of carbon from surface waters to the deep ocean largely occurs by the slow ocean basin circulation and takes about 500-1000 years.

Methane is an extremely effective absorber of radiation, though its atmospheric concentration is less than CO_2 and its lifetime in the atmosphere is brief (10-12 years), compared to some other greenhouse gases (such as CO_2, N_2O). Concentrations of nitrous oxide also began to rise at the beginning of the industrial revolution and is understood to be produced by microbial processes in soil and water, including those reactions which occur in fertilizer containing nitrogen. Ultraviolet radiation and oxygen interact to form ozone in the stratosphere. Existing in a broad band, commonly called the ozone layer, a small fraction of this ozone naturally descends to the surface of the Earth. The potential radiative impact of the ozone change on the stratosphere–troposphere exchange was investigated in (Xie et al., 2008). This analysis shows that a 15% global O_3 decrease can cause a maximum cooling of 2.4 K in the stratosphere. Carbon monoxide (CO) is not considered a direct greenhouse gas, mostly because it does not absorb terrestrial thermal IR energy strongly enough. However, CO is able to modulate the production of methane and tropospheric ozone.

Water vapor and atmospheric gases, such as CO_2, N_2O, CH_4, and others, have a decisive influence on the formation of thermal radiation fields. However, according to the Le Chatelier principle, there are opposite compensating processes in the atmosphere. Clusterization of greenhouse gases can be considered as one of these processes. Water molecules bind easily to one another by forming hydrogen bond complexes - clusters. These weak interactions (approximately 0.21 eV) perturb the rovibronic and electronic states of individual molecules, and alter the spectroscopy of water vapor (Vaida et al., 2001). The possible involvement of water clusters in these processes has been proposed for a long time (Bignell et al., 1970; Lee et al., 1973; Coffey, 1977; Wolynes & Roberts, 1978; Carlon, 1979; Dias-Lalcaca et al., 1981). The abundance and the vibrational spectrum of the water dimer and trimer were evaluated both theoretically (Shillings et al., 2010) and in an experiment (Goss et al., 1999). The temperature contributions of the greenhouse gases of the Earth's atmosphere to the greenhouse effect are determined in (Barrett, 2005) according to their volume fraction. So for water vapor this contribution should make 37.4 K. However, correction of 13.4 K is necessary to take into account the effect of water evaporation. Therefore effectively the contribution of water vapor decreases to 24.0 K.

Nitrogen forms the following series of oxides that formally correspond to all possible oxidation states from +1 to +5: N_2O, NO, N_2O_3, NO_2, N_2O_5; however, only two of them, nitrogen(II) oxide and nitrogen(IV) oxide, are stable under ordinary conditions and actively participate in the natural and industrial nitrogen cycles. Nitrogen oxides (NO_x) are of great importance for many atmospheric reactions. NO_3, radical is the most reactive of them; it is formed in the dark and is regulated by the chemical reactions occurring in the night atmosphere. When interacting with water, nitrogen oxides form nitric acid (HNO_3), which is not only the main component of acid rain, but also governs the main process of removing nitrogen oxides from atmosphere. Nitric acid plays the key role in the chemistry of polar stratosphere. It is involved in the formation of ozone holes. Nitrogen oxides directly participate in the formation and decomposition of the tropospheric ozone. Under the action of sunlight, nitrogen dioxide is transformed into monoxide NO, which, in turn, forms NO_2 by interacting with oxygen. Ozone and peroxide radicals are also involved in this cycle.

More carbon monoxide is emitted to the atmosphere every year than any other pollutant except for CO_2, and the amount of emitted CO increases every year. However, the data on the rise of CO concentration in the troposphere are quite contradictory. The results of measurements of CO content in samples of Arctic and Antarctic ice indicate that its concentration hardly varied for centuries (Platt & Stutz, 2008). There exist at least five varieties of CO, namely, two varieties with light oxygen (O_{16}) and three varieties with heavy oxygen (O_{18}). The first two varieties of CO may form directly in the atmosphere the year round. The much less common varieties of CO enriched with O_{18} are seasonal. Carbon monoxide is liberated from the ocean surface, similar to methane. Depending on the concentration of CO, the soil may be both the source and the sink of this gas. The main anthropogenic source of CO is the incomplete combustion of hydrocarbon fuel, including automobile fuel. The resident time of CO in the troposphere is from 0.1 to 0.2 year. As a result of the effect of hydroxyl, CO oxidizes to carbon dioxide. Because of the existing transverse gradient of CO concentration in the troposphere, the stratosphere serves as the sink of carbon monoxide. Monoxide NO stands out from among nitrogen oxides; its resident time in the troposphere is approximately one hundred years. At present, its concentration in the atmosphere increases by 0.25% a year. The transition of NO to the stratosphere causes a reduction of stratospheric ozone: nitrogen monoxide reacts with ozone to form nitrogen dioxide and oxygen.

The study of the properties of clusters is usually considered as a way of investigating the transition from gas to condensed liquid. In the case of water clusters, it is also possible to estimate the effect of hydration on the dynamics of chemical reactions in a regulated form. For example, the reaction between SO_3 and water vapor consisting of both monomers and clusters is extremely important from the standpoint of heterogeneous atmospheric chemistry (Akhmatskaya, et al., 1997), as it represents the final stage in SO_2 tropospheric oxidation, which eventually leads to the formation of sulfuric acid aerosols and acid rains. These reactions also result in the depletion of ozone, thus weakening the absorption of incident solar radiation and influencing the Earth's climate.

Atomic oxygen plays an important part in the formation of atmospheric ozone. Oxygen atoms formed in the dissociation of an O_2 molecule either associate again in the presence of another particle M necessary to withdraw the energy from a formed molecule according to the equation

$$\mathrm{O} + \mathrm{O} + M \rightarrow \mathrm{O}_2 + M, \qquad (1)$$

or interact with an O_2 molecule also in the presence of another particle and form an ozone molecule as follows:

$$\mathrm{O}_2 + \mathrm{O} + M \rightarrow \mathrm{O}_3 + M. \qquad (2)$$

Physically, an ozone molecule is stable, but the decomposition rate of gaseous ozone grows appreciably with an increase in the temperature and amounts of some gases, for example, NO, Cl_2, Br_2 and I_2, and also under the action of different radiations and particle fluxes.

The autoregulation of the atmosphere composition is due to the formation of water clusters and their subsequent capture of greenhouse gas molecules is, as a rule, ignored in the estimation of the Earth's radiation balance. The characteristics of infrared (IR) radiation absorption by water clusters incorporating molecules of the most abundant atmospheric gases - nitrogen, oxygen, and argon - were studied in (Galashev et al., 2005; Novruzova et al., 2007a, 2007b; Novruzova & Galashev, 2008).

The objective of this study is to investigate the absorption of different greenhouse gases by ultradisperse water medium with determining the spectra of infrared (IR) absorption and emission by $(H_2O)_n$ and $X_i(H_2O)_n$ (where X= CO_2, N_2O, CH_4, C_2H_2, C_2H_6, NO_2, NO, CO and O_3) systems under conditions typical of the troposphere. Also, in the light of the molecular dynamics calculations, the impact of atmospheric gases' clustering on greenhouse effect is examined.

2. Computer model

This work deals with the molecular dynamics study of the influence of clusterization of H_2O vapor and atmospheric gases such as CO_2, N_2O, CH_4, C_2H_2, C_2H_6, NO_2, NO, CO and O_3 on the greenhouse effect. The IR absorption spectra were calculated for systems formed by water clusters and molecules of the above greenhouse gases at 233 K. Due to clusterization a homogeneous single-phase system represented by a mixture of gases transforms into a "two-phase" or even "three-phase" state since, depending on the ambient conditions, clusters constituting a fine "phase" can be in both the liquid and solid states.

Water is among the most studied of chemicals, owing, in part, to its ubiquity and its necessity for all life. In addition to these "natural" reasons, water is an interesting compound because it has unique physical properties and is a model hydrogen-bonded liquid. Most of the available water interaction potentials are parametrized to reproduce the thermodynamic and structural properties of bulk water. The polarizable model allowed us to examine the changes in the dipole moment of the individual water molecules as a function of their environment. This can provide insight into many-body effects in water clusters at a molecular level. Dang and Chang (Dang & Chang, 1997) have developed a polarizable potential model for water that behaves reasonably well with changes in the environments (i.e.., cluster, liquid, and liquid/vapor). The simulation of water clusters was performed using a refined TIP4P interaction potential for water and the rigid four-center model of H_2O molecule (Jorgensen, 1981). The total interaction energy of the system can be written as

$$U_{\text{tot}} = U_{\text{pair}} + U_{\text{pol}} , \quad (3)$$

where the pairwise additive part of the potential is the sum of the Lennard-Jones and Coulomb interactions,

$$U_{\text{pair}} = \frac{1}{2}\sum_i \sum_j \left[4\varepsilon^{(\text{LJ})} \left\{ \left(\frac{\sigma_{ij}^{(\text{LJ})}}{r_{ij}} \right)^{12} - \left(\frac{\sigma_{ij}^{(\text{LJ})}}{r_{ij}} \right)^{6} \right\} + \frac{q_i q_j}{r_{ij}} \right] . \quad (4)$$

Here, r_{ij} is the distance between site i and j, q is the charge, $\sigma^{(\text{LJ})}$ and $\varepsilon^{(\text{LJ})}$ are the Lennard-Jones parameters.

The nonadditive polarization energy is given by

$$U_{pol} = \frac{1}{2}\sum_i \mathbf{d}_i \cdot \mathbf{E}_i^0 , \quad (5)$$

where $\mathbf{E}_i^0$ is the electric field at site i produced by the fixed charges in the system

$$\mathbf{E}_i^0 = \sum_{j \neq i} \frac{q_j \mathbf{r}_{ij}}{r_{ij}^3}, \tag{6}$$

$\mathbf{d}_i$ is the induced dipole moment at atom site i, and is defined as

$$\mathbf{d}_i = \alpha_i^p \mathbf{E}_i, \tag{7}$$

where

$$\mathbf{E}_i = \mathbf{E}_i^0 + \sum_{j \neq i} \mathbf{T}_{ij} \cdot \mathbf{d}_j. \tag{8}$$

In the above, $\mathbf{E}_i$ is the total electric field at atom i, α_i^p is the atomic polarizability, and $\mathbf{T}_{ij}$ is the dipole tensor

$$\mathbf{T}_{ij} = \frac{1}{|r_{ij}|^3}\left(3\hat{\mathbf{r}}_{ij}\hat{\mathbf{r}}_{ij} - \mathbf{1}\right). \tag{9}$$

In (9), $\hat{\mathbf{r}}_{ij}$ is the unit vector in the direction $\mathbf{r}_i - \mathbf{r}_j$, where $\mathbf{r}_i$ and $\mathbf{r}_j$ are the positions of the centers of mass of molecules *i* and *j*, and **1** is the 3 × 3 unit tensor.

The modification of interaction potential for water by Dang and Chang (Dang & Chang, 1997) concerned the variation of the parameters of the Lennard–Jones part of potential and localization of negative charge. As a result, the value of permanent dipole moment for water molecule was taken to be equal to its experimentally obtained value of 1.848 D. The geometry of this molecule corresponds to the experimentally obtained parameters of the molecule in the gas phase: r_{OH} = 0.09572 nm and angle HOH of 104.5° (Benedict et al., 1956). Fixed charges (q_H= 0.519 *e*, q_M= −1.038 *e*) are ascribed to H atoms and to point M lying on the bisectrix of angle HOH at a distance of 0.0215 nm from oxygen atom. The values of charges and the position of point *M* are selected so as to reproduce the experimentally obtained values of dipole and quadrupole moments (Xantheas, 1996; Feller & Dixon, 1996), as well as the ab initio calculated energy of dimer and the typical distances in the dimer (Smith & Dang, 1994). The stabilization of short-range order in water clusters is largely attained owing to the short-range Lennard-Jones potential with the center of interaction ascribed to the oxygen atom. Related to point *M* in addition to the electric charge is the polarizability which is required for the description of nonadditive polarization energy. The standard iterative procedure is used at every time step for calculating induced dipole moments (Dang & Chang, 1997). The accuracy of determination of $\mathbf{d}_i$ is given in the range of 10^{-5} –10^{-4} D.

The atom-atom impurity (carbon, oxygen, hydrogen, nitrogen)–water interactions were preassigned in terms of the sum of repulsion, dispersion, and Coulomb contributions,

$$\Phi(r_{ij}) = b_i b_j \exp[(c_i + c_j) r_{ij}] - a_i a_j r_{ij}^6 + \frac{q_i q_j}{r_{ij}}, \tag{10}$$

where the parameters a_i, b_i, and c_i of the potential describing these interactions were borrowed from Spackman (Spackman, 1986a, 1986b). Experimental polarizability values have been used (Lide, 1996).

The added linear CO_2, N_2O, CO, NO and C_2H_2 molecules were placed along the beams, connecting $(H_2O)_n$ cluster's center of the mass to those of these molecules. At first NO_2, O_3, CH_4 and C_2H_6 molecules were placed in the knots of imaginary BCC lattice piercing the cluster. In all cases the admixture molecules were situated outside the cluster.

The Gear method of fourth order was used for the integration of the motion equations of the molecules' centers of the mass (Gear, 1971). The analytical solution of motion equations for molecules' rotation was carried out with a help of the Rodrigues - Hamilton parameters (Petrov & Tikhonov, 2002). Scheme of the equations of motion integration at presence of rotation corresponded with the approach offered by Sonnenschein (Sonnenschein, 1985). The duration of calculation for each cluster lasted not less than $3 \cdot 10^6$ Δt, where time step was $\Delta t = 10^{-17}$c.

The overall dipole moment of a cluster $\mathbf{d}_{cl}$ was calculated via the formula

$$\mathbf{d}_{cl}(t) = Z_+ \sum_{i=1}^{N_{tot1}} \mathbf{r}_i(t) + Z_- \sum_{j=1}^{N_{tot2}} \mathbf{r}_j(t), \tag{11}$$

where $\mathbf{r}_i(t)$ is a vector denoting the position of atom i or point M at time moment t; Z_+ and Z_- are the electric charges located at positively and negatively sites; and N_{tot1} and N_{tot2} are the numbers of positively and negatively charged atoms in a cluster, respectively.

Static permittivity ε_0 was calculated from the fluctuations of overall dipole moment $\mathbf{d}_{cl}$ (Bresme, 2001) as follows:

$$\varepsilon_0 = 1 + \frac{4\pi}{3VkT}\left[\left\langle \mathbf{d}_{cl}^2 \right\rangle - \left\langle \mathbf{d}_{cl} \right\rangle^2\right], \tag{12}$$

where V is the cluster volume and k is Boltzmann's constant.

Permittivity $\varepsilon(\omega)$ was represented as a function of frequency ω by a complex value $\varepsilon(\omega) = \varepsilon'(\omega) - i\varepsilon''(\omega)$, which was estimated through the following equation (Bresme, 2001; Neumann, 1985):

$$\frac{\varepsilon(\omega) - 1}{\varepsilon_0 - 1} = -\int_0^\infty \exp(-i\omega t)\frac{dF}{dt}dt = 1 - i\omega\int_0^\infty \exp(-i\omega t)F(t)dt, \tag{13}$$

where function $F(t)$ is the normalized autocorrelation function of the overall dipole moment of the cluster

$$F(t) = \frac{\langle \mathbf{d}_{cl}(t)\mathbf{d}_{cl}(0)\rangle}{\left\langle \mathbf{d}_{cl}^2 \right\rangle}. \tag{14}$$

Let us consider the scattering of nonpolarized light when the molecule path length l is much less than the light wavelength λ. The extinction (attenuation) coefficient ξ of an incident ray can be determined both by the Rayleigh formula (Landau & Lifshits, 1982)

$$\xi = \frac{2\omega^4}{3\pi c^4}\frac{(\sqrt{\varepsilon_0} - 1)^2}{N}, \tag{15}$$

and through the scattering coefficient μ by the expression ($\xi = \frac{16\pi}{3}\mu$) (Przhibel'sky, 1994) in the approximation of a scattering at an angle 90°. Here, N is the concentration of scattering centers, c is the light velocity, ε is the dielectric permittivity of the medium, and ω is the incident wave frequency.

Taking into consideration that $\xi = \alpha + \mu$, where α is the absorption coefficient, we obtain

$$N_{i,n} = \frac{2\omega^4}{3\pi c^4}\frac{(\sqrt{\varepsilon_0} - 1)^2}{\alpha}(1 - \frac{3}{16\pi}) \cdot \tag{16}$$

Let us form system in such a way that the statistical weight of a cluster containing i admixture molecules and n water molecules can be expressed as follows:

$$W_{in} = \frac{N_{in}}{N_{\Sigma}}, \tag{17}$$

where N_{in} is the concentration of clusters with i admixture molecules and n water molecules, $N_{\Sigma} = \sum_{i=1}^{n_i} N_i$, where n_i is the maximal value of i. Similar weights were used for $(H_2O)_n$ clusters constituting system. The further calculation of all the spectral characteristics was performed with consideration for the assumed statistical weights W_{in}. The average concentration of each type of cluster in the systems under consideration was 12–13 orders of magnitude lower than the Loschmidt number.

The absorption coefficient α of incident IR radiation can be expressed through the imaginary component of the frequency-dependent dielectric permittivity $\varepsilon(\omega)$ as follows (Neumann, 1986):

$$\alpha(\omega) = 2\frac{\omega}{c}\mathrm{Im}[\varepsilon(\omega)^{1/2}]. \tag{18}$$

In the case of depolarized light, the Raman spectrum $J(\omega)$ is determined as follows (Bosma et al., 1993):

$$J(\omega) = \frac{\omega}{(\omega_L - \omega)^4}\left(1 - e^{\hbar\omega/kT}\right) \times \mathrm{Re}\int_0^{\infty} dte^{i\omega t}\langle \Pi_{xz}(t)\Pi_{xz}(0)\rangle \tag{19}$$

where

$$\Pi(t) \equiv \sum_{j=1}^{N}\left[\alpha_j^p(t) - \langle \alpha_j^p \rangle\right], \tag{20}$$

$\hbar = h/2\pi$, h is the Planck constant, ω_L is the exciting laser frequency, Π_{xz} is the xz component of the $\Pi(t)$ value, the x axis is directed along the molecular dipole, and xy is the molecular plane.

The frequency dispersion of the dielectric permittivity determines the frequency dependence of dielectric losses $P(\omega)$ according to the equation (Levanuk & Sannikov, 1988)

$$P = \frac{\varepsilon'' < E^2 > \omega}{4\pi}, \tag{21}$$

where $< E^2 >$ is the average squared intensity of an electrical field, and ω is the frequency of an emitted electromagnetic wave.

The rigid model of molecules used in this work does not permit consideration of intramolecular vibrations, which, as a rule, have frequencies $\omega > 1200$ cm^{-1}. In this work, we study the frequency range $0 \le \omega \le 1200$ cm^{-1}, which characterizes vibrational and rotational motions of molecules.

3. Water clusters and main greenhouse gases absorption

To determine the influence of absorbed polyatomic molecules on the greenhouse effect, we considered different types of ultradisperse systems: $(H_2O)_n$ (I), $(CO_2)_i(H_2O)_{10}$ (II), $(CH_4)_i(H_2O)_{10}$ (III), and $(N_2O)_i(H_2O)_{10}$ (IV), where n = 10–20 and i = 1–10. Investigation of clusterization of the GHG was executed in works (Galashev et al., 2006a, 2006b, 2007; Novruzov et al., 2006; Galasheva et al., 2007; Chukanov & Galashev, 2008; Galashev, 2011). Here we shall consider influence of capture of admixture molecules by water clusters on some spectral characteristics of cluster systems. In analyzing frequency-dependent characteristics we restrict ourselves to frequency range $0 \le \omega \le 1200$ cm^{-1} because no intramolecular vibrations are present in the model employed by us for systems I-IV. The structure of equimolecular heteroclusters is shown in Fig. 1, where configurations of $(CO_2)_{10}(H_2O)_{10}$ and $(CH_4)_{10}(H_2O)_{10}$ clusters are represented. One may observe that both clusters have a significantly irregular structure. Moreover, CO_2 molecules intermix with H_2O molecules and CH_4 molecules gather in one group forming heterocluster's surface. The correlation in the orientation of CO_2 and H_2O molecules can be observed, while CH_4 molecules are disoriented.

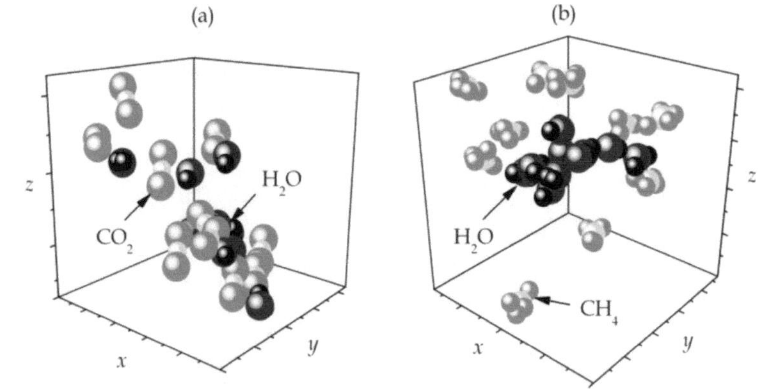

Fig. 1. Configurations of clusters: (a) - $(CO_2)_{10}(H_2O)_{10}$, (b) - $(CH_4)_{10}(H_2O)_{10}$, corresponding to the moment of time 30 ps.

The greenhouse effect caused by atmospheric gases is in fact the absorption of the Earth's thermal radiation by them and a subsequent dissipation of the absorbed energy. The spectrum of Earth's thermal radiation, together with experimental IR radiation absorption spectrum of liquid water (Goggin & Carr, 1986), is shown in Fig. 2(a). The spectrum for water overlaps practically all of the Earth's radiation frequency range and indicates the greatest significance of atmospheric moisture in greenhouse effect creation. The

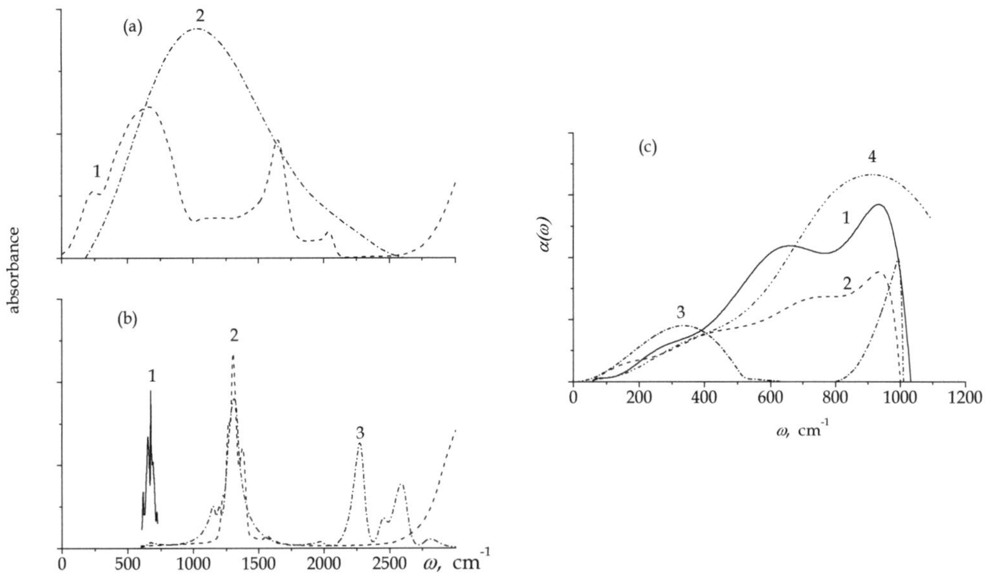

Fig. 2. IR absorption spectra, (a) *(1)* experimental spectrum for liquid water (Goggin & Carr, 1986), *(2)* spectrum of thermal radiation of the Earth at $T = 280$ K; (b) *(1)*, *(2)*, *(3)* experimental spectra for gaseous CO_2, CH_4, and N_2O (Kozintsev et al., 2003) correspondingly; (c) *(1)* system I, *(2)* II, *(3)* III, *(4)* IV.

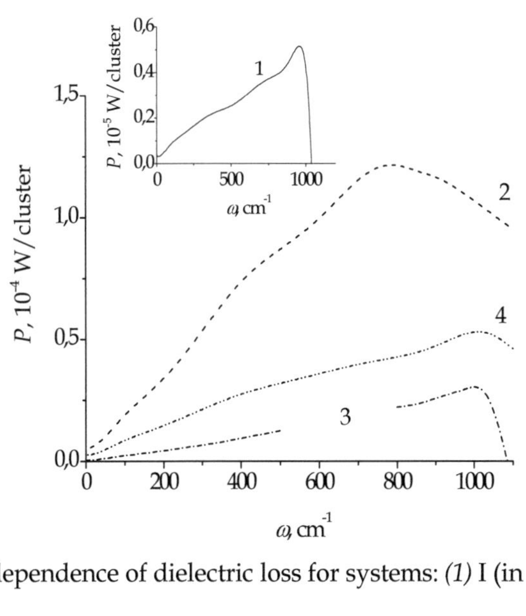

Fig. 3. The frequency dependence of dielectric loss for systems: *(1)* I (in the inset), *(2)* II, *(3)* III, *(4)* IV.

experimental IR spectra of gaseous CO_2, N_2O and CH_4 (Kozintsev et al., 2003) absorption are represented in Fig. 2(b). In the investigated frequency range the locations of the main spectrum peaks for gaseous N_2O and CH_4 coincide. These spectra represent a landmark for location of changes of relevant IR spectra peaks during transition from clusters of pure water to heteroclusters. IR spectra of systems I, II, III and IV presented in Fig. 2(c) are calculated by the method described in (Galashev et al., 2006b). The spectra for ultra-disperse systems of pure water and $CO_2)_i(H_2O)_{10}$ clusters system have two peaks, the main of which situates at ω = 974 (I) and 960 cm^{-1} (II), and the second peak at 661 cm^{-1} (I) and 724 cm^{-1} (II), it is less expressed. The IR spectra in the frequency range of $0 \le \omega \le 1200$ cm^{-1} related with aqueous systems containing N_2O and CH_4 molecules are characterized by one peak. These peaks are situated respectively at the 911 and 340 cm^{-1} frequency. In the observed frequency range the integral intensity of IR radiation absorption by systems II and III significantly decreases. On the contrary, for the system IV this value slightly increases in comparison with the corresponding characteristic of system I.

Let us consider energy exchange between photons representing falling electromagnetic wave and phonons - collective oscillations of molecules in clusters. The most probable events of this process are (Poulet & Mathieu, 1970):

1. the absorption of photon with frequency ω (or $\hbar\omega$ energy) with a birth of two phonons with the same ω_1 frequency extending in opposite directions under the law of energy conservation $\hbar\omega = 2\hbar\omega_1$ or $\omega = 2\omega_1$,
2. the absorption of photon with frequency ω causes appearance of two phonons with different frequencies ω_1 and ω_2, where $\omega = \omega_1 + \omega_2$,
3. the absorption of photon, the collapse of one phonon and the emergence of another with $\omega = \omega_1 - \omega_2$, $\omega_1 \neq \omega_2$.

The exchange of electromagnetic radiation energy with clusters is an essentially unharmonious process. Due to this the emerged phonon can be characterized not only by one frequency but by a number of frequencies from the appointed interval.

The most probable result of IR radiation interaction with clusters, as well as with crystals (Poulet & Mathieu, 1970), is the appearance of two phonons of the same frequency (event 1). We especially refer to this event at ω = 974 cm^{-1} frequency of the main IR spectrum peak of the system I. Then the frequency expected for appearing phonons is defined by ω_1 = 487 cm^{-1} value. The event 2 should be the second one according to the frequency occurrence where absorbed photon energy distributes among excited phonons in uneven portions. To this event we can attribute the emergence of the second IR spectrum peak in the system I at ω = 661 cm^{-1} frequency. The frequency $\omega_2 = \omega - \omega_1$ = 174 cm^{-1} can be thus derived. The third event of the system I, where the frequency of IR spectrum peak localization is defined by $\omega_1 - \omega_2$, is the least probable. In this case the location of expected peak is given by ω = 313 cm^{-1}. In the IR spectrum of the system I in the vicinity of this frequency (at 348 cm^{-1}) there is only one inflection point of dependence $\alpha(\omega)$. It is possible to estimate the influence of admixture molecules on clusters' phonons, and consequently, on corresponding IR spectra by inharmonic contributed by them. In the case of CO_2 molecules the interaction of clusters with IR radiation gives events 1 and 2, and the inharmoniousness is characterized by the quantities $\Delta\omega_1 = \omega_1(II) - \omega_1(I)$ = -7 cm^{-1}, $\omega_2(II) - \omega_2(I)$ = 70 cm^{-1} (II:ω_1 = 480 cm^{-1} and ω_2 = 280 cm^{-1}). The event 1 with phonon frequency ω_1 = 455.5 cm^{-1} takes place at interaction of IR radiation with $(N_2O)_i(H_2O)_{10}$ clusters, and the inharmoniousness is defined by the value $\Delta\omega_1(IV)$ = 31.5 cm^{-1}. From the minimum unharmonious effect estimation it follows that for

$(CH_4)_i(H_2O)_{10}$ clusters the realization of event 3 with phonon frequencies ω_1 = 650 cm^{-1} and ω_2 = 310 cm^{-1} leading to a quantity of inharmoniousness $\Delta\omega_2$(III) = 136 cm^{-1}, is more probable. Thus, according to the quantity of contributed inharmoniousness (from greater to smaller) the admixture molecules locate as CH_4, CO_2, N_2O.
The frequency distribution of power dissipated by cluster systems under consideration is given in Fig. 3. One can see that the addition of CO_2, CH_4 and N_2O molecules to water clusters causes a significant increase in the rate of energy dissipation. The energy of absorbed IR radiation is dissipated most rapidly by system II with the maximum of dissipation at a frequency 800 cm^{-1}. The system IV has next rate of energy dissipation with a maximum of spectrum $P(\omega)$ at frequency 1036 cm^{-1}. The system of water clusters with absorbed CH_4 also strengthens IR radiation emission power. Water clusters exhibit the highest rate of dissipation of stored energy at frequency ω = 974 cm^{-1}, and clusters which absorbed CH_4 molecules (system III) - at 1014 cm^{-1}.

4. Hydrocarbon molecules' absorption by ultra disperse water medium

We consider some types of ultra disperse systems: $(C_2H_2)_j(H_2O)_n$, $(C_2H_6)_j(H_2O)_n$, $(C_2H_2)_i(H_2O)_{20}$, $(C_2H_6)_i(H_2O)_{20}$, j = 1, 2; $10 \le n \le 20$, $1 \le i \le 6$. Configurations of (a) $(C_2H_2)_6(H_2O)_{20}$ and (b) $(C_2H_6)_6(H_2O)_{20}$ clusters corresponding to the moment of time 30 ps show the absence of stirring of H_2O molecules with C_2H_2 and C_2H_6 molecules even when the number of admixture molecules reaches six (Fig. 4). Acetylene molecules are attracted by water clusters. Finally, they get the orientation of a tangent to the $(H_2O)_{20}$ cluster surface. It is reached due to the attraction of C atoms to H atoms of water molecules oriented mainly outwards to the cluster. Thus, H atoms, which are on the ends of the C_2H_2 molecule, feel the repulsion from the surface of the water cluster. In this case, acetylene molecules act as proton acceptors (Allouch, 1999). Also, the bond energy with water clusters is estimated as - 13.8, –15.4 and –12.9 kJ/mol for systems $C_2H_6(H_2O)_n$, $(C_2H_6)_2(H_2O)_n$ and $(C_2H_6)_i(H_2O)_{20}$, respectively. These values are coordinated with the estimation of bond energy (–15.4 kJ/mol) for the acetylene acting as a proton acceptor to amorphous ice (Silva & Devlin, 1994).

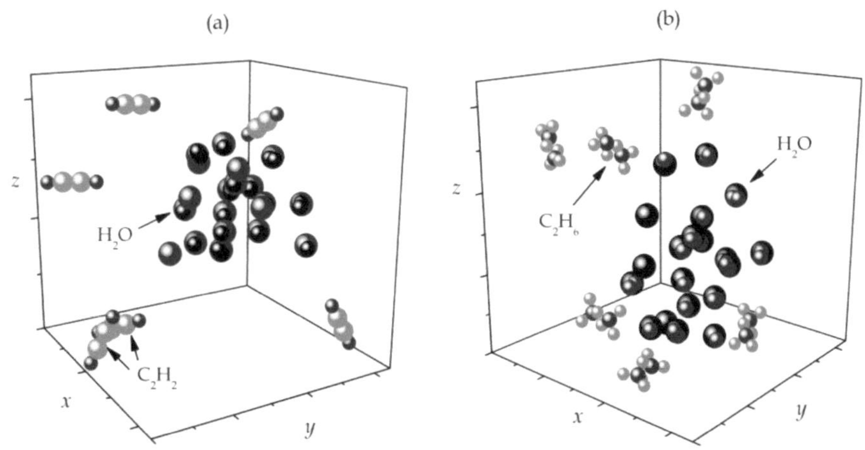

Fig. 4. Configurations of (a) $(C_2H_2)_6(H_2O)_{20}$ and (b) $(C_2H_6)_6(H_2O)_{20}$ clusters, corresponding to the moment of time 30 ps.

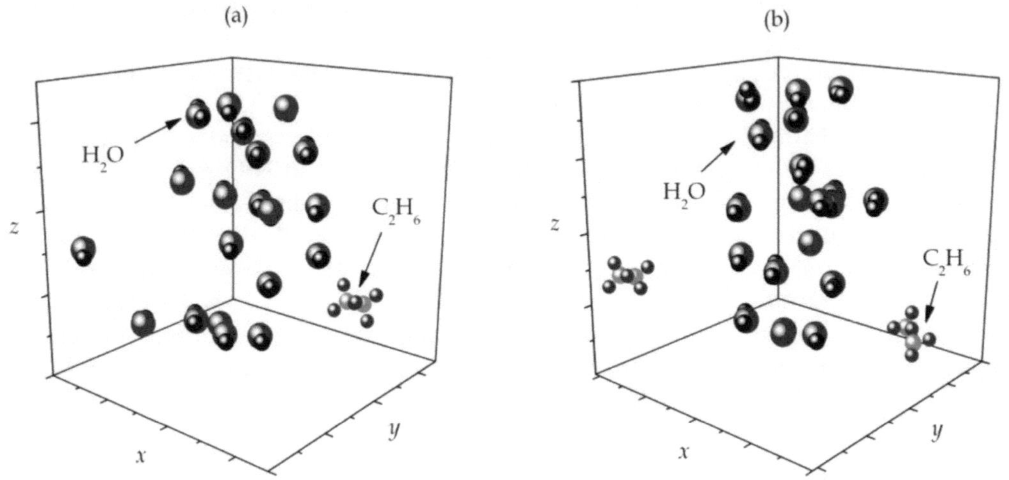

Fig. 5. Configurations of the clusters corresponding to a time of 30 ps: (a) C_2H_6 $(H_2O)_{20}$ and (b) $(C_2H_6)_2(H_2O)_{20}$ clusters.

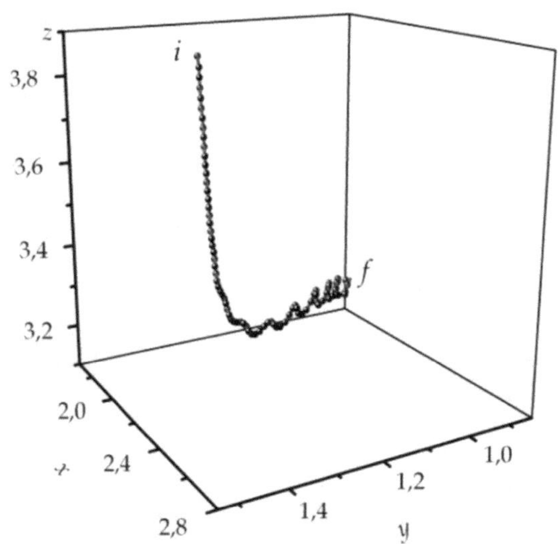

Fig. 6. Trajectory of motion for the center of mass of the C_2H_2 molecule within 30 ps: i and f are the beginning and the end of the trajectory, respectively. The coordinates are given in units σ_w = 0.3234 nm.

The bond energy for acetylene-water dimer is determined at -(5.0–7.9) kJ/mol (acetylene-proton acceptor) and -(8.7–13.3) kJ/mol (acetylene-proton donor) (Frish et al., 1983). Interaction of ethane molecules with water clusters is characterised by higher values of bond energy. These values for systems of $C_2H_6(H_2O)_n$, $(C_2H_6)_2(H_2O)_n$ and $(C_2H_6)_i(H_2O)_{20}$ are -11.7, -11.7 and -6.7 kJ/mol, respectively. If, in general, for the first two systems, the ethane acts as a proton acceptor, then for the last system, it acts as a proton donor (oxygen-hydrogen of ethane pairs gives the lowest energy in the ethane–water interaction). Thus, with the growth of ethane concentration and the amplification of interaction between C_2H_6 molecules, the switching of the bond type from the proton acceptor to the proton donor is observed. Bond energy in a complex amorphous ice–ethylene (proton acceptor) is estimated as –15.8 kJ/mol (Silva & Devlin, 1994). At atmospheric pressure, the ethylene has the higher solubility in water than ethane. The C_2H_6 molecules are not rejected by water cluster but also do not approach it too close being symmetrically on opposite sides of the cluster.

Theoretical determination of the ground-state geometry of water clusters is a difficult task. As the number of local minima grows exponentially with the number of atoms, finding the global minimum is a real challenge. Exact definition of the cluster configurations with the minimal energy is complicated due to energies, corresponding to the different equilibrium configurations, that are close enough. These configurations frequently differ only in the way of arrangement of atoms of hydrogen in the system of bonds. Already for polytetrahedral clusters from eight molecules, the number of isomers formed due to the hydrogen disorder makes 450, and it is already more than 40,000 of such isomers for similar clusters from 14 molecules (Kirov, 1993). Another difficulty is that the use of various models of water results in clusters with the minimal energy, having various forms. In addition, their structures differ by symmetry (Sremaniak et al., 1996). Even with the use of the same model of water, different forms of the $(H_2O)_{20}$ clusters with the minimal energy were received. With the help *ab initio* calculations, using TIP4P potential, Tsai and Jordan (Tsai & Jordan, 1993) have defined such cluster in the form of the fused cube, and by a method of molecular dynamics with the same potential Wales and Ohmine (Wales & Ohmine, 1993) have found an even lower energy structure for $(H_2O)_{20}$, which consists of three pentagonal prisms sharing three faces. Extensive *ab initio* calculations have been performed for several possible structures of water clusters $(H_2O)_n$, n=8–20 (Stern & Berne, 2001). It is found that the most stable geometries arise from a fusion of tetrameric or pentameric rings. As a result, $(H_2O)_n$, n = 8, 12, 16 and 20, are found to be cuboids, while $(H_2O)_{10}$ and $(H_2O)_{15}$ are fused pentameric structures. It is necessary to notice that the adding polarisability in an explicit manner has the effect favouring a reduction in strain energy at the expense of hydrogen bonding (Sremaniak et al., 1996). Many-body intermolecular interaction expansions provide a promising avenue for the efficient quantum mechanical treatment of molecular clusters and condensed-phase systems, but the computationally expensive three-body and higher terms are often non-trivial (Maheshwary et al., 2001).

The neighbourhood already with one ethane molecule changes the structure of water cluster. The C_2H_6 molecules are not detached by the water core of a cluster; however, at the same time, they do not approach the core too closely (Fig. 5). In the case of both clusters, every C_2H_6 molecule is adjacent to three water molecules. Also, every C_2H_6 molecule is oriented arbitrarily to the water core of a cluster, i.e. its C–C axis is directed neither at the centre of a cluster mass nor tangentially to the core "surface". The presence of a second C_2H_6 molecule changes the shape of the water skeleton of a cluster. This is due to the fact that every ethane molecule adjusts the nearest water molecules of a cluster to its conformation. The C_2H_6 molecules are quite remote from water molecules comparing to the distance between the adjacent H_2O molecules. In general, the surface of the $C_2H_6(H_2O)_{20}$ cluster turned out to be looser than the surface of the $(C_2H_6)_2(H_2O)_{20}$ cluster. The structures shown in Fig. 4 and 5 correspond to the energy close to the minimal one for these clusters under considered conditions.

The motion of the center of mass of one of the C_2H_2 molecules during the formation of the cluster $(C_2H_2)_2(H_2O)_{20}$ is reflected by the trajectory depicted in Fig. 6. The trajectory was followed over the time span of 30 ps. The initial and final positions of the center of mass of the molecule are denoted by the symbols *i* and *f*, respectively. It is seen that the C_2H_2 molecule first moves strictly rectilinearly toward the cluster. Covering a certain distance, the molecule turns toward the tangent to the cluster surface and follows in this direction, experiencing small-scale oscillation.

The frequency dependence of $\alpha(\omega)$ IR radiation absorption coefficient of investigated systems is shown in Fig. 7. The $\alpha(\omega)$ coefficient for disperse systems containing C_2H_2 molecules is higher than that for the disperse system of pure water (Fig. 7(a)). The intensity of the spectrum is increased with the growth of C_2H_2 molecules' number in the water system. The principal maximum of the $\alpha(\omega)$ distribution for the system of pure disperse water falls in the frequency of 780 cm^{-1}, and that for the similar water system with one and two acetylene molecules in each cluster is at 970 cm^{-1}. After absorption from one up to six acetylene molecules by the monodisperse water system, the $\alpha(\omega)$ spectrum becomes strongly oscillating with the principal maximum at $\omega = 920$ cm^{-1}. The $\alpha_w(\omega)$ spectrum of bulk liquid water (Goggin & Carr, 1986) has two maxima at frequencies of $\omega = 200$ and 690 cm^{-1}. The higher integrated intensity of IR radiation absorption for bulk water is caused by its density, which is higher, on average, than the density of water clusters by a factor of 1.4. However, in the presence of hydrocarbon molecules in these clusters, the density is not a determining factor in the absorption of IR radiation. The expansion of the system owing to the attachment of hydrocarbon molecules results in an increase in the number of vibration modes including the $0 \leq \omega \leq 1000$ cm^{-1} frequency range of the $\alpha(\omega)$ spectrum. The orientation of acetylene molecules on a tangent to the water cluster gives a stable amplification of the integrated intensity of the $\alpha(\omega)$ spectrum with the growth of acetylene concentration. This is promoted by the repulsion of the positive charges of C_2H_2 molecules. The bending band of the $\alpha(\omega)$ spectrum of gaseous acetylene is located at a frequency of ω =730 cm^{-1} (Kozintsev et

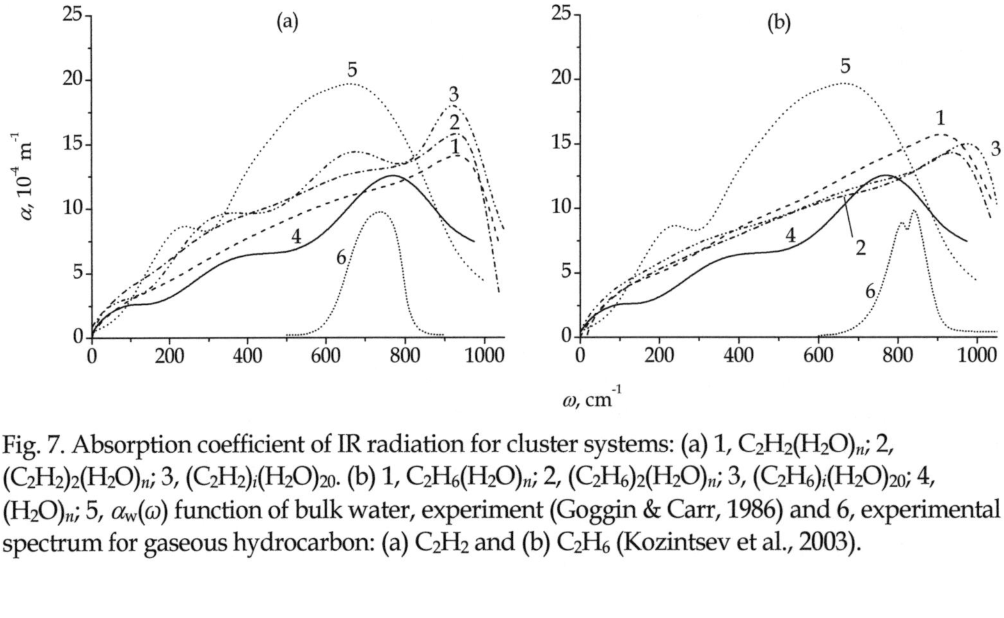

Fig. 7. Absorption coefficient of IR radiation for cluster systems: (a) 1, $C_2H_2(H_2O)_n$; 2, $(C_2H_2)_2(H_2O)_n$; 3, $(C_2H_2)_i(H_2O)_{20}$. (b) 1, $C_2H_6(H_2O)_n$; 2, $(C_2H_6)_2(H_2O)_n$; 3, $(C_2H_6)_i(H_2O)_{20}$; 4, $(H_2O)_n$; 5, $\alpha_w(\omega)$ function of bulk water, experiment (Goggin & Carr, 1986) and 6, experimental spectrum for gaseous hydrocarbon: (a) C_2H_2 and (b) C_2H_6 (Kozintsev et al., 2003).

al., 2003). Thus, the attachment of acetylene molecules by water clusters strengthens the integrated I_{tot} absorption intensity.

Ethane molecules have the more chaotic orientation. This results in a more random change of the vibrations' intensity with the growth of the number of C_2H_6 molecules in clusters. As a consequence, the behaviour of the IR absorption spectrum for these systems is less stable when the number of C_2H_6 molecules changes. After absorption of one ethane molecule, the absorption of external IR radiation in the investigated frequency range by ultradispersed water systems is also amplified (Fig. 7(b)) and the form of the curve becomes smoother. However, the following addition of C_2H_6 molecules results in some decrease of the IR radiation absorption, at least at frequencies of $\omega > 310$ cm^{-1}. The values of the α coefficient for the systems consisting of heteroclusters are higher than that for a cluster system of pure water almost in all frequency ranges (except for the area of $690 \leq \omega \leq 800$ cm^{-1}). The principal maximum of the $\alpha(\omega)$ frequency spectrum appears at $\omega = 910$ cm^{-1} ($C_2H_6 (H_2O)_n$ system) and $\omega = 973$ cm^{-1} ($(C_2H_6)_2(H_2O)_n$ and $(C_2H_6)_i(H_2O)_{20}$ systems). The part of the $\alpha(\omega)$ spectrum determined by the bending vibrations of gaseous ethane molecules has the doubled maximum in the $810 \leq \omega \leq 840$ cm^{-1} frequency range (Kozintsev et al., 2003).

Water clusters including clusters absorbing acetylene or ethane molecules are capable of re-emitting the falling IR radiation. Calculations show that the disperse system consisting of pure water clusters has low values of P radiation power, in comparison with the system enriched with acetylene (Fig. 8(a)). For the disperse system formed by pure water, two characteristic frequencies of IR radiation are observed: $\omega_1 = 657$ cm^{-1} and $\omega_2 = 973$ cm^{-1}. The maximal value of emitted radiation power corresponds to a frequency of 970 cm^{-1} when only one C_2H_2 molecule is present in the clusters, and to a frequency of 910 cm^{-1} at the presence of two acetylene molecules in every cluster of the system. After adsorption of the second C_2H_2 molecule by clusters, the radiation power of disperse systems is increased. Moreover, the absorption of ethane molecules by water clusters causes the essential decrease of clusters' P radiation power (Fig. 8(a), insert). Arrangement of C_2H_2 molecules on a tangent to a surface of water clusters makes their structure more dense (Novruzov et al., 2008a). Due to

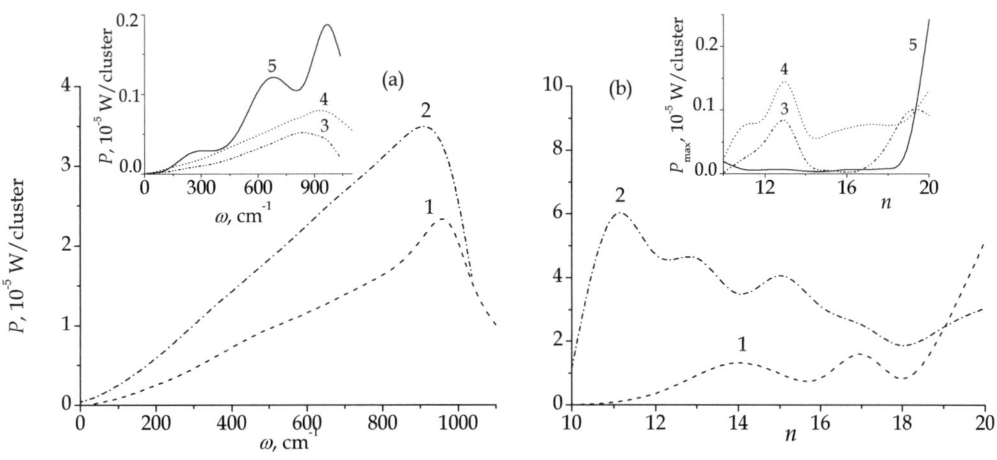

Fig. 8. Frequency dependence of (a) $P(\omega)$ IR radiation power and (b) maximum P quantity on n number of water molecules in clusters: 1, $C_2H_2(H_2O)_n$; 2, $(C_2H_2)_2(H_2O)_n$; 3, $C_2H_6(H_2O)_n$; 4, $(C_2H_6)_2(H_2O)_n$; 5, $(H_2O)_n$.

C atoms, the acetylene molecule is attached to the cluster. Due to H atoms, the molecule keeps a tangent direction to the water cluster and makes it denser. As a rule, $\overline{n}_b$, an average number of hydrogen bonds per molecule, decreases, and $\overline{L}_b$, the H-bond length, increases during the addition of both acetylene and ethane molecules to water clusters (Novruzov et al., 2008a, 2008b). When the number of admixture molecules $i < 5$, the value of $\overline{n}_b$ for water clusters with C_2H_2 molecules is higher and $\overline{L}_b$ is lower than those for clusters with C_2H_6 molecules. The volume of water clusters with C_2H_2 molecules appears to be 5–10% less than the volume of similar clusters with C_2H_6 molecules. As a result, the D_w self-diffusion coefficient of water molecules in $(C_2H_2)_i(H_2O)_n$ clusters is on average 20–30% lower than that in $(C_2H_6)_i(H_2O)_n$ clusters. The D_a self-diffusion coefficient of C_2H_2 molecules is higher than the D_a value of C_2H_6 molecules by no more than 1%. The decrease of the D_w value causes the amplification of collective vibrations, which, in turn, causes an appreciable amplification of dissipation of absorbed energy. An average $\overline{d}_{cl}$ dipole moment of $(C_2H_2)_i(H_2O)_n$ clusters exceeds the appropriate $\overline{d}_{cl}$ value of $(C_2H_6)_i(H_2O)_n$ clusters on an average by 5%. The ε'' imaginary part of dielectric permittivity of $(C_2H_2)_i(H_2O)_n$ clusters can increase by an order in comparison with the ε'' value of pure water clusters. This causes a sharp amplification of emitted radiation power for the system formed from these clusters. For $(C_2H_6)_i(H_2O)_n$ clusters, the opposite picture is observed, i.e. the ε'' value significantly reduces (up to five times), which causes an appreciable easing of the power of emitted IR radiation by an appropriate cluster system. The $P(\omega)$ spectra for the systems consisting of heteroclusters become smoother than those for the clusters system of pure water. The doubling of C_2H_6 molecules' number in ultra disperse system results in some amplification of radiation power. The principal maxima of the $P(\omega)$ spectra for the systems containing one and two ethane molecules in clusters are located at $\omega = 847$ and 910 cm^{-1}, respectively.

The behaviour of the P_{max} maximal value of clusters in dependence on a number of water molecules containing in clusters is shown in Fig. 8(b). Pure water clusters are characterised by an extremely low rate of energy dissipation up to the size $n = 18$ (an insert in Figure 8(b)). For $(H_2O)_n$ clusters with $n > 18$, the radiation power is sharply increased, but still remains low enough in comparison with the similar characteristic of clusters containing C_2H_2 molecules. The maximal values of the radiation power of water clusters adding one C_2H_2 molecule up to the size $n = 18$ are considerably lower than the P_{max} quantity of $(C_2H_2)_2(H_2O)_n$ clusters. Almost everywhere, the P_{max} values for clusters of the system where each aggregate contains two ethane molecules exceed the appropriate characteristics of aggregates of the system with one C_2H_6 molecule. At the same time, clusters containing ethane molecules, as a rule, have the higher P_{max} values than pure water clusters. Water clusters with $n = 10$, 16 and 20 are an exception here.

5. Characteristics of water clusters in the presence of nitrogen dioxide

For a system consisting of $(NO_2)_i(H_2O)_{50}$, heteroclusters, the $\alpha(\omega)$ spectrum becomes more rough than for clusters system of pure water (Fig. 9(a)). The integral intensity ratio between the absorption spectra for the system of pure water clusters (I_1) and the system of water heteroclusters containing i number of NO_2 molecules (I_2) is $I_1 : I_2 = 1 : 0.95$. For the system of pure water clusters, the spectrum of the absorption coefficient demonstrates one maximum at 838 cm^{-1} (curve *1*), whose localization disagrees with that of the main peak in the experimental absorption spectrum of bulk water, $\omega = 690$ cm^{-1} (curve *3*). The IR absorption spectrum of liquid water also comprises a slightly pronounced peak in the vicinity of $\omega \approx$

200 cm^{-1} (Goggin & Carr, 1986). The discrepancy in the positions of the main peaks may be related to the fact that the collective vibrations of the dipole moments of molecules in the clusters as nanosized objects may be characterized by frequencies different from the corresponding characteristics of liquid water. In the $\alpha(\omega)$ spectrum of heteroclusters (curve 2), four peaks are distinguished at $\omega = 88, 338, 713$, and 901 cm^{-1} (the main maximum). In the examined frequency range, a local maximum is present at $\omega = 780$ cm^{-1} in the experimental absorption spectrum of gaseous nitrogen(IV) dioxide (curve 4). The high-frequency maxima are observed in the absorption spectrum of NO_2 at ≈1300, 1650, and 1900 cm^{-1} (Kozintsev et al., 2003).

The IR absorption spectra obtained for single clusters of the $(NO_2)_i(H_2O)_{50}$ system, $0 \leq i \leq 6$ are illustrated in Fig. 9(b). The maximum integral intensity is inherent in the clusters containing 3 nitrogen (IV) dioxide molecules. On average, the spectra comprise three to four peaks, with the most intense peak being most often located in the frequency range $800 \leq \omega \leq 1000$ cm^{-1}. The exception is the $\alpha(\omega)$ spectrum of the pure water cluster whose main maximum is observed at 680 cm^{-1}. As awhole, the spectral intensities are comparable; the ratio between the maximum I_{max} and the minimum integral intensity I_{min} is 1.12. Even the addition of one impurity molecule can markedly change the pattern of an absorption spectrum.

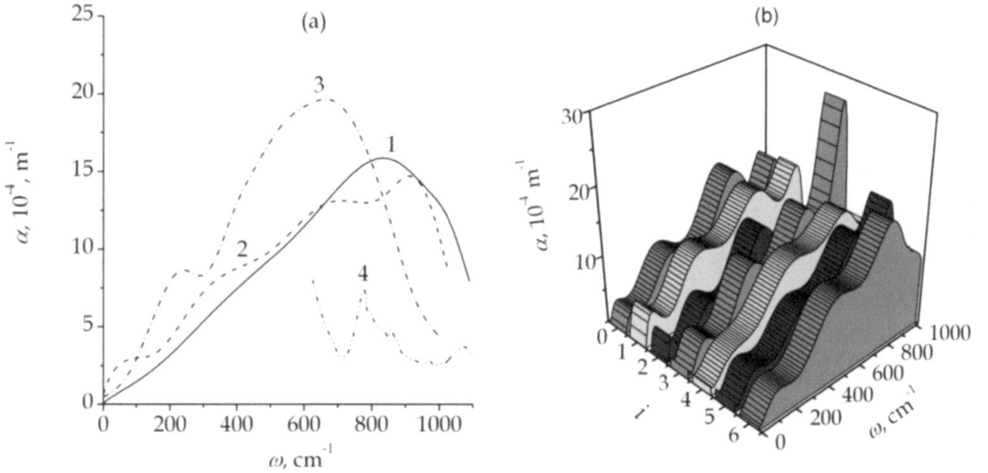

Fig. 9. (a) Frequency dependences of the IR absorption coefficients for different systems: (*1*) $(H_2O)_n$ at $10 \leq n \leq 50$, (*2*) $(NO_2)_i(H_2O)_{50}$ at $1 \leq i \leq 6$, (*3*) the experimental spectrum of bulk liquid water (Goggin & Carr, 1986), and (*4*) experimental spectrum of gaseous NO_2 (Kozintsev et al., 2003); (b) IR absorption coefficients for $(NO_2)_i(H_2O)_{50}$ clusters at $0 \leq i \leq 6$.

The scattering of electromagnetic waves by molecules of a substance takes place due to the fact that the electric field of an electromagnetic wave induces a dipole moment in a molecule and this moment vibrates at the frequency of the field. At weak electric fields, the induced dipole moment is proportional to the field strength, with the molecule polarizability playing the role of the proportionality coefficient. According to the classical physics, the mechanism of the scattering of incident electromagnetic waves by molecules lies in the fact that molecules emit secondary electromagnetic waves with the frequencies of the primary incident waves. However, because the polarizability of a molecule depends on the interatomic distances in it and varies with time during its vibration, the spectrum of a scattered light is transformed. In addition to the lines corresponding to the frequencies of the incident light, lines corresponding to the combinations (sums and differences) of these frequencies and normal vibration frequencies arise in the spectra (Günzler, H. & Gremlich, 2002).

A Raman spectrum consists of a most intense line (the exciting line) with the frequency equal to that of the incident radiation and less intense lines on each side of the former. The lines on the side of higher frequencies (the anti-Stokes lines) are noticeably less intense than those on the side of lower frequencies (the Stokes lines) (Wilson et al., 1980). Figure 10(a) depicts the anti-Stokes regions of the Raman spectra for $(H_2O)_{50}$ cluster (curve *1*) and $(NO_2)_i(H_2O)_{50}$ system (curve *2*), which comprise peaks with close positions but different intensities. The shown pattern of the spectra reflects the shifts relative to the exciting frequency $\omega_L = 19436.3$ cm^{-1} (the green line of the argon laser, $\lambda = 514.5$ nm).

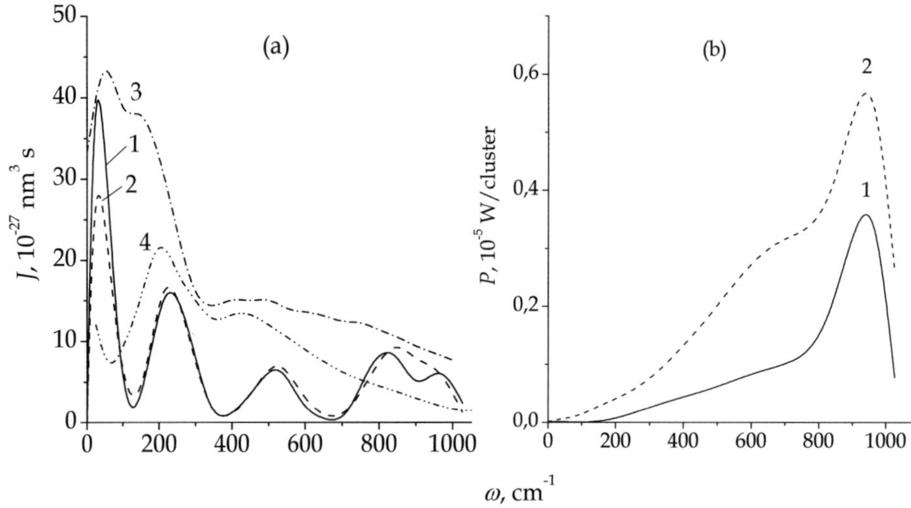

Fig. 10. (a) Raman spectra for (1) $(H_2O)_{50}$ cluster and (2) $(NO_2)_i(H_2O)_{50}$ system with $1 \le i \le 6$, (3) experimental spectrum of bulk water (Murphy, 1977), (4) spectrum calculated for a $(H_2O)_{50}$ cluster by molecular dynamic method using SPC/E three-site rigid model of water molecule (Bosma et al., 1993); (b) IR emission spectra for different systems: (1) $(H_2O)_n$ at $10 \le n \le 50$ and (2) $(NO_2)_i(H_2O)_{50}$ at $1 \le i \le 6$.

A distinct line with the frequency 30 cm^{-1} is the first reflection of the exciting frequency. A fainter line ($\omega = 220$ cm^{-1}) located in the immediate proximity to the exciting line is related to a change in the state of the rotational energy within the limits of the same vibrational state (Wilson et al., 1980). In the vibrational IR and Raman spectra of liquid water, the low-frequency region ($30 \le \omega \le 300$ cm^{-1}) resulting from the displacements of oxygen atoms (intermolecular motions) and the high-frequency region (higher than 3000 cm^{-1}) resulting from the modes of hydrogen atoms (intramolecular stretching vibrations of O–H bonds) are distinguished between (Agmon, 1996). The short-range order with the tetrahedral symmetry is created by complexes composed of five water molecules. Four normal modes are distinguished between in the tetrahedral symmetry, i.e., the breathing mode (ω_1) which describes completely nondegenerate symmetric stretching; the twice degenerate torsional mode (ω_2); two three-times degenerate modes, i.e., asymmetric stretching (ω_3) and bending

(ω_4). In the Raman spectrum of water, all four modes must be active, with the following values attributed to them: ω_4 = 50 cm^{-1}, ω_2 = 70 cm^{-1}, ω_1 = 150 cm^{-1}, and ω_3 = 180 cm^{-1} (Agmon, 1996). The first peak (ω = 30 cm^{-1}) in the Raman spectrum obtained for water clusters may be interpreted as the confluence of modes ω_2 and ω_4, whereas the second peak (at 220 cm^{-1}) may be considered as the combination of modes ω_1 and ω_3. Hence, the even (ω_2 and ω_4) and odd (ω_1 and ω_3) tetrahedral modes exhibit the red and blue shifts, respectively.
Proton mobility in water is characterized by jump time τ which, at room temperature, is $\approx$ 1.5 ps (Agmon, 1996). Time τ is determined from the data on the narrowing of NMR signals (Halle & Karlstrom, 1983). This time reflects the abnormal mobility of protons in water upon a jump 0.25–0.26 nm long. The time τ of the fast reorientation in water may be related to the intensity of the Raman peak assigned to the asymmetric stretching of the tetrahedron, i.e. the ω_3 frequency. For a system of clusters, this peak corresponds to a frequency of 220 cm^{-1}. Under the four-coordinated water approximation, hydrogen bonds are reoriented at a constant rate of $I_{\omega 3}/\tau$. It can be assumed that this rate is independent of temperature and, at equilibrium, we have $\tau \sim I_{\omega 3}$. For the examined clusters, the half-width of the Raman peak at 220 cm^{-1} is nearly 130 cm^{-1} and slightly varies with changes in their composition. A proton mobility time of $\approx$ 0.26 ps corresponds to this half-width. It is seen that the evaluation time of the reorientation of hydrogen bonds in clusters containing 50 water molecules is decreased by a factor of five to six as compared to τ of liquid water. This time, which is predetermined by the number of water molecules in a cluster and temperature, slightly varies as the cluster uptakes NO_2 molecules. Note that the estimation of τ is of an approximate character, because the peak at 220 cm^{-1} is predetermined by both the mode and the symmetric location of the tetrahedron of water molecules with frequency ω_1. Therefore, the real time τ of the clusters must be longer.
The lines located farther from the exciting line, ω = 526 and 838 cm^{-1}, are attributed to simultaneous changes in the vibrational and rotational energy states. The Raman spectra comprise overtones which, ignoring anharmonicity, represent the combinations of the exciting frequency and frequencies divisible by the frequency (~200 cm^{-1}) of vibrational-rotational transitions. In the Raman spectrum of $(H_2O)_{50}$ clusters (curve *1*), the peak at 838 cm^{-1} can be considered to be an overtone of the peak at 220 cm^{-1}. The experimental spectrum of bulk water (curve *3*) demonstrates three peaks at ω=53, 143 and 447 cm^{-1} (Murphy, 1977). According to our calculations, the first of these peaks is somewhat shifted to the left and is located at ω = 30 cm^{-1}. Curve *4* corresponds to the Raman spectrum of a pure water cluster consisting of 50 molecules and obtained in terms of the molecular dynamic model using the SPC/E rigid three-site polarizable model of a water molecule (Bosma et al., 1993). In this case, two peaks are observed, i.e., one at ω = 200 cm^{-1} that is in good agreement with our calculations, and another at 450 cm^{-1}. The latter peak appears to be somewhat shifted to the red relative to the position of the corresponding peak in the Raman spectrum that we calculated for the cluster, for which ω = 526 cm^{-1}. At frequencies higher than 550 cm^{-1}, the spectrum of the SPC/E cluster is a long damped tail, while, in the $J(\omega)$ spectrum of a modified TIP4P model, vibrations with maxima at 838 and 963 cm^{-1} are observed.
The power of scattering the energy accumulated in the systems of clusters is presented in Fig. 10(b). The rate of the energy scattering for heteroclusters containing nitrogen(IV) oxide is 2.3-fold higher than that for pure water clusters, with the most intense peaks being located at a frequency of 963 cm^{-1} in both cases. The $P(\omega)$ spectra are, in both cases, unimodal. However, in the $P(\omega)$ spectrum of the heteroclusters, a shoulder arises in the vicinity of $\omega \approx$ 700 cm^{-1}.

6. Spectral characteristics of the ozone-water cluster system

Let us consider the dielectric properties for two cluster systems: $(H_2O)_n$, $10 \leq n \leq 50$ (I') and $(O_3)_i(H_2O)_{25}$, $0 \leq i \leq 6$ (II'). The configuration of an $(O_3)_6(H_2O)_{25}$ cluster corresponding to the time moment of 30 ps is shown in Fig. 11. It can be seen that the cluster central part (skeleton) consisting of water molecules has an elongated form. Ozone molecules evenly surround the skeleton formed by water molecules and do not strive to get inside it. The orientation of ozone molecules is caused by the character of their contact with the water skeleton. An ozone molecule is bonded with one or, more rarely, two water molecules. As a rule, the side atoms of O_3 molecules are located more closely to the hydrogen atoms of H_2O molecules. A nonlinear O_3 molecule had one positive (q_{cen}) and two negative (q_{side}) charges placed at the localization points of the central and side atoms, respectively. Since the hydrogen atoms are mainly oriented outward from the water skeleton, the central atom of three of six O_3 molecules is oriented outwards. The molecular planes of two O_3 molecules are positioned tangentially to the skeleton, and the central atom of one O_3 molecule is closer to the water skeleton than its side atoms.

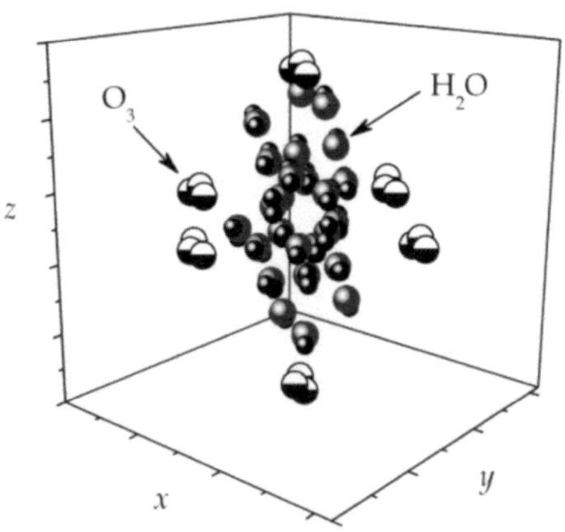

Fig. 11. Configuration of the $(O_3)_6(H_2O)_{25}$ cluster at a time moment of 30 ps.

The IR absorption spectrum $\alpha(\omega)$ of the disperse water system is appreciably changed after ozone has been absorbed (Fig. 12(a)). The main peak in the $\alpha(\omega)$ spectrum of the pure water system at a frequency of 843 cm^{-1} is transformed into two peaks corresponding to the frequencies of 690 and 970 cm^{-1}. The location of the first of these peaks coincides with that of the main peak in the IR absorption spectrum of bulk water (Goggin & Carr, 1986), and the second one is positioned close to the peak typical of the atmospheric ozone spectrum $\alpha(\omega)$ obtained by a satellite at a height of 83 km (Fichet et al., 1992). The absorption of ozone leads to a 14.1% decrease in the integral intensity of the IR absorption spectrum of the disperse water system.

The IR absorption spectra of individual $(O_3)_i(H_2O)_{25}$ clusters are illustrated in Fig. 12(b). It can be seen that their shapes and the intensities of peaks are substantially changed with addition of new ozone molecules to a water cluster. The $\alpha(\omega)$ spectra of the clusters forming system II' have two to four peaks. The integral intensity I_{tot} of the $\alpha(\omega)$ spectrum is minimal for the $(O_3)_6(H_2O)_{25}$ cluster and maximal for the $(O_3)_4(H_2O)_{25}$ aggregate. The value of I_{tot} of the clusters of system II' changes within 25.0%, and that of the water clusters, which have captured from two to five ozone molecules, becomes ~10.8% higher on average in comparison with I_{tot} of the $(H_2O)_{25}$ cluster.

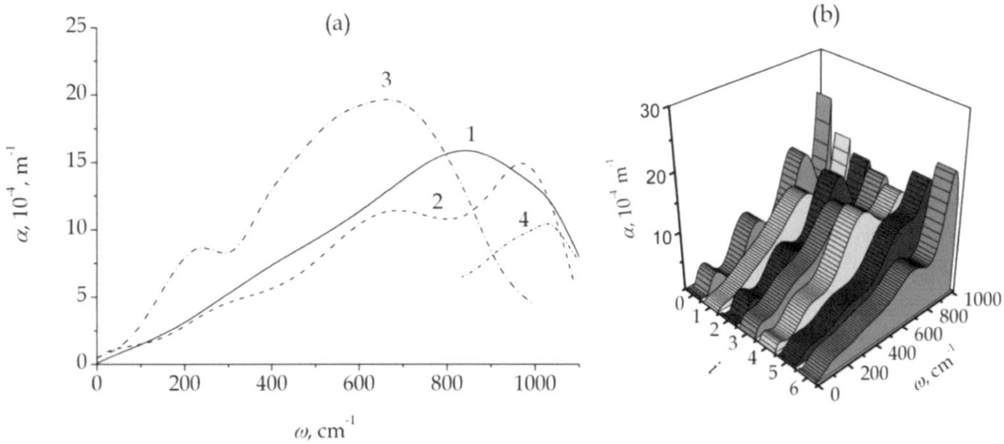

Fig. 12. (a) IR absorption spectra of (1) $(H_2O)_n$, $10 \leq n \leq 50$ and (2) $(O_3)_i(H_2O)_{25}$, $1 \leq i \leq 6$ systems; (3) the experimental $\alpha(\omega)$ function of bulk liquid water (Goggin & Carr, 1986); and (4) experimental spectrum of gaseous O_3 (Fichet et al., 1992); (b) IR absorption spectra of $(O_3)_i(H_2O)_{25}$ clusters, $0 \leq i \leq 6$.

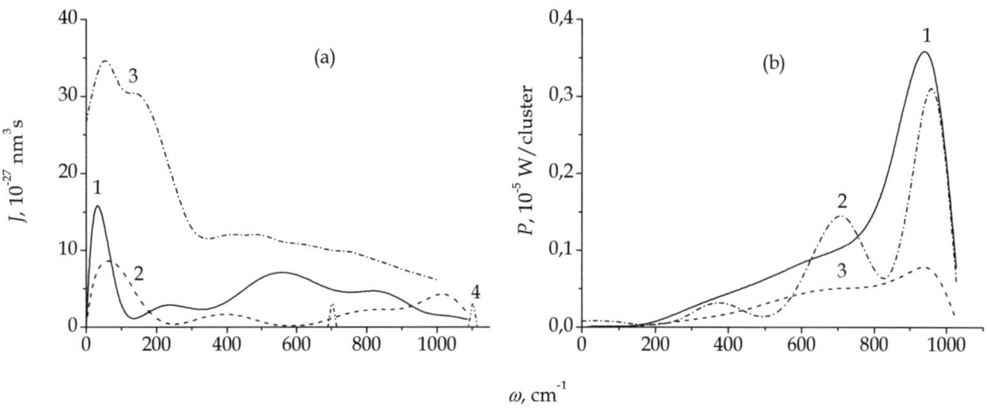

Fig. 13. (a) Raman spectra of (*1*) $(H_2O)_n$, $10 \leq n \leq 50$ and (*2*) $(O_3)_i(H_2O)_{25}$, $1 \leq i \leq 6$ systems; (*3*) bulk water (experimental) (Murphy, 1977); and (*4*) gaseous O_3 (experimental) (Andrews, L. & Spiker, 1972); (b) IR emission spectra of (*1*) $(H_2O)_n$, $10 \leq n \leq 50$; (*2*) $(H_2O)_{25}$; and (*3*) $(O_3)_i(H_2O)_{25}$, $1 \leq i \leq 6$ systems and cluster .

The difficulties in studying the Raman spectra $J(\omega)$ of ozone are connected with its high sensitivity to photodecomposition. Nevertheless, in (Andrews & Spiker, 1972), the Raman spectrum of gaseous ozone was obtained with the use of a He–Ne laser at a pressure of 0.2–0.4 MPa. It has been established that the basic $^{16}O_3$ transition frequencies ω_1 and ω_2 are 1103.3 and 702.1 cm^{-1}, respectively, and correspond to individual weak spectral lines shown in Fig. 13(a). The calculated Raman spectra of systems I' and II' and also the Raman spectrum of bulk water (Murphy, 1977) are illustrated in the same figure. It can be seen that the absorption of ozone leads to an appreciable change in the Raman spectrum of the disperse water system. The integral intensity of the Raman spectrum of system II' consisting of heterogeneous clusters was reduced by 1.8 times. The peaks at 402, 830, and 1016 cm^{-1} were formed instead of the peaks at 240, 561, and 821 cm^{-1} in the $J(\omega)$ spectrum of system I'. The main peak of the $J(\omega)$ spectrum reduced its intensity by 1.9 times and had a blue shift from 30 to 65 cm^{-1}. The combination principle consists in that the transitions with the frequencies equal to the combinations (sums or differences) of the frequencies of other transitions can be observed. Taking into account anharmonicity, the weak peak at 830 cm^{-1} can be considered as an overtone of the peak at 402 cm^{-1}. In the anti-Stokes region of the Raman spectrum of bulk water, the Raman frequency shifts correspond to 53, 143, and 447 cm^{-1} (Fig. 13(a), curve *3*). The study of a Raman spectrum makes it possible to determine the lifetime of a system in an excited state, or the radiation time τ. After absorbing a photon, a system transfers to a higher energy level, i.e., becoming excited. In the absence of external actions, an excited system loses its energy in the form of emitted photons, so its lifetime is finite. The time τ is reciprocal to the change in the collective vibration frequency (mode difference $\Delta\omega$). To estimate the upper boundary of τ, we choose the minimum $\Delta\omega$ value, which corresponds to the width of the first Raman spectrum peak at half height $\Delta\omega_1$ equal to 71.4 cm^{-1} for system I' and 133.3 cm^{-1} for system II'. Hence, the minimum radiation time τ is 0.51 ps for system I' and 0.25 ps for system II'. Therefore, the absorption of ozone molecules by the disperse water system reduces appreciably the observed photon emission time. As the lifetime τ_{cl} of the clusters studied in the given work exceeds the time of computational experiments, i.e., the clusters are not decomposed during the calculations, the radiation time τ of $(O_3)_i(H_2O)_{25}$ clusters, where $1 \le i \le 6$, is at least an order of magnitude less than τ_{cl}.

The radiation source power indicates how quickly the intensity of this radiation is changed. The density of radiation power emitted by the particles determines how clearly these particles are visible. The frequency dependence of the cluster emitted IR radiation power $P(\omega)$ is shown in Fig. 13(b). In system I', the radiation power per cluster (curve *1*) is ~1.3 times higher than the value of P for the $(H_2O)_{25}$ cluster (curve 2). The $P(\omega)$ spectrum is characterized by a single maximum at 939 cm^{-1} for system I' and by three maxima at 369, 709, and 956 cm^{-1} for the $(H_2O)_{25}$ cluster. The scattered IR radiation power is appreciably reduced in the system of $(H_2O)_{25}$ clusters with absorbed ozone molecules (system II'). In this case, the $P(\omega)$ spectrum is also characterized by a single maximum at 942 cm^{-1}. The integral intensity of the emitted radiation power for system II' is 3.4 times lower than that for system I'.

7. Absorption of carbon and nitrogen monoxides by ultradisperse aqueous system

Three systems of clusters were investigated, namely, I'' – a monodisperse system consisting of $(H_2O)_{20}$ clusters, II'' – $(CO)_i(H_2O)_{20}$ clusters, and III'' – $(NO)_i(H_2O)_{20}$ clusters, $i = 1, \ldots, 10$. In the frequency range under consideration, the absorptance of IR radiation decreases after absorption of CO molecules by water clusters and increases as a result of absorption of NO molecules by these aggregates (Fig. 14(a)). The integral intensities of IR absorption spectrum for systems I''–III'' are correlated as 1 : 0.71 :1.03. In the case of monodisperse system of $(H_2O)_n$, the $\alpha(\omega)$ spectrum has four peaks at frequencies of 103, 356, 744, and 1023 cm^{-1}. The respective spectrum for systems II'' and III'' is unimodal. For system II'', the peak falls on the

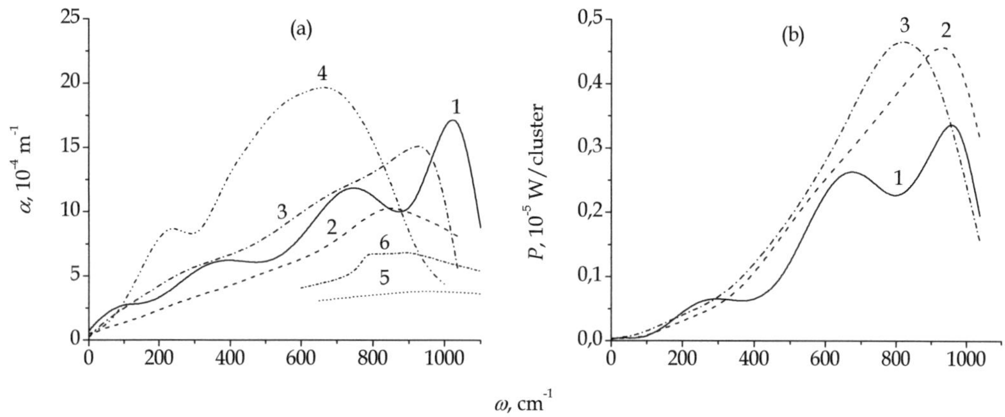

Fig. 14. (a) The absorptance of IR radiation of disperse systems: (*1*) I'', (*2*) II'', (*3*) III''; (*4*) experimentally obtained coefficient α for liquid water (Goggin & Carr, 1986); (*5, 6*) coefficient α of gaseous CO and NO, experiment (Kozintsev et al., 2003); (b) the radiation power generated under conditions of dissipation of thermal energy by systems (*1*) I'', (*2*) II'', (*3*) III''.

frequency of 845 cm^{-1}, and for system III'' - on 925 cm^{-1}. Note that the main peak of $\alpha(\omega)$ spectrum of liquid water falls on the frequency of 690 cm^{-1} (Goggin & Carr, 1986); for gaseous CO and NO, the extremely slightly pronounced peaks of deformation vibrations correspond to frequencies of 950 and 800 cm^{-1} (Kozintsev et al., 2003). The weak frequency dependence of the function $\alpha(\omega)$ of gaseous CO and NO shows up in the smoothing of the absorption spectrum of IR radiation for systems II'' and III'' with respect to system I''.

The electromagnetic radiation may be treated as the process of generation of free electromagnetic field under conditions of nonuniform motion and interaction of electric charges. The validity of this approach is supported by the fact that the field of moving electric charge is given by the sum of intrinsic field and field extending to infinitely long distances from the charge. Referred to as intrinsic field is the field which is concentrated in the vicinity of the charge and moves along with this charge. The frequency dependence of the power of radiation of systems I''–III'' is given in Fig. 14(b). The integral power of radiation increases both when the disperse aqueous system absorbs carbon monoxide and when it absorbs nitrogen monoxide. The integral intensities of emission power for systems I''–III'' are correlated as 1 : 1.36 : 1.38. A monodisperse aqueous system has maxima of power of IR radiation on frequencies of 276, 672, and 955 cm^{-1}, and systems II'' and III''—only on frequencies of 930 and 820 cm^{-1}, respectively.

8. Variation of integrated absorption during the growth of clusters

The anti-greenhouse effect can be developed due to a decrease in the number of absorbing centers upon the formation of water clusters. Figure 15(a) (curve 1) shows the relative change in the integrated IR absorption intensity I_{tot} upon the formation of the $(H_2O)_{20}$ cluster by means of successive addition of Δn water molecules to the $(H_2O)_2$ dimer. Curve 2 reflects the total relative intensity of the IR radiation absorption by the $(H_2O)_{10}$ cluster and five successively added dimers, which combine to produce the $(H_2O)_{20}$ cluster.

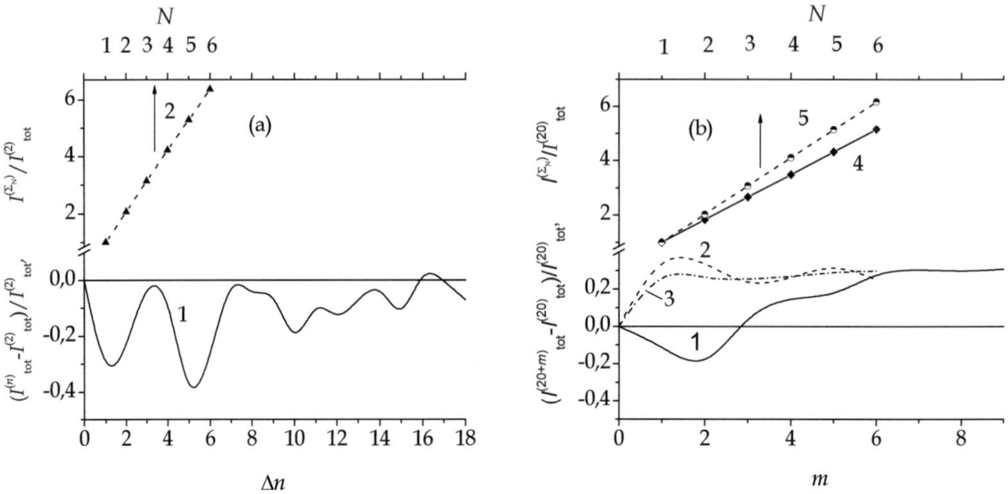

Fig. 15. (a) Relative integrated intensity of the IR absorption by (1) the water dimer that grows by adding Δn H_2O molecules and (2) total relative integrated intensity of absorption of N clusters: N_1 is $(H_2O)_{10}$ and (N_2–N_6) are $(H_2O)_2$ dimers; (b) relative integrated intensity of the IR absorption by the $(H_2O)_{20}$ cluster adding m molecules of (1) CH_4, (2) C_2H_2 and (3) C_2H_6 and total relative integrated intensity of absorption of N clusters: (4) N_1 is $(CH_4)_8(H_2O)_{10}$ and (N_2–N_6) are $(H_2O)_2$ dimers; (5) N_1 is $(CH_4)_3(H_2O)_{10}$ and (N_2–N_6) are $CH_4(H_2O)_2$ clusters.

In the case of water, the formation of the cluster leads to a twofold reduction in the greenhouse effect. The major contribution to the anti-greenhouse effect is made by the decrease in the number of absorbing centers. In addition, the buildup of the water cluster is accompanied by a decrease in its absorption capacity (curve 1 is in the range of negative values almost everywhere) due to a change in the frequency and amplitude characteristics of the total dipole moment. The introduction of hydrocarbon molecules can enhance the IR absorption by growing clusters (Figure 15(b)). The increase in I_{tot} for water cluster combined with C_2H_2 and C_2H_6 molecules is 0.2-0.4 of the I_{tot} value even at $m > 1$, whereas, when water cluster is combined with CH_4 molecules, such I_{tot} values are achieved at m > 6. When one or two CH_4 molecules are added to the $(H_2O)_{20}$ cluster, the anti-greenhouse effect is enhanced since the relative integrated intensity of the IR absorption by the heterocluster decreases. However, finally, the incorporation of hydrocarbon molecules in the water clusters reduces the greenhouse effect due to a decrease in the number of absorbing centers. The anti-greenhouse effect caused by clustering is tens of times stronger than the effect due to a change in the characteristics of vibrations of the total dipole moment of clusters caused by absorption of greenhouse gas molecules.

For estimating the contribution made by the absorption of CO and NO molecules by water clusters, it is important to know the variation of integral absorption during the growth of $X(H_2O)_n$ cluster owing to attachment of X molecules (where X = CO or NO), as well as the integral effect of absorption of IR radiation produced by sets of clusters. In so doing, the combined $X_{10}(H_2O)_n$ cluster is formed of the entire set of clusters.

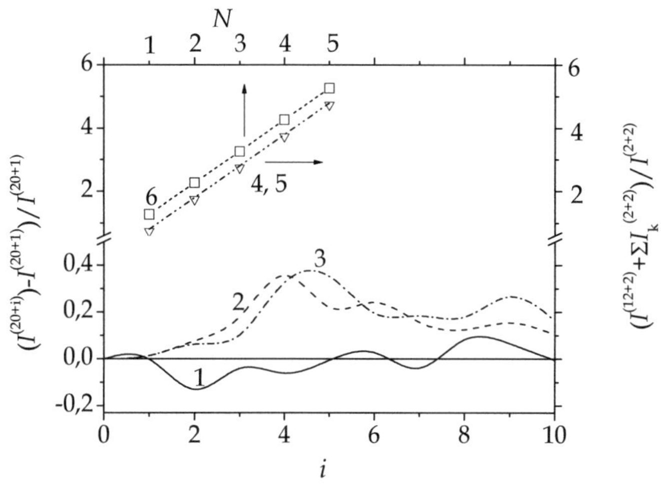

Fig. 16. Relative variation of integral intensity of absorption of IR radiation for clusters: (1) $(H_2O)_{10+i}$, (2) $(CO)_i(H_2O)_{20}$, (3) $(NO)_i(H_2O)_{20}$ as a result of attachment of *i* molecules of impurity; combined relative intensity of absorption of IR radiation, produced by *N* clusters: (4) water clusters, (5, 6) clusters of water with CO and NO molecules, respectively.

We will first consider the clustering of water molecules. If water molecules are successively added one-by-one to $(H_2O)_{10}$ cluster, the relative variation of integral intensity of absorption $(I^{(10+i)} - I^{(10)})/ I^{(10)}$ will be defined by curve 1 (Fig. 16). One can see that the quantity under consideration is a sign-variable function of *i*, i.e., each subsequent, larger, cluster may exhibit both higher and lower integral intensity of IR absorption than $(H_2O)_{10}$ cluster. The integral absorption of the resultant $(H_2O)_{20}$ cluster is 99.5% of IR absorption of the initial $(H_2O)_{10}$ cluster or 92.8% of the value of $I^{(2)}$ for water dimer.

A somewhat different pattern of variation of $(I^{(20+i)} - I^{(20+1)})/ I^{(20+1)}$ is observed in the case of attachment of *i* X (CO or NO) molecules to $X(H_2O)_{20}$ cluster. In both cases, i.e., where X = CO or NO, the reduced integral intensity of absorption increases relative to the initial value. However, this rise is not monotonic, i.e., each subsequent attachment of X molecule may make the absorptance of cluster both higher and lower than that the cluster had when its size was one X molecule less, if its previous state does not correspond to the initial one. The ratio between the combined values of $(I^{(k+i)} - I^{(k+1)})/ I^{(k+1)}$ (k = 10 for clusters of pure water and k = 20 for clusters of water with CO or NO molecules) for systems I''–III'' is 1 : 20.5 : 20.8. Therefore, the maximal impact on the variation of integral part of absorption of IR radiation caused by the variation of amplitude-frequency characteristics of total dipole moment of cluster is made by the capture of NO molecules by clusters.

We will further consider clusters of three types. Those of the first type are formed by $(H_2O)_{12}$ cluster and successively added $(H_2O)_2$ dimers of water. The dependence of combined intensity of absorption of IR radiation of successive set of clusters on the number of such clusters *N* is given by points 4. Similarly, $X_2(H_2O)_2$ complexes (miniclusters) were added to the $X_2(H_2O)_{12}$ cluster. In the case of X = CO, the dependence of intensity of absorption of IR radiation of combined cluster on the number *N* is given by points *5* which coincide with points *4*; at X = NO, the analogous function is given by points *6*. The relative combined intensity of absorption, given by points 4 and 5, increased by a factor of six as a result of variation of *N* from 1 to 5, and the analogous characteristic given by points 6 increased by a factor of four with the same variation of *N*. The integral intensity of absorption of IR radiation of $(CO)_{10}(H_2O)_{20}$ cluster is 1.16 times lower, and the value of $I^{(20+10)}$ of $(NO)_{10}(H_2O)_{20}$ cluster is 1.18 times that of the similar characteristic for $(CO)_2(H_2O)_2$ and $(NO)_2(H_2O)_2$ complexes, respectively. Of most importance here is the fact that the combined intensity of absorption of initial systems, i.e., of $(H_2O)_{12}$ cluster plus a set of *N* $(H_2O)_2$ dimers and of $X_2(H_2O)_{12}$ cluster plus a set of *N* $X_2(H_2O)_2$ complexes, is several times higher than the intensity of absorption of clusters formed of those systems. This effect will be stronger if individual molecules will be attached rather than dimers and complexes. The decrease in combined absorption due to reduction of the number of absorbing centers in the case of molecules uniting into a cluster exceeds many times over the increase in absorption associated with the variation of vibrational characteristics of the total dipole moment. This is the essence of anti-greenhouse effect caused by clustering.

The absorption of IR radiation by a disperse water system insignificantly increases when it incorporates N_2O, C_2H_6, and NO molecules and decreases when CO_2, CH_4, C_2H_2, NO_2, O_3, and CO molecules are incorporated. When methane molecules are incorporated in water clusters, the system becomes IR transparent. Formation of water clusters and subsequent incorporation of greenhouse gases in these clusters are accompanied by a decrease in the number of absorbing centers. Remarkably, the Earth's surface receives on average more radiation from the atmosphere and clouds than direct radiation from the Sun. On the whole, the absorption of greenhouse gases by the disperse water system results in the anti-greenhouse effect. This effect is latent because it can be defined in comparison with situation when clusters are absent. It appears that the same number of free molecules absorbs IR radiation more powerfully than when they are incorporated in a cluster.

The variation of relative combined power of emission upon addition of molecules to clusters is given by Fig. 17. In the case of clusters of pure water, we consider the quantity $(P^{(10+i)}-P^{(10)})/P^{(10)}$, and for water clusters containing CO or NO molecules—the ratio $(P^{(20+i)} - P^{(20+1)})/P^{(20+1)}$. Here, *P* denotes the combined power of IR radiation generated by cluster.

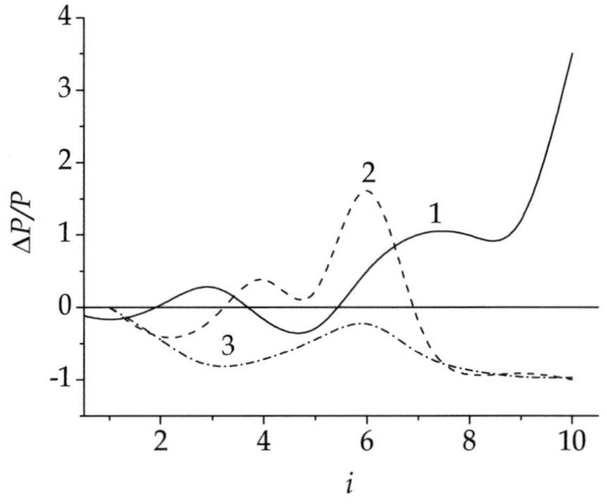

Fig. 17. Relative variation of combined intensity of emission of IR radiation for clusters: (1) $(H_2O)_{10+i}$, (2) $(CO)_i(H_2O)_{20}$, (3) $(NO)_i(H_2O)_{20}$ as a result of attachment of *i* molecules of impurity.

The first term in the superscript indicates the number of water molecules in the cluster, and the second term—the number of molecules of impurity. For clusters of system I'', water molecules are considered instead of molecules of impurity. In all cases, the dependence of relative combined intensity of emission power on the number of molecules i exhibits a fluctuating pattern. For clusters of pure water up to size $n = 16$ ($i = 6$), alternation of decrease and increase in emission power after attachment of individual molecules is observed, i.e., the quantity $(P^{(10+i)} - P^{(10)})/ P^{(10)}$ assumes both negative and positive values. However, at $i >$ 6, this quantity becomes especially positive and rapidly increases in the region of $i \geq 9$. In the case of absorption of CO molecules by clusters, the quantity $(P^{(20+i)} - P^{(20+1)})P^{(20+1)}$ at $i < 7$ may both decrease and increase; at $i \geq 7$, it decreases. When the clusters attach NO molecules, the quantity $(P^{(20+i)} - P^{(20+1)})/P^{(20+1)}$ is always negative at $i > 1$, i.e., the emission power decreases relative to its value corresponding to $NO(H_2O)_{20}$ cluster. Therefore, the power of IR radiation emitted by clusters depends on both the number and the sort of molecules of impurity absorbed by the clusters.

9. Estimation of the greenhouse and anti-greenhouse effect

Atmospheric water clusters absorb IR radiation going from the Earth, as well as free (not incorporated in clusters) molecules of greenhouse gases create a greenhouse effect. However the quantity of absorption by clusters of radiation is not proportional to the number of molecules forming it. Moreover, clusters absorb energy of IR radiation in quantity comparable, and sometimes even smaller, than that absorbed by free water molecule (Fig. 18). As a result free molecules (before they have been incorporated into cluster) make much stronger absorption of IR radiation (the total intensity $\sum I_i$), than clusters (I_{cl}). The difference of values $\sum I_i$ and I_{cl} is defined as an anti-greenhouse effect of the cluster. This value can be expressed in degrees. The estimation of the anti-greenhouse effect that exists now, created by atmospheric water clusters, can be made as follows. The distribution of the atmospheric moisture at altitude submits to the empirical Ghan dependence (Ghan et al., 1997):

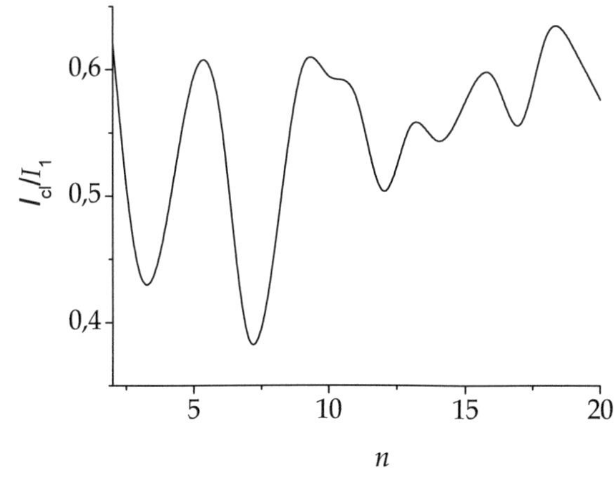

Fig. 18. The relation of integrated intensities of IR radiation absorption by water clusters, formed by n molecules, to the appropriate characteristic of a free water molecule at average surface humidity of 11 g/m³.

$$\rho = \rho_0 \times 10^{h/6.3}, \tag{22}$$

where ρ_0 is the humidity at certain altitude (as a rule, direct at the Earth's surface), h is an altitude in km. On one hand, the formula (22) can be used for calculation of the distribution of water vapor monomer above the Earth's surface, and on another hand - for definition of high-altitude distribution of total moisture in the atmosphere. In the first case, constant ρ_0 is a fraction of water monomers near to the Earth's surface. It can be established by extrapolation of the Boltzmann distribution of clusters on the sizes to the $i = 0$ value. Thus the value determined equals ~ 66.3 % from known value of humidity (~11 g/m^3) which at the surface is defined by the number of water monomers and clusters. In the second case value ρ_0 corresponds to the humidity at the altitude of 3 km, established on the basis of spectroscopic measurements (Kebabian et al., 2002). Experimental measurements are executed in the presence of clouds at the altitude from 1 up to 2 km. The spectroscopic sensitive element allows to measure spatially divided profiles of moisture density both around tropospheric clouds, and inside them. Integration on concentric layers of thickness Δh of the first and second distributions gives value of mass M_{vap} of vapor monomers and total quantity M_{tot} of moisture in the atmosphere. The mass of droplets and crystals in the atmosphere was found by the formula

$$M_{\text{drop(cryst)}} = \sum_{n=1}^{100} \left(\rho^{(n)} - \rho_{\text{sv}}^{(n)} \right) \times V_n, \tag{23}$$

where $\rho^{(n)}$ and $\rho_{\text{sv}}^{(n)}$ are densities of moisture and saturated water vapor in n layer thickness of 1 km, V_n is the volume of a layer.
We consider particles which size exceeds 0.5 μm (the minimal size of a drop observable in clouds) as drops. Water drops and ice crystals are formed in clouds when T < 273 K, but in the first instance it is necessary for air to be supersaturated in relation to water, and in the second instance it should be supersaturated in relation to ice. Curves of saturation of water vapor taking place above surfaces of water and ice are situated close enough, so both drops and ice crystals can be in the same cloud. Obviously, the full weight of ice in clouds is proportional to concentration of crystal nucleus in clouds. It was accepted, that relative concentration of crystals $c_{\text{crys}} = c'_{\text{crys}} / (c'_{\text{crys}} + c'_{\text{liq}})$ linearly changes from 0 to 1 at decrease of temperature from 273 to 230 K (Fletcher, 2011). The mass of ice crystals was determined in accordance with dependence $c_{\text{crys}}(T)$ given in Fig. 19. The $\rho_{\text{svl}}^{(n)}$ density of saturated water vapor over the supercooled water was designed as $\rho_{\text{sv}}^{(n)}$ at the definition of droplets' mass in the atmosphere, and for calculation of crystals' mass the $\rho_{\text{svc}}^{(n)}$ density of saturated water vapor over ice at temperature of n layer was used. Masses M_{drop} and M_{cryst} thus were established. The mass of clusters in the atmosphere was defined as

$$M_{\text{cl}} = M_{\text{tot}} - M_{\text{vap}} - M_{\text{drop}} - M_{\text{cryst}}. \tag{24}$$

Similarly, through the appropriate distributions, the high-altitude profile of clusters' density was defined. The designed dependences are given in Fig. 20a. It is visible that the great bulk of water clusters in the atmosphere settles up to an altitude of 2 km. Overwhelming quantity of drops is concentrated within the limits of the same altitude, and crystals are formed, from the altitude of 3 km. The ratio between quantities of monomer vapor, clusters, drops

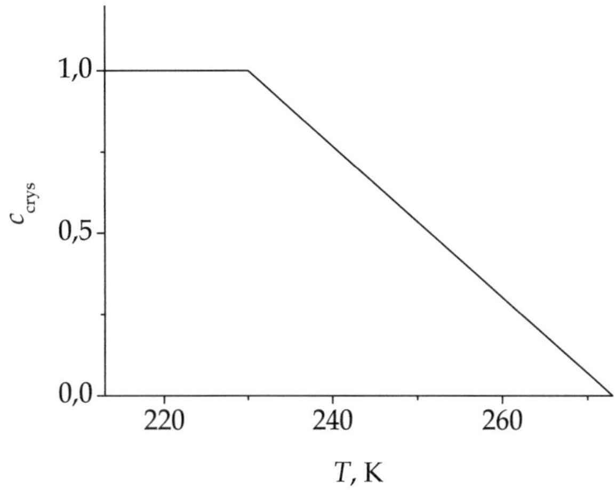

Fig. 19. The fraction of ice crystals in the condensed phase of clouds, determined in accordance with temperature dependences of ice concentration in (Fletcher, 2011).

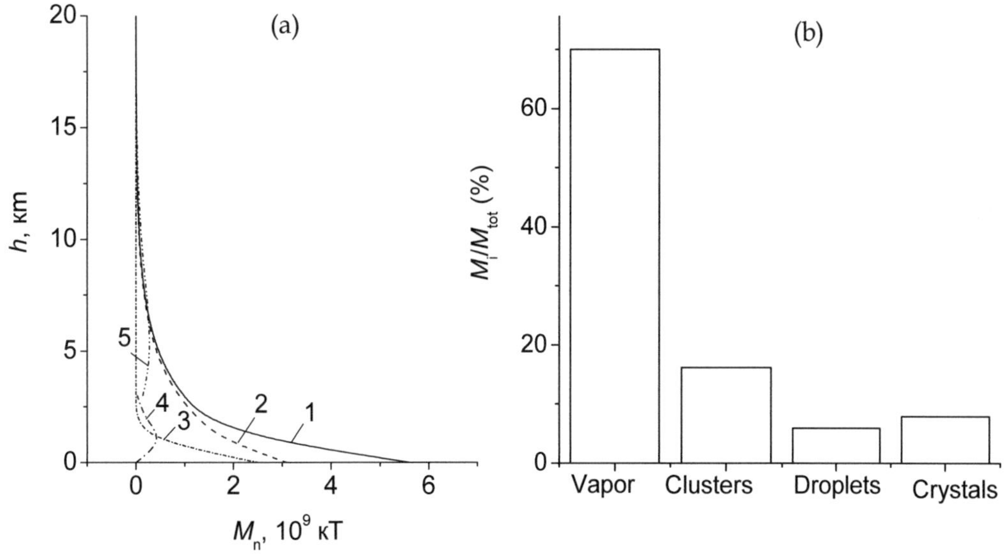

Fig. 20. (a) Contributions to a total mass of the atmosphere moisture: 1 – all moisture, 2 – monomers vapor, 3 – clusters, 4 – droplets, 5 – crystals; (b) the ratio between monomer vapor, clusters, droplets, and crystals, obtained using the linear temperature dependence for the number of crystals in accordance with (Fletcher, 2011).

and crystals looks as 70.0 : 16.2 : 5.9 : 7.9, if total magnitude of the moisture in the atmosphere is 100 % (Fig. 20b). On average, the integrated intensities of IR radiation absorption of water clusters are lower than the intensity of separate molecules. According to the estimation the number of clusters in the atmosphere at time 2.38 is less than the number of molecules forming them. Created by clusters and corresponding free molecules greenhouse effects are ~ 1.0 and 4.3 K accordingly. Thus, the average temperature of the planet could rise by 4.3-1.0 =3.3 K in the absence of clusters, that would result in a significant change of climate. During the last 100 years the average global temperature of the Earth has gone up by 0.6 K (Halmann & Steinberg, 1999).

10. Conclusion

As the temperature of the atmosphere rises, more water is evaporated from ground storage (rivers, oceans, reservoirs, soil). Because the air is warmer, the absolute humidity can be higher, leading to more water vapor in the atmosphere. In other words, the air is able to "hold" more water when it's warmer. As a greenhouse gas, the higher concentration of water vapor is then able to absorb more thermal IR energy radiated from the Earth, thus further warming the atmosphere. The warmer atmosphere can then hold more water vapor and so on. However as water vapor increases in the atmosphere, more of it will eventually also condense into clouds, which are more able to reflect incoming solar radiation and thus allowing less energy to reach the Earth's surface and heat it up.

Besides water vapor, many other feedback mechanisms operate in the Earth's climate system and impact the sensitivity of the climate response to an applied radiative forcing. The relative strengths of feedback interactions between clouds, aerosols, snow, ice, and vegetation, including the effects of energy exchange between the atmosphere and ocean, create a great influence on climate. For the greenhouse effect to work efficiently, the Earth's atmosphere must be relatively transparent to sunlight at visible wavelengths so that significant amounts of solar radiation can penetrate to the ground. Also, the atmosphere must be opaque at thermal wavelengths to prevent thermal radiation emitted by the ground from escaping directly to space. Without the presence of water vapor, carbon dioxide, and other gases in the atmosphere, too much heat would escape and the Earth would be too cold to sustain life. Carbon dioxide, methane, nitrous oxide, and other greenhouse gases absorb the infrared radiation rising from the Earth and hold this heat in the atmosphere instead of reflecting it back into space. Although greenhouse gases make up only about 1 percent of the Earth's atmosphere, they regulate our climate by trapping heat and holding it in a kind of warm-air blanket that surrounds the planet.

Until recently there is no complete picture of all the factors influencing climate change. The reliability of forecasts depends on an understanding of cycles that result in climate change. In the present work we have shown that atmospheric water clusters are also capable of influencing the climate. Accumulation of CO_2 in the atmosphere has increased since 1900. By 2000 the volumetric concentration of CO_2 has increased by 1.25 times. The anti-greenhouse effect created by clusterization of the atmospheric water vapor limits the growth of global temperature.

When water evaporates from the surface of the Earth, it cools the surface. This keeps the surface from getting too hot. The total cooling effect of water evaporation from the surface of seas and reservoirs, is 13.4 K. Transition of the part of atmospheric water vapor into clusters causes a decrease of atmospheric temperature by 3.3 K, that is the effect of 24.6% cooling of the heat received during water evaporation. The thermal energy of the lower atmosphere reaches the upper atmosphere by radiative transfer, thermal conduction, convection and by the release of energy when water vapor condenses. From the higher atmosphere radiative transfer becomes overwhelmingly predominant as the loss of energy to space can only take place radiatively. The potential absorption by the combination of water vapor and CO_2 is almost complete due to the logarithmic relationship between concentration and radiance/absorption. Thus, the content of CO_2 in the atmosphere of the Earth is not an absolute criterion of the efficiency of the greenhouse effect made by this gas which now is estimated at 9.3 K, i.e. constitutes about 19.6% of the effect given by all greenhouse gases. In the absence of reliable thermal protection from the ozone layer of the lower stratosphere and depletion of atomic oxygen in the thermosphere, CO_2 can partially or completely carry out the functions of anti-greenhouse gas.

The bulk of clusters are concentrated up to the altitude of 2 km, and the droplets are largely located at altitudes up to 2.5 km. The crystals are, as a rule, located above 3 km. Water vapor is most widespread among other components of atmospheric moisture. Water clusters are injected into the upper atmosphere during flights of space vehicles. Such clusters may exist for a long time; however, their fraction is small at present, and the greenhouse effect they produce is low.

It is well known that condensed phase reactions occurring in the Earth's atmosphere play a role in atmospheric phenomena, for example, in the formation of the Antarctic ozone hole. The presence of the XH_2O complexes (X is a molecule of greenhouse gases) in the Earth's atmosphere is confirmed both experimentally and theoretically. Weakly bound complexes, considered as precursors to the condensed phase, are important because perturbations and interactions between the monomer units can alter the spectroscopy and reactivity compared with the constituent molecules. Water is a major component in the absorption of radiation in the atmosphere and is a key component in the Earth's radiative balance. The significant abundances of H_2O and high absorbability of greenhouse gases (including water) in the atmosphere could affect the Earth's climate.

Water vapour is itself too strong a greenhouse gas. However the greenhouse effect will decrease at presence of the clusters formation in water vapour instead of to amplify. The increase of the water vapour content in atmosphere at the present condition of an atmosphere results in amplification of a greenhouse effect, but not in direct ratio to quantity added vapour. Now clusters are formed and they reduce a greenhouse effect. However there can come the moment of such warming in atmosphere, that clusters will not be formed. It will result to significant jump of global temperature and its further very fast growth, i.e. to catastrophe.

Clouds modulate Earth's radiation and water balances, namely:

clouds cool Earth's surface by reflecting incoming sunlight;

clouds warm Earth's surface by absorbing heat emitted from the surface and re-radiating it back down toward the surface; clouds warm or cool Earth's atmosphere by absorbing heat emitted from the surface and radiating it to space; clouds warm and dry Earth's atmosphere and supply water to the surface by forming precipitation;

clouds are themselves created by the motions of the atmosphere that are caused by the warming or cooling of radiation and precipitation.

If the climate should change, then clouds would also change, altering all of the effects mentioned above. What is important is the sum of all these separate effects, the net radiative cooling or warming effect of all clouds on Earth. If Earth's climate should warm due to the greenhouse effect , the weather patterns and the associated clouds would change. However it is not known whether the resulting cloud changes would diminish the warming (a negative feedback) or enhance the warming (a positive feedback). Moreover, it is not known whether these cloud changes would involve increased or decreased precipitation and water supplies in particular regions. Improving our understanding of the role of clouds in climate is crucial to understanding the effects of global warming.

11. References

Agmon, N. (1996) Tetrahedral Displacement: The Molecular Mechanism behind the Debye Relaxation in Water. *Journal of Physical Chemistry*, Vol. 100, No.3, pp. 1072–1080, ISSN 0022–3654.

Akhmatskaya, E.V.; Apps, C.J.; Hillier, I.H.; Masters, A.J.; Watt, N.E. & Whitehead, J.C. Formation of H_2SO_4 from SO_3 and H_2O, Catalysed in Water Clusters. *Chemical Communications*, 1997, Issue 7, pp. 707–708, ISSN 1364–548X.

Allouch, A. (1999) Quantum Study of Acetylene Adsorption on Ice Surface. *Journal of Physical Chemistry A*, Vol. 103, No.45, pp. 9150–9153, ISSN 1538–6414.

Andrews, L. & Spiker, R.C. (1972) Argon Matrix Raman and Infrared Spectra and Vibrational Analysis of Ozone and the Oxygen_18 Substituted Ozone Molecules. *Journal of Physical Chemistry*, Vol. 76, No. 22, pp. 3208–3213, ISSN 0022–3654.

Barrett, J. (2005) Greenhouse Molecules, Their Spectra and Function in the Atmosphere. *Energy & Environment*, Vol.16, No.6, pp. 1037-1045, ISSN 0958–305X.

Benedict, W.S.; Gailar, N. & Plyler, E.K. (1956) Rotation–Vibration Spectra of Deuterated Water Vapor. *Journal of Chemical Physics*, Vol.24. No.6, pp. 1139-1165, ISSN 0021–9606.

Bignell, J. (1970) The Water Vapour Infrared Continuum. *Quarterly Journal of the Royal Meteorological Society*, Vol.96, Issue 409, 390–403, ISSN 1477–870X.

Bosma, W.B.; Fried, L.E. & Mukamel, S.J. (1993) Simulation of the Intermolecular Vibrational Spectra of Liquid Water and Water Clusters, *Journal of Chemical Physics*, Vol.98, No.6, pp. 4413-4421.

Bresme, F. (2001) Equilibrium and Nonequilibrium Molecular-Dynamics of the Central Force Model of Water. *Journal of Chemical Physics*, V.115, No. 16, pp. 7564–7574, ISSN 0021–9606.

Carlon, H.R. (1979) Do Clusters Contribute to the Infrared Absorption Spectrum of Water Vapor? *Infrared Physics*, Vol.19, No.5, pp. 549–557, ISSN: 0020-0891.

Chukanov, V. N. & Galashev, A. Y. (2008) Cluster Mechanism of the Anti-Greenhouse Effect. *Doklady Physical Chemistry*, Vol.421, Part 2, pp. 226–229, ISSN 0012-5016.

Coffey, M.T. (1977) Water Vapour Absorption in 10-12 μm Atmospheric Window. *Quarterly Journal of the Royal Meteorological Society*, Vol.103, Issue 438, pp. 685–692, ISSN 1477–870X.

Dang, L.X. & Chang, T.M. (1997) Molecular Dynamics Study of Water Clusters, Liquid and Liquid–Vapor Interface of Water with Many-Body Potentials. *Journal of Chemical Physics*, Vol.106, No.19, pp. 8149-8159, ISSN 0021–9606.

Dias-Lalcaca, P.; Packham N.J.C. & Gebbie H.A. (1981) The Effect of Ultraviolet Radiation on Water Vapour Absorption Between 5 and 50cm^{-1}. *Infrared Physics*, Vol.24, No.5, pp. 437-441, ISSN: 0020-0891.

Essenhigh, R.H. (2009) Potential Dependence of Global Warming on the Residence Time (RT) in the Atmosphere of Anthropogenically Sourced Carbon Dioxide. Energy & *Fuels*, Vol. 23, No.5, pp. 2773–2784, ISSN: 0887-0624.

Feller, D. & Dixon, D.A. (1996) The Hydrogen Bond Energy of the Trimer. *Journal of Physical Chemistry*, Vol.100, No.8, pp. 2993–2997, ISSN 0022-3654.

Fichet, P., Jevais, J.R., Camy-Peyret, C. & Flaud, J.M. (1992) Calculation of NLTE Ozone Infrared Spectra, *Planetary Space Science*, Vol. 40, No.9, p. 1289–1297, ISSN 0032–0633.

Fletcher, N.H. (2011) *The Physics of Rainclouds*. University Press, ISBN-13:9780521154796, Cambridge, UK.

Frish, M.J.; Pople, J.P. & Del Bene, J.E. (1983) Hydrogen Bonds Between First-Row Hydrides and Acetylene. *Journal of Chemical Physics*, Vol.78, No.6, pp. 4063–4065, ISSN 0021-9606.

Galashev, A.Y.; Rakhmanova, O.R. & Chukanov, V.N. (2005) Absorption and Dissipation of Infrared Radiation by Atmospheric Water Clusters, *Russian Journal of Physical Chemistry A*, Vol.79, No.9, pp. 1455-1159, ISSN 0036-0244.

Galashev, A. Y.; Chukanov, V. N.; Novruzov, A. N. & Novruzova, O. A. (2006a) Molecular-Dynamic Calculation of Spectral Characteristics of Absorption of Infrared Radiation by $(H_2O)_j$ and $(CH_4)_i(H_2O)_n$ Clusters. *High Temperature*, Vol. 44, No.3, pp. 364–372, ISSN 0018-151X.

Galashev, A. Y.; Rakhmanova, O. R.; Galasheva, O. A. & Novruzov, A. N. (2006b) Molecular Dynamics Study of the Greenhouse Gases Clusterization. *Phase Transitions*, Vol.79, No.11, pp. 911–920, ISSN 0141-1594.

Galashev, A. Y.; Chukanov, V. N.; Novruzov, A. N. & Novruzova, O. A. (2007) Simulation of Dielectric Properties and Stability of Clusters $(H_2O)_i$, $CO_2(H_2O)_i$, and $CH_4(H_2O)_i$. *Russian Journal of Electrochemistry*, Vol.43, No.2, pp. 136–145, ISSN 1023-1935.

Galashev, A. Y. (2011) Greenhouse Effect of Clusterization of CO_2 and CH_4 with Atmospheric Moisture. *Environmental Chemistry Letters*, Vol.9, No.1, pp. 37–41, ISSN 1610-3653.

Galasheva, A. A.; Rakhmanova, O. R.; Novruzov, A. N. & Galashev, A. Y. (2007) Spectral Effects of the Clusterization of Greenhouse Gases: Computer Experiment. *Colloid Journal*, Vol.69, No.1, pp. 56–65, ISSN 1061-933X.

Gear, C.W. (1971) The Automatic Integration of Ordinary Differential Equations, *Communications of the* ACM, Vol.14, Issue 3, pp. 176-179, ISSN 0001-0782.

Ghan, S.J.; Leung, L.R.; Easter, R.C. & Abdul-Razzak, H. (1997) Prediction of Cloud Droplet Number in a General Circulation Model, *Journal of Geophysical Research,* Vol.102, D18, pp. 777–794, ISSN 0148-0227.

Goggin, P.L. & Carr, C. (1986) Far Infrared Spectroscopy and Aqueous Solutions, In: *Water and Aqueous Solutions,* Vol. 37, G.W. Neilson & J.E. Enderby, (Ed.), 149-161, Adam Hilger, ISBN 0-85274-576-1, Bristol, USA.

Goss, L.M.; Sharpe, S.W.; Blake, T.A.; Vaida, V. & Brault, J.W. (1999) Direct Absorption Spectroscopy of Water Clusters. *Journal of Physical Chemistry A 103,* No.43, 1999, 8620-8624, ISSN 1538-6414.

Günzler, H. & Gremlich, H.-U. (2002) *IR Spectroscopy. An Introduction,* Wiley-VCH, ISBN 3-527-28896-1, Weinheim, Germany.

Halle, B. & Karlstrom, G. (1983) Prototropic Charge Migration in Water. Part 1.—Rate Constants in Light and Heavy Water and in Salt Solution from Oxygen-17 Spin Relaxation. *Journal of the Chemical Society, Faraday Transactions,* Vol. 79, No.7, pp. 1031–1046, ISSN 0956-5000.

Halmann, M.M. & Steinberg, M. (1999) *Greenhouse Gas Carbon Dioxide Mitigation. Science and Technology,* Lewis publishers, ISBN 1-56670-284-4, Roca Raton, London, New York, Washington, USA.

Jorgensen, W.L. (1981) Transferable Intermolecular Potential Functions for Water, Alcohols and Ethers. Application to Liquid Water. *Journal of American Chemical Society*, Vol.103. No.2, pp. 335-340, ISSN 0002–7863.

Kebabian, P.L.; Kolb, C.E. & Freedman, A. (2002) Spectroscopic Water Vapor Sensor for Rapid Response Measurements of Humidity in the Troposphere. *Journal of Geophysical Research*, Vol.107, D23, 4670, 14 pp, ISSN 0148-0227.

Kirov, M.V. (1993) F-structure of polyhedral water clusters, *Journal of Structural Chemistry*, Vol. 34, No. 4, pp. 557-561, ISSN 0022–4766.

Kozintsev, V.I.; Belov, M.L.; Gorodnichev, V.A. & Fedotov, Yu.V. (2003) *Lazernyi Optiko-Akusticheskii Analiz Mnogokomponentnykh Gazovykh Smesei [Laser Optical Acoustic Analysis of Multicomponent Gas Mixtures]*, Izd. MGTU im. N.E. Baumana, ISBN 5–7038–2134–7, Moscow, Russia.

Landau, L.D. & Lifshits E. M. (2001) *Elektrodinamika Sploshnykh Sred [Electrodynamics of Continuous Media]*, Vol. 8, Fizmatlit, ISBN 5–9221–0125–0, Moscow, Russia.

Lee, A.C.L. (1973) A Study of the Continuum Absorption Within the 8-13 μm Atmospheric Window. *Quarterly Journal of the Royal Meteorological Society*. Vol.99, Issue 421, pp. 490–505, ISSN 1477–870X.

Levanuk, A.P. & Sannikov, D.G. (1988) Dielectric Loses, In: *Fizicheskaya Entsiklopediya [Physical Encyclopedia]*, Vol. 1, A.M. Prokhorov, (Ed.), 702, Sovetskaya Entsiklopediya, ISBN 5–85270–034–7, Moscow, USSR.

Lide D.R. (Ed.) (1996) *CRC Handbook of Chemistry and Physics*. 77th edition, CRC Press, ISBN-13: 978-0849304774, Boca Ration-New York-London-Tokyo, USA.

Maheshwary, S.; Patel, N.; Sathyamurthy, N.; Kulkarni, A.D. & Gadre, S.R. (2001) Structure and Stability of Water Clusters $(H_2O)_n$, n=8–20: An Ab initio Investigation. *Journal of Physical Chemistry A*, Vol.105, No.46, pp. 10525–10537, ISSN 1538–6414.

Murphy, W.F. (1977) The Rayleigh Depolarization Ratio and Rotational Raman Spectrum of Water Vapor and the Polarizability Components for the Water Molecule, *Journal of Chemical Physics*, Vol. 67, No.12, pp. 5877– 5882, ISSN 0021–9606.

Neumann, M. (1985) The Dielectric Constant of Water. Computer Simulations with the MCY Potential, *Journal of Chemical Physics*, Vol.82, No.12, pp. 5663-5672, ISSN 0021–9606.

Neumann, M. (1986) Dielectric Relaxation in Water. Computer Simulations with the TIP4P Potential. *Journal of Chemical Physics*, Vol. 85, No. 3, pp. 1567–1679, ISSN 0021–9606.

Novruzov, A. N.; Chukanov, V. N.; Rakhmanova O. R. & Galashev, A. Y. (2006) A Computer Study of the Absorption of Infrared Radiation by Systems of Molecular Clusters. *High Temperature*, Vol.44, No.6, pp. 932–940, ISSN 0018–151X.

Novruzov, A.N.; Rakhmanova, O.R.; Novruzova, O.A. & Galashev, A.Y. (2008a) The Structure of Water Clusters Interacting with Gaseous Acetylene. *Russian Journal of Physical Chemistry B*, Vol. 2, No. 1, pp. 115–122, ISSN 1990-7931.

Novruzov, A.N.; Rakhmanova, O.R. & Galashev, A.Y. (2008b) Computer Simulation of the Structure of Water Clusters Containing Absorbed Ethane Molecules. *Colloid Journal*, Vol. 70, No. 1, pp. 64–70, ISSN 1061–933X.

Novruzova, O.A.; Galasheva, A.A. & Galashev, A.Y. (2007a) IR Spectra of Aqueous Disperse Systems Adsorbed Atmospheric Gases: 1. Nitrogen. *Colloid Journal*, Vol.69, No.4, pp. 474-482, ISSN 1061–933X.

Novruzova, O.A.; Galasheva, A.A. & Galashev, A.Y. (2007b) IR Spectra of Aqueous Disperse Systems Adsorbed Atmospheric Gases: 2. Argon. *Colloid Journal*, Vol.69, No.4, pp. 483-491, ISSN 1061-933X.

Novruzova, O.A. & Galashev, A.Y. (2008) Numerical Simulation of IR Absorption, Reflection, and Scattering in Dispersed Water-Oxigen Media. *High Temperature*. Vol.46, No.1, pp. 60-68, ISSN 0018-151X.

Petrov, K.G. & Tikhonov A.A. (2002) Equations of rotary motion of a rigid body based on the utilization of quaternion parameters. Mechanics of solid, Vol.37, No.3, pp. 1-12, ISSN 0025-6544.

Platt, U. & Stutz, J. (2008) *Differential Optical Absorption Spectroscopy. Principles and Applications*. Springer, ISBN 978-3-540-21193-8, Verlag-berlin-Heidelberg, Germany.

Poulet, H.& Mathieu, J.-P. (1970) *Specters de Vibration et Symetrie des Cristaux*, Gordon & Breach, ISBN 0677501803, New York, USA.

Przhibel'sky S.G. (1994) The molecular scattering of light, In: *Fizicheskaya Entsiklopediya [Physical Encyclopedia]*, Vol. 4, A.M. Prokhorov, (Ed.), 280-283, Sovetskaya Entsiklopediya, ISBN5-85270-087-8, Moscow, Russia.

Shillings, A. J. L.; Ball, S. M.; Barber, M. J.; Tennyson, J. & Jones R. L. (2010) A Upper Limit for Water Dimer Absorption in the 750nm Spectral Region and a Revised Water Line List. *Atmospheric Chemistry and Physics Discussions*, Vol.10, No.5, pp. 23345-23380, ISSN 1680-7316.

Silva, S.C. & Devlin, J.P. (1994) Interaction of Acetylene, Ethylene, and Benzene with Ice Surfaces. *Journal of Physical Chemistry*, Vol.98, No.42, pp. 10847-10852, ISSN 0022-3654.

Smith, D.E. & Dang, L.X. (1994) Simulations of NaCl Association in Polarizable Water. *Journal of Chemical Physics*, Vol.100, No.5, pp. 3757-3762, ISSN 0021-9606.

Sonnenschein, R. (1985) An Improved Algorithm for Molecular Dynamics Simulation of Rigid Molecules. *Journal of Computational Physics*, Vol.59, No.2, pp. 347-350, ISSN 0021-9991.

Spackman, M.A. (1986a) Atom-Atom Potentials via Electron Gas Theory. *Journal of Chemical Physics*, Vol.85, No.11, pp. 6579-6585, ISSN 0021-9606.

Spackman, M.A. (1986b) A Simple Quantitative Model of Hydrogen Bonding. *Journal of Chemical Physics*, Vol.85, No.11, pp. 6587-6601, ISSN 0021-9606.

Sremaniak, L.S.; Perera, L. & Berkowitz, M.L. (1996) Cube to Cage Transitions in $(H_2O)_n$ (n=12, 16, and 20), *Journal of Chemical Physics*, Vol.105, No.9, pp. 3715-3721, ISSN 0021-9606.

Stern, H.A. & Berne, B.J. (2001) Quantum Effects in Liquid Water: Pathintegral Simulations of a Flexible and Polarizable Ab initio Model. *Journal of Chemical Physics*, Vol.115, No.16, pp. 7622-7628, ISSN 0021-9606.

Tsai C.J. & Jordan, K.D. (1993) Theoretical Study of Small Water Clusters: Low-Energy Fused Cubic Structures for $(H_2O)_n$, n=8, 12, 16, and 20. *Journal of Physical Chemistry*, Vol.97, No. 20, pp. 5208-5210, ISSN 0022-3654.

Vaida, V.; Daniels, J.S.; Kjaergaard, H.G.; Goss, L.M. & Tuck, A.F. (2001) Atmospheric Absorption of Near Infrared and Visible Solar Radiation by the Hydrogen Bonded Water Dimmer, *Quarterly Journal of the Royal Meteorological Society*, Vol.127, Issue 575, pp. 1627-1643, ISSN 1477-870X.

Wales, D.J. & Ohmine, I. (1993) Structure, Dynamics, and Thermodynamics of Model $(H_2O)_8$ and $(H_2O)_{20}$ Clusters. *Journal of Chemical Physics*, Vol.98, No.9, pp. 7245–7256, ISSN 0021–9606.

Wilson, E.B.Jr.; Decius, J.C. & Cross, P.C. (1980) *Molecular Vibrations. The Theory of Infrared and Raman Vibrational Spectra*, Dover Publications, ISBN 0–486–63941–X, New York, USA.

Wolynes, P.G. & Roberts, R.E. (1978) Molecular Interpretation of the Infrared Water Vapour Continuum. *Applied Optics*, Vol.17, No.10, pp. 1484–1485, ISSN 0003–6935.

Xantheas, S. (1996) The Hamiltonian for a Weakly Interacting Trimer of Polyatomic Monomers. *Journal of Chemical Physics*, Vol.104. No.21. pp. 8821–8824, ISSN 0021–9606.

Xie, F.; Tian, W. & Chipperfield, M.P. (2008) Radiative Effect of Ozone Change on Stratosphere-Troposphere Exchange. *Journal of Geophysical Research*, Vol.113, D00B09, 15 pp., ISSN 0148–0227.

9

California Coastal-Cooling a Reverse Reaction from Global Warming General Circulation and Mesoscale Effects

Bereket Lebassi-Habtezion[1], Jorge González[2] and Robert Bornstein[3]
[1]*Department of Environmental Earth System Science*
Stanford University, Stanford, California
[2]*Department of Mechanical Engineering, the City College of New York, New York*
[3]*Department of Meteorology, San José State University, San José, California*
USA

1. Introduction

California complex-topography coastal-areas are influenced by mesoscale coastal flows, initiated by near-surface (herein referred to as "surface") temperature differences between land and ocean that produce onshore daytime sea breezes and off-shore nighttime land breezes (Pielke 1984). Related phenomena include return flows aloft, MBLs (all acronyms are defined in Appendix) and associated inversions, LLJs, upslope and down slope winds, and topographic channeling (Mahrer and Pielke 1977). These phenomena depend on time of year, latitude, SSTs, PBL-depth and -stability, and factors that alter land-surface energy balances (e.g., clouds, land use, albedo [α], and soil moisture). SoCAB sea breezes have been extensively documented, e.g., a shallow coastal MBL that deepens inland (Edinger 1959) and cool marine air that warms as it advances inland, until its frontal inversion weakens and finally erodes (Edinger 1963).

These SoCAB mesoscale phenomena have also been modeled, e.g., Ulrickson and Clifford (1990) found that summer daytime upslope flows ventilate the Basin by merging with and strengthening sea breeze flows. This flow converges at inland mountain tops, producing strong vertical motions, which create return offshore-directed flows. Polluted air is prevented from exiting the SoCAB (in the pass between the San Gabriel and San Bernardino Mountains) by opposing upper level easterly-flows associated with mesoscale highs north of that gap (Boucouvala et al. 2003).

Observational analyses have indicated that cooling of JJA maximum temperature (T_{max}) values) in coastal California over the last several decades could be due to increased: irrigation (Bonfils and Lobell 2007), coastal upwelling (Bakun 1990), stratus cover (Nemani et al. 2001), and/or urban cool-islands (LaDochy 2007). Lebassi et al. (2009), however, showed that this cooling from 1970-2005 was due to a "reverse reaction" to GHG induced-warming of inland areas, which increases sea breeze activity that overwhelms the warming in low-elevation SFBA and SoCAB coastal areas (see Fig. 1 for SoCAB). Irrigation is not a factor in these coastal areas, but increased upwelling, cold air advection, and stratus changes probably contribute to this observed cooling. Their results for all 253 California COOP sites together showed increased

(0.15°C decade^{-1}) T_{ave}-values; asymmetric warming, as minimum temperature (T_{min}) values increased faster than T_{max}-values (0.27 vs. 0.04°C decade^{-1}); and thus decreased DTR values (-0.23°C decade^{-1}). While JJA nighttime SoCAB and SFBA minimum temperatures T_{min} showed expected GHG-induced warming, their T_{max}-values exhibited a complex spatial pattern, with cooling (-0.30°C decade^{-1}) in low-elevation coastal-areas open to marine air penetration and warming (0.32°C decade^{-1}) at inland areas, which suggests that the inland warming resulted in increased coastal sea-breeze activity. Lebassi et al. (2010) showed that per capita energy consumption for summer cooling in the two coastal-cooling areas decreased during the period, while it increased in inland warming areas.

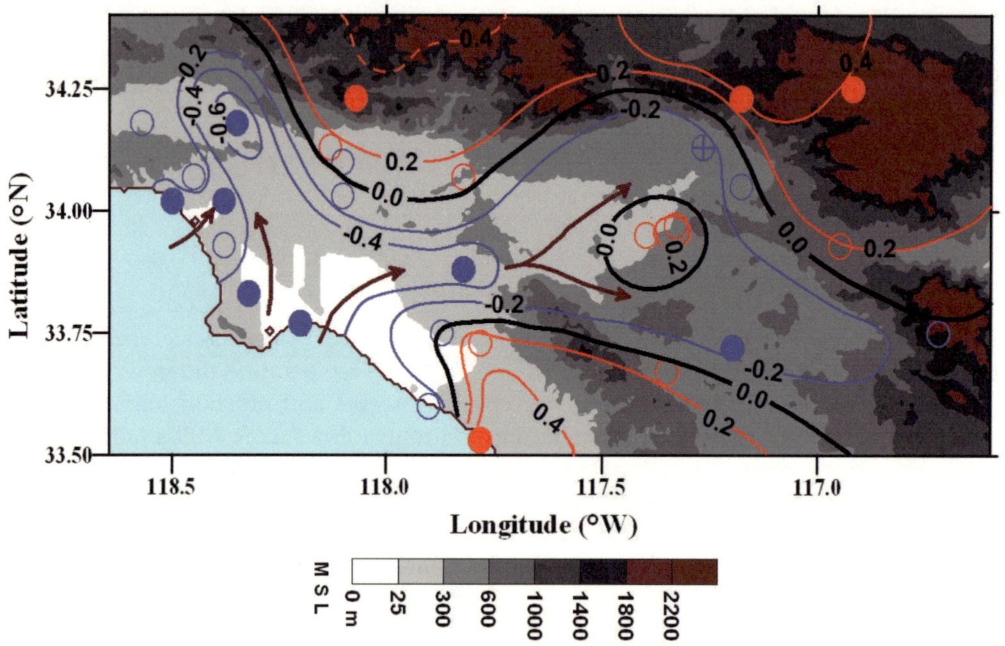

Fig. 1. Spatial distribution of observed SoCAB 2-m summer maximum-temperature trends (°C decade-1) for 1970-2005; arrows indicate predominant summer daytime surface-flow patterns; blue, red, and black colors indicate cooling, warming, and no-change isopleths and station locations; dashed isopleths are extrapolated; and statistical significance values of >99%, between 95 and 99%, between 90 and 95%, and < 90% are represented, respectively, by full-colored, half-colored, plus sign in circles, and open circles; from [Lebassi et al., 2009].

GCM simulations by Cayan et al. (2008) project a total 21st century warming of 2-5°C for California surface temperatures. Downscaled regional climate modeling by Kueppers et al. (2007) showed western-US irrigation lowering average temperature (T_{ave}) and T_{max} values at rates comparable to the increases from GHG warming. Similar modeling by Miller et al. (2003) showed California T_{ave} warming rates of 2.1-4.5°C, while ensemble downscaling by Maurer (2007) of 1950-99 JJA median temperatures showed warming rates that decreases from 0.13°C decade^{-1} in inland California to 0.08°C decade^{-1} at coastal areas north of the SoCAB. While this showed coastal influences, its coarse spatial resolution could not sufficiently resolve local topographic features that affect mesoscale circulations and thus local temperature trends. Local modeling by Lobell et al. (2006) showed California temperature changes generally determined by GHG warming, but with large land-use changes locally dominant.

Mesoscale climate-change modeling by Snyder et al. (2003) and Duffy et al. (2006) showed increased GHG-warming strengthens onshore pressure and temperature gradients, as land areas warm faster than ocean areas due to thermal differences. Thus enhance alongshore winds strengthened upwelling, which further increased onshore temperature gradients. McGregor et al. (2007) observed this effect over coastal northwest Africa, while Bakun (1990) hypothesized a similar scenario to explain observed 30 year increases of California upwelling.

IPCC (2001) annual temperature trends for 1976-2001 do, in fact, show such cooling (up to 0.6°C decade^{-1}) at all the west coast areas listed above, except Portugal. Recent local observational studies have also detected coastal cooling, e.g., Falvey et al. (2009) analyzed 1979-2006 NCDC and gridded observed (annual-average and maximum) temperature trends over coastal Chile, and found coastal-cooling. Coastal sediment-cores along Peru, analysed by Gutiérrez et al. (2011), also showed cooling SSTs for the latter part of the 20th century.

Oglesby et al. (2010) analyzed 4 km WRF modeled-differences (between 2000-2004 and 2050-2054), forced with CCSM output, for Meso-America; coastal cooling was predicted along the Pacific coasts of Mexico and Central America. This result indicates that the currently discovered SoCAB coastal cooling might likewise continue over the next 50 years, although a similar series of WRF simulations by Zhao et al. (2011) for all of California did not show coastal-cooling, perhaps because they forced their simulations with output from the older Parallel Climate Model (PCM) model.

While previous studies generally attributed observed JJA asymmetric warming in coastal California during the last three decades to increases in cloud cover, SSTs, upwelling, land use changes (i.e., urbanization, and/or irrigation); Lebassi et al. (2011) used the RAMS mesoscale meteorological model to quantify global warming effects on SoCAB summer sea breeze patterns and thus on T_{max} trends. Thus, this study analyzed large scale input data used by Lebassi et al. 2011) for a difference simulations between summertime past (1966-'70) and present (2001-'05) atmospheric fields of surface air temperature, sea surface temperature and pressure over California. The changes in the vertical profiles of temperatures, wind, marine boundary layer height and subsidence are quantified. This will increase the understanding of the changes in the vertical structure of the boundary layer and possibly impacts of these changes to coastal cooling.

2. Methodology

2.1 Model description

The overall goal of the current high-resolution RAMS mesoscale meteorological numerical-model simulations of current (i.e., 2001-05) and past (i.e., 1966-70) SoCAB summer sea breeze wind and temperature patterns is to analyzed the large scale forcings of the RAMS simulations and the vertical structures of coastal cooling.

An outer grid with a horizontal grid spacing of 16 km, and an inner grid of 4 km, are specified to capture larger scale (i.e., GC and synoptic) forcings and to further resolve more complex smaller mesoscale and urban phenomena, respectively. Simulations of 35 years of summer months are not feasible on such high resolution grids, because of the required large computational (i.e., one hour CPU time for one-day simulation-time with a 24 processor cluster) and storage resources. The current simulations are thus only for five consecutive summers during both a current (2001-05) and a past (1966-70) period. Resulting temperature and wind difference-fields are then quantified.

RAMS (Version 6.0), developed at CSU, solves the Reynolds-averaged quasi- Boussinesq, non-hydrostatic, primitive equations on a Polar stereographic map projection (Tripoli and Cotton 1986). It uses terrain-influenced sigma coordinates and an Arakawa-C staggered grid on which thermodynamic and moisture variables are defined at grid-volume centers, with velocity components on grid-face centers perpendicular to each component (Mesinger and Arakawa 1976).

Two fixed two-way interactive grids are centered over the SoCAB, with 30 vertical layers, stretched from a spacing of 0.03 to 1.2 km in the first 7.5 km (to increase near-surface uses time-split differencing (Pielke 1984), as well as a variable-field model initialization and update process, in which gridded 3-D fields of wind, potential temperature, and relative humidity values from a large-scale, i.e., NCEP in this application, model are assimilated at 6 h intervals via Newtonian relaxation (Davies 1983).

The cloud microphysics formulation of Walko et al. [1995] and Meyers et al. [1997] was used; as it includes complex parameterization of many ice species. A cumulus parameterization was not used, as both grid resolutions were at or below the 9-km resolution needing such formulations. Vertical diffusion coefficients were computed from the 2.5-level closure in Mellor and Yamada (1982), which employs a prognostic TKE equation. Short- and long-wave radiation were calculated following Mahrer and Pielke (1977), in which simulated water vapor and specified CO_2 concentrations affect radiative flux divergence optical-path calculations via their effects on atmospheric-emissivity.

The CO_2 concentration is assumed uniform in both domains at 330 ppm from the surface to 40 km, while down-welling long-wave flux at the model top is set to zero. The CO_2 value is comparable with the summer 329 ppm value observed by Keeling and Whorf (2004) at La Jolla, California in the early 1970s, while it underestimates the 2001 value of 369 ppm. Radiative effects from long-term global CO_2-increases, accounted for via the NCEP BCs, are ingested into RAMS every 6 hrs. Local inner-domain radiative effects from CO_2 trends are not accounted for, but should be small over the five-summer simulation period.

The 1966-1970 and 2001-2005 periods were selected for simulation, as they showed similar large-scale PDO climate-variability factors. Strong winter El Niños produce decreased coastal upwelling in southern California, while strong winter La Niñas produce the reverse (LadoChy et al. 2007). The more important multi-decaldal index influencing California summer climate, however, is the spring (MAM) PDO, evaluated from north Pacific SSTs. The JJA past- and present-periods were thus selected based on similarities between the trends of the anomalies of this index and those of coastal California JJA T_{ave}-values; i.e., both periods generally show small upward trends in both variables [*http://jisaao.washington. edu/ pdo/PDO.latest*].

2.2 Initialization, BCs, and statistical evaluation

RAMS was initialized with NCEP gridded data-sets, interpolated (offline) to its grid by its internal isentropic analysis package. The NCEP data contains the following 4-D fields on horizontal pressure surfaces at horizontal increments of 0.5 deg: horizontal velocity components, temperature, geopotential height, and relative humidity. These fields generate 3-D assimilation fields every 6 h during the execution (started at 2400 UTC; 1700 LT in summer) that nudge the coarse-grid lateral boundary-regions. Lateral BCs on the outer grid follow Klemp-Lilly (1978), a variant of Orlanski (1976), in which gravity-wave propagation speeds computed for each model-cell are averaged vertically, with the single average-value applied over the entire vertical column. In this scheme, horizontal diffusion coefficients are computed as the product of horizontal deformation rate and length-scale squared (Smagorinsky 1963).

Initialization of RAMS also requires characterization of its surface BCs, via the following four input data sets, constructed "off line" and then interpolated to its 2-D surface grid-points:

- *Topography:* USGS topographic heights at a resolution of 30 arc-sec (about 1 km), obtained from the RAMS web site.
- *SSTs:* Monthly JJA ERSST, originally produced by statistical methods that allow for stable reconstructions from sparse data (Smith and Reynolds 2003) by use of ICOADS data. Values are down-loaded from NCDC at a resolution of 1-deg (about 111 km) for current-condition runs and at 2-deg for past–condition runs.
- *NDVI:* Monthly vegetation distributions, obtained from the RAMS web site on a 30 arc-sec grid.
- *LULC:* AVHRR data are used in the 1-km resolution Olson (1994) Global OGE 94-class LULC classification scheme, available from the USGS EROS Data Center (Lee 1992). OGE LULC data are input into the RAMS BATS scheme, which condenses its 94 classes into 21. OGE data are used in the outer-domain for the current-period simulations, but the inner-domain uses the 30 m resolution, 38-class, 2000 data from the NOAA C-CAP (http://www.csc.noaa.gov/digitalcoast/data/ccapregional/index.html). OGE LULC data are normally condensed to one urban and 20 rural classes in RAMS, but for the current simulations, two SoCAB urban LULC classes dominate: commercial and low-density residential; the one additional urban class was thus added to LEAF-3.

The LEAF-3 model (Walko et al. 2005) is used to calculate time-varying 2-D surface-temperature and -humidity BCs from linked prognostic surface-energy and -moisture balance equations for each of the following basic LULC type (i.e., group of similar classes): bare soil, plant-canopy covered soil, and urban; water surface temperature is assumed constant. Within any surface grid cell, 30-m patches are used to represent the heterogeneity of LULC classes. Once moisture and temperature values for each LULC category within each surface grid are calculated, the heat and moisture fluxes are area-averaged within each cell, used within SBL-parameterizations as (constant with height) fluxes, and finally used as the lower BCs for the sub-grid diffusion schemes in the finite-differenced prognostic PBL equations.

As the inner domain is mainly urbanized, the RAMS surface BCs were modified, to better account for urban processes, by the addition of a time- and spacing-varying Q_A for each urban sub-grid 30 m "patch." In US cities, Q_A is typically 60% from traffic sources; 40% from residential and industrial activities; and only a few percent from human metabolism (Sailor and Lu 2004). The current effort implements the time-varying daily profile, which shows morning and afternoon peaks associated with rush hour traffic.

RAMS look-up tables provide literature-values (as a function of LULC class) for the radiative, physical, vegetative, and thermal parameters associated with the surface-BC formulation. Values for several of key parameters for urban and rural LULC classes show increased values α and Vegetation height (H_v) values in urban areas.

As is the spatial distribution of vegetation fraction (F_v) is an important input-parameter, a technique was developed to determine it from high resolution Google™ earth-visible images as follows: (a) start with visible color Google map, (b) reduce map colors to 16, (c) count pixel values of each color in selected "typical" area for each urban class, (d) calculate fraction of green-color pixels to determine F_v, (e) reduce map colors to two (black and white), (f) count both the black and white pixels, (g) calculate fractions of white pixels to determine rooftop fractions, and (h) calculate street fractions as black fraction minus vegetation fraction.

Only the F_v–value for each urban class can currently be input in the lookup table in RAMS. Further details can be found in Lebassi (2009).

An evaluation of RAMS-results at the 19-m level (i.e., lowest RAMS half-grid wind and temperature level) against 1-10 June 2002 (first 10-days of the "present" simulation) hourly observed 2-m temperatures and 10-m wind speeds from 12 SoCAB METAR weather stations was carried out by use of RAMS-values interpolated to those levels at the grid-point closest to each site. Daily-averaged surface temperatures from 15 COOP station had to be used to evaluate the past (1970) simulation results, as no hourly observations are available for that period. Both the RAMS 19-m level values and all observed values are hereafter referred to as "surface" values.

Spatial distributions of statistical significance-levels of RAMS-produced temperature and wind- differences (i.e., present minus past) in Domain-2 were also calculated. Mean-values and standard deviations of RAMS surface winds and temperatures were calculated for each Domain-2 grid-point at 1200, 1400, and 1600 LT over each five- or one-year (as appropriate) past and present simulation-period for use in Student-t significance tests of differences. The study focused on these hours, as they are the period of strongest sea-breeze impacts on the coastal temperatures.

3. Results

3.1 Model evaluation

A validation of present (1-10 June 2002) and past (JJA, 1970) model results against available surface observations was carried out. Figure 2 compares the time series of hourly average surface temperature and wind speed from the 12 SoCAB METAR stations for the period of the present simulations and daily surface temperatures from15 COOP station for the period of the past simulations with values from the RAMS grid point closest to each site. Due to the unavailability of observational data, only the daily maximum temperature is used to validate the past simulations.

Present RAMS temperatures (Fig. 2a) generally compare well with the observations [r^2=0.87, Fig. (2c)], as they capture diurnal cycles and day to day trends in peak-values. The 10 day average observed value was 19.1°C, while the modeled value was 20.3°C. RAMS captured the large-scale cooling trend over the first three days, the warming trend over the next four days, and the final three-day cooling trend. The largest discrepancies occurred on warm days, with overestimations of maxima by 2.5°C and minima by 2.0°C.

RAMS wind speeds (Fig. 2b) also compare well with observations, as they again generally capture diurnal cycles and daily peak values [r^2=0.8, Fig. (2c)]. The 10 days average observed value was 2.9 m s^{-1}, while the modeled value was 3.1 m s^{-1}. Minor discrepancies exist during the three hottest nights, which had observed near calm winds, when RAMS minima were about 1 m s^{-1}.

Past JJA daily maximum surface RAMS temperatures (Fig. 3) also generally compare well with the observations [r^2=0.7, Fig. (3a), blue], as they capture average JJA cycles and day to day trends in peak values. The JJA summer average T_{max} observed value was 26.5°C, while the modeled value was 27.3°C. RAMS captured the large-scale cooling trend over the first 13 days, followed by a warming trend over the next 13. The largest discrepancies again occurred on the warm days, with overestimations of maxima by 2.5°C and minima by 2.0°C.

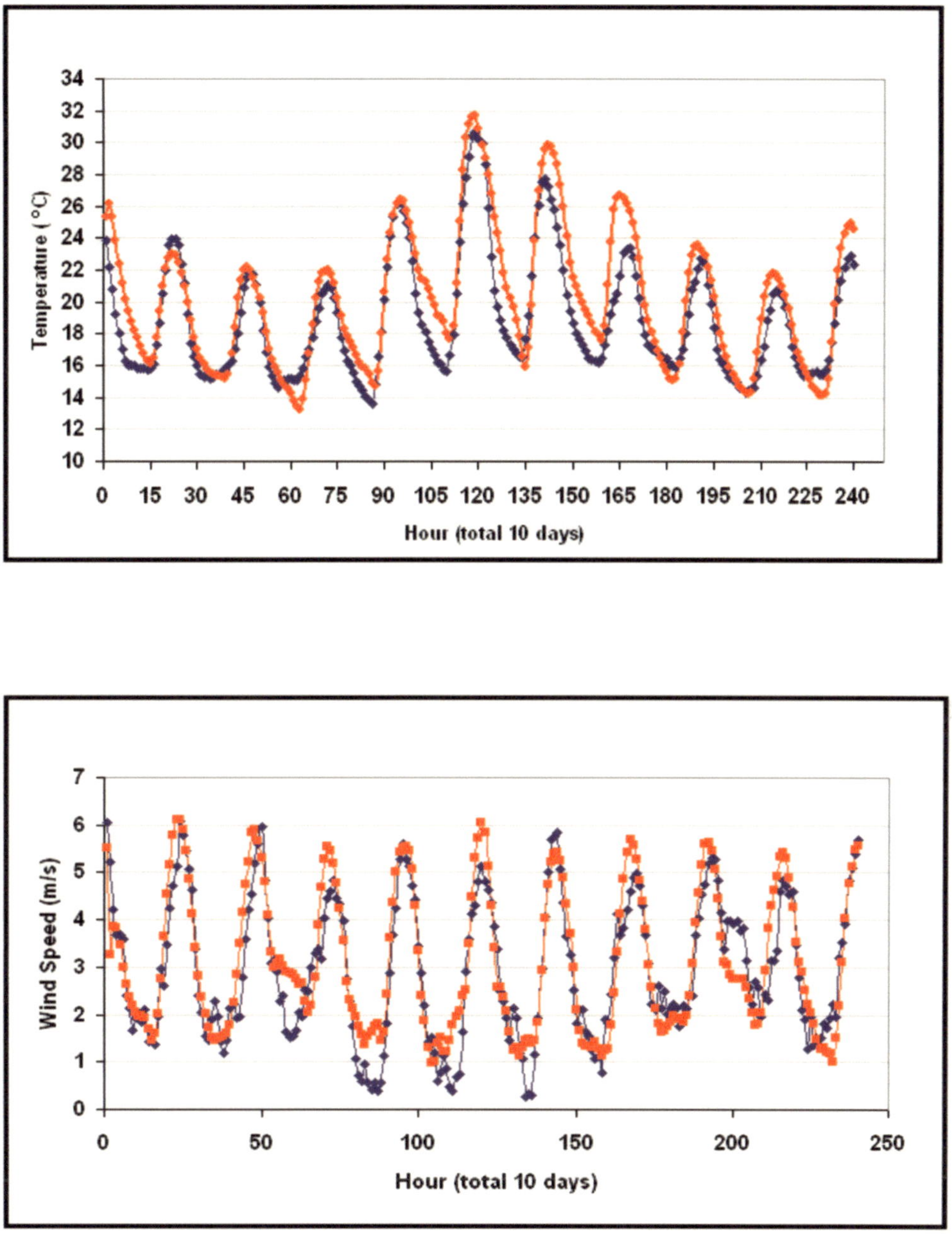

Fig. 2. a, b. (a) Modeled (red) vs. observed (blue) hourly 2-m temperatures (°C) averaged over 12 METAR stations for 1-10 June 2002; (b) same as Fig. 1a, but for wind speed (m s^{-1}).

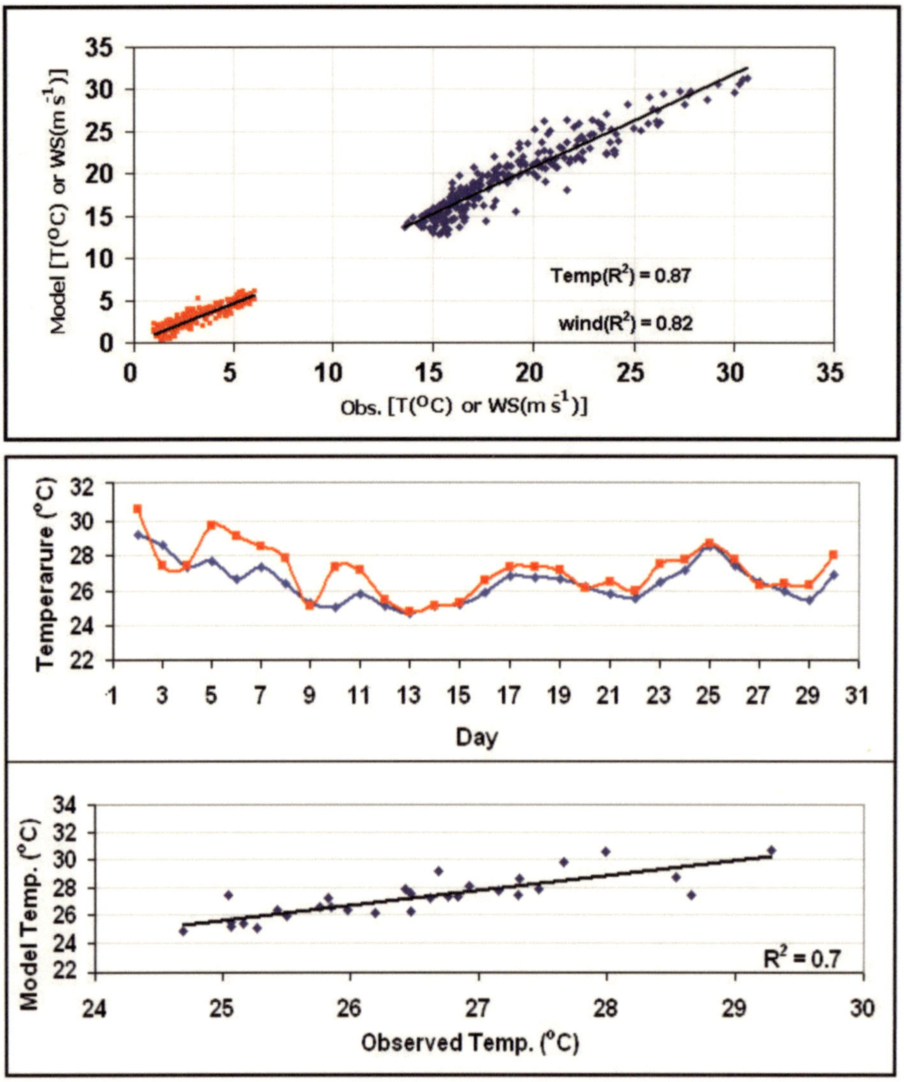

Fig. 3. a, b, c. (c) correlation of modeled vs. observed METAR temperatures (°C, blue) and wind speeds (red, m s^{-1}); (d) modeled (red) vs. observed (blue) daily 2-m summer temperatures (°C) averaged over 15 COOP stations; and (e) correlation plot of data in (d).

3.2 General circulation effects

GC forcing, the main external driver of the mesoscale RAMS model, enters through its initialization and BCs procedures. During summer, the semi-permanent Pacific High pressure system is located on its eastern edge, while a thermal low is over southeastern California. The resulting horizontal pressure gradient creates an onshore sea breeze flow over the SoCAB. This flow peaks in summer, when SSTs are coldest along the California coast within the southward flowing coastal California-current, as upwelling water reaches the surface as Ekman transport carries surface waters away from the coast.

NCEP 2.5 deg spatial-resolution sea level pressure p_s-values are interpolated from the input geopotential heights that initialize RAMS and provide its updated lateral BCs at 6 h intervals. The JJA-averaged 1000 hPa input NCEP p_s-values at 1700 LT for the five-year past-period of Run-2 shows the Pacific High centered west of California (Fig. 4a), with a central pressure center of 1024.3 hPa. The axis of the thermal low (central pressure of 1006 hPa) extends northward from Mexico to the California-Nevada border.

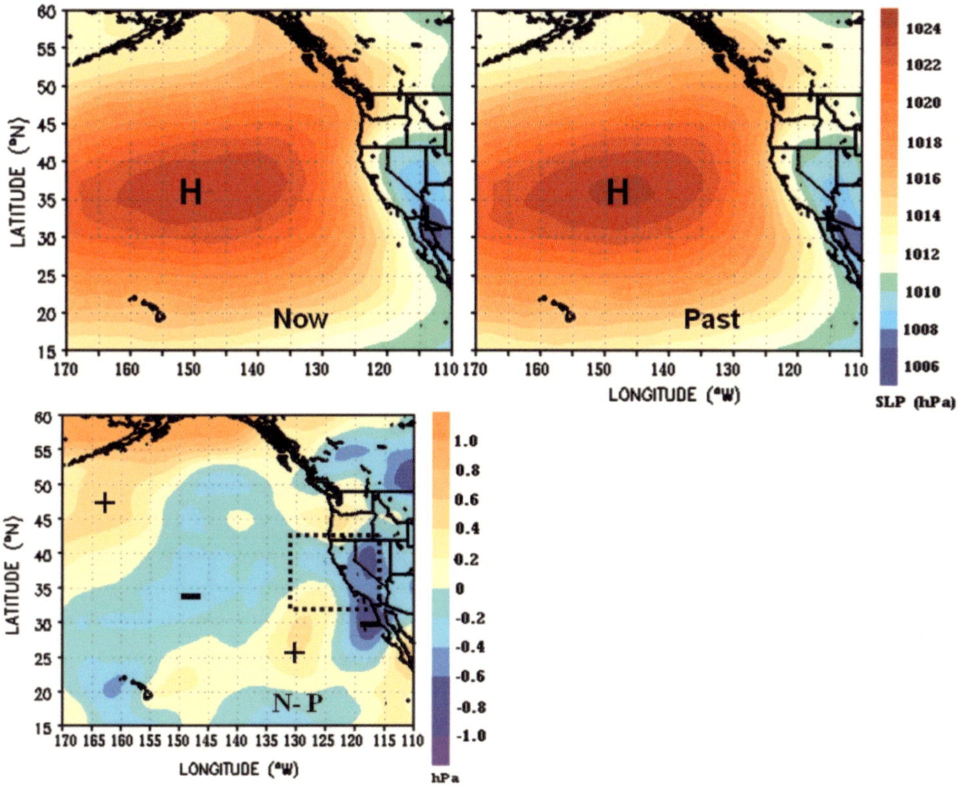

Fig. 4. Summer-averaged input NCEP 2.5-deg sea level pressures (hPa) at 17 LT, averaged over five-year periods of past and present simulations: (a) top right for past period, (b) top left for present period, and (c) lower for present minus past values (where pluses and minus denote increases and decreases, respectively).

The analogous input p_s field for the five-year present-period of Run-1 (Fig. 4b) shows the High basically unchanged in position and orientation, while the thermal low (as expected from the greater global warming over the continent) has extended northward (further into Nevada) and westward beyond the southern California-Mexico coast. Changes in p_s magnitude between the two simulations are best seen in Fig. 4c, in which the difference (herein defined as present minus past) pattern shows a broad region of decreased values of up to -0.4 hPa across the center of the High, north of a smaller area of increased values (up to 0.6 hPa). The two areas of decreased p_s in Nevada and along the coast (from the SoCAB to Mexico) are due to the expanding thermal low; over-water decreases peak at -1.0 hPa along the Mexican coast. The thermal low, however, has expanded northward and southward, producing two areas of maximum pressure decreases (up to -1.0 hPa). It has also intensified and the pressure at its center has decrease by only -0.6 hPa, a smaller amount than both to the north (over Nevada) and south (over the SoCAB-Mexico coast). These input NCEP p_s changes produce an increased west to east large-scale pressure gradient over the area that will strengthen the onshore large-scale background flow, onto which RAMS-produced mesoscale thermally-driven flows will be superimposed.

The monthly-averaged ICOADS SSTs used as BCs in the RAMS Run-1 (present) and Run-2 (past) simulations have 2- and 1-deg spatial resolutions, respectively. Five-year average Domain-1 SSTs from the past-period Run-2 (Fig. 5a) show cold values (14°C) northwest of Domain-2 that increase southward (to 18°C).

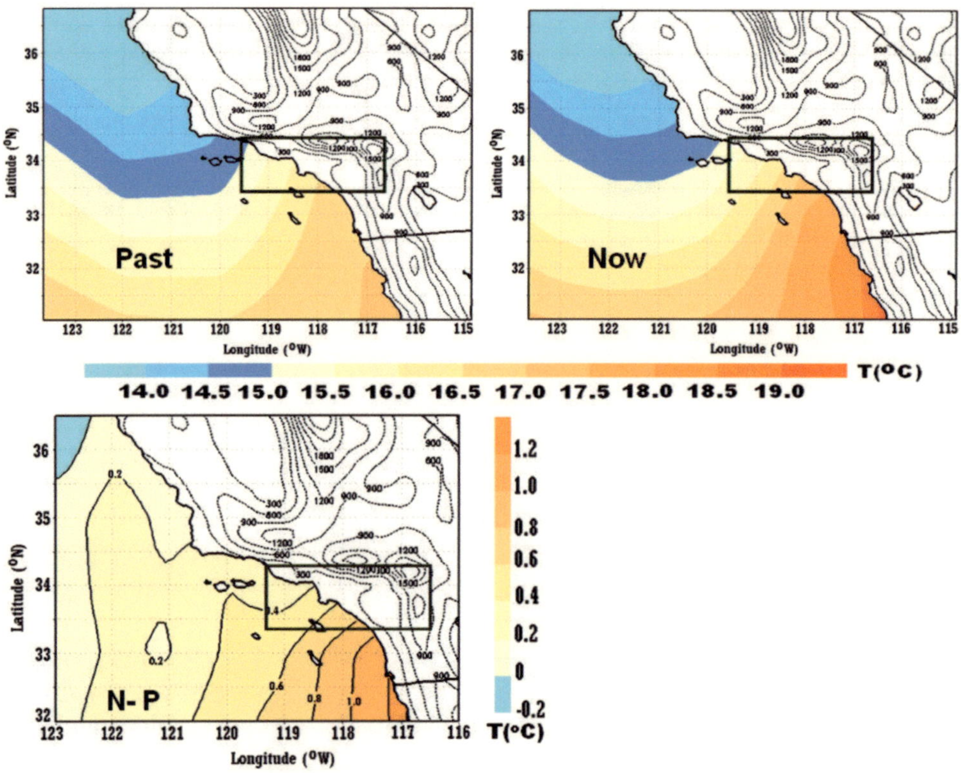

Fig. 5. Summer-averaged input ICOADS sea surface temperatures (°C) at 17 LT averaged over five-year periods of past and present simulations: (a) top left for past period (at 2 deg input resolution), (b) top right for present period (at 1 deg input resolution), and (c) lower for present minus past values; box represents Domain-2 area.

Present-period Run-1 results show cool-area values slightly reduced (Fig. 5b), with warm-area values increased. Differences (Fig. 5c) show the peak cooling at only -0.2°C, while the warming increased southward (up to 1.2°C). Global warming is expected, as others have observed increased SSTs over this period (e.g., Hansen 2006).

Averaged input JJA NCEP 1000-hPa temperatures 1700 LT for Run-2 (past) at a 2.5 deg spatial resolution (Fig. 6a) show cool ocean values (11 to 25°C), with warm land areas (25 to 41°C). The coldest area is west of SoCAB, while the warmest (41°C) is over Nevada. Run-1 (present) values show that the coldest ocean temperatures moved north-westward, while the hottest inland values moved southeast towards Arizona (Fig. 6b).

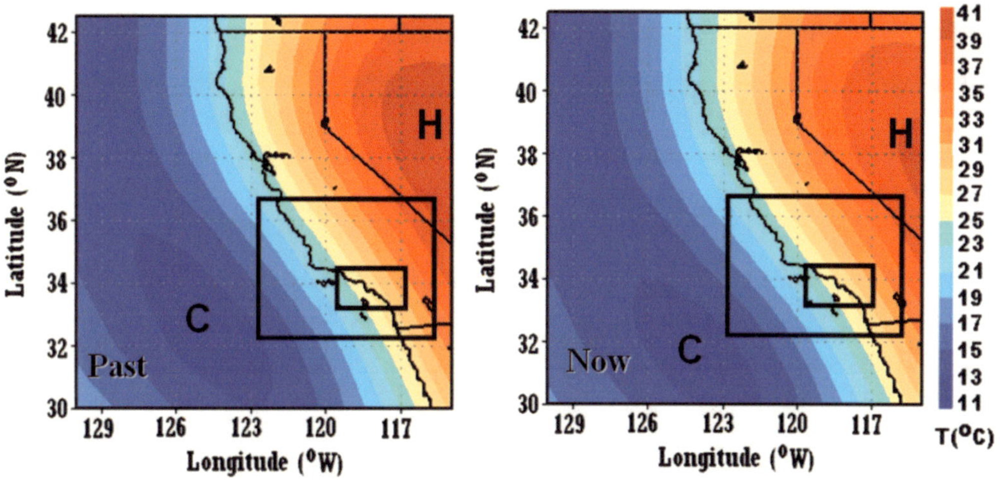

Fig. 6. Summer-averaged input NCEP 2.5 deg 1000-hPa temperature (°C) at 17 LT, averaged over five-year periods of past and present simulations: (a) right for past period and (b) left for present period; boxes indicate Domain 1 and Domain 2.

Differences (Fig. 7a) show cooling (up to -1.0°C) over the central ocean area, surrounded by a narrow band of warming ocean areas with values increasing (up to 0.6°C) southward along the coast from the SoCAB to Mexico. Over land, a narrow band of coastal areas (north SoCAB) showed slight cooling (up to -0.2°C), while inland areas warmed, with a maximum of 1.8°C over central Nevada. A vertical east-west cross section of these NCEP temperature changes at 34.0 N (Fig. 7b) shows the cooling over the coastal waters (up to -1.0°C) contained within the MBL (up to 925 hPa or 400 m); it also shows a local maximum global warming (1.2°C) at 900 hPa and an absolute peak (1.8°C) at 850 hPa. While global warming is expected, the cooling has not been found in previous observational or modeling studies (to the best of our knowledge), but is here-in hypothesized could be due to the southward movement of the Thermal Low in Fig. 4.

In summary, the large scale Pacific High is generally unchanged in position and orientation, but shows a diminished central pressure. JJA average 1700 LT NCEP temperatures at 1000 hPa showed cooling over the central California coast that extends up-wards to about 925 hPa. Finally, SSTs generally showed warming, which increased to the south. Vertical Profile of the NCEP data did not resolve the base of the marine boundary layer over the California coast. This large scale results are input for the model as ICs and BCs, and RAMS results in the next section will be analyzed to evaluate the vertical structure of GHG-induced coastal cooling.

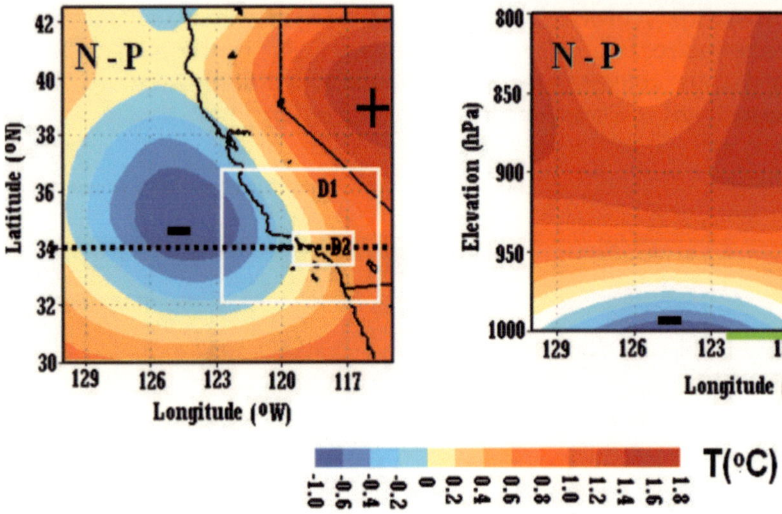

Fig. 7. Temperature changes (°C, past minus present) for values in plots in Fig. 6 in: (a) horizontal plane (right) and (b) vertical plane, shown by dashed line in (a).

3.3 MesoScale effects: Modeled vertical structure of coastal cooling

A cross-section at 33.83 N (through Domain-2) shows the vertical extents of MBL-changes and coastal cooling, i.e., the 1600 LT Run-2 (present) temperature results shows cold marine air (ex post facto defined, see below, as having temperatures ≤ 21°C) coming onshore and up against the coastal mountains (Fig. 8a).

The subsidence inversion top from the High is at about 750 m and at a temperature between 22 and 23°C. Its base is at the surface (i.e., at 19-m, the lowest RAMS temperature grid level), as opposed to several hundred meters. According to Thompson (2010, personal communication), this results as the RAMS TKE scheme does not generate sufficient mechanical turbulence over ocean areas. This occurred in the Coupled Ocean/ Atmosphere Mesoscale Prediction System (COAMPS) model (developed for over-ocean forecasting) until recent changes were made to its TKE formulation. Air temperatures above the inversion-top decreases to 20°C at 2 km.

The Run-1 (present) temperature cross-section (Fig. 8b) also shows a similar structure, but the inversion-base is now elevated to 60 m (i.e., second lowest temperature grid level), while the inversion-top has grown to 900 m, which could be due to the decreased subsidence discussed below. Near-surface temperatures at the higher inland-elevations have warmed by about 1°C, while those at 2 km have cooled by a similar amount, both of which are expected from global warming theory.

The corresponding difference plot (Fig. 8c) more accurately visualizes these relatively small temperature changes, i.e., a shallow layer (50 to 120 m) of over-ocean temperature increase (up to 0.4°C) exists, due to the increased SSTs discussed above and which causes formation of the shallow marine sub-inversion layer in Fig. 8b. Above this, a 450 m layer that has cooled (up to -1.0°C) resulted due to the large-scale cooling in the NCEP BCs (Fig. 6). Inland global warming peaks (at 1.4°C) at 2 km, while coastal cooling (up to -0.6°C) extends inland to the top of the coastal hills. The cooling is due to the increased sea breeze activity, as unconnected at 19 m to the NCEP-introduced cooling. If the inversion-base was accurately reproduced, then the SST induced near-surface warming-layer in Fig. 8c would extent upwards by several hundred more meters, which would thus also show the disconnected-nature of these two cooling areas. The shallow over-ocean SST warming, large scale elevated-cooling, coastal cooling, and global warming where all significant at 99%, while the narrow transition layer was less significant (<90%) due to cancellation effects of coastal cooling and inland warming.

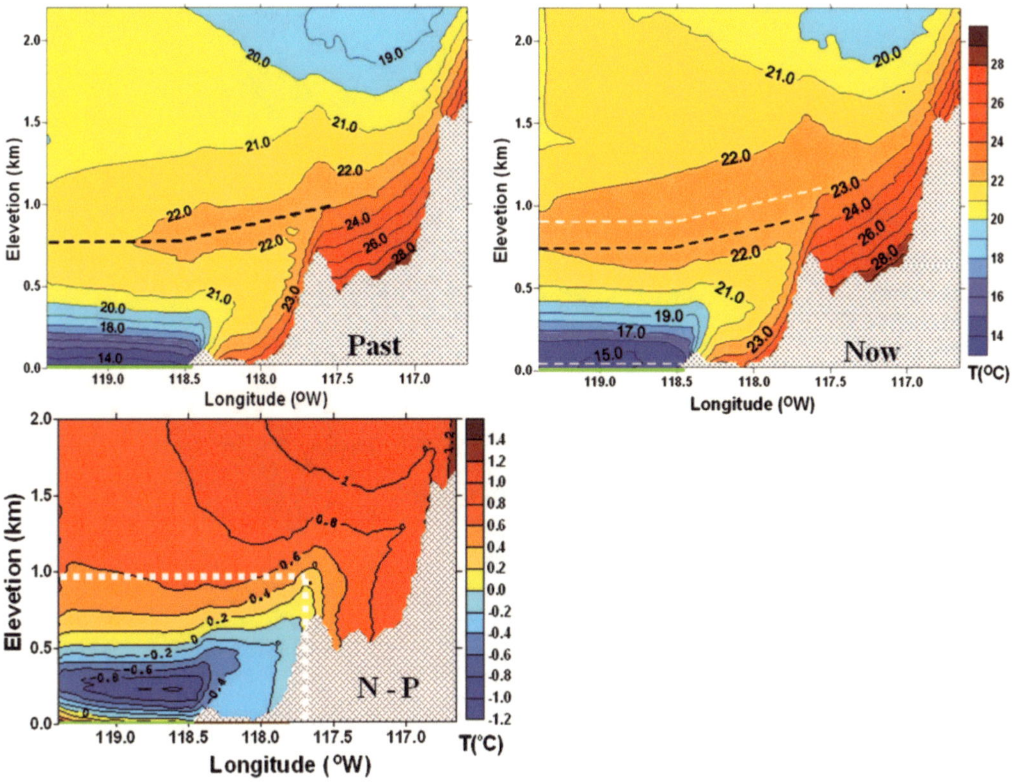

Fig. 8. Summer-averaged (over five-year periods of past and present simulations) Domain-2 temperature (°C) cross-section at 14 LT for Run-2 (past, top left) period, Run-1 (present, top right) period, and (c) present minus past (lower), where dashed lines represents inversion base and top and white dashed box represents area of D2 cross section at 33. 83 N (see Fig. 7 for location).

The vertical velocities in the vertical plane of Fig. 9 also showing the disconnected nature of the two cooling areas, i.e., the 1400 LT Run-2 (past) section (Fig. 9a) shows a peak subsidence from the High (up to -1.6 cm s^{-1}) over the ocean at 800 m, with peak upward motion areas over both the coastal plane and coastal topography (up to 3.6 cm s^{-1} above the

latter). Similar results are found for Run-1 (present) (Fig. 9b), but with a weaker subsidence (only up to -0.6 cm s^{-1}) over the ocean and decreased upward motion over coastal areas.
The "difference" vertical-velocity field (Fig. 9c) shows these changes more clearly, i.e., a maximum decreased of over-ocean subsidence of 0.6 cm s^{-1} at 500 m. The decreased upward motion over both land areas is also clearly shown, i.e., over the coastal basin and coastal mountains (both up to -0.6 cm s^{-1}), the former associated with the increased stability of the stronger sea-breeze flow. The decreased over-ocean subsidence is due to the extension of the Thermal Low in Fig. 4c.

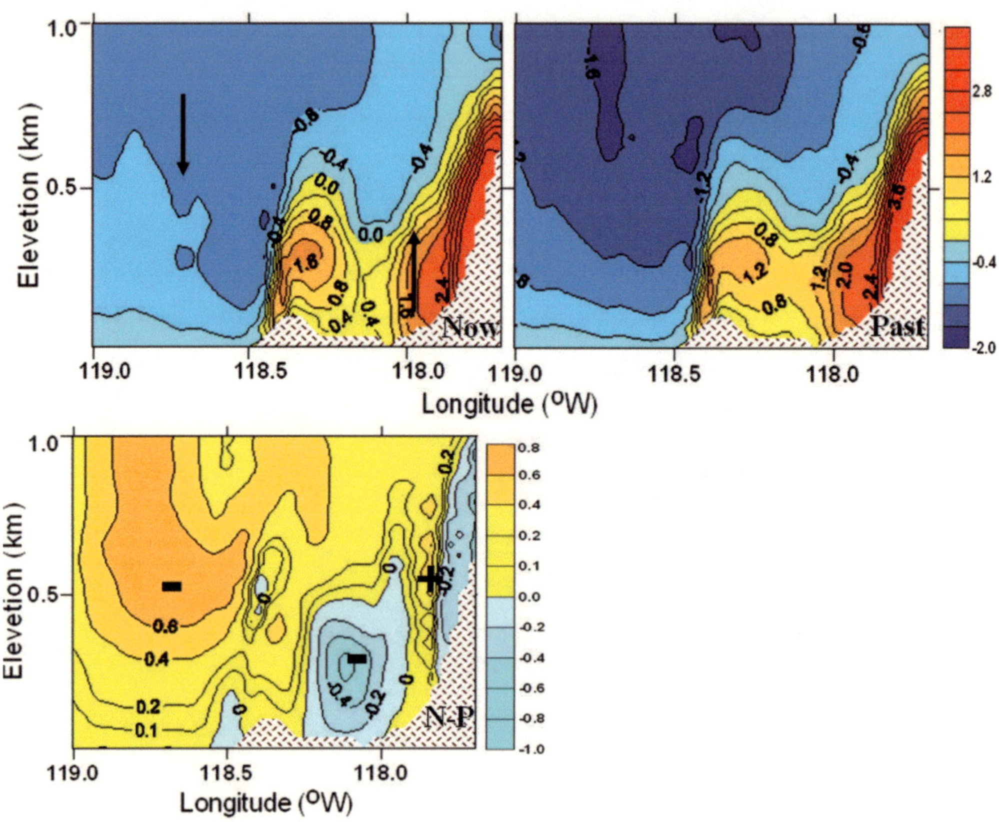

Fig. 9. Same as Fig. 8, but in sub-area of its white box and for vertical velocity values (cm s^{-1}) fields, where up/down arrows represent up/down motions and plusses/minuses represent increases/decreases, respectively.

4. Discussion and conclusion

The meso-met RAMS model was used to further investigate local (on 16 and 4 km grids) climate changes in the SoCAB, with simulations designed to quantify impacts from both global warming and urbanization. Results from a simulation with both present (2001-05) summer climate-conditions and LCLU-patterns were compared to one with present-LCLU and past (1966-1970) climate.

Evaluation of model results against hourly surface temperature and wind speed observations during a current 10-day summer period showed that they captured diurnal cycles and day to day trends in peak and minimum values, with an exception on several warm days; average errors were only 1.2°C and 0.2 m s^{-1}, respectively. Evaluation of model results against past daily maximum temperatures averaged over 15 COOP stations also showed good agreement in trend and magnitude, with an average error of 0.7°C.

The shallow SST warming layer over the ocean, large scale elevated cooling, coastal cooling, and global warming were all significant at 99%. All transition-layers, in which two opposite effects cancel, showed less-significant (<90%) results.

NCEP input BCs to RAMS of summer 1000-hPa at 1700 LT showed that during the 35-years between the past and present simulation periods, the Pacific High was generally unchanged in position and orientation, but slightly diminished in central pressure (up to -0.4 hPa). The over-land Thermal Low expanded northward and southward, producing two areas of maximum pressure-decrease (up to -1.0 hPa) over Nevada and the SoCAB-Mexico coast. These input NCEP SLP-changes produced increased west to east large-scale pressure gradients that thus strengthened onshore large-scale background flows, onto which RAMS-produced mesoscale thermally-driven flows were superimposed.

ICOADS SSTs generally showed warming the 35-year period, which increased southward (up to 1.2°C) and which resulted from global warming. Concurrent input NCEP air temperatures generally showed a maximum global warming of 1.8°C at an elevation of 2-km over land areas. The warming extended downward to the surface over land areas, while a surface large-scale cooling over central California coastal waters (up to -1.0°C) extended upwards to about 925 hPa (or 400 m); effects from this imposed-cooling on RAMS results are discussed below.

Lebassi et al. 2009 in their observational study analyzed that average-JJA California land-surface min and max temperatures from 1970-2004 showed expected asymmetric-warming, with increases of 0.27 and 0.06°C decade^{-1}, respectively. The larger min-temperature increase was significant at the 93% level and was comparable with (and correlated to) the coastal daily-mean SST trend of 0.24°C decade^{-1} (significant at the 92% level). Different rates of max and min temperature increase is consistent with previously observed asymmetric global-warming from increased greenhouse-gasses. The spatial distribution of comparable observed SoCAB JJA max-temperatures, however, showed a more complex pattern, with both cooling at low-elevation coastal-areas open to marine air penetration and warming at inland and high-elevation coastal areas. The scarcity of observational sites in some parts of both regions meant for additional modeling study of the region. Lebassi et al. 2011 further modeled the SoCAB coastal cooling using the RAMS model and results captured many important aspects of the observed surface inland-warming and concurrent coastal-cooling that developed during the 35-years between the present and past summer-daytime simulation periods, i.e., sequentially: weaker global warming over the ocean than over inland areas, increased surface temperature gradients, strengthened sea breezes, and thus coastal cooling.

This study showed the vertical structure of coastal cooling in the finer RAMS domain. The model captured the vertical and horizontal extent of GHG-induced coastal cooling while there is some limitations discussed below, which are introduced from the meso scale and forcing model, in this case NCEP. Fine-domain temperatures at 1400 LT showed an over-ocean Pacific High subsidence inversion-top that increased by 150 m (to a height of 900 m) over the 35-year period. Its base, however, was much below a reasonable MBL height (i.e., of several hundred meters). According to Thompson (2010, personal communication), these too low values resulted as the TKE scheme in RAMS does not generate sufficient mechanical turbulence over ocean surfaces, which also occurred in the COAMPS meso-met model (developed for over-ocean forecasting) until recent changes were made to its TKE formulation.

In addition, vertical profiles of NCEP temperatures showed that the model does not resolve the inversion base due the few number of grid point in the boundary layer. Even though further future research is needed to analyze the impacts of the large scale input on the mesoscale model, this could underestimate the base of the marine boundary layer which is several hundred meters in the coast of California. This study showed a too-shallow warming (over the 35-year period) layer of only 50 to 120 m thus developed over the ocean, as the increasing input-SSTs were unable to warm (and thus destroy) the lowest inversion layers and thus produce a realistically deep MBL. Above the lower warming layer, a 450 m layer cooled over the 35-years due to the large scale cooling within the input NCEP BCs. Over-land coastal cooling (to the same height) extended to the first coastal hill-tops due to the increased sea breeze activity. This cooling is unconnected to NCEP-introduced large-scale cooling, because the coarse-domain results showed separate areas for each effect.

The past vertical-velocity pattern showed peak subsidence (-1.6 cm s^{-1}) from the High over the ocean at 800 m, with peak upward motion areas over both the coastal plain (1.2 cm s^{-1} at 300 m) and first coastal topographic peak (>3.6 cm s^{-1} at 500 m). Similar present-simulation results were found, but with a smaller (reduction of up to -0.6 cm s^{-1}) subsidence over the ocean, in association with the decreased intensity of the High and the increased elevation of its inversion top, discussed above. The (present minus past) difference vertical-velocity field showed decreased upward motion over both the coastal basin and first coastal peaks, with the former associated with the increased stability from the more intense present-case cool stable marine-air sea breeze flow. These distinct vertical velocity difference-areas also demonstrate the disconnected nature of the two cooling areas.

In summary, the coarse domain RAMS results captured many important aspects of the observed surface and upper-level coastal cooling and inland warming that developed during the 35-years between the present and past simulation summer-daytime periods, i.e., sequential weaker global warming over the ocean than over inland areas, increased surface temperature gradients, strengthened sea breezes, and coastal cooling.

Various significance impacts may result from coastal cooling, e.g., summer daytime max ozone levels are decreased due to reduced rates of: fossil-fuel usage for cooling, natural hydrocarbon production, and photochemical photolysis. Agricultural production, water supply, and human thermal-stress and mortality are also effected by max summertime temperatures. Similar effects could exist in other coastal regions, while other regional reverse-reactions may be found in a variety of topo-climatic areas, possibilities that require further study. Future modelling efforts could enhance the current simulations by inclusion of finer scale horizontal grid resolutions and better large scale that resolved the marine boundary layer base.

5. Acknowledgments

The authors would like to thank Profs. Edwin Maurer and Drazen Fabris of Santa Clara University (SCU), and Dr. Cristina Milesi of the California State University Foundation, Monterey Bay, at NASA Ames Research Center, for their insightful comments. We also thank the School of Engineering, at SCU and the National Science Foundation Grant No. 0933414 for funding the lead author. We also acknowledge the CCNY and San Jose State University for providing the computational time.

6. References

Baklanov, A., B. Grisogono, R. Bornstein, L. Mahrt, S. Zilitinkevich, P. Taylor, S. Larsen, M. Rotach, and H. Fernando, 2010: On the nature, theory, and modelling of atmospheric planetary boundary layers. Accepted, *Bull. Amer. Met. Soc., 92,* 123-128.

Bakun, A., (1990), Global climate change and intensification of coastal ocean upwelling, *Science, 247,* 198–201, doi:10.1126/science.247.4939.198.

Bonfils, C., and D. Lobell, (2007), Empirical evidence for a recent slowdown in irrigation-induced cooling, *Proc. Natl. Acad. Sci., 104,* 13582-87, doi:10..1073/pnas.0700144104.

Bornstein, R. D., (1975), The two-dimensional URBMET urban boundary layer model, *J. Appl. Meteor., 14,* 1459-1477.

Bornstein, R. D., S. Klotz, U. Pechinger, R. Salvador, R. Street, L. J. Shieh, F. Ludwig, and R. Miller, (1986), Application of linked three-dimensional PBL and dispersion models to New York City, *Air Pollution Modeling and its Applications V,* Plenum Press, 543-564.

Boucouvala, D., R. Bornstein, J. Wilkinson, and D. Miller, (2003), MM5 simulations a SCOS97-NARSTO episode, *Atmos. Environ., 37,* S95–S117.

Cayan, D. R., E. P. Maurer, M. D. Dettinger, M. Tyree, and K. Hayhoe, (2008), Climate Change Scenarios for the California Region, *Climatic Change,* published online, doi:10.1007/s 10584-007-9377-6

Davies, H. C., (1983), Limitations of some common lateral boundary schemes used in regional NWP models, *Mon. Wea. Rev., 111,* 1002–1012.

Duffy, P. B., C. Bonfils, and D. Lobell (2007), Interpreting recent temperature trends in California, *Eos Trans. AGU, 88,* 409-410.

Edinger, J. G., (1959), Changes in the depth of the marine layer over the Los Angeles Basin, *J. of Meteor., 16,* 219-226.

Edinger, J. G., (1963), Modification of the marine layer over coastal southern California, *J. of App. Meteor.,* 16, 706-712.

Falvey, M., and R. D. Garreaud (2009), Regional cooling in a warming world: Recent temperature trends in the southeast Pacific and along the west coast of subtropical South America (1979–2006), *J. Geophys. Res., 114,* D04102, doi:10.1029/ 2008JD010519.

Gutiérrez, D., and et al. (2011), Coastal cooling and increased productivity in the main upwelling zone off Peru since the mid twentieth century, *J. Geophys. Res., 38,* L07603, doi:10.1029/2010GL046324, 2011.

Intergovernmental Panel on Climate Change (IPCC). (2001), IPCC Third Assessment Report - Climate Change 2001.

Keeling, C. D. and T. P. Whorf, (2004), Atmospheric CO_2 concentrations derived from flask air samples at sites in the SIO network. In Trends: A Compendium of Data on Global Change. Carbon Dioxide Information Analysis Center, Oak Ridge National Laboratory, U.S. Department of Energy, Oak Ridge, Tennessee, USA, 123 pp.

Klemp, J. B., and D.K. Lilly, (1978), Numerical simulation of hydrostatic mountain waves. *J. Atmos. Sci., 35,* 78-107.

Kueppers, L. M., M. A. Snyder, and L. C. Sloan, (2007), Irrigation cooling effect: Regional climate forcing by land-use change, *Geophys. Res. Lett.,* 34, L03703, doi:10.1029/2006GL028679.

LaDochy, S., R. Medina, and W. Patzert, (2007), Recent California climate variability: spatial and temporal patterns in temperature trends, *Climate Research, CR 33*, 159–169.

Lebassi, B., J. E. González, D. Fabris, E. Maurer, N. L. Miller, C. Milesi, P. Switzer, and R. Bornstein, (2009), Observed 1970-2005 cooling of summer daytime temperatures in coastal California, *J. Clim., 22,* 3558-3573, doi:10.1175/2008JCLI2111.1.

Lebassi, B. (2010), Observational and Modeling Study of Global Warming and Urbanization Impacts on Coastal California Climate, Ph.D. thesis, Dep. of Mech. Eng., Santa Clara Univ., Santa Clara, California.

Lebassi, B., J. E. Gonzalez, R. Bornstein, and, D. Fabris, (2010), Impacts of Climate Change in Degree Days and Energy Demand in Coastal California, *J. Sol. Energy Eng, 132,* 031005, doi:10.1115/1.4001564

Lebassi-Habtezion, B., J. E. Gonzalez, and R. Bornstein (2011), Modeled Large-Scale Warming Impacts on Summer California Coastal-Cooling Trends, *J. Geophys. Res., Accepted.*

Lee, R. L., and D. B. Olfe, (1974), Numerical calculations of temperature profiles over an urban heat island. *Bound.-Layer Meteor., 7,* 39-52.

Lobell, D. B., G. Bala, C. Bonfils, and P. B. Duffy, (2006), Potential bias of model project-ed greenhouse warming in irrigated regions, *Geophys. Res. Lett., 33,* L13709, doi:10.1029/ 2006GL026770.

Lobell, D., and C. Bonfils, (2008), The effect of irrigation on regional temperatures: A spatial and temporal analysis of trends in California, 1934–2002, *J. Clim., 21,* 2063-2071.

Mahrer, Y., and R. A. Pielke, (1977), The effect of topography on sea and land breezes in a two dimensional numerical model, *Mon. Wea. Rev., 105,* 115–1162.

Martilli, A., (2002), Numerical study of urban impact on the boundary layer structure: sen-sitivity to wind speed and urban morphology, and rural soil moisture, *J. Appl. Meteor., 41,* 1247-1267.

Masson, V., (2000), A physically-based scheme for the urban energy budget in atmospheric models. *Bound.-Layer Meteor., 94*, 357-397.

Maurer, E. P., (2007), Uncertainty in hydrologic impacts of climate change in the Sierra Nevada, California under two emissions scenarios, *Climatic Change, 82*, 309-25.

McGregor, H. V., M. Dima, H. W. Fischer, and S. Mulitzal, (2007), Rapid 20th-Century Increase in Coastal Upwelling off Northwest Africa, *Science, 315*, 637–639.

Mellor, G. L., and T. Yamada (1982), Development of a turbulence closure model for geophysical fluid problems, *Rev Geophys. Space Phys., 20*, 851-75.

Meyers, M. P., R. L. Walko, J. Y. Harrington, and W. R. Cotton (1997), New RAMS cloud microphysics parameterization. The two-moment scheme. *Atmos.* Res., 45, 3–39.

Mesinger, F., and A. Arakawa, (1976), *Numerical methods used in atmospheric models*, GARP Publication, No. 14, WMO/ICSU Joint Organizing committee, 64 pp.

Miller, N. L., K. Hayhoe, J. M. Jin, and M. Auffhammer, (2007), Climate change, extream heat, and electricity demand, *J. Appl. Meteor. and Climatology., 47, 6*, 1834-44.

Myrup, L., (1969), A numerical model of the urban heat island. *J. Appl. Meteor.*, 8, 908-18.

Nemani, R. R., M. A. White, D. R. Cayan, G. V. Jones, S. W. Running, and J. C. Cough-lan, (2001), Asymmetric warming over coastal California and its impact on the premium wine industry, *Climate Research, 19*, 25–34.

Oglesby, R. J., C. M. Rowe, C. Hays, (2010), Using the WRF Regional Model to Produce High Resolution AR4 Simulations of Climate Change for Mesoamerica, A23F-07, Dec 14, 2010, San Francisco, CA.

Pielke, R. A., (1984), *Mesoscale Meteorological Modeling.* Academic Press, San Diego, California, 612 pp.

Sailor D. J., and L. Lu, (2004), A top-down methodology for developing diurnal and seasonal anthropogenic heating profiles for urban areas, *Atmos. Env., 38*, 2737–2748.

Smith, T. M., and R. W. Reynolds, (2003), Extended reconstruction of global sea surface temperature based on ICOADS data, *J. Climate, 16,* 1495–1510.

Snyder, M., L. Sloan, N. Diffenbaugh, and J. Bell, (2003), Future climate change and upwelling in the California current, *Geophys. Res. Lett.,* 30, 1823, doi:10.1029/2003GL017647.

Walko, R. L., W. R. Cotton, M. P. Meyers, and J. Y. Harrington (1995), New RAMS cloud microphysics parameterization. Part I: The single-moment scheme. *Atmos. Res., 38,* 29–62.

Taha, H., (1999), Modifying a mesoscale meteorological model to better incorporate urban heat storage: a bulk-parameterization approach, *J. Appl. Meteor.*, 38, 466-473.

Taha, H., (2007), Urban surface modification as a potential ozone air-quality improvement strategy in California: A mesoscale modeling study, *Bound. –Layer Meteor., 127,* 219-239, doi:10.1007/s10546-007-9259-5.

Tripoli, G. J., and W. R. Cotton, (1982), The Colorado State University three-dimensional cloud/mesoscale model – 1982. Part I: General theoretical framework and sensitivity experiments, J. *de Rech. Atmos., 16,* 185-220.

Ulrickson, B. L., and C. F. Mass (1990), Numerical investigation of mesoscale circulations in the Los Angeles Basin, Part I: verification study. *Mon. Wea Rev., 118,* 2238-61.

Walko, R. L., and C. J. Tremback, (2001), RAMS technical description, http://www.atmet.com/html/docs/rams/rams_techman.pdf.

10

Causes and Consequences of the Late 1960s Great Salinity Anomaly

Mihai Dima[1] and Gerrit Lohmann[2]
[1]*University of Bucharest, Faculty of Physics*
[2]*Alfred Wegener Institute for Polar and Marine Research*
[1]*Romania*
[2]*Germany*

1. Introduction

The second half of the 20th century showed a series of decadal-scale anomalies of salinity, temperature and sea ice cover in the northern North Atlantic. One pronounced event, the 'great salinity anomaly' (GSA) is observed in the late 1060s and the early 1970s (Dickson et al., 1988). This anomaly can be linked to the sea ice volume out of the Arctic through Fram Strait, which represents a major source of freshwater (Aagaard & Carmack, 1989; Schmith & Hansen, 2003). Increased sea ice export from the Arctic Ocean stabilizes the upper water column in the North Atlantic, diminishes the production of intermediate and deep water masses through ocean convection and can influence the large-scale ocean circulation (Häkkinen, 1999). An important part of sea ice export from the Arctic is forced by specific atmospheric structures (Hilmer et al., 1998; Häkkinen and Geiger, 2000; Cavalieri 2002). These can be parts of coupled ocean-atmosphere modes of climate variability with large-scale projections.

The large-scale Atlantic meridional overturning circulation (AMOC) is a key component of the climate system. It can be defined as the part of the global ocean circulation, which is driven by density differences. The relatively dense surface waters in the North Atlantic generate downward mixing and southward movement of deep-water masses. This is partially balanced by a transport of saline waters by surface ocean currents from the tropics to midlatitudes. Changes in the AMOC can be very abrupt and have a worldwide climatic impact as demonstrated in theoretical and numerical studies (Stommel, 1961; Rahmstorf, 1994). There is significant evidence that the past abrupt climate changes (Dansgaard et al., 1993; Heinrich, 1988) were linked to AMOC variations (EPICA, 2006). One of the largest climate changes during the last 10.000 years occurred about 8.200 years ago (Alley, 2001) and it had a large impact over the Northern Hemisphere. It was proposed that this event included a freshwater pulse entering the North Atlantic (Barber et al., 1999) and modeling simulations suggest that it was associated with AMOC reduction (Renssen et al., 2002). Given the prominence of the GSA and its potential worldwide impact through the AMOC, it is of significant importance to understand the physical mechanisms associated to this outstanding climatic anomaly. Therefore, the goal of this study is to investigate the causes and the consequences of the GSA.

The "climate mode" represents an useful concept to reduce the complexity of climate variations and can be defined as a set of physical processes with coherent spatial variability and quasi-periodic variations in time. Based on North Atlantic sea ice and sea level pressure

(SLP) data for the 1903-1994 period, it was noted that the constructive inference between four interannual-to-interdecadal modes may be responsible for the GSA (Venegas & Mysak, 2000). This observation is further investigated here based on a reconstruction of the sea ice export through Fram Strait (Fram Strait Sea Ice Export - FSSIE), which extends back in time to 1820. Further more, a possible impact of the GSA on the AMOC is also presented. The identification of the quasi-periodic components in the FSSIE time series is presented in sections 2 and 3, and their physical relevance is addressed in section 4. The possible GSA impact on AMOC is presented in section 5. The 6th part of this study contains a summary and the conclusions.

2. Possible causes of the GSA

The Singular Spectrum Analysis (SSA) method (Ghil et al., 2002) is applied on a reconstruction of Fram Strait Sea-Ice Export (FSSIE) (Schmith & Hansen, 2003) in order to investigate the GSA origin. The SSA technique is used to extract information from short and noisy time series by providing data-adaptive filters that help separate the time series into statistically independent components like trends, oscillatory signals and noise (Allen & Smith, 1997). The analyzed record had an annual resolution and extends over the 1820-2000 period (Fig. 1). Since significant salinity anomalies in the North Atlantic are separated by decadal scale time intervals, here the interest is in decadal to interdecadal components. Therefore, before SSA, a 7-year running mean filter is applied to the initial time series, to obtain the record shown in Fig. 2a. This index is used in the following analyzes.

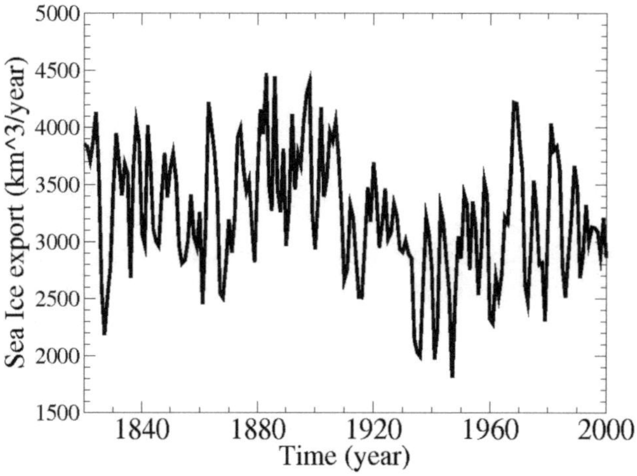

Fig. 1. Fram Strait Sea Ice Export (FSSIE) reconstruction for the 1820-2000 period (Schmith & Hansen, 2003).

In order to identify the long-term component, an SSA is performed on the FSSIE record using an 80-year window. This analysis emphasizes a dominant centennial signal (Fig. 2b) explaining 48% of variance. The next components have interdecadal time scales but these are overshadowed by this centennial signal. While here the focus is on decadal and interdecadal variations, the centennial component is subtracted from the record to obtain the residual time series shown in Fig. 2c. A key feature of this record is the pronounced maximum in the late 1960s, which represents the highest values in the last century and a half. One notes that around the 1970s, the removed centennial component (Fig. 2b) is in transition phase and therefore has no significant contribution to the maximum of the late 1960s.

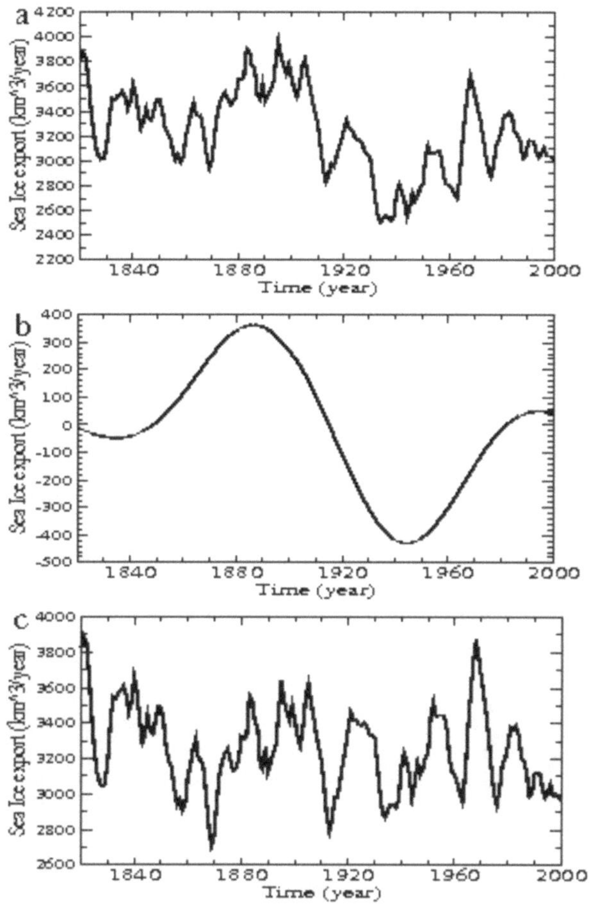

Fig. 2. a) 7-year running mean filtered Fram Strait sea ice export reconstruction (km^3/year) for the 1820-2000 period; b) the centennial component identified through Singular Spectrum Analysis, using a 80-year window; c) residual record obtained after subtracting the centennial component; the 1960s maximum is unprecedented in the last 130 years.

The dominant quasi-periodic components contained in this residual record are identified through SSA. The first three periodic signals identified in the analysis have the properties of robust oscillatory components: the associated eigenvalues are above the noise level (Fig. 3a) and the corresponding time-EOFs are in quadrature (Fig. 3b,c,d).

The first one (Fig. 4a) has a multidecadal time scale with a period of approximately 70-years and explains 29% of the variance. The second (Fig. 4b) and the third (Fig. 4c) signals are characterized by periods of 14-years and 20-years and explain 20%, respectively 15% of the variance. An important common feature of all these three periodic components is the maximum around 1968. The signal reconstructed from these three components has the highest value during the late 1960s (Fig. 4d). Accordingly, here it is argued that the maximum of the FSSIE observed in the late 1960s can be considered as resulting from the synchronization of the multidecadal, decadal and bidecadal quasi-periodic components.

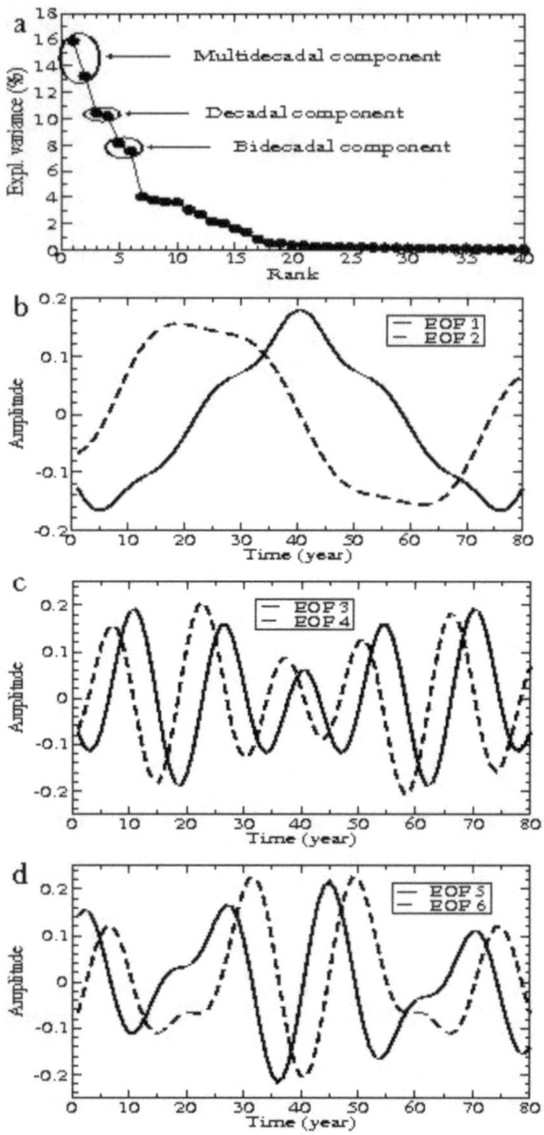

Fig. 3. Singular Spectrum Analysis of the residual record shown in Fig. 2c, using a 80-year window. a) the eigenvalues spectrum; b) the time-EOFs 1 and 2; c) the time-EOFs 3 and 4 d) the time-EOFs 5 and 6. The robustness of the first three quasi-periodic components is supported by their distinction from the noise level in the eigenvalues spectrum and by the constant phase-lag of their associated time-EOFs.

3. Robustness of the quasi-periodic components

The significance of these components is further investigated through Monte Carlo simulations (Allen & Smith, 1997; Ghil et al., 2002). An ensemble of surrogate data is generated and a covariance matrix is computed for each realization. These matrices are projected onto the eigenvector basis of the original data to derive the corresponding eigenvalues, which measure the resemblance between the original data set and the surrogate data. The statistical distribution of the eigenvalues gives confidence intervals outside which a time series can be considered significantly different from a random realization of the process. If an eigenvalue lies outside a 90% noise percentile, then the red noise null hypothesis for the associated EOF can be rejected with this level of confidence.

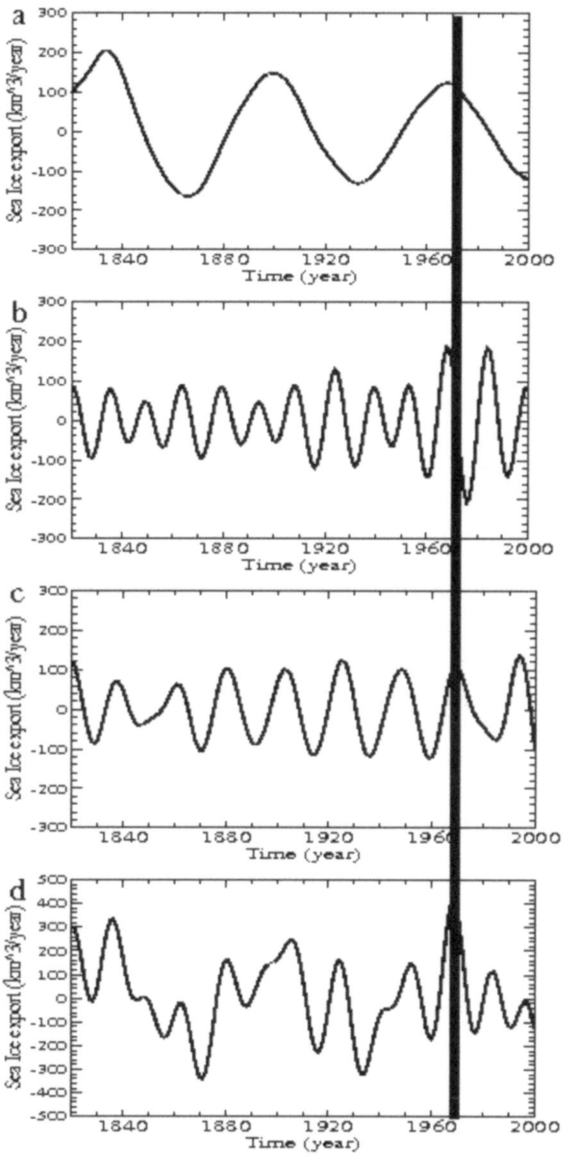

Fig. 4. Singular Spectrum Analysis of the residual record shown in Fig. 2c emphasizes three quasi-periodic components, with periods of a) ~70-year; b) ~14-year; c) ~20-year. All three quasi-periodic signals show a maximum around 1968, concurrent with the Great Salinity Anomaly. d) reconstruction based on these components, which has the highest value in the whole record in the late 1960s.

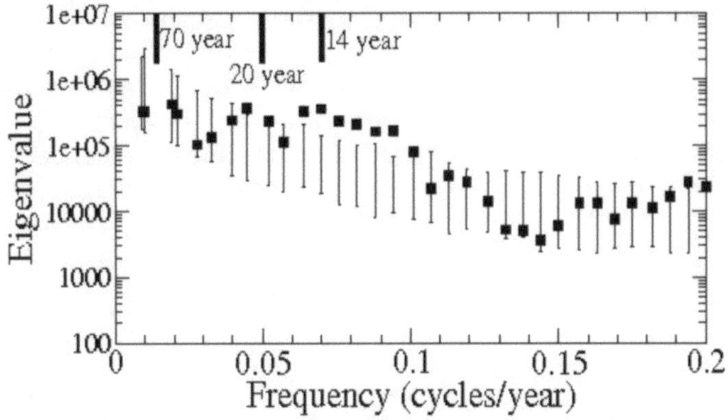

Fig. 5. Monte Carlo singular spectrum of the FSSIE record. The rectangles indicate projections of the noise eigenvectors on the data correlation matrix. The lower and upper thicks on the error bars indicate the 5th and 95th noise percentile. A characteristic frequency was estimated for each EOF, by maximizing its correlation with a sinusoid. Therefore, the frequency at each rectangle is just an average estimate.

The statistical distribution of the eigenvalues derived from simulations is shown in Fig. 5, as a function of frequency. The rectangles are lying outside the error bars for the periodicities of ~14-yr and ~20-yr, supporting the significance of these quasi-periodic components identified through SSA. Although the ~70-yr component is the dominant one in terms of explained variance (Fig. 3a), the rectangles associated to multidecadal variability are not located outside the error bars (Fig. 5). This is likely due to the non-harmonicity of the EOFs (Fig. 4b), which makes the estimation of the dominant frequency uncertain, mainly for low frequency (Ghil et al., 2002).

The causal relation between the peak in the late 1960s and the three quasi-periodic components can be considered in two directions. Any significant peak can be decomposed in quasi-periodic signals while synchronization of several oscillatory components can generate a maximum. One notes that when a peak is decomposed into quasi-periodic components, their amplitudes are decreasing away from its position. However, the quasi-constant amplitude of the three peaks (Fig. 4) suggests that these signals do not result from the decomposition of the peak. On the contrary, their robustness and quasi-constant amplitude suggest that their synchronization generates the peak.

4. Physical relevance of the oscillatory modes

The wind field is an important factor for the sea ice dynamics in the Fram Strait (Hilmer et al., 1998). In order to construct the atmospheric structures responsible for the decadal and interdecadal variations of FSSIE, three sea level pressure composite maps are constructed. These are computed based on the quasi-periodic components resulted from SSA (Fig. 4) and the winter sea level pressure field for the 1899-2000 period (Trenberth & Paolino, 1980), to which a 7-year running mean filter is applied. Because the sea-ice dynamical response lags the wind forcing by several years (Hilmer et al., 1998), the composite maps are constructed so that the detrended SLP field leads the FSSIE by four years. The SLP composite map corresponding to the multidecadal component (Fig. 6a) has a wave-number-1 structure at the hemispheric scale, including positive values over the North Pole and the North Pacific,

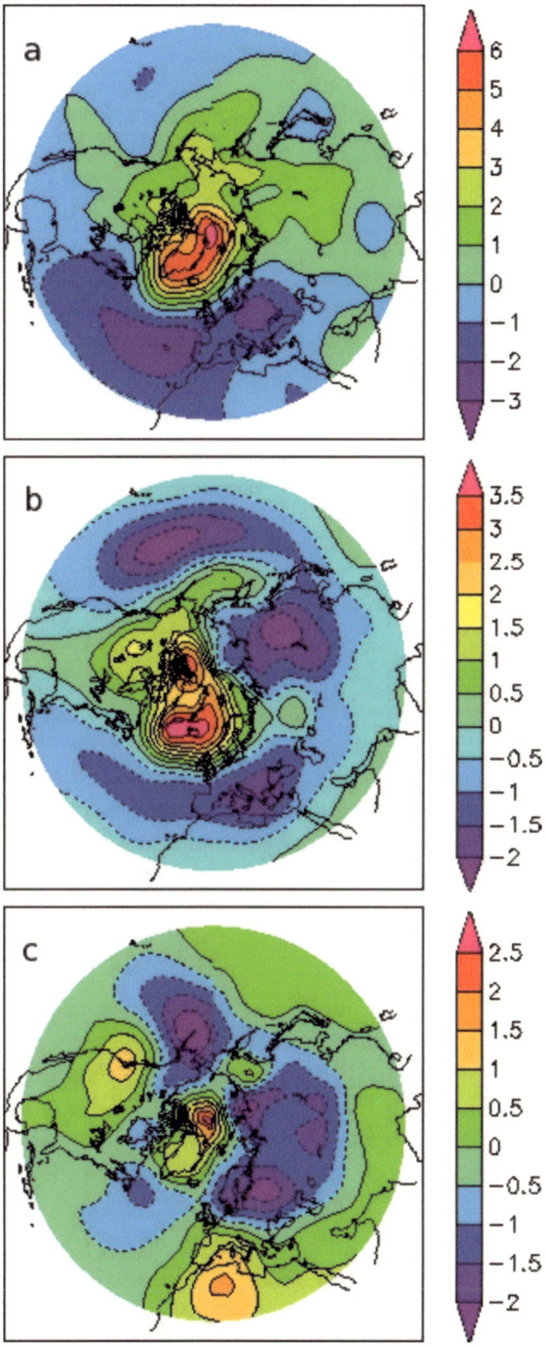

Fig. 6. Winter SLP composite maps constructed based on the ~70-year (a), ~14-year (b) and ~20-year (c) components identified through SSA; in calculating the composite maps the SLP field leads the time series by 4 years. In all the composite maps the centers over Greenland, Western Europe and North Atlantic are significant at 95% confidence level according to a t-test, considering 14 degrees of freedom.

and negative values in the North Atlantic and over Europe. The highest values are concentrated in a high pressure system centered over Greenland. The associated wind field has a strong projection on FSSIE and contributes to an intensification of freshwater and sea ice export from the Arctic to the North Atlantic. This SLP structure was associated with the Atlantic Multidecadal Oscillation (Dima & Lohmann, 2007).

The SLP composite map associated to the decadal mode (Fig. 6b) has an annular structure with strong projection on the negative phase of the Arctic Oscillation (Thompson et al, 2000). The positive center extends over Greenland and the North Pole and favors freshwater and sea ice export from the Arctic. A similar influence on the sea ice dynamics is suggested by the SLP structure corresponding to the bidecadal mode (Fig. 6c), which includes also a positive center over the North Pole. This SLP structure is similar to that presented by White & Cayan (1998) in association with bidecadal variability (Figure 2 in their study). One notes that the axis of maximum wind field from the Arctic toward the North Atlantic for each SLP composite map aligns better with Fram Strait, as the mode explains more variance in the FSSIE record (Fig. 6). Similarly, the SLP gradient across the Fram Strait associated to each mode is higher, as the mode explains more variance in the reconstruction. The composite maps show that the quasi-periodic decadal and interdecadal variations identified in the FSSIE reconstruction are forced by specific structures of atmospheric circulation.

In order to investigate possible associations between the quasi-periodic signals identified in the FSSIE reconstruction and decadal modes of climate variability, correlation maps are calculated between these time components and the sea surface temperature (SST) field from the HADISST1 dataset (Rayner al., 2003). A maximum influence of a quasi-periodic climate mode on FSSIE is generated for a specific phase of the modes. In general, this is different from that phase of the mode which maximizes its influence on other fields, because different physical processes could be involved. In order to consider these phase-shifts between modes' projection on different fields, specific lags are considered when calculating correlation maps, so that the obtained patters corresponds to the canonical SST structures of these modes, presented in previous studies.

The SST map associated to the multidecadal component has maximum values in the North Atlantic (Fig. 7a) and its monopolar structure is very homogeneous. These properties of the correlation map and the characteristic time scale of the multidecadal component identified in the FSSIE reconstruction were associated with the Atlantic Multidecadal Oscillation in several studies (Schlesinger & Rammankutty, 1994; Enfield & Mestaz-Nunes, 1999). The SST correlation map corresponding to the decadal component does also have maximum values in the North Atlantic (Fig. 7b). Such a tripolar SST structure is specific for a dominant mode of North Atlantic variability, having a 12-14-year characteristic time scale (Deser & Blackmon, 1993; Dima et al., 2001). For the bidecadal component, the correlation map includes high values in both, the Pacific and Atlantic basins (Fig. 7c). This structure was associated with a global bidecadal mode (Zhang et al., 1997; White & Cayan 1998). The SLP and SST maps suggest that the quasi-periodic components identified in the FSSIE are associated to specific modes of climate variability. In view of these results and along the same line with Venegas & Mysak (2000) we argue that the GSA in 1970s was generated through synchronization between three decadal climate modes, which have simultaneous maximum projection on the FSSIE, in the late 1960s.

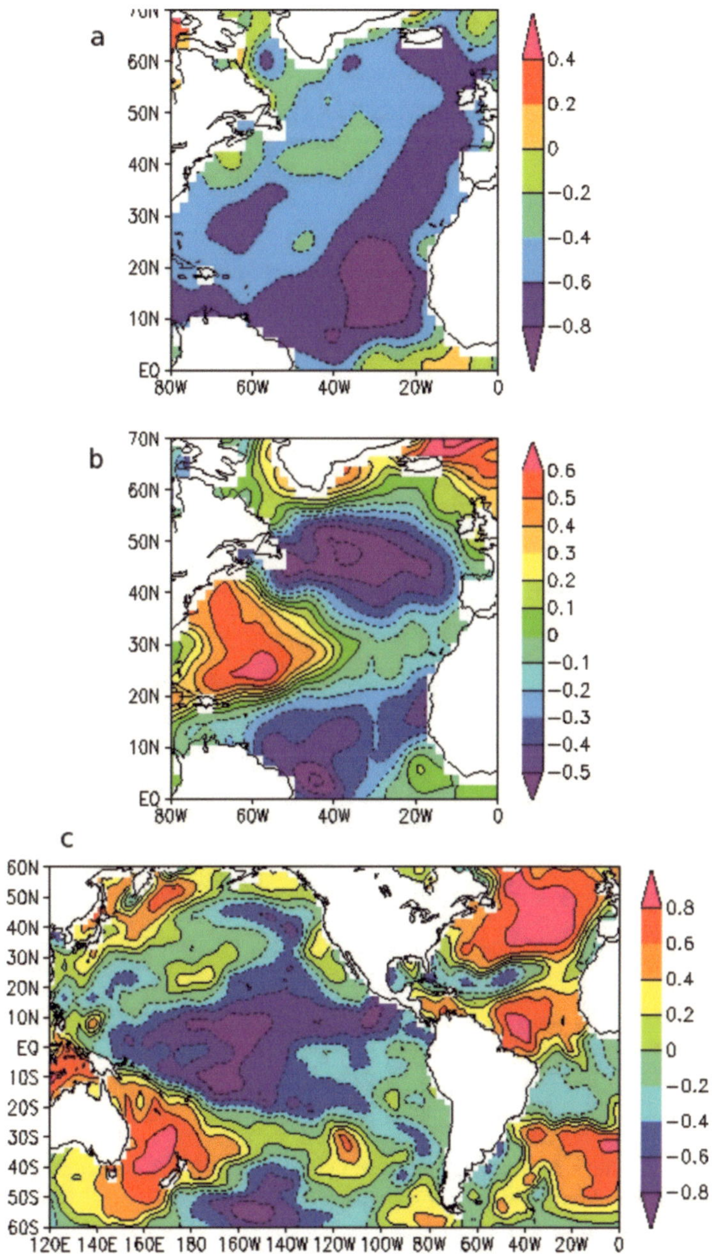

Fig. 7. Correlation maps derived based on the SST field (Rayner et al., 2003) and the quasi-periodic signals obtained through SSA: a) the multidecadal (the component leads the field by 15-year), b) decadal (the component leads the field by 6-year), c) bidecadal (the component leads the field by 9-year). Before calculating the maps the warming trend was removed, by subtracting the annual global average from each grid point, and the SST field was filtered in the 10-17-year band (for the decadal component) and 17-26-year band (for the bidecadal component). For the multidecadal component a 21-year running mean filter was applied to the SST field.

5. Possible consequences of the GSA

It is conceivable that the freshwater pulse associated with the GSA in the 1960s altered the convection pattern and the Atlantic meridional overturning circulation (Lenderink et al., 1994). Modeling studies show that significant changes of the AMOC mean state project on the surface temperature as a dipole between hemispheres (Schiller et al., 1999; Velinga & Wood, 2002), suggesting that the interhemispheric sea surface temperature gradient can be used as a proxy for AMOC variations. Following this indication, we used annual means of SST for the 1870-2006 period (Rayner et al., 2003) to calculate an interhemispheric index, as a difference between mean temperatures of the Northern and Southern Hemispheres. By construction, this time series captures the global AMOC signature but preserves its interannual variations. The resulted index (Fig. 8) shows interannual and decadal variability and is marked by a pronounced and abrupt jump around 1970. The index value dropped by 0.39 from 1967 to 1972. The climatological values (marked by horizontal line in Fig. 8) have dropped from 0.16°C, for the 1950-1968 period, to -0.08°C, for the 1971-1987 time interval. This abrupt change in the interhemispheric SST index is significant at the 99% confidence level according with a t-test. The amplitude and rapidness of the change suggest that MOC suffered a jump to a weaker state. In the interhemispheric index, this regime starts in the late 1960s and ends in the late 1980s, when the MOC seems to recover to its previous warm state.

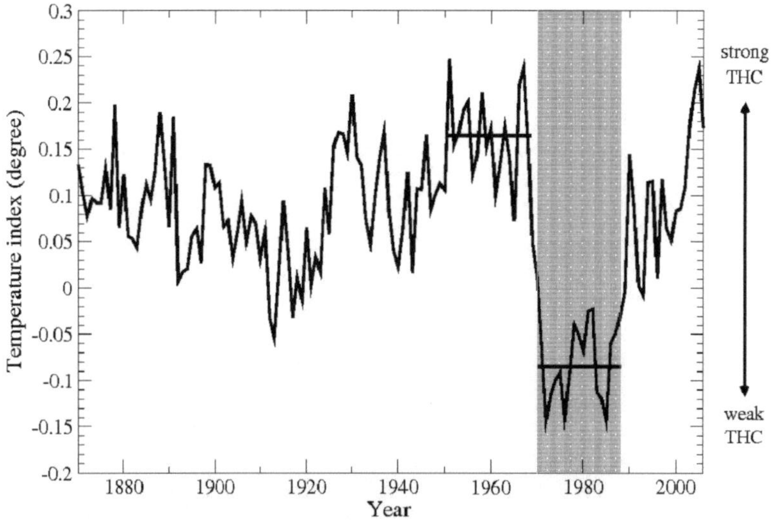

Fig. 8. An interhemispheric SST index (degree) calculated as a difference between the Northern and Southern Hemisphere averages, based on Hadley Center data set. Based on previous modeling studies, it is considered a proxy for AMOC variations. The climatological values for the 1950-1968 and 1971-1987 periods are marked by horizontal lines.

In order to identify the large-scale spatial projection of the AMOC shift, difference maps are constructed for the annual SST (Rayner et al., 2003) and winter SLP fields (Trenberth & Paolino, 1980). For each of these quantities, average maps are calculated for the linearly detrended fields over the 1950-1968 and 1971-1987 periods. The shift patterns are then obtained as a difference between the last and the former period means. The SST projection of the shift (Fig. 9a) emphasizes the interhemispheric dipole structure. Extended regions of statistically significant changes are located in the North Atlantic, the South Atlantic, the

Indian Ocean and the North Pacific. Together, the negative anomalies south of Greenland and the anomalies of opposite signs extending eastward from the South America indicate a weakened AMOC (Lohmann, 2003).

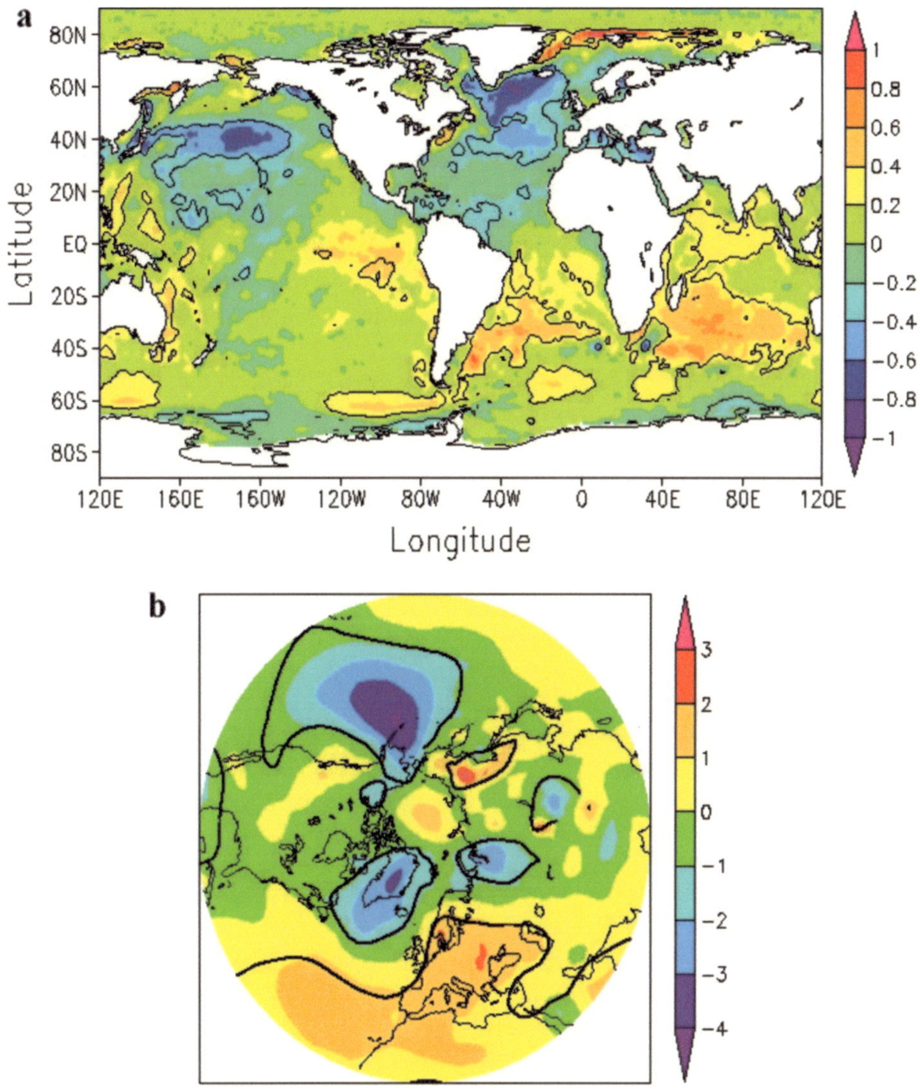

Fig. 9. a) The SST (°C) pattern of the abrupt MOC change around 1970, obtained as a difference between the climatological maps for the 1971-1987 and 1950-1968 periods. The average values were calculated based on annual detrended SST fields for these periods. The contours indicate regions were the SST difference is significant at the 90% confidence level according with a two-sided t-test. The pattern is consistent with an abrupt AMOC weakening during the early 1970s; b) The winter sea level pressure (hPa) pattern of the abrupt AMOC change around 1970, obtained as a difference between the climatological maps for the 1971-1987 and 1950-1968 periods. The average values were calculated based on annual detrended winter sea level pressure fields from an updated version of the Trenberth and Paolino data set.

The Northern Hemisphere winter SLP projection of the shift includes positive anomalies over the North Atlantic and Europe, and centers of negative values over the North Pacific and Greenland (Fig. 9b). The quasi-uniform positive SLP anomalies disposed over the negative surface temperatures in the North Atlantic basin resembles an atmospheric thermal response to a weakening oceanic overturning, as was also derived from numerical simulations (Lohmann, 2003). Both SST and SLP patterns were linked to AMOC variations in relation to the Atlantic Multidecadal Oscillation (Dima & Lohmann, 2007).

In order to derive the temporal fingerprints of the shift in the key regions emphasized by the SST pattern (Fig. 9a), three indexes were constructed by averaging the SST anomalies over the larger significant regions shown in Fig. 10a: North Atlantic (50°W-10°W, 30°N-70°N), South Atlantic (60°W-0, 50°S-20°S) and South Indian oceans (40°E-100°E, 50°S-30°S). In all three indexes the AMOC jump is marked by abrupt changes around 1970 (Fig. 10). The shifts manifests first in the North Atlantic time series in the late 1960s, followed in the South Atlantic and in the South Indian Ocean in the early 1970s. The weak state extends over 2-to-3 decades.

One interesting feature of the SLP fingerprint of the AMOC jump is the dipolar structure in the North Atlantic sector, which projects strongly on the North Atlantic Oscillation (NAO). Consistent with an atmospheric thermal response to weak oceanic overturning induced SSTs, the NAO index shows a shift toward a regime of positive values in the early 1970s, which ends during the 1990s (Fig. 11b). In this view, the NAO is modulated through MOC variations, and is most likely not directly related to anthropogenic effects.

Long-term climate data suggest that AMOC regimes and Sahel precipitation are linked (Street-Perrott & Perrott, 1990). A reduced AMOC state is associated with significant changes in the precipitation pattern in the tropical Atlantic, manifested as an interhemispheric dipole (Vellinga & Wood, 2002; Lohmann, 2003). Consistent with the weak overturning regime, the Sahel precipitation index (Janowiack 1988) (Fig. 11c) shows a pronounced change towards a regime of low values in early 1970s, which ends in the early 1990s. This indicates a southward shift of the Intertropical Convergence Zone (ITCZ) which was linked to significant precipitation anomalies in the sub-Saharian western Africa. This regime is unprecedented in the 20th century.

A climate shift during the mid 1970s was reported also for the Pacific basin (Trenberth & Hurrel, 1994; Stephens et al., 2001), with a characteristic SST pattern, which is similar to the SST fingerprint of the AMOC abrupt weakening (Fig. 9a). From the perspective of our results the global climate regime shift in the 1970s reported in different climate variables (Baines & Folland, 2007), including the NAO index (Fig. 11b), Sahel precipitation time series (Fig. 11c), tropical and North Pacific indexes, represent surface manifestations of an abrupt AMOC weakening. Consequently, the regions in which the shift presents large amplitudes could be used to monitor future oceanic overturning changes.

The weak oceanic overturning regime ended in the early 1990s with a recovery to a strong AMOC state, at a level similar with that observed before 1970 (Figs. 8 and 10a,b). This is supported also by the observed oceanic overturning strengthening since the 1980s (Latif et al., 2006) and by the record-high salinities observed in the Atlantic inflow to the Nordic Seas and the Arctic Ocean (Hatun et al., 2005). As the early 1970s shift, the AMOC resumption from the weak state is detected in all presented climate indexes, like the North Atlantic, South Atlantic and South Indian SST time series (Fig. 10), the NAO index (Fig. 11b), the Sahel precipitation time series (Fig. 11c) and the Pacific index (Trenberth & Hurrel, 1994; Stephens et al., 2001). Given the amplitude and the abruptness of the resumption (Figs. 8

and 10a,b) in the early 1990s, one can consider that a climate regime shift, associated with a rapid strengthening of Atlantic meridional overturning, was experienced in the last decade.

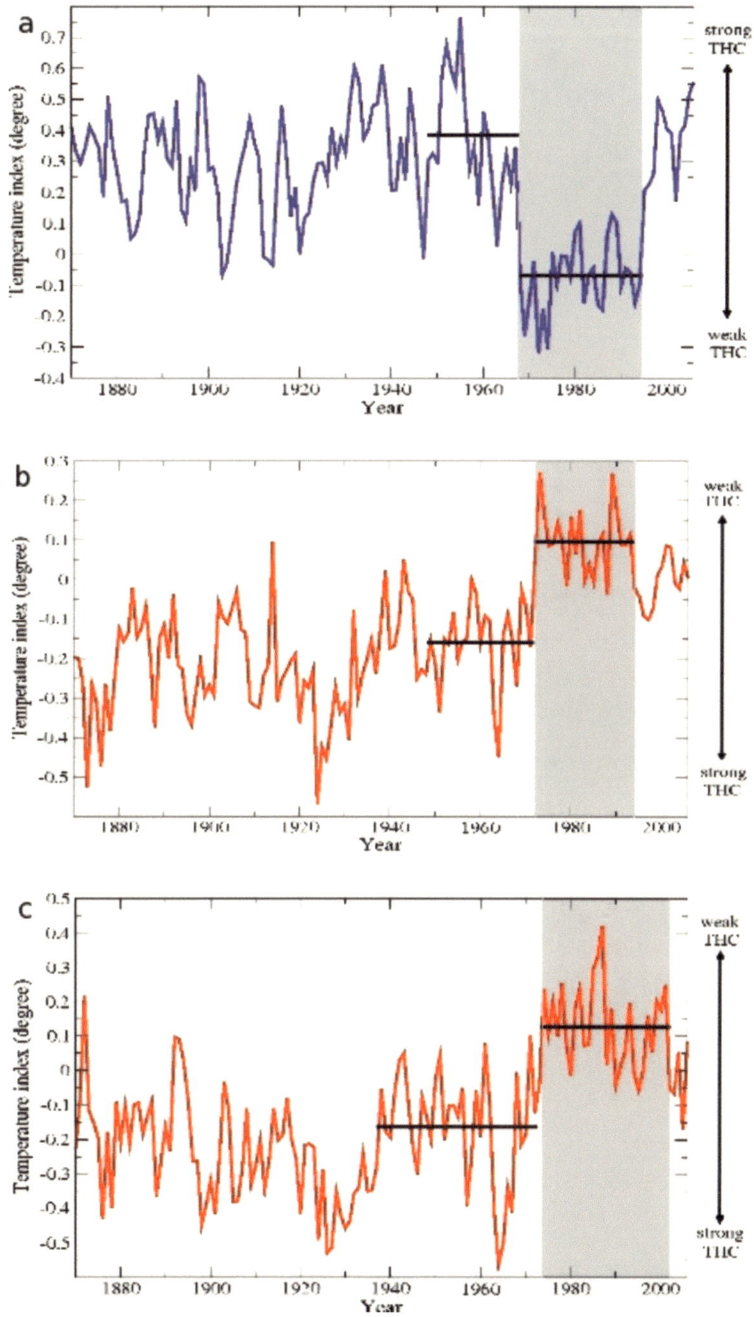

Fig. 10. SST indexes (degree) derived as averages over the a) North Atlantic (50°W-10°W, 30°N-70°N); b) South Atlantic (60°W-0, 50°S-20°S); c) South Indian oceans (40°E-100°E, 50°S-30°S). All three indexes emphasize an AMOC jump to a weaker state during 1970s.

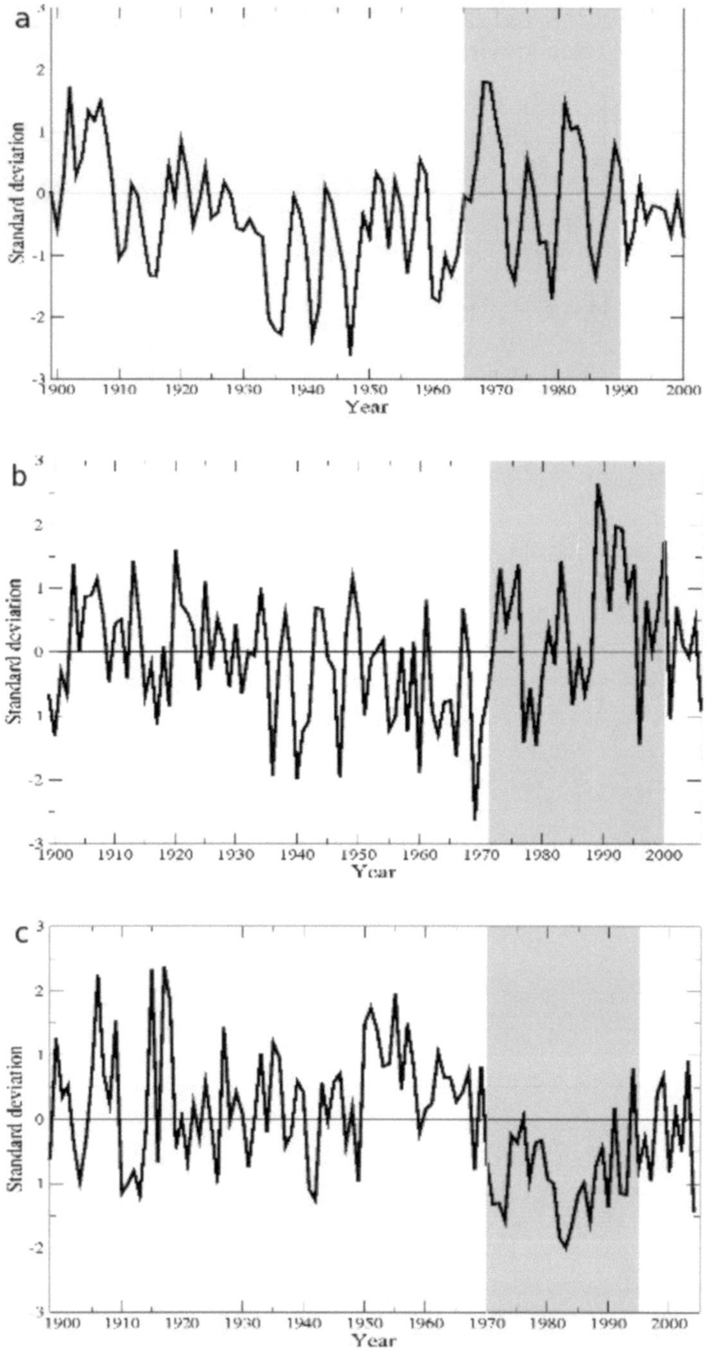

Fig. 11. a) Normalized Fram Strait Sea Ice Export for the 1899-2000 period; b) Principal Component based December-January-February-March average North Atlantic Oscillation index for the 1899-2006 period; the PC analysis was performed on the SLP field over the Atlantic sector (20°-80°N, 90°W-40°E); c) Normalized Sahel precipitation index for the 1898-2004 period

Due to its prominence and spreading path along east Greenland, the GSA in the 1960s represented a significant forcing reaching the Labrador Sea convection region (Dickson et al., 1988). It was the first event from a series of several significant salinity anomalies with decreasing amplitudes, which ended in the late 1980s. Anomalous strong sea-ice export from the Arctic through Fram Strait (Fig. 11a), as that associated with the GSAs (Schmith & Hansen, 2003), would increase the fresh water fluxes in the North Atlantic (Hilmer et al., 1998) and can modulate the AMOC (Häkkinen, 1999). Numerical experiments show that a fresh water pulse applied in the North Atlantic can generate an AMOC shift on a weak state, which is manifested at the surface as a center of pronounced negative anomalies south of Greenland (Rahmstorf, 1994), as seen in Fig. 9a. A similar fingerprint on the surface temperature was attributed to a weakening oceanic overturning in response to a freshwater pulse associated with the 8.2-kyr event (Renssen et al., 2002). For preindustrial climate the SST and SLP response patterns to an AMOC collapse are similar to the corresponding patterns of the 1970s shift, although with a larger amplitude (Vellinga & Wood, 2002) (Fig. 9a,b). A multi model ensemble of numerical experiments related to the ocean response to fresh water forcing reproduces well the general features of the SST pattern of the shift (Fig. 9a) and accurately the dipole anomaly pattern in North Atlantic, which includes cooling south of Greenland and warming over the Barents and Nordic Seas (Stouffer et al., 2006).
It is conceivable that the freshwater pulse associated with the GSA in the 1960s altered the convection pattern and the AMOC (Lenderink & Haarsma, 1994). The changed convection pattern is possibly responsible for the jump to a weaker state. According to model simulations (Lohmann, 2003), the associated ~1.5°C change in SST can be translated to a moderate AMOC reduction of 10%-20%. This results in a freshening of the North Atlantic surface layers in the last decades (Dickson et al., 2002; Curry & Mauritzen, 2005). The AMOC attends the new weak state several years after the GSA in the late 1960s, consistent with a fast response of the meridional overturning to North Atlantic forcing (Eden & Willebrand, 2001). From this perspective, the 1960s GSA represents the cause for the AMOC jump to a new weaker state around 1970. The significant fresh water forcing related to Fram Strait Sea Ice Export starts in the late 1960s and ends in the late 1980s (Fig. 11a), while the weak state extends from the early 1970s to the early 1990s (Figs. 8 and 10a). It appears therefore that the AMOC responded in a direct way to the forcing so that the weak state persisted as long as the anomalous forcing acted.

6. Summary and conclusions

Following the hypothesis that the GSA in the late 1960s has an Arctic origin, the SSA method is applied to identify the quasi-periodic components in the FSSIE reconstruction (Schmith & Hansen, 2003). The analysis emphasizes that the maximum in the late 1960s in the FSSIE record resulted from a synchronization between time-components with periods of ~70-yr, ~14-yr and ~20-yr, consistent with a previous study based on a shorter data set (Venegas & Mysak, 2000). The statistical significance of time-components was tested with Monte Carlo simulations. The time components are associated with known climate cycles (Enfield & Mestaz-Nunes, 1999; Deser & Blackmon, 1993; White & Cayan, 1998), suggesting that modes synchronization can represent an important deterministic mechanism for generating large-scale climate anomalies. In view of the restricted number of analyzed fields and time series, and their limited time extension, more investigations based on different data sets are necessary in order to validate these results.

As we suggest that the late 1960s GSA was caused by synchronization between three climatic modes, one may speculate that the ~70-years, ~14-years and ~20-years periods of the climate modes, which synchronized in the late 1960s, point to a ~140-yr recurrence time of the GSA. Accordingly, a similar synchronization is observed in 1830s (Fig. 4c) and the next one is anticipated for the 2110s, if the cycles continue to exist.

An interesting point concerns a possible factors of the anthropogenic influences on the GSA. The quasi-periodic nature of the synchronized time-components which generate the GSA excludes a potential anthropogenic origin of this freshwater anomaly because this evolves in time as a trend. However, the anthropogenic factor could provide a relatively warm background on which the amplitude and impact of a potential future GSA can be much larger.

Using information derived from numerical experiments we synthesize a large amount of observational data and increase the signal-to-noise ratio related to AMOC changes. Based on an inter-hemispheric and on regional indexes we present evidence and argue that the freshwater pulse associated with the late 1960s GSA modified the convection pattern in the northern North Atlantic, generating an abrupt AMOC weakening, which manifested at the surface as a climate regime shift. This weak state lasted 2 to 3 decades and ended in the early 1990s, with a rapid resumption of the AMOC on a strong state. Our study demonstrates the possibility of monitoring large-scale oceanic changes using observational data and points to key surface locations on which such variations have optimal projections. The identification of abrupt AMOC changes over the modern period and its association with climate regime shifts suggest that future rapid climate changes are also likely in a globally warming world. The fact that an abrupt but moderate MOC reduction caused a climate regime shift in surface variables points to a larger climatic impact generated by possibly more significant AMOC changes.

It was shown based on numerical experiments that prominent climatic events can be generated in the climate system without variable external forcing (Hall & Stouffer, 2001). Our results based on observational and reconstructed data show that a significant climate anomaly, like GSA, can be generated without need of anomalous external forcing, but just through a constructive interference of quasi-periodic time-components. Consequently, the climate shift in the 1970s appears as a result of a resonance between several climate modes.

7. References

Aagard, K. & Carmack, E. C. (1989). The role of sea ice and other freshwater in the Arctic circulation, *Journal of Geophysical Research*, Vol.94, pp. 305-311.

Allen, M. & Smith, L. A. (1997). Optimal filtering in Singular Spectrum Analysis, *Physical Letters*, Vol.234, pp.419-428.

Alley, R. B.; Anandakrishman, S. & Jung, P. (2001). Stochastic resonance in the North Atlantic, *Paleoceanography*, Vol.16, pp. 190-198.

Baines, P. G., & Folland, C. K. (2007). Evidence for a Rapid Global Climate Shift across the Late 1960s, *Journal of Climate*, Vol.20, pp. 2721-2744.

Barber, D. C.; Dyke, A; Hillaire-Marcel, C.; Jennings, A. E.; Andrews, J. T.; Kerwin, M. W.; Bilodeau, G;, McNeely, R.; Southon, J.; Morehead, M. D. & Gagnon, J. M. (1999). Forcing of the cold event of 8,200 years ago by catastrophic drainage of Laurentide lakes, *Nature*, Vol.400, pp. 344-348.

Cavalieri, D. J. (2002). A Link Between Fram Strait sea ice export and atmospheric planetary wave phase, *Geophysical Research Letters*, Vol.29, No.12, pp. 56-1 - 56-4.

Curry, R., & Mauritzen, C. (2005). Dilution of the Northern North Atlantic Ocean in Recent Decades, *Science*, Vol.308, pp. 1772-1774.

Dansgaard, W.; Johnsen, S. J.; Clausen, H. B.; Dahl-Jensen, D.; Gundestrup, N. S.; Hammer, C. U.; Hvidberg, C. S.; Steffensen, J. P.; Sveinbjomsdottir, A. E.; Jouzel, J. & Bond, G. (1993). Evidence for general instability of past climate from a 250-kyr ice-core record, *Nature*, Vol.364, pp. 218-220.

Deser, C. & Blackmon, M. L. (1993). Surface Climate Variations over the North Atlantic Ocean during Winter: 1900-1989, *Journal of Climate*, Vol.6, pp. 1743-1753.

Dickson, R. R.; Meinecke, J.; Malmberg, S.-A. & Lee, A. J. (1988). The "great salinity anomaly" in the northern North Atlantic 1968-1982, *Progress in Oceanography*, Vol.20, pp. 103-151.

Dickson, R.R.; Yashayev, I.; Meincke, J.; Turrell, W.; Dye, S. & Holfort, J. (2002). Rapid freshening of the deep North Atlantic Ocean over the past four decades, *Nature*, Vol.416, pp. 832-837.

Dima, M; Rimbu, N.; Stefan, S. & Dima, I. (2001). Quasi-Decadal Variability in the Atlantic Basin involving Tropics-Midlatitudes and Ocean-Atmosphere Interactions, *Journal of Climate*, Vol.14, pp. 823-832.

Dima, M. & Lohmann, G. (2007). A Hemispheric Mechanism for the Atlantic Multidecadal Oscillation, *Journal of Climate*, Vol.20, pp. 2706-2719.

Eden, C., & Willebrand, J. (2001). Mechanism of Interannual to Decadal Variability of the North Atlantic Circulation, *Journal of Climate*, Vol.14, pp. 2266-2280.

Enfield, D. B. & Mestas-Nunes, A. M. (1999). Multiscale variabilities in global sea surface temperatures and their relationship with tropospheric climate patterns, *Journal of Climate*, Vol.12, pp. 2719-2733.

EPICA Community Members (2006). One-to-one coupling of glacial climate variability in Greenland and Antarctica, *Nature*, Vol.444, pp. 195-198.

Ghil, M.; Allen, M. R.; Dettinger, M. D.; Ide, K.; Kondrashov, D.; Mann, M. E.; Robertson, A. W.; Saunders, A.; Tian, Y.; Varadi, F. & Yiou, P. (2002). Advanced spectral methods for climatic time series, *Reviews of Geophysics*, 8755-1209/02/2000RG000092, 1-1-1-41.

Hall, A., & Stouffer, R. J. (2001). An abrupt climate event in a coupled ocean-atmosphere simulation without external forcing, *Nature*, Vol.409, pp. 171-174.

Hatun, H., Sande, A. B., Drange, H., Hansen, B. & Valdimarsson, H. Influence of the Atlantic Subpolar Gyre on the Thermohaline Circulation, *Science*, 309, 1841-1844 (2005).

Häkkinen, S. (1999). A simulation of thermohaline effects of a Great Salinity. Anomaly, *Journal of Climate*, Vol.12, pp. 1781-1795.

Häkkinen, H. & Geiger, C. A. (2002). Simulated low-frequency modes of circulation in the Arctic, *Journal of Geophysical Research*, Vol.105, No.C3, pp. 6549-6564.

Heinrich, H. (1988). Origin and consequence of cyclic ice rafting in the northeast Atlantic Ocean during the past 130,000 years, *Quaternary Research*, Vol.29, pp. 3359-3362.

Hilmer, R.; Harder, M. & Lemke, P. (1998). Sea ice transport: A highly variable link between Arctic and North Atlantic, *Geophysical Research Letters*, Vol.25, pp. 3359-3362.

Janowiak, J. E. (1988). An investigation of interannual rainfall variability in Africa, *Journal of Climate, Vol.*1, pp. 240-255.

Latif, M. et al. (2006). Is the Thermohaline Circulation Changing ?, *Journal of Climate*, Vol.19, pp. 4631-4637.

Lenderink, G. & Haarsma, H. (1994). Variability and multiple equilibria of the thermohaline circulation associated with deep water fromation, Journal of Physical Oceanography, Vol.24, pp. 1480-1493.

Lohmann, G. (2003). Atmospheric and oceanic freshwater transport during weak Atlantic overturning circulation, *Tellus*, Vol.55, pp. 439-448.

Rahmstorf, S. (1994) Rapid climate transitions in a coupled ocean-atmosphere model, *Nature*, Vol.372, pp. 82-85.

Rayner, N. A. et al. (2003). Global analyses of sea surface temperature, sea ice, and night marine air temperature since the late nineteenth century, *Journal of Geophysical Research,* Vol.108, 10.1029/2002JD002670.

Renssen, H.; Goosse, H. & Fichelet, T. (2002). Modeling the effect of freshwater pulses on the early Holocene climate: the influence of high-frequency climate variability, *Paleoceanography*, Vol.17, No.2, 1020, DOI: 10.1029/2001PA000649.

Schiller, A.; Mikolajewicz, U. & Voss, R. (1997). The stability of the North Atlantic thermohaline circulation in a coupled ocean-atmosphere general circulation model. *Climate Dynamics,* Vol.13, pp. 325-347.

Schlesinger, M. E. & Ramankutty, N. (1994). An oscillation in the global climate system of period 65-70 years, *Nature,* Vol.367, pp. 723-726.

Schmith, T. & Hansen, C. (2003). Fram Strait ice export during the nineteenth and twentieth centuries reconstructed from a multiyear sea ice index from Southwestern Greenland, *Journal of Climate,* Vol.16, pp. 2782-2792.

Stephens, C., Levitus, S., Antonov, J. & Boyer, T. P. (2001). On the Pacific Ocean regime shift, *Geophysical Research Letters,* Vol.28, No.19, pp. 3721-3724.

Stommel, H. (1961). Thermohaline convection with two stable regimes of flow, *Tellus,* Vol.13, pp. 2782-2792.

Stouffer, R. J. et al. (2006). Investigating the Causes and Response of the Thermohaline Circulation to Past and Future Climate Changes, *Journal of Climate,* Vol.19, pp. 1365-1387.

Thompson, D. W. J.; Wallace, J. M. & Hegerl, G. C. (2000). Annular modes in the extratropical circulation. Part II: Trends, *Journal of Climate,* Vol.13, pp. 1018-1036.

Trenberth, K. A. & Paolino, D. A. (1980). The Northern Hemisphere sea-level pressure, *Monthly Weather Review,* Vol.108, pp. 855-872.

Vellinga, M. & Wood, R. A. (2002). Global climatic impacts of a collapse of the Atlantic thermohaline circulation, *Climate Change,* Vol. 54, pp. 251-267.

Venegas, S. A. & Mysak, L. A. (2000). Is there a Dominant Timescale of Natural Climate Variability in the Arctic?, Journal of Climate, Vol.13, pp. 3412-3434.

White, W. B. & Cayan, D. R. (1998). Quasi-periodicity and global symmetries in interdecadal upper ocean temperature variability, *Journal of Geophysical Research,* Vol.103, pp. 21335-21354.

Zhang, Y.; Wallace, J. M. & Battisti, D. S. (1997). ENSO-like interdecadal variability: 1900-1993, *Journal of Climate,* Vol.10, pp. 1004-1020.

11

Tropical Dry Forests in the Global Picture: The Challenge of Remote Sensing-Based Change Detection in Tropical Dry Environments

Betsabé de la Barreda-Bautista[1], Alejandra A. López-Caloca[1], Stephane Couturier[2] and José Luis Silván-Cárdenas[1]
[1]*Centro de Investigación en Geografía y Geomática "Ing. Jorge L. Tamayo"*
[2]*Universidad Nacional Autónoma de México (UNAM). Instituto de Geografía*
Mexico

1. Introduction

Global environmental change has recently pushed the scientific community in the quest for more comprehensive spatial information on the continental biosphere. In terms of climate change, ecosystem monitoring has become one of the priorities to better understand the evolution of terrestrial carbon stocks, as well as to foster conservation policies for these carbon stocks. According to IPCC (2002), deforestation and land clearing activities, mostly from sub-tropical regions, contributed with one fifth of the greenhouse gas emission during the 1990s.

The tropical dry forests are one of the most extended tropical forested ecosystems, and yet have received only recent attention from the scientific community. This ecosystem is also scarcely represented in the international protection schemes, which perhaps causes increased vulnerability of this ecosystem to the tropical fingerprint of global human development. Additionally, the climatic conditions are relatively attractive for human settlement and the ecosystem has historically supported dense agriculture activity. In megadiverse Mexico for example, these forests extend up to 60% of tropical forests, and an estimated 30% of this extent is considered as highly modified under anthropic pressure. The annual deforestation rate of the deciduous tropical dry forest in Mexico has been evaluated at around 1.4 - 2 %. The contribution of the latter to climate change is manifolds, including carbon emissions, increased albedo and regional hydrographic cycle alteration. Moreover, the very loss of biodiversity derived from the conversion of forest to grassland for pasture is considered as a triggering factor for future forest fires and conversion to more grassland.

The monitoring and analysis of the forest distribution pattern, including phenological and anthropogenic modifications, contributes to the uneasy task of slowing down the tendency of forest loss. Remote sensing has proved a fundamental tool for such monitoring, owing to its contribution to the study and understanding of the global environment through time, and the calibration of models which help building environmental scenarios in the future.

However, in the case of tropical dry forests, remote-sensing based studies report high levels of error for its detection and mapping. One of the reasons for this is associated with the seasonality of the vegetation: During the wet season, cloud-free satellite imagery is difficult to acquire, while during the dry season when the imagery is more available, the leaf off configuration of the forest causes misclassification with savanna, shrubland or grassland ecosystems. Other sources of uncertainty are related to tropical dry forest definitions and classification methodology. For example, classification schemes at regional scales have mainly been calibrated on other, better studied, ecosystems, and have led to poor accuracy results for tropical dry forests and there is no widely accepted, standard methodology so far for the monitoring of this forest type.

This chapter focuses on the specificity of the tropical dry forest ecosystem and its related challenge for remote-sensing based mapping and monitoring. An emphasis is laid on the exploration of change detection techniques in tropical dry environments. The first section of the chapter relates to the global importance and specificity of the tropical dry forest, in terms of biophysical features, ecological function, and role in climate change. In the second section, remote sensing as a tool to monitor forested ecosystems is reviewed. We discuss the difficulties for monitoring tropical dry forests, among which methodological difficulties of change detection techniques, and the potential of the Multivariate Alteration Detection (MAD) as an efficient technique in this case. In the third section, a comparative study of change detection techniques is presented for Landsat ETM+ imagery on a portion of the tropical dry forest along the western coast of Mexico. Finally, a discussion is proposed on the perspectives learned for dry tropical forest monitoring.

2. Global importance and specificity of tropical dry forest

Most tropical ecosystems have been subjected to a diminishment of their extension due to the impact of human being's colonization. Murphy & Lugo (1986) report that about 40% of the earth's subtropical landmass is dominated by open or closed forest, and, out of this landmass, 42% would be tropical dry forest (TDF), 33% would be moist forest, and only 25% would be wet and rain forest. Large patches of TDF in the Americas are located in Northeastern Brazil and Southeastern Bolivia, Paraguay, and Northern Argentina (52.4%). The rest of the existing TDF is found in highly fragmented landscapes in Mexico and Central America (Sánchez-Azofeifa *et al.*, 2009).

Since there is a need to generate scientific knowledge on the tropical dry forest, a series of studies regarding its structure (plant height, density), functioning and composition have already been reported. Several efforts have been made at a local and regional level in Central America and the Caribbean (Murphy & Lugo, 1995; Griscom & Ashton, 1995)), Brazilian caatinga (Sampaio, 1995), Africa (Menaut, 1995) and Thailand (Rundel & Boonpragob, 1995), Costa Rica (Sánchez-Azofeifa *et al.*, 2009), islands in Lago Guri, Venezuela (Feeley *et al.*, 2005) and Mexico (Maas, 1995; Martínez-Yrízar, 1995; Trejo & Dirzo, 2000). In order to monitor their extent, global distribution maps of TDF were built based on the MODIS Vegetation Continuous Fields (VCF) product; an estimated 1,048,700 km^2 of tropical dry forest remains distributed throughout the three sub-tropical regions. More than half of the forest area (54.2%) is located within South America, the remaining area being almost equally divided among North and Central America, Africa and Eurasia, with a relatively small proportion (3.8%) within Australasia and Southeast Asia (Miles *et al.*, 2006).

The TDF is one of the most predominant types of tropical vegetation in Mexico (Trejo & Dirzo, 2000). It is distributed in Mexico along the Pacific Coast where it covers large, practically uninterrupted extensions from the South of Sonora and Southeast Chihuahua all the way down to Chiapas, continuing to Central America. It also exists in patches on the Southern most point of Baja California. It takes up a large portion of Chiapas in the Isthmus of Tehuantepec, also including the Gulf' side in Southern Tamaulipas, Southeastern San Luis Potosí, Northern Veracruz, Yucatán and Campeche (Rzedowsky, 1978). This type of vegetation develops in Mexico in regions of warm sub-humid weather, with an annual average temperature above 20 to 29°C, an annual rainfall of maximum 1200 mm and between 0 and 1700 m.a.s.l and its spatial distribution is characterized by a heterogeneous matrix of topographic, aspect, climatic and edaphic conditions (Rzedowsky, 1978; Trejo & Dirzo, 2000). These types of forests generally present only one arborous stratum, the height of the trees generally oscillates between 5 and 15 m, the treetops are convex or flat and their width sometimes matches or surpasses the height of the trees (Rzedowsky, 1978).

According to Rzedowsky (1978) the impact of human activities on this type of vegetation had been considerably less than in other types of vegetation due to the shallow, rocky floors, stating that they are not good for the development of agriculture even though there were regions of TDF which were exploited for livestock farming, selective woodcutting and affected by forest fires. Still, TDF covered around 14% of the Mexican territory. However, Janzen (1988) assured that the TDF is the most threatened of the major tropical forest types and that this type of vegetation has declined from an original area of 550,000 km^2 and only 0.09% has been granted a conservation status. Later in the 80's and 90's Flores & Gerez (1994) reported a reduction of 5.42% of its surface between 1981 and 1992. Ceballos & García (1995) stated that there is no consensus about the rates of deforestation in the Neotropics. However, large dry forest patches had disappeared mainly to support agriculture and cattle ranching activities. Trejo & Dirzo (2000) found that in Mexico 73% of the TDF has been altered, degraded or converted mainly into agricultural or livestock grasslands.

Quadri de la Torre (2000) states that this ecosystem took up an area of approximately 16 million hectares in the 1950's, of which less of a third remained in that time due to deforestation. One of the last studies made and one of the first to use remote sensing (Miles *et al*, 2006) demonstrates that in Latin America TDF experienced the greatest decreases between 1980 and 2000 in percentage of forested area, with an estimated figure of 12%, and the main causes for change and loss in these type of ecosystems were mainly due to climatic changes, fires, forest fragmentation, conversion to agriculture and population growth.

2.1 Ecological functions and ecosystem services

The relatively simple structure of dry forests makes them more suitable to unravel ecological functions, patterns and processes than their wet counterparts, and, indeed, some of the most significant insights into ecosystem structure and function in the tropics have had their basis in research conducted in dry rather than in wet forests (Bullock *et al.*, 1995).

The distinctive characteristic of TDF is the availability of humidity. It creates a very clear physiognomic contrast between the dry and rainy seasons. The dry season has a length of 7 to 8 months and the vegetation formations lose their leaves (Rzedowsky, 1978). This creates a vegetation's hydric and available mineral nutrients deficit, giving as a result a highly complex and diverse ecosystem (Bullock *et al.*, 1995). The seasonality that the TDF presents becomes a dominant ecological force when temporal patterns of biological activity such as growth or reproduction become synchronized with the availability of water or when the geographic distributions of plant or animal taxa are constrained by moisture limitations during certain times of the year (Murphy & Lugo, 1986).

Ecosystem services refer to the biological properties or processes in forests. They also represent human population benefits derived directly or indirectly from ecosystem functions. Ecosystem services consist of flows of materials, energy, and information from natural capital stocks combined with manufactured and human capital services to produce human welfare (Constanza *et al.*, 1987). For example, forest ecosystems may absorb significant quantities of carbon dioxide. As a result of this, in the last decades a considerable amount of interest has appeared to increase the amount of carbon content in the land vegetation by means of forest preservation, reforestation, agroforestry and other methods of land use.

Tropical dry forests are less diverse than wet forests; total biomass of trees and net annual primary productivity (NPP) are also lower, but the diversity of plant life forms is greater. Improved measurements of NPP have included annual litterfall and have suggested high values of forest biomass (Martínez-Yrízar *et al.*, 1996; Bullock *et al.*, 1995). TDF presents well defined phenological cycles, which make them potential proxies for the biological response to climate change and specifically to drought effects. According to Maass *et al.* (2005) little research has been conducted therefore, little is known about the ecosystem services provided by the TDF.

Kalacska *et al.* (2008) point out the need of characterizing, as accurately as possible, the initial state and extent of the forest to estimate the payments for environmental services. On his own, it studies one of the most important ecosystem services of the TDF: Carbon sequestration.

Maass *et al.* (2005) provide a conceptual framework to construct an integrated understanding of ecosystem services delivered by tropical dry forest of the Pacific Coast of Western Mexico. In order to explain this, the structural and functional aspects of the natural and transformed ecosystems were considered, as well as a more recent understanding regarding social dimensions of ecosystem management. Maass *et al.* (2005) identified nine main services in the Chamela Region, Jalisco, Mexico, which are fresh water, agricultural and pastoral goods, diverse resource provisioning, biodiversity, climate regulation, soil fertility maintenance, flood control, bio-regulation and scenic beauty. Additionally, Pielke *et al.* (2002) also claim that the climate-related ecosystem services that tropical forests provide include the maintenance of elevated soil moisture and surface air humidity, reduced sunlight penetration, weaker near-surface winds and the inhibition of anaerobic soil conditions.

2.2 Tropical dry forest removal in the context of global warming

The human activity is causing deforestation at global scale. The main driver of deforestation is land use change, including conversion to grasslands and to agricultural land, forest fires and irrational woodcutting, among others. The conversion through shifting cultivation and clearing of secondary vegetation makes significant contributions to global emissions of greenhouse gases today, and has the potential for large additional emissions in future decades. Additionally, the non-sustainable exploitation of timber where extraction exceeds regeneration and reforestation implies a surplus of emissions of greenhouse gases (Intersecretarial Commission of Climatic Change 2009).

The consequences of deforestation are multifold: Firstly the production of considerable carbon emissions due to the combustion and decomposition of vegetable biomass and also to the loss of organic carbon located in the soil.

Deforestation also impacts on regional and global climates, not only because of emissions of the greenhouse gases, but also because of the increased albedo and evapotranspiration derived (IPCC, climate change and biodiversity, 2009). The regional consequences of the future climate change are not yet known, but it is highly probable that the changes in variables induced by climate change may modify in turn the regional climates and may induce unexpected effects over the weather conditions in different regions. Local and regional climate changes may consist of alterations in the length of the agricultural seasons, in the availability of water or in a greater incidence of perturbation regimes (extreme values of high temperatures, flooding, droughts, forest fires or plagues), which will, in turn, notably impact the structure and function of the natural and artificial environments (IPCC, 1997).

Strictly in terms of CO_2 emissions in the world which seem to prevail in the factors of global warming (Mintzer, 1992, as cited in Ordoñez, 1999), deforestation contributes roughly one fourth of the emissions while fossil fuel consumption by humans contributes roughly three fourths (IPCC, 2009). However, deforestation has consequences beyond global warming, much further into the multi-threatening global change: for example, the fragmentation of habitats and conversion of extensive land uses to intensive ones cause a severe loss in biodiversity. In fact, changes in ecosystems may have important consequences within global warming. For example, the alteration of tropical landscapes, primarily the conversion of forests to agriculture or pasture, changes the partitioning of insolation into sensible and latent turbulent heat forms. Less transpiration associated with the agricultural and pasture regions results in less thunderstorm activity over this landscape (Pielke et al., 2002). Each change in the quantity or quality of types of natural capital and ecosystem services may impact human welfare. These are part of both small changes at large scales and large changes at small scales; in other words changes in particular forms of natural capital and ecosystem services which both alter the costs or benefits of maintaining human welfare (Constanza *et al.*, 1987).

From the strict global warming perspective, tropical ecosystems in general, and forests in particular, can accumulate a significant amount of the global carbon in biomass (Pfaff, 2000). As a consequence, there is a great need to estimate, map and monitor with greater precision the accumulated carbon in the forests (Goetz, 2009). The amount of biomass is generally much greater in tropical rainforests than in tropical dry forests. However, worldwide many tropical rainforests are replaced by TDF (William, 1985), especially in South America and Africa. Rainforests are converted to Savanna and other dry tropical forest types, often irreversibly, because of fire events. This phenomenon justifies the importance of studying TDF carbon stocks in spite of their lesser content with respect to rainforests.

In Mexico, TDF, described as a low size tropical deciduous forest, has been one of the ecosystems most affected by land use change. It is also likely that in the future, an increasing land use change trend will continue due to the increasing demand over ecosystem services and natural resources. Burgos and Maass (2004) unveil the causes and consequences of tropical dry forest conversion into other types of land use in Mexico. They show how much of the TDF has already been converted to pasture and agricultural land. Human settlements surrounding the reserve of Chamela (Jalisco, Mexico) are organized under the Ejido land tenure system where peasants have tended to clear TDF for slash and burn cropping and pasture for cattle grazing. Ejidos are considered land use administrative units under the control of local peasant communities. Decisions on land use change at the Ejido level are mostly driven by the needs to fill local market expectations. The decisions are nevertheless controlled and/or motivated by Mexican agrarian and environmental regulations (Sánchez-Azofeifa, 2009).

Ordoñez & Masera (2001) report that one of the main emitters of greenhouse gases in Mexico is land use change: 30.2 million tons of carbon released each year; the total emissions produced in Mexico represent 1.4% of the total carbon emissions around the globe. The greenhouse gas which contributes the most to global warming is CO_2 and tropical forests are said to play a very important role in the cycle of this gas. They claim deforestation produced 69.5% of the reported CO_2 emissions in Mexico in 2006.

In order to quantify the TDF ecosystem's biomass, carbon (C) and nitrogen (N) pools of both intact forests and converted sites, Jaramillo *et al* (2003) sampled the total aboveground biomass (TAGB), and the N and C pools of two floodplain forests, three upland dry forests, and four graze lands that had been dry forest extents. The lack of a consistent decrease in soil pools caused by land-use change suggests that C and N losses result from the burning of aboveground biomass. It is estimated that in México dry forest landscapes store approximately 2.3 Pg C, which is about the same C stored by the evergreen forests in that country (approximately 2.4 Pg C). Potential C emissions to the atmosphere from the burning of biomass in the dry tropical landscapes of Mexico may be higher (708 Tg C), as compared with from evergreen forests (569 Tg C). Efforts to determine the status of TDFs in Mexico are particularly critical in certain regions with the highest diversity and endemism of TDF in the world. These regions include the Yucatan Peninsula, the coast of the State of Jalisco, the Balsas and Santiago river basins and scattered areas along the Gulf coast (Trejo & Dirzo, 2000, 2002).

3. Remote sensing as a tool to monitor forested ecosystem

Given the need to create a strategy to go deeper into the behavior of TDF, it is necessary to go further into characterization through the dominion of a spatial and temporal scale. In this work, the resources of geospatial information such as satellite images are used. The identification of places is essential in the determination of spatial patterns. The related images are connected to the geographical characteristics of the place. Nowadays, satellite images are tools which help us to identify geographic features or places that are associated to a certain type of application; provide a context of the place as well as helping us identify natural and environmental aspects of the geographical space. However, there are different ways to approach the geospatial problem in a more integral way, such as using more technological, geospatial tools that allow organizing and optimizing the use in the handling of the information. This is presented within a conceptual framework in Geomatics where the interaction of the changes in landscape is motivated with the appropriate models in order to simulate identified social, economic and ecological processes, and their dynamics and interactions given by the social actors as well as the government itself, from a territorial point of view (Reyes *et al.*, 2006).

The identification, follow-up and notification of the changes experienced by the terrestrial cover may be made through field information, geographic statistics, models based on forest inventories, remote sensing techniques, flow measurements, soil sampling and ecological studies (IPCC, 2002). In Mexico, as in the world, the ecosystems are in permanent change due to natural causes and the pressure exercised upon them by human beings. Remote sensing and geographic information systems have been fundamental tools in monitoring these changes throughout the years.

These sciences allow the pinpointing and following-up of these changes from a historical perspective due to their temporal resolution. Remote sensing has been a very valuable instrument in the observation and follow-up of Earth and one of its most relevant contributions is the study of the weather and the following-up of the processes happening on it (Chuvieco, 2002). However, there are several challenges associated with developing the remote sensing inputs needed to understand landscape dynamics. Land use and land cover change studies are often based on information about landscape structure and composition at different spatial and temporal resolutions derived from remotely sensed data.

There are two change detection approaches according to Chuvieco (2002) considering a temporal aspect: a multiseasonal and a multiannual approach. The first one refers to following patterns by seasons in the same year where, for example, the changes in the phenology of the vegetation can be analyzed. The multiannual approach refers to the detection of changes happening between two or more dates. This last approach is the one this work is focussed on in order to monitor the changes in soil cover and its implication in climatic change.

The changes in the terrestrial cover and the changes in land use have important implications for the future climatic changes on Earth. This is why the need to understand the interactions between human activity and natural resources appears. These same interactions are seen reflected in the change patterns of land use/terrestrial cover. These changes are a process widely distributed, accelerated and significant, mainly caused by human actions and, in many cases, also bring about changes which impact on humanity (Rosete *et al.*, 2008). This is why remote sensing proves to be fundamental in the evaluation of changes in the patterns of deforestation and changes in land use, because this allows a qualitative and quantitative analysis of the changes happening in the ecosystems.

Remote sensing has been used for the extraction of information from the environment. Within the most common applications we find coverage mapping, changes in land use, weather monitoring, biomass accounting, vegetation phenology, desertification, deforestation, monitoring of forest fires and natural disaster prevention, all of which are closely linked to climatic change.

There are several sensors which have made possible all the studies above mentioned. However, it is noteworthy to mention that the Landsat program has a compilation of images dating 40 years, which represents the lengthiest registry of information of the Earth's surface obtained in a global and repetitive way from space (USGS, 2010). The availability of periodic data obtained through satellite observation has made possible for climatic change detection to be one of the most important applications in remote sensing. Since 2009, the information obtained from this program has become available providing a lot of material to make scientific studies, as well as the advance and implementation of new methodologies to study Earth's surface from different fields.

3.1 Importance of change detection for ecosystem monitoring

Mapping TDF distribution remains a difficult challenge in spite of the diversity and potential of the remote sensing data that have been employed for this task so far. Parts of the challenge are because most of the studies have been performed in tropical evergreen forest, these studies are successful in the characterization of this forest and they obtained good results in accuracy (e.g. user's accuracy of MODIS derived global map for broadleaf evergreen forests: MODIS team, 2003). TDF has received considerably less attention and the extraction of TDF information from remote sensing imagery has been hampered by at least three factors:

1. The difference in the classification method for TDF cover extraction.
2. The difference in definitions of TDF among countries and scientists.
3. The difficulty in mapping or characterizing TDF with remote sensing data.

Regarding points 1 and 2, the extraction of the distribution of ecosystems is in general a complex task because vegetation classification systems differ (e.g. bioclimatic types, global biogeographic classifications, vegetation types according to the Marrakech's accords, etc.) along with natural boundaries of these ecosystems (Miles *et al.*, 2006; Kalacska *et al.*, 2008). Another interesting difference reported by Lucas *et al.* (2004) is related to natural boundaries of tropical ecosystems, due to excessive loss and fragmentation.

With regard to point 3, the noise in the remote sensing data due to atmospheric conditions causes more problems than for evergreen forests (Lucas *et al.* 2004), because of the seasonal phenology of the forest. During wet season, the acquisition of the imagery is hampered by cloud cover. During dry season, when more imagery are available, a varying percentage of trees is leafless which causes the forest to be possibly misinterpreted, because of a similar spectral signature, with non treed ecosystems or with highly modified evergreen forests.

Couturier (2010) assessed the accuracy of the MODIS derived global map MOD12Q1 for the major forest biomes in the total extent of mega-diverse Mexico. He found that the greatest amount of error occurred for the broadleaf deciduous forests while evergreen forests were relatively well mapped; oak forests tended to be confused with grassland and TDF tended to be confused with evergreen forests. Only 39% of the TDF in Mexico were correctly mapped. His findings corroborated Kalacska et al. (2007), who found only 36% accuracy for the TDF class in Mexico's Chamela Natural reserve and Costa Rica. The TDF was significantly underrepresented in the land cover maps. In the case of confusion with evergreen forests, this means that carbon stock estimations derived from the maps are optimistically flawed by a significant amount.

In the perspective of precise ecosystem service tracking and global warming mitigation models, there seems to be a need for enhancing classification methods or/and optimizing remote sensing inputs for the estimation of TDF extent.

3.2 Difficulties for monitoring dry forest

In order to estimate the total value of ecosystem services, it is necessary to estimate the total global extent of the ecosystem (Constanza *et al.*, 1987) and the rate of changes of the ecosystem extent (Kalacska *et al.*, 2008). In the last few years efforts have been focused in the creation of models that help to create future scenarios which can help in the management of issues related to the preservation of the biodiversity and sustainable management of natural resources (Lambin, 1997).

The changes in vegetation cover and land use are dynamic processes. Therefore, the transitions are produced in different ways and in different places within its own limitations, as a response to social, economic or environmental factors. Change analyses provides fundamental information for the managing and planning of territory, the comparison in landscaping dynamics, in environmental impact studies (Serra *et al.*, 2001), the detection of modifications in the conditions, abundance and distribution of species and the updating and systematization of natural resources inventories, among others.

This information, in order to determine significant changes in those critical components in an ecosystem, is one of the initial conditions and it will assist as a basis for the development of managing and preservation plans for ecosystems and the preservation throughout time of the natural resources sustaining these ecosystems (Ramírez-Bojórquez *et al.*, 2006). Balance between the natural habitat and human landscape could determine the future of the conservation of biological diversity in large areas of the planet. For this reason, it is important to monitor these conversions in the habitats. (Lee *et al.*, 1995 as cited in López *et al.*, 2001).

Transformation in ecosystems should be handled taking into account three main questions: Which environmental and cultural variables are mainly contributing to the changes on Earth's cover? Which are the places affected by these type of changes? How is the progress of these changes? The challenge, in order to model and monitor the terrestrial cover, is answering these questions in such a way that these studies become crucial for any decision-making (Lambin, 1997).

Because of all the above mentioned, the importance of monitoring natural environments and the research and improvement of change detection methods are an issue under constant development in order to insert itself in the decision-making processes and investigation lines which interest the scientific community and society as well as is the case of climatic change.

It is important to note that a better change detection is based on the better understanding of the process and driving force of change (Sui *et al.*, 2008). This is why it is interesting to know the behavior of the TDF because proposing efforts towards the conservation of the tropical dry forest would help alleviate, among other things, the emission of greenhouse gases. This is important because it has been demonstrated that changes in these kinds of ecosystems have caused changes in the carbon deposits (Estrada-Salvador & Navar, 2009).

In order to analyze and understand the ecosystem services like the capture of carbon that is in the TDF, it is necessary to know the intrinsic characteristics of the forests such as temperature, rainfall, soil characteristics and —of course and most importantly for the information within this chapter— the changes in land cover as well as the change in land use. With this, the number of biomes is calculated as well as the changes they have suffered throughout time in order to make estimations in these ecosystem services. The most significant impact of human activities on C storage comes through deforestation (Houghton *et al.*, 1999, as cited in Pfaff *et al.*, 2000). It is necessary for this to know the baselines for land use and translate observable land use and physical characteristics of sites into measures of carbon (Pfaff *et al.*, 2000).

Land-cover change provides an additional major forcing of climate, through changes in the physical properties of the land surface (Pielke *et al.*, 2002).

To estimate the change in cover and to understand the forces and factors related to these changes can be important measures to decrease these practices and in this way minimize the carbon emissions which would help to alleviate climatic change, biodiversity and endemic species conservation, keep land in natural conditions, regulate the water cycle with direct benefits in a better microclimate and a better recharge of aquifers, among others (Estrada-Salvador & Navar, 2009).There are many ways to tackle a change detection study and they should work mainly towards the proposed study's objectives, without leaving aside technical matters such as the equipment's availability, images and processing software. This line of investigation through remote sensing has become essential for several applications. Table 1 (below) shows the relationship between the most explored applications in change detection and the methods being used for each one of them.

Typical applications	Used Methods
Change in land use or landscape cover	Post-classification comparison, indirect classification of multi-temporal images, visual interpretation
Change in vegetation	Image differencing, images rationing, post-classification comparison
Monitoring of natural disasters	Image differencing, images rationing, regression, fire models, visual interpretation
Changes in urban zones and SIG updates	Image differencing, images rationing, object-oriented methods, visual interpretation
Object follow-up and monitoring	Image differencing, images rationing, post-classification comparison, object-oriented methods, visual interpretation
Research and monitoring for military applications	Image differencing, images rationing, object-oriented methods, visual interpretation, time series analysis
Other applications (e.g. changes in wetlands and landscape, crop monitoring, etc.)	Image differencing, images rationing, post-classification comparison, object-oriented methods, visual interpretation

Table 1. Typical applications in change detection studies and used methods.

4. Potential monitoring techniques

In order to analyze the dynamic characteristics of the land covers, a great quantity of techniques has been developed for the extraction of information to allow visualization and understanding of the terrestrial cover through time. Change detection methods have been developed over the last decades and have been improved through time. The techniques being used to make change detection studies are focused on showing the transformations or modifications that have happened in a determined area as a consequence of natural phenomena or man-made actions.

Digital change detection analysis is affected by spatial, spectral and temporal factors (Coppin *et al.*, 2004). The type of method used may give different final results; therefore choosing an effective technique becomes crucial depending on the study's objectives, the case to be studied, the phenological characteristics of the area, etc. The limitations, advantages and disadvantages of each technique must be important points to consider in the analysis, so that the right choice is made with the one that covers the necessary characteristics to get a good result in the change analysis.

Several existing methodologies have been proposed for change detection analyses, which have been modified with the making of different works. They are not exclusive methodologies, meaning that some of them use concepts from others and in this way they have evolved in the improvement and development of them. Whichever combination of change detection algorithms and classification routine is applied, it is evident that several alternatives exist and all of them have different degrees of flexibility and availability (Coppin *et al.*, 2004).

New methods and techniques are in constant development and invention (Lu *et al.*, 2004), and it is sought that they are widely used in fields such as the detection, follow-up and modeling of the changes in land use and land cover, forests, ecosystems, natural disasters and urban areas (Sui *et al.*, 2008).

The techniques for the change detection analysis, through digital analysis, have been divided or categorized and described by different authors (Chuvieco, 1998; Coppin *et al.*, 2004; Pacifici, 2007; Niemeyer *et al.*, 2008; Lu *et al.*, 2004) according to the involved processes and the algorithms' characteristics. Each method has a different degree of robustness, flexibility and significance. However, since the change detection techniques require making a complete and complex analysis of multi-temporal, multispectral images with several spatial resolutions and geographic data, nowadays no classification is enough to frame all the algorithms' bases or the developing methods.

Singh (1989) classified the methods in two categories: post-classification change detection and direct comparison with multi-temporal data. Lu *et al.* (2003) grouped the methods into seven categories: algebra, transformation, classification, advanced models, geographical information system approaches, visual analysis and other approaches. Wyatt (2000) basically groups them in two: post-classification methods and algorithms for quantifying change. Coppin *et al.* (2004) classified the methods in two large groups: bi-temporal change detection and temporal trajectory analysis. Sui *et al.* (2008), in turn, classified the methods in seven categories: direct comparison, classification methods, object-oriented methods, time series analysis, visual methods and the hybrid method. These seven categories are classified within the Coppin *et al.* (2004) classification.

The direct comparison methods use algebraic operations, trying to compare arithmetically bands derived from two different dates (Chuvieco, 1998). These methodologies include image differencing and image rationing, correlation coefficient methods, biophysical variables used to compare values from two images from different years; Coppin *et al.* (2004) calls this univariate image differencing. The most widely used biophysical variables for this type of analysis are the vegetation indexes, e.g. Normalized Difference Vegetation Index (NDVI), Principal components analysis (PCA), etc. (Coppin *et al.*, 2004).

The research in change detection techniques is still very active as new techniques should be developed to effectively use both the available data obtained through remote sensing, which is increasingly diverse and complex, and expected data from airborne and space-borne sensors under development.

Even a glimpse of the remote sensing literature provides enough evidence to conclude that the images obtained through satellites at different dates may be used to detect and monitor changes happening in the ecosystems (Coppin *et al.*, 2004). However, the efforts towards developing a better understanding of the change detection process should not decrease so that a better understanding of the applications and methodologies on this matter continue to be made.

It is also important to mention that the change detection techniques may focus on three fundamental aspects: "if it changed", which means whether there has been a change in the images or not; "what is changed", which determines which characteristics are changing, meaning that it identifies the "from-to" change; and lastly "how it changed", which identifies the change process and trajectories (Sui *et al.*, 2008).

4.1 Post-classification

One of the most frequently used methodologies is the post-classification change detection method which involves two steps: In the first step, images acquired at two times comprising the time period of interest are classified independently. In the second step, a pixel-by-pixel or segment-by-segment comparison is carried out in order to determine the type of changes in land cover class. This method has the advantage of minimizing impacts of atmospheric, sensor and environmental differences between multi-temporal images; it provides complete matrix of change information, whereas its disadvantages are the need of a great amount of time and expertise to create accurate classification products (Lu *et al.*, 2004). By adequately coding the classification results, a complete change matrix is obtained, and change classes can be defined by the analyst (Coppin *et al.*, 2004). This method has been applied in studies on land cover change, forest or vegetation change, and urban change (Sui *et al.*, 2008). It is important to mention that the accuracy of this method depends on the accuracy of the classification. The final accuracy resembles very closely to that resulting from the multiplication of the accuracies of each individual classification and may be considered intrinsically low (Coppin *et al.*, 2004).
In order to make image classification, the spectral information is used by a pattern recognition technique. This means that the objects' spectral response is analyzed and separated into classes. The objective of this process is to assign a class to all the image pixels (e.g. bodies of water, types of vegetation, bare soil, and so forth). This way a thematic map is obtained, which means that we get an image with a mosaic of pixels, each one belonging to a particular class. (Ortiz, 2005).
One of the advantages of this method is that the classification of multitemporal images avoids the need to normalize for atmospheric conditions and sensor differences, between the acquisitions. However, the biggest disadvantage is that the performances of the post-classification comparison technique critically depend on the accuracies of the classification maps (Jensen, 1996). In particular, the final change detection map exhibits an accuracy close to the product of the accuracies yielded at the two times. This is due to the fact that post-classification comparison does not take into account the dependence existing between two images of the same area acquired at two different times (Pacifici, 2007). The effectiveness of this alternative depends on how the change classes are spectrally differentiated from the no change classes.

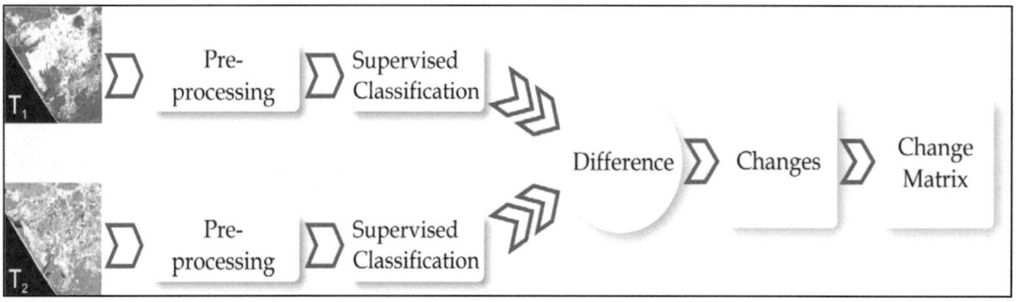

Fig. 1. Scheme of post-classification techniques.

4.2 Multivariate Alteration Detection

The Multivariate Alteration Detection (MAD) (Nielsen and Conradsen, 1997) procedure is based on a classic statistical transformation, referred to as canonical correlation analysis, to enhance the change information in the different images. The property of the Multivariate Alteration Detection transformation is the linear scale invariance. Therefore, with the use of MAD, the preprocessing by linear radiometric normalization becomes unnecessary, meaning that this implies insensitivity to linear differences in atmospheric conditions or sensor calibration at the two different image times.

Canonical correlation analysis investigates the relationship between two groups of several variables. It finds two sets of linear combinations of the original variables, one for each group. Canonical correlation analysis (CCA) is used to measure the strength of association between two sets of variables. The first two linear combinations are the ones with the largest correlation. This correlation is called the first canonical correlation and the two linear combinations are called the first canonical variates. The second two linear combinations are the ones with the largest correlation subject with the condition that they are orthogonal to the first canonical variates. This correlation is called the second canonical correlation and the two linear combinations are called the second canonical variates. Higher order canonical correlations and canonical variates are defined similarly. Nielsen (1998), looks for canonical variates that are as similar as possible as measured by correlation, so he requests positive canonical correlations.

The essence of this method is to transform two groups of multispectral data (two images in time 1 and time 2 in such a way that they show the maximum change (variance) in all the bands in a new set of images (MAD components), so that the change may be seen at the same time in various resulting MADs showing the changes of the different characteristics of the image (Fig. 2).

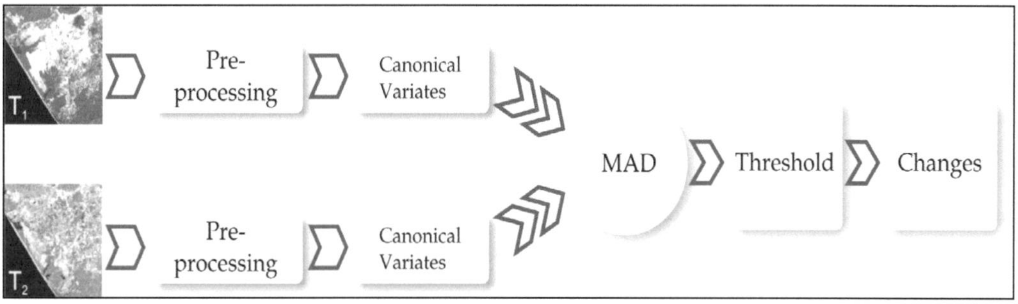

Fig. 2. Scheme of Multivariate Alteration Detection technique.

The mathematical expression of the MAD transformation is:

$$Z = a^T X - b^T Y \tag{1}$$

where X and Y are the two multivariates of two images, a and b are coefficient vectors of linear combinations, and Z is the difference variable. Variance of Z is to be the maximized subject to constrain that both $a^T X$ and $b^T Y$ have unit variances. (Zhang, 2007)

Vectors a and b could be solved by employing the established Canonical Correlation Analysis in multivariate statistics (Thompson, 1984). The correlation between $a^T X$ y $b^T Y$ is called canonical correlation. The lower canonical correlation is the larger variance of Z. Therefore, the core idea of MAD is to find out maximum differences between two images by removing correlations between them as much as possible.

$$a^T X = a_1 X_1 + ... + a_p X_p$$
$$b^T Y = b_1 Y_1 + ... + b_q Y_q \qquad (2)$$

To find a and b is to define a set of a and b simultaneously, to let us maximize the variance with $Var\{a^T X - b^T X\}$. A multiplication of a and b with a constant c will multiply the variance with c^2. Therefore a choice concerning a and b must be made and natural choices in this case are requesting unit variance of $a^T X$ and $b^T Y$. The criterion then is: maximize $Var\{a^T X - b^T X\}$ with $Var\{a^T X\} = Var\{b^T X\} = 1$. By choosing this, we have

$$Var\{a^T X - b^T X\} = Var\{a^T X\} + Var\{b^T X\} - 2Cov\{a^T X, b^T X\} \qquad (3)$$

$$Var\{a^T X - b^T X\} = -2(1 - Cov\{a^T X, b^T X\}) \qquad (4)$$

Also, to find a and b a principal component (PC) analysis can be used. Doing PC transformation generates linear combinations of multiespectral pixel intensities which have maximum variance. The difference between MAD and PC is that PC does not intend to retain the common information between dates, but actually the one that changes. The first PC gathers the stability dimension of the image (stable brightness). The secondary components offer the change which is what interests us (Chuvieco, 2002). The second corresponds to the change between dates with clarity, since the coefficients that are negative represent the change in brightness. The third shows the contrast between the IRC and the other bands and it can be defined as stable greenery. The fourth, according to Chuvieco (2002) may refer to the transition from dry vegetation to a vigorous one.

Theoretically, among all MAD components, the first one with largest variance ought to contain most of the change information, while others contain change information in decreasing order. In practice, in order to understand the physical content of the MAD components, it is possible do correlations between the change areas of the MADs and the original bands (wavelength regions) (Nori *et al.*, 2008). However in this study a supervised classification of MAD components was done in order to understand better those components.

However, this is not the case in practice by visual inspection due to noise in remotely sensed images. Noise may often be modeled as random variables uncorrelated with signals. During MAD transformation, this noise will be concentrated in those MAD components with low canonical correlations, which will result in a deviation of change information distribution from theoretical model, so as to hinder subsequent interpretations of change detection results (Zhang *et al.*, 2007). To solve such a problem, Canty (2010) introduced another orthogonal multivariate linear transformation, called Minimum Noise Fraction (MNF), into change detection as a post-processing step for MAD transformation (Canty, 2010). The objective of MNF transformation is to separate signal from noise as much as possible.

An important problem in satellite images change detection is the one produced by the variations introduced by the observation conditions, atmospheric conditions or the sensor's calibration. These effects modify the spectral signatures of land cover classes; therefore it is necessary to homogenize the images through conversion models to physical parameters incorporating the parameters of atmospheric correction and lighting. The MAD components are robust in this aspect, which means they are invariant under any linear transformation including any linear atmospheric corrections that may be made to the data. Therefore such corrections are not necessary (Nielsen, 1998; Canty *et al.*, 2004). Other advantages of this algorithm are that it doesn't require parameters to adjust, it can be applied at an object level, and it can be iterated on no change probability (Canty & Nielsen, 2008).

4.2.1 MAD & thresholding

The MADs are separated into change and no change classes (the change and no-change), and change thresholds are established. There are different methods to implement the change thresholds. The main idea in the choice of thresholds is to detezct significant changes thrown by the results in the change detection. In order to exploit the change information yielded by the MAD components, statistic criteria have been frequently used, assuming a normal distribution of the MAD components, based on the average standard deviation (Figure 3).

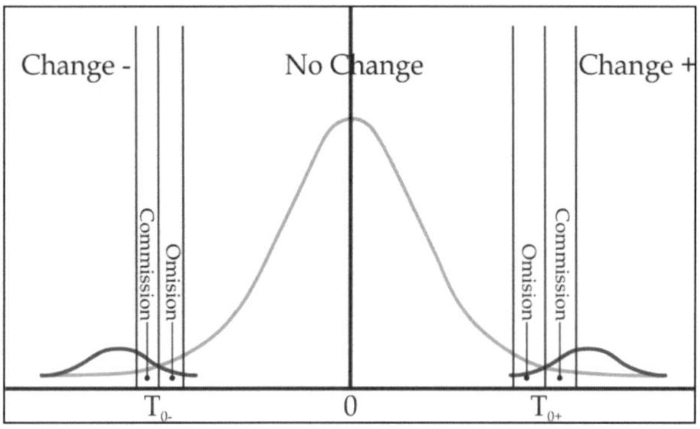

Fig. 3. Definition of Threshold (T). The curve shows MAD components distribution. The centre of the curve represents the "stable areas" or "areas of no change", whilst the areas from the sides represent the "dynamic areas" or "areas of changes detected". The omission and commission areas of the curve represent "over or under estimations" of possible changes detected.

There are several authors who suggest improvements to the threshold by applying unsupervised classification (Canty *et al.*, 2006), or applying different algorithms such as the expectation-maximization algorithm (EM) (Bruzzone & Fernandez-Prieto 2000). One method is to determine a threshold based on standard deviations (σ). The value of the thresholds is placed separately for each MAD component, considering that the intensity values are at ± 2σ of zero, which corresponds to the unchanged pixels.

4.3 Comparative study of change detection techniques

To develop models that allow observing the change dynamics is necessary to understand the causes and consequences of processes linked to global warming. In this study, two different change detection techniques were applied in order to assess land cover changes in the Jalisco Coast (Pacific Ocean). First, we focused on change detection based on post-classification comparison. Second, we investigated the usefulness of Multivariate Alteration Detection technique.

The Jalisco Coast, Mexico, has a surface of 1,451,465.85 hectares, corresponding to 17.5% of the State of Jalisco surface (Fig. 4). Up to 18 types of vegetation can be found in the area, being the most important and representative the tropical dry forest (Castillo et al., 2007).

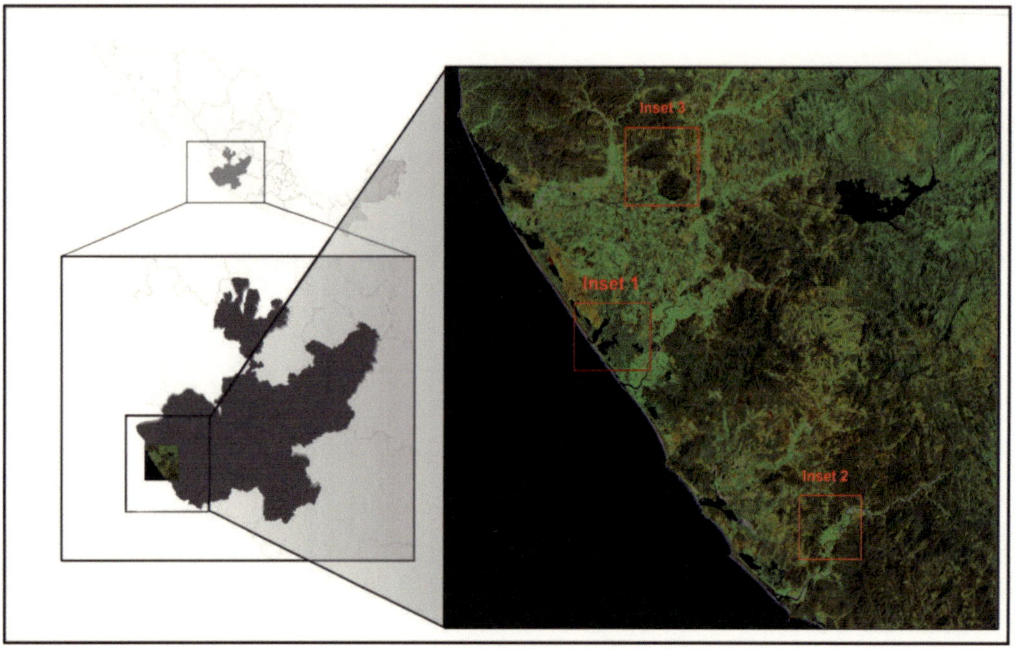

Fig. 4. Research area.

For a complete land cover change detection analysis, four basic remote sensing processes were used. First, a phase is allocated for satellite definition (for its orbital perception and repetition time) and for the sensor (for the desired properties in the image). Afterwards, the pre-processing takes place for the image correction. Thirdly, analysis techniques are used to outline the areas with significant alterations, which is what strictly concerns change detection. Finally, the analysis and validation should be performed with the use of the ancillary data, which can be field or official statistical data, and using techniques to estimate the precision of the whole process. (Singh *et al.*, 1989; Lu *et al.*, 2004; Chuvieco, 1998; Coppin *et al.*, 2004).

In this study, two Landsat ETM+ images acquired in 2000 and 2003 were used. In order to minimize the effects from atmospheric effects and seasonal differences (phenotype differences), images from the same time of the year, or as close as possible, were obtained. In the Figure 5 is shown the methodological approach. Two approaches were compared in order to assess differences between both methodologies. The steps followed were: 1) the acquisition and pre-processing; 2) changes analyses made through MAD and Post-classification, 3) MAD classification in order to compare with Post-classification technique; 4) A change matrix acquisition was undertaken followed by; 5) Result analysis and qualitative comparison.

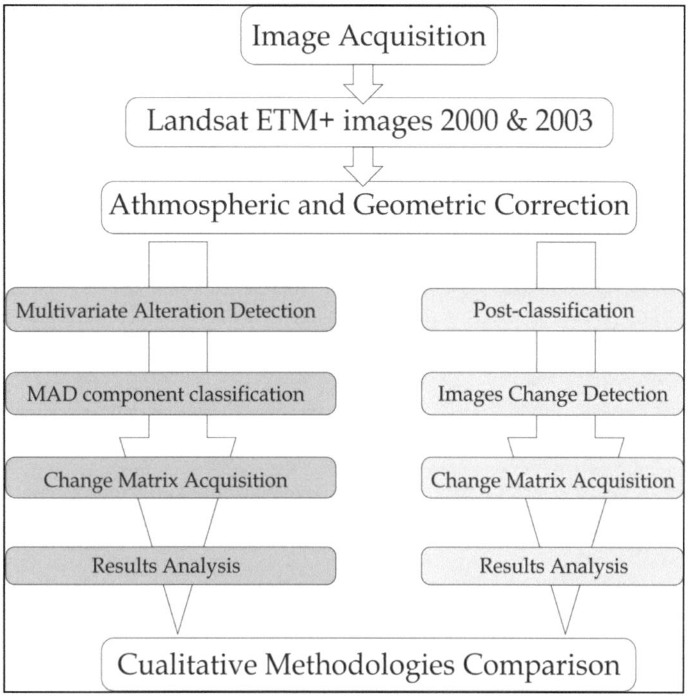

Fig. 5. Methodological approach. The figure shows the methodology followed for MAD in the left and Post classification in the right.

4.3.1 Results of the comparative study

We selected a set of three areas where notorious land cover change had occurred. The change map was extracted from the 6 components of the MAD transformation and the Chi square probability image, using a maximum likelihood algorithm and a threshold method. The change classes were the following:

1. TDF to bare soil and agriculture crop

Other vegetation to bare soil and agriculture crop

2. Agriculture crop to other vegetation

First of all, the Landsat imagery shows greater greenness in the year 2000 than in the year 2003. However, this trend, which holds for the bulk of TDF, may be explained by a difference in forest phenology stages between March 2000 and February 2003, because of greater precipitations earlier in year 2000 than in year 2003. Indeed, the rivers and water bodies are wider in 2000 than in 2003.

Most changes due to deforestation seemed to have occurred in easily accessible sites, within few pixels of human settlements, water bodies and agricultural areas in 2000. This indicated deforestation processes associated with an expansion of agricultural activity.

Three image subsets are shown in figures 6, 7 and 8, where a visual inspection helped the interpretation of the change detection. Red, orange and green colors referred respectively to deforestation of the TDF, removal of other vegetation types to a land use, and conversion of agriculture crop to a type of vegetation cover.

A visual inspection reveals that the post-classification approach yielded a much larger change surface than the MAD approach.

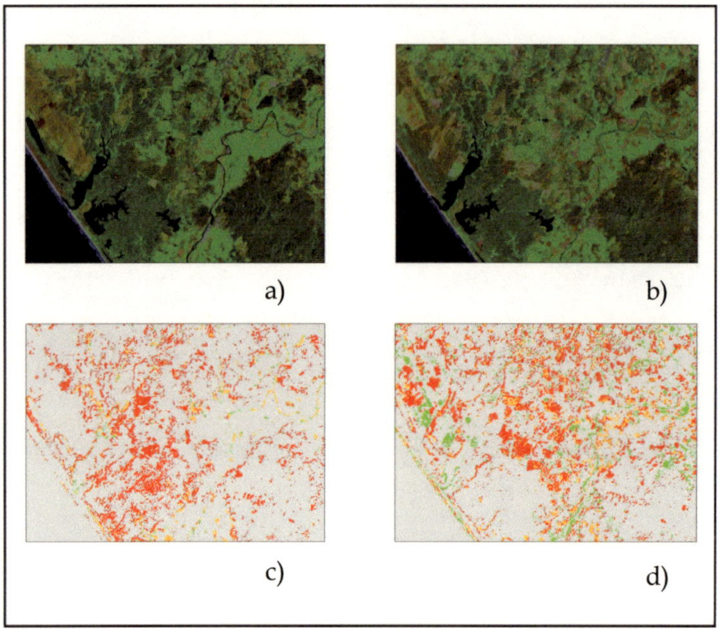

Fig. 6. Inset 1: a) Original Landsat ETM+ 2000; b) Original Landsat ETM+ 2003; c) Post-classification change map; d) Multivariate Alteration Detection change map.

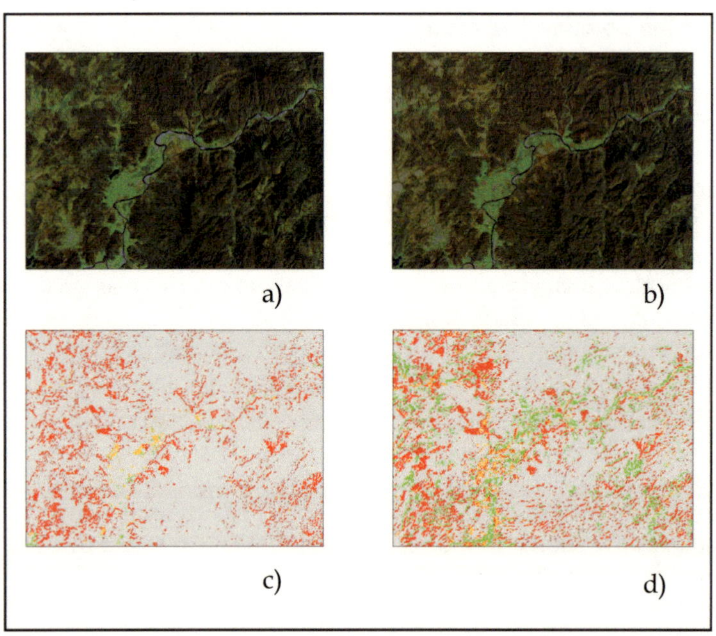

Fig. 7. Inset 2: a) Original Landsat ETM+ 2000; b) Original Landsat ETM+ 2003; c) Post-classification change map; d) Multivariate Alteration Detection change map.

A close qualitative analysis of the MAD change map indicated a relatively good match between large red patches with actual occurrences of removal of TDF. By contrast, there seem to be a few inconsistencies for classes indicated in orange and green. Much green occurrence corresponded to a change of soil moisture or vegetative state of a crop field (agricultural crop in 2000 and 2003). The spectral signatures of some of these crop fields are indeed close to that of a TDF, which makes it difficult to detect by any of the methods. In figure 8, many patches detected as deforestation by the post-classification method are in fact changes in crop conditions.

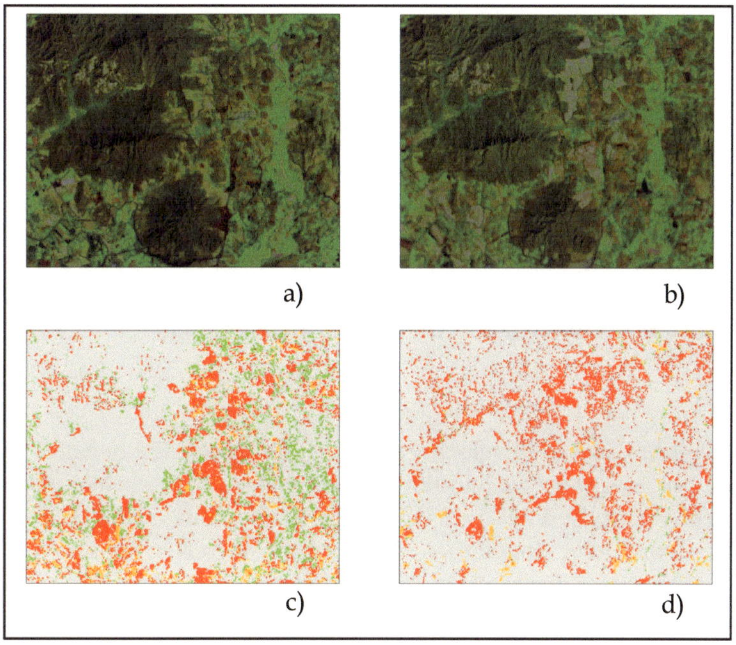

Fig. 8. Inset 3: a) Original Landsat ETM+ 2000; b) Original Landsat ETM+ 2003; c) Post-classification change map; d) Multivariate Alteration Detection change map.

In the mountainous area, where TDF prevails (see figures 7), red spots would indicate many small change patches according to the post-classification method, and less so according to the MAD approach. These red spot most likely correspond to false changes due to misregistration errors of pixels located in steep terrain or near to a river bank. Overall, it is likely that the post-classification method largely overestimates total land cover change.

The directionality of changes is observed on the change matrices (Tables 2 and 3). From the first method (MAD), the results showed a total area of 247.2 km^2 of deforestation, whereas the post-classification method resulted in a higher change with 294.7 km^2. Regarding to the areas with the opposite process ("forestation"), the matrix showed an area of 89 km^2 and 29 km^2 for MAD and for post-classification, respectively.

MAD	Other vegetation	Agriculture	Bare land	Deforested area [km^2]	Regenerated area [km^2]
Tropical Dry Forest	NA	144145	55641	179.8074	
Other vegetation	NA	41773	33218	67.4919	
Agriculture	99021	NA	NA		89.1189
			Total	**247.2993**	**89.1189**

Table 2. Change matrix resulted from MAD technique.

Post-classification	Other vegetation	Agriculture	Bare land	Deforested area [km2]	Regenerated area [km2]
Tropical Dry Forest	NA	274459	3211	249.903	
Other vegetation	NA	49551	316	44.8803	
Agriculture	32698	NA	NA		29.4282
			Total	294.7833	29.4282

Table 3. Change matrix resulted from Post-Classification technique.

5. Conclusion

Society faces the challenge of developing strategies to reduce the negative environmental impacts of land use change across multiple services and scales while maintaining social and economic benefits. This is why it is important to measure changes and find optimal techniques to measure them. Remote sensing is a valuable tool for monitoring land use and land cover. In the case of Tropical Dry Forests, the change in land cover is challenging because of its unsteady reflectance characteristics. Within a context of climatic change, the TDF is an important ecosystem for its preservation due to the ecosystem services it presents as well as for its high value in carbon capturing. It is necessary to diminish the land use and land cover changes in the TDF because of its high value in content of carbon. One of the necessary tasks to diminish climatic change is the creation of studies that focus on specific ecosystems where the changes in land use dynamics are made at a local and regional level.

This research contributes by exploring change detection techniques in the challenging settings of tropical sub-humid conditions, towards an optimization of tropical land use change detection. The Multivariate Alteration Detection technique was tested against a more conventional technique (Post-classification), on a pilot area where TDF prevails in Mexico.

In the case of post-classification detection, the high amount of detected change could be interpreted as a large overestimation of change. In both approaches, classes of potential change included false change, such as pixels affected by small misregistration errors in heterogeneous landscapes and next to steep terrain, and signal change within a particular class (change of soil moisture conditions within the agriculture class for example). A possible interpretation was that many categories of false changes were difficult to handle with post-classification whereas they could be handled better with the MAD technique.

Indeed, post-classification is a three step process (classification of the first image, classification of the second image, and then comparison of two layers) where the interpreter has to 'foresee' potential detection problems at each step. By contrast, each component of the MAD data represented a potential change type. This change type could have been a kind of false change, or a change within a particular class. In each case, the interpreter could directly handle the potential problem of the change type in a supervised manner. Instead of this, in post-classification, the change within a particular class would have to be detected ('foreseen') at the classification step.
In the case of TDF landscapes, where the phenology causes forests to be confounded with grassland or cropland, the characterization of each class is in itself a difficult task, which needs a certain amount of supervision. In such case, it is worth concentrating the efforts of the interpreter directly on change classes rather than first on thematic classes and then on change classes.
The global environmental challenge includes maintaining land use surfaces below a reasonable level, intending to reach a trade-off between immediate human needs and the capacity of ecosystems to provide goods and services in the future. Studies in deforestation and land use/land cover change made at a regional and local level have more and more importance every day creating models that help understand the climatic variations.

6. References

Bullock, S. Mooney, H. & Medina, E. (1995). *Seasonally dry tropical forests.* Cambridge: Cambridge University Press. 450 pp.

Bruzzone, L., Fernandez Prieto, D. (2000). A minimum-cost thresholding technique for unsupervised change detection, *International Journal of Remote Sensing*, Vol. 21, No. 18, 2000, pp. 3539 - 3544 . DOI: 10.1080/014311600750037552

Burgos A. & Maass M.J. (2004). Vegetation change associated with land-use in tropical dry forest areas of Western Mexico, *Agriculture, Ecosystems and Environment*,104 pp. 475–481.

Canty M.J., Nielsen A.A. and Schmidt, M. (2004). Automatic radiometric normalization of multitemporal satellite imagery. *Remote Sensing of Environment* Vol. 91(3-4), pp.441-451.

Canty M.J and Nielsen A.A.(2006). Visualization and unsupervised classification of changes in multispectral satellite imagery. *International Journal of Remote Sensing* Vol.27, No.(18), pp.3961-3975.

Canty M.J. and Nielsen A.A. (2008). Automatic Radiometric Normalization of Multitemporal Satellite Imagery with the Iteratively Re-weighted MAD Transformation. *Remote Sensing of Environment* Vol. 112, No (3), pp.1025-1036.

Canty, M. J., (2010), Image Analysis, Classification and Change Detection in Remote Sensing: With algorithms for ENVI/IDL, 2nd edition, pp. 249-253., ISBN: 978-1-4200-8713-0

Castillo, A., Dominguez, García, A., Quesada, M. & Vega, J, (2007). Proyectos de desarrollo turísticos "La Huerta"(clave:14JA2006T0018) y "TAMBORA " (Clave:14JA20-

06T0011) en áreas vecinas a la reserva de la biosfera Chamela-Cuixmala. Panel técnico de análisis. Universidad Nacional Autónoma de México.

Ceballos, G & García, A. (1995). Conserving Neotropical Biodiversity: The Role of Dry Forests in Western Mexico. *Conservation Biology*, Vol. 9, No (6), pp. 1349-1353.

Chuvieco, E. (1998). El factor temporal en teledetección: evolución fenomenológica y análisis de cambios. *Revista de Teledetección*, No.10, pp. 1-9.

Chuvieco, E. (2002). Teledetección ambiental: la observación de la Tierra desde el espacio. Ariel. España. pp. 439-441. ISBN: 8434480476

Comisión Intersecretarial de Cambio Climático, México (2009). Cuarta Comunicación Nacional ante la Convención Marco de las Naciones Unidas sobre el Cambio Climático. Secretaría de Medio Ambiente y Recursos Naturales & Instituto Nacional de Ecología, México.

Constanza, R., d'Arge, R., de Groot, R., Farber, S., Grasso, M., Hannon, B., Limburg, K., Naeem, S., O'Neill, R., Paruelo, J., Raskin, R., Sutton, P. & Van den Belt, M. (1987). The value of the world's ecosystem services and natural capital. *Nature*, Vol. 387, No. 15, pp. 253-260.

Coppin, P. R. & Bauer, M. E. (1996). Change detection in Forest Ecosystems with Remote Sensing Digital Imagery. *Remote Sensing Reviews*, No. 13, pp. 207-234.

Couturier, S. (2010). A fuzzy-based method for the regional validation of global maps: the case of MODIS-derives phonological classes in a mega-diverse zone, *International Journal of Remote Sensing*, Vol.31, No. 22, pp. 5797-5811.

Estrada-Salvador & Navar, (2009). Flujos de carbono por deforestación en la selva baja caducifolia del estado de Morelos, México. *XIII Congreso Forestal Mundial*, Buenos Aires, Argentina.

Feeley, K.J., Gillespie, T.W., Terborgh, J.W. (2005). The Utility of Spectral Indices from Landsat ETM+ for Measuring the Structure and Composition of Tropical Dry Forests1, *Biotropica* Vol. 37, No. 4, pp. 508-519, ISSN 1744-7429.

Flores V, O. & Gerez, P. (1994). Biodiversidad y conservación en México: vertebrados, vegetación y uso de suelo. CONABIO y UNAM. México.

Goetz, S., Baccini, A., Laporte, N., Johns, T., Walker, W., Kellndorfer, J., Houghton, R., and Sun, M. (2009) Mapping and monitoring carbon stocks with satellite observations: a comparison review, Carbon Balance and Management, 4:2.

Griscom, H. P. Ashton, M.S., (2010), Restoration of dry forest in Central America: A review of pattern and process, *Forest Ecology and Management*, Vol. 261, No. 10, pp.1564-1574, ISSN 0378-1127.

Houet T. and Verburg P.H. (2010), Monitoring and modelling landscape dynamics, Landscape Ecol. 25.pp.163–167, DOI 10.1007/s10980-009-9417-x

Intergovermental Panel on Climate Change (IPCC). (1995). Climate Change. The Suplementary Report to the IPCC Scientific Assessment. Cambridge University Press. Cambridge, Inglaterra.

Intergovermental Panel on Climate Change (IPCC), (1997). Estabilización de los gases atmosféricos de efecto invernadero: implicaciones físicas, biológicas y socioeconómicas. Documento técnico III.

Intergovermental Panel on Climate Change (IPCC). (2002). Climate change and biodiversity. IPPCC technical paper V. WMO y UNEP.

Janzen, D.H., (1988). Tropical dry forests: the most endangered major tropical ecosystem. In: Wilson, E.O. (Ed.), *Biodiversity*, National Academy Press, Washington, pp. 130–137.

Jaramillo J. V., Boone Kauffman J., Rentería-Rodríguez L.,Cummings L. D. and Ellingson L. J . (2003). Biomass, Carbon, and Nitrogen Pools in Mexican Tropical Dry Forest Landscapes, Ecosystems 6. pp.609–629, DOI: 10.1007/s10021-002-0195-4

Jensen, J. (1996). Introductory Digital Image Processing A Remote Sensing Perspective. Prentice Hall, New Jersey.

Kalacska, M.. Sanchez-Azofeifa, G.A., Rivard, B., Caelli, T., White, P. & Calvo-Alvarado, M. (2007). Ecological fingerprinting of ecosystem succession: Estimating secondary tropical dry forest structure and diversity using imaging spectroscopy. *Remote Sensing of Environment*, Vol. 108, No 1, pp. 82-96.

Kalacska, M., Sanchez-Azofeifa, G.A., Rivard, B., Calvo-Alvarado, M. & Quesada, M. (2008). Baseline assessment for environmental services payments from satellite imagery: a case study from Costa Rica and Mexico. *Journal of environmental management*, Vol. 88, No. 2, pp. 348-59.

Lambin, E.F. (1997). Modelling and monitoring land-cover change processes in tropical regions. *Progress in Physical Geography*. No. 21, pp. 375–393.

López Granados, E.M., G. Bocco & M.E. Mendoza. (2001). Cambio de cobertura vegetal y uso de suelo. En: Mendoza, M.E., G. Bocco y E. López 2001. Regionalización ecológica, conservación de recursos naturales y ordenamiento territorial en la cuenca del lago de Cuitzeo, Michoacán. Informe Técnico. Proyecto No. 98306024. Programa SIMORELOS - CONACYT, México 273 pp

Lu, D., Mausel, P., Brondízio, E. & Moran, E. (2004). Change detection techniques. *International Journal of Remote Sensing*. Vol. 25, No. 12, pp. 2365-2401

Lucas, R.M., Held, A.A., Phinn, S.R. and Saatchi, S., (2004), Tropical forests. In *Remote Sensing for Natural Resource Management and Environmental Monitoring*, S.L. Ustin (Ed.), pp. 239–315 (Hoboken, NJ: John Wiley).

Maass, J.M., Balvanera, P., Castillo, A., Daily, G.C., Mooney, H.A., Ehrlich, P., Quesada, M., Miranda, A., Jaramillo, V.J., García-Oliva, F., Martínez-Yrizar, A., Cotler, H., López-Blanco, J., Pérez-Jiménez, A., Búrquez, A., Tinoco, C., Ceballos, G., Barraza, L., Ayala, R.& Sarukhán, J. (2005). Ecosystem Services of Tropical Dry Forests : Insights from Long- term Ecological and Social Research on the Pacific Coast of Mexico. *Ecology And Society, Vol.* 10, No. 1, pp. 17.

Maas, J.M. (1995). Conversion of tropical dry forest to pasture and agriculture. In: *Seasonally dry tropical forests*, Bullock, S. H., Mooney, H.A. & Medina, E. pp. 326-345. Cambridge University Press, ISBN 0521435145, Great Britain.

Martínez-Yrízar, A. (1995). Biomass distribution and primary productivity of tropical dry forests. En: Bullock, S. H., Mooney, H. & Medina, E. (eds), Seasonally dry tropical forests. Cambridge University Press. pp. 326-345.

Martínez-Yrízar, A., Maass, M., Pérez-Jiménez, L. A. & Sarukhán, J. (1996). Net primary productivity of a tropical deciduous forest ecosystem in western Mexico. *Journal of Tropical Ecology. Vol.* 12, No. 1, pp. 169-175.

Menaut, J.C., Lepage, M. & Abbadie, L. (1995). Savannas, woodlands and dry forest in Africa. In: *Seasonally dry tropical forests,* Bullock, S. H., Mooney, H.A. & Medina, E. pp. 64-88. Cambridge University Press. ISBN 0521435145, Great Britain.

Miles, L., Newton, A.C., DeFries, R. S., Ravilious, C., May, I., Blyth, S., Kapos, V. & Gordon, J. (2006). A global overview of the conservation status of tropical dry forests. *Journal of Biogeography,* Vol. 33, No. 3, pp. 491-505.

Murphy, P.G. & Lugo, A.E.,(1986). Ecology of tropical dry forest. *Annual Review of Ecology and Systematics* 17: 67–88.

Murphy, P.G. (1995). Dry forest of Central America and the Caribbean. In: *Seasonally dry tropical forests,* Bullock, S. H., Mooney, H.A. & Medina, E. pp. 9-29. Cambridge University Press. ISBN 0521435145, Great Britain.

Nielsen, A.A. and Conradsen, K. (1997). Multivariate Alteration Detection (MAD) in Multispectral, Bi-temporal Image Data: A New Approach to Change Detection Studies. IMM Technical Report No. 1997-11.

Nielsen, A.A., Conradsen, K. and Simpson, J.J. (1998). Multivariate Alteration Detection (MAD) and MAF Post-Processing in Multispectral, Bi-temporal Image Data: New Approaches to Change Detection Studies. Remote Sensing of Environment Vol. 64, No.1, pp. 1-19.

Niemeyer, I., Marpu, P.R. & Nussbaum, S. (2008). Change detection using object features. In: object-based image analysis. Spatial concepts for knowledge-driven Remote sensing applications. Thomas Blaschke, Stefan Lang and Geoffrey J. Hay (eds).

Nori, W., Elsidding, E. and Niemeyer, I., (2008), "Detection of Land Cover Changes Using Multi-Temporal Satellite Imagery", The International Archives of the Photogrammetry, Remote Sensing and Spatial Information Sciences. Vol.37, No. 7, pp. 947-951.

Ordoñez, B. (1999). Captura de carbon en un bosque templado: el caso de San Juan Nuevo, Michoacán. INE-SEMARNAT, México.

Ordóñez, B. & Masera, O. (2001). Captura de carbono ante el cambio climático. *Madera y bosques*. Vol. 7, No.1, pp. 3-12. Instituto de Ecología, Xalapa.

Ortiz, R. V. 2005. Hyperspectral Change Detection Using Temporal Principal Component Analysis. Master thesis. Universidad de Puerto Rico. Mayaguez Campus

Pacifici, F., Del Frate, F., Solimini, C. & Emery, W. J. (2007). An Innovative Neural-Net Method to Detect Temporal Changes in High-Resolution Optical Satellite imagery, *IEEE Transactions on Geoscience and Remote Sensing,* Vol. 45, No. 9, pp. 2940-2952.

Pielke, R. A. Sr, Gregg Marland, Richard A. Betts, Chase, T., Eastman, J., Niles, J., Niyogi, D. & Running, S. (2002). The Influence of Land-use Change and Landscape Dynamics on the Climate System: Relevance to Climate-change Policy Beyond the Radiative Effect of Greenhouse Gases. *Philosophical Transactions of the Royal Society* A, No. 360, pp. 1705-1719.

Pfaff, A., Kerr, S., Flint Hughes, R., Liu, S., Sánchez-Azofeifa, A., Schimel, D., Tosi, J. & Watson, V. (2000). The Kyoto protocol and payments for tropical forest: An interdisciplinary method for estimating carbon-offset supply and increasing the feasibility of a carbon market under the CDM. *Ecological Economics*, Vol. 35, No. 2, pp. 203-221

Quadri de la Torre, G. (2000). Incendios forestales y deforestación en México: una perspectiva analítica. Comisión de estudios del sector privado para el desarrollo sustentable (CESPEDES), México.

Ramírez Bojórquez P.R. (2006). Análisis espacial de la cobertura de manglar en los municipios de Tecuala y Santiago Ixcuintla, Nayarit en el periodo de 1973-2005. Tesis de Maestría. CIAD, A.C. México.

Reyes C., Taylor F., Martínez E. y López F. (2006). Geo-Cybernetics: A new Avenue of Research in Geomatics, *Cartographica*, 41, Issue 1.DOI. 10.3138/C034-6P5T-W322-1G72

Rosete, A., Pérez, J.L. & Bocco, F. (2008). Cambio de uso de suelo y vegetación en la Península de Baja California, México. Investifaciones Geográficas, *Boletín del Instituto de Geografía*, UNAM. No. 67, pp 39-58.

Rundel, P.W., Boonpragob, K. (1995). Dry forest of Thailand. In: *Seasonally dry tropical forests*, Bullock, S. H., Mooney, H.A. & Medina, E. pp. 93-119. Cambridge University Press. ISBN 0521435145, Great Britain.

Rzedowsky, J. (1978). *Vegetación de México*. Ed. Limusa. Mexico

Sánchez-Azofeifa A.G., (2009). Land cover and conservation in the area of influence of the Chamela-Cuixmala Biosphere Reserve, Mexico. *Forest Ecology and Management* Vol. 258, pp.907–912.

Sampaio, E.V.S.B. (1995). Overview of the Brazilian caatinga. In: *Seasonally dry tropical forests*, Bullock, S. H., Mooney, H.A. & Medina, E. pp. 35-58. Cambridge University Press. ISBN 0521435145, Great Britain.

Serra, P., Pons, X. & Saurí, D. (2001). Protocolo para la detección de cambios a través de diferentes sensores. *Revista de teledetección*. No. 16, pp 17-24.

Sui, H., Zhou, Q., Gong, J. & Ma, G. (2008). Processing of multitemporal data and change detection. In: Advances in Photogrammetry, Remote Sensing and Spatial Information Science: 2008 ISPRS Congress Book. Li, Z., Chen, J. & Baltsavias (Eds).

Thompson, B. (1984). Canonical Correlation Analysis. Uses and Interpretation. Series: Quantitative Applications in the Social Sciences. Sage University. USA. pp.8-25. USA.ISBN 0-8039-2392-9. OK

Trejo, I. & Dirzo, R. (2000). Deforestation of seasonally dry tropical forest a national and local analysis in Mexico. *Biological Conservation*, Vol. 94, No. 2, pp.133-142.

U.S. Geological Survey (USGS), (2010), In: http://landsat.usgs.gov/ about_project_ descriptions.php.

William R. E. (1985), Climatic change and the broad scale distribution of terrestrial ecosystem complexes, Climatic Change, No, 7 pp. 29-43.

Wyatt, B.K., 2000. Remote sensing of land cover and land cover change. In: Observing Land from Space: Science, Customers and Tecnology. Verstraete *et al.* (Eds) pp. 127-133.

Zhang, L., Liao, M., Yang, L., & Lin. H. (2007). Remote Sensing Change Detection Based in Canonical Correlation Analysis and Contextual Bayes Decision. *American Society for Photogrammetry and Remote Sensing* Vol.73, No, 3. pp. 311-318.

12

Cirrus Clouds and Climate Engineering: New Findings on Ice Nucleation and Theoretical Basis

David L. Mitchell[1], Subhashree Mishra[1] and R. Paul Lawson[2]
[1]*Desert Research Institute, Reno, Nevada*
[2]*SPEC, Inc., Boulder, Colorado*
USA

1. Introduction

Geo-engineering, henceforth referred to as climate engineering, can be described as an intentional intervention on the Earth's climate system for the purpose of temporarily reducing the increase in surface temperatures due to global warming. While it is generally acknowledged that climate engineering cannot solve the problem of global warming resulting from unrestricted greenhouse gas (GHG) emissions (Launder and Thompson, 2010), it may "buy time" for non-carbon energy systems to dominate global energy production and for GHG emissions to be reduced to safe levels. The longer meaningful global policy decisions on GHG emissions are delayed, the stronger the demand is likely to be for climate engineering research and development. As described by Anderson and Bows (2010), even if an effective global treaty on GHG emissions were to take effect now, it would still be extremely difficult to prevent mean global surface temperatures from rising above 2°C (above which it is argued that the consequences of climate change would be unacceptable). It is thus defensible to argue that some combination of GHG mitigation and climate engineering may be necessary to limit the mean increase in global surface temperatures to 2°C.

A modeling study by Arora et al. (2011) also finds that a mean global warming exceeding 2°C may be unavoidable. Using updated GHG emission scenarios and an upgraded Earth system model (which accounts for aerosol effects and represents the carbon-cycle more realistically), the study finds that even under the lowest, most optimistic GHG emission scenario, the global average temperature will still increase more than 2°C (the limit agreed to by various governments in the Copenhagen accord) by 2100. To limit warming to 2°C by 2100, carbon dioxide emissions would need to be reduced to zero over the next 50 years followed by ongoing carbon sequestration (removal of CO_2 from the atmosphere) through the end of this century. While few scientists advocate tinkering with the earth's climate (Schneider 2010), climate engineering of some kind may become a necessity rather than an option.

Two types of climate engineering have been recognized: carbon dioxide removal (CDR) strategies and solar radiation management (SRM). CDR approaches can be biological, such as iron fertilization of the oceans to increase phytoplankton uptake of CO_2 (Lampitt et al., 2010; Smetacek and Naqvi, 2010), or they can be physically based, such as the direct capture of CO_2 from the atmosphere using chemically treated scrubbers with subsequent carbon sequestration or conversion into carbon-neutral fuels (e.g. Keith et al., 2010). CDR is arguably the best climate engineering approach as it could potentially stabilize atmospheric CO_2 concentrations at safe levels and prevent ocean acidification. Unfortunately, iron fertilization is unlikely to remove CO_2 from the atmosphere at rates sufficient to prevent future warming, and air capture would likely require decades before such devices are numerous enough throughout the world to make a significant impact.

SRM refers to efforts that increase the reflection of solar radiation to space, thus reducing surface temperatures. A major drawback common to all SRM methods is that they do not address the problem of ocean acidification. The most studied SRM method is to reflect incoming sunlight with sulfuric acid aerosol injected into the stratosphere (similar to how volcanic aerosols cool climate). Although this method may have serious drawbacks such as a likely interference with the hydrological cycle (producing drought in some regions), ozone destruction and turning the sky white, this SRM method appears capable of rapidly neutralizing the warming due to a CO_2 doubling (Lenton and Vaughan, 2009), and appears relatively easy and quick to deploy. Another SRM method that might neutralize the warming from a CO_2 doubling is the brightening of marine stratus clouds (Latham 1990, 2002; Latham et al., 2008, Lenton and Vaughan 2009). These clouds are made more reflective by enriching the marine boundary layer with abundant cloud condensation nuclei (CCN) produced by a fine spray of sea water from specially designed boats (Salter, 2008). Potential drawbacks include regional changes in sea surface temperatures producing changes in ocean circulation patterns and modification of regional weather systems (The Royal Society, 2009).

A conceptual shortcoming of SRM-based methods is that they implicitly address a problem of reduced heat output (GHGs trapping thermal radiation) by reducing radiation input (reflecting sunlight to space). A third climate engineering approach recently considered (Mitchell and Finnegan, 2009; henceforth MF2009) is to directly address the problem of increased troposphere heat content by increasing its heat output (i.e. shifting its thermal equilibrium in favor of cooler temperatures). If it were possible to modify the troposphere's heat balance to reduce surface temperatures, this would offset the effect of heat-trapping GHGs in a more direct fashion. This third approach might be called earth radiation management (ERM). Under this definition, reducing GHG emissions and CDR are in effect ERM methods.

The purpose of this chapter is to describe an ERM approach pertaining to cirrus clouds and to test this hypothesis against new, more reliable cirrus cloud measurements. Like the SRM methods, this ERM method does not address ocean acidification, and effective GHG mitigation strategies along with possibly CDR methods would be needed to responsibly address the problem of global warming. This chapter is organized as follows. Section 2 provides an overview of the cirrus cloud climate engineering idea, which is described in greater detail in MF2009. Section 3 presents new research findings that test the conceptual foundation of this ERM approach. These findings are not only relevant to climate engineering, but are relevant to basic research on cloud-aerosol-climate interactions as well. Section 4 addresses the potential of over-cooling the planet and the need for climate monitoring. Some possible social, political and economic ramifications of climate engineering are discussed in Section 5, and a summary with conclusions is given in Section 6.

2. Modification of cirrus clouds to reduce global warming

In a study involving 1000's of "perturbed physics" global climate model (GCM) simulations using the U.K. Hadley Centre GCM, Sanderson et al. (2008) identified physical processes responsible for changing climate sensitivity (i.e. the equilibrium response of global-mean surface temperature to CO_2 doubling). The most important process affecting climate sensitivity was entrainment associated with deep convection, which strongly affects water vapor amounts in the upper troposphere. The second most important process was the ice fall speed from cirrus clouds, which affects the cirrus cloud life cycle and cloud coverage, the cirrus cloud optical depth and the upper troposphere water vapor. These findings imply that upper troposphere water vapor and cirrus clouds are having the greatest impact on climate sensitivity. Thus, an effective ERM strategy might target these components of the climate system which regulate outgoing longwave radiation, or OLR.

A modeling study by Mitchell et al. (2008), using an atmospheric GCM known as the Community Atmosphere Model version 3 or CAM3, also found that the CAM3 climatology was sensitive to the ice fall speed which had a profound influence on cirrus cloud cover, ice water path, optical depth and related radiative fluxes and heating rates. In addition, this study demonstrated a substantial dependence of the ice fall speed on the concentrations of relatively small ice crystals in the ice particle size distribution (PSD), thus suggesting a dependence of the representative fall speed on ice crystal nucleation rates. This suggestion will be explored further in this chapter.

Cirrus clouds are similar to GHGs in that they trap OLR by absorbing/emitting upwelling thermal radiation back towards the surface, which has a warming effect on climate. This "greenhouse" effect of cirrus clouds becomes stronger the higher they get since the effective temperature at which the Earth radiates thermal energy to space depends on the temperature of the overlying clouds. An abundance of cirrus will cause this effective temperature to be relatively low (similar to the temperature of the cold cirrus clouds). This reduces the OLR and traps thermal radiation below the cirrus clouds that would otherwise escape to space. In addition, cirrus clouds also reflect sunlight back to space (a cooling effect on climate), but this albedo effect is less efficient than their "greenhouse" effect. Thus, cirrus clouds have a net warming effect on the earth's climate (Hartmann et al., 1992; Chen et al., 2000). It follows then that the most effective way to increase OLR may be to reduce the cloud cover of the highest, coldest cirrus clouds.

One method for how this goal might be accomplished is described in MF2009. They suggested "seeding" cirrus clouds with efficient ice nuclei that would out-compete the natural ice nuclei for water vapor. Aerosol particles that serve as ice nuclei often "activate" to form ice crystals at some threshold supersaturation with respect to ice. Natural aerosols that serve as ice nuclei activate at relatively high thresholds compared to those required for efficient ice nuclei such as silver iodide (used in cloud seeding operations). Introducing efficient ice nuclei into cirrus clouds at relatively low concentrations should then preclude the activation of most natural ice nuclei since the efficient ice nuclei (i.e. the cloud seeding aerosol) will activate and grow ice crystals at lower supersaturations and their growth will prevent water vapor concentrations from ever reaching the threshold supersaturation needed to activate the natural ice nuclei. This has a two-fold effect of depressing water vapor concentrations in supersaturated environments and producing fewer ice crystals. While lower in concentration than the natural ice nuclei, the concentration of the cloud seeding aerosol should be sufficient to prevent the ice supersaturation from rising significantly as an air parcel ascends. This will insure that primarily only the seeded ice nuclei will produce ice crystals and that these crystals will grow to larger sizes than ice crystals produced by the more numerous natural ice nuclei. That is, a given amount of cloud condensate distributed among fewer ice crystals results in larger ice crystals. These larger, more massive ice crystals will have larger fall velocities than the natural ice crystals. As noted above, higher fall speeds will result in shorter cloud lifetimes, less cloud coverage, lower ice water paths and lower cloud optical depths. This would allow more OLR to escape to space.

2.1 Ice nucleation in cirrus clouds

Ice nucleation mechanisms can be grouped into two categories: (1) heterogeneous nucleation processes and (2) homogeneous freezing nucleation. The first category includes the mechanisms of deposition nucleation, contact nucleation, immersion freezing nucleation, condensation freezing nucleation and others (for a description of these mechanisms, see Pruppacher and Klett, 2010). These processes are referred to as heterogeneous since they involve a substrate other than water (vapor or liquid) to initiate freezing. The substrate could be mineral dust, black carbon aerosol or virtually any substance that promotes the freezing of supercooled liquid water. The rate of ice crystal production from these heterogeneous mechanisms is markedly slower than the production rate from homogeneous freezing nucleation (Pruppacher and Klett, 2010). Thus ice crystal concentrations produced through heterogeneous nucleation processes are lower than those produced through homogeneous freezing nucleation. In contrast to heterogeneous nucleation, homogeneous freezing nucleation does not require a substrate to initiate freezing. It occurs when a soluble aerosol particle (e.g. ammonium sulfate or nitrate, sulfuric acid, or any soluble hydrophilic substance) accretes water vapor in a humid environment to form a haze droplet (below water saturation), whereby the particle is dissolved resulting in a homogeneous solution. When the temperature is less than ~ -38°C and the relative humidity with respect to ice (RH_i) is greater than ~ 140%, these haze droplets spontaneously freeze forming ice crystals. For the details of this homogeneous nucleation process, see, for example, Sassen and Dodd (1988), Heymsfield and Sabin (1989) and Koop et al. (2000). Kärcher et al. (2007) provide a discussion of ice nucleation processes relevant to cirrus clouds.

2.2 Reducing cloud cover for the coldest cirrus clouds

Cirrus clouds might be modified by seeding aerosol at any temperature, provided the natural ice nuclei, whether heterogeneous or homogeneous, activates at a RH_i well above the RH_i threshold of the seeding aerosol. For example, if the RH_i threshold for natural ice nuclei is ~ 120% and the threshold for seeding aerosol is 105% RH_i, the cirrus modification process described above should be possible. For homogeneous freezing nucleation, the RH_i threshold is greater than 140% (e.g. Kärcher et al 2007), making it even more likely that the seeding aerosol will out-compete the natural ice nuclei for water vapor. Moreover, homogeneous freezing nucleation requires temperatures colder than -37.5°C (Rosenfeld and Woodley 2000), making this temperature regime ideal for cirrus modification provided that homogeneous freezing is an important nucleation mechanism in that temperature regime. On this question there is considerable controversy, and one of the aims of this chapter is to provide new evidence towards resolving this controversy. If homogeneous nucleation is an important nucleation process for temperatures less than -40°C, then cirrus clouds at such temperatures should be most susceptible to modification as described above. Forming in an air mass conditioned with seeding aerosol at the right concentration, cirrus ice crystals should grow larger and fall faster, depleting the air mass of moisture and reducing cirrus cloud coverage over time. As noted above, the most effective way to increase OLR may be to reduce the cloud cover of the highest, coldest cirrus clouds, and it appears that these cirrus clouds could be the most susceptible to modification.

2.3 Reducing upper tropospheric water vapor

To increase ice crystal fall-speeds, the direct aircraft seeding of cirrus may have the opposite effect (to decrease fall-speeds), since the seeding aerosol is initially concentrated upon release from an aircraft and may produce many relatively small ice crystals. Rather, seeding aerosol should be released in the upper troposphere under clear conditions and allowed to become well dispersed, conditioning a vast air mass with seeding aerosol that has a concentration range that promotes large ice crystal growth. Then under the right conditions cirrus clouds will form in this clear air having relatively large ice crystals and thus a relatively short life-cycle.

Vast regions in the upper troposphere are supersaturated with respect to ice yet remain cloud-free (Helten et al. 1998; Read et al. 2001; Spichtinger et al. 2003, 2004). The above seeding strategy would thus initially produce more cirrus clouds since the seeding aerosol would have a much lower RH_i threshold than the natural ice nuclei, forming cirrus clouds in these supersaturated regions. However, this effect should be transient since over time, the relatively large ice crystals would sediment to lower levels and warmer temperatures where the cirrus greenhouse effect is less. Water vapor concentrations in the upper troposphere should decrease as this moisture is exported to lower levels, decreasing the water vapor greenhouse effect. Indeed, this is what was observed in two GCM experiments when the ice fall-speed was increased (Lohmann et al. 2008, MF2009). The globally averaged clear-sky OLR was decreased by ~ 0.7 W m^{-2} in the Lohmann et al. study (MF2009) due to small changes in the ice fall-speed.

2.4 GCM studies

In MF2009, two GCM studies (Lohmann et al. 2008 and Mitchell et al. 2008) are discussed that lend credibility to this cirrus engineering idea. The study by Lohmann et al. contrasted ECHAM5 GCM simulations based on homogeneous freezing nucleation with simulations based on either heterogeneous nucleation or a combination of homo- and heterogeneous nucleation. This study indicated that a heterogeneous nucleation threshold of RH_i = 130% would increase effective ice crystal size by ~ 11%, and that this would increase the ice fall-speed enough to produce a net global cooling of ~ -2.7 W m^{-2}, with 2 W m^{-2} coming from cirrus cloud reduction and 0.7 W m^{-2} due to water vapor reduction. This compares with the radiative forcing due to a CO_2 doubling of 3.7 W m^{-2}. No GCM experiments have been conducted using very efficient ice nuclei (RH_i ~ 105%), and thus greater increases in ice crystal size and fall speed, and greater cooling effects, might well be possible.

The study by Mitchell et al. (2008) did not examine the impact of different ice nucleation mechanisms on cirrus cloud radiative forcing, but it did look at the impact of varying the concentration of small ice crystals in the PSD, which affects the mass-weighted ice fall-speed. Ice fall-speed differences between simulations were most significant at temperatures less than -45°C. The higher fall-speed simulation (lower concentrations of small ice crystals) had 5.5% less cirrus cloud coverage on a global average. In the mid-latitudes and polar regions, shortwave cloud forcing was changed very little while longwave cloud forcing decreased up to 6 W m^{-2} (i.e. OLR increased by up to 6 W m^{-2}) in the higher fall-speed simulation.

2.5 Possible seeding strategies

This seeding could be done continuously on a global scale using commercial airliners to sustain a quasi-constant concentration of seeding aerosol in regions frequented by airliners. Commercial airliners typically fly at temperatures between -45 and -60°C, which is the temperature region of interest. If the seeding material was soluble in jet-fuel, the jet-fuel could be doped with seeding material, but extensive tests would need to be conducted to insure that burning the doped fuel had no detrimental effect on the jet engines. Alternatively, a flammable solution of seeding material could be injected into the hot engine exhaust.

Another approach would be to develop a fleet of unmanned drone aircraft to disperse the seeding material. While perhaps more expensive, this approach would not face the many legal and technical issues dealing with public safety on commercial aircraft. The lightweight drone aircraft would be relatively fuel-efficient and could be directed wherever seeding was needed, remaining in flight for extensive periods.

Bismuth tri-iodide (BiI_3) has been suggested as a possible seeding material (MF2009) since it is as effective as silver iodide (AgI, the most effective seeding agent known) at temperatures colder than -20°C, is $1/12^{th}$ the cost of AgI, and is non-toxic. Note that the active ingredient in Pepto Bismol, commonly used for treating diarrhea, is bismuth subsalicylate (525 mg per dose).

2.6 Advantages and drawbacks

Since the cirrus cloud engineering idea is new and untested in climate models, it is unknown what undesirable environmental consequences may exist regarding its implementation. As stated above, it does not address the problem of ocean acidification. However, a number of advantages are apparent: (1) it appears potentially capable of neutralizing the warming due to a doubling of CO_2; (2) if negative environmental consequences resulted from implementing the idea, the climate system should return to its pre-engineered state within months after termination of seeding (due to the fact that the seeding aerosol have an atmospheric residence time of 1-2 weeks); (3) unlike marine stratus clouds, cirrus clouds are generally present everywhere over the earth, making preferential cooling of certain regions less likely; (4) if GCM simulations indicate seeding only certain regions (such as polar and/or mid-latitude regions) renders the most favorable global climate conditions, then such seeding strategies could be adopted; (5) The use of non-toxic seeding aerosol does not appear to present an environmental hazard since conventional cloud seeding does not, with AgI concentrations in snow being less than 10 parts per trillion (Super 1986; Warburton et al. 1995). Moreover, this climate engineering approach does not appear to have the problems associated with stratospheric injection of sulfuric acid aerosol: (1) likely interference in the hydrological cycle, enhancing the frequency of drought in some regions; (2) ozone destruction and (3) turning the sky white. Clearly more research is needed to better understand the pros and cons of this approach.

2.7 Needed research

To determine whether this approach is viable, targeted research will be needed in the following areas:

1. Improved understanding of ice nucleation mechanisms in cirrus clouds and potential seeding aerosol
2. Small-scale field campaigns with research aircraft to provide limited testing of the climate engineering approach
3. Plume dispersion studies to determine the feasibility of establishing a background concentration of seeding aerosol suitable for favorable modification of cirrus clouds
4. GCM studies to determine the potential effectiveness and limitations of the climate engineering method, especially in regards to unwanted environmental consequences. The credibility of the results will depend on how realistically the cirrus clouds are represented in the GCM, including consistency between predicted and observed cirrus cloud coverage.

There are other areas of research and engineering that need to be explored for assessing the viability of this approach, but progress regarding the above research topics would be a good start.

A major uncertainty regarding this climate engineering approach is whether homogeneous freezing nucleation is an important process by which ice crystals are formed in cirrus clouds. Before GCM experiments can be taken very seriously, a basic understanding of the dominant ice nucleation mechanisms at temperatures less than -38°C is needed. While it appears possible to modify cirrus clouds when heterogeneous nucleation processes dominate (provided the difference in RH_i threshold is appreciable regarding natural and seeded ice nuclei), cirrus may be more easily modified if homogeneous freezing nucleation is active and an important process, since this indicates the RH_i threshold for natural ice nuclei is more than 140%, substantially above that of the seeding aerosol ($RH_i \sim 105\%$). In this case it should be relatively easy to modify cold cirrus clouds since the seeding aerosol should easily out-compete the natural ice nuclei for water vapor.

There is a large body of literature dealing with hetero- and homogeneous freezing nucleation in cirrus clouds with no clear verdict on what process dominates at temperatures less than -38°C. A recent study by Krammer et al. (2009) suggests that heterogeneous nucleation processes may dominate at temperatures less than -40°C based on RH_i measurements and ice particle concentration measurements in cirrus clouds. A common problem to this and many other studies addressing this nucleation issue from a measurement angle is that the ice PSD has only recently been measured with reasonable accuracy. Earlier measurements were plagued with the problem of ice particle shattering, whereby ice particles sampled by a probe impact and shatter on the inlet tube, producing anomalously high concentrations of small ice crystals (McFarquhar et al. 2007, Jensen et al. 2009, Lawson et al. 2010, Mitchell et al. 2010, Zhao et al. 2011). The subsequent sampling of these artifact ice fragments along with natural ice particles made it virtually impossible to say anything about ice nucleation processes in the atmosphere based on aircraft probe measurements. Recently PSD measurement probes have been improved to greatly reduce the shattering problem (Lawson et al. 2006a, Jensen et al. 2009, Lawson et al. 2010, Korolev et al. 2011, Lawson 2011). These relatively recent PSD measurements are discussed below in relation to several field campaigns sampling cirrus clouds. Moreover, these measurements provide greater insight into the role of homogeneous freezing nucleation in cirrus clouds and the viability of this climate engineering idea.

3. New findings

This section presents new in situ measurements regarding the role of homogeneous freezing and heterogeneous nucleation processes at temperatures less than -38°C. For tropical cirrus clouds, homogeneous freezing nucleation may dominate ice production under the higher updraft conditions associated with anvil cirrus clouds, based on a change in the PSD shape, ice particle number concentration/ice water content (IWC) ratio and the mean ice particle size near -40°C. For mid-latitude synoptic cirrus, homogeneous freezing nucleation appears to dominate ice production based on a marked change ~ -40°C regarding the PSD shape, the ice particle number concentration/IWC ratio, the mean ice particle size, the ice particle shape and the mass-weighted ice fall-speed. These results suggest that ice production in fresh anvil and mid-latitude synoptic cirrus clouds occurs at RH_i levels generally exceeding 140%, which should make these clouds easily susceptible to modification from seeding with efficient ice nuclei.

3.1 Field campaigns and measurement methods

The new findings reported here deal with the role of homogeneous freezing nucleation in cirrus clouds, based on field observations from primarily two field campaigns: (1) the Tropical Composition, Cloud and Climate Coupling (TC4) campaign, sampled in August 2007 near Costa Rica (funded by the USA's National Aeronautical and Space Administration or NASA), and (2) the Small PARTicles In CirrUS (SPARTICUS) campaign, with six months of sampling over the continental USA during the winter and spring of 2010 (funded by the U.S. Department of Energy, Atmosphere Radiation and Measurement (ARM) program). During TC4, the NASA DC-8 aircraft was used to sample fresh anvil cirrus (anvil cloud connected to column of deep convection), aged anvil cirrus (anvil cloud detached from column of deep convection), and in situ cirrus (not directly related to convection). The in situ cirrus appeared thin and tenuous in satellite imagery. This study uses TC4 flight data obtained on 22 July, 24 July, 5 August and 8 August 2007. On August 5th, the NASA WB57 joined with the DC-8 to vertically profile a deep anvil cirrus deck. To complement the in situ cirrus measured during TC4, in situ cirrus data from the NAMMA (NASA African Monsoon Multidisciplinary Analysis) campaign of 2006 were also used in this study for the following flight days: 19 and 26 August, 3 September 2006. Constant temperature transects or legs were flown through the clouds during TC4 and NAMMA, with one measured PSD representing one flight leg. No evidence of liquid water was found in the TC4 cirrus clouds.
The SPARTICUS PSD used here are from 1-2 minute periods of flight time where the cirrus microphysical properties (median mass dimension and extinction coefficient) were not changing rapidly over time. A total of 174 SPARTICUS PSD were processed from 9 different days and 13 different flights, where only synoptic (non-anvil) cirrus was sampled. Using measurements from the Cloud Particle Imager (CPI), the relative humidity sensor, Forward Scattering Spectrometer Probe (FSSP) and the 2D-S probe, no evidence of liquid water was detected in these clouds.
In both field campaigns a relatively new probe was used for measuring the PSDs; the 2D-Stereo or 2D-S probe (Lawson et al., 2006a, Lawson 2011). The 2D-S directly measures ice particle length and projected area, and indirectly measures ice particle mass. The ice particle mass in each size bin was determined from 2D-S measurements of ice particle area using the mass-area relationship described in Baker and Lawson (2006). The IWC was calculated by integrating the PSD ice particle masses, and this method was shown to produce IWCs

consistent with those measured directly by the Counterflow Virtual Impactor (CVI) during TC4, with differences less than ~ 50% and an r^2 of 0.88 (Lawson et al. 2010; Mitchell et al., 2010). The CVI uncertainty in IWC has been estimated to be 13% at water contents of 0.05 to 1.0 g m^{-3} increasing to 16% at 0.010 g m^{-3} and to 40% at 0.0025 g m^{-3} (Heymsfield et al., 2007; Twohy et al., 1997, 2003). Thus the 2D-S yields direct or indirect measurements of the size distributions for ice particle concentration, area and mass. These three types of size distributions were used to calculate the mass-weighted fall speed, V_m, as described in Mitchell et al. (2011). As noted in Section 2, V_m is a critical parameter controlling the cloud life cycle and coverage, and associated radiative forcing.

As noted, historical PSD measurements in cirrus clouds have suffered from ice particles shattering on the probe inlet tube which artificially enhanced the concentration of small ($D < 100$ µm) ice crystals. The problem of ice particle shattering is greatly reduced in the 2D-S probe due to both probe design and the removal of shattered ice particles based on ice particle interarrival times (Lawson et al., 2006a; Jensen et al., 2009; Baker et al., 2009a,b; Lawson et al., 2010). Thus the PSDs measured during these field programs appear to be much more realistic than previous PSD measurements and render more realistic estimates of ice particle concentrations and V_m.

An additional capability of the 2D-S to make improved measurements of ice particle size and projected area is its ability to provide true 10-µm pixel resolution at jet aircraft speeds. In laboratory experiments the 2D-S probe was shown to accurately image an 8-µm fiber rotating at 233 m s^{-1} (Lawson et al. 2006), which is greater than the true airspeeds that aircraft experienced during these field campaigns. Ice crystals smaller than this appear to have a negligible impact on most anvil cirrus optical properties (Jensen et al. 2009). This higher resolution improves the measurement of particle projected area.

3.2 Calculation of V_m

Perhaps for the first time, the mass weighted ice fall-speed or V_m was estimated directly from in situ measurements of ice particle area and mass, as described in Mitchell et al. (2011). This reduces uncertainty in V_m by eliminating intermediate steps or approximations used in its calculation. Since the ice particle mass is estimated from the 2D-S probe measurements (Baker and Lawson 2006), and the 2D-S IWCs compare favorably with the CVI IWCs, it appears that ice particle masses based on the Baker and Lawson area-mass relationship are realistic regarding the TC4 cirrus anvils. CVI measurements of IWC were made during the last week of the SPARTICUS campaign and have not yet been compared against 2D-S probe IWCs. Thus for now we assume that ice particle masses derived from the Baker and Lawson (2006) area-mass relationship are valid for SPARTICUS cirrus. The violation of this assumption could impose a systematic bias on our results regarding V_m.

With measurements of the concentration of ice particle area, estimated mass and number (to calculate mean bin values of area and mass) for each size bin of the 2D-S probe, these properties can be used directly to calculate the ice particle fall velocity in each size bin, using the fall speed formulation of Heymsfield and Westbrook (2010). The formulation of Heymsfield and Westbrook is more accurate for ice particles having aspect ratios far from unity, such as long columns and needles or "stellar" dendritic ice crystals.

The treatment of the mass-weighted PSD fall velocity, V_m, is calculated as:

$$V_m = \Sigma v(D) IWC(D) / IWC, \quad (1)$$

where v(D) is the fall speed calculated for a given bin size D and IWC(D) is the measurement derived ice mass concentration in a given 2D-S size bin. To calculate v(D), the mean area A_m(D) and mean mass m(D) of the size-bins are needed, which are calculated from the measurements as:

$$A_m(D) = A(D) / N(D), \tag{2}$$

$$m(D) = IWC(D) / N(D), \tag{3}$$

where N(D) is the bin number concentration and A(D) is the measured projected area concentration in a given bin. In this way V_m is calculated as directly as possible from measurements. The V_m calculations reported in this study assume a reference temperature of -20 °C and a pressure of 500 hPa.

3.3 Calculation of area ratios

The area ratio AR is the projected area of an ice particle divided by the area defined by a circle having a diameter equal to the particle's maximum dimension D. The AR is thus a metric for ice particle shape, with more branched or columnar ice crystals tending to have relatively low ARs whereas compact or quasi-spherical ice particles have relatively high ARs near unity. If there is a change in ice crystal nucleation mechanism ~ -40°C, then this could be accompanied by a change in ice crystal shape. For this reason the AR was calculated for each size bin of the 2D-S probe. All measurements of A_m(D) were checked to insure that AR ≤ 1.0. In rare cases when AR > 1.0, then A_m(D) was set to the area of a corresponding circle.
Since the AR is a function of ice particle shape, ice particle shape differences among PSDs were evaluated using an AR representing the entire PSD. This was done by summing A(D) over the PSD for ice particles greater than 60 μm. While the 2D-S probe has 10-μm pixel resolution, ARs are more accurate when estimated for size D > 60 μm. Moreover, smaller ice crystals tend to be quasi-spherical (Korolev and Isaac 2003; Mitchell et al. 2010; 2011). The same was done using ice spheres (i.e. ice particle maximum dimension defines a sphere from which the area cross-section is calculated). Then the area ratio of the PSD was determined as

$$AR_{PSD} = \frac{\sum_{D=60\mu m}^{D=D_{max}} A(D)}{\sum_{D=60\mu m}^{D=D_{max}} A_s(D)}, \tag{4}$$

where A_s(D) is the area of a sphere having diameter equal to the ice particle maximum dimension and D_{max} is the maximum particle size of the PSD. Probability distribution functions (PDFs) of AR_{PSD} were produced for each type of cirrus cloud evaluated here to determine whether ice crystal shapes change near -40°C.

3.4 Tropical cirrus clouds: TC4 results

As noted, cirrus clouds sampled during the TC4 campaign were classified as either fresh anvils, aged anvils or as in situ cirrus. The PSDs from these clouds were grouped into temperature intervals of 5°C and mean PSDs for each temperature interval were calculated as shown in Fig. 1, which shows the temperature dependence of these mean PSD for each cloud type. Since updraft speeds in deep convective clouds are relatively high (meters per second), ice crystals in fresh anvil cirrus were likely produced under relatively high updraft conditions. In aged anvil cirrus, no longer attached to their parent cumulonimbus cloud, updrafts should be much weaker or possibly not present. In situ cirrus clouds are formed through the gradual ascent of an air mass and thus ascending air motions in these clouds should be relatively weak. It is ascending air motions that cause air to become supersaturated and to produce a cloud, and the higher the updraft, the higher the supersaturation level. For this reason, homogeneous freezing nucleation may become more likely as the updraft increases since it requires $RH_i > 140\%$.

Figure 1 shows that for the fresh anvil cirrus there is a clear transition in PSD shape ~ -40°C, with PSD colder than -40°C being approximately mono-modal in shape, and PSD warmer than -40°C being more bimodal in shape. This can be interpreted in terms of nucleation processes, with homogeneous freezing nucleation affecting the PSDs at temperatures less than -40 C. Homogeneous nucleation will produce ice crystals at a faster rate than

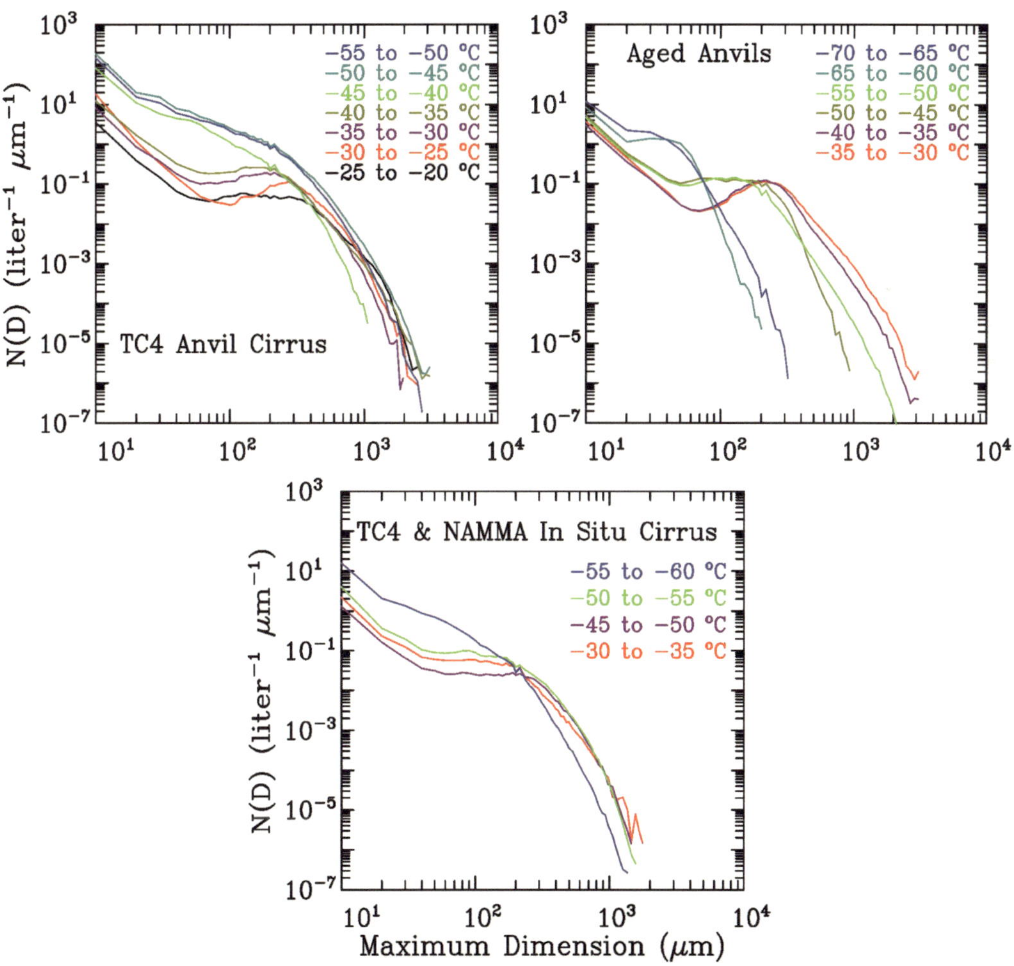

Fig. 1. Temperature dependence of mean PSD for three types of tropical cirrus clouds sampled during the TC4 and NAMMA field campaigns.

heterogeneous processes, thus elevating the concentrations of the smaller ice crystals relative to heterogeneous nucleation conditions (T > -40°C). Ice particle growth processes also affect the PSD shape, such as aggregation. Aggregation is a process where ice particles collide with each other and combine to form larger "aggregates" comprised of individual ice crystals. It has been demonstrated that aggregation is an important growth process for ice particles in cirrus clouds (Mitchell et al. 1996). Aggregation will act on all PSD, depleting the concentrations of small ice crystals to form larger ones and possibly producing a secondary maximum in concentration ~ 200 μm particle size when ice crystals are produced only through heterogeneous processes. This bimodal behavior apparently disappears when ice crystals are nucleated at a higher rate, building up the small crystal concentrations to produce approximately mono-modal PSD. Thus there is a counter-balancing effect between nucleation producing new small ice crystals and the removal of small crystals due to aggregation. When nucleation rates are relatively low (i.e. heterogeneous processes), aggregation effects should be more apparent, with bimodality developing. This explanation appears to fit the fresh anvil observations in Fig. 1 where updrafts are higher.

The temperature dependence of the fresh anvil cirrus could also be explained by a process known as size sorting, where larger ice particles fall faster than smaller ones, leaving higher concentrations of smaller ice crystals at higher cloud levels and building up the concentration of larger ice particles at lower levels. This process could explain the observed bimodality at warmer temperatures, or it could be a contributing factor, but this process alone cannot explain the transition observed at -40°C. Some may argue this transition at -40°C is a mere coincidence.

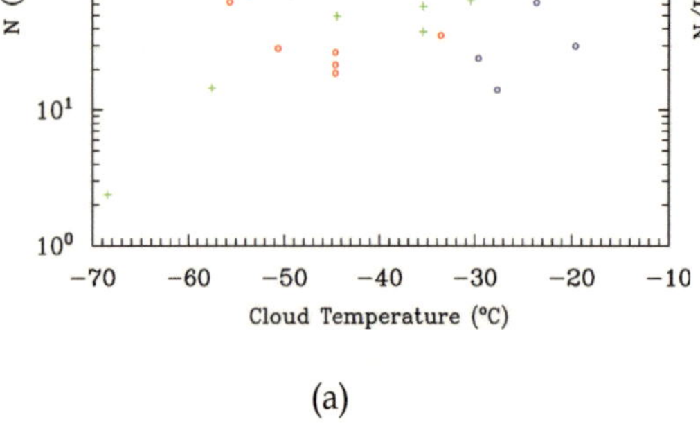

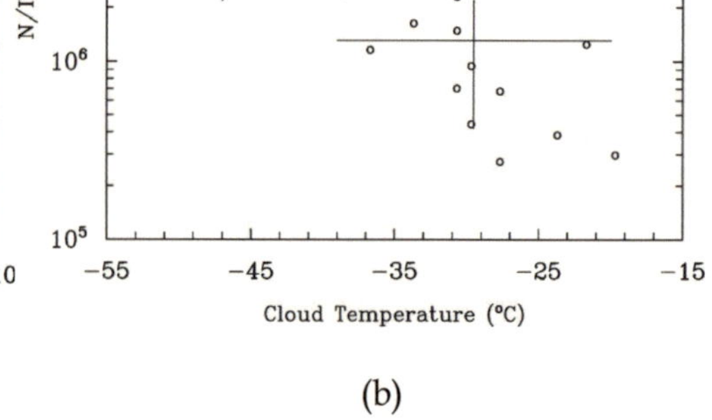

Fig. 2. a: Total number concentration for the three types of cirrus clouds sampled during the TC4 campaign. b: For the fresh anvil cirrus cases, shown is the total number concentration divided by the corresponding PSD IWC related to PSD temperature.

The aged anvil PSDs in Fig. 1 exhibit a similar temperature dependence, but without a distinct transition near -40°C. It is not clear whether this behavior is a "memory effect" from an earlier period when the anvil was "fresh" or whether homogeneous freezing nucleation was having an ongoing influence on the PSD temperature dependence. The PSD associated with in situ cirrus do not show a transition from bimodal to mono-modal size-spectra near -40°C, which appears consistent with their lower updrafts. The mono-modal PSD at the coldest level for in situ cirrus may reflect a reduced impact from aggregation near cloud top (aggregation influences on PSDs increase with distance from cloud top (Mitchell 1988, Mitchell et al. 1996)).

The above information can be presented in terms of the bulk properties of the PSD; total number concentration N and mean ice particle size (maximum dimension), $\bar{D}$. Figure 2a shows N for the three cirrus cloud types. The highest five N values are associated with an IWC greater than 0.3 g m^{-3} (very high for cirrus clouds), due to sampling closer to the center of convection. Since N partly depends on RH_i (related to the condensation rate and IWC), N was divided by the corresponding IWC to remove this dependence in Fig. 2b for the fresh anvil cirrus. This ratio is related to the nucleation rate coefficient J, which is the number of ice germs formed per unit volume (related to mass) of condensate per unit time. N/IWC can be thought of as a time integration of J in a cloud parcel, from ice nucleation onset to sampling time, although this is oversimplified as it ignores other processes such as aggregation and ice fallout. Since J for homogeneous freezing is known to be higher than J for heterogeneous processes, N/IWC should also be higher for temperatures less than -40°C if homogeneous freezing is active. It can be seen in Fig. 2b that N/IWC values can be described as two groups; one for T < -40°C and one for T > -40°C. The horizontal bars indicate the range of temperature values in each group while the vertical bars indicate the standard deviation for each group of points. The intersection point gives the mean ratio value for each group. Figure 3 shows $\bar{D}$ for the fresh anvil cirrus related to temperature. Like Fig. 2b, the $\bar{D}$ values cluster into two groups: one for T < -40°C and one for T > -40°C. The horizontal and vertical bars follow the same convention as Fig. 2b. Higher nucleation rates should produce smaller values of $\bar{D}$.

Figures 1, 2b and 3 collectively suggest that a change in ice nucleation mechanism occurs near -40°C when updrafts are sufficiently high (corresponding to fresh anvil cirrus). Figure 1 shows that it is the concentration of the smaller more recently nucleated ice particles in fresh anvils that increase at temperatures below -40°C. For cloud updrafts between 1 and 4 m s^{-1} (common for marine deep convection at cirrus levels), homogeneous freezing predicts N between ~ 1 and 10 cm^{-3} (or 1000 to 10,00 $liter^{-1}$) for relatively low aerosol particle concentrations (< ~ 150 cm^{-3}) between -40 and -50°C (Barahona and Nenes 2008). Most of the N values for fresh anvil cirrus in Fig. 2a fall into this approximate range of expected values. The close grouping of N/IWC values in Fig. 2b for T < -40°C indicates that the two N values ~100 $liter^{-1}$ in Fig. 2a are associated with lower IWCs and perhaps lower updrafts and thus still consistent with homogeneous freezing nucleation. Moreover, entrainment of drier air into the anvils with subsequent sublimation, inhomogeneous mixing and dilution will reduce N. Overall, this analysis indicates that homogeneous freezing appears to be an important process, perhaps the main nucleation process, for fresh anvil cirrus clouds. Therefore RH_i levels should be more than 140% in the region where anvil ice crystals are formed.

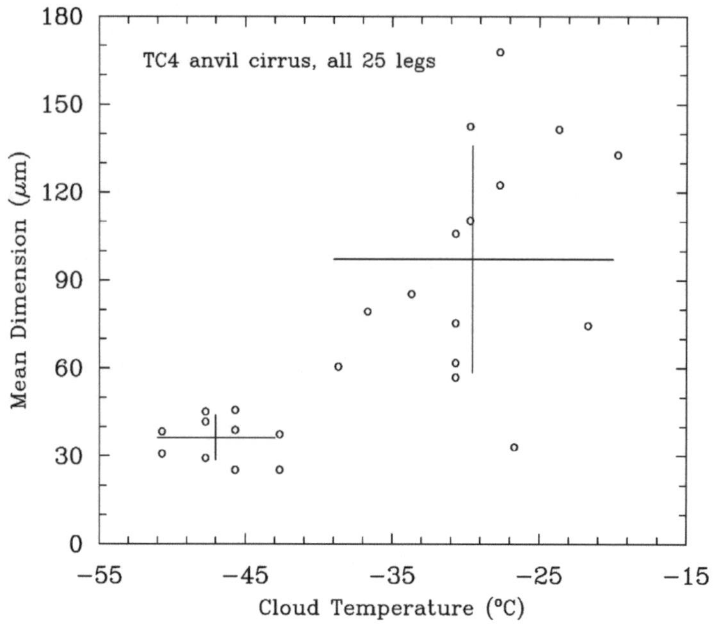

Fig. 3. Mean maximum dimension (i.e. mean ice particle size) related to temperature for the fresh anvil PSDs sampled during TC4.

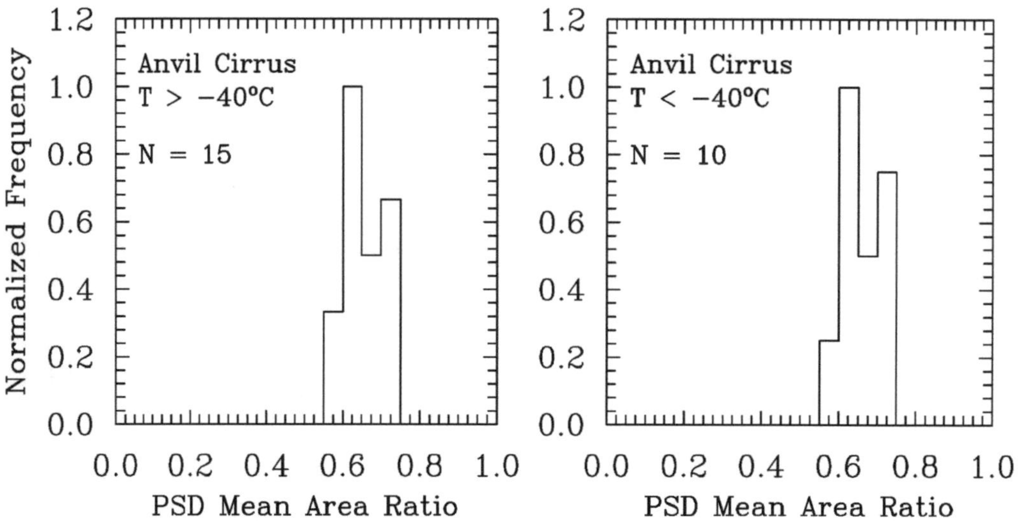

Fig. 4. Normalized frequency distributions of the PSD area ratio for fresh anvil cirrus in two temperature regimes. N refers to the number of PSD sampled.

3.4.1 Temperature dependence of ice particle shape

As mentioned, a change in ice nucleation mechanism may be accompanied by a change in ice crystal habit (i.e. shape). A change in ice particle shape can be detected through evaluating the area ratio (AR) as described in section 3.3. Figure 4 shows the temperature dependence of the AR for fresh anvil cirrus for temperatures more than -40°C and for temperatures less than -40°C. Since there is no significant change in the normalized frequency distribution between temperature regimes, there is no evidence for a change in ice particle shape. Similar results were obtained for aged anvil and in situ tropical cirrus, although aged anvil cirrus exhibited a frequency value 0.5 in the largest AR bin (corresponding to unity) for T < -40°C (this was not present for T > -40°C).

For the fresh anvil cirrus case, the results might be explained if the geometry of the ice nuclei (e.g. irregular mineral dust particle vs. spherical haze or cloud droplets) influences subsequent ice crystal growth patterns. It is reasonable to expect the freezing of supercooled cloud droplets (e.g. via immersion freezing nucleation or contact nucleation (Pruppacher and Klett 2010)) to contribute substantially to new ice production for temperatures greater than -38°C. These same supercooled cloud droplets may experience flash freezing at the -38°C level where all liquid water freezes via homogeneous freezing, as observed by Rosenfeld and Woodley (2000) in strong convective updrafts. This would increase N above the -38°C level but may not result in a change in ice crystal shape.

3.4.2 Application to climate engineering

The above analysis indicates that fresh anvil cirrus may be susceptible to modification by seeding aerosol. But due to the formation mechanism of anvil cirrus, it is less clear just how seeding could be accomplished, since anvil cirrus represents the outflow of deep convection where boundary layer air (typically within 1 km of the surface) has been vertically advected into the anvil cirrus. Thus if the seeding aerosol were dispersed at anvil cirrus levels, and ice crystal production occurs in the convective updraft, there may be little opportunity for the seeding aerosol to produce new ice crystals (scavenging by wet deposition processes could remove most of the aerosol before it sediments to boundary layer levels). Alternatively, the boundary layer could be seeded, but some portion of the seeding aerosol would be scavenged and removed through wet deposition processes in the convective updraft. Fridlind et al. (2004) present evidence indicating mid-tropospheric aerosols contribute substantially to ice nucleation in anvil cirrus, and if this is true, then seeding at temperatures less than -45°C may be effective. These factors make the seeding of tropical anvil cirrus a more complicated proposition relative to synoptic cirrus clouds that are commonly found in the subtropics, the mid-latitudes and the polar regions. Synoptic cirrus clouds are formed in gradually ascending air masses such as lifting motions due to a warm front. Thus seeding the region of atmosphere where they typically form under clear conditions could modify these cirrus clouds once they form in the treated air mass.

To counteract the worst effects of global warming, it may not be necessary to seed anvil cirrus clouds. Both observational evidence and GCM predictions show that global warming due to GHG emissions will affect the polar regions much more than the tropics (Soloman et al., 2007), thus reducing the temperature gradient between the poles and tropics. To counteract this effect, seeding could be applied to only the mid-latitudes and polar regions, increasing OLR from those regions. This would cool the planet where it needs it most, and also act to restore the "normal" temperature gradient between the tropics and polar regions.

3.5 Synoptic cirrus clouds: SPARTICUS results

Synoptic cirrus clouds are characterized by ice formation conditions distinctly different from those of anvil cirrus; updrafts are much weaker (typically ~ 10-30 cm s^{-1}) and heterogeneous processes like deposition and condensation freezing nucleation may play a larger role. The

SPARTICUS field campaign, designed in part to evaluate the concentrations of small ice crystals in cirrus clouds using improved instruments much less vulnerable to artifacts from ice particle shattering, was conducted over the continental USA, especially over the Southern Great Plains ARM site. Although both anvil and synoptic cirrus clouds were sampled during SPARTICUS, this study considers only the synoptic cirrus clouds. The cirrus clouds were identified as synoptic based on flight notes and satellite observations. There were 13 synoptic cirrus clouds analyzed that met the criteria established for good data quality. This resulted in 174 PSD that were grouped into 5°C temperature intervals and mean PSD were calculated for each temperature interval, producing 11 mean PSD. The number of PSD found in each temperature interval is shown in Fig. 5.

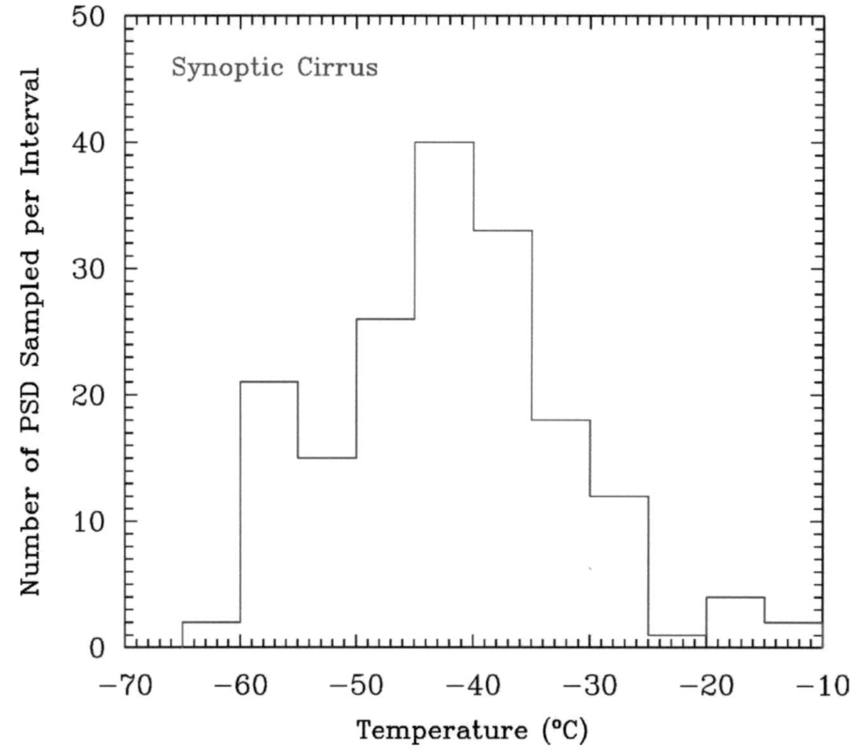

Fig. 5. PSD sampling statistics for SPARTICUS synoptic cirrus, based on 13 cirrus flights during 9 days.

The PSD data processing criteria established for good data quality are as follows: A sampled flight segment begins when the extinction coefficient β_{ext} (twice the measured particle projected area) exceeds 0.1 km^{-1} over a 5 second period. The mean value of β_{ext} and the median mass dimension D_m are calculated were the sample time resolution is 1 second (i.e. 1 Hz). If the maximum values of β_{ext} and D_m do not exceed their mean value by a factor of two, and their minimum values are not less than 0.4 times the mean value, then the process continues. Sampling time is then increased in 1-second increments; every second the value of β_{ext} and D_m are calculated and compared against their respective cumulative mean values. If they do not exceed two times the cumulative mean and are not less than 0.4 times the cumulative mean, then the segment passes and another second is added. A segment must reach 60 seconds in length to be kept and when an acceptable segment reaches 120 seconds in length the first half is cut off and kept as a PSD measurement and the second half continues adding seconds provided the above criteria are met. The process goes on in this way such that all PSD measurements have sampling times between 60 and 120 seconds. The method makes use of a large portion of the data while insuring relatively stable cloud conditions.

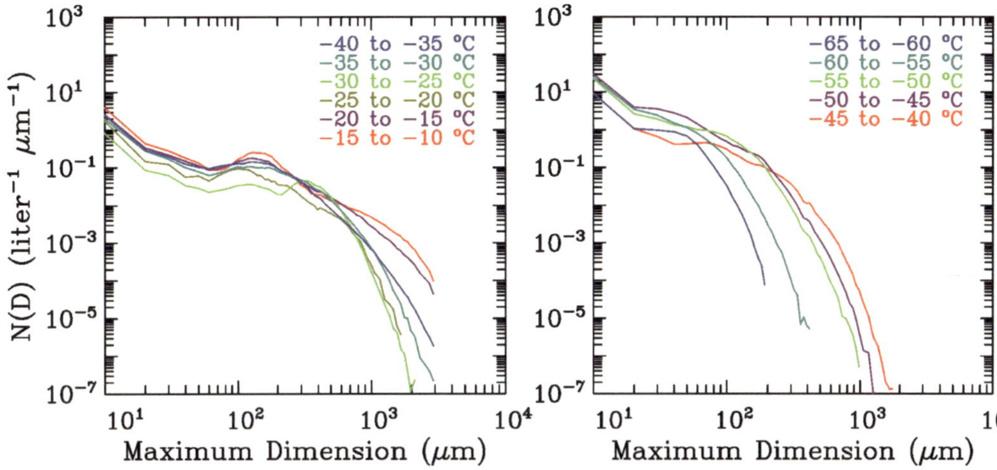

Fig. 6. Temperature dependence of the synoptic PSD during SPARTICUS.

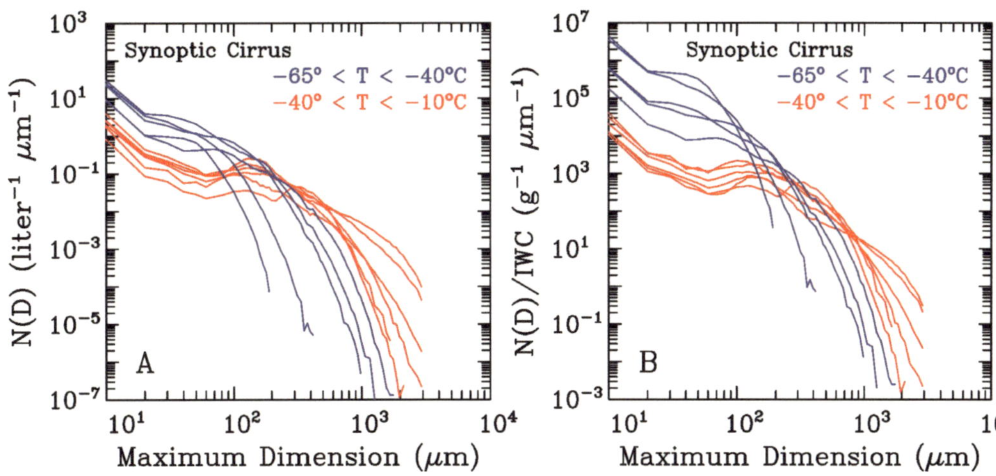

Fig. 7. A: As in Fig. 6 but with all 11 mean PSD shown together and color-coded by temperature regime. B: The same PSD as in panel A, but divided by their corresponding IWC to express the impact of ice nucleation rates on PSD.

The temperature dependence of the PSD that were processed and selected in this way is shown in Fig. 6, with PSD warmer than -40°C on the left and PSD colder than -40°C on the right. A clear difference in PSD shape is evident, with the warmer PSD being bimodal and the colder PSD being approximately mono-modal. As with the anvil cirrus, this could be explained by differences in ice crystal nucleation rates, with higher rates regarding the mono-modal PSD. Lower nucleation rates would change the balance between small ice crystal production rates and small ice crystal removal rates from aggregation, perhaps resulting in bimodal PSD as shown in Fig. 6 and discussed in Section 3.4.

The mean PSD are again shown in Fig. 7a, but are color-coded to show the transition in PSD shape at -40°C. A transition in ice nucleation rates at -40°C is better expressed by dividing each PSD by its corresponding IWC, as shown in Fig. 7b, showing the number of ice particles per gram of condensate per μm size interval. Differences in nucleation rates are now easily recognized by the concentration differences of the smaller ice crystals. This is similar in principle to the N/IWC ratio as described in relation to Fig. 2b. Figures 6 and 7 demonstrate that at -40°C, there are distinct changes in PSD shape, and for small ice crystals (D < 100 μm), distinct changes in N and N/IWC.

The total number concentration N and the N/IWC ratio are plotted for the mean PSD, as shown in Fig. 8. When less than three PSD occurred in a given temperature interval, no standard deviation was calculated (see Fig. 5 for sampling statistics). Merely relating N to PSD temperature may not reveal a change in nucleation rate unambiguously, but the N/IWC ratio reveals a change in nucleation rate near -40°C more clearly. As predicted by homogeneous freezing nucleation theory (Barahona and Nenes 2008), J should be most sensitive to changes in temperature and cloud updraft (RH_i). The increase in N/IWC with colder temperatures (< -40°C) appears consistent with homogeneous freezing theory. Moreover, the N values shown in Fig. 8a also appear consistent with those predicted for updrafts typical of synoptic cirrus clouds (10-30 cm s^{-1}) at temperatures between -40 and -60°C and for an aerosol particle concentration of 200 cm^{-3}, where the vapor deposition coefficient for ice growth ranges from 0.1 to 1.0 (Barahona and Nenes 2008). However, continental aerosol concentrations measured over Florida between 5 km and 15 km ranged from 3000 cm^{-3} to 100 cm^{-3} (Fridlind et al. 2004). If typical aerosol concentrations during SPARTICUS were on the order of ~ 1000 cm^{-3} at cold cirrus levels, then the observed Ns may be explained by entrainment and dilution processes.

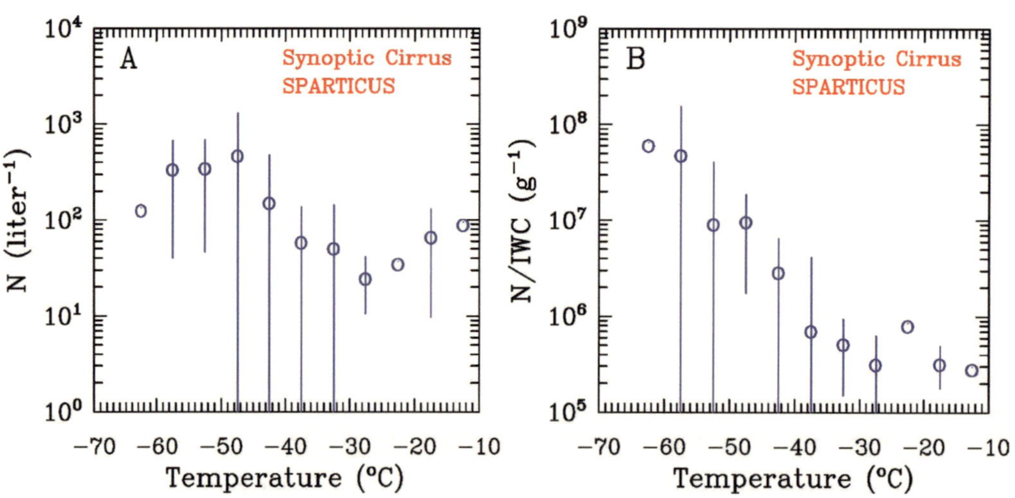

Fig. 8. Temperature dependence of the total ice particle number concentration and the N/IWC ratio based on mean PSDs. Vertical bars are standard deviations. Data points based on less than 3 PSD have no standard deviation.

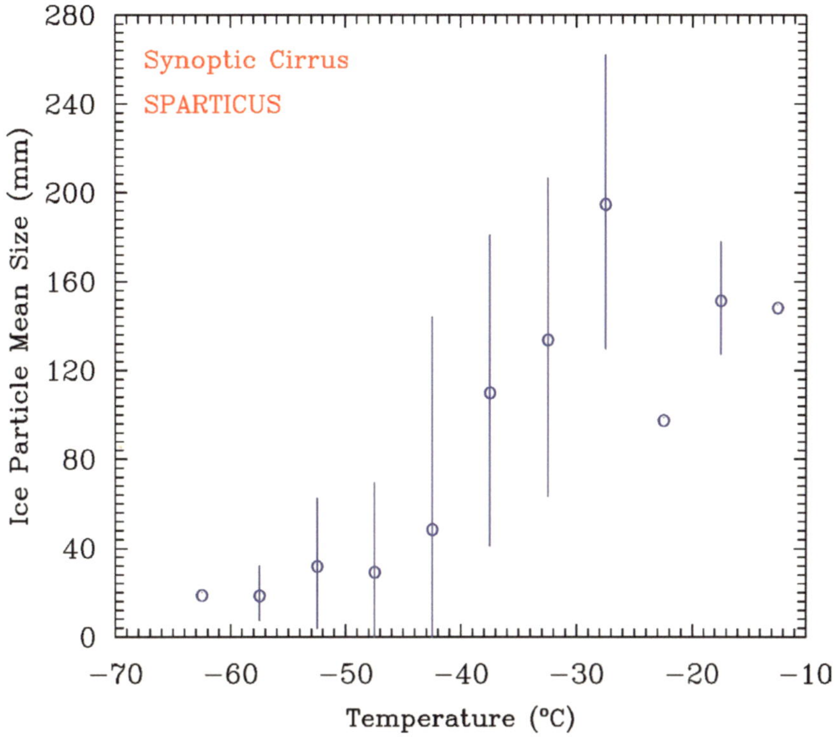

Fig. 9. The temperature dependence of the mean maximum dimension from the mean SPARTICUS PSDs. Points derived from less than 3 individual PSD have no standard deviation.

The same basic pattern shown by the N/IWC ratio is repeated by plotting the mean size of the mean PSD against temperature, as shown in Fig. 9. As ice nucleation rates increase, the mean ice particle size generally decreases.

3.5.1 Temperature dependence of ice particle shape

As mentioned, a change in nucleation rate could be accompanied by a change in ice crystal habit (i.e. shape), depending upon the ice nucleation mechanisms involved. To explore this possibility, normalized frequency distributions of the mean area ratio of a PSD, for all of the synoptic PSDs processed, were developed for two temperature regimes: more than -40°C and less than -40°C. This is shown in Fig. 10. When the size threshold was changed from 60 μm to 100 μm (i.e. excluding particle sizes less than 100 μm from the analysis), the results were almost the same. This is clearly an unambiguous signal that ice particle shapes generally differ between these temperature regimes. Figure 10 is strikingly different than the similar analysis of area ratios for fresh anvil cirrus shown in Fig. 4, suggesting that the ice growth physics in mid-latitude synoptic cirrus clouds differs appreciably from that of anvil cirrus. The solitary peak near unity for temperatures less than -40°C indicates that these ice crystals are generally either quasi-spherical or isometric (i.e. having nearly equal length and width) in geometry. Since homogeneous freezing nucleation commences at about -38°C, these results strongly suggest that this nucleation mechanism is the reason for the observed change in ice particle morphology. Moreover, they also indicate that almost all of the ice particles less than -40°C have the same shape. For almost all the ice particles to have the same shape, a common origin may be implied, suggesting that homogeneous freezing produced almost all of the ice particles.

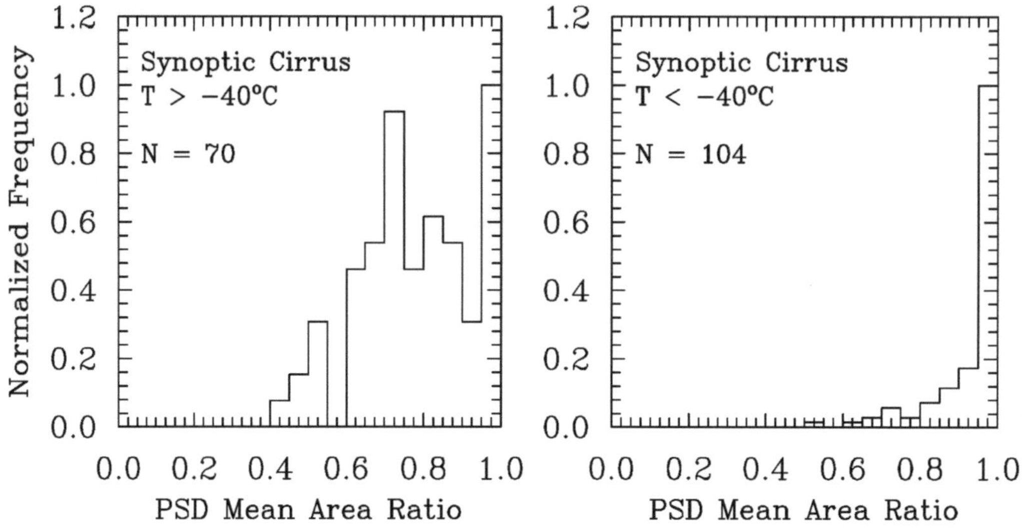

Fig. 10. Normalized probability distribution functions of the mean area ratio for all 174 synoptic cirrus PSDs, divided into two temperature regimes. N refers to the number of PSDs evaluated in each temperature regime.

To evaluate how rapidly this transition in ice particle shape occurs, the PSD area ratio was determined for each of the mean PSD and related to the mean temperature of each interval. This is shown in Fig. 11. The PSD area ratio is lowest at temperatures warmer than -20°C where dendritic ice crystals can grow and ice particles tend to be more aggregated. Between -20 and -40°C, the mean area ratio is virtually flat, but increases abruptly for temperatures less than -40°C. If we attribute this abrupt change near -40°C to homogeneous freezing nucleation, then the onset of homogeneous freezing with decreasing temperature appears rapid.

At temperatures warmer than -40°C, the gradually ascending air motions characterizing synoptic cirrus may make droplet freezing a less likely source of ice crystals since the RH_i levels will be lower with lower updrafts. That is, RH_i may never exceed the water saturation threshold that would allow cloud droplets to form and freeze. This would mean that heterogeneous nucleation processes would more likely occur through deposition nucleation and condensation freezing nucleation rather than droplet-freezing, contact nucleation on cloud droplets and immersion freezing within cloud droplets. If true, the ice crystal shapes produced through nucleation mechanisms not involving cloud droplets might be different than for ice crystals beginning from a frozen cloud droplet. At temperatures colder than -40°C, homogeneous freezing occurs on spherical haze droplets and ice germs begin growing as spheres. On this basis there could be a change in ice crystal shape occurring ~ -40°C.

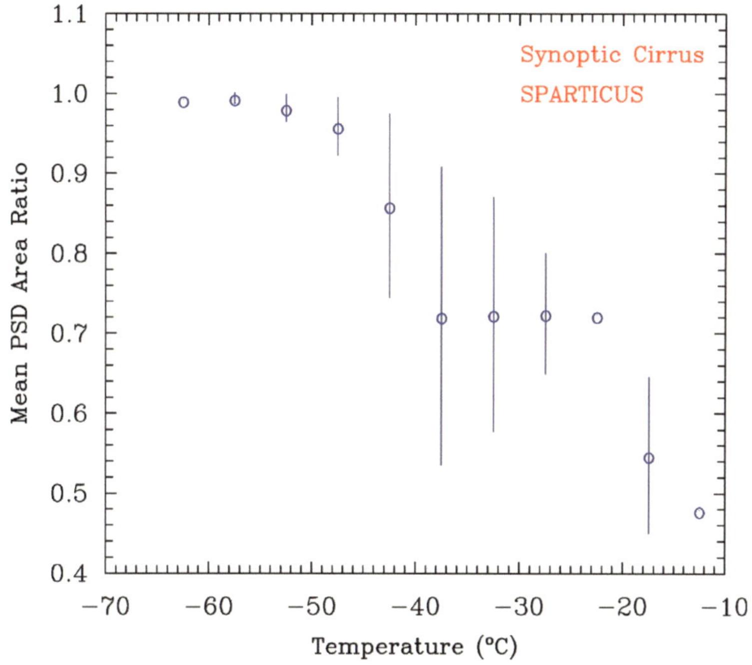

Fig. 11. Temperature dependence of the area ratio calculated from the mean PSD. Vertical bars are standard deviations. Points without such bars have less than 3 PSD samples.

In a dynamically, microphysically and radiatively complex system like the upper troposphere, things are often not what they appear to be. An alternate explanation for the observed ice particle morphology is that when moist layers of the upper troposphere gradually ascend, they become supersaturated and ultimately exceed the threshold RH_i for homogeneous freezing nucleation. Relatively high concentrations of ice crystals are then produced throughout this layer and grow in the supersaturated environment, rapidly drawing down RH_i levels close to ice saturation. The chance of an aircraft sampling regions near the homogeneous freezing threshold would be relatively small since such regions would be transient and short-lived. This appears consistent with the extensive observations in Kramer et al. (2009), who found similar maximum RH_i levels in both clear-sky and inside-cirrus conditions, with maximum RH_i near the homogeneous freezing threshold. For inside-cirrus conditions between -33 and -90°C, RH_i was most commonly found between 85% and 115%. Under such conditions, ice will either grow slowly or slowly sublimate. Ice crystals grown in the laboratory between -40° and -70°C at RH_i between 101% and 110% exhibit isometric growth tendencies, with short columns, thick plates and irregular compact crystals common between -20° and -70°C (Bailey and Hallett 2004, 2009). This is shown in Fig. 12, which is an adaption of Fig. 5 in Bailey and Hallet (2009), courtesy of Dr. Matthew Bailey. This suggests that the results in Fig. 10 and 11 at temperatures less than -40°C might be due to a low RH_i growth environment. The Krämer et al. observations also suggest that heterogeneous ice nucleation processes may not always be active for RH_i well above ice saturation since peak RH_i's were similar for both clear-sky and inside-cirrus conditions. Overall, this explanation attempts to broach the paradox of high RH_i required for homogeneous nucleation and low RH_i commonly found in cirrus clouds and associated with ice particle shapes at temperatures less than -40°C. This description of cirrus cloud

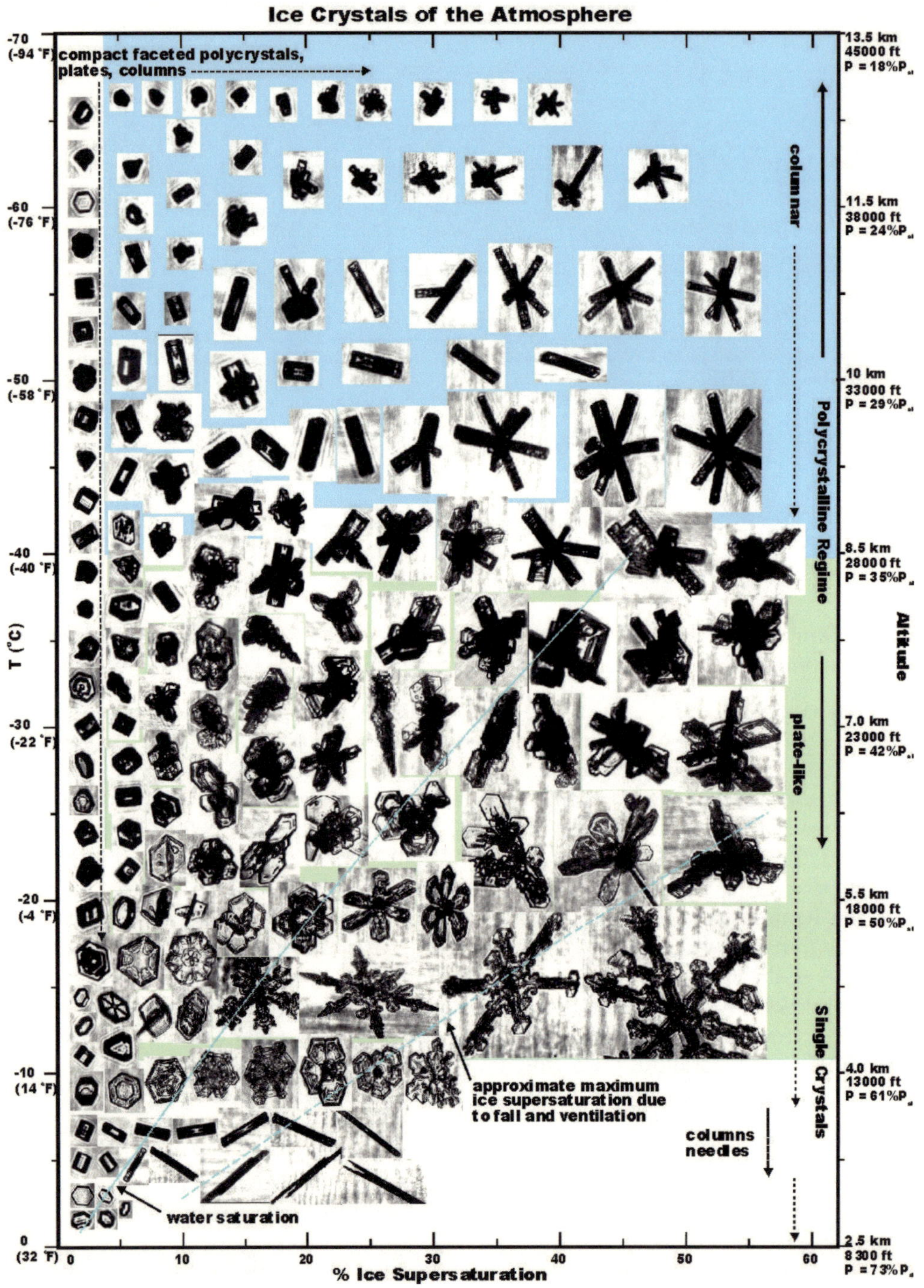

Fig. 12. Ice crystal shape as a function of formation temperature and supersaturation with respect to ice. Right axis gives approximate height and pressure of crystal formation where P_o is standard atmosphere pressure. Figure adapted from Bailey and Hallett (2009), courtesy of Dr. Bailey.

formation is similar to that of others (e.g. Haag et a. 2003; Spichtinger et al. 2004) except it offers an explanation of ice particle shape. It may not apply in cases where cirrus emanate from a convective "seeder" region, such as cirrus uncinus, where ice crystals nucleate and grow in more convective air, resulting in fall streaks as these ice crystals descend from this region. Rather, in this description, ice is produced throughout the whole moist layer without a specific "source region" involved, which appears consistent with most visual observations of cirrus clouds. In summary, two hypotheses have been described for explaining the change in ice crystal shape; one based on the morphology of the original ice germ and another based on the average supersaturation level within the cirrus cloud. These need not be mutually exclusive, and both processes could be relevant in determining ice particle shape.

Similar observations are reported in Lawson et al. (2006b), where extensive measurements in mid-latitude cirrus clouds were made using the Cloud Particle Imager (CPI) probe. The CPI images ice particles with a resolution of 3 µm, yielding very detailed images of their structure (Lawson et al. 2001). The CPI analysis in Lawson et al. (2006b) showed that for temperatures warmer than -40°C and sizes greater than 50 µm in length, number concentrations were dominated by rosette-shaped ice crystals (~ 57%), whereas at colder temperatures, irregular ice crystals dominated, increasing in abundance with decreasing temperature. The irregular crystal category in Lawson et al. (2006b) corresponds to ice particle shapes that do not fit into "traditional" shape categories such as hexagonal columns and plates, bullet rosettes, aggregates, etc. These irregular ice crystals were generally isometric, having "blocky" and sometimes quasi-spherical shapes. Examples of ice crystals in mid-latitude cirrus clouds, including these blocky and bullet rosette ice crystals that appear to dominate the homogeneous freezing and heterogeneous nucleation zones, respectively, are shown in Fig. 13.

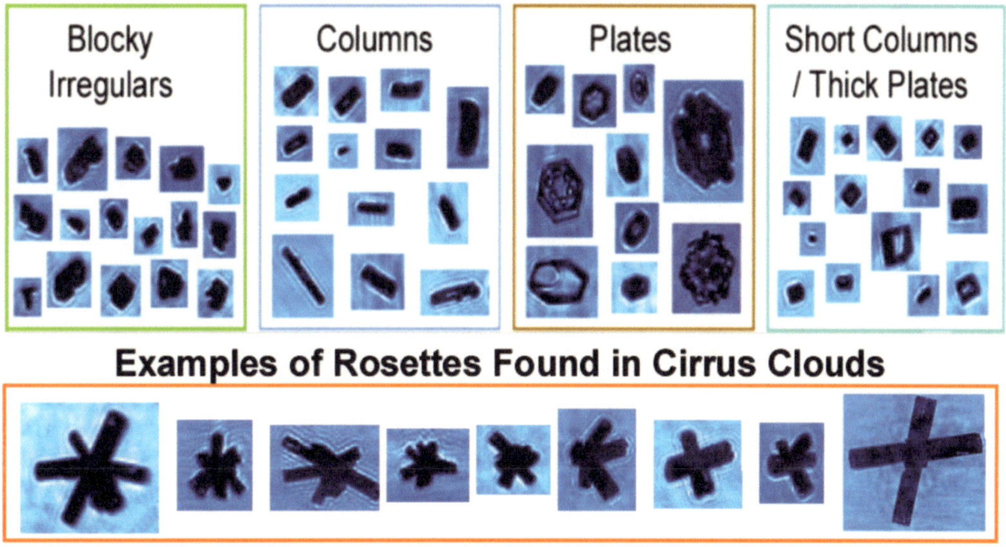

Fig. 13. CPI images of ice crystals sampled in mid-latitude cirrus clouds, with blocky irregulars being most common at temperatures below -40°C and rosette-shaped crystals being more common at warmer temperatures.

3.5.2 Temperature dependence of the ice fall-speed

As discussed in Section 2, the ice fall-speed has a strong impact on climate sensitivity in some GCMs. It is therefore critical to understand the processes determining the ice fall-speed, such as ice nucleation and ice particle morphology. The way these processes conspire to determine the mass-weighted ice fall-speed V_m is shown in Fig. 14. While V_m is roughly constant between -20° and -40°C, V_m abruptly decreases at colder temperatures. Ice particle shape impacts V_m through the ice particle mass/area ratio, with higher ratios increasing V_m. These results illustrate how changes in ice particle size (a product of nucleation rates) have a greater impact on V_m than ice particle shape does, since the mass/area ratio should be greater for compact, high density irregular ice particles relative to side planes and rosette-shaped crystals common to temperatures warmer than -40°C (Bailey and Hallett, 2009; Lawson et al. 2006b). Thus it appears that the onset of homogeneous freezing nucleation produces a rather dramatic decrease in V_m in spite of an expected increase in the ice particle mass/area ratio. As discussed in Section 2, this should increase the lifetime and cloud coverage of cirrus at temperatures below -40°C relative to warmer cirrus clouds. In addition, these results indicate that V_m might be substantially increased by seeding aerosol for temperatures colder than about -45°C. This would reduce cirrus coverage and increase OLR, cooling the planet.

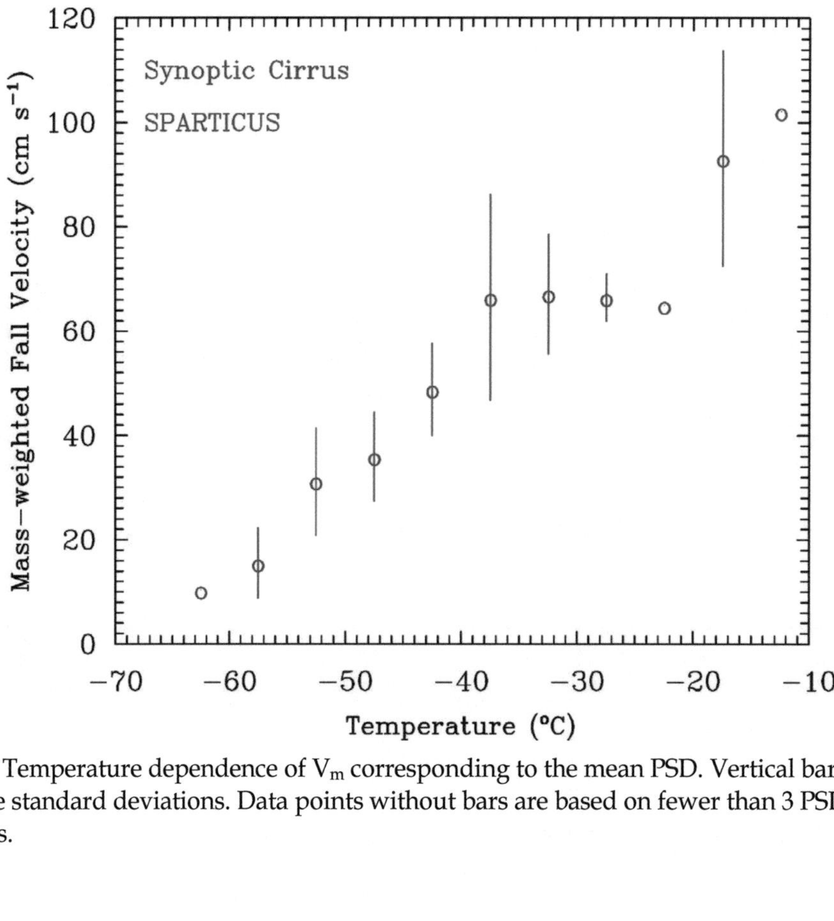

Fig. 14. Temperature dependence of V_m corresponding to the mean PSD. Vertical bars indicate standard deviations. Data points without bars are based on fewer than 3 PSD samples.

Similar V_m results were not found for tropical anvil cirrus clouds, and the temperature dependence of V_m for these clouds is reported in Mitchell et al. (2011). The PSD in these clouds for temperatures less than -40°C were generally characterized by relatively high IWCs, indicating strong updrafts and condensation rates. As a result, the concentrations of larger ice particles in these PSD were relatively high, increasing V_m, and a significant decrease in V_m at temperatures below -40°C was not observed. However, only 25 PSD were sampled for fresh anvil cirrus, and a larger dataset could reveal a temperature dependence for V_m similar to mid-latitude synoptic cirrus.

4. Over-cooling the planet

In section 2.4 a GCM study by Lohmann et al. (2008) was discussed that demonstrated that a substantial global cooling effect might be possible through seeding with heterogeneous ice nuclei that activate at ~ RH_i = 130%. This produced an 11% increase in effective ice particle size and a corresponding increase in the ice fall-speed, which resulted in a mean global net cooling of ~ 2.7 W m^{-2}. This compares with a net radiative forcing due to a CO_2 doubling of 3.7 W m^{-2}. Since we are suggesting the use of very efficient ice nuclei having RH_i's ~ 105% or lower, the impact of cirrus seeding on the cirrus cloud ice fall-speed V_m and hence global cooling could be considerably greater than in the Lohmann et al. study. This presents the possibility of "over-cooling" the planet.

As mentioned, cirrus cloud climate engineering could be accomplished by only seeding cirrus clouds outside the tropics. As temperatures in the mid-latitudes and polar regions decrease, snow-cover in those regions would likely increase, with increases in Arctic sea-ice and polar ice sheets. This would reflect more sunlight to space. The climate system is very sensitive to these climate feedback effects (Thompson and Sieber, 2010), which could substantially amplify the direct radiative cooling produced by the cirrus cloud climate engineering.

Since currently the treatment of cirrus clouds and climate feedbacks in GCMs can vary considerably, the cooling effect predicted by different GCMs could vary substantially regarding this climate engineering approach. Thus it may be difficult to determine at this time through GCM simulations how much and how rapidly the planet would cool, although a range of estimates would be useful. To know whether the climate is responding at an appropriate rate, it will be necessary to closely monitor Earth's radiation budget and the global climate.

GCM simulations indicate there is considerable time lag (e.g. decades) for a change in globally averaged net radiative forcing to fully manifest a corresponding change in global mean temperature. Thus temperature is not an appropriate variable to use for monitoring the effects of climate engineering. Rather shortwave and longwave radiative forcing should be monitored, as these will respond immediately to changes in the climate system (Evans 2010). This should involve both satellite and ground-based monitoring networks (Evans 2010).

5. Social, ethical and political implications of climate engineering

Global warming is just one of many challenges facing humanity, but it is unique in that it could produce a severe and irreversible change in the global climate if the global mean surface temperature exceeds 2°C (relative to the beginning of the industrial era) and trigger key tipping points in the climate system, such as the melting of the polar ice caps (Hansen et al., 2008). Whether humankind could survive such a climate change is an open question.

There appears to be a growing consensus among scientists that, while no one wants to geo-engineer the climate, some kind of geo-engineering will be needed in combination with severe GHG mitigation to avoid triggering climate system tipping points that would result in catastrophic climate disasters, irreversibly locking the climate into a much warmer state (e.g., Thompson and Sieber, 2010; Arora et al. 2011; Greene et al., 2010; Hansen et al., 2008). And if the Earth's climate sensitivity is 2 to 4 times more than GCMs have predicted based on paleoclimate research (Kiehl 2011), it may already be too late to avoid passing the 2°C threshold.

From another point of view, perhaps our fixation with technological solutions is distracting us from larger ontological questions regarding our relationship to Nature and our spiritual being. What if the root of the problem lies in Man's consciousness and the relentless pursuit of insatiable desires? Can the planet handle this? Is it possible that global warming is Nature's way of getting us "back on track" to live in harmony with Nature and with each other? Perhaps climate engineering would at best merely postpone and then intensify an apocalyptic tragedy if human consciousness is not transformed. Perhaps these are the real issues that climate change is presenting to us. A more extensive exploration of this viewpoint is given in an essay by Prof. Clive Hamilton (2011).

If climate engineering in combination with resource conservation, renewable energy systems and GHG reduction are all needed for our survival, then it behooves us to explore what new opportunities climate engineering presents for manifesting positive social and political changes in the world. Since it would affect the entire world, climate engineering should be internationally organized and executed, requiring the cooperation of all the nations of the world. Seen in this way, global warming may bring about a situation mandating the cooperation of the entire human race, asking people and nations to go beyond their immediate self-interest and act for the good of the whole planet. The future climate of the planet may depend on whether nations can cooperate in a spirit of shared-sacrifice, and for democratic nations, it depends on whether the people themselves can act in this way. As it has always been, our collective destiny depends on our collective consciousness and our ability to transform it to meet the challenges of our time.

6. Summary and conclusions

A new approach to climate engineering, described in Mitchell and Finnegan (2009), has been reviewed in Section 2. A key principle determining the success of this approach is the ability of cloud seeding aerosol to out-compete the natural ice nuclei for water vapor when cirrus cloud temperatures are below -40°C. This vapor competition effect will be most pronounced, with cloud modification due to seeding very likely, if homogeneous freezing nucleation plays an important role in ice crystal production. This chapter provides new findings regarding the importance of homogeneous freezing in cirrus clouds relative to heterogeneous nucleation processes. These findings were made possible through recent improvements in ice particle measuring probes consisting of (1) a dramatic reduction in small ice particle artifacts due to the shattering of natural ice particles at the probe inlet tube, (2) the ability to accurately measure ice crystal concentrations down to 10 μm in size and (3) 10 μm resolution for imaging the shapes of ice crystals, resulting in more accurate ice particle area ratios as described in Section 3.3.

For mid-latitude synoptic cirrus around -40°C, these new measurements show that clear changes in the ice particle size distribution (PSD) and its properties occur regarding (1) PSD shape, (2) total number concentration-to-ice water content ratio (N/IWC), (3) PSD mean size, (4) PSD mean area ratio and (5) the mass-weighted fall velocity (V_m). These five changes are consistent with a change in ice nucleation mechanism, with heterogeneous nucleation processes active at temperatures warmer than -40°C and homogeneous freezing nucleation at temperatures colder than -40°C. The change in V_m implies that synoptic cirrus colder than -40°C will have longer lifetimes and greater cloud coverage than warmer cirrus clouds, all other relevant factors remaining equal.

Regarding tropical anvil cirrus clouds near -40°C, changes were observed in the first 3 of the 5 attributes listed above. For these fresh anvil cirrus clouds, homogeneous freezing nucleation also appears to be an important process for ice crystal production. The lack of a change in ice particle shape at -40°C may be due different dominant heterogeneous ice nucleation processes (e.g. droplet freezing) and less of a change in RH_i between temperature regimes. The lack of a change in V_m behavior near -40°C may be due to relatively high condensation rates associated with PSDs sampled at temperatures less than -40°C, producing higher concentrations of both small and larger ice particles. Note that V_m depends mostly on the larger ice particle concentrations (Mitchell et al. 2011). A larger PSD dataset for tropical anvil cirrus would help determine whether this result was general or not.

In regards to synoptic cirrus clouds, we feel that the evidence for homogeneous freezing nucleation being a major source of ice crystals is exceptionally strong. While it is entirely possible that heterogeneous processes are also an important source of ice crystals at temperatures below -40°C, the dramatic change in ice particle shape near -40°C suggests that the vast majority of ice crystals came from homogeneous freezing. These ice crystals below -40°C had area ratios near unity, indicating compact-irregular, isometric and/or quasi-spherical ice crystal shapes. Such ice crystals are associated with growth conditions having low RH_i (< 110%), presenting the following paradox: how is it possible for ice crystal growth below -40°C to correspond to low RH_i while these same ice crystals were formed at high RH_i needed for homogeneous nucleation? This was explained by postulating that moist layers in the upper troposphere experience the RH_i threshold for homogeneous freezing nucleation, which then produces sufficiently high concentrations of ice crystals to vigorously compete for water vapor immediately after nucleation. This rapidly lowers the RH_i throughout the layer, creating a low RH_i environment for subsequent growth. Heterogeneous processes can still contribute new ice crystals at these lower RH_i but at lower rates. This explanation is similar to that proposed by Haag et al. (2003) and others, and appears consistent with the RH_i measurements in Krämer et al. (2009), noting that high RH_i regions corresponding to homogeneous nucleation should be transient and short-lived, making it unlikely that research aircraft would sample such regions. Overall, these results indicate that synoptic cirrus clouds could be modified relatively easily through seeding aerosol in the right concentration range to increase V_m, which should decrease cirrus cloud cover, ice water path and optical depth, and thus increase outgoing longwave radiation (OLR).

Due to the nature of global warming, with surface temperatures increasing in the polar regions and mid-latitudes considerably more than in equatorial regions, it makes sense to enhance OLR primarily in the mid-latitudes and polar regions. This would also stabilize the temperature gradient between the equatorial and polar regions and thus limit changes to the storm track (i.e. jet stream) that might otherwise be altered due to global warming.

Although the seeding of cirrus clouds to reduce global warming is referred to as a climate engineering approach, this is somewhat of a misnomer since "engineering" generally refers to the application of well established scientific principles. In the case of cirrus cloud engineering, the science is still in the process of being worked out. Moreover, the scientific issues involved are not only relevant to climate engineering; they are of fundamental importance to our understanding of cirrus clouds, their role in the climate system, and their representation in climate models. Therefore this research should also be viewed in the broader context of basic research on the climate system, having the potential to improve the ability of climate models to predict future climates.

GCMs used to study this climate engineering approach should have a physically realistic treatment of cirrus clouds, including a realistic cirrus cloud coverage and treatment of supersaturation with respect to ice. Moreover, realistic representations of climate feedback effects are needed, such as those associated with Arctic sea-ice and polar ice sheets. The latter is needed since the direct radiative cooling from cirrus cloud modification may enhance these feedbacks in the climate system and thus amplify the direct cooling.

While the social and political ramifications of climate engineering are many, its potential role as a catalyst for increased global cooperation should not be discounted. It appears that cirrus cloud climate engineering would be relatively inexpensive to implement.

While this climate engineering approach may appear promising, it is absolutely critical to understand that this is not a "silver bullet" for the global warming problem. To think in these terms would be a prescription for sudden, catastrophic climate change, since there is only so much "wiggle room" in the climate system. That is, a successful modification of cirrus clouds at temperatures below -40°C might cool the planet for a limited period of time, but not indefinitely, and at current GHG emission rates it would not take long for any cirrus OLR benefit to be neutralized. A "business as usual" GHG emission scenario would result in a sudden and dramatic increase in global temperatures once any cirrus OLR benefit was exhausted, and such a temperature increase might make human life on Earth unbearable. Therefore this cirrus climate engineering method may "buy time" for societies to transition to renewable non-carbon fuels and to implement carbon sequestration methods that address ocean acidification, but it should never be viewed as a solution to global warming.

7. Acknowledgment

This research was supported by the Office of Science (BER), U.S. Department of Energy and by discretionary funds from the Desert Research Institute. Dr. Matt Bailey is gratefully acknowledged for providing Figure 12 for this chapter. We also wish to thank Dr. Cedric Francois for coining the term "Earth Radiation Management" in relation to this climate engineering approach, and for his tireless efforts to secure funding for this research.

8. References

Anderson, K. & Bows, A. (2010). Reframing the climate change challenge in light of post-2000 emission trends. In: *Geo-Engineering Climate Change*, B. Launder & J. M. T. Thompson, (Eds.), 27-49, Cambridge University Press, ISBN 978-0-521-19803-5, Cambridge, United Kingdom.

Arora, V. K., Scinocca, J. F., Boer, G. J., Christian, F. R., Denman, K. L., Flato, G. M., Kharin, V. V., Lee, W. G., & Merryfield, W. J. (2011). Carbon emission limits required to satisfy future representative concentration pathways of greenhouse gases. *Geophys. Res. Lett.*, Vol.38, L05805, (March 2011), pp. 1-6, ISSN 0094-8276.

Baker, B., & Lawson, R. P. (2006). Improvement in determination of ice water content from two-dimensinal particle imagery. Part I: Image-to-mass relationships. *J. Appl. Meteorol. and Climatol.*, 45, No.9, pp. 1282-1290, ISSN 1558-8432.

Barahona, D. & Nenes, A. (2008). Parameterization of cirrus cloud formation in large-scale models: Homogeneous nucleation. *J. Geophys. Res.*, Vol.113, No.D11, D11211, (June 2008), pp. 1-15, 0148-0227.

Bailey, M. & Hallett, J. (2004). Growth rates and habits of ice crystals between -20° and -70°C. *J. Atmos. Sci.*, Vol.61, No. 5, pp. 514-544, ISSN 0022-4928.

Bailey, M. & Hallett, J. (2009). A comprehensive habit diagram for atmospheric ice crystals: Confirmation from the laboratory, AIRS II, and other field studies. *J. Atmos. Sci.*, Vol.66, No.9, pp. 2888-2899, ISSN 0022-4928.

Caldeira, K. & Wood, L. (2010): Global and arctic climate engineering: numerical model studies. In: *Geo-Engineering Climate Change*, B. Launder & J. M. T. Thompson, (Eds.), 181-204, Cambridge University Press, ISBN 978-0-521-19803-5, Cambridge, United Kingdom.

Chen, T., Rossow, W. & Zhang, Y. (2000). Radiative effects of cloud-type variations. *J. Climate*, Vol.13, No.1, pp. 264-286, ISSN 1520-0442.

Evans, W. F. J. (2010). Personal communication. North West Research Associates, Washington, U.S.A.

Fridlind, A. M., and coauthors (2004). Evidence for the predominance of mid-tropospheric aerosols as subtropical anvil cloud nuclei. *Science*, Vol.304, No. 5671, (April 2004), pp. 718-722, ISSN 0036-8075.

Greene, C. H., Monger, B. C., Huntley, M. E. (2010). Geoengineering: The inescapable truth of getting to 350. *Solutions*, Vol.1, Issue 5, (September-October 2010), pp. 57-66.

Haag, W., Kärcher, B., Strom, J., Minikin, A., Lohmann, U., Ovarlez, J., & Stohl, A. (2003): Freezing thresholds and cirrus cloud formation mechanisms inferred from in situ measurements of relative humidity. Atmos. Chem. Phys., Vol.3, (October 2003), pp. 1791-1806, ISSN 1680-7316.

Hamilton, C. (2011): The ethical foundations of climate engineering. Available at the following URL: http://www.clivehamilton.net.au/cms/media/ethical_foundations_of_climate_engineering.pdf , Centre for Applied Philosophy and Public Ethics, Canberra, Australia.

Hansen, J., and coauthors. (2008). Target atmospheric CO2: where should humanity aim? *Open Atmospheric Science Journal*, Vol.2, pp. 217-231, ISSN 1874-2823.

Hartmann, D. L., Ockert-Bell, M. E. & Michelsen, M. L. (1992). The effect of cloud type on Earth's energy balance: Global analysis. *J. Climate*, Vol.5, No. 11, pp. 1281-1304, ISSN 1520-0442.

Helten, M., Smit, H. G. J., Strater, W., Kley, D., Nedelec, P., Zoger, M. & Busen, R. (1998). Calibration and performance of automatic compact instrumentation for the measurement of relative humidity from passenger aircraft*J. Geophys. Res.*, Vol.103, No.D19, pp. 25643-25652, ISSN 0148-0227.

Heymsfield, A. J., & Sabin, R. M. (1989). Cirrus Crystal Nucleation by Homogeneous Freezing of Solution Droplets *J. Atmos. Sci.*, Vol.46, No. 14, pp. 2252-2264, ISSN 0022-4928

Heymsfield, A. J., Bansemer, A. & Twohy, C. (2007). Refinements to ice particle mass dimensional and terminal velocity relationships for ice clouds. Part I: Temperature dependence. *J. Atmos. Sci.*, Vol.64, No.4, pp. 1047-1067, ISSN 0022-4928.

Heymsfield, A. J. & Westbrook, C. (2010). Advances in the estimation of ice particle fall speeds using laboratory and field measurements. *J. Atmos. Sci.*, Vol.67, No. 8, pp. 2469-2482, ISSN 0022-4928.

Jensen, E. J. & coauthors. (2009). On the importance of small ice crystals in tropical anvil cirrus. *Atmos. Chem. Phys.*, Vol.9, No. 15, pp. 5519-5537, ISSN 1680-7316.

Kärcher, B., Möhler, O., DeMott, P. J., Pechtl, S. & Yu, F. (2007). Insights into the role of soot aerosols in cirrus cloud formation. *Atmos. Chem. Phys.*, Vol.7, No. 16, (August 2007), pp. 4203-4227, ISSN 1680-7316.

Keith, D., Heidel, K. & Cheery, R. (2010). Capturing CO_2 from the atmosphere: rationale and process design considerations. In: *Geo-Engineering Climate Change*, B. Launder & J. M. T. Thompson, (Eds.), 107-126, Cambridge University Press, ISBN 978-0-521-19803-5, Cambridge, United Kingdom.

Kiehl, J. (2011). Lesson's from Earth's past. Science, Vol.331, (January 2011), pp. 158-159, ISSN 1095-9203.

Koop, T., Luo, B., Tsias, A. & Peter, T. (2000). Water activity as the determinant for homogeneous ice nucleation in aqueous solutions. Nature, Vol.406, (August 2000), pp. 611-614, ISSN 0028-0836.

Korolev, A. & Isaac, G. A. (2003). Roundness and aspect ratios of particles in ice clouds. *J. Atmos. Sci.*, Vol.60, No. 15, pp. 1795-1808, ISSN 0022-4928.

Korolev, A., Emery, E. F., Strapp, J. W., Cober, S. G., Isaac, G. A., Wasey, M. & Marcotte, D. (2011). Small particles in tropospheric clouds: fact or artifact? Airborne Icing Instrument Evaluation Experiment. Accepted in *Bull. Amer. Meteorol. Soc.*

Krämer, M., and coauthors. (2009). Ice supersaturations and cirrus cloud crystal numbers. *Atmos. Chem. Phys.*, Vol.9, No. 11, pp. 3505-3522, ISSN 1680-7316.

Lampitt, R. S. & coauthors. (2010). Ocean fertilization: a potential means of geo-engineering? In: *Geo-Engineering Climate Change*, B. Launder & J. M. T. Thompson, (Eds.), 149-180, Cambridge University Press, ISBN 978-0-521-19803-5, Cambridge, United Kingdom.

Latham, J. (1990). Control of global warming. *Nature*, Vol.347, No. 6291, pp. 339–340. (doi:10.1038/347339b0), ISSN 0028-0836.

Latham, J. (2002). Amelioration of global warming by controlled enhancement of the albedo and longevity of low-level maritime clouds. *Atmos. Sci. Lett.* Vol.3, No.2-4, pp. 52. (doi:10.1006/asle.2002.0048), ISSN 1530-261X.

Latham, J., Rasch, P., Chen, C.-C. J., Kettles, L., Gadian, A., Gettelman, A., Morrison, H., Bower, K. & Choularton, T. (2008). Global temperature stabilization via controlled albedo enhancement of low-level maritime clouds. *Phil. Trans. R. Soc. A*, Vol.366, No. 1882, pp. 3969–3987. (doi:10.1098/rsta.2008.0137), ISSN 1471-2962.

Launder, B. & Thompson, J. M. T. (Eds.). (2010). *Geo-Engineering Climate Change*, Cambridge University Press, ISBN 978-0-521-19803-5, Cambridge, United Kingdom.

Lawson, R. P., Baker, B. A., Schmitt, C. G., Jensen, T. L. (2001). An overview of microphysical properties of Arctic cirrus clouds observed in May and July 1998 during FIRE ACE. *J. Geophys. Res.*, Vol.106, No. D14, (July 2001), pp. 14,989-15,014, ISSN 0148-0227.

Lawson, R. P., O'Connor, D., Zmarzly, P., Weaver, K., Baker, B., Mo, Q. & Jonsson, H. (2006a). The 2D-S (Stereo) Probe: Design and preliminary tests of a new airborne, high-speed, high-resolution particle imaging probe. J. Atmos. Oceanic Technol., Vol.23, No.11, pp. 1462-1477, ISSN 1520-0426.

Lawson, R. P., Baker, B., Pilson, B. & Mo, Q. (2006b). In situ observations of the microphysical properties of wave, cirrus and anvil clouds. Part II: Cirrus clouds. *J. Atmos. Sci.*, Vol.63, No. 12, pp. 3186-3203, ISSN 0022-4928.

Lawson, R. P., Jensen, E., Mitchell, D. L., Baker, B., Mo, Q. & Pilson, B. (2010). Microphysical and radiative properties of tropical clouds investigated in TC4 and NAMMA. *J. Geophys. Res.*, Vol.115, No. D10, D00J08, (August 2010), pp. 1-16, ISSN 0148-0227.

Lawson, R. P. (2011): Effects of ice particles shattering on the 2D-S probe. Atmos. Meas. Tech., Vol.4, pp. 1361–1381, ISSN 0022-4073.

Lenton, T. M. & Vaughan, N. E. (2009). The radiative forcing potential of different climate geoengineering options. *Atmos. Chem. Phys.*, Vol.9, No.15, (August 2009), pp. 5539-5561, ISSN 1680-7316.

Lohmann, U., Spichtinger, P., Jess, S., Peter, T. & Smit, H. (2008). Cirrus cloud formation and ice supersaturated regions in a global climate model. *Environ. Res. Lett.*, Vol.3, No.4, pp. 1-11, ISSN 1748-9326.

McFarquhar, G. M., Um, J., Freer, M., Baumgardner, D., Kok, G. & Mace, G. (2007). Importance of small ice crystals to cirrus properties: Observations from the Tropical Warm Pool International Cloud Experiment (TWP-ICE). *Geophys. Res. Lett.*, Vol.34, No.13, L13803, pp. 1-6, ISSN 0094-8276.

Mitchell, D. L. (1988). Evolution of snow-size spectra in cyclonic storms. I: Snow growth by vapor deposition and aggregation. *J. Atmos. Sci.*, Vol.45, No. 22, pp. 3431-3451, ISSN 0022-4928.

Mitchell, D.L., Chai, S., Liu, Y., Heymsfield, A. J., & Dong, Y. Y. (1996). Modeling cirrus clouds. Part I: Treatment of bimodal size spectra and case study analysis. *J. Atmos. Sci.*, Vol.53, No. 20, pp. 2952-2966, ISSN 0022-4928.

Mitchell, D. L., Rasch, P. J., Ivanova, D., McFarquhar, G. M., Nousiainen, T. (2008). Impact of small ice crystal assumptions on ice sedimentation rates in cirrus clouds and GCM simulations. *Geophys. Res. Lett.*, Vol.35, No.9, L09806, pp. 1-5, ISSN 0094-8276.

Mitchell, D. L. & Finnegan, W. (2009). Modification of cirrus clouds to reduce global warming. *Environ. Res. Lett.*, Vol.4, 045102, (October 2009), pp. 1-8, ISSN 1748-9326.

Mitchell, D.L., d'Entremont, R. P., & Lawson, R. P. (2010). Inferring cirrus size distributions through satellite remote sensing and microphysical databases. *J. Atmos. Sci.*, Vol.67, No. 4, pp. 1106-1125, ISSN 0022-4928.

Mitchell, D. L., Lawson, R. P., & Mishra, S. (2011). Representing the ice fall speed in climate models: Results from TC4 and ISDAC. Accepted for publication in *J. Geophys. Res.* (ISDAC special issue) in June 2011.

Pruppacher, H. & Klett, J. (2010). Microphysics of Clouds and Precipitation (2nd edition), Springer, ISBN 978-0-7923-4211-3, New York, N.Y., U.S.A.

Read, W. G., Waters, J. W., Wu, D. L., Stone, E. M. & Shippony, Z. (2001). UARS Microwave Limb Sounder upper tropospheric humidity measurement: Method and validation*J. Geophys. Res.*, Vol.106, No. D23, pp. 32207-32258, ISSN 0148-0227.

Rosenfeld, D. & Woodley, W. L. (2000). Deep convective clouds with sustained supercooled liquid water down to -37.5°C. *Nature,* (May 2000), Vol.405, No. 6785, pp. 440-442, ISSN 0028-0836.

Salter, S., Sortino, G. & Latham, J. (2008). Sea-going hardware for the cloud albedo method of reversing global warming. *Phil. Trans. R. Soc. A*, Vol.366, (August 2008), pp. 3989–4006, ISSN 1471-2962.

Sanderson, B. M., Piani, C., Ingram, W. J., Stone, D. A., & Allen, M. R. (2008). Towards constraining climate sensitivity by linear analysis of feedback patterns in thousands of perturbed physics GCM simulations. *Clim. Dyn.*, Vol.30, pp. 175-190, ISSN 0930-7575.

Sassen, K. & Dodd, G. C. (1988). Homogeneous nucleation rate for highly supercooled cirrus cloud droplets. *J. Atmos. Sci.*, Vol.45, No. 8, pp. 1357-1369, ISSN 0022-4928.

Schneider, S. H. (2010). Geo-engineering: could we or should we make it work?, In: *Geo-Engineering Climate Change*, B. Launder & J. M. T. Thompson, (Eds.), 3-26, Cambridge University Press, ISBN 978-0-521-19803-5, Cambridge, United Kingdom.

Smetacek, V. & Naqvi, S. W. A. (2010): The next generation of iron fertilization experiments in the Southern Ocean. In: *Geo-Engineering Climate Change*, B. Launder & J. M. T. Thompson, (Eds.), 181-204, Cambridge University Press, ISBN 978-0-521-19803-5, Cambridge, United Kingdom.

Solomon, S., Qin, D., Manning, M., Chen, Z., Marquis, M., Averyt, K.B., Tignor, M. & Miller, H.L. (Eds.), (2007). IPCC Fourth Assessment Report (AR4). Climate Change 2007: The Physical Science Basis. Contribution of Working Group I to the Fourth Assessment Report of the Intergovernmental Panel on Climate Change, Cambridge University Press, ISBN 978 0521 70596-7, Cambridge, United Kingdom and New York, NY, USA.

Spichtinger, P., Gierens, K. & Read, W. (2003). The global distribution of ice-supersaturated regions as seen by the microwave limb sounder. *Q. J. Roy. Meteor. Soc.*, Vol.129, pp. 3391-3410, ISSN 0035-9009.

Spichtinger, P., Gierens, K., Smit, H. G. J., Ovarlez, J., & Gayet, J. F. (2004). On the distribution of relative humidity in cirrus clouds. *Atmos. Chem. Phys.*, Vol.4, No.3, (April 2004), pp. 639-647, ISSN 1680-7324.

Super, A. B. (1986). Further Exploratory Analysis of the Bridger Range Winter Cloud Seeding Experiment. *J. Clim. Appl. Meteorol.*, Vol.25, Issue 12, (December 1986), pp. 1926-1933, ISSN 1520-0450.

The Royal Society (2009). Geoengineering the Climate. The Royal Society, ISBN 978-0-85403-773-5, London, United Kingdom.

Thompson, J. M. T. & Sieber, J. (2010). Predicting climate tipping points, In: *Geo-Engineering Climate Change*, B. Launder & J. M. T. Thompson, (Eds.), 3-26, Cambridge University Press, ISBN 978-0-521-19803-5, Cambridge, United Kingdom.

Twohy, C. H., Schanot, A. J. & Cooper, W. A. (1997). Measurement of condensed water content in liquid and ice clouds using an airborne counter-flow virtual impactor. *J. Atmos. Oceanic Technol.*, Vol.14, No.1, pp. 197-202, ISSN 1520-0426.

Twohy, C. H., Strapp, J. W., & Wendisch, M. (2003). Performance of a counterflow virtual impactor in the NASA icing research tunnel. *J. Atmos. Oceanic Technol.*, Vol.20, No. 6, pp. 781-790, ISSN 1520-0426.

Warburton, J. A., L. G. Young & Stone, R. H. (1995). Assessment of seeding effects in snowpack augmentation programs: Ice nucleation and scavenging of seeding aerosols. *J. Appl. Meteorol.*, Vol.34, No. 1, (January 1995), pp. 121-130, ISSN 1520-0450.

Zhao, Y., Mace, G. G. & Comstock, J. M. (2011). The occurrence of particle size distribution bimodality in midlatitude cirrus as inferred from ground-based remote sensing data. *J. Atmos. Sci.*, Vol.68, No.6, pp. 1162-1177, ISSN 0022-4928.

13

Destruction of Fluorinated Greenhouse Gases by Using Nonthermal Plasma Process

Young Sun Mok
Department of Chemical & Biological Engineering, Jeju National University
Republic of Korea

1. Introduction

Most of fluorinated compounds such as hydrofluorocarbons (HFCs), perfluorocarbons (PFCs) and sulfur hexafluoride (SF_6) are considered as significant greenhouse gases due to their chemical stability, long atmospheric lifetimes, and high infrared absorption capacity. The 100-year global warming potentials (GWPs) of them are reported to be from a few thousand times to tens of thousand times that of carbon dioxide (CO_2). Fluorinated compounds have been widely used in semiconductor and polymer industries, commercial refrigeration and air conditioning. SF_6 that is the most potent greenhouse gas with a GWP of 22,200 has been used as a common gaseous dielectric medium in electrical power equipments and in the etching and cleaning processes of the semiconductor industry. Up to now, several technologies for abating such gaseous fluorinated compounds have been developed, which include incineration, catalytic decomposition, thermal or nonthermal plasma destruction, and so forth (Bickle et al., 1994; Futamura & Yamamoto, 1997; Lee & Choi, 2004; Ogata et al., 2004; Föglein et al., 2005; Mizeraczyk et al., 2005). Even though the incineration is the only field-proven technology so far, it necessarily requires a lot of energy and long preheating time to reach high temperature enough to destroy chemical bonds in fluorinated compounds. The use of catalyst can largely lower the operation temperature, but still high temperature above 700°C is needed to achieve sufficient catalytic activity. The application of thermal or nonthermal plasma to the destruction of fluorinated compounds as an emission control technology is a relatively new research area. Previous studies have shown that low-pressure plasma processes like inductively coupled plasma (ICP) can destroy fluorinated compounds effectively (Kuroki et al., 2005). However, they inevitably require high investment and operation cost for vacuum. For this reason, atmospheric nonthermal plasma systems can be more desirable from a practical point of view.

The non-thermal plasma has been created in different plasma reactors such as microwave, pulsed streamer corona and dielectric barrier discharge (DBD) reactors, offering an innovative approach for the abatement of fluorinated compounds. The pulsed corona discharge is induced by the application of fast-rising narrow high voltage pulse to non-uniform electrode geometry (Mok et al., 1998). It develops by forming a number of streamers, the starting points of which are discrete and distributed over the surface of discharging electrode. Free electrons produced by the discharge can be accelerated by an imposed electric field to gain energy. During their drift, they can collide with various molecules and lose energy. The collisions of energetic electrons with gas molecules result in the formation of various reactive species. The DBD is a kind of gaseous electrical discharge occurring between two electrodes separated by at least one insulating layer. Typically, DBD reactors use high voltage alternating current (AC) and operate in atmospheric pressure range. The DBD plasma reactor becomes increasingly very attractive due to its capability of producing abundant reactive species by enormous energetic electron collisions with running gas molecules at atmospheric pressure. In addition, dielectric barrier discharge plasma reactors are generally compact and operates stably with high average power.

The drawbacks associated with the nonthermal plasma destruction of fluorinated compounds may be high energy consumption, slow destruction rate and unwanted byproducts formation. In order to get over these problems, recent studies have been focused on the combination of plasma with catalysis (Kim et al., 2005). The use of catalysis together with nonthermal plasma can promote the oxidative destruction of fluorinated compounds through C-F bond cleavage and prevent the recombination of decomposed fragments, thereby remarkably depressing unwanted byproducts formation. As well, the nonthermal plasma can assist catalytic reactions by triggering additional activation of catalyst. As demonstrated in many studies conducted elsewhere, the combination of nonthermal plasma and catalysis has provided a broad range of applications.

In this work, the destruction of fluorinated compounds, including trifluoromethane (CHF_3 or HFC-23), sulfur hexafluoride (SF_6), 1,1,1,2-tetrafluoroethane ($C_2H_2F_4$ or HFC-134a) and hexafluoroethane (C_2F_6), has been investigated over a wide temperature range up to 500°C with a dielectric-packed nonthermal plasma reactor and a simulated exhaust gas consisting of fluorinated compounds, oxygen and nitrogen. The GWP values and atmospheric lifetimes of these greenhouse gases are summarized in Table 1 (Intergovernment Panel on Climate Change Third Assessment Report, 2011).

Chemcial formula	Code name	GWP (-)	Atmospheric lifetime (yr)
CHF_3	HFC-23 or R-23	12,000	260
$C_2H_2F_4$	HFC-134a or R-134a	1,300	13.8
SF_6	-	22,200	3,200
C_2F_6	R-116	11,900	10,000

Table 1. Global warming potentials and atmospheric lifetimes of fluorinated compounds

The nonthermal plasma can be used not only to produce a variety of reactive species capable of destroying gaseous pollutants, but also to improve catalytic activity through various actions. In this context, proper selection of a packing material that can in turn affect the performance of the plasma reactor is of great importance. There have been several previous articles of our research group, where the characteristics of three different packing materials such as alumina, zirconia and glass beads were comparatively examined (Kim et al., 2010a, 2010b; 2010c; Kim & Mok, 2011a). In those articles, the destruction behaviors of the fluorinated compounds were characterized with respect to electric power and reactor temperature, and the effects of several other variables on the destruction were evaluated. In this paper, various aspects of the nonthermal plasma destruction of fluorinated compounds were discussed and plausible destruction mechanisms were illustrated with experimental results, referring to the previous articles of our research group. Two key parameters controlling the performance are the electric power and the reactor temperature, because they dominate reactive species generation and catalytic reaction rate, respectively.

2. Theoretical background

Electrical discharge plasma - that is the DBD plasma - produces various reactive spcies including excited molecules (ex. N_2 ($A^3\Sigma_u^+$)), excited atoms (ex. O(^{1}D) and N(^{2}D)), radicals (ex. OH and HO_2), ions and energetic electrons, all of which can contribute to the destruction of gaseous pollutants more or less. When it comes to the destruction of fluorinated compounds with strong carbon-fluorine bond, however, the reactive species that are mainly responsible for the destruction are considered to be excited atoms, radicals and energetic electrons rather than excited molecules and ions. The destruction mechanisms of the fluorinated compounds dealt with in this work are elucidated below.

2.1 Basic processes

High-energy electrons, i.e., energetic electrons created by nonthermal electrical discharge plasma are in the range of 5~10 eV (1 eV = 1.6 × 10^{-19} J) on the average. The production of reactive species associated with the destruction of fluorinated compounds is initiated by collisions between energetic electrons and background molecules like H_2O, N_2 and O_2. The electron-molecule collision processes are

$$e + N_2 \rightarrow e + N_2(A^3\Sigma_u^+) \tag{1}$$

$$e + N_2 \rightarrow e + N + N \tag{2}$$

$$e + O_2 \rightarrow e + O + O \tag{3}$$

$$e + O_2 \rightarrow e + O + O\left(^1D\right) \tag{4}$$

$$e + H_2O \rightarrow e + H + OH \tag{5}$$

The rates of reactions (1)-(5) are a function of electron energy, namely, imposed electric field, which can be estimated from the solution of Boltzmann equation and the appropriate collision cross-section data. The excited atomic oxygen O(^{1}D) generated from reaction (4) can produce additional OH radicals as a result of rapid quenching with H_2O as follows:

$$O\left(^1D\right) + H_2O \rightarrow OH + OH \tag{6}$$

The rate constant for reaction (6) is reported to be 2.6×10^{-10} cm^3 $molecule^{-1}$ (Li et al., 1995; Chang et al., 1991). In the case of electron-beam irradiation process utilizing fast electrons in the range of 300~750 keV, most of OH radicals are formed through charge-exchange reactions. On the other hand, the OH formation by nonthermal plasma is dominated by hydrogen abstraction from water vapor by O(^{1}D) under most conditions.

2.2 Destruction pathways

2.2.1 Trifluoromethane

In most cases, the major constituent of contaminated gas to be treated is nitrogen. Regarding the reaction of excited state N_2 ($A^3\Sigma_u^+$) with CHF_3, Piper et al. (1985) identified the process as

$$N_2(A^3\Sigma_u^+, v=1) + CHF_3 \rightarrow N_2(A^3\Sigma_u^+, v=0) + CHF_3 \; 6.0\times10^{11}cm^3mol^{-1}s^{-1} \quad (7)$$

Tao et al. (1992) also reported the dominant process to be vibrational relaxation. Thus, it is reasonable to consider the CHF_3 destruction by excited state $N_2(A^3\Sigma_u^+)$ to be negligible. On the contrary, electronically excited atomic nitrogen $N(^2D)$ can play an important role in destroying CHF_3 by abstracting hydrogen atom (Herron, 1999)

$$N(^2D) + CHF_3 \rightarrow NH + CF_3, \; 6.0\times10^{10}cm^3mol^{-1}s^{-1} \quad (8)$$

The rate coefficients of reactions (7) and (8) are the values at 298 K, and other rate coefficients given below are also at 298 K. Energetic electrons can also initiate the destruction of CHF_3. The destruction products formed by the energetic electrons may be inferred by referring to mass spectroscopy, since the electron impact dissociation has some analogy with fragmentation reactions in a mass spectrometer. The cracking pattern of CHF_3 indicates that the most abundant destruction products are CF_3 and CHF_2. Thus, the electron impact dissociations can be written as

$$e + CHF_3 \rightarrow CF_3 + H + e \quad (9)$$

$$e + CHF_3 \rightarrow CHF_2 + F + e \quad (10)$$

The bond dissociation energies of C-H and C-F are 4.3 and 5.1 eV, respectively. Although C-F has larger bond energy than C-H, strong electro-negativity of F atom can lead to reaction (10). Both reactions (9) and (10) depend on the electric field controlling the electron energy. H and F radicals from reactions (9) and (10) can react as follows (Barker, 1995; NIST, 1998):

$$H + CHF_3 \rightarrow H_2 + CF_3, \quad 2.2\times10^{4}cm^3mol^{-1}s^{-1} \quad (11)$$

$$F + CHF_3 \rightarrow HF + CF_3, \quad 9.4\times10^{10}cm^3mol^{-1}s^{-1} \quad (12)$$

The small rate coefficient of reaction (11) implies that the contribution of H radical to the CHF_3 destruction is insignificant. In the presence of oxygen, reactions (3)~(6) form O and OH radicals, which participate in the following reactions (Barker, 1995; NIST, 1998)

$$O + CHF_3 \rightarrow OH + CF_3, \quad 7.6\times10^{4}cm^3mol^{-1}s^{-1} \quad (13)$$

$$OH + CHF_3 \rightarrow H_2O + CF_3, \; 2.9\times10^{8}cm^3mol^{-1}s^{-1} \quad (14)$$

Compared to reactions (8), (12) and (14), the rate of reaction (13) is much slower, implying that the contribution of atomic oxygen to the destruction is trivial.

The processes for the primary destruction steps (reactions (8)~(14)) produce CF_3 and CHF_2, which can react with oxygen to form peroxy radicals

$$CF_3 + O_2 \rightarrow CF_3O_2, \quad 2.4\times10^{12}cm^3mol^{-1}s^{-1} \quad (15)$$

$$CHF_2 + O_2 \rightarrow CHF_2O_2, \; 2.4\times10^{12}cm^3mol^{-1}s^{-1} \quad (16)$$

where the rate coefficient for reaction (16) was assumed to be equal to that for reaction (15) (Barker, 1995). The peroxy radicals, i.e., CF_3O_2 and CHF_2O_2, further react to form alkoxy radicals as (Barker, 1995; NIST, 1998)

$$CF_3O_2 + CF_3O_2 \rightarrow 2CF_3O + O_2, \quad 1.36 \times 10^{12} mol^{-1}s^{-1} \tag{17}$$

$$CHF_2O_2 + CHF_2O_2 \rightarrow 2CHF_2O + O_2, \quad 3.0 \times 10^{12} mol^{-1}s^{-1} \tag{18}$$

$$CF_3O_2 + NO \rightarrow CF_3O + NO_2, \quad 9.6 \times 10^{12} mol^{-1}s^{-1} \tag{19}$$

$$CHF_2O_2 + NO \rightarrow CHF_2O + NO_2, \quad 7.8 \times 10^{12} mol^{-1}s^{-1} \tag{20}$$

It should be noted that NO involved in reactions (19) and (20) is formed by the reaction between excited atomic nitrogen $N(^2D)$ and O_2 (Ricketts et al., 2004; Harling et al., 2005). The alkoxy radicals CF_3O and CHF_2O formed by the reactions (17)~(20) are further degraded to give carbonyl fluoride (COF_2) as follows (Barker, 1995; NIST, 1998):

$$CF_3O + O_2 \rightarrow COF_2 + FO_2, \quad 1.0 \times 10^{6} mol^{-1}s^{-1} \tag{21}$$

$$CHF_2O + O_2 \rightarrow COF_2 + HO_2, \quad 1.1 \times 10^{9} cm^3 mol^{-1}s^{-1} \tag{22}$$

$$CF_3O + CO \rightarrow COF_2 + COF, \quad 1.2 \times 10^{9} cm^3 mol^{-1}s^{-1} \tag{23}$$

$$CF_3O + NO \rightarrow COF_2 + NOF, \quad 3.2 \times 10^{13} cm^3 mol^{-1}s^{-1} \tag{24}$$

$$CF_3O + NO_2 \rightarrow COF_2 + NO_2F, \quad 1.9 \times 10^{12} cm^3 mol^{-1}s^{-1} \tag{25}$$

Although the rate coefficient of reaction (21) is small, its reaction rate is not negligible because oxygen is often one of the main constituents in the gas to be treated. Besides the reactions above, other channels for COF_2 formation are as follows:

$$CF_3 + O \rightarrow COF_2 + F, \quad 1.9 \times 10^{13} cm^3 mol^{-1}s^{-1} \tag{26}$$

$$CF_3 + NO_2 \rightarrow COF_2 + NOF, \quad 1.5 \times 10^{13} cm^3 mol^{-1}s^{-1} \tag{27}$$

$$COF + COF \rightarrow COF_2 + CO, \quad 8.0 \times 10^{12} cm^3 mol^{-1}s^{-1} \tag{28}$$

Under electrical discharge plasma, a part of COF_2 leads to carbon oxides such as CO_2 and CO (Herron, 1999; NIST, 1998)

$$COF_2 + O(^1D) \rightarrow CO_2 + F_2, \quad 1.3 \times 10^{13} cm^3 mol^{-1}s^{-1} \tag{29}$$

$$CO_2 + N(^2D) \rightarrow CO + NO, \quad 2.2 \times 10^{11} cm^3 mol^{-1}s^{-1} \tag{30}$$

Energetic electrons can also decompose CO_2 into CO by direct bond cleavage. The reaction pathways regarding CHF_3 destruction are described in Fig. 1.

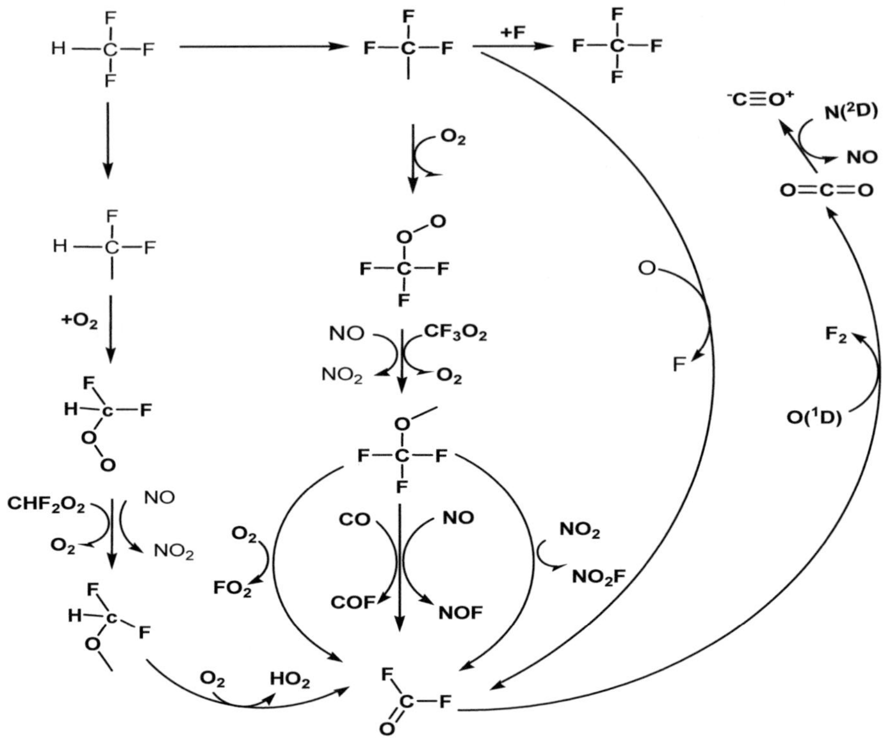

Fig. 1. Reaction pathways of CHF_3 destruction in the presence of N_2 and O_2

2.2.2 Hexafluoroethane

Hexafluoroethane (C_2F_6) is chemically very stable, which is attributed to the strength of C-F bond in it and the shielding effect of the fluorine atoms. Hence, the reactivity of C_2F_6 with the species generated by the plasma is extremely low. The most likely process for initiating the destruction of C_2F_6 is believed to be the dissociation by the energetic electrons generated by the plasma, which can be written as (Motlagh & Moore, 1998)

$$e + C_2F_6 \rightarrow CF_3 + CF_3 + e \tag{31}$$

$$e + C_2F_6 \rightarrow C_2F_5 + F + e \tag{32}$$

The bond dissociation energies of C-C and C-F are 3.6 and 5.1 eV, respectively, indicating that reaction (31) is superior to reaction (32). According to the mass spectrum of C_2F_6 that illustrates the cracking pattern, the most abundant fragment is CF_3, which supports that reaction (31) is the predominant electron impact dissociation process. Once CF_3 is produced somehow, succeeding reactions that convert CF_3 into carbon oxides can be explained by reactions (15)~(30). The plausible reaction pathways responsible for the destruction of C_2F_6, leading to the formation of CO and CO_2 are illustrated in Fig. 2 (Kim et al., 2010b). In the destruction of fluorinated compounds, the overall rate is determined by its initial fragmentation by energetic electrons, i.e., electron impact dissociation process acts as the rate-determing step, and succeeding reactions are considered to be a lot faster.

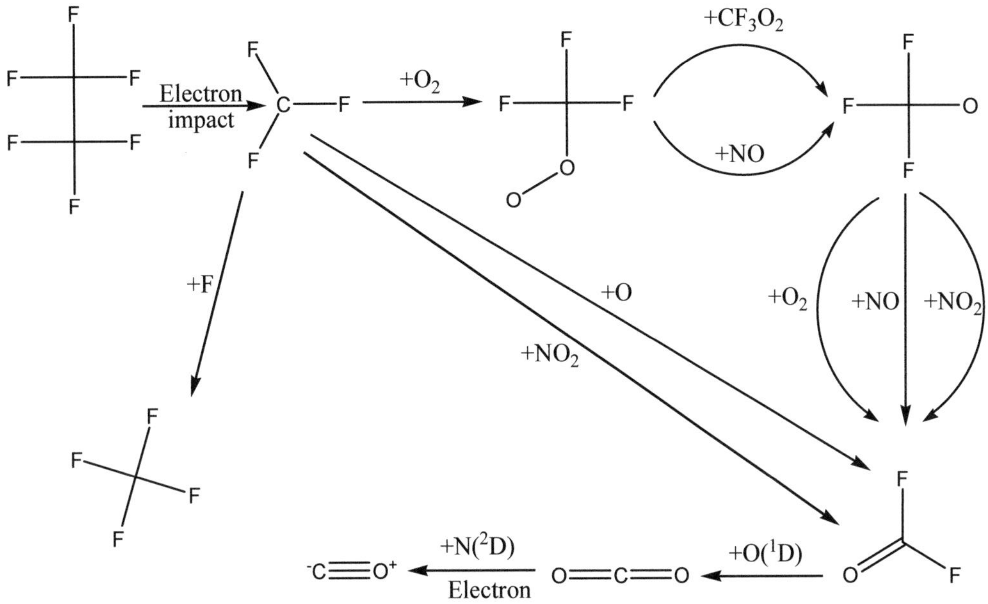

Fig. 2. Plausible reaction pathways responsible for the destruction of C_2F_6

2.2.3 1,1,1,2-tetrafluoroethane

In the same way, the destruction of $C_2H_2F_4$ can be explained with the bond dissociations followed by subsequent reactions leading to the formation of CO and CO_2. The initial destruction fragments of $C_2H_2F_4$ that are produced through collisions with energetic species such as electrons, $N(^2D)$ and N_2 $(A^3\Sigma_u^+)$ include CH_2F, CF_3 and $CHF{=}CF_2$ (Mok et al., 2008):

$$C_2H_2F_4 + \text{energetic species} \rightarrow CH_2F + CF_3 \tag{33}$$

$$C_2H_2F_4 + \text{energetic species} \rightarrow CHF - CF_3 + H \rightarrow CHF = CF_2 + HF \tag{34}$$

Likewise, the fragments CH_2F and CF_3 further react with O, O_2 and O_3 to form carbon oxides, as described in reactions (15)~(30), which may be summarized as follows:

$$CF_3 \rightarrow CF_3O_2 \rightarrow CF_3O \rightarrow COF_2 \rightarrow CO,\ CO_2 \tag{35}$$

$$CH_2F \rightarrow CH_2FO_2 \rightarrow CH_2FO \rightarrow COF_2 \rightarrow CO,\ CO_2 \tag{36}$$

$$C_2H_2F_3 \rightarrow CHF = CF_2 \rightarrow HCOF\ +\ COF_2 \rightarrow CO,\ CO_2 \tag{37}$$

Being more specific about the reaction scheme (37), $CHF{=}CF_2$ formed by dehydrofluorination yields HCOF and COF_2 as a result of the addition of ozone to its double bond, which is typical in the reaction of alkene compounds with ozone:

$$CHF = CF_2 + O_3 \rightarrow HCOF + COF_2 + O,\ 1.1\times10^{6} cm^{3} mol^{-1} s^{-1} \quad (38)$$

HCOF successively reacts as below to form COF_2

$$HCOF + F \rightarrow COF + HF,\ 5.3\times10^{11} cm^{3} mol^{-1} s^{-1} \quad (39)$$

$$COF + F_2 \rightarrow COF_2,\quad 2.7\times10^{10} cm^{3} mol^{-1} s^{-1} \quad (40)$$

Even though the $C_2H_2F_4$ destruction can systematically be elucidated with the above reactions, there are many other possible reactions that simultaneously occur.

2.2.4 Sulfur hexafluoride

Finally, the electron impact dissociation of SF_6 can generally be expressed as (Nanjo & Ohyama, 2005)

$$SF_6 + e \rightarrow SF_x + (6-x)F,\ x \le 5 \quad (41)$$

Here, SF_x stands for intermediate decomposition products such as SF_5, SF_4, SF_3, etc, and they can be oxidized to form SO_2 and SO_2F_2. The reaction schemes involving oxygen have been proposed by different authors in order to explain the SO_2 and SO_2F_2 formation (Khairallah et al., 1994; Nanjo & Ohyama, 2005). The intermediate decomposition products of reaction (41) further react with atomic or molecular oxygen as follows:

$$SF_4 + O \rightarrow SOF_2 + 2F \quad (42)$$

$$SF_3 + O \rightarrow SOF_2 + F \quad (43)$$

$$SOF_2 + F \rightarrow SOF_3 \quad (44)$$

$$SOF_3 + O \rightarrow SO_2F_2 + F \quad (45)$$

$$SOF_2 + O \rightarrow SO_2F_2 \quad (46)$$

$$SF_3 + O_2 \rightarrow SO_2F_2 + F \quad (47)$$

$$SF_2 + O_2 \rightarrow SO_2F_2 \quad (48)$$

$$SO_2F_2 \rightarrow SO_2 + F_2 \quad (49)$$

These consecutive reactions lead to the formation of SO_2 and SO_2F_2.

3. Experimental section

DBD reactors can be constructed in many configurations, for instance, planar type using parallel-plate metal electrodes separated by a dielectric layer or cylindrical type consisting of two coaxial electrodes separated by a tubular dielectric layer between them. In this work, a cylindrical-type DBD reactor packed with catalyst pellets or dielectric beads was employed for the destruction of fluorinated compounds. The DBD nonthermal plasma reactor of this work, referred to as plasma reactor, was made up of a ceramic tube serving as a dielectric barrier, a concentric stainless steel screw and a copper foil wrapping around the ceramic tube. The experimental details including apparatus, feed gas preparation, methods and analyses are described below.

3.1 Apparatus

Fig. 3 depicts the schematic diagram of the cylindrical plasma reactor of this work. The inner and outer diameter of the ceramic tube were 24.5 mm and 28 mm, where the stainless steel screw with a thickness of 6 mm was coaxially placed. In this reactor configuration, the stainless steel screw and the copper foil acted as the discharging and ground electrode, respectively. The effective reactor length for creating nonthermal plasma was about ~150 mm. The plasma reactor prepared as above was packed with 3-mm α-alumina beads (Sigma-Aldrich Co.), 3-mm glass beads (Sigmund-Lindner, Germany) or 3-mm zirconia beads (Daihan Scientific, Korea) to a volume of 127 cm^3. Unlike alumina widely used as a catalyst or a catalyst support, zirconia and glass beads have no catalytic activity, thus the contribution of the nonthermal plasma to the destruction of fluorinated compounds can solely be evaluated with these packing materials. The plasma reactor was also operated without any packing materials to contrast the results with those obtained in the presence of packing materials. So as to change the reactor temperature to a desired value, the plasma reactor was covered with a heating tape and the temperature was controlled by a proportional-integral-derivative (PID) controller. The reactor temperature was measured at the midpoint of the reactor wall by using a K-type thermocouple.

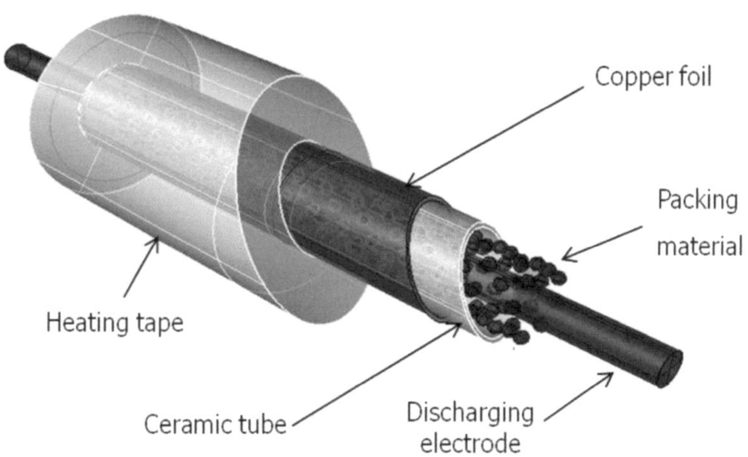

Fig. 3. Schematic diagram of the nonthermal plasma reactor

3.2 Methods

The schematic representation of the experimental setup is shown in Fig. 4. An alternating current (AC) high voltage power supply (operating frequency: 400 Hz) was used to energize the plasma reactor. The voltage applied to the discharging electrode of the plasma reactor was varied in the range of 7~16 kV (rms value) to change the electrical power delivered to the plasma reactor. The fluorinated compounds dealt with in this work were CHF_3, $C_2H_2F_4$, SF_6 and C_2F_6. The behavior of destruction of these compounds was separately examined one by one. The feed gas consisted of three gases, i.e., a fluorinated compound, nitrogen and oxygen, whose flow rates were regulated by mass flow controllers (MKS Instruments, Inc., USA). The concentration of the fluorinated compound at the reactor inlet was typically at 2,000 ppm (parts per million, volumetric). The overall flow rate of the feed gas was 1.0 L min^{-1} or 60 L h^{-1} on the basis of room temperature. The reactor temperature was changed up to 500°C by applying heat to the reactor using a heating tape. The simulated exhaust gas processed in the plasma reactor was directed to the Fourier transform infrared (FTIR) Spectrometer (Bruker IFS 66/S, Germany) for analyzing fluorinated compounds and destruction products. The fluorinated compounds and byproducts were assigned in the spectra and the measured absorbance of each compound was converted into concentration units. The decomposition efficiency was defined as $100 \times (C_0\text{-}C)/C_0$, where C_0 and C are the concentrations at the inlet and outlet of the reactor, respectively. The electrical power (input power) was measured by a digital power meter (Model WT200, Yokogawa, Japan) and the voltage was monitored using a digital oscilloscope (TDS 3032, Tektronix, USA) equipped with a 1000 : 1 high voltage probe (P6015, Tektronix, USA).

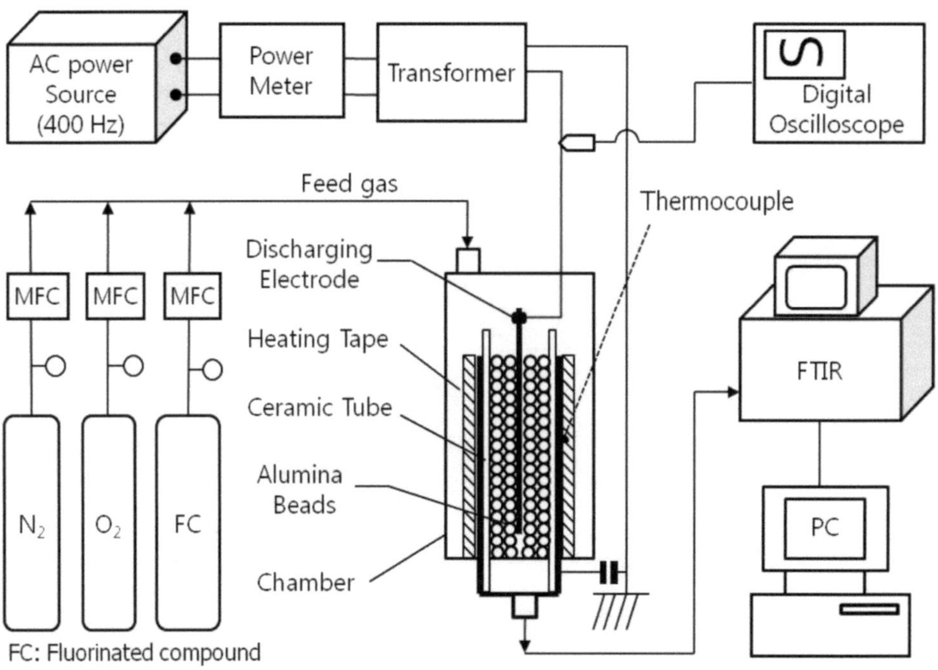

Fig. 4. Schematic of the experimental setup for destroying fluorinated compounds

Discharge power that is actually consumed in the plasma reactor was determined by using the so-called Lissajous charge-voltage curve (Rosocha, 2005). The Lissajous curve was obtained by measuring the voltages across the electrodes of the plasma reactor and across the capacitor (0.43 μF) connected to the plasma reactor in series. The voltage across the capacitor multiplied by its capacitance corresponds to the charge, which is, in principle, equal to the charge accumulated on the electrodes of the plasma reactor because the capacitor and the plasma reactor are a series circuit. Fig. 5 (a) shows example waveforms of the voltage applied to the plasma reactor and the charge deposited, and Fig. 5 (b) shows the Lissajous curve for the voltage versus the charge. The energy per cycle consumed in the plasma reactor is equal to the enclosed area, and the discharge power can be calculated by multiplying the energy per cycle by the operating frequency (400 Hz).

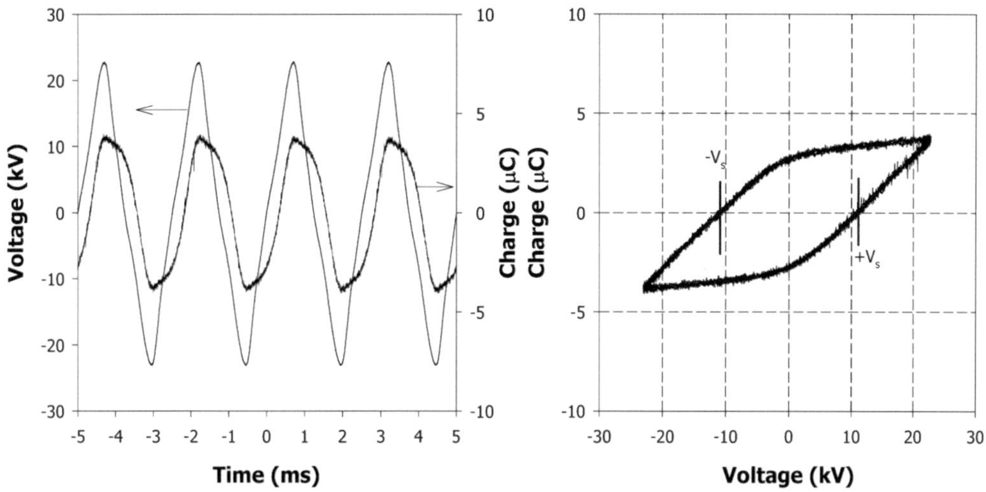

Fig. 5. (a) Waveforms of the voltage applied to the plasma reactor and the charge deposited and (b) the corresponding Lissajous figure

4. Results and discussion

4.1 Discharge power

The performance of the plasma reactor is a strong function of temperature, especially when it is packed with catalyst pellets, and accordingly, it is necessary to recognize the relationship between the reactor temperature and the discharge power. The DBD plasma is characterized by numerous short lifetime microdischarges, which are generated when the applied voltage exceeds the breakdown voltage of the gas between the electrodes. The microdischarges form conduction paths between the electrodes, and self-extinguish as the charge accumulated on the dielectric reduces the local electric field. Fig. 6 shows the dependence of the discharge power on the reactor temperature and the type of packing materials (alumina, zirconia and glass beads), when the input power was 60~100 W. The plasma reactor can electrically be treated as a capacitor, and the discharge power was measured by using the Lissajous curve (Rosocha, 2005). As presented in Fig. 6, regardless of the packing material used, the temperature-discharge power relationships were similar to one another. Regarding the temperature effect for 60 W and 80 W input power, the discharge power gradually increased with increasing the temperature up to 200°C, and then stabilized with further increase in the temperature above 200°C. As well known, gaseous molecules can be more easily ionized at higher temperatures, which results in increasing the discharge power. For 100 W input power, the temperature effect on the discharge power was not significant. The efficiency of power transfer, defined as the ratio of discharge power to input power, was calculated to be about 70%.

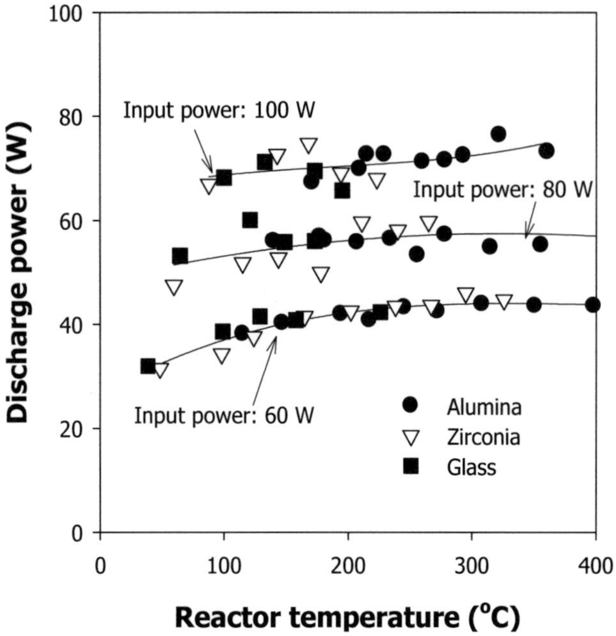

Fig. 6. Dependence of the discharge power on the reactor temperature and the type of packing material

4.2 Effect of reactor temperature

The plasma reactor packed with alumina beads showed different behaviors according to the reactor temperatures. Below a certain threshold temperature, the destruction efficiency slowly increased with increasing the reactor temperature, while on the other hand there was a steep increase in the destruction efficiency as the reactor temperature was further increased beyond the threshold temperature (Kim et al., 2010c; Kim & Mok, 2010; Kim and Mok, 2011). The threshold temperature was found to vary with the fluorinated compounds investigated. Details are given below.

The effect of the reactor temperature on the destruction of CHF_3 is shown in Fig. 7. The input power was fixed at 80 W (discharge power: ~56 W) ove a reactor temperatures range up to 300°C. The destruction efficiency was observed to increase with increasing the reactor temperature, implying that the reactions responsible for the CHF_3 destruction are advantageous at elevated temperatures. In addition, the increase in the discharge power with increasing the temperature (see Fig. 6) may partly explain why higher destruction efficiency was observed at higher temperature. Meanwhile, at temperatures below 150°C where the catalytic activity may be neglected, the difference in the destruction efficiency between the alumina and glass beads was inconsiderable. On the other hand, the difference became pronounced when the reactor temperature was further increased over 150°C, because the catalytic destruction played an important role in this temperature region and the plasma possibly assisted the catalytic reactions.

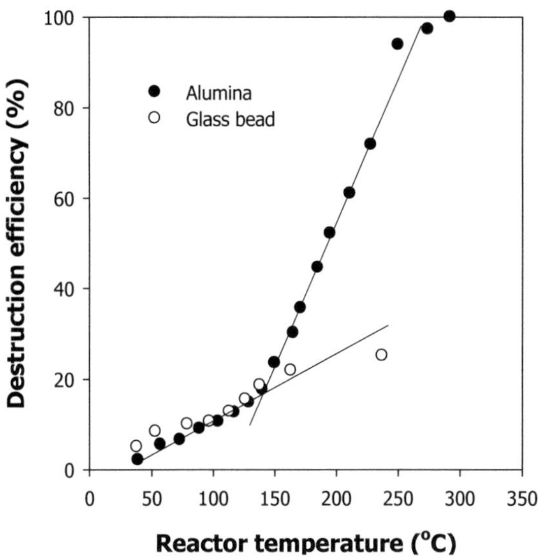

Fig. 7. Variations of the HFC decomposition efficiency as a function of reactor temperature (CHF_3: 2,000 ppm; O_2: 1.0%(v/v); input power: 80 W)

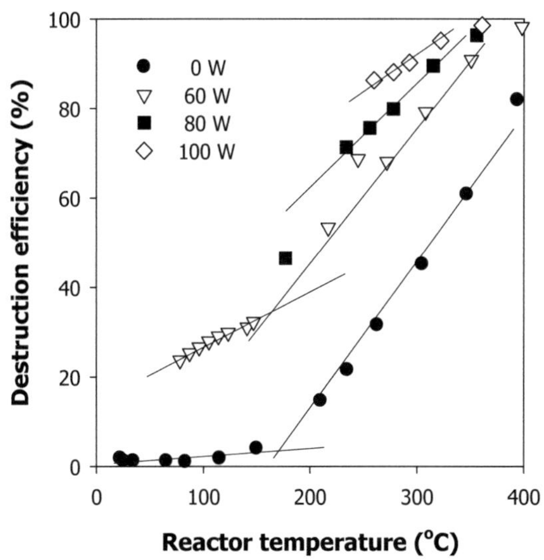

Fig. 8. Comparison of $C_2H_2F_4$ destruction efficiencies between catalysis alone and plasma-catalysis over a temperature range up to 400°C ($C_2H_2F_4$: 2,000 ppm; O_2: 2.0%(v/v)

Fig. 8 presents a comparison of $C_2H_2F_4$ destruction efficiencies between without plasma and with plasma over a temperature range up to 400°C at input powers of 60~100 W (discharge power: 42~60 W), which was obtained with alumina beads as the packing material. In this figure, the catalysis-alone case corresponding to the results obtained at 0 W indicates that the alumina catalyst was thermally activated without applying high voltage to the reactor, and the plasma-catalysis (60~100 W) represents that both thermal and plasma activation of the catalyst worked upon the $C_2H_2F_4$ destruction.

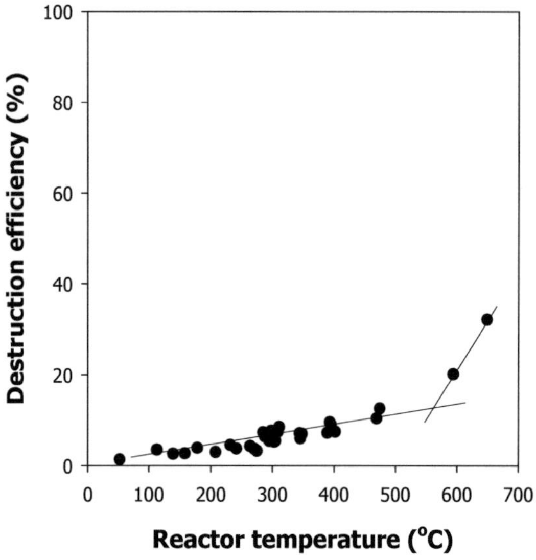

Fig. 9. Effect of reactor temperature on the destruction of C_2F_6 (C_2F_6: 2,000 ppm; O_2: 2.0%(v/v); input power: 100 W)

As can be seen in Fig. 8, the catalyst alone started destroying the fluorinated compound from about 150°C, exhibiting negligible destruction efficiency at temperatures below 150°C. When the reactor temperature was gradually increased from 150°C to 400°C at 60 W, the destruction efficiency proportionally increased, reaching 80% at 400°C. In the case of the plasma plus catalysis, it is apparent that the $C_2H_2F_4$ destruction can be divided into two different regions by the steepness of the temperature-destruction efficiency relationship. In the lower temperature region below 150°C, the $C_2H_2F_4$ destruction efficiency slowly increased with increasing the reactor temperature, while in the higher temperature region the destruction efficiency rapidly increased with increasing the reactor temperature, approaching complete destruction at around 400°C. The lower temperature region where the catalyst has no activity is understood to have been dominated by gas-phase reactions resulting from various actions of the plasma. On the other hand, above 150°C, both the plasma and the catalysis must have affected the $C_2H_2F_4$ destruction, thereby leading to a steeper increase in the destruction efficiency. The threshold temperatures were around 150°C for $C_2H_2F_4$, but as shown in Fig. 9, the threshold temperature for C_2F_6 destruction was found to be much higher around 600°C, despite similar molecular structure to one another. This result can be attributed to the fact that C_2F_6 does not have relatively weak C-H bonds, i.e., since C_2F_6 consists only of strong C-F bonds, the high bond dissociation potential and and the shielding effect of the fluorine atoms make the reactivity of C_2F_6 extremely low.

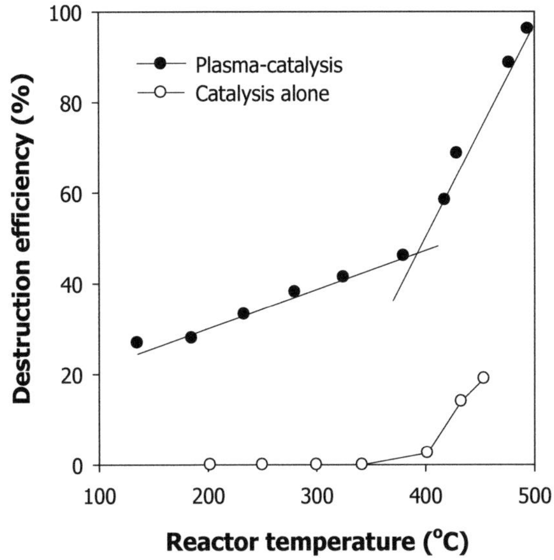

Fig. 10. Effect of reactor temperature on the destruction of SF_6 (SF_6: 2,000 ppm; O_2: 2.0%(v/v); input power: 100 W)

Fig. 10 shows the effect of the reactor temperature on the destruction of SF_6. It can be seen that the SF_6 destruction efficiency increased slowly from 27% to 46% with increasing the reactor temperature from 135°C to 380°C, and then abruptly increased with further increasing the reactor temperature, reaching 96% destruction efficiency at 495°C. From this result, the destruction of SF_6 in the plasma-catalysis combined system can be divided into two different regions, according to the reactor temperature. The threshold temperature for the nonthermal plasma-assisted catalysis is seen to be around 400°C. Below 400°C is the region of no catalytic activity, and above 400°C is the region dominated by the nonthermal plasma-assisted catalysis. In the low temperature region below 400°C, the slow increase in the destruction efficiency with increasing the reactor temperature can be explained by the decrease in the gas density. As the gas density decreases, electrons generated in the plasma reactor can be accelerated more efficiently to destroy SF_6 molecules because the mean free path correspondingly increases. In the high temperature region above 400°C, there was a precipitous increase in the destruction efficiency. It is the region dominated by the nonthermal plasma-assisted catalysis, where processing the simulated exhaust gas produced over 95% decomposition efficiency at temperatures higher than 490°C, compared with 27~46% in the low temperature region. In Fig. 10, the data resulting from the catalysis alone are also presented. In the absence of plasma, the catalyst exhibited no SF_6 decomposition efficiency at temperatures below 400°C. It was observed that the catalyst began decomposing SF_6 from about 400°C, which is in agreement with the threshold temperature mentioned above. Consequently, it can be said that the temperature starting to show an abrupt increase in the decomposition efficiency is the one for the catalyst to begin exhibiting its activity. Once the catalyst begins exhibiting its activity, the nonthermal plasma can remarkably enhance the decomposition, as shown in Fig. 10. From such a large enhancement in the destruction efficiency in the presence of the nonthermal plasma, it is apparent that the plasma-assisted catalysis is a very effective way to improve the catalytic activity at relatively low temperatures.

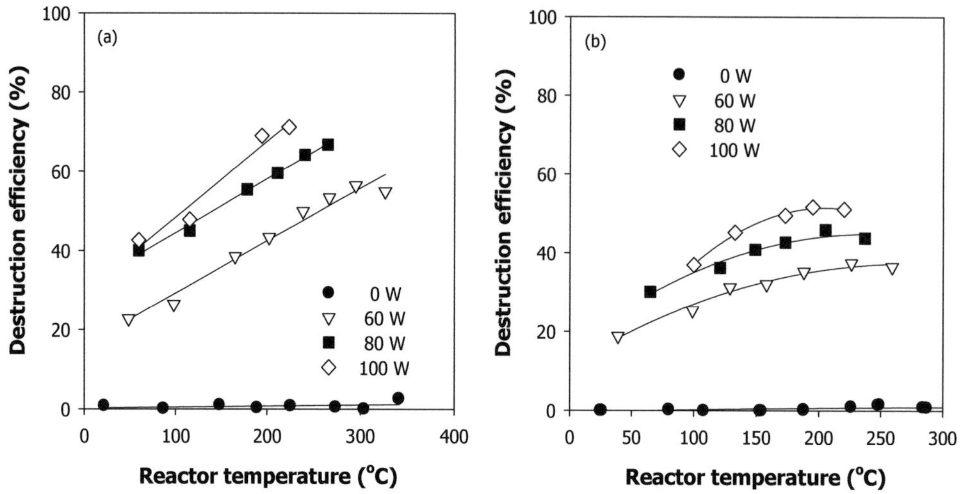

Fig. 11. Comparison of $C_2H_2F_4$ destruction efficiencies between (a) zirconia and (b) glass beads ($C_2H_2F_4$: 2,000 ppm; O_2: 2.0%(v/v))

4.3 Effect of packing material type

The destruction of fluorinated compounds by the plasma can strongly depend on the type of packing material, especially at high temperatures that the contribution of catalysis can be significant. In Fig. 11, a comparison between zirconia and glass beads was made under the same input power conditions as a function of the reactor temperature. The results obtained with alumina are presented in Fig. 8 above. It should be noted that zirconia and glass beads have negligible catalytic activity for the destruction of fluorinated compounds. In all cases, increasing the reactor temperature enhanced the destruction efficiency. The increase in the destruction efficiency with increasing the temperature can be attributed to the decrease in the gas density. Namely, the decrease in the gas density increases the mean free path of electrons, which in turn accelerates electrons more efficiently to increase the generation of reactive species and the electron-impact dissociation of $C_2H_2F_4$ molecules. Meanwhile, as observed in Fig. 8 and Fig. 11, the difference in the destruction efficiencies between the three types of packing materials was inconsiderable at lower temperature region. Such a phenomenon was also shown in the destruction of CHF_3 as described in Fig. 7. However, at temperatures above 150°C, the difference in the destruction efficiencies became pronounced, obviously because the catalytic action of alumina significantly contributed to the $C_2H_2F_4$ destruction. Despite both having negligible catalytic activity, the behavior of $C_2H_2F_4$ destruction with zirconia beads was different from that with glass beads, which may be explained by the difference in their dielectric constants. The discharge characteristics of a packed-bed type plasma reactor largely depend on the dielectric constant of packing material, and larger dielectric constant is generally more advantageous to the performance of plasma reactor. The dielectric constants of zirconia and silica glass are 15~22 and 3.8, respectively, and it is reasonable that the reactor packed with zirconia beads produced higher destruction efficiency than that with glass beads.

Fig. 12 presents the FTIR spectra of the gas processed in the plasma reactor, which were obtained with alumina and zirconia as the packing materials (Mok & Kim, 2011). With

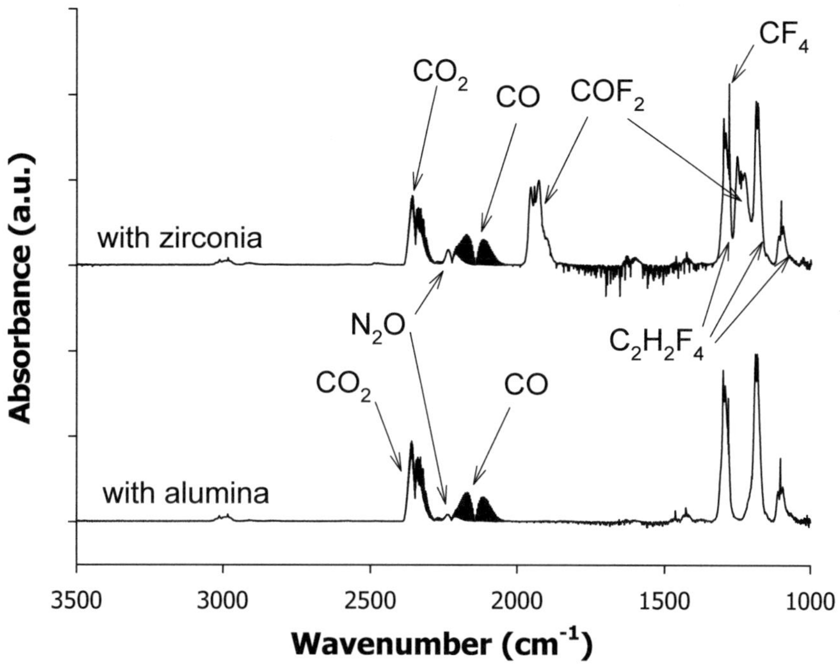

Fig. 12. FTIR spectra of the gas processed in the plasma reactor ($C_2H_2F_4$: 2,000 ppm; O_2: 2.0%(v/v); input power: 60 W)

zirconia, the carbon-containing byproducts identified included CO_2, CO, carbonyl fluoride (COF_2) and carbon tetrafluoride (CF_4). According to the reactions schemes explained above, COF_2 serves as a key intermediate compound to convert fluorinated compounds into CO and CO_2. Different from the result obtained with zirconia, however, COF_2 was disappeared in the FTIR spectrum when alumina was used as the packing material. It has been shown above that alumina exhibits it catalytic activity for destroying $C_2H_2F_4$, starting at around 150°C. Besides such a capability for destroying $C_2H_2F_4$, the FTIR spectrum suggests that alumina can also move the process towards the total oxidation, oxidizing COF_2 to CO and CO_2. Since COF_2 is a highly toxic compound, the use of alumina is more advantageous.

Figs. 13 (a) and (b) show the concentrations of the carbon-containing destruction products that were obtained with alumina and zirconia as the packing material, respectively. Although data are not shown, nitrous oxide (N_2O) was also identified in the effluent gas stream. The formation of N_2O is common in plasma processing of air-like gases containing both N_2 and O_2 (Fitzsimmons et al., 2000; Ricketts et al., 2004; Harling et al., 2005). Properly, the amounts of destruction products were more in the presence of alumina than zirconia, because of higher destruction efficiency. In both cases of packing materials, the abundant destruction products were CO_2 and CO. The concentrations of CO_2 and CO tended to increase with increasing the reactor temperature. Even though the amounts were small, the plasma reactor packed with zirconia beads produced COF_2 and CF_4. The main reaction for the formation of CF_4 is the recombination of CF_3 and F.

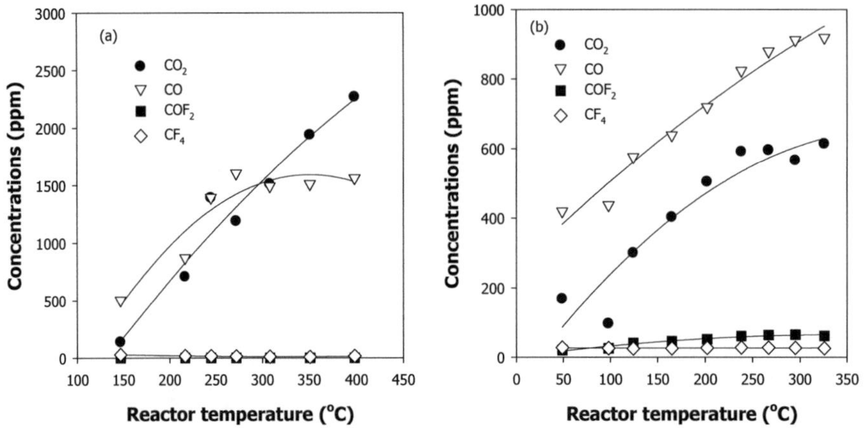

Fig. 13. Byproducts distribution: (a) with alumina and (b) with zirconia ($C_2H_2F_4$: 2,000 ppm; O_2: 2.0%(v/v); input power: 60 W)

4.4 Synergistic effect of plasma-catalysis

In Fig. 14, the destruction efficiencies obtained with the nonthermal plasma-alone case, the catalyst-alone case and the plasma-catalyst case are compared for input powers of 60 W and 80 W. In the case of "the plasma alone", the reactor was packed with the glass beads, because the glass beads do not have any catalytic activity for destroying fluorinated compounds. The cases of "the plasma-catalyst" and "the catalyst alone" represent the catalytic CHF_3 decomposition performed with and without the plasma, respectively. In Fig. 14, an interesting aspect of the plasma-catalyst is the so-called synergistic effect. At an input power of 80 W, the plasma-alone case decomposed about 24% and 28% of CHF_3 at 200°C and 250°C, respectively. The respective destruction efficiencies obtained by the catalyst-alone case at the corresponding temperatures were 10% and 33%. The arithmetic sum of

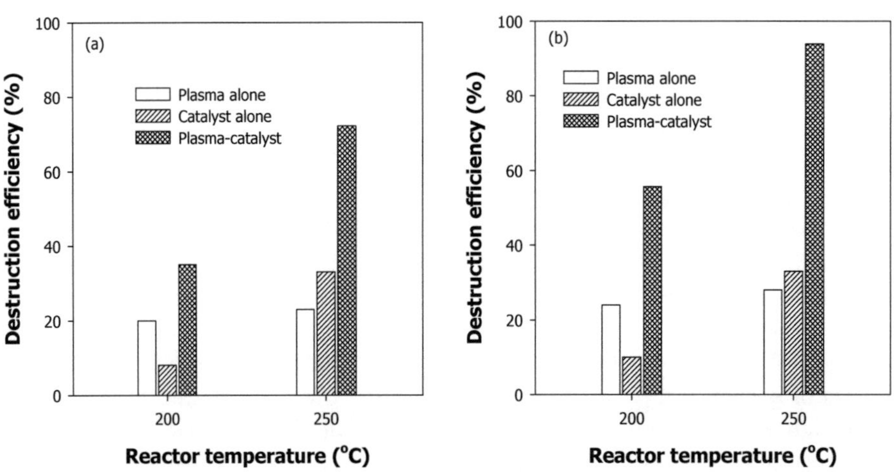

Fig. 14. Comparison of decomposition efficiency between the plasma alone, the catalyst alone and the plasma-catalyst combination at input powers of (a) 60 W and (b) 80 W (CHF_3: 2,000 ppm; O_2: 1.0%(v/v)).

the plasma and catalyst decompositions individually adds to 34% and 61% at 200°C and 250°C, respectively. However, in the case of the plasma-catalyst, the destruction efficiencies at the respective temperatures were 56% and 94%, clearly showing that there was some synergy. The synergistic effect implies that the plasma can provide some additional activation of the alumina catalyst (Kim et al., 2005; Kim et al., 2010c). In this context, it would be advisable to combine the plasma and the catalyst rather than use them separately. When the input power was 60 W, the destruction efficiencies obtained by the plasma-alone case were 20% and 23% at 200°C and 250°C, and those obtained by the plasma-catalyst were 35% and 72 %, respectively. The sum of the destruction efficiencies individually obtained by the plasma and catalyst is 30% at 200°C and 56% at 250°C, which are smaller values when compared to 35% and 72 % obtained by the combination of plasma and catalyst at identical temperatures.

In the destruction of fluorinated compounds, the overall destruction rate is determined by the electron impact dissociation, and succeeding reactions for oxidizing the destruction fragments are much faster. The enhancement of the destruction efficiency with the plasma-catalyst case may be explained by the acceleration of the rate-determining step. The C-F bond strength for a gaseous fluorinated compound is 5.1 eV, but it gets weak when adsorbed on the catalyst surface. As a result, the energetic electrons generated by the plasma can more easily break the C-F bond of the adsorbed molecule through direct electron impact, speeding up the rate-determining step.

4.5 Effect of electric power

In Fig. 15, the effect of the electrical power on the destruction of CHF_3 at different temperatures in the range of 150~250°C. The input power was changed up to 100 W. The discharge power was about 70% of the input power. In this figure, the results at 0 W that the CHF_3 destruction efficiencies obtained with the catalysis alone. As can be seen in Fig. 15 (a), the alumina exhibited negligible catalytic destruction efficiency below 150°C, but its catalytic activity was gradually enhanced with increasing the reactor temperature. On the contrary, the reactor packed with the glass beads did not decompose CHF_3 at all temperatures explored. Moreover, the effect of the electric power on the destruction was even more significant for

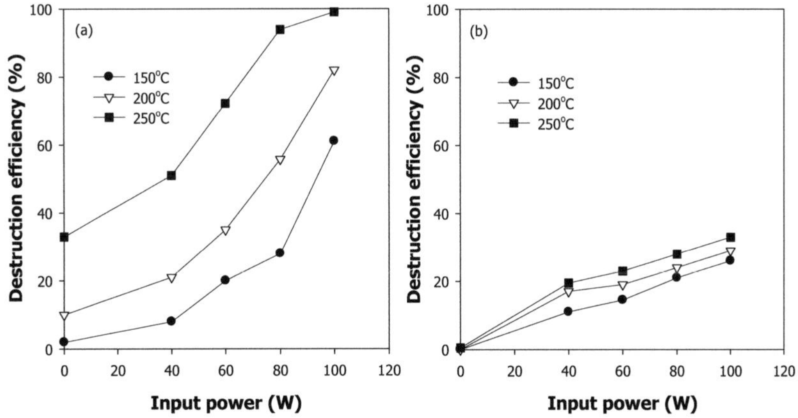

Fig. 15. Effect of the electric power on the decomposition of CHF_3 (a) with alumina and (b) with glass beads (CHF_3: 2,000 ppm; O_2: 1.0%(v/v))

alumina beads than for glass beads, obviously because various actions of the plasma assisted catalytic reactions. The destruction efficiency versus the input power for alumina-packed case (Fig. 15 (a)) showed an exponential growth whereas that for glass-packed case (Fig. 15 (b)) showed a linear increase. With glass beads, only gas-phase reactions induced by the plasma are responsible for the CHF_3 destruction. As shown in Fig. 15 (a), the higher the reactor temperature, the less the input power required to destroy CHF_3, due to the improved catalytic activity. When the reactor temperature was 250°C, the destruction efficiency with the alumina approached 100% at an input power of 100 W. This is neary three times higher destruction efficiency, when compared to a little more than 30% in the presence of glass beads.

5. Conclusions

The destruction of several fluorinated compounds such as trifluoromethane (CHF_3), sulfur hexafluoride (SF_6), 1,1,1,2-tetrafluoroethane ($C_2H_2F_4$) and hexafluoroethane (C_2F_6) was investigated in the plasma reactor packed with alumina, zirconia or glass beads. This work was concentrated on the effects of reactor temperature and electric power, and the conclusions drawn are as follows. Operating the plasma reactor at elevated temperatures was advantageous, owing to increased rates of destruction reactions. Particularly, the favorable effect of the elevated temperature on the destruction was remarkable in the presence of alumina, because it acted as a catalyst.

From several sets of catalyst-alone experiments, it was found that the threshold temperature from which the destruction efficiency began rapidly rising corresponds to the minimum activation temperature of the alumina catalyst. The threshold temperatures were around 150°C for CHF_3 and $C_2H_2F_4$, around 400°C for SF_6, and around 600°C for C_2F_6. On the contrary, with zirconia or glass beads as the packing material, the temperature dependence of the destruction efficiency did not show such a threshold, indicating that the destruction was mainly caused by gas-phase reactions. Even though the temperature dependence on the destruction with zirconia or glass beads was not as remarkable as with alumina beads, the reactor packed with zirconia beads having larger dielectric constant produced higher destruction efficiency than that with glass beads.

This study has shown that the combination of plasma and catalyst may be an effective method to destroy fluorinated compounds. The plasma-catalyst combination showed higher destruction efficiency than the sum of those individually obtained by the plasma and the catalyst. This synergistic effect indicates that the nonthermal plasma created in the catalytic reactor can provide some additional activation of the catalyst.

6. Acknowledgment

The author would like to thank members of the Jeju National University Plasma Applications Laboratory whose work is referred to this article: D. H. Kim, S. B. Lee, J. H. Oh, E. J. Jwa, and M. Gandhi. Financial support from the National Research Foundation of Korea (Grant number 2010-0021672) is greatly acknowledged.

7. References

Barker, J. R. (1995). *Progress and Problems in Atmospheric Chemistry*, World Scientific, ISBN 981-02-1868-0, London, UK

Bickle, G.M.; Suzuki, T. & Mitarai, Y. (1994). Catalytic Destruction of Chlorofluorocarbons and Toxic Chlorinated Hydrocarbons. *Applied Catalysis B: Environmental*, Vol. 4, (1994), pp. 141-153, ISSN 0926-3373

Chang, M.B.; Balbach, J.H.; Rood, M.J. & Kushner, M.J. (1991). Removal of SO_2 from Gas Streams Using a Dielectric Barrier Discharge and Combined Plasma Photolysis. *Journal of Applied Physics*, Vol.69, No. 8, (April 1991), pp. 4409-4417, ISSN 0021-8979

Futamura, S. & Yamamoto, T. (1997). Byproducts Identification and Mechanism Determination in Plasma Chemical Decomposition of Trichloroethylene. *IEEE Transactions on Industry Applications*, Vol. 33, No. 2, (March & April 1997), pp. 447-453, ISSN 0093-9994

Fitzsimmons, C.; Ismail, F.; Whitehead, J.C. & Wilman, J.J. (2000). The Chemistry of Dichloromethane Destruction in Atmospheric-Pressure Gas Streams by a Dielectric Packed-Bed Plasma Reactor. *Journal of Physical Chemistry A*, Vol. 104, (2000), pp. 6032-6038, ISSN 1089-5639

Föglein, K.A.; Szabó, P.T.; Babievskaya, I.Z. & Szépvölgyi, J. (2005). Comparative Study on the Decomposition of Chloroform in Thermal and Cold Plasma. *Plasma Chemistry and Plasma Processing*, Vol. 25, No. 3, (2005), pp. 289-302, ISSN 0272-4324

Futamura, S. & Yamamoto, T. (1997). Byproducts Identification and Mechanism Determination in Plasma Chemical Decomposition of Trichloroethylene. *IEEE Transactions on Industry Applications*, Vol. 33, No. 2, (March & April 1997), pp. 447-453, ISSN 0093-9994Harling, A.; Whitehead, J.C. & Zhang, K. (2005). NO_x Formation in the Plasma Treatment of Halomethane. *Journal of Physical Chemistry A*, Vol. 109, (2005), pp. 11255-11260, ISSN 1089-5639

Herron, J. T. (1999). Evaluated Chemical Kinetics Data for Reactions of $N(^2D)$, $N(^2P)$, and $N_2(A^3\Sigma_u^+)$ in the Gas Phase. *Journal of Physical and Chemical Reference Data*, Vol.28, No.5, (1999), pp. 1453-1483, ISSN 0047-2689

Kim, D.H.; Mok, Y.S.; Lee, S.B. & Shin, S.M. (2010a). Nonthermal Plasma Destruction of Trifluoromethane Using a Dielectric-Packed Bed Reactor. Jornal of Advanced Oxidation Technologies, Vol. 13, (2010), pp. 36-42, ISSN 1203-8407

Kim, D.H.; Mok, Y.S.; Lee, S.B. & Shin, S.M. (2010b). Destruction of Hexafluoroethane in a Dielectric-Packed Bed Plasma Reactor. *Journal of Zhejiang University-Science A*, Vol.11, No.7, (July 2010), pp. 538-544, ISSN 1673-565X

Kim, D.H.; Mok, Y.S. & Lee, S.B. (2010c). Effect of Temperature on the Decomposition of Trifluoromethane in a Dielectric Barrier Discharge Reactor. *Thin Solid Films*, (November 2010) doi:10.1016/j.tsf.2010.11.060, ISSN 0040-6090

Kim, H.H. Ogata, A. & Futamura, S. (2005). Atmospheric Plasma-Driven Catalysis for the Low Temperature Decomposition of Dilute Aromatic Compounds. *Journal of Physics D: Applied Physics*, Vol. 38, No.8, (2005), pp. 1292-1300, ISSN 0022-3727

Kim, D.H. & Mok, Y.S. (2010). Decomposition of Sulfur Hexafluoride by Using Nonthermal Plasma-Assisted Catalytic Process, *3rd Euro-Asian Pulsed Power Conference & 18th International Conference on High-Power Particle Beams*, Jeju, Korea, October 10-14, 2010

Kim, D.H. & Mok, Y.S. (2011). Destruction of Tetrafluoroethane with Atmospheric Nonthermal Plasma Created in Dielectric-Packed Bed Reactors, *15th International Congress of Chemistry and Environment* (Organizer: Research Journal of Chemistry and Environment, India), Negeri Senbilan, Malaysia, May 27-29, 2011

Kim, H.H. Ogata, A. & Futamura, S. (2005). Atmospheric Plasma-Driven Catalysis for the Low Temperature Decomposition of Dilute Aromatic Compounds. *Journal of Physics D: Applied Physics*, Vol. 38, No.8, (2005), pp. 1292-1300, ISSN 0022-3727.

Khairallah, Y.; Khonsari-Arefi, F. & Amouroux, J. (1994). Decomposition of Gaseous Dielectrics (CF_4, SF_6) by a Non-Equilibrium Plasma. Mechanisms, Kinetics, Mass Spectrometric Studies and Interactions with Polymeric Targets. *Pure and Applied Chemistry*, Vol.66, No.6, (1994), pp. 1353-1362, ISSN 1365-3075

Kuroki, T.; Mine, J.; Okubo, M.; Yamamoto, T. & Saeki, N. (2005). CF_4 Decomposition Using Inductively Coupled Plasma: Effect of Power Frequency. *IEEE Transactions on Industry Applications*, Vol. 41, No. 1, (January & February 2005), pp. 215-220, ISSN 0093-9994

Lee, M.C. & Choi, W. (2004). Development of Thermochemical Destruction Method of Perfluorocarbons (PFCs). *Journal of Industrial and Engineering Chemistry*, Vol. 10, No. 1, (2004), pp. 107-114, ISSN 1226-086X

Li, J.; Sun W.; Pashaie, B. & Dhali, S. K. (1995). Streamer Discharges Simulation in Flue Gas. *IEEE Transactions on Plasma Science*, Vol.23, No.4, (August 1995), pp. 672-678, ISSN 0093-3813

Mizeraczyk, J.; Jasinski, M. & Zakrzewski, Z. (2005). Hazardous Gas Treatment Using Atmospheric Pressure Microwave Discharges. *Plasma Physics and Controlled Fusion*, Vol. 47, (2005), pp. B589–B602, ISSN 1361-6587

Mok, Y.S.; Ham, S.W. & Nam, I.-S. (1998). Mathematical Analysis of Positive Pulsed Corona Discharge Process Employed for Removal of Nitrogen Oxides. *IEEE Transactions on Plasma Science*, Vol. 26, No. 5, (October 1998), pp. 1566-1574, ISSN 0093-3813

Mok, Y.S.; Demidyuk, V. & Whitehead, J.C. (2008). Decomposition of Hydrofluorocarbons in a Dielectric-Packed Plasma Reactor. *Journal of Physical Chemistry A*, Vol. 112, (2008), pp. 6586-6591, ISSN 1089-5639

Motlagh, S. & Moore, J.H. (1998). Cross Sections for Radicals from Electron Impact on Methane and Fluoroalkanes. *Journal of Chemical Physics*, Vol.109, No.2, (July 1998), pp. 432-438, ISSN 0021-9606

Nanjo, Y. & Ohyama, R. (2005). An Experimental Study on Vacuum-Ultraviolet Photochemical Reaction to Non-Thermal Plasma Oxidized SF_6 Gases, *IEEE Conference on Electrical Insulation and Dielectric Phenomena*, pp. 689-692, Nashville, Tennessee, USA, October 16-19, 2005

National Institute of Standards and Technology (NIST) Chemical Kinetics Database: Version 2Q98, 1998

Ogata, A.; Kim, H.H.; Futamura, S.; Kushiyama, S. & Mizuno, K. (2004). Effects of Catalysts and Additives on Fluorocarbon Removal with Surface Discharge Plasma. *Applied Catalysis B: Environmental*, Vol. 53, (2004), pp. 175-180, ISSN 0926-3373

Piper, L.G.; Marinelli, W.J.; Rawlins, W.T. & Green, B.D. (1985). The Excitation of IF($B^3\Pi_{0+}$) by $N_2(A^3\Sigma_u^+)$. *Journal of Chemical Physics*, Vol. 83, No. 11, (December 1985), pp. 5602-5609, ISSN 0021-9606

Ricketts, C.L.; Wallis, A.E.; Whitehead, J.C. & Zhang, K. (2004). A Mechanism for the Destruction of CFC-12 in a Nonthermal, Atmospheric Pressure Plasma. *Journal of Physical Chemistry A*, Vol. 108, (2004), pp. 8341-8345, ISSN 1089-5639

Rosocha, L.A. (2005). *IEEE Transactions on Plasma Science*, Vol.33, No.1, (February 2005), pp. 129-137, ISSN 0093-3813

Tao, W.; Golde, M.F. & Ho, G.H. (1992). Experimental Study of the Reactions of $N_2(A\ ^3\Sigma^+_u)$ with CH_3CN and HCN: The Effect of Vibrational Energy in $N_2(A)$. *Journal of Chemical Physics*, Vol.96, No. 1, (January 1992), pp. 356-366, ISSN 0021-9606

The Intergovernmental Panel on Climate Change (IPCC) Third Assessment Report: Climate Change. (2001)

14

The Choice of Biofuels to Mitigate Greenhouse Gas Emissions

Rogério Cezar de Cerqueira Leite[1], Manoel Regis Lima Verde Leal[2] and Luís Augusto Barbosa Cortez[3]
[1]*Interdisciplinary Center of Energy Planning / University of Campinas*
[2]*Brazilian Bioethanol Science and Technology Laboratory*
[3]*Faculty of Agricultural Engineering / University of Campinas*
Brazil

1. Introduction

The tentatives to use biofuels can be traced back to the years around 1900 when Henry Ford and Rudolf Diesel used ethanol and vegetable oil in their Otto cycle and Diesel cycle engines, respectively. With the introduction of oil in the energy scenario as a very cheap option, the interest in biofuels dwindled down rapidly and the transport sector was fully dominated by gasoline and diesel. However, the idea did not die and biofuels came and went a few times; in the mid 1920s ethanol fuel was used in Brazil and in 1931 a Federal Law mandated the blending of 5% ethanol in all imported gasoline used in the country. Since then, several biofuels options have been produced and tentatively used in some countries: ethanol, methanol, higher alcohols, vegetable oils, fat acid methyl/ethyl esters, biogas and dimethyl ether (DME) just to mention the main ones.

The driving forces behind the use of biofuels are many, but can be separated in four groups: environmental benefits (local and global), high oil prices, energy security and support to local agriculture. Different countries in different times were drawn by different motivations that changed in time in each case. Looking into the main biofuel programs today it can be seen that USA alcohol program was originally intended to mitigate local pollution problems due to vehicle tail pipe emissions but today is driven by the support to local agriculture, energy security and only very recently, with the Energy Independence and Security Act of 2007 (EISA 2007), it has shown some interest in global warming mitigation, with the introduction of minimum greenhouse gas (GHG) emission reduction limits for different alternative of biofuels (Renewable Fuel Standard - RFS2); ethanol dominates the first generation technologies (1G) with biodiesel playing a minor but important role. The Brazilian Alcohol Program launched in 1975 was aimed at reducing the oil imports (due to high oil prices and energy security), but also at improving the sugarcane industry conditions, badly hit by the low sugar prices and overproduction; after the decline in oil prices in the mid 1980s the focus became the reduction of local pollution in the large cities resulting from vehicle tail pipe emissions; more recently, in 2004,the National Biodiesel Production and Use Program was initiated in Brazil with a strong focus on social inclusion and support to small producers, but also with the justification to eliminate diesel imports. Some countries in the European Union (EU), notably Germany, introduced biodiesel to support local agriculture with surplus production problems, however, the Renewable Energy Directive (RED) introduced in April 2009 and the revision of Fuel Quality Directive (FQD) present some sustainability requirements, including GHG emission reduction minimum threshold values, tightening along the time, and fuel quality standards. Since the motivations are many and variable in time, with the changes in context both in local and global scale, it is critically important that a biofuels program, to be launched by any country, should have very clear objectives and a long term view to reduce the risk of supporting inadequate alternatives that will prove unsustainable, or at least inefficient, in the future.

Besides meeting the objectives of the main driving forces, the biofuel alternative to be sustainable and come to represent a meaningful positive impact on the performance of the transport sector it must have some characteristics such as environmental benefits (both local and global), be able to be produced in large quantities without negatively impacting food and feed production, have a good positive energy balance and, last but not least, be competitive in the long run with fossil fuels and other renewable energy alternatives. Although there were many alternatives studied and developed in the past decades, today ethanol and biodiesel from first generation technologies (1G) dominate the biofuels scene and the escalating oil prices have demonstrated to be a very strong driver, as can be seen in Figure 1 for the case of Brazil. The low oil prices between 1986 and 2001 are responsible for the stagnation of the ethanol use in Brazil and the escalating oil prices after 2002 (that peaked above US$ 140/barrel in 2008) can be blamed for the very fast growth in biofuels production and use in the world seen after 2004, when the global biofuel production increased from 32 billion liters (30 billion liters of ethanol and 2 billion liters of biodiesel) in 2004 to 93 billion liters (76 billion liters of ethanol and 17 billion liters of biodiesel) in 2009 (REN21, 2010); fortunately the major players have noticed the danger of embarking in wrong options and introduced legislations establishing some requirements to differentiate the alternatives in terms of feedstocks, local producing conditions, processing paths and, most importantly, GHG emission reduction potential.

The global energy market is several times larger than the agricultural commodities market and, therefore, the question is how much the world in general, and each country in particular, can or should produce before the demand for natural resources for biofuels becomes a problem. The International Energy Agency (IEA, 2009) forecasts in the Low Carbon Scenario (Scenario 450, to keep the temperature increase at no more than 2 °C above the pre-industrial age values) that biofuels will represent some 11% of world transport fuel consumption by 2030; this means around 278 Mtoe of biofuels by 2030. Second generation technologies (2G) will start to be significant around 2020 and will dominate after 2030. IEA also points out that, although 2G technologies will dominate after 2030, sugarcane ethanol will be the only 1G biofuel to survive in the long term (IEA, 2008). In comparison the Reference Scenario estimates the biofuels global production in 2030 as 132 Mtoe representing 4% of the transport fuel demand. To play an important role in GHG mitigation, biofuels should come to represent at least 10% in the world transport fuel pool, with 20% representing a more ambitious, but probably achievable, target for the long term. Setting a tentative target the question now is if can we do it, in terms of resources availability, investments required and how much would cost with respect to GHG mitigation effect (US$/t$CO_2$). It is quite clear that the options should be carefully chosen and that only a handful of countries can play a significant role in this endeavor; this does not rule out the use of biofuel alternatives for niche and specific applications and the participation of several countries in the global biofuel production.

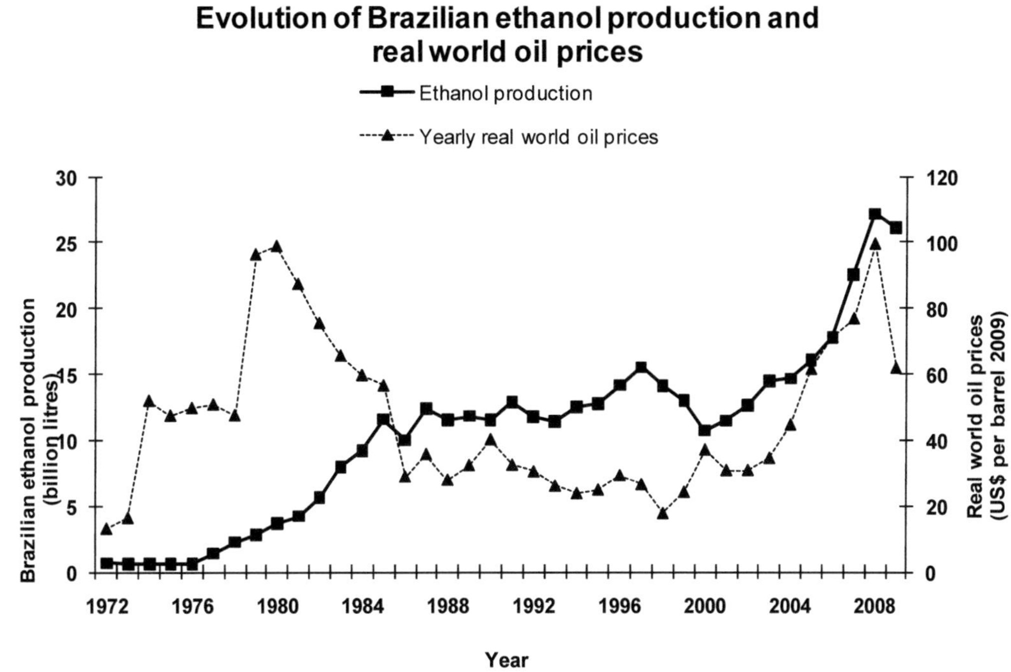

Source: BP, 2009 and EIA, 2010 (oil prices); Energy Research Company (EPE, 2010) and Brazilian Ministry of Agriculture, Livestock and Food Supply (MAPA, 2009) (Brazilian ethanol production)

Fig. 1. Evolution of Brazilian ethanol production and real world oil prices

Therefore, the aim of this chapter is to analyze several options of the most promising biofuels in terms of GHG mitigation potential, taking into consideration the demand for natural resources, GHG emission reduction, and technology availability in time.

The state of the art of lifecycle analysis (LCA) methodology to estimate de GHG emissions in the production chain of the biofuels are evaluated based on a selected literature review, aiming at the identification of the key issues in terms of reliability and reproducibility of the results, unresolved problems and comparing the biofuel alternatives with respect to their GHG emission reduction potential. The two major legislations related to biofuels (European Renewable Energy Directive and US Renewable Fuel Standard) are analyzed to identify the main points relative to GHG emission reduction requirements and the listed default values for the various alternatives; this is a key issue in this chapter as it offers a good indication on which ones are likely to survive in the long run.

Considered by many to be the most important unresolved problem, the determination of the GHG emissions derived from land use changes, both direct and indirect (LUC and ILUC), the theme is discussed based on the most recent works in this area that are or have already been submitted to public consultation process. The impacts estimated are analyzed to supplement the data presented in the previous section, indicating that they can be significant and that more work is certainly required to improve the confidence of the results to an acceptable level and to bring them to a broader range.

The land requirement to produce biofuels is discussed with respect to the potential availability, possible long term biofuels targets and the importance on the sustainability; it is stressed the importance of biofuel yields in all aspects of sustainability.
To be able to produce practical data, the work will concentrate in a case study, taking sugarcane ethanol produced in Brazil as an alternative to displace significant amounts of gasoline worldwide as transport fuel. In this way, the impacts of resource demand, energy balance and GHG lifecycle analysis, including LUC and ILUC effects, can be assessed. The introduction of 2G technologies after 2020 using the sugarcane fiber as feedstock is also investigated as well as the introduction of improvements in the 1G technology of sugarcane ethanol production in the future.

2. Characteristics of biofuels

There are some important characteristics of the biofuels that indicate how well they will perform in terms of meeting the objectives of mitigating the GHG emissions, improving energy security and strengthening the rural economy, without causing meaningful negative impacts on the local environment, food/feed production and prices, biodiversity, social-economic conditions of the local community and, probably the most important of all, be economically viable in the long term without subsidies. For the sake of maintaining the focus on the main issues, only the aspects of GHG mitigation potential and land demand will be considered.

2.1 GHG mitigation potential

The technical and scientific literature is rich in articles and reports dealing with biofuels GHG Life Cycle Analysis (LCA). Larson (2006) reviewed more than 30 publications on this subject covering a broad spectrum of biofuels such as first generation (1G) ethanol, biodiesel, pure vegetable oil, esters, ethers from different feedstocks and second generation (2G) ones including lignocellulosic ethanol and ETBE, Fischer-Tropsch diesel and dimethyl ether (DME) from crop residues and woody crops and grasses; a wide variation in the results was observed in terms of net energy balances and GHG mitigation potentials, even for the same type of biofuel. The reasons for this high uncertainty were attributed to possible range of the input values and variability of the assumptions related to GHG included, N_2O emissions, soil carbon dynamics and allocation method used to give credit to co-products. All but two of the works reviewed were related to developed countries conditions and the two exceptions referred to Brazil and India. In this work it is stressed the importance to refer the GHG emissions and net energy balances to the land used in the feedstock production (tonnes of CO_2eq/ha) and not only to the energy content of the biofuel (gCO_2eq/MJ). Another review of biofuels energy balances and GHG LCA covering 30 relevant papers and reports (Manichetti and Otto, 2008), screened from a set of 60 works, has also indicated a wide variation in the results for the different types of biofuels and feedstocks, even among analyses of the same biofuel and feedstock. First generation (1G) ethanol from maize, wheat, sugarcane and sugar beet and biodiesel from rapeseed, soy beans, sunflower and palm oil, as well as second generation (2G) cellulosic ethanol and Biomass to Liquid (BtL) biofuels, were included in the works reviewed. The agricultural phase is appointed as responsible for most of the GHG emissions and for the adverse local environmental impacts, while the processing phase had the largest contribution to energy use, with the exception of sugarcane ethanol. The main uncertainties are related to N_2O emissions, due to its complex process and dependence on site specificities, agricultural inputs and co-products allocation methodology with its many alternatives. The emissions due to land use change, both direct and indirect (LUC and ILUC), were not included in the analyses but their importance was emphasized. Among other causes of uncertainties and variation in the results it was mentioned the temporal, geographical and technological representativeness of the life cycle inventory data, derived mainly from the use of different data sources for the same unit process. The use of best values or average values for a specific production path, differences in yields and inputs have a strong impact on the final results. The integration of different inputs to produce different products (the biorefinery concept) and the technological evolution impacts were suggested as topic to be considered in future work in this area. Once again, the data and results were mainly related to developed countries pointing to the necessity to know better the performance of biofuels in terms of GHG LCA and energy balance in developing countries.

There are many more publications dealing with this theme, but they generally lead to the same conclusions of the two works discussed above:

- To be able to compare biofuels in terms of energy balance and GHG emissions, more precise procedures and methods need to be developed and reliable data from the same or similar sources should be used. System boundaries, GHG species considered, co-product impact allocation methods, yields and inputs data, non energy GHG emissions calculation procedures and assumptions are the key issues.
- The agricultural phase is the key area in terms of GHG emissions; non energy related GHG emissions, such as soil emissions due to fertilizer use and soil carbon stock dynamics, result from very complex processes that depend on the local soil and climate conditions and agricultural practices and even the past history of land use.
- The processing path is very important in terms of energy balance, especially for ethanol, and can be critical also with respect to GHG emissions if high carbon footprint fuels (such as coal) are chosen.
- Land use change derived emissions, both direct and indirect (LUC and ILUC), need to be considered, but the methodology and tools necessary for this task are not yet established properly; the dynamics of LUC and ILUC evaluated using econometric models need to evolve a lot more to be widely accepted, and soil emission data from world wide data basis that does take into account the local conditions in many countries and regions need to be produced and properly organized.

Recently, two major biofuels programs were launched supported by specific legislation in USA (Renewable Fuel Standard - RFS2, defined in the Energy Independence and Security Act of 2007) and in the EU (the 10% share of renewable energy in transport by 2020 mandated by the Directive 2009/28/EC Of the European Parliament and of the Council of 23 April 2009). Both legislations establish requirements to qualify the biofuels to be counted to meet the targets and the potential GHG emission reduction is a key parameter in this qualification process.

The EU Directive requires a minimum threshold limit for GHG emission reduction compared with the replaced fossil fuel of 35% starting in 2013 (biofuels produced in new installations are already required to meet this limit); this limit will be increased to 50% by 2017 and to 60% by 2018 (in this last case it applies to biofuels produced in new plants). Second generation biofuels (2G) and those produced from wastes and residues will count twice toward the targets; biofuels produced from feedstocks cultivated in restored degraded land will receive a bonus of 29 gCO_2eq/MJ to be discounted from the biofuel LCA emissions. To facilitate the qualification of the biofuels according to this criterion the rules to calculate the GHG impacts of biofuels are presented in the Directive (Annex V) and default values (without LUC and ILUC effects) are included to be optionally used instead of values obtained from a formal calculation procedure. Table 1 presents some of these typical and default values for different biofuels and production pathway.

Biofuel production pathway	Typical GHG emission saving (%)	Default GHG emission saving (%)
Wheat ethanol (lignite as process fuel in CHP plant)	32	16
Wheat ethanol (natural gas as process fuel in CHP plant)	53	47
Wheat ethanol (straw as process fuel in CHP plant)	69	69
Corn ethanol , Community produced (NG in CHP plant)	56	49
Sugar beet ethanol	61	52
Sugarcane ethanol	71	71
Rapeseed biodiesel	45	38
Sunflower biodiesel	58	51
Soybean biodiesel	40	31
Palm oil biodiesel (process not specified)	36	19
Waste vegetable oil or animal oil	88	83
Pure vegetable oil from rapeseed	58	57
Wheat straw ethanol (2G)	87	85
Farmed wood ethanol (2G)	76	70
Waste wood Fischer-Tropsch diesel (2G)	95	95
Farmed wood Fischer-Tropsch diesel (2G)	93	93

Source: Directive 2009/28/EC

Table 1. Typical and default values for GHG emission reduction for biofuels not including LUC/ILUC derived emissions

From the Table 1 above some conclusions can be drawn:

- The GHG emission reduction performances of the biofuels are very much dependent on the feedstock and production pathway, especially on the process fuel used.
- Waste materials, either as feedstock or process fuel, offer superior performance in terms of GHG abatement potential.
- Second generation biofuels (2G) promise the highest GHG emission reductions but the production technologies have yet to be demonstrated at commercial scale.
- Among the first generation alternatives (1G), sugarcane ethanol has the highest impact on GHG emission reduction, even considering the emissions derived from the transport of the biofuel from the producing country to EU.
- Some biofuel alternatives will not meet the Directive GHG emission reduction threshold limits, but nevertheless are widely used today and their productions are still expanding.
- The inclusion of the GHG emissions derived from LUC and ILUC will bring several biofuel alternatives, that seems to meet the Directive limits, to the non attainment area, especially those that exhibit lower yields such as grain ethanol and oil seed biodiesel (except palm oil).

In the USA, the Renewable Fuel Standard (RFS2) has taken a slightly different approach in the sense that it established four different types of biofuels with different GHG emission reduction threshold values, minimum volume to be used in 2022 and phase in time schedule: renewable fuel (essentially corn starch ethanol), cellulosic biofuels, biomass based biodiesel (excludes vegetable oil and animal fat co-processed with petroleum) and other advanced biofuels (including co-processed biodiesel). The minimum annual volumes in 2022 and minimum threshold limits for the life cycle GHG emission reduction compared with the fossil fuel displaced are: corn starch ethanol – maximum volume of 15 billion gallons (56.8 billion liters) to be reached in 2015 and minimum GHG emissions reduction of 20% (for the new plants); cellulosic biofuels – 16 billion gallons (60.6 billion liters) and minimum GHG emission reduction of 60%; biomass-based biodiesel – 1 billion gallons (3.8 billion liters) and minimum GHG emission reduction of 50%; and other advanced biofuels – 4 billion gallons (15.1 billion liters) and GHG emission reduction of 50% (EPA, 2010).

Likewise the EU Directive, EPA presents default values for lifecycle GHG emissions reduction for different biofuels and production pathway; in the EPA case the LUC and ILUC derived emissions are included. Some of these default values are shown in Table 2.

Here also, there are significant differences among biofuels production pathways and fuel used in the process. When the LUC and ILUC effects are included, corn ethanol does not qualify as an advanced biofuel even in the case where biomass is used as fuel; with natural gas as the fuel it barely qualifies as a renewable biofuel and with coal fueled plants there is practically no GHG emission reduction in the lifecycle. Sugarcane ethanol qualifies nicely as an advanced biofuel even in the case where the residues are not collected and used; when this is done, there is a considerable benefit for GHG emission reduction. Second generation ethanol offer a considerable advantage in terms of GHG reductions. The values above 100% are the result of co-product credits allocated in favor of the biofuel.

Biofuel	Production pathway	GHG emission reduction (%)
Corn ethanol	Dry mill with NG	21
Corn ethanol	Dry mill with coal	1
Corn ethanol	Dry mill with biomass	38
Sugarcane ethanol	No residue collection	61
Sugarcane ethanol	With residues collection and surplus power generation	91
Cellulosic ethanol	Switchgrass - biochemical	110
Cellulosic ethanol	Switchgrass - thermochemical	72
Cellulosic ethanol	Corn stover - biochemical	129
Cellulosic ethanol	Corn stover - thermochemical	92
Corn butanol	Dry mill with NG	31
Soy bean biodiesel	FAME	57
Waste grease biodiesel	FAME	86

Source: EPA, 2010

Table 2. Lifecycle GHG emissions reduction default values for different biofuels and production pathways in 2022 (LUC and ILUC derived GHG emissions are included)

In a quick comparison between the two major pieces of biofuels legislation some observations can be made:

- The amount of biofuels required in the next 10 years will be considerable, but looking only at the additional volumes needed above today's production the goals seem achievable in the total. However, the expectations on second generation (2G) biofuels may not materialize in such a short time. The USA ethanol and biodiesel productions in 2009 were 41 billion liters and 2.1 billion liters, respectively, and in EU these volumes were 3.6 and 8.9, respectively (REN21, 2010).
- The installed capacity in USA was already 12 billion gallons/year of ethanol (EPA, 2010). In the EU, the 10% renewable energy participation in transport is estimated to be divided in 5.6% for biofuels (in 2008 biofuels already represented 3.3% of transport fuel use) and 4.4% for other renewable alternatives, mainly electricity, what will demand a significant improvement in the electric vehicle (EV) technologies to reduce the costs from the present values.
- The threshold limit of 50% for the GHG emission reduction, including the LUC/ILUC impacts, will not be easily satisfied by most grain ethanol and oil seed biodiesel with pathways using fossil fuels. Biomass fuel will help to improve the GHG performance of the biofuels, but it is constrained by cost and availability in large scale and there will be competition for feedstock for 2G plants. Sugarcane has a tremendous benefit in this respect since there are large amounts of residues in the distillery (bagasse that is the residue from the juice extraction operation) quite enough to supply all the energy needed to operate the distillery and to generate surplus electricity for sale; with the collection of the agricultural residues (trash: sugarcane tops and leaves) and the surplus bagasse it is possible to have a 2G plant operating integrated with the 1G distillery with synergies that will result in lower investment and operating costs as well as higher GHG emission reduction.

- Some type of biofuel sustainability certification criteria will be formally introduced in the two regions including other aspects beyond GHG emission reduction capabilities such as local environmental impact (air, soil and water), protection of biodiversity, avoided use of land with high carbon stock, social impacts and others.

3. LUC/ILUC impacts

An important component of the total lifecycle emission of GHG of biofuel production/use chain is likely to come from the impacts of LUC and ILUC. To put the subject under a right perspective the differences between the two effects must be made clear: direct land use change is the change in land use that occurs within the system boundary when the feedstock is planted replacing an existing land use (pasture, other crop, forest, etc.); indirect land use change takes place when agricultural production displaced by the biofuel feedstock crop will take place in other area, or even in other country, in its turn, displacing an existing land use. This concept exists for many years but it came to the spotlights with two articles published in 2008 in *Science* (Searchinger and al. and Fargione and al., 2008); these articles had the merit to bring the concept to a broader discussion, but on the negative side they presented results for LUC/ILUC GHG emissions that were unreasonably high as a consequence of inadequate data and assumptions used (see Table 4 at the end of this section). It was a case similar to the famous publication of Thomas Malthus in 1798 *An Essay on the Principle of Population* where he predicted that the world population would starve in the future because it increased in a geometric rate while the food supply grows according to an arithmetic rate; his calculations were corrected, but his hypotheses were not, since he did not considered the agricultural yield growth and other technological improvements.

GHG emissions resulting from land use change are related to soil carbon stocks loss (or gain when the impact is positive) and N_2O soil emissions due to fertilizer decomposition, change in soil carbon stock, residues decomposition and other. There are emission factors suggested by the IPCC, but real values are difficult to estimate because they depend on soil and climate conditions, agricultural practices (type of fertilizer and way it is applied) and previous use of the soil, since soil carbon stocks change slowly. The modeling of land use change uses normally economic based model of the computable general equilibrium or partial equilibrium concepts, sometimes coupled with optimization models, but several other types of model are available for this application in spite the fact that they were developed for other uses and have to be adapted to analyze the land use dynamics (CBES, 2009). A literature review concerning the impact of land use change on greenhouse gas emissions from biofuels was prepared by the DG Energy for the European Commission (DG Energy, 2010) and has shown that, although scientific progress has been made, consensus is still far from being reached. Some of the critical issues identified are: land use data, a fundamental part of the LUC modeling, is very poor and unreliable; there are some confusion with respect to handling crop yields variation and multi-cropping intensity; elasticities between increase in demand and improving yields are difficult to quantify empirically; rotation of land in and out of crop production leads to erroneous classification of land use type; how the biofuel feedstock is determined and the co-products credits are allocated and the corresponding impacts on land demand are not clearly explained; and, last but not least, it is not only how much land will be converted that matters, but also what type of land since this has a strong implication on the emissions due to the fact the carbon stock (above and below the ground) vary with the type and location of soil and present land use.

In this last issue, it is critical the share of forest/woodland converted to crops, considering the high carbon stocks and the impacts on biodiversity and other environmental services. It was also observed from the review that carbon stocks had significant variation among the studies, even for the same type of land (sometimes by a factor of 15) and the dynamics of pasture use for livestock production is poorly understood.

Considering the complexity of the land use impacts and the lack of consensus on how they can be estimated, the US Department of Energy (USDOE) Biomass Program sponsored a workshop on May 11 to 14, 2009, with more than 50 experts from around the world, to review the state of science, identify opportunities for collaboration, prioritize the next steps for research and discuss the data needed in terms of availability and quality. The focus was selected to be the interface between land use changes and global economic models; the main finding was that there was a need to improve current generation of land use change models and the central limiting problems were the historical data on land use (not land cover) that are frequently nonexistent or available only in a very coarse scale, and the poor understanding of the driving factors of LUC. Initial land use change drivers (cultural, technical, biophysical, political, economic and demographic) usually change in time and location, a condition not handled by the models; data from different sources with varying quality and high level of aggregation just add more uncertainties to the modeling. Other important drivers are governance capacity, population growth, land tenure regimes, macroeconomic and trade policy, environmental policy, infrastructure, land suitability, domestic and international agricultural and energy markets, and climate conditions; it seems unlikely that a single model, or even a combination of different models, could handle quantitatively all these drivers and produce consistent and replicable results. The most recent important tentatives to estimate the LUC/ILUC impacts have used economic equilibrium models (general or partial equilibrium) oriented to agriculture associated with spatially explicit land use models and optimization models. A combination of models tends to increase the scope of the analyses, but bears the risk of increasing the uncertainties due to error propagation from one model to the other. A final conclusion of the workshop was that there was a strong agreement among the participants regarding the uncertainties surrounding current use of global economic models to project the land use change effects of biofuels.

Nevertheless, both the EISA and EU Renewable Energy Directive require that the ILUC impacts be included in the lifecycle analysis of GHG emissions of the biofuels, immediately in the case of EISA and in a near future in the EU Directive.

EPA was in charge of managing the EISA mandate and produced the necessary studies and analyses leading to the RFS2 Final Rule in terms of default values for different biofuels and production pathways shown in Table 2 above (EPA, 2010). The international ILUC GHG emission values for corn ethanol and sugarcane ethanol were established as 30.3 gCO_2eq/MJ and 3.8 gCO_2eq/MJ, respectively.

The European Commission has not come to a final decision about the values of GHG emissions resulting from ILUC effects from the production of biofuels, but several reports have been submitted to public consultation in 2010 (IFPRI, DG Energy, 2010 and JRC, 2010a and 2010b) covering several aspects of the problem. The work of the International Food Policy Research Institute (IFPRI, 2010), that seems to be the main document, used a modified version of the MIRAGE model (Modeling International Relationships in Applied General Equilibrium) to analyze the impact of the increase of biofuel consumption in the EU due to the requirements of the EU Renewable Energy and Fuel Quality Directives. The baseline was determined assuming the biofuel share in transport fuel in 2008 (3.3%) would remain constant from 2009 to

2020, and the Directive scenario assumes that in 2020 the first generation biofuel will represent 5.6% of the total transport fuel demand. The modeling considered that biofuels will compete in the international market considering three alternatives: same situation as today in terms of import duties and other barriers (Business as Usual Scenario), global free trade regime and free trade with the MERCOSUR. Some results are presented in Table 3 for the Business as Usual Scenario (BAU), without considering the peatland drainage for the production of palm oil, for the marginal (considers the effects of the new production of biofuels disregarding the past average values) Indirect Land Use and marginal Net Emission Reductions by the production and use of ethanol and biodiesel from different feedstocks.

Biofuel	ILUC Emissions	Net Emissions Reductions (*)
Ethanol - sugar beet	16.1	-35.9
Ethanol - sugarcane	17.8	-54.0
Ethanol – maize	54.1	3.6
Ethanol - wheat	37.3	-7.0
Biodiesel – palm oil	46.4	-22.0
Biodiesel – rapeseed	53.0	8.8
Biodiesel – soy bean	74.5	24.1
Biodiesel - Sunflower	59.9	8.7

Source: IFPRI, 2010
(*) Negative values mean emission reduction with respect to the fossil fuel displaced

Table 3. Marginal indirect land use emissions and marginal net emission reductions from the production and use of biofuels (gCO2eq/MJ, 20 years lifecycle)

Analyzing the data presented in Tables 2 and 3 for the first generation biofuels it can be concluded that:

- Only sugarcane ethanol qualifies as an advanced biofuel according to RFS2 rules (50% minimum emission reduction) and can meet the requirements of the EU Directive for 2017 (50% minimum emission reduction), and with a little improvement can meet also the 2018 requirement (60% minimum emission reduction).
- Palm oil biodiesel is the only alternative of this biofuel that can reduce the GHG emissions in the replacement of fossil diesel; however, it does not meet the emission reduction minimum threshold value of the EU Directive even for the initial value of 2013 (35%) for this value is only 24% (fossil fuel lifecycle emissions of 92 gCO2eq/MJ).
- All other biodiesel alternatives will increase the GHG emission compared with mineral diesel.
- Maize ethanol is the worst alternative of this biofuel in terms of GHG emission reduction, and even increases slightly the emissions compared with fossil gasoline, for the European Union case.
- Sugar derived ethanol have a better GHG abatement performance than the grain ethanol alternatives, but sugar beet ethanol can meet only the 2013 requirement.
- Ethanol, in general, is a better option to reduce GHG emission in transport than biodiesel.

With that said, it remains the question why so many countries are persisting with the idea to develop programs to promote biofuels with such a poor performance in terms of GHG abatement potential (grain ethanol and oil seed biodiesel)? The possible explanation is the intention to help the local agricultural sector and to reduce a little the oil imports. Although there is a large amount of uncertainties in the LCA of GHG emissions of biofuels in general, and the LUC/ILUC derived emissions in particular, these results are at least a qualitative indication that biofuels are not equal. Table 4 presents results from different sources including the extremely high value from Searchinger and co-authors (Searchinger et al., 2008), that are out of the range of the results from the other studies by the California Air Resources Board (CARB), US Environmental Protection Agency (EPA) and International Food Policy Research Institute (IFPRI).

Reference	Maize ethanol	Soybean biodiesel
Searchinger et al., 2008	156	165-270
CARB, 2009	45	63
EPA, 2010	47	54
IFPRI, 2010	54	75

Source: EC, 2010

Table 4. Land use change GHG emissions results from different works (gCO_2eq/MJ)

3.1 Land requirement for biofuel production

It is interesting to start to look the land availability situation around the world today and in the future to have a clear picture of how much and where there is land availability for this purpose. The second step will be to look what are the possible targets for biofuel production in the long term.

Doornbosch and Steenblik (2007) have made a good assessment of the land use and availability worldwide based on the work developed by the Food and Agriculture Organization of the United Nations (FAO) and the International Institute for Applied System Analysis (IIASA). The results indicate that around 440 million hectares (Mha) will be available by 2050 for rain-fed cultivation of energy crops. This figure considers that the land needed to feed the additional population (from 6.5 to 9 billion people), estimated in 200 Mha, 100 Mha to accommodated population growth (housing and infrastructure) and the preservation of forests is discounted from the total land available. Nearly all this land availability is concentrated in South and Central America and Africa, and is presently being used as grassland for livestock production; therefore, land use change will take place and pasture intensification will be needed, a fact that is already taking place in many regions in the world. It is important to notice that this 440 Mha represents less than 10% of the 5,000 Mha of land under management (1,500 Mha arable and 3,500 Mha grassland), but should be considered as an upper limit for land available for energy crops by 2050. Deforestation is a major concern, but the causes are very complex and poorly understood, varying in space and time, deserving the attention of the scientific community to develop science based cause-effect relationships; today, the problem is being treated more on the emotional and subjective basis. Nogueira, 2008, presented some data on deforestation rates in Brazil between 1988 and 2006 and the variation does not seem to correlate well with the increase in agricultural production. In summary, land for energy crops production is not unlimited and, therefore, the biofuel options that present higher yields have a clear advantage in this aspect. Table 5 presents the estimated yields of different biofuels/feedstocks where a wide variation can be observed, even for the same biofuel/feedstock produced in different regions.

Region-Biofuel	Feedstock	Yields 2005 (l/ha)	Yields 2050 (l/ha)
EU-ethanol	Wheat	2 500	3 420
EU-ethanol	Sugar beet	5 000	6 750
EU-biodiesel	Rapeseed	1 200	1 640
US/Canada-ethanol	Corn	3 000	4 040
US/Canada-biodiesel	Soybean/rapeseed	800	1 100
Brazil-ethanol	Sugarcane	6 800	9 160
Brazil-biodiesel	Soybean	700	1 100
Rest of world-ethanol	Sugarcane	5 500	8 480
Rest of world-ethanol	Grain	2 000	3 090
Rest of world-biodiesel	Oil palm	2 500	3 910
Rest of world-biodiesel	Soybean/rapeseed	1 000	1 570
World-ethanol	Ligno-cellulose	4 300	7 580
World-BtL biodiesel	Biomass	3 000	5 960

Source: IEA, 2008

Table 5. Biofuels yields for different feedstocks and production regions

Even considering that the effects of the co-products are not included in Table 5, it can be seen that there are significant differences among biofuels and feedstocks, and among different regions in the world, in terms of land requirement for biofuels production, that must be taken into consideration in deciding which alternatives should be implemented. Besides the competition with land for food/feed production, the biofuel yields affects heavily the LUC/ILUC derived GHG emissions, as seen in session 3 above, and have a significant impact on production costs (agricultural inputs and field operations are related to cropped area and not to crop production quantity) and biodiversity. Sugarcane appears again as the best option, now and in the future, and can compete in equal terms with second generation alternatives with respect to land demand, with the advantage that the technology is ready now and not in the future. The superiority of ethanol compared with biodiesel is also demonstrated and sugarcane ethanol is the winner; considering that sugarcane is produced in more than 100 countries it seems reasonable to expect that the dissemination of this biofuel alternative has some probability to succeed if the right approach and policies are used. Molasses, the byproduct of sugar production, seems to be the cheapest feedstock for bioethanol, although there are some uncertainty about its availability for this application since it is already widely used for several applications, such as, beverage production, cattle feed, other products from fermentation (lysine, glutamates, solvents, etc.).

4. Sugarcane ethanol: A case study for Brazil

It is interesting at this stage to use the information described in the previous section to make some simulations, using the Brazilian current and future conditions, to get a feeling of the impacts of the production and use of biofuels in general, and ethanol in particular, on the GHG emission reduction potential and land demand.

The Brazilian Government prepared and released the Agroecological Zoning (AEZ) of sugarcane (EMBRAPA, 2009) identifying 64.7 Mha of land available for rain-fed cultivation of sugarcane without significant impacts on food production, deforestation, biodiversity and protected areas. It is important to point out that these 64.7 Mha represent only 7.5% of the country's area, meaning that 92.5% of Brazil surface will not be used to produce sugarcane. A recent study by the Interdisciplinary Center of Energy Planning of the University of Campinas (Leite, 2009, Leite et al., 2009) tried to indentify the land demand and availability for the production of a volume of ethanol sufficient to displace 5% or 10% of the projected gasoline consumption in 2025; the socioeconomic impacts, necessity of investments in distilleries, cane fields and infrastructure were estimated. The assessment of the land needs for sugar production for internal and external markets was also included. These works will be the reference for resources demand calculations and technology improvements with impact on ethanol yields.

For the estimate of future biofuels consumption the values projected by IEA (IEA, 2009) for the Reference scenario for 2030 will be used instead of the original estimates made in the two works above (104 and 205 billion liters in 2025). That means 132 Mtoe of total biofuels of which ethanol represents 79%, resulting in an ethanol demand around 200 billion liters in 2030, comparable to the estimated value used in the studies by Leite and Leite et al., 2009.

To estimate the land required to produce that amount of first generation ethanol in 2030 it is necessary to estimate the yields for that date. Using IEA data as shown in Table 5 above, which represents 0.7%/year yield improvement for sugarcane ethanol, the yield in 2030 would be around 8 100 liters/ha, demanding some 25.7 Mha. These figures are very conservative since sugarcane yields have increased at a rate of approximately 1.6%/year in the recent past; Landell et al., a group of sugarcane breeders, have drawn a roadmap for sugarcane quality improvement resulting in average sucrose yield per hectare increasing at a rate a little above 1.4%/year starting from 12,150 kg/ha/yr in 2010; considering also gains in efficiency in the distillery leading to a global distillery efficiency of 90% by 2030, up from 85% in 2010, the resulting ethanol yield would be around 12,000 liters per hectare. The land required to produce the 200 billion liters in 2030 would be reduced to 17 Mha, representing only 26% of the 64.7 Mha indicated in the sugarcane AEZ and just a little more than 1% of the current world arable land (1 500 Mha).

The 200 billion liters of ethanol in 2030 would be displacing 134 billion liters of gasoline that would produce, using EPA data, some 400 million tonnes of CO_2. Using also the EPA estimate of 61% GHG emission reduction potential for the Brazilian ethanol indicated in Table 2 the GHG emission reduction would be 244 Mt CO_2eq (14.4 t CO_2eq/ha), or 3.5% of the total emissions in road transport estimated for that year (IEA, 2009). In the future, if the sugarcane residues (bagasse and straw) were better used in a 2G plant integrated with the 1G distillery an additional 3000 to 4000 liters of ethanol would be obtained (Leite et al., 2009), reducing the land demand to no more than 13 Mha and the saved GHG emissions would increase to 364 Mt CO_2eq or 5% of the road transport emissions.

Just to make a quick comparison with the alternative of US corn ethanol, using the IEA yields of 3 600 liters/ha (IEA, 2008) and the GHG emission reduction default value from EPA of 21% (EPA, 2010) the required area would be 55 Mha and the GHG emission savings of only 84 Mt CO_2eq, or just a little over 1% of the projected road transport emissions. It must be said that these estimates are good for qualitative comparison only, since there are many uncertainties that need to be resolved in the LCA GHG emissions of the biofuels production/use chain, specially related to the ILUC effects. Another point is that the ILUC derived emissions calculated by econometric models are not linear with respect to biofuel volume produced; therefore the use of EPA values for other volumes is a simplified approach.

More data on Brazilian sugarcane ethanol LCA GHG emissions can be found on Macedo et al, 2008 and Macedo and Seabra, 2008.

Other considerations concerning the sustainability of ethanol production in Brazil are not included here, but they can be found in several publications dealing specifically with this subject such as, Smeets et al., 2006, Macedo, 2007, Walter et al., 2008, Zuurbier and van de Vooren, 2008, Goldemberg et al., 2008, Oliveira, 2011.

5. Final comments

The presentation of a plentiful of data obtained from studies made by well recognized and reputable institutions and researchers had the aim of indicating significant differences among the biofuels alternatives considering only two of their main characteristics: GHG abatement potential and land demand. A third very important characteristic, the production cost, was not included in the effort to compare biofuels because it was outside of the scope of this chapter, but nonetheless it is the key characteristic for the long term survival of the biofuel option without subsidies.

GHG emission savings is a fundamental characteristic for attainment of the qualification status of the biofuel according to the two major legislations in effect today: the EU Renewable Energy Directive and the US Renewable Fuel Standard (RFS2), and therefore the LCA GHG emissions of a biofuel is a crucial characteristic to be taken into account in the process of selecting the best alternative. Needless to say that it should be the "go no go" test if the biofuel production and use is intended to mitigate the global warming effect when displacing fossil fuels. In spite all that, the methodology and procedures to perform the LCA of the GHG emissions in the production path still have several points that need improvements and definitions: climate active gases included, allocation methods to divide the LCA emissions among co-products, N_2O emissions, soil carbon dynamics and the indirect land use change impacts (ILUC). The soil emissions, one of the most complex point in the analyses, are highly dependent on the local conditions (climate, soil, agricultural practices, past history of land use) and there is an urgent need to improve and extend the few existing data bases on soil characteristics, land use past dynamics, agricultural practices (fertilizer use, tillage types, crop rotation, double cropping, etc.). Besides the improvement of the input data, the determination land use change indirect effects is another area that needs more research and development of the models to make them able to simulate the driving forces of land use (highly variable in time and space), the cause/effect relationship of crop dynamics (where the displaced crops really go and what caused the occupation of native vegetation), cattle grazing and many other things.

Land for agriculture is a finite resource and, therefore, the demand for biofuels production must be carefully considered and in this process the yields are the main point. Besides, the ILUC impact on the LCA GHG emissions are highly dependent on the land, as well as the production costs and energy demand are more related to the area cultivated than to the volume of feedstock produced (fertilizer, herbicide, land preparation, agriculture operations and land rental). This reasoning should lead to the selection of biofuels alternatives with higher yields, such as sugarcane and sugar beet ethanol and palm oil biodiesel, but the reality is quite different with the domination of ethanol from grains and biodiesel from rape seed. In 2008, according to UNEP, 2010, to world biofuel crop production used around 36 Mha, or 2.3% of the arable land, to produce 67 billion liters of ethanol and 12 billion liters of biodiesel (REN21, 2009). Using the yield values indicated in Table 5 and assuming all ethanol from US corn and all biodiesel from EU rapeseed the total area required would be 32 Mha, very close to UNEP value; in the case where all ethanol is produced from sugarcane in Brazil and all biodiesel from oil palm the total area required would be 15 Mha, or less than half of the previous case. The 36 Mha estimated by UNEP is an indication that very low yield options are being widely used around the world, in spite the dominance of USA and Brazil in ethanol and EU in biodiesel. In the IEA projections land demand projections for biofuels in 2050 an average value of 160 GJ/ha is used, including second generation biofuels; Leal, 2007, projected for 2020 the yield gains for Brazilian sugarcane ethanol to 7,900 l/ha for 1G ethanol and 11,700 l/ha (245 GJ/ha) for an integrated 1G and 2G production using sugarcane sugars (1G) and fibers (2G).

Other important characteristics were not included in the evaluations due to the limitation necessary to keep the chapter at a reasonable size and scope. It is important to point out that several other characteristics such as impacts on the local environment and biodiversity, as well as some of the main socioeconomic impacts are strongly related to the extension of the land required to produce the biofuel feedstocks. Therefore, biofuel yield, energy balance and GHG emission reduction potential are critical issues for most of the situations around the world, but there are some specific local conditions that take the priorities to other areas such as job creation, local energy supply and development and creation of outlet for some local production potential constrained by lack of market access due non existence of storage and distribution infrastructure. Different driving forces may lead to different optimal solutions.

The Brazilian experience with the efficient and economic production of sugarcane ethanol is available as a reference for countries interested to deploy a biofuel program, but it cannot be expected to be readily transferable to some of the more than 100 sugarcane producing countries due to significant differences in the local conditions, including technology access, land tenure issues, human resources, cultural aspects and strength of the different drivers.

In summary, to make sure that the negative impacts on land demand are minimized and the positive impacts on GHG emission reductions are maximized it is crucial to make the proper choices if biofuels are to play an important role in the future world energy scenario.

6. References

BP, 2010, British Petroleum Statistical Review of World Energy June 2009, www.bp.com/ statisticalreview.

CBES, 2009, Center for BioEnergy Sustainability, Oak Ridge National Laboratory, Land-use change and bioenergy: Report from the 2009 workshop, 76p.

Doornbosch, R. and R. Steenblik, 2007, Biofuels: Is the cure worse than the disease?, Organization for Economic Co-operation and Development Report SG/SD/RT (2007)3, Paris, 57p.

EC, 2010, European Commission, Report from the Commission on the indirect land-use change related to biofuels and bioliquids, COM(2010)811 final, Brussels, December 22, 2010.

EIA, 2010, US Energy Information Administration, EIA - Short Term Energy Outlook - April 2010.

EMBRAPA, 2009, Empresa Brasileira de Pesquisa Agropecuária, Zoneamento Agroecológico da Cana-de-Açúcar, Rio de Janeiro, RJ, 2009, 56p.

EPA, 2010, US Environmental Protection Agency, Renewable Fuel Standard Program (RFS2) Regulatory Impact Analysis, EPA-420-R-10-006, 1120p.

EPE, 2010, Energy Research Company (Empresa de Pesquisa Energética) - EPE, Brazilian Energy Balance 2010, year 2009, Rio de Janeiro, EPE, http://ben.epe.gov.br/ BENRelatorioFinal2010.aspx.

Fargione, J., J. Hill, D. Tillman, S. Polasky, and P. Hawthorne, 2008, Land Clearing and the biofuel carbon debt, Science 319, p 1235-1238.

Goldemberg, J., F.E.B. Nigro and S.T. Coelho, 2008, Bioenergia no Estado de São Paulo: Situação Atual, Perspectivas, Barreiras e Propostas, São Paulo Official Press, 2008, 151p.

IEA, 2008, Energy Technology Perspectives 2008 - Scenarios and Strategies to 2050, International Energy Agency, Paris, 2008.

IEA, 2009, World Energy Outlook 2009, International Energy Agency, Paris, 2009.

IFPRI, 2010, International Food Policy Research Institute, Global Trade and Environmental Impact Study of the EU Biofuels Mandate, Final Report prepared by Perrihan Al-Riffai, Betina Dinamaran and David Laborde, March 2010, 125p.

JRC, 2010a, Impacts of the EU biofuel target on the agricultural markets and land use: a comparative modeling assessment, Scientific and Technical Report by the Joint Research Council

JRC, 2010b, Indirect Land Use Change from increased biofuels demand, Scientific and Technical Report by the Joint Research Council

Larson, E.D., 2006, A review of lifecycle analysis studies on liquid biofuel systems for the transport sector, Energy for Sustainable Development, Vol. X, No. 2, June 2006, p 109-126.

Leal, M.R.L.V., 2007, The potential of sugarcane as an energy source, Proceedings of the XXVI International Society of Sugar Cane Technologists (ISSCT) Congress, July 31st to August 3rd, 2007, Durban, South Africa, p 23-34.

Leite, R.C.C, 2009, Bioetanol combustível: uma oportunidade para o Brasil (in Portuguese), CGEE , Brasilia, DF, Brazil, Coordinated by Rogério C.C. Leite, Brasília, DF, 536p.

Leite, R.C.C., M.R.L.V. Leal., L.A.B. Cortez, M. G. Griffin and M.I.G. Scandiffio, 2009, Can Brazil replace 5% of the 2025 gasoline world demand with ethanol?, Energy 34 (2009), 655-661.

Macedo, I.C. (organizer), 2007, Sugar cane's energy - Twelve studies on Brazilian sugar cane agribusiness and its sustainability, 2nd edition, São Paulo: Berlendis & Vertecchia: UNICA, 2007.

Macedo, I.C.,J.E.A. Seabra, and J.E.A.R Silva,2008, Green house gases emissions in the production and use of ethanol from sugarcane in Brazil: the 2005/2006 averages and a prediction for 2020, Biomass & Bioenergy 32, 582-595, 2008.

Macedo, I.C. and J.E.A. Seabra, 2008, Mitigation of GHG emissions using sugarcane bioethanol, in Sugarcane ethanol: Contributions to climate change mitigation and the environment, Wageningen Academic Publishers, The Netherlands, 95-111, 2008.

Menichetti, E. and M. Otto, 2009, Energy balance and greenhouse gas emissions of biofuels from life-cycle perspective in R.W. Howarth and S. Bringezu (eds) Biofuels: Environmental Consequences and Interactions with Changing Land Use, Proceedings of the Scientific Committee on Problems of the Environment (SCOPE) International Biofuels Project Rapid Assessment, 22-25 September 2008, Gummersbach, Germany, Cornell University, Ithaca, NY, USA, Ch. 5, p 81-109.

Nogueira, L.A.H., 2008, Sugarcane-Based Ethanol: Energy for Sustainable Development, report prepared for the Banco Nacional de Desenvolvimento Econômico e Social (National Bank for the Economic and Social Development) and Centro de Gestão e Estudos Estratégicos (Center for Strategic Studies and Management in Science, Technology and Innovation), Rio de Janeiro, 2008, 304p.

Oliveira, J.G., 2011, Indicadores Socioeconômicos em Estados Produtores de Cana-de-Açúcar: Análise Comparativa Entre Municípios, Doctoral Thesis, College of Mechanical Engineering, University of Campinas, Campinas, 2011, 202p.

Searchinger, T., R. Heimlich, R.A. Houghton, R.A., F. Dong, A. Elobeid, J. Fabiosa, S. Tokgoz, D. Heyes, and T.H. Yu, 2008, Use of US croplands for biofuels increases greenhouse gases through emission from land use change, Science Express, February 7th, 2008.

Smeets, E., M. Junginger, A. Faaij, A. Walter and P. Dozan, 2006, Sustainability of Brazilian Bio-ethanol, Report of the Copernicus Institute, University of Utrecht, for SenterNovem, The Netherlands, 2006,107p.

UNEP, 2010, United Nations Environment Programme, Towards Sustainable Production and Use of Resources: Assessing Biofuels, key authors Stefan Bringezu, Helmut Schütz, Meghan O´Brien, Lea Kauppi, Robert W. Howarth and Jeff McNeely, Paris, 2010, 119p.

Walter, A., P. Dozan, O. Quilodrán, J. Garcia, C. da Silva, F. Piacente and A. Sergerstedt, 2008, A Sustainability Analysis of the Brazilian Ethanol, Report of the University of Campinas to the UK Embassy in Brasilia, Campinas, November 2008, 167p.

Zuurbier, P. and J. van de Vooren (editors), 2008, Sugarcane ethanol: Contributions to climate change mitigation and the environment, Wageningen Academic Publishers, The Netherlands, 2008, 255p.

15

Contribution of the Atmospheric Chlorine Reactions to the Degradation of Greenhouse Gases: CFCs Substitutes

Iván Bravo[1], Yolanda Díaz-de-Mera[2], Alfonso Aranda[2], Elena Moreno[2] and Ernesto Martínez[2]
[1]*Instituto de Ciencias Ambientales (ICAM), University of Castilla-La Mancha, Toledo*
[2]*Departamento de Química Física, Chemistry Faculty*
University of Castilla-La Mancha, Ciudad Real
Spain

1. Introduction

During the last few decades it has been shown that the use and dispersion of chemical compounds emitted from anthropogenic sources, firstly considered as innocuous, have dramatic effects on the global Atmosphere. The adverse environmental impacts of chlorinated hydrocarbons on the Earth's ozone layer have focused attention on the effort to replace these compounds by non-chlorinated substitutes with environmental acceptability. Although new materials have been developed for a large number of applications, a comprehensive solution remains to be found. Therefore, many provisional applications, using chemicals with unknown effects, are still currently found such as, refrigerants, foam agents, flame inhibitors, solvents, propellants, anaesthetics, etc [see for example: 3M; EPA; IPCC; Shine, 2010].

Hydrofluoroethers (HFEs) have been introduced as ozone friendly alternatives in many instances such as, refrigeration, electronic equipment, carrier fluids for lubricant deposition, and fire suppression (EPA). HFEs contain no chlorine and, thus, have ozone depletion potentials of essentially zero. One of the principal advantages of the HFE structure has been determined to be the significantly shorter atmospheric lifetimes, when compared to HFCs (hydrofluorocarbons) and PFCs (perfluorocarbons) (IPCC). However, the presence of the C-O bond, together with C-F bonds in the hydrocarbon molecule, enhance the absorption features in the atmospheric infrared window. In other words, HFEs are absorbers of infrared radiation, thus raising concern about their possible roles as greenhouse gases. Thus, it is necessary to improve our knowledge about lifetimes and global warming potentials (GWP) of these compounds in order to get a complete evaluation of their environmental impact.

To provide an accurate evaluation of the global warming potentials, the lifetimes must first be obtained. The atmospheric lifetimes of pollutants is generally calculated on the basis of the reaction rates with OH only (Kurylo & Orkin, 2003), assuming that the reaction rates are independent of temperature. This is not suitable for chemicals with low reactivity. As a relatively homogeneous vertical distribution in the troposphere is expected, to a large extent, the losses of such chemicals take place at temperatures which are significantly lower than 298K. Thus, for reactions with relatively high activation energy, E_a, neglecting the temperature dependence of the kinetic rate constants may lead to an underestimation of the corresponding lifetimes. In this regard, lifetimes 2.5 times longer were found for several hydrofluoro(poly)ethers when the temperature dependence was considered (Myhre et al., 1999). When calculating OH-based lifetimes, the use of 272K as an average tropospheric temperature, and methyl chloroform (CH_3CCl_3), as a chemical of well known sources and sinks, has been suggested (Spivakovsky et al., 2000) to minimize the errors resulting from neglecting the specific temperature dependences.

Generally, HFEs show low surface sticking coefficients and low water solubility. Thus, primary removal of HFEs in the troposphere will mainly be initiated by reaction with OH radicals. Although global atmospheric abundance of OH radicals is around 2 orders of magnitude greater than that of chlorine atoms, Cl reactions are generally faster than OH reactions, $k_{Cl}/k_{OH} \geq 10$, so their contribution to the degradation of organic compounds may be not negligible compared to the role of OH (Finlayson-Pitts & Pitts, 2000). The contribution of Cl to the oxidation of HFEs could be significant in areas where the concentration of Cl precursor species has been reported to be high, such as the coastal boundary layer (Spicer et al., 1998).

The influence of the tropospheric temperature profile on the Cl rate constants has been studied and reviewed for many halocarbons (IUPAC, NASA). Recently, this has been done for HFEs as well. As it has been shown for OH reactions, the understanding of the kinetic rate constants as a function of temperature is required to properly evaluate the contribution of Cl reactions to the degradation of HFEs. The use of the rate constants at only 298K tends to overestimate the global degradation rates of both OH and Cl reactions, given the decrease of T with altitude. The degree of overestimation may be different for OH and Cl depending on the specific value of E_a. The data on temperature dependence are thus crucial to quantify the absolute roles of OH and Cl, and their relative contributions.

In this work we will report the results obtained in the absolute kinetic study of the reactions of Cl atoms with different CFC substitutes (four segregated HFEs), at temperatures ranging from 234-343K, thus providing useful data to simulate the temperature profile characteristic of the troposphere.

$$C_4F_9OC_2H_5(\text{HFE}-7200) \rightarrow \text{Products} \tag{1}$$

$$C_4F_9OCH_3(\text{HFE}-7100) \rightarrow \text{Products} \tag{2}$$

$$C_3F_7OCH_3(\text{HFE}-7000) \rightarrow \text{Products} \tag{3}$$

$$n-C_3F_7CF(OC_2H_5)CF(CF_3)_2(\text{HFE}-7500) \rightarrow \text{Products} \tag{4}$$

To conclude, we will discuss some different strategies that can be used to design CFC substitutes with low environmental impact. For this, computational chemistry offers an alternative to the experimental procedures currently used to assess environmental compatibility parameters such as, lifetimes, reaction mechanism or GWP. In the present work, we will evaluate the radiative ability, and hence the contribution to Global Warming, of the HFE-7500, using a recently reported theoretical method based on computational techniques.

Thus, in the Experimental Section we describe the experimental method used in this work. In the Results Section we describe the experimental conditions and we obtain the values for the rate constants for all the studied reactions at different temperatures, driving to the Arrhenius' expression for each compound. Furthermore, we present a study of the products of the reactions, obtaining the branching ratio for the abstraction channel for each one. In the Discussion Section, we compare the results obtained in this work with previous studies, we discuss the reactivity of the studied compounds taking into account the number of $-CF_2-$ groups in the structure, and, finally, we compare the ionization potential versus the k values for segregated and no segregated HFEs. In the Atmospheric Implications Section, we discuss the atmospheric implications of the studied reactions from the calculus of the lifetimes and GWP for the CFCs substitutes. Finally, in Section 6 (Strategies to design CFC alternatives with low environmental impact: The scope of the computational chemistry), we show and discuss the results obtained for the radiative efficiency of HFE-7500 using new computational techniques.

2. Experimental section

The experimental method used in this work (figure 1) (Aranda et al., 2006; Díaz-de-Mera et al., 2008, 2009), is the absolute discharge flow-mass spectrometry. It incorporates a dual-stage molecular beam system for the sampling. The mass spectrometer was equipped with an electron-impact ion source and a Chaneltron electron-multiplier. The energy level of the ionizing electrons was E_e=40 eV. Typical pressures in the first chamber and in the chamber hosting the mass spectrometer were below $1x10^{-6}$ and $\approx 1x10^{-8}$ Torr, respectively. Both radical and molecular species were fed from the reactor to the first high vacuum chamber through a stainless steel cone (250 μm orifice diameter). They were then channelled through a second stainless steel cone (1000 μm hole diameter) into the mass spectrometer vacuum chamber, as a molecular beam.

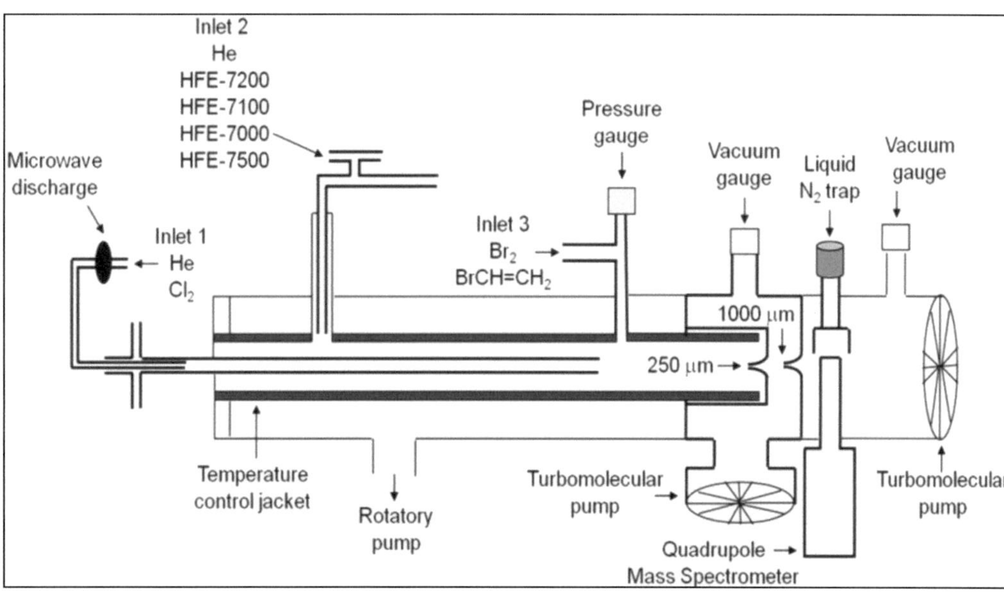

Fig. 1. Schematic view of the experimental set-up.

Cl atoms were produced by flowing mixtures of Cl_2 and He through a microwave discharge joined to the main reactor, inlet 1. The discharge tube was coated with phosphoric acid to increase the Cl_2 dissociation yield. To reduce the wall losses of Cl atoms, the inner surfaces of the reactor and the injector were coated with halocarbon wax. HFEs were added through inlet 2 and the reactions with the Cl radicals were observed downstream at the end of the axial injector.

All reactants were diluted in helium and stored in bulbs of known volume. For some experiments, where concentrations of HFEs (7200 y 7100) had to be enhanced, the reactants were used without dilution in helium, directly from the storage bulb. In order to assure constant and accurate HFE concentrations, their flows were regulated with mass flow controllers. The direct detection of the organic compounds was not possible since the mass spectrometer is only able to detect masses just below 200 amu. However, the signals found at *m/e*=131, 69, 120, and 69 for HFE-7200, HFE-7100, HFE-7000, and HFE-7500, respectively, showed good intensity and no overlap with the peaks of the rest of the species.

Molecular chlorine was detected at its parent peak *m/e*=70 and the absolute concentration of Cl atoms was measured by titration with $BrCH{=}CH_2$ in excess, and subsequent mass spectrometric detection of $ClCH{=}CH_2$ at *m/e*=62 and $BrCH{=}CH_2$ at *m/e*=106 (Park et al., 1983):

$$Cl + BrCH = CH_2 \rightarrow ClCH = CH_2 + Br \qquad k = 1.4x10^{-10} cm^3 molecule^{-1} s^{-1} \qquad (5)$$

During a kinetic run (for HFE-7200, HFE-7100, and HFE-7000), the remaining chlorine was observed indirectly as BrCl at *m/e*=116 by scavenging Cl atoms with Br_2 (Aranda et al., 2003). Br_2 was introduced in excess at the end of the reactor through inlet 3, to ensure the complete consumption of Cl atoms by Br_2 (Bedjanian et al., 1998):

$$Cl + Br_2 \rightarrow BrCl + Br \qquad k = (2.3 \pm 0.4)x10^{-10} exp[-135 \pm 60) / T]\ cm^3 molecule^{-1} s^{-1} \quad (6)$$

For HFE-7500, the remaining chlorine was followed indirectly as $ClCH{=}CH_2$ (at *m/e*=62) by scavenging Cl atoms with $BrCH{=}CH_2$ (reaction 5). Following both procedures, the detection limit for Cl atoms was $9x10^{11}$ molecule cm^{-3}.

Reagents

Liquid compounds were purified by trap-to-trap distillation. The chemical used were: He (Praxair, 99.999%), Cl_2 (Praxair, >99.8%), Br_2 (Fluka, ≥99.5%), $BrCH{=}CH_2$ (Aldrich, 98%), $ClCH{=}CH_2$ (Fluka, ≥99.5%), HCl (Aldrich, >99%), HFE-7200 (3M Novec, >99%), HFE-7100 (Fluka, ≥99%), HFE-7000 (3M Novec, >99%), HFE-7500 (3M Novec, >99%).

3. Results

The experimental conditions for four HFEs studied reactions are shown in table 1.

All the kinetic runs were carried out at 1 Torr total pressure in the reactor and under pseudo-first order conditions with the organic compound in excess over Cl atoms. Preliminary experiments were conducted in which the reactions between Cl_2 and Br_2 (or $BrCH{=}CH_2$) with HFEs were evaluated. No reaction was observed within the time used in the experiments. Homogeneous losses (Cl-self reaction) did not contribute to the observed temporal profiles because of the low radical concentration (from $0.8x10^{11}$ to $3.0x10^{11}$ molecule cm^{-3}).

Heterogeneous wall losses of chlorine atoms were checked in additional experiments at all the studied temperatures. These experiments were carried out in the absence of organic compounds, but under similar conditions to those of a kinetic run. In such experiments, with Cl atoms entering the main tube from the injector and Br_2 (or $BrCH=CH_2$) entering from inlet 3, the formation of BrCl (or $ClCH=CH_2$) was observed at different contact times. The mean value obtained for the wall loss rate constants was k_w = 8, 6, 8, and 7 s^{-1} for HFE-7200, HFE-7100, HFE-7000, and HFE-7500, respectively.

Experimental conditions	HFE-7200	HFE-7100	HFE-7000	HFE-7500
T(K)	234-333	234-315	266-333	253-343
P (Torr)		1		
Flow velocity (m s^{-1})	600-850	600-800	700-900	650-950
Reaction time (ms)	0-25	0-45	0-43	0-35
[Cl_2] (10^{11} molecule cm^{-3})	6-25	2-20	5-6	2.5-7
[Br_2] (10^{13} molecule cm^{-3})	0.9-50	0.9-50	2.5-8	
[$BrCH=CH_2$](10^{13}molecule cm^{-3})				3-7
[HFE](10^{14}molecule cm^{-3})	0.04-2.5	0.8-7.0	0.15-2.1	0.03-0.45
[Cl] (10^{11} molecule cm^{-3})	0.8-1.8	0.8-1.8	2-3	1.1-2
Mixing time for Cl/He (ms)		1.1-0.6		

Table 1. Experimental conditions in the kinetic study of HFEs with Cl atoms

For the bimolecular reaction between Cl and HFEs, the integrated rate constant that applies to our experimental conditions is

$$\text{Ln}\,[\text{Cl}]_t = \text{Ln}\,[\text{Cl}]_0 - k'\,t \qquad \text{(I)}$$

where k' is the pseudo-first-order kinetic rate constant, k' = k [HFE] + k_w. k_w again, represents the heterogeneous wall losses of Cl in the injector. Typical pseudo-first-order decays of Cl (measured as BrCl or $ClCH=CH_2$), against time are shown in figure 2 for reaction (3) and (4). Similar plots are obtained for HFE-7200 and HFE-7100 reactions.
The pseudo first-order constant values, k', obtained for the slope, were corrected to take into account the axial and radial diffusion of Cl atoms (Kaufman, 1984) by:

$$k' = k'_{exp}\left(1 + \frac{k'_{exp}D}{v^2} + \frac{k'_{exp}r^2}{48D}\right) \qquad \text{(II)}$$

where v is the linear flow velocity of the gas mixture in the reactor (cm s^{-1}), r is the radius of the reactor (cm) and D is the effective diffusion coefficient (cm^2 s^{-1}). The effective diffusion coefficients of the Cl in He mixture were calculated from the atomic diffusion volumes (Perry et al., 2001). The values obtained within the temperature range used (234-343K) were (330-680), (330-615), (415-680), and (380-650) $cm^2\ s^{-1}$ for HFE-7200, HFE-7100, HFE-7000, and HFE-7500, respectively. Corrections in k′ from diffusion were less than 20, 10, 5, and 8% for HFE-7200, HFE-7100, HFE-7000, and HFE-7500, respectively.

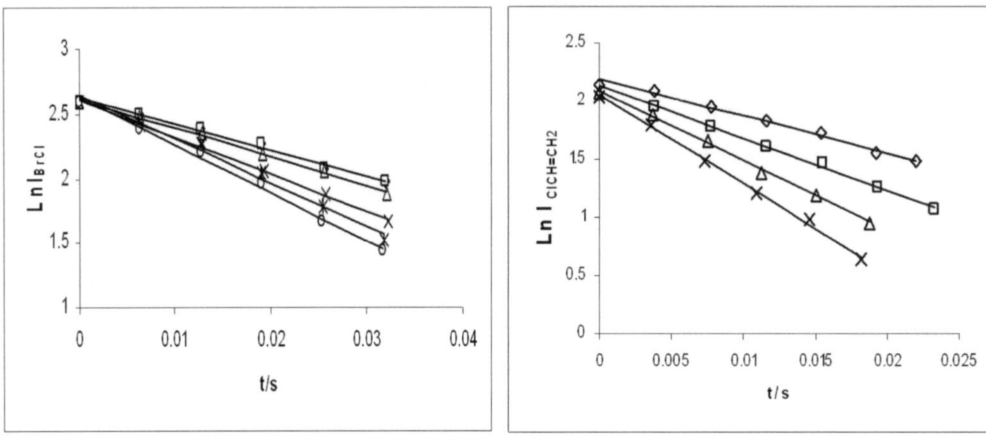

(a) (b)

Fig. 2. Typical pseudo first-order decays for Cl for the reaction of a) HFE-7000+Cl at 298K and 1 Torr: [HFE-7000]=0.66 (□); 0.95 (▵); 1.35 (x); 1.85 (∗); 2.10 (○) x10^{14} molecule cm^{-3}. b) HFE-7500+Cl at 298 K and 1 Torr: [HFE-7500]=0.53 (◊); 0.73 (□); 1.42 (▵); 2.09(x) x10^{13} molecule cm^{-3}

HFE-7200		HFE-7100		HFE-7000		HFE-7500	
T(K)	k	T (K)	k	T(K)	k	T (K)	k
234	1.0 ± 0.1	234	1.7 ± 0.7	266	11.7 ± 2.2	253	1.1 ± 0.1
266	1.5 ± 0.1	263	4.6 ± 1.1	273	12.2 ± 2.3	273	1.5 ± 0.1
298	2.1 ± 0.1	273	5.5 ± 0.9	285	13.1 ± 2.2	298	2.2 ± 0.3
315	2.5 ± 0.2	285	7.3 ± 1.5	298	12.4 ± 2.5	307	2.4 ± 0.2
333	3.1 ± 0.4	298	14.3 ± 2.8	315	13.8 ± 2.3	324	2.9 ± 0.3
		315	17.6 ± 2.0	333	16.8 ± 2.6	333	3.3 ± 0.4
						343	3.9 ± 0.5

Table 2. Summary of the second-order rate constants at different temperatures for HFEs+Cl reactions. k in units of 10^{-12} cm^3 $molecule^{-1}$ s^{-1} for HFE-7200 and HFE-7500, and of 10^{-14} cm^3 $molecule^{-1}$ s^{-1} for HFE-7100 and HFE-7000. Errors are 2σ

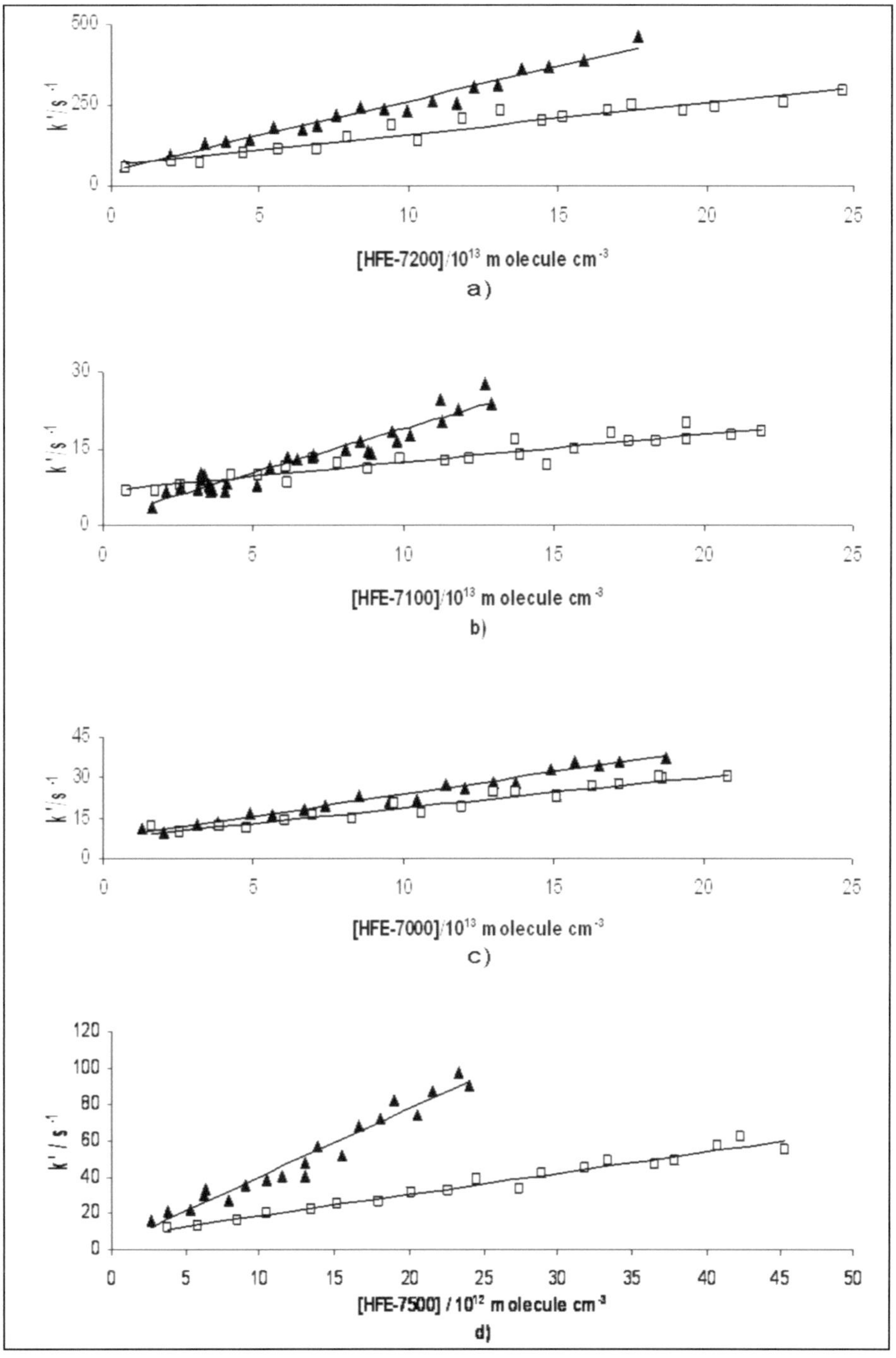

Fig. 3. Plots of the pseudo-first-order rate constants, k′, against segregated HFEs concentrations, at 1 Torr. a) HFE-7200+Cl = (▲) 298 K and (□) 234 K; b) HFE-7100+Cl = (▲) 315 K and (□) 263 K; c) HFE-7000+Cl = (▲) 333 K and (□) 266 K; d) HFE-7500+Cl = (▲) 333 K and (□) 253 K

The second order rate constant was calculated by plotting the pseudo-first-order constant against the HFE concentration and applying weighted least-squares fittings as shown in figure 3. At all temperatures the intercepts agree well with the Cl wall losses measured in the absence of HFEs. Table 2 summarizes the results for all the experimental conditions.
The reaction rate constants were found to increase with increasing temperature for reactions (1) to (4). The Arrhenius equation has been used to fit the rate constant-temperature data:

$$k = A\,e^{-\frac{E_a}{RT}} \tag{III}$$

Using logarithms:

$$Ln\,k = Ln\,A - \frac{E_a}{RT} \tag{IV}$$

Plotting Ln k vs. 1/T, the linear weighted, least-squared analyses of the data, yields the activation energy, the pre-exponential factor (errors are 2σ), and allows the calculation of the kinetic rate constant in the studied temperature range at 1 Torr total pressure as shown in figure 4:

$$k(1) = (3.7 \pm 0.5)\text{x}10^{-11}\exp[-(852 \pm 38)/T]\ \text{cm}^3\text{molecule}^{-1}\text{s}^{-1} \quad T = 234 - 333\text{K}$$
$$k(2) = (2.3 \pm 1.4)\text{x}10^{-10}\exp[-(2254 \pm 177)/T]\ \text{cm}^3\text{molecule}^{-1}\text{s}^{-1} \quad T = 234 - 315\text{K}$$
$$k(3) = (6.1 \pm 3.8)\text{x}10^{-13}\exp[-(445 \pm 186)/T]\ \text{cm}^3\text{molecule}^{-1}\text{s}^{-1} \quad T = 266 - 333\text{K}$$
$$k(4) = (1.2 \pm 0.4)\text{x}10^{-10}\exp[-(1186 \pm 88)/T]\ \text{cm}^3\text{molecule}^{-1}\text{s}^{-1} \quad T = 253 - 343\text{K}$$

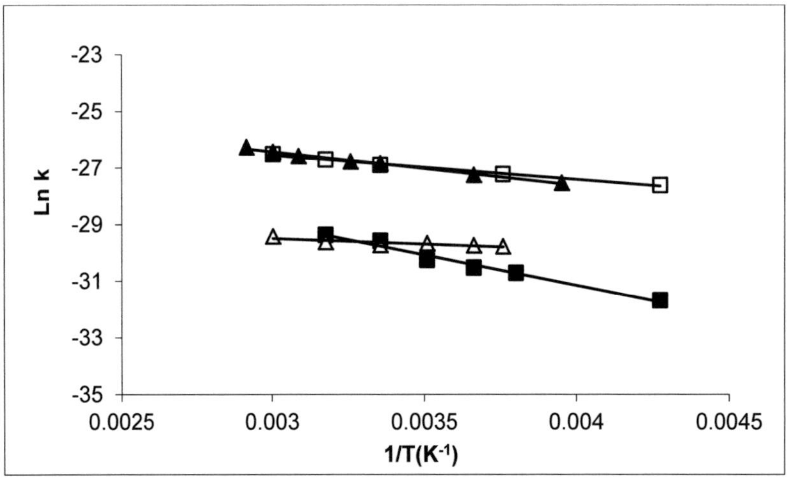

Fig. 4. Temperature dependence of the rate constant for: Cl+HFE-7200 (□), Cl+HFE-7100 (■), Cl+HFE-7000 (△), and Cl+HFE-7500 (▲) reactions at 1 Torr total pressure.

Further experiments were also conducted to identify the products of reactions (1) to (4) using higher concentrations of the reactants, in order to enable the detection of possible weak signals. These experiments were carried out at 298K, 1 Torr, and in the absence of Br_2

(or $BrCH=CH_2$) to avoid secondary chemistry. The expected reaction mechanism is the abstraction of an H atom to form HCl and the corresponding radical:

$$C_4F_9OC_2H_5(HFE-7200) + Cl \rightarrow C_4F_9OC_2H_4 + HCl \quad (1)$$

$$C_4F_9OCH_3(HFE-7100) + Cl \rightarrow C_4F_9OCH_2 + HCl \quad (2)$$

$$C_3F_7OCH_3(HFE-7000) + Cl \rightarrow C_3F_7OCH_2 + HCl \quad (3)$$

$$\begin{aligned} &n-C_3F_7CF(OC_2H_5)CF(CF_3)_2(HFE-7500) + \\ &Cl \rightarrow n-C_3F_7CF(OC_2H_4)CF(CF_3)_2 + HCl \end{aligned} \quad (4)$$

The masses of the expected radicals, $C_4F_9OC_2H_4$, $C_4F_9OCH_2$, $C_3F_7OCH_2$, and n-$C_3F_7CF(OC_2H_4)CF(CF_3)_2$ exceed the mass range of the mass spectrometer so they could not be confirmed. The scan for masses up to 200 amu only revealed the formation of HCl whose signals (*m/e*=36 and 38) increased with the time of reaction. No other new peak was observed probably due to the fact that radicals $C_4F_9OC_2H_4$, $C_4F_9OCH_2$, $C_3F_7OCH_2$, and n-$C_3F_7CF(OC_2H_4)CF(CF_3)_2$ may undergo ionization patterns similar to those of their preceding species. The detection of HCl and the positive activation energy obtained for reactions (1) to (4) are consistent with t he expected reaction mechanism, the hydrogen atom abstraction.
Additional experiments were carried out at 298 K and 1 Torr total pressure to measure the yield on HCl of reactions (1) to (4) following the next procedure. To avoid residual contributions present in signals at *m/e*=36 and 38, first, Br_2 (or $BrCH=CH_2$) in excess was added together with the corresponding HFE (inlet 2), completely removing Cl atoms (giving BrCl or $ClCH=CH_2$) and avoiding the reaction between Cl and HFE. The residual signals at *m/e*=36 and 38 were obtained under such conditions. Then, Br_2 (or $BrCH=CH_2$) was changed from inlet 2 to inlet 3, enabling the Cl+HFE reaction and the formation of HCl. Under such conditions Cl is also lost in the reactor's wall. Commercial HCl was used to prepare samples of known concentration to obtain the corresponding calibration plot signal intensity /concentration. The absolute HCl concentrations obtained during the Cl-HFE reactions were, thus, calculated from the HCl signals and the calibration data. The commercial mixtures of HCl were prepared and flowed from time to time testing the signal for a constant flow to the reactor. The intensity of the *m/e* signal remained constant showing the stability of HCl in the storage bulb and glass tubing. No observable heterogeneous wall losses of HCl in the reactor were found. Table 3 shows experimental conditions for these experiments.

Conditions	HFE-7200+Cl	HFE-7100+Cl	HFE-7000+Cl	HFE-7500+Cl
[HFE] (molecule cm^{-3})	$1.3x10^{14}$	$1.2x10^{15}$	$6.6x10^{14}$	$3.5x10^{13}$
[Br_2] / [$BrCH=CH_2$] (molecule cm^{-3})	$(0.7-1.5)x10^{14}$	$(0.7-1.5)x10^{14}$	$(0.96-1.4)x10^{13}$	$\approx 3x10^{13}$
Reaction Time (ms)	>45	>45	>50	>50
$[Cl]_0$ (molecule cm^{-3})	$(1-9)x10^{11}$	$(0.5-7)x10^{11}$	$(1.3-9.1)x10^{11}$	$(2.5-8.5)x10^{11}$

Table 3. Experimental conditions for the quantification experiments. $[Cl]_0$ was determined by the titration reaction with $BrCH=CH_2$ as described in the Experimental Section. Reaction Time is the time used in the experiments, corresponding to >99% of conversion.

For different Cl initial concentrations ($[Cl_0]$) introduced into the reactor and in presence of the HFE, HCl signal produced in reactions (1) to (4) were followed at *m/e*=36 and 38. Figure 5 shows, as an example, the plots for HFE-7200+Cl and HFE-7100+Cl reactions.

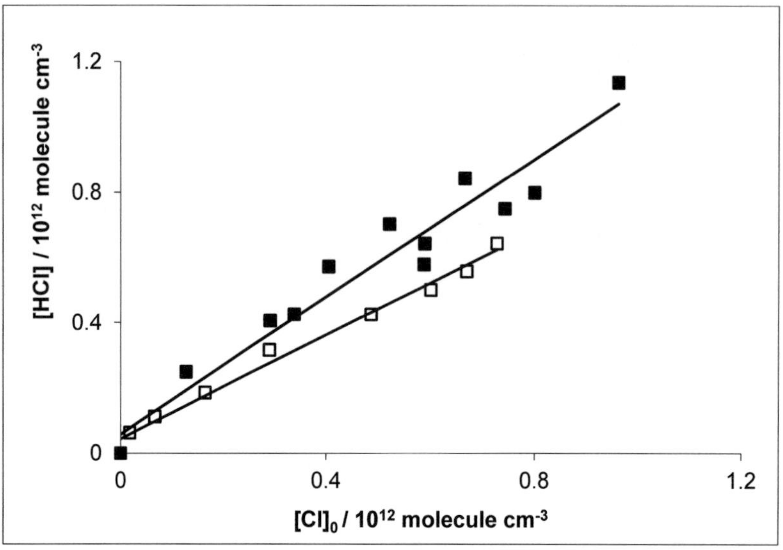

Fig. 5. Yield on HCl. HCl produced against initial Cl atoms concentration at 298 K and 1 Torr total pressure for Cl+HFE-7200 (■) and Cl+HFE-7100 (□) reactions.

The branching ratio for HCl formation was obtained from the slope and taking into account the competitive losses of Cl in the reactor's wall:

$$\frac{[HCl]}{[Cl]_0} = \frac{k_{abstraction}\,[HFE]}{k\,[HFE] + k_w} \tag{V}$$

Where k is the global kinetic rate constant reported in table 2 (considering the total losses of Cl due to reactions with the HFEs), $k_{abstraction}$ is the rate constant for the pathway giving HCl, and k_w is the Cl wall losses constant. The obtained results at 298 K and 1 Torr were $k_{abstraction}/k$ = 0.95±0.10 for reaction (1), 0.88±0.09 for reaction (2), 0.95±0.38 for reaction (3), and 0.98±0.02 for reaction (4) (errors are 2σ). These results confirm that the studied reactions quantitatively proceed through H-abstraction mechanism to form HCl and the corresponding radical. Thus, reactions (1) to (4) are expected to be independent of pressure conditions and the results obtained in this work may apply also to atmospheric pressure conditions.

During the kinetic studies, organic radicals $C_4F_9OC_2H_4$, $C_4F_9OCH_2$, $C_3F_7OCH_2$, and n-$C_3F_7CF(OC_2H_4)CF(CF_3)_2$ could contribute to regenerate Cl atoms through the reaction with Cl_2 present in the reactor (the dissociation efficiency in the microwave discharge was <100% leaving undissociated Cl_2):

$$C_4F_9OC_2H_4 + Cl_2 \rightarrow C_4F_9OC_2H_4Cl + Cl \tag{7}$$

$$C_4F_9OCH_2 + Cl_2 \rightarrow C_4F_9OCH_2Cl + Cl \tag{8}$$

$$C_3F_7OCH_2 + Cl_2 \rightarrow C_3F_7OCH_2Cl + Cl \tag{9}$$

$$n-C_3F_7CF(OC_2H_4)CF(CF_3)_2 + Cl_2 \rightarrow n-C_3F_7CF(OC_2H_4Cl)CF(CF_3)_2 + Cl \tag{10}$$

If reactions (7) to (10) were very fast and Cl_2 concentrations in the experiments were high, the net regeneration of Cl would be important and would drive to measured rate kinetic constants lower than real value. To check this possible influence some experimental runs were carried out under the experimental conditions show previously and introducing additional Cl_2 through inlet 2 together with HFEs. The measured rate constants were the same as those obtained in the experiments without additional Cl_2 from inlet 2. Furthermore, as shown previously, the yields for HCl remained constant for large reactions times, also supporting the conclusion that regeneration of Cl through reactions (7) to (10) must be negligible under our experimental conditions.

4. Discussion

In table 4 we report the previous studies for reactions (1) to (4) with the obtained results in this work. Taking into account error limits, the results obtained under low-pressure conditions in this work are in good agreement with those obtained in relative experiments at atmospheric pressure (700 Torr) and room temperature for all the studied HFEs.

Our samples of both HFE-7200 and HFE-7100 were a mixture of two isomers, however in the studies of Christensen et al. (1998) and Wallington et al. (1997), the authors had access to pure samples of n-HFE-7200, n-HFE-7100, i-HFE-7200, and i-HFE-7100 and could study their reactions with chlorine atoms separately. They found no discernible difference in reactivity showing that kinetic rate constant for these isomer mixtures are expected to be independent of composition. These results are important in order to study different commercial mixtures of HFEs.

Comparing the reactivity with Cl for different HFEs of the same series, for example, HFE-7000 ($CF_3CF_2CF_2OCH_3$), and HFE-7100 ($CF_3CF_2CF_2CF_2OCH_3$) belonging to $C_nF_{2n+1}OCH_3$ series, we can conclude that there is almost no difference in the reactivity when a $-CF_2-$ group is introduced in the chain [$(1.24\pm0.25)x10^{-13}$ and $(1.43\pm0.28)x10^{-13}$ cm^3 $molecule^{-1}$ s^{-1} for HFE-7000 and HFE-7100, respectively]. Besides the studies showed in table 4, Christensen et al. (1999) studied the reaction with Cl for n=1 [$(1.4\pm0.2)x10^{-13}$ cm^3 $molecule^{-1}$ s^{-1}] and reactions for n=2, 3, and 5 were studied by Nohara et al. (2001) [$(1.1\pm0.14$, 1.18 ± 0.14, and $1.03\pm0.14)x10^{-13}$ cm^3 $molecule^{-1}$ s^{-1}, respectively]. Taking into account the error limits, the results presented in this work are in good agreement with previous studies for this series of ethers, and show that the kinetic rate constant are almost independent of the number of $-CF_2-$ groups in the perfluorated chain. However, the kinetic rate constants are very sensitive to the length of the hydrocarbon chain [$(1.43\pm0.28)x10^{-13}$, and $(2.1\pm0.1)10^{-12}$ cm^3 $molecule^{-1}$ s^{-1} for HFE-7100 ($C_4F_9OCH_3$) and HFE-7200 ($C_4F_9OC_2H_5$), respectively]. For the same perfluorated chain, the kinetic rate constant increases with the number of $-CH_2-$ groups in the HFE molecule, what is expected because increases the number of H atoms which can be attacked by Cl.

Furthermore, if we compare the reactivity for HFE-7500 and HFE-7200, considering both of them as ROC_2H_5 [$(2.3\pm0.7)x10^{-12}$, and $(2.1\pm0.1)x10^{-12}$ cm^3 $molecule^{-1}$ s^{-1}, respectively], we can see that there is almost no change in the reactivity of this series when the number of $-CF_2-$ is changed, even if the per-fluorinated chain is ramified (HFE-7500). This conclusion can be very useful because changing R, we can obtain fluorinated compounds with a particular physico-chemical properties for a specific use without altering their atmospheric reactivity.

Reaction	T (K)	k(298 K) (cm³ molecule⁻¹s⁻¹)	E_a/R (K)	P (Torr)	References
HFE-7200 + Cl	234-333	$(2.1 \pm 0.1)x10^{-12}$	852±38	1	This work
	296	$(2.7 \pm 0.6)x10^{-12}$		700	[Christensen, 1998] Relative
HFE-7100 + Cl	234-315	$(1.43 \pm 0.28)x10^{-13}$	2254±177	1	This work
	298	$(0.97 \pm 0.14)x10^{-13}$		700	[Wallington, 1997] Relative
HFE-7000 + Cl	266-333	$(1.24 \pm 0.25)x\ 10^{-13}$	445±186	1	This work
	298	$(1.18 \pm 0.14)x10^{-13}$		700	[Nohara, 2001] Relative
	295	$(0.91 \pm 0.13)x10^{-13}$		700	[Ninomiya, 2000] Relative
HFE-7500 + Cl	253-343	$(2.22 \pm 0.28)x10^{-12}$	1186 ± 88	1	This work
	298	$(2.3 \pm 0.7)x10^{-12}$		700	[Goto, 2002] Relative

Table 4. Rate constants for the reaction between the Cl radical and the title compounds. Errors are 2σ

Any H atom in the aliphatic chain is susceptible to an oxidant attack. Generally, the radicals (OH, NO_3, Cl, etc.) will tend to abstract the most weakly bound hydrogen atom in the molecule (Seinfeld & Pandis, 1998). Unfortunately, our mass spectrometer was not able to provide the experimental direct evidence of the Cl reaction on the CH_2 or the CH_3 group of HFE-7200, and HFE-7500. In the study of Christensen et al. (1999) they showed that Cl predominantly attack the CH_2 group.

The reactivity on the CH_3 terminal group is, thus, well isolated from changes in the perfluorated chain by the ether link. This is also clear if we compare $k_{HFE\text{-}7100}$ (298K) with the result obtained for the reaction of Cl with CH_3CF_3, k_{298K}=2.6x10^{-17} cm³ molecule⁻¹ s⁻¹ (IUPAC). Thus, the presence of the fluorinated chain gives them the physico-chemical behaviour for industrial or domestic use while the CH_3- or -CH_2CH_3 groups remain relatively reactive toward the atmospheric radicals (due to the ether linkage), so driving to relatively low lifetimes and mitigating the contribution as greenhouse gases.

The results obtained for E_a/R (852, 2254, 445, and 1186 K for reactions (1) to (4), respectively) are very similar to the obtained for other reactions of Cl with HFEs (Kambanis et al., 1998; Papadimitriou et al., 2004). For example, Papadimitriou et al. (2004) obtained E_a/R values of 929 and 1112 K for reactions of Cl with $CHF_2CF_2OCH_3$ and $CF_3CHFCF_2OCH_3$, respectively. Also, for many halocarbons with high Cl or F substitution, E_a values are the same order of magnitude to that obtained in this work (IUPAC): E_a/R = 2000, 2420, and 3720 K has been obtained for reactions of Cl with CH_3CFCl_2, CH_3CF_2Cl, and CH_3CF_3 (IUPAC).

Segregated and no segregated HFEs

The significant decreases in the lifetimes of segregated HFEs compared to non segregated HFEs is attributed to the direct activating effect of the oxygen on the contiguous carbon with CH bonds. In general, the strength of the C-H bonds in hydrofluoroethers depends on the interplay of two counteracting electronic effects: a) the strengthening due to the electron-withdrawing inductive effects of F and/or O atoms through σ-bonds, and b) the weakening of the adjacent C-H bonds due to the π-electron transfer from F or O atom to the central C atom (Papadimitriou et al., 2004).

On the other hand, the inductive effect caused by an F atom is usually more important than their conjugative effect, contrary to the O atom. Taking into account both aspects, we can say that if the O atom is directly bonding to C of C-H bond, this O atom will produce a conjugative effect very important, decreasing the bond strength and so, enabling the abstraction of the H atom. In terms of stability we can say that the conjugative effect of O becomes more stable the C supporting the odd electron after the H abstraction. The inductive effect of F atoms directly bonding to C-H removes electronic density destabilizing the possible radical, and so, causing more difficult the breaking of C-H bond.

The strength of the more labile bond in a molecule can be indirectly measured by means of their ionization potential (IP) as shown in figure 6, in which we can distinguish three different behaviours. In the first group (left in the plot), we found the no fluorinated ethers, like $CH_3CH_2OCH_2CH_3$. In these ones, the absence of F atoms and the weakness C-H bonds drives to low IP and high k values.

In the right side of the plot are located the no segregated HFEs because they present high IP and very low k values. This behaviour is in agreement with the described above about the inductive and conjugative effects. Finally, the segregated HFEs are located in the intermediate zone of the plot. This type of HFEs presents all the F atoms in one side of the ether group and the H atoms to the other side. This situation is similar to the hydrogenated ethers, where the conjugative effect of the O is very important in order to make the breaking of the C-H bonds easier. Furthermore, the inductive effect of $-CF_2-$ and $-CF_3$ groups over the C-H bonds are attenuated by the fact that these groups are separated of the hydrocarbon chain. This attenuation of the fluorocarbon chain can be appreciated comparing HFE-7000 ($C_3H_7OCH_3$) and HFE-7100 ($C_4H_9OCH_3$). These compounds have similar k values with a different fluorocarbon chain and the same hydrocarbon chain. The HFE-7200 presents an IP value similar to the rest of HFEs segregated and a higher k value because it has a high number of H atoms. The abstraction of H from $-CH_2-$ group will be easier than from the $-CH_3$ group due to the conjugative effects of O over the $-CH_2-$ group.

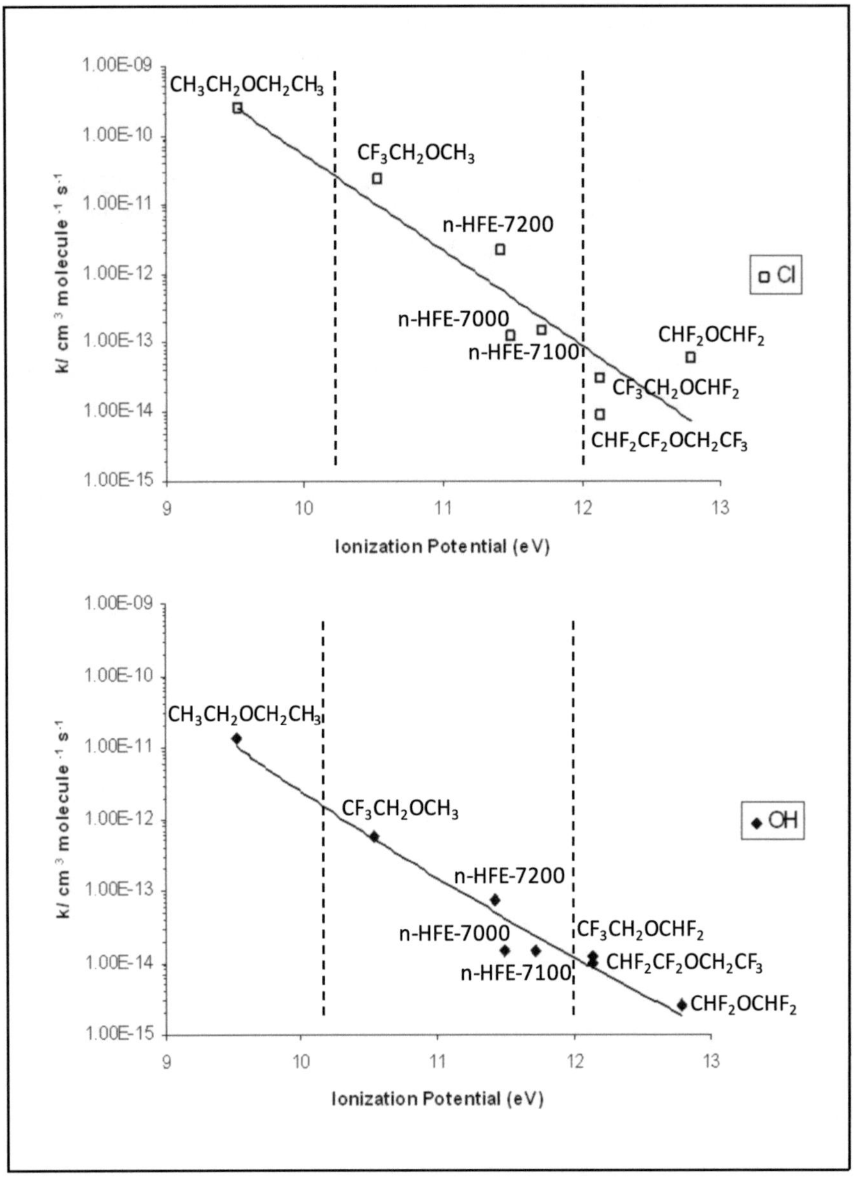

Fig. 6. a) Kinetic rate constants of some HFEs with Cl atoms versus their IP. k(HFE-7200), k(HFE-7100), and k(HFE-7000) from this work; k($CH_3CH_2OCH_2CH_3$) from Notario et al. (2000); k($CF_3CH_2OCH_3$), k(CHF_2OCHF_2), and ($CF_3CH_2OCHF_2$) from Kambanis et al. (1998); k($CHF_2CF_2OCH_2CF_3$) from Papadimitriou et al. (2004). b) Kinetic rate constants of some HFEs with OH atoms versus their IP. k(HFE-7200), k(HFE-7100), and k(HFE-7000) from this work; k($CH_3CH_2OCH_2CH_3$) from Mellouki et al. (1995); k($CF_3CH_2OCH_3$) from Oyaro & Nielsen (2003); k(CHF_2OCHF_2) from Orkin et al. (1999); k($CF_3CH_2OCHF_2$) from Zhang et al. (1992); k($CHF_2CF_2OCH_2CF_3$) from Chen et al. (2003). IP of $CH_3CH_2OCH_2CH_3$ and $CF_3CH_2OCH_3$ are experimental data from Bowen & Maccoll (1984) and Molder et al. (1983). The rest of IP is from Papadimitriou et al. (2004).

5. Atmospheric implications

An estimation of the gas-phase lifetime for organic compounds may be obtained for their reactions towards the tropospheric agents. Since oxidative processes against OH radical are the major route of elimination in most of cases, normally the lifetimes are calculated against this OH radical by means of:

$$\tau_{OH} = \frac{1}{k_{OH}[OH]} \quad \text{(VI)}$$

However, this equation does not take into consideration the errors due to the vertical temperature profile of the troposphere. Thus, lifetimes estimations for CFCs substitutes are generally calculated on the basis of gas-phase removal by OH only and with methyl chloroform (MCF) as reference:

$$\tau_{OH} = \frac{k_{OH}^{MCF}(272K)}{k_{OH}(272K)} \tau_{OH}^{MCF} \quad \text{(VII)}$$

where τ_{OH} and τ_{OH}^{MCF} [τ_{OH}^{MCF} =5.99 year (Kurylo & Orkin (2003))] are the lifetimes of a given compound and MCF, respectively, due to the reactions with hydroxyl radical in the troposphere only. $k_{OH}(272K)$ and $k_{OH}^{MCF}(272K) = 6.0 \times 10^{-15} cm^3$ $molecule^{-1}$ s^{-1} (Kurylo & Orkin, 2003) are the rate constants for the reactions of these compounds with OH at 272K. The use of 272K in place of 298K overcomes the problems associated with the use of temperature dependent OH reaction and the errors are minimized compared to estimates using 298K.
Reactions with Cl atoms, and their dependence with temperature, can be especially relevant because, as described in the Introduction, Cl reactions are generally faster than OH reactions and high Cl atoms concentrations have been observed in the marine boundary layer. This fact can significantly affect the mean lifetimes. However, for Cl the transport models are not so developed and its vertical distribution in the troposphere remains rather uncertain, so, an equation similar to (VII) is not available for Cl based lifetimes, τ_{Cl}. Thus, to enable a comparison with OH, τ_{Cl} is estimated from the data obtained in this work at 272K using:

$$\tau_{Cl} = \frac{1}{k_{Cl}[Cl]} \quad \text{(VIII)}$$

In table 5 are shown the lifetimes for the studied CFCs substitutes in this work.
In the context of estimating the climate impact of the emissions of these gases, a fundamental parameter is the radiative forcing per unit concentration change, or radiative efficiency (RE); this measures the change in the Earth's radiation balance for a 1 ppbv increase in concentration of the gas. RE values for the studied compounds are included in table 5. The global warming potential (GWP) is one method for calculating the carbon-dioxide equivalent of a 1 kg emission of a gas—it takes into account both the lifetime and the RE of a gas. It is the radiative forcing of an emission of 1 kg at time zero, integrated over some given time horizon, divided by the same value for a 1 kg emission of carbon dioxide. The 100 year GWP is used within the Kyoto Protocol of the United Nations Framework Convention on Climate Change to place emissions on a common scale and IPCC (IPCC) regularly reports 20, 100 and 500 year GWP values for a large number of gases. Table 5 includes the values for these parameters obtained from the global lifetimes (considering OH and Cl degradation) calculated in this work.

As can be seen in table 5, it is clear that the use of rate coefficients determined at 298 K leads to underestimates of the lifetimes by up to a factor of two, which have knock-on effects on the determination of GWPs.

Considering the atmospheric lifetimes, we can see that the HFEs studied would be scavenged mainly by OH radicals. However, it is necessary take into consideration the degradation via Cl radicals because τ_{global} can be considerably modified when we take into account the Cl reactions. Also, under local conditions as in coastal regions or in the marine boundary layer in the early hours where Cl concentrations can be high, the elimination of these compounds via Cl reactions can be even more important than OH reactions.

τ_{global} for HFEs studied are small compared with CFCs. Thus, their degradation processes take place mainly in the troposphere and their transport to the stratosphere is lower than for the CFCs. However, τ_{global} are large enough (>0.5 years) to ensure proper vertical distribution in the troposphere and to minimize the possible isolated smog episodes due to rapid oxidation in the lower troposphere.

Despite having high values for RE, due to the lots of C-F bounds present in these compounds, the obtained values for GWP_{100} are relatively low (<500) compared with CFCs (several miles). Thus, the four HFEs could affect the radiative balance because of the high RE values, but their short lifetimes lead to short-term effect only.

	$C_4F_9OCH_2CH_3$ *HFE-7200*	$C_4F_9OCH_3$ *HFE-7100*	$C_3F_7OCH_3$ *HFE-7000*	*n*-$C_3F_7CF(OC_2H_5)CF(CF_3)_2$ *HFE-7500*
[a]τ_{OH}(298K)	[g]0.4	[h]2.1	[i]2.1	[j]1.2
[b]τ_{OH}(272K)	0.9	4.2	4.6	
[c]τ_{Cl}(272K)	3.9	109.5	53.4	4.32
[d]τ_{global}(272K)	0.73	4.08	4.21	1.05
[e]RE	[f]0.43	[f]0.38	[f]0.34	[k]0.37
[f]GWP_{20}	269	1403	1618	213
[f]GWP_{100}	76	401	463	61
[f]GWP_{500}	23	122	141	18

[a]τ_{OH} (298K) has been calculated by means equation (VI) using k(298K), where [OH] = 1×10^{6} molecule cm^{-3} (Prinn et al., 2001).

[b]τ_{OH} (272K) has been calculated by means equation (VII) with τ_{OH}^{MCF} =5.99 year and $k_{OH}^{MCF}(272K)$ =6.0×10^{-15} cm^3 $molecule^{-1}$ s^{-1}.k values from Bravo et al. (2010)

[c]τ_{Cl} (272K) has been calculated by means equation (VIII) using k(272K) obtained in this work, where [Cl] = 5×10^{3} molecule cm^{-3} (Pszenny et al., 1993; Wingenter et al., 1996; Spicer et al., 1998).

[d]$\tau_{global}^{-1} = \tau_{OH}^{-1} + \tau_{Cl}^{-1}$

[e]In units of W m^{-2} $ppbv^{-1}$

[f]From Bravo et al. (2010)

[g]k(298K) = 7.3×10^{-14} cm^3 $molecule^{-1}$ s^{-1} (Bravo et al., 2010)

[h]k(298K) = 1.5×10^{-14} cm^3 $molecule^{-1}$ s^{-1} (Bravo et al., 2010)

[i]k(298K) = 1.5×10^{-14} cm^3 $molecule^{-1}$ s^{-1} (Bravo et al., 2010)

[j]k(298K) = 2.6×10^{-14} cm^3 $molecule^{-1}$ s^{-1} (Goto et al., 2002)

[k]From Goto et al. (2002)

Table 5. Several atmospheric parameters for the HFEs studied.

6. Strategies to design CFC alternatives with low environmental impact: The scope of the computational chemistry

The availability of the relationship between molecular structure and the atmospheric oxidation mechanism is the main key in order to determine and design environmentally innocuous materials. The molecular structure can be easily modified to get the desired physical and chemical properties such as thermodynamic behaviour, stability, toxicity, lifetime or radiative properties. For instance, the inclusion of H atoms in the molecular structure is an environmentally advantage since it makes the molecule more reactive against the atmospheric oxidants like OH, Cl, or NO_3. However, an increase of the number of H atoms increases the flammability of the species. On the other hand, an increase on the number of Cl or F atoms increases the lifetimes. Besides, F atoms drive to negligible ozone depletion potential (ODP) parameters compared to Cl atoms. At the meantime, F atoms promote the ability of the molecule to absorb infrared radiation in the atmospheric windows (800-1400 cm^{-1}) what increase the GWPs.

Generally, the studies of the environmental parameters that determine the compatibility of the new CFC alternatives have been undertaken by direct measurement of the compounds' infrared (IR) absorption spectra, kinetic behavior against the tropospheric oxidants (lifetimes) or product distribution and mechanistic studies [see for example: Sihra et al., 2001; Bravo et al., 2010] . From these measurements the radiative forcing of the species is determined, which together with the atmospheric lifetime, then allows an assessment of its GWP. But there are a huge number of molecules which may have industrial or other uses, and it would require a massive investment in time and money to carry out all the measurements required.

Recent studies have shown up that the correct use of computational techniques might be the key to sort out this problem, being a very important tool for the design of CFC alternatives with low environmental impact. In this way, recent researches have indicated that it is possible to calculate infrared spectra using *ab initio* and DFT (Density Functional Theory) methods with useful accuracy, and that radiative transfer models can then be applied to these spectra to determine radiative efficiencies and hence GWPs [Papasavva et al., 1997; Blowers et al., 2007; Bera et al., 2010; etc..].

In the method performed by Bravo et al. (2010b) theoretical spectra for a set of perfluorocarbons were determined using DFT methods. Then, the radiative efficiencies (REs) were determined using the method of Pinnock et al. (1995) and combined with atmospheric lifetimes from the literature to determine global warming potentials (GWPs). Theoretically-determined absorption cross sections were within 10% with experimentally determined values. They found that the calculated RE is extremely sensitive to the exact position of the C-F stretch at around 1250 cm^{-1} and the raw calculated frequencies cannot be used directly in radiative transfer models. Thus, they used a combination of theoretical and experimental results to obtain a very precise correction to the band position generated directly from the DFT calculations.

As an example, here we used this method to predict the RE of HFE-7500, which experimentally-determined value is summarized in table 5. In figure 7 we can see an schematic view of this procedure, where the cross section spectra of HFE-7500 has been performed using Gaussian 03 software package at B3LYP/6-31G** level of theory.

The computed wavenumbers were corrected following the expression $\bar{\nu}_{scal} = 0.977\ \bar{\nu}_{calc} + 11.664\ cm^{-1}$ to obtain scaled wavenumbers, where $\bar{\nu}_{calc}$ is the calculated vibrational mode wavenumber and $\bar{\nu}_{scal}$ is the empirically-corrected value. These modes can be assumed to be Gaussian in shape and here we use a full with of 14 cm^{-1} to simulate the complete infrared spectrum of HFE-7500. Broadly speaking, the wavenumber position and integrated cross sections are then used to calculate the (instantaneous) REs using the simple Pinnock et al. (1995) method. In this method the raditaive forcing function describes the radiation able to get the Earth's surface evaluated over the tropopause. Using this approaching we found a REs of 0.55 and 0.43 W m^{-2} $ppbv^{-1}$ for the 0-2500 and 900-1900 cm^{-1} wavenumber intervals, respectively. The previous literature value measured for the 900-1900 cm^{-1} interval was reported by Goto et al. (2002), and it is 16% lower than the predicted here, 0.37 vs 0.43. However, differences within 14-25% of existing experimental values provide a valuable data for the REs in order to calculate accurate GWPs values (Blowers et al., 2007; Bravo et al., 2010b). Another advantage of using computational techniques to predict REs and hence GWPs, is the possibility of evaluate the cross-section spectrum over the overall infrared spectral interval, 0-2500 cm^{-1}, since the range of 0-700 cm^{-1} is difficult to measure using commercial infrared spectrometers. This wavenumber range is particularly important due to the radiative forcing function has a maximum there as is illustrated in Figure 7. This effect can be observed in our calculation over the HFE-7500 where the RE increase around 28% whether we include the 0-900 cm^{-1} interval in the RE calculation, 0.55 vs 0.44 W m^{-2} $ppbv^{-1}$. Apart from the used on the prediction of radiative properties of molecules, computational techniques have successfully been used to establish reaction pathway in chemical mechanisms along with the predictions of atmospheric kinetic rates and hence lifetimes with relatively good accuracy [see for example: Rodríguez et al, 2010; Garzón et al., 2010].

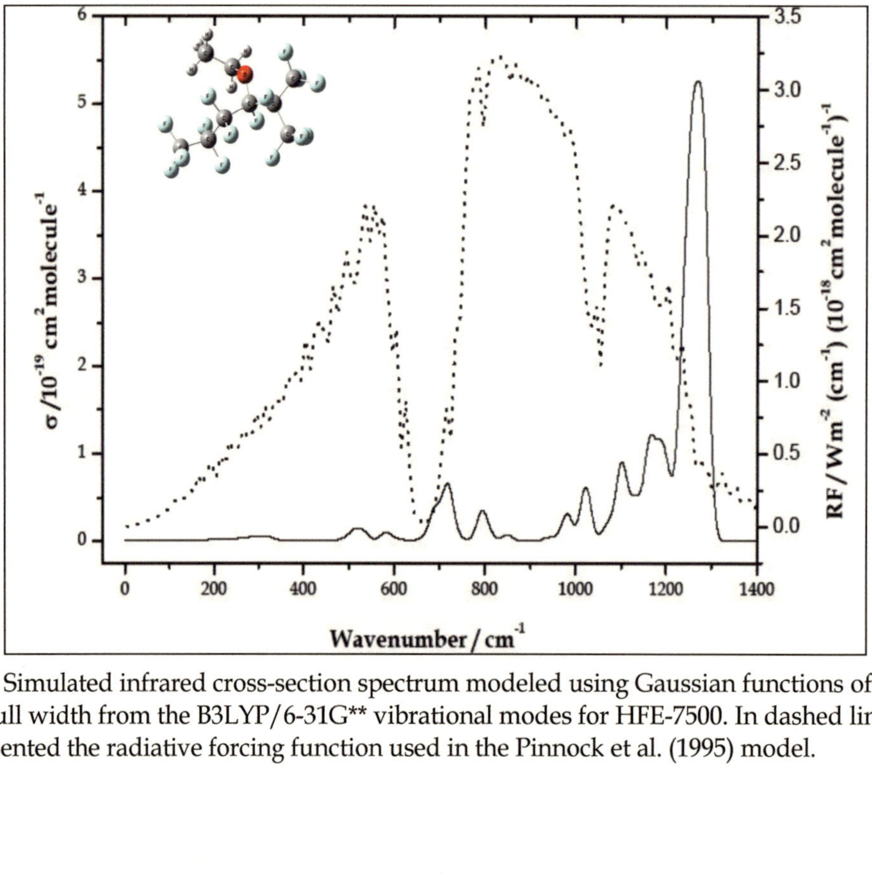

Fig. 7. Simulated infrared cross-section spectrum modeled using Gaussian functions of 14 cm^{-1} full width from the B3LYP/6-31G** vibrational modes for HFE-7500. In dashed lines is represented the radiative forcing function used in the Pinnock et al. (1995) model.

7. Conclusion

Rate coefficients as a function of temperature have been determined for the reactions of Cl with a range of HFEs. The room-temperature data are in good agreement with previous measurements obtained using different techniques and under different conditions. The branching ratio for the abstraction channels of the studied reactions has been determined showing that these reactions proceed almost exclusively via this channel. Using the RE values for these compounds and combining these data with the kinetic data (k values) allows the determination of their GWPs, which are considerably smaller than those for the CFCs that they have been manufactured to replace.

Taking into account the atmospheric aspects and leaving aside the health aspects, we can conclude that segregated HFEs with chemical structures similar to those studied in this work present *a priori* an acceptable environmental compatibility and they can be good substitutes for CFCs: They have a nule contribution to the ozone depletion, a minimum contribution to the smog formation and a low contribution to the greenhouse effect both medium and long term, and a moderate contribution to a short term.

On the other hand, computational techniques are an important and handy key to predict the environmental behavior of new compounds in the atmosphere. Combining different methodology that include the use of physical and chemical software, levels of theory and basic sets, we will be able to calculate environmental parameters such as REs, GWPs or lifetimes. As an example, in this work we have determined a theoretical RE value for HFE-7500, which is in good agreement with previous experimental measurements. This means that when a new compound is proposed to replace a CFC in a determined application because of they have similar physicochemical properties, a right use of these techniques will warns us important information about their environmental behavior. Such information might be very useful for the industry in order to go through with the manufacturing processes. There are several examples where apparently environmentally-safe species have been manufactured and then wrongly used in industrial application. For instance, this is the case of several perfluorocarbons and some hydrofluorocarbons which have been used to replace CFCs in several applications since they do not contain Cl atoms in the structure but they contribute strongly to the global warming. The use of these computational techniques might avoid such wrong uses.

8. Acknowledgment

This work was supported by the Spanish Ministerio de Ciencia e Innovación (project CGL2007-62479/CLI) and Junta de Comunidades de Castilla La Mancha (Project PEII09-0262-2753). The authors also thank Krystle Ince for her assistance and words of advises through the writing process.

9. References

3M™ Novec™ Engineered Fluids. http://www.3M.com

Aranda, A.; Díaz-de-Mera, Y.; Rodríguez, A.; Rodríguez, D. & Martínez, E. (2003). A kinetic and mechanistic study of the reaction of Cl atoms with acrolein: temperature dependence for abstraction channel. *Journal of Physical Chemistry A*, Vol. 107, pp. 5717-5721, ISSN 1089-5639

Aranda, A.; Díaz-de-Mera, Y.; Bravo, I.; Rodríguez, D.; Rodríguez, A. & Martínez, E. (2006). Atmospheric HFEs degradation in the gas phase: Reactions of HFE-7100 and HFE-7200 with Cl atoms at low temperatures. *Environmental Science & Technology,* Vol. 40, pp. 5971-5976, ISSN 0013-936X

Bedjanian, Y.; Laverdet, G. & Le Bras, G. (1998). Low-pressure study of the reaction of Cl atoms with isoprene. *Journal of Physical Chemistry A,* Vol. 102, pp. 953-959, ISSN 1089-5639

Bera, P.P.; Francisco, J. S. & Lee, T. J. (2010). Design strategies to minimize the radiative efficiency of global warming molecules. Proceeding of the National Academy of Sciences of the United States of America, vol. 107, pp. 9049-9054, ISSN 0027-8424

Blowers, P.; Moline, D. M.; Tetrault, K. F.; Wheeler, R. R. & Tuchawena, S.L. (2007). Prediction of radiative forcing values for hydrofluoroethers using density functional theory methods, *Journal of Geophysical Research,* Vol. 112, article number D15108, ISSN 0148-0227

Bowen, R.D. & Maccoll, A. (1984). Low-energy, low-temperature mass-spectra of some small saturated alcohols and ethers. *Organic Mass Spectrometry,* Vol. 19, pp. 379-384, ISSN 0030-493X

Bravo, I.; Díaz-de-Mera, Y.; Aranda, A.; Smith, K.; Shine, K.P.; Marston, G. (2010). Atmospheric chemistry of $C_4F_9OC_2H_5$ (HFE-7200), $C_4F_9OCH_3$ (HFE-7100), $C_3F_7OCH_3$ (HFE-7000) and $C_3F_7CH_2OH$: temperature dependence of the kinetics of their reactions with OH radicals, atmospheric lifetimes and global warming potentials. *Physical Chemistry Chemical Physics,* Vol. 12, pp. 5115-5125, ISSN 1463-9076

Bravo, I.; Aranda, A.; Hurley, M. D.; Marston, G.; Nutt, D. R.; Shine, K. P.; Smith, K. & Wallington, T. J. (2010b). Infrared absorption spectra, radiative efficiencies, and global warming potentials of perfluorocarbons: Comparison between experiment and theory. *Journal of Geophysical Research,* Vol. 115, article number D24317, ISSN 0148-0227

Chen, L.; Kutsuna, S.; Tokuhashi, K.; Sekiya, A.; Takeuchi, K. & Ibusuki, T. (2003). Kinetics for the gas-phase reactions of OH radicals with the hydrofluoroethers $CH_2FCF_2OCHF_2$, $CHF_2CF_2OCH_2CF_3$, $CF_3CHFCF_2OCH_2CF_3$, and $CF_3CHFCF_2OCH_2CF_2CHF_2$ at 268-308K. *International Journal of Chemical Kinetics,* Vol. 35, pp. 239-245, 0538-8066

Christensen, L.K.; Sehested, J.; Nielsen, O.J.; Bilde, M.; Wallington, T.J.; Guschin, A.; Molina, L.T. & Molina M.J. (1998). Atmospheric chemistry of HFE-7200 ($C_4F_9OC_2H_5$): Reaction with OH radicals and fate of $C_4H_9OCH_2CH_2O(\cdot)$ and $C_4F_9OCHO(\cdot)CH_3$ radicals. *Journal of Physical Chemistry A,* Vol. 102, pp. 4839-4845, ISSN 1089-5639

Christensen, L.K.; Wallington, T.J., Guschin, A.; Hurley, M.D. (1999). Atmospheric degradation mechanism of CF_3OCH_3. *Journal of Physical Chemistry A,* Vol. 102, pp. 953-959, ISSN 1089-5639

Díaz-de-Mera, Y.; Aranda, A.; Bravo, I.; Rodríguez, D.; Rodríguez, A. & Moreno, E. (2008). Atmospheric chemistry of HFE-7000 ($CF_3CF_2CF_2OCH_3$) and 2,2,3,3,4,4,4-heptafluoro-1-butanol ($CF_3CF_2CF_2CH_2OH$): kinetic rate coefficients and temperature dependence of reactions with chlorine atoms. *Environmental Science & Pollution Research*, Vol. 15, pp. 584-591, ISSN 0944-1344

Díaz-de-Mera, Y.; Aranda, A.; Bravo, I.; Moreno, E.; Martínez, E. & Rodríguez, A. (2009). Atmospheric HFEs degradation in the gas phase: Reaction of HFE-7500 with Cl atoms at low temperatures. *Chemical Physics Letters,* Vol. 479, pp. 20-24, ISSN 0944-1344

EPA, U.S. Environmental Protection Agency. Clean Air Act. http://www.epa.gov/ ozone/snap/index.html

Finlayson, B.J. & Pitts, J.N.Jr. (2000). *Chemistry of the upper and lower atmosphere: theory, experiments, and application,* Academic Press, ISBN 978-0-12-257060-5, San Diego, CA

Garzón, A.; Moral, M.; Ceacero-Vega, A.A.; Fernández-Gómez, M. & Albaladejo, J. (2010). Atmospheric reactions of (H)- and (D)-fluoroalcohols with chlorine atoms. *ChemPhysChem,* Vol. 11, pp. 442-451, ISSN 1439-4235

Gaussian03. Frisch, J.; Schlegel, H.B.; Scuseria, G.E.; Robb, M.A.; Cheeseman, J.R.; Montgomery, J.A. Jr.; Vreven, T.; Kudin, K.N.; Burant, J.C.; Millam, J.M.; Iyengar, S.S.; Tomasi, J.; Barone, V.; Mennucci, B.; Cossi, M.; Scalmani, G.; Rega, N.; Petersson, G.A.; Nakatsuji, H.; Hada, M.; Ehara, M.; Toyota, K.; Fukuda, R.; Hasegawa, J.; Ishida, M.; Nakajima, T.; Honda, Y.; Kitao, O.; Nakai, H.; Klene, M.; Li, X.; Knox, J.E.; Hratchian, H.P.; Cross, J.B.; Bakken, V.; Adamo, C.; Jaramillo, J.; Gomperts, R.; Stratmann, R.E.; Yazyev, O.; Austin, A.J.; Cammi, R.; Pomelli, C.; Ochterski, J.W.; Ayala, P.Y.; Morokuma, K.; Voth, G.A.; Salvador, P:; Dannenberg, J.J.; Zakrzewski, V.G.; Dapprich, S.; Daniels, A.D.; Strain, M.C.; Farkas, O.; Malick, D.K.; Rabuck, A.D.; Raghavachari, K.; Foresman, J.B.; Ortiz, J.V.; Cui, Q.; Baboul, A.G.; Clifford, S.; Cioslowski, J.; Stefanov, B.B.; Liu, G.; Liashenko, A.; Piskorz, P.; Komaromi, I.; Martin, R.L.; Fox, D.J.; Keith, T.; Al-Laham, M.A.; Peng, C.Y.; Nanayakkara, A.; Challacombe, M.; Gill, P. M. W.; Johnson, B.; Chen, W.; Wong, M.W.; Gonzalez, C. & Pople, J.A. (2004). Gaussian 03, Revision C.02; Gaussian, Inc: Wallingford CT

Goto, M.; Inque, Y; Kawasaki, M.; Guschin, G.; Molina L.T.; Molina, M.J., Wallington, T.J. & Hurley, M.D. (2002). Atmospheric chemistry of HFE-7500 [n-$C_3F_7CF(OC_2H_5)CF(CF_3)_2$]: Reaction with OH radicals and Cl atoms and atmospheric fate of n-$C_3F_7CF(OCHO)CF(CF_3)_2$ and n-$C_3F_7CF(OCH_2CH_2O)CF(CF_3)_2$ radicals. *Environmental Science & Technology,* Vol. 36, pp. 2395-2402, ISSN 0013-936X

IPCC. Forster, P.M.D.; Ramaswamy, V.; Artaxo, P.; Berntsen, T.; Betts, R.; Fahey, D.W.; Haywood, J.; Lean, J.; Lowe, D.C.; Myhre, G.; Nganga, J.; Prinn, R.; Raga, G.; Schulz, M. & Van Dorland, R. (Ed.: Solomon, S.) (2007). *Fourth Assessment Report of the Intergovernmental Panel on Climate Change,* Cambridge.

IUPAC. *Summary of Evaluated Kinetic and Photochemical Data for Atmospheric Chemistry.* IUPAC Subcommittee on Gas Kinetic Data Evaluation for Atmospheric Chemistry. http://www.iupac-kinetic.ch.cam.ac.uk/summary/IUPACsumm_ web_ latest. pdf

Kambanis, K.G., Lazarou, Y.G & Papagiannakopoulos, P. (1998). Kinetic study for the reactions of chlorine atoms with a series of hydrofluoroethers. *Journal of Physical Chemistry A*, Vol. 102, pp. 8620-8625, ISSN 1089-5639

Kaufman, F. (1984). Kinetics of elementary radial reactions on the gas phase. *Journal of Physical Chemistry*, Vol. 88, pp. 4909-4917, ISSN 0022-3654

Kurylo, M.J. & Orkin, V.L. (2003). Determination of atmospheric lifetimes via the measurement of OH radical kinetics. *Chemical Reviews*, Vol. 103, No. 12, pp. 5049-5076, ISSN 0009-2665

Mellouki, A.; Teton, S. & Le Bras, G. (1995). Kinetics of OH radical reactions with a series of ethers. *International Journal of Chemical Kinetics*, Vol. 27, pp. 791-805, ISSN 0538-8066

Molder, U.H., Pikver, R.J. & Koppel I.A. (1983). Photoelectron-spectra of molecules. 2. Ethers. *Organic Reactivity*, Vol. 20, pp. 208-229, ISSN 0206-4766

Myhre, G.; Nielsen, C.J.; Powel, D.L. & Stordal, F. (1999). Infrared absorption cross section, radiative forcing, and GWP of four hydrofluoro(poly)ethers. *Atmospheric Environment*, Vol. 33, pp. 4447-4458, ISSN 1352-2310

NASA. *Chemical Kinetics and Photochemical Data for Use in Atmospheric Studies*. Evaluation Number 14. February 1, 2003. http://jpldataeval.jpl.nasa.gov/

Ninomiya, Y.; Kawasaki, M.; Guschin, A.; Molina L.T.; Molina M.J. & Wallington, T.J. (2000). Atmospheric chemistry of n-$C_3F_7OCH_3$: Reaction with OH radicals and Cl atoms and atmospheric fate of n-$C_3F_7OCH_2O$ radicals. *Environmental Science & Technology*, Vol. 34, pp. 2973-2978, ISSN 0013-936X

Nohara, K.; Toma, M.; kutsuna, S.; Takeuchi, K. & Ibusuki, T. (2001). Cl Atom-initiated oxidation of three homologous methyl perfluoroalkyl ethers. *Environmental Science & Technology*, Vol. 35, pp. 114-120, ISSN 0013-936X

Notario, A.; Mellouki, A. & Le Bras, G. (2000). Rate constants for the gas-phase reactions of Cl atoms with a series of ethers. *International Journal of Chemical Kinetics*, Vol. 32, pp. 105-110, ISSN 0538-8066

Orkin, V.L.; Villenave, R.E. ; Huie, R.E. & Kurylo, M.J. (1999). Atmospheric lifetimes and global warming potentials of hydrofluoroethers: Reactivity toward OH, UV spectra, and IR absorption cross sections. *Journal of Physical Chemistry A*, Vol. 103, pp. 9770-9779, ISSN 0022-3654

Oyaro, N. & Nielsen, C.J. (2003). Atmospheric chemistry of hydrofluoroethers: Reaction of a series of hydrofluoroethers with OH radicals and Cl atoms, atmospheric lifetimes, and global warming potentials. *Journal of Physical Chemistry A*, Vol. 109, pp. 337-346, ISSN 0022-3654

Papadimitriou, V.C.; Kambanis, K.C.; Lazarou, Y.G. & Papagiannakopoulos, P. (2004). Kinetic study for the reactions of several hydrofluoroethers with chlorine atoms. *Journal of Physical Chemistry A*, Vol. 108, pp. 2666-2674, ISSN 0022-3654

Papasavva, S.; Tai, S.; Illinger, K.H. & Kenny, J.E. (1997). Infrared radiative forcing of CFC substitutes and their atmospheric reaction products. *Journal of Geophysical Research*, Vol. 102, pp. 13643–13650, ISSN 0148-0227

Park, J.Y.; Slagle, I.R. & Gutman, D. (1983). Kinetics of the reaction of chlorine atoms with vinyl bromide and its use for measuring chlorine-atom concentrations. *Journal of Physical Chemistry,* Vol. 87, pp. 1812-1818, ISSN 0022-3654

Perry, R.H.; Green, D.W. & Maloney, J.O. (Eds.) (2001). *Handbook of air pollution analysis* (7th ed.), McGraw-Hill, ISBN 84-481-3008-1, Madrid

Pinnock, S.; Hurley, M. D.; Shine, K. P.; Wallington, T. J. and Smyth, T. J. (1995). Radiative forcing of climate by hydrochlorofluorocarbons and hydrofluorocarbons. *Journal of Geophysical Research,* Vol. 100, pp. 23227-23238, ISSN 0148-0227

Prinn, R.G.; Huang, J.; Weiss, R.F.; Cunnold, D.M.; Fraser, P.J.; Simmonds, P.G.; McCulloch, A.; Harth, C.; Salameh, P.; O'Doherty, S.; Wang, R.H.J., Porter, L. & Miller, B.R. (2001). Evidence for substantial variations of atmospheric hydroxyl radicals in the past two decades. *Science,* Vol. 292, pp. 1882-1888, ISSN 0036-8075

Pszenny, A.A.P.; Keene, W.C.; Jacob, D.J.; Fan, S.; Maben, J.R.; Zetwo, M.P.; Springer-Young, M. & Galloway, J.N. (1993). Evidence of inorganic chlorine gases other than hydrogen chlorine in marine surface air. *Geophysical Research Letters,* Vol. 20, pp. 699-702, ISSN 0094-8276

Rodríguez, A.; Rodríguez, D.; Garzón, A.; Soto, A.; Aranda, A. & Notario, A. (2010). Kinetics and mechanism of the atmospheric reactions of atomic chlorine with 1-penten-3-ol and (Z)-2-penten-1-ol: an experimental and theoretical study. *Physical Chemistry Chemical Physics,* Vol. 12, pp. 12245-12258, ISSN 1463-9076

Seinfeld, J.H. & Pandis, S.N. (1998). *Atmosperhic chemistry and physics: From air pollution to climate change,* John Wiley & sons, ISBN 978-471-17816-3, New York

Shine, K.P. (2010). Climate effect of inhaled anaesthetics. *British Journal of Anaesthesia,* Vol. *105,* pp. 731-733, ISSN 0007-0912

Sihra, K.; Hurley, M. D.; Shine, K. P. & Wallington, T. J. (2001) Updated radiative forcing estimates of 65 halocarbons and nonmethane hydrocarbons. *Journal of Geophysical Research,* Vo. 106, pp. 20493-20505, ISSN 0148-0227

Spicer, C.W.; Chapman, E.G.; Finlayson-Pitts, B.J.; Plastridge, R.A.; Hubbe, J.M.; Fast, J.D. & Berlowitz, C.M. (1998). Unexpectedly high concentrations of molecular chlorine in coastal air. *Nature,* Vol. 394, pp. 353-356, ISSN 0028-0836

Spivakovsky, C.M.; Logan, J.A.; Montza, S.A.; Balkanski, Y.J.; Foreman-Fowler, M.; Jones, D.B.; Horowitz, L.W.; Fusco, A.C.; Brenninkmeijer, C.A.; Prather, M.J.; Wofsy, S.C. & McWlroy, M.B. (2000). Three-dimensional climatological distribution of tropospheric OH: Update and evaluation. *Journal of Geophysical Research,* Vol. 105, pp. 8931-8980, ISSN 0148-0227

Wallington, T.J.; Schneider, W.F.; Sehested, J.; Bilde, M.; Platz, J.; Nielsen, O.J.; Christensen, L.K.; Molina M.J.; Molina, L.T. &Wooldridge, P.W. (1997). Atmospheric chemistry of HFE-7100 ($C_4F_9OCH_3$): Reaction with OH radicals, UV spectra and kinetic data for $C_4F_9OCH_2$ and $C_4F_9OCH_2O_2$ radicals, and the atmospheric fate of $C_4F_9OCH_2O\cdot$ radicals. *Journal of Physical Chemistry A,* Vol. 105, pp. 5146-5154, ISSN 1089-5639

Wingenter, O.W.; Kubo, M.K.; Blake, M.K.; Sith, T.W. & Rowland, F.S. (1996). Hydrocarbon and halocarbon measurements as photochemical and dynamical indicators of atmospheric hydroxyl, atomic chlorine, and vertical mixing obtained during

Lagrangian flights. *Journal of Geophysical Research,* Vo. 101, pp. 4331-4340, ISSN 0148-0227

Zhang, Z.; Saini, R.D.; Kurylo, M.J. & Huie, R.E. (1992). Rate constants for the reactions of the hydroxyl radical with several partially fluorinated ethers. *Journal of Physical Chemistry,* Vol. 96, pp. 9301-9304, ISSN 0022-3654

16

The Importance of Advective Fluxes to Gas Transport Across the Earth-Atmosphere Interface: The Role of Thermal Convection

Uri Nachshon[1], Noam Weisbrod[1], Maria I. Dragila[2] and Yonatan Ganot[1]

[1]Ben Gurion University of the Negev
[2]Oregon State University, Oregon
[1]Israel
[2]USA

1. Introduction

Understanding of gas exchange between the Earth's upper crust and the atmosphere is vital, as it affects many important processes which concern the water cycle, agricultural activities, greenhouse gas emissions, and more. From a hydrological aspect, water vapor transport is the most important process related to Earth-atmosphere gas exchange, since it affects above-land water vapor concentration; soil water content; and soil salinity. These three important hydrological parameters respectively affect the global water cycle (Hillel, 1998); water management and agricultural practices; and the formation of salt crusts at and near land surface - which can lead to soil salinization (Weisbrod et al., 2000; Nachshon et al., 2011), an important process from an agricultural point of view. In addition to soil salinization, with respect to agriculture, gas transport in the upper soil profile, i.e., the root zone, is important for soil aeration or movement of oxygen within the soil. Soil aeration is critical for plant root growth, as plants generally cannot get enough oxygen from their leaves (Lambers et al., 2008). Oxygen is not always readily available in the soil pores, since respiration of plants and other organisms and microbial degradation of organic compounds in the ground emit high volumes of CO_2 into soil pores, while consuming O_2 (Brady, 1999). The exchange rate of air between soils and the atmosphere is crucial to maintain the needed soil aeration and oxygen concentration for plant growth.

Since most underground biological activity takes place in the upper parts of the soil profile, the majority of CO_2 is formed from the ground surface down to shallow depths of a few meters (Amundson, 2005). As soil temperature and water content increase, the CO_2 production increases (Fang & Moncrieff, 1999; Rastogi et al., 2002; Buyanovsky et al., 1986). The increase of organic matter availability will also lead to an increase in CO_2 production (Amundson, 2005). For example: Buyanovsky et al. (1986) calculated CO_2 production in soil surface cultivated with wheat. Values varied from 4 to 8 g/m d in spring, but in winter as soil temperature dropped below 5°C, CO_2 production was reduced to less than 1 g/m d. As for organic matter availability; soil CO_2 concentrations at 1 m depth in Tundra, temperate grassland and tropical rain forest are 1000, 7000 and 20,000 ppm, respectively (Amundson, 2005), corresponding to the richness of these soils in organic matter. CO_2 concentration in the pores of unsaturated soils, in the range of 3000 ppm is very common for agricultural and grasslands areas (Brady,

1990; Reicosky et al., 1997). Even though this concentration seems minor, it is eight times greater than atmospheric CO_2 concentration (391.48 ppm in 2011, (NOAA, 2011)). This difference emphasizes the importance of understanding gas exchange mechanisms between terrestrial environments and atmosphere as it affects CO_2 atmospheric concentration.

Migration of gases in unsaturated rocks is also very important in the context of Earth-atmosphere gas exchange. Therefore, the general term 'vadose zone', which includes both unsaturated soils and rocks above the water table, is often used. For example, the world's largest carbon reservoir are carbonate rocks, containing about $6.1*10^7$ billion tons of carbon, which is 1694 and $1.1*10^5$ fold more than the carbon content in oceans and world vegetation, respectively (Houghton & Woodwell, 1989). Chemical interactions between these carbonate rocks and atmosphere may be a source or a sink for large volumes of CO_2 (Liu & Zaho, 2000).

Emission of gases from the Earth's subsurface has an important role from an environmental aspect. For example, water vapor and CO_2, beside being important for agricultural and hydrological concerns, are important components of the global warming process, as they are major green house gases, together with N_2O and CH_4 (Weihermüller et al., 2011). Thus, it is important to understand the transport of these gases across the Earth-atmosphere interface. Additionally, understanding of gas transport is important for environmental and public health concerns. Movement of volatile radionuclides, such as ^3H, ^{14}C and Rd from radioactive waste disposal facilities, as well as natural emission of Rn from natural sources and industrial volatile organic components such as chlorinated volatile organic compounds (e.g., Lenhard et al., 1995; Conant et al., 1996; Smith et al., 1996; Choi et al., 2002; Ronen et al., 2010) can greatly affect public health when emissions occur in buildings or populated areas (Nazaroff, 1992; Scanlon et al., 2001).

For the reasons mentioned above and more, it is vital to understand gas exchange processes and gas flow rates between the Earth crust and atmosphere. Traditionally, diffusion was considered as the main mechanism of gas exchange between the atmosphere and the vadose zone, driven by gas concentration gradients (Hirst & Harrison, 1939; Penmann, 1940a, 1940b; Marshall 1958, 1959; Millington & Quirk, 1961; Cunningham & Williams, 1980; Amali & Rolston, 1993). In the last few decades several advective gas transport mechanisms, corresponding to pressure gradients, were introduced, resulting in faster gas exchange rates between terrestrial environments and atmosphere. Moreover, while diffusion impacts the transport of each gas independently, according to its concentration gradient, advective mechanisms impact the migration of the bulk assembly of gases within pores. Advective mechanisms that could result in gas flux across the Earth-atmosphere interface include: (1) wind pumping (Fukuda, 1955; Weeks, 1993, 1994); (2) atmospheric barometric changes (Pirkle et al., 1992; Wyatt et al., 1995; Rossabi, 2006); and (3) thermal convection flows in fractures and other cavities (Weisbrod et al., 2005; Weisbrod & Dragila, 2006; Weisbrod et al., 2009).

This chapter first presents a general overview of relatively known gas transfer mechanisms between the vadose zone and atmosphere, which include diffusion, wind driven advection and advection driven by atmospheric barometric changes. Subsequently, we describe recent exploration and findings regarding thermal convection flow within surface exposed fractures and its important effect on fracture ventilation. This is followed by a description of field and laboratory investigations conducted to study thermal convection in rock fractures. Lastly, results are presented with an aim to answer the following scientific questions:

- Is there a temporal dependence on development of thermal convections in surface-exposed fractures?
- Do fracture aperture and thermal conditions affect convection flow velocities and to what extent?

- Do fracture aperture and thermal conditions affect convection flow cell dimensions and geometry?
- Do fracture aperture and thermal conditions affect total mass transfer rates between fractures and atmosphere and to what extent?

Deciphering these issues enables quantitative estimation of gas exchange rates between the vadose zone and atmosphere for fractured media. Moreover, while the studies presented here focus on thermal convection developing within fractured media, these same processes are likely pertinent for other surface exposed cavities in the vadose zone such as animal burrows, boreholes and karstic structures.

2. Gas transfer mechanisms

As aforementioned, two gas transport mechanisms, diffusion and advection, control gas movement between the Earth and the atmosphere. While the former occurs spontaneously, whenever there is a concentration gradient of a certain substance (gas or liquid, even under isothermal and isobaric conditions), the latter occurs only if there is a thermal or a pressure gradient. Diffusion mass transfer rates are usually several orders of magnitude lower than advection mass transfer rates. If a total pressure gradient exists in a soil in response to natural phenomena such as barometric changes or surface wind, gases will flow from points of higher to those of lower pressure. It has been shown that relatively small gradients of total pressure can result in advective gas fluxes that are much higher than diffusive gas fluxes (Thorstenson & Pollock, 1989; Massmann & Farrier, 1992, Scanlon et al., 2001).

2.1 Diffusion

Diffusion has long been considered the dominant process by which gases move from the vadose zone to the atmosphere. Penman (1940a, 1940b) explored soil properties and their effect on water vapor and CO_2 diffusion across the vadose zone-atmosphere boundary. This work showed that diffusion alone can support CO_2 transfer rates needed for plant respiration. Fick (1855) formulated gas and liquid diffusive fluxes as a function of their concentration field, with flow resulting from regions of high concentration to regions of low concentration. Fick's law is:

$$J_D = -D\nabla C \tag{1}$$

where J_D (kg/m^2 s) is the mass flux, D (m^2/s) is the gas diffusion coefficient in free air, and C is gas (e.g., water vapor) concentration (kg/m^3). Table 1 introduces D (in free air) for common vadose zone gases.

Gas	Temperature (°C)	Diffusion coefficient x10^{-5} (m^2/s)
H_2O	16	2.82
CH_4	9	1.96
O_2	0	1.76
CO_2	9	1.02

Table 1. Diffusion coefficients in free air for common vadose zone gases (from: Cussler, 1997)

A few relationships have been proposed to describe gas diffusion in porous media (e.g., Marshall, 1958, 1959; Millington & Quirk, 1961; Moldrup et al., 2000); in all of them, the diffusion coefficient in porous media (D_{eff}) is low compared with that in free air and is proportional to the air-filled porosity ϕ_a. As per Millington & Quirk (1961):

$$D_{eff} = D \cdot \phi_a^{\frac{4}{3}} \tag{2}$$

For example, assuming sandy soil (ϕ_a = 0.35) with the evaporation front at a depth of 1 m below surface, soil pores just above the evaporation front have water vapor concentration of 0.03 kg/m^3, corresponding to 100% relative humidity at 30°C (Ho, 2006). For an atmospheric relative humidity of 50% the vapor concentration is 0.015 kg/m^3, hence ∇C between soil surface and water table is 0.015 kg/m^3 over a distance of 1 m. From Table 1, D for water vapor is 2.82x10^{-5} m^2/s. Using Equation 2 (D_{eff} is equal to 6.95x10^{-6}) and Equation 1, the water vapor flux from the water table toward the atmosphere is 1.04x10^{-7} kg/m^2 s. The constraining effect of pores is removed for the diffusion of gases from the vadose zone towards the atmosphere when the diffusion path is via fractures, caves, karstic holes or animal burrows. In these cases, or the same physical conditions but for diffusion in free air, the vapor flux would be four times higher than vapor flux through the soil, which would be equal to 4.23 x10^{-7} kg/m^2 s.

2.2 Advective gas transport

Non-turbulent advective gas transport in porous media is described by Darcy's law which correlates gas fluxes J_A (m/s) to matrix permeability, K(m^2), gas viscosity, μ (Kg/m s) and pressure gradient, ∇P:

$$J_A = -\frac{K}{\mu} \nabla P \tag{3}$$

Small pressure gradients can result in substantial advective gas fluxes because the resistance to flow is small, due to the low viscosities of gases (Granger, 1995). When comparing advective to diffusive gas mass transport it is convenient to use the Sherwood number (Sh) which is a dimensionless number that gives the ratio of convective to diffusive mass transport. The Sherwood number is defined as: (Weast, 1980),

$$\mathrm{Sh} = \frac{J_A L}{D} \tag{4}$$

where L is the length scale of interest. For Sh>1 advection is the dominant transport mechanism, while for Sh<1, diffusion is the main transport mechanism.

2.2.1 Barometric pumping

One mechanism which can generate pressure gradients between the air within vadose zone pores and the above ambient atmosphere is related to barometric pressure changes at the order of 1 kPa; these pressure changes are caused by diurnal thermal and gravitational fluctuations, and larger fluctuations corresponding to regional scale weather patterns, which cause pressure changes of tens of kPa (Auer et al., 1996; Scanlon et al., 2001; Rossabi, 2006). The penetration depth of barometric pressure fluctuations increases with the permeability of the medium. In very high permeability fractures and boreholes, this mechanism is more prominent and the pressure wave can penetrate to a large depth, in some cases down to depth of several hundred meters (Nilson et al., 1992; Holford et al., 1993).

Rossabi (2006) presented a comprehensive review of this mechanism. It was shown that the barometric pumping mechanism results in very high mass transfer rates of air through boreholes crossing the vadose zone. For a 32 m deep, 1 inch diameter borehole, and an average atmospheric pressure fluctuation of 1 kPa per day, the maximal measured air flow velocities were in the range of 0.7 m/s, much faster than typical diffusion mass transfer rates. Massmann & Farrier (1992) computed gas transport in the unsaturated zone resulting from atmospheric pressure fluctuations and found that atmospheric "fresh" air can migrate several meters into the subsurface during a typical barometric pressure cycle. Auer et al. (1996) included the effects of barometric pumping in an airflow and transport model and found that as little as 13% of a simulated original contaminant mass remained after 50 years, compared with 49% of the original mass remaining for the case of diffusion only.

2.2.2 Wind effect

Another mechanism which results in ventilation of the vadose zone, and has been explored mainly in connection with fractured media (Fukuda, 1955; Weeks, 1993, 1994) and animal burrows (Vogel et al., 1973; Fenton & Whitford, 1978; Kleineidam et al., 2001), is related to wind flow above the ground surface. It follows naturally from the conservation of energy law that for any fluid (including gas) the balance between its pressure, velocity and elevation on a single stream line is described by the Bernoulli equation: (Schaschke, 1998),

$$P + \frac{\rho V^2}{2} + \rho g h = CONSTANT \tag{5}$$

where V is the gas flow velocity (m/s), ρ is density (kg/m^3), g is gravitational acceleration (m/s^2) and h (m) is height over a reference. The first term in the equation, P, is called the static pressure (kg/m s^2) and is the pressure induced by the gas that can be measured by a barometer. The second term in the equation is known as the dynamic pressure and the third term is the gravitational potential. As can be seen from Equation 5, as gas velocity increases the dynamic pressure increases and induces a static pressure decrease. For a stagnant bulk of air (V = 0 m/s) at sea level (h=0 m), Equation 5 can be written as P=$CONSTANT$=ATM (atmospheric pressure). Now, if this bulk air moves at constant elevation with velocity U (m/s), the static pressure becomes: $P = ATM - \frac{\rho U^2}{2}$.

Therefore, when wind blows over porous or fractured media a pressure gradient is induced between the stagnant air below ground surface and the moving atmospheric air, resulting in advective gas fluxes from the vadose zone towards the atmosphere. As per Darcy's Law (Equation 3), an increase in matrix permeability increases gas fluxes, therefore this mechanism is expected to be most prevalent in fractures and other cavities exposed to atmosphere that have much higher permeability, compared to porous media (Weeks, 1993). Assuming air density of 1.3 kg/m^3 (Evett, 2001) and wind velocity of 0.5 m/s at ground surface (Morgan, 2005), the pressure difference between stagnant air below soil surface and atmospheric air is 0.1625 Pa (Equation 5). Using Equation 3, this pressure difference along a fracture with 0.5 m depth and aperture of 2 cm results in a gas flux of 0.58 m/s towards the atmosphere (values used where k = (2b)^2/12 and 2b is the fracture aperture (Shemin, 1997) and μ=1.85x10^{-5} (kg/m s)).

Assuming the fracture is saturated with water vapor, which diffused into it from the surrounding porous media, the induced gas exchange between the fracture and atmosphere will increase water vapor extraction from the fracture, resulting in a net vapor flux into the atmosphere. The Sh number (Equation 4) can give an estimate of the wind effect on mass transfer rates. For example, the ratio between advective velocity of 0.58 m/s to diffusion of H_2O vapor with D=2.82x10^{-5} m^2/s, over a 0.5 m fracture depth results in Sh>10,000, indicating a potential mass transfer for advection that is 10,000 times greater than diffusion, demonstrating the important role of wind in ventilation of the fracture.

Another passive ventilation mechanism, occurring mainly for surface exposed cavities in the ground, is called "viscous entrainment". Viscous entrainment is caused by the resistance of fluids to rapid shear rates and the consequent attraction of stagnant fluid by adjacent rapidly moving fluid. For surface exposed cavities this shear stress develops between the fast moving surface wind and the stagnant air in the cavity (Vogel et al., 1973; Kleineidam et al., 2001). In addition, several works (Kimball & Lemon, 1971, 1972; Ishihara et al., 1992) have shown that turbulent wind above ground surface might result in turbulent diffusion of gases below ground surface, hence increasing the overall gas exchange rates. Apart from this, when wind blows over irregular surfaces, the formed eddies increase pressure on the windward side of an obstruction and reduce pressure in the leeward side (Don scott, 2000). This mechanism is more relevant to flow in high porosity and permeability matrices, with its effect limited to a depth of several cm (Ishihara et al., 1992). Kimball & Lemon (1971) reported an increase of up to 2 orders of magnitude in fluxes of Heptane vapors from 2 cm below surface to the atmosphere due to turbulent winds with an average velocity of 0.65 m/s.

3. Thermal convection mechanism

Besides wind and barometric pumping mechanisms, a third mechanism was recently explored, relevant to increasing gas exchange between vadose zone and atmosphere. Weisbrod et al. (2005) and Weisbrod & Dragila (2006) proposed that thermally-driven convection might be a primary mechanism for venting of surface-exposed fractures and cracks open to the atmosphere. While the above mentioned barometric pumping and the wind mechanisms require specific meteorological conditions to occur, conditions necessary for thermal convections occur every night in fractures and cavities exposed to the atmosphere, in response to daily thermal fluctuations (Weisbrod et al., 2009).

The authors suggested that during cold nights, especially in arid climates, when the ambient air becomes colder and consequently denser than the fracture air, density-driven air convection may develop within the surface exposed fractures. Exchange of warm fracture air with cooler atmospheric air effectively ventilates fracture air. Conversely, during the day, atmospheric air is warmer than fracture air, generating a stable condition in which no appreciable convective movement driven by air-density differences is expected. Theoretical calculations have suggested that convection could markedly increase gas exchange between fractures and atmosphere and increase evaporation rates from fractures with apertures of more than a few millimeters and thermal differences of more than a few degrees Celsius between fracture and atmospheric air (Weisbrod & Dragila, 2006; Weisbrod et al., 2005).

Several experimental studies (Nachshon et al., 2008; Kamai et al., 2009) as well as field measurements (Weisbrod et al., 2009) have demonstrated the validity of this concept and quantified the thermally-driven convection mechanism. A thorough review of this mechanism will be presented in this chapter.

3.1 Density-driven gas fluxes

In contrast to barometric pumping and the wind mechanisms, which are associated with pressure gradients, thermal convection in fractures responds to density differences between fracture and atmospheric air. The fundamental assumption is that because of the high geometric aspect ratio and low heat capacity of air, fracture air temperature is similar to the temperature of surrounding fracture walls. Consequently, the daytime thermal profile of fracture air consists of warm air in the upper parts of the fracture and colder air in the fracture's deeper parts, in accordance with the typical thermal profile of the vadose zone matrix (Don scott, 2000, Weisbrod & Dragila 2006). At night atmospheric air temperature drops dramatically, especially in arid areas and the vadose zone's thermal profile is inverted (Evett, 2001). The ground surface, which is exposed to the relatively cold atmosphere and from where energy is radiated, becomes colder than the deeper parts of the matrix. Thus, the fracture's thermal profile is in an unstable state, with cold, dense air on top of warmer, less dense air.
Unstable thermal gradients of sufficient magnitude lead to buoyancy-driven convection flows. A conceptual model for day and night thermal conditions and their associated mass transport mechanisms are depicted in Figure 1.

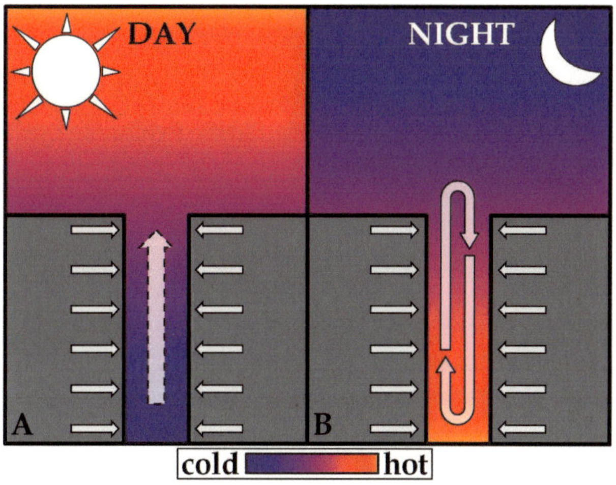

Fig. 1. Conceptual model for fracture venting, consisting of (A) diffusive venting of fracture air during the day; and (B) thermally driven convection at night. Horizontal arrows indicate diffusive gas flux from the porous matrix towards the fracture.

3.1.1 Rayleigh number

In a density-driven gas transport process, two forces oppose one the other: buoyancy and viscosity. The onset of convection in fractures has been theoretically proposed to occur when the Rayleigh number (Ra), which is a dimensionless number that compares buoyant and viscous forces, exceeds a critical value of $4\pi^2$ (Lapwood, 1948; Nield, 1982). Ra is defined as:

$$\mathrm{Ra} = \frac{\Delta \rho g \kappa L}{\mu k} \tag{6}$$

where $\Delta\rho$ is the density (kg/m^3) difference between the air at the top and bottom of the fracture over a length scale L (m), μ is the dynamic viscosity of the air (kg/m s), g is the gravitational constant (m/s^2), κ is the thermal diffusivity (m^2/s), and k is the fracture permeability (m^2) (k = (2b)^2/12, where 2b is the fracture aperture (Shemin, 1997). Within the range of 0 to 80°C, air density can be assumed to be a linear function of temperature and expressed by the thermal expansion coefficient (α = 0.00367(1/°C)). Therefore, Equation 6 can be recast as a function of temperature, which can be measured easily,

$$\mathrm{Ra} = \frac{\Delta T \alpha g \kappa L}{\nu k} \tag{7}$$

where ΔT (°C) is the temperature difference between fracture air at the top and bottom of the fracture over the length scale L, and ν is the kinematic viscosity (m^2/s). As an example, for a discrete fracture of 1 m depth and 2 cm aperture, Ra would predict a minimum necessary temperature difference of 0.01°C. For a 0.5 cm aperture, a temperature difference of 0.17°C is needed. Such thermal differences are much lower than typical thermal differences along a 1 m profile in the vadose zone (e.g., Brady, 1990; Don scott, 2000) and in surface exposed fractures (Weisbrod et al., 2009), which are in the range of 2-15°C, at least for temperate and semi-arid environments. These differences were measured on a nightly basis, thus, theoretically convection flows are expected to be formed in the surface exposed fractures, every night.

3.2 Quantifying thermal convection flows

Weisbrod et al. (2009) and Nachshon et al. (2008) explored and quantified thermal convection *in situ* within a natural fracture and in laboratory experiments. The former used a two dimensional grid of thermocouples to monitor spatial thermal distribution within a natural fracture in the field. The fracture, within a Chalk rock unit, is located in the south of Israel, in a semi-arid environment. The fracture depth is up to 120 cm, with average aperture of 1.5 cm over a lateral width of 150 cm. The installed 25-thermocouple grid occupied part of the fracture, penetrating to a depth of 60 cm and a 60 cm lateral width. Measurements were taken at a spatial resolution of 15 cm every 10 minutes. In addition a vertical line of thermocouples inserted to a depth of 120 cm, measured matrix temperatures (Figure 2).

Nachshon et al (2008) simulated a natural fracture exposed to atmosphere using a 50 cm x 50 cm Hele-Shaw cell with apertures of 1 and 2cm. The Hele-Shaw cell walls were made of glass, enabling viewing of the fracture volume. Thermocouples were installed along the glass walls, as well as a two dimensional grid of thermocouples, similar to the one used in the field that was inserted into the simulated fracture. The entire setup was placed on heating plates in a climate control room. By controlling heating plates and room ambient air temperatures, thermal gradients similar to those measured in nature were imposed. No wind was blowing in the climate control room to eliminate wind effect on convection dynamics. Smoke filled the simulated fracture to enable tracing of air movement. A video camera (24 frames per second) was used to film the migration of smoke within the Hele-Shaw cell and to the atmosphere above. Images were analyzed frame-by-frame to quantify tracer transport.

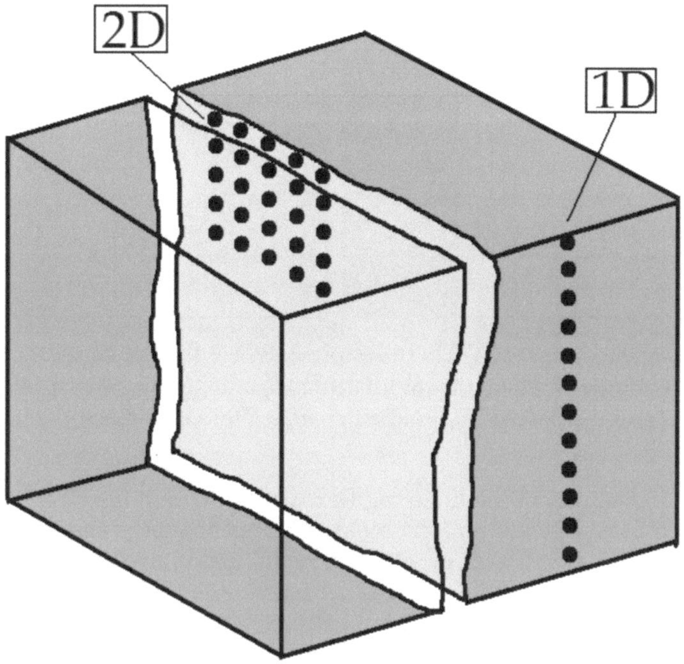

Fig. 2. Schematic of thermocouples installation as two dimensional grid within the fracture and the vertical line within the rock. Two dimensional grid dimensions are 60x60 cm. Fracture aperture is ~1-2 cm.

3.2.1 Field measurements and experimental results

Field measurements indicate diurnal appearance of convection flows every night. The thermal signature of the convection cells was measured by the two dimensional thermocouples grid. It was found that every night, unstable conditions of hot air below cold air develop within the fracture, sustained through the night until the next morning. Calculated Ra numbers for the measured thermal conditions are in the range of few thousands, much larger than the critical Ra necessary for the onset of convection.

3.2.2 Thermal convection patterns

Convection patterns between parallel ducts fall into two categories, Thermo-siphon (Figure 3A; Balaji & Venkatesehan, 1995; Lanchao & Amir, 1998) and finger flow (Figure 3B). For the case of natural convection in natural fractures with apertures in the range of a few cm, it was shown both experimentally and *in situ* in the field that the convection flows form a fingering pattern.

A convection cell is defined as the region which includes one down and one up air stream line. For example; in Figure 3B there are two convection cells. For typical thermal conditions in the field, with a thermal gradient in the range of 2-15°C over a vertical distance of less than 1 m, convection cell dimensions ranged from 20 to 50 cm in width, reaching a maximal depth of ~60cm. As the thermal gradient increases, cell aspect ratio increases as cells become taller and narrower, i.e. more convection cells occupied the fracture volume, increasing the effectiveness of the system for mass and energy exchange with the atmosphere.

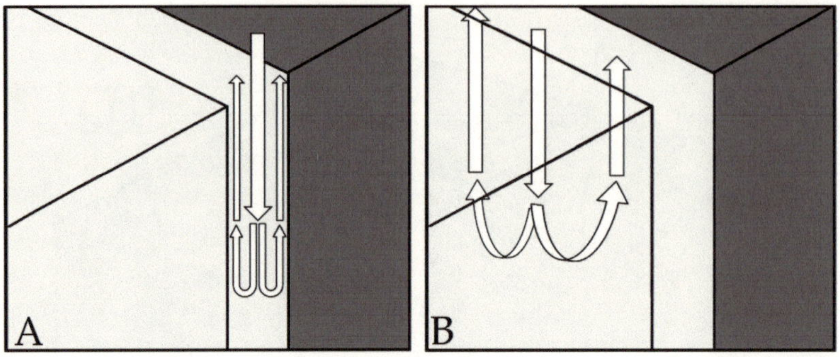

Fig. 3. Convection cell patterns: (A) Thermo-siphon, where the hot air ascends along the hot walls and the cold air descends in the middle; and (B) a fingering pattern, where the ascending hot air and the descending cold air change alternately along the fracture width dimension.

Figure 4 presents examples of the fingering flow pattern within the Hele-Shaw cell (Figure 4A), for a ΔT of 5°C over 50 cm, and thermal signatures of a large and small fingering flow pattern in the natural fracture (Figures 4B and 4C, respectively). Comparison between Figure 4B and 4C demonstrates the effect of an increase in the thermal difference between fracture bottom and upper boundaries on the convection cells' dimensions. In Figure 4B a single, wide convection cell is observed for ΔT of less than 6°C, while for ΔT of 8°C (Figure 4C) two narrow convection cells are observed, enabling more efficient heat removal.

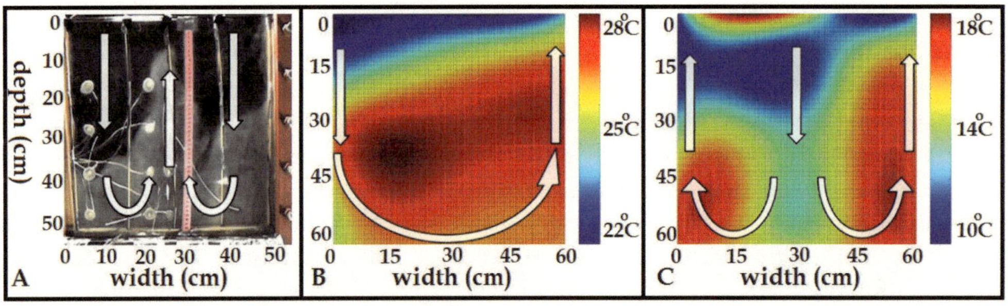

Fig. 4. Experimental and field observations of the fingering flow pattern. Arrows indicate air flow directions. (A) snapshot of the flowing air in the Hele-Shaw chamber, visualized by the smoke, for a thermal difference of 5°C between fracture bottom and atmosphere. (B), (C) thermal measurements from the natural field fracture. The convection cell's thermal footprint is observed. (B) and (C) are for different thermal gradients of 5°C and 8°C respectively.

3.2.3 Thermal convection durations

As observed from the field measurements, the convection flows started as the atmospheric air temperatures started to decline, every evening. The convection cells were observed throughout the night until the following morning. During winter, when the nights are longer and atmosphere is cold the duration of the convection flows were longer than in summer (Figure 5). During December-January the average duration for convection conditions was ~18 hours, starting at ~17:00 and ending at ~11:00 the following day. During July-August, convection started at ~19:00 and lasted until ~07:00 the following day.

A diurnal cycle of thermal distribution, resulting from thermal convection within a natural fracture in the field is presented in Figure 6 (during September 28–29, 2006). At this time, it was observed that the convection cells commenced at 19:00 and ceased around 09:00 the following day.

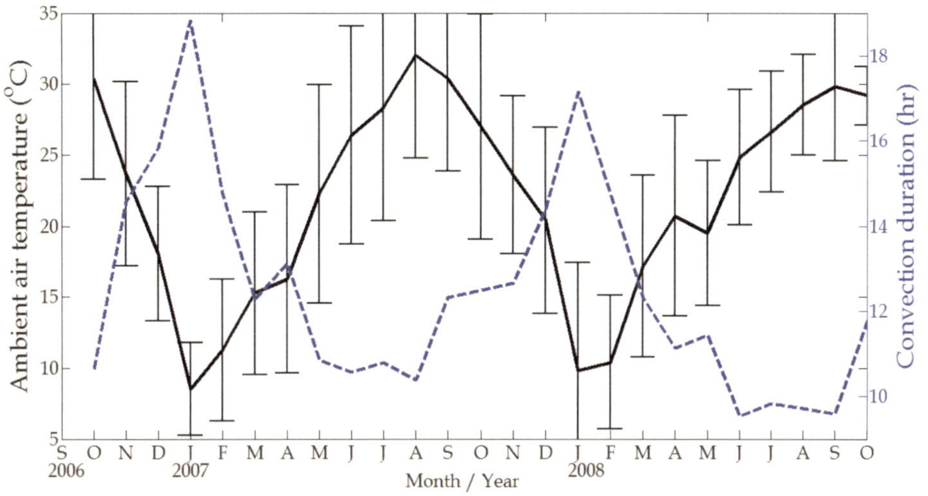

Fig. 5. Variability in daily duration of convective venting. Average monthly atmospheric air temperature (solid line) and monthly average of the daily duration of convection (dashed line) for a 24-month period from October 2006 to October 2008. Vertical bars depict daytime-nighttime temperature range.

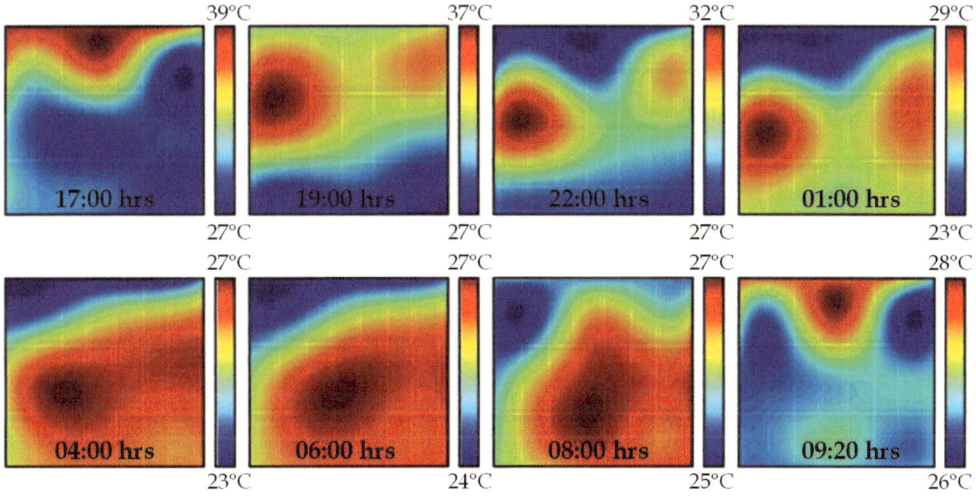

Fig. 6. Thermal map of air temperature within the natural fracture. Thermal convection cells were observed to form in a natural fracture exposed to the soil surface. Note that temperature scale is unique to each map to maximize image range: blue (cool) to red (warm). Dimensions of each map are 60 cm by 60 cm. Time corresponding to each map is shown at each map box. Date: Sep 28–29, 2006.

3.2.4 Mass transfer

Hele-Shaw experiments enabled quantification of the mass transfer rates between the simulated fracture and atmosphere, as well as providing a method for which to measure convection flow velocities for various Ra values which correspond to different thermal gradients and apertures. Nachshon et al. (2008) defined two velocities to describe a convection flow: (1) the maximal velocity at the center of the ascending hot air (U_{max} (m/s)) and (2) the average up flow velocity for the entire fracture (Hele-Shaw) volume (U_{ave} (m/s)). The maximal velocity was measured by analyzing video camera sequences of the smoke flow within the Hele-Shaw cell. The average velocity was calculated based upon the time it took to empty the chamber from smoke. Table 2 presents physical conditions and measured U_{max} and U_{ave} for various Ra values. It is shown that U_{max} is ~50 times faster than U_{ave} for the lower Ra values and only ~15 times faster for higher Ra values. This disparity reflects the linear increase in U_{max} (U_{max}=4x10^{-6}(Ra)) against a logarithmic increase of U_{ave} as Ra increases.

Aperture (cm)	ΔT	Ra	Total width of upward flow (cm)	U_{ave} (m/s)	U_{max} (m/s)
1	5	2973	10	0.00038	0.019
1	10	5947	12	0.00055	0.0231
1	13	7731	15	0.00065	0.0265
2	5	11893	10	0.00093	0.051
2	7	16658	--------	0.00119	--------
2	8	19037	--------	0.00152	--------
2	9	21417	--------	0.00333	--------
2	10	23796	20	0.00439	0.0678
2	12	28556	--------	0.00556	--------
2	13	30923	25	0.00725	0.124

Table 2. Hele-Shaw experimental conditions and results (Nachshon et al., 2008).

The mass transfer rate of air from the fracture to the atmosphere is comprised of two components: (1) the upward air velocity; and (2) the fraction of the fracture which participates in the upward flow. As can be seen from Table 2 and from the convection cell pattern, as Ra increases, more fracture areas participate in upward flow. Therefore, even though U_{max} increases linearly, the overall mass transfer rates and the subsequent U_{ave} both increase exponentially. Figure 7 displays U_{max} and the mass transfer rates for various Ra values, showing a linear increase in U_{max} and the exponential increase in the mass transfer rate of $\approx$ 4.3x10^{-9}(Ra)^2. Figure 8 displays the Sh number for various Ra values, indicating the importance of thermal convections to gas exchange between vadose zone fractures and atmosphere. Thermal convections increase mass transfer rates by almost two orders of magnitude for the higher Ra values and by an order of magnitude for lower Ra values. The thermal conditions in the range of these Ra values were measured every night in the natural fracture.

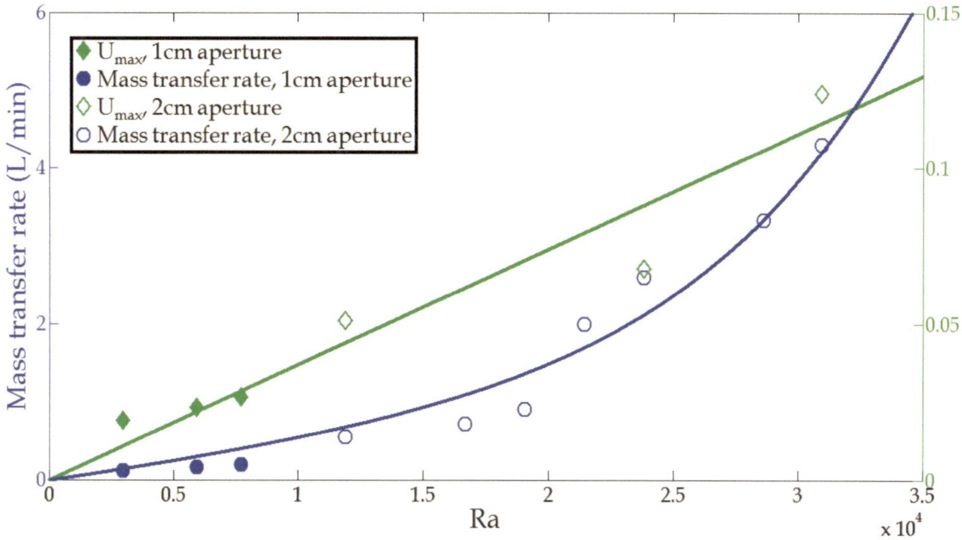

Fig. 7. Mass-transfer rates (blue symbols) from the smoke-replacement experiments increase exponentially with Rayleigh number (Ra) (mass transfer ≈ 4.3 × 10−9(Ra)^2, R^2 = 0.97), while maximum velocity (U_{max}) (green symbols) increases linearly (U_{max} = 4 × 10^{-6}(Ra), R^2 = 0.94). Full and empty polygons represent data from the 1 and 2 cm apertures, respectively.

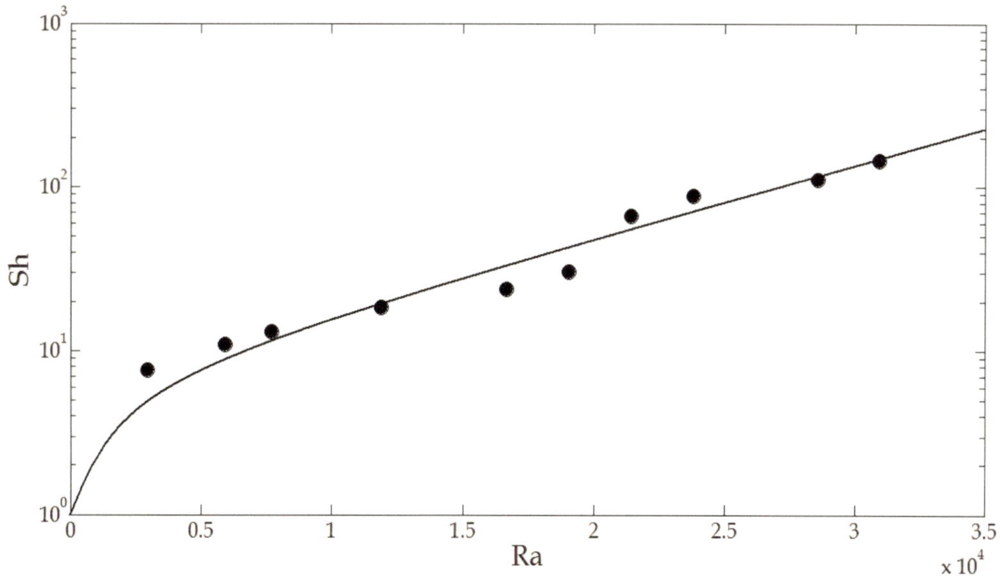

Fig. 8. Sherwood number (Sh) as a function of Rayleigh number (Ra). Black marks are calculated from the smoke-replacement experiment. Black line is a numeric fitting of the results: Sh ≈ 1.5E^{-7}(Ra)^2 + 1.

4. Conclusions

Both barometric variability and surface winds drive gas and vapor exchange between the vadose zone and the atmosphere. However, these two mechanisms are infrequent and depend upon the capriciousness of local weather patterns. On the other hand, thermally driven free convection, occurs on a nightly basis and should therefore be considered as a primary transfer mechanism in areas where the surface is cracked or fractured.

Evidence presented here categorically shows that convective venting of cracks and fractures is a natural and pervasive process that may have a pronounced impact on earth-atmosphere gas exchange in areas where surface cracks and fractures prevail (e.g. desert playas, cracked soils in agricultural regions or rock fractures). Soil cracks and fractures on the Earth's surface are not rare; they are ubiquitous features that can be commonly found in arid, moist and frigid climatic settings. However, thermally driven convection may also prevail in the fractures and cracks of karst systems, boreholes, and animal burrows. Karstic systems provide almost 25% of the world's potable groundwater supply (Ford & Williams, 1989), demonstrating their importance and abundance and subsequently the potential impact of advective gas flux to Earth-atmosphere gas and vapor exchange rate. Preliminary measurements within an ancient borehole (52 m deep, 3.5 m in diameter, Negev Desert, Israel) indicated unstable thermal conditions to a depth of 40 m throughout most of the year nights (data not shown) suggesting the deep reach of the mechanism into the vadose zone.

The studies reviewed herein have shown the diurnal cycle appearance of thermal convection flows in surface exposed fractures. On a yearly scale it was shown that during winter, duration of convection flows is longer compared to summer time, in accord with winter cold atmospheric temperatures and long nights. Velocities of thermal convection flows were found to increase linearly with increasing thermal gradient and fracture aperture (higher Ra). In addition, as Ra increased, convection cells' geometry changed. For high Ra values, convection cells were narrower and taller, resulting in more convection cells per unit fracture volume. Consequently, as Ra increased, the mass transfer rates between the fracture and ambient air increased with a quadratic relationship, reflecting the combined effect of the linear increase in velocity and changes in convection cell geometry. It was shown that for the natural thermal conditions under which measurements were made, surface exposed fractures with aperture in the range of 1-2 cm will accommodate gas transport due to the thermal convection that is two orders of magnitude greater than non convective conditions. These findings demonstrate the importance of the thermal convection in fractures to atmosphere gas exchange processes and hence, vadose zone ventilation.

It is vital to understand gas transfer mechanisms between vadose zone and atmosphere, particularly with respect to greenhouse gases emissions, since vadose zone gas composition is in many cases markedly different than atmospheric gas composition (having high concentrations of greenhouse gases such as H_2O, CO_2, CH_4 and N_2O) (Reicosky et al., 1997; Hillel, 1998; Brady, 1999; Scanlon et al., 2001). Convection can lead to large mass exchange rates of the subsurface and atmospheric air via fracture and other discontinuities of the porous media. Nevertheless, vadose zone ventilation may be limited by gas diffusion through the matrix, as gases need to diffuse through the porous media toward fractures. Since diffusion fluxes are several orders of magnitude lower than convection fluxes within fractures, the diffusion rate from the matrix to the fracture will be the limiting factor in vadose zone-atmosphere gas exchange rates. Yet, even with this limit, fracture and other discontinuities in the porous media ventilation by thermal convection have an important effect on gas emission from the vadose zone as they expose deeper parts of the vadose zone to fresh atmospheric air, increasing the total atmospheric-matrix boundary area, consequently increasing overall diffusion fluxes.

Figure 9 presents an estimate of diffusive fluxes from a CO_2 point source, at 1 m depth, for conditions of unfractured porous media and fractured media with thermal convection flows. Diffusion rates were calculated using Equations 1 and 2 and Table 2 (for CO_2 diffusion coefficient in free air), for various porosities, ranging from 0 to 50%, to simulate the variation in soil or rock porosity due to texture and/or morphology. CO_2 concentration was considered as 3000 ppm and 400 ppm for the source and atmosphere respectively. For the fracture conditions the following assumptions were made: a spacing of 1 m between fractures, and CO_2 concentration equal to atmospheric concentration, due to the high mass transfer rates between fractures and atmosphere. Figure 9 shows that as porosity increases, diffusion fluxes increase too, due to increase of the effective diffusion coefficient of the porous media (Equation 2). The important contribution of the thermal convection within fractures to the overall diffusion of CO_2 from the vadose zone is thus shown, with total convective mass flux being five fold higher than for unfractured conditions, regardless of porosity. Convective venting, besides increasing gas exchange rates between vadose zone and atmosphere, may also affect greenhouse gases production within the vadose zone. The increase in vadose zone ventilation and subsequently the increase in oxygen supply for deeper parts of the vadose zone may result in increasing biological activity and organic matter decomposition, which are the major producers of CO_2 and CH_4 in the upper parts of the vadose zone (Jenkinson et al., 1991; Ryan & Law, 2005).

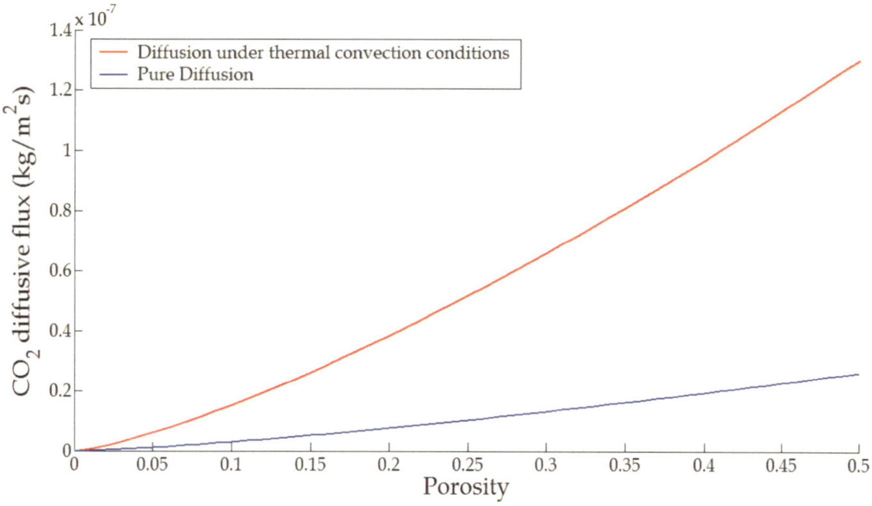

Fig. 9. Calculated diffusive mass fluxes for CO_2 from a point source 1 m deep for fractured-thermal convection conditions (red curve) and unfractured, pure diffusion conditions (blue curve).

Convective venting from cracks and fractures was found to make a marked contribution to vadose zone-atmosphere water vapor flux. Kamai et al. (2009) found that the contribution of nighttime thermal convection to vapor flux from a fractured porous media region, assuming a 1 m fracture spacing, is of similar order of magnitude as the daytime soil evaporative flux observed in field studies (Cahill & Parlange, 1998). Consequently, to ignore this mechanism is to ignore potentially half of the water vapor flux to the atmosphere in regions of cracked soil or fractured rock surface. Salt deposition data also indicate a significant contribution of the thermal convection mechanism for water evaporation. Within a 1 m deep, 2 cm aperture fracture in the arid Negev Desert, Weisbrod & Dragila (2006) calculated an accumulated salt-crust mass corresponding to a vapor venting rate of up to 200 times greater than that which would be predicted by diffusive venting alone. This explains some field observations in the Israeli Negev desert, where enhanced salt-loading inside fractures occurs (Weisbrod et al., 2000). As fractures and cracks are well-known to serve as a bypass for groundwater recharge, enhanced accumulation of evaporates within fractures could pose a risk to groundwater salinization and contamination, especially in arid and semiarid zones with deep vadose zones (Nativ & Nissim, 1992; Scanlon et al., 1997; Weisbrod et al., 2000), hence, advective air fluxes also play a determinant role in groundwater salinization.

In summary, the work presented here demonstrates that advective venting should be incorporated into predictive models in which gas flux across the earth-atmosphere boundary is being considered and quantified. The recently explored ventilation mechanism through thermal convection flows in surface exposed fractures was introduced. It was shown that this mechanism plays an important role in ventilation of fractured porous media. Most likely, it is also very important in other cases where the porous media is contains discontinuities such as karstic systems, shafts, caves, animal burrows etc. It was shown that advective mechanisms can significantly enhance gas flux across this interface. These fluxes, which were briefly mentioned above, have relevance to greenhouse gases migrating into and out of the earth, the water cycle, and soil and groundwater salinization.

5. Acknowledgement

Supported by the Israeli Science Foundation (ISF), grant 70/06, the Binational Science foundation (BSF), grant 2002058, and the National Science Foundation (NSF), grants 0208384 and 0510825.

6. References

Amali, S. & Rolston D.E. (1993). Theoretical investigation of multicomponent volatile organic vapor diffusion: steady-state fluxes. *Journal of Environmental Quality*, Vol. 22, pp. 825–831.

Amundson, R. (2005). Soil formation, in: *Surface and ground water, weathering, and soils*, Drever, J. I., (Ed.), 1-37, Elsevier, London.

Auer, L.H., Rosenberg, N.D., Birdsell, K.H. & Whitney, E.M. (1996). The effect of barometric pumping on contaminant transport. *Journal of Contaminant Hydrology*, Vol. 24, pp. 145-166.

Balaji, C. & P., Venkatesehan. (1995). Combined conduction, convection and radiation in a slot. *International Journal of Heat and Fluid Flow*, Vol. 16, pp.139–144.

Brady, C. N. (1990). *The nature and properties of soils*, 10th ed., Macmillan publishing company, New York.

Buyanovsky, G.A., Wagner, G. H. & C. J., Gantzer (1986). Soil respiration in a winter wheat ecosystem. *Soil Science Society of America Journal*, Vol. 50, pp. 338-344.

Cahill, A. T. & M. B., Parlange. (1998). On water vapor transport in field soils, *Water Resour. Res.*, 34, 731–739.

Choi, J.W., F.D., Tillman & J.A., Smith. (2002). Relative importance of gas phase diffusive and advective trichloroethene (TCE) fluxes in the unsaturated zone under natural conditions. *Environmental Science & Technology*, Vol. 36, pp. 3157–3164.

Conant, B.H., R.W., Gillham & C.A., Mendoza. (1996).Vapor transport of trichloroethylene in the unsaturated zone: field and numerical modeling investigations, *Water Resources Research*, Vol. 32, No. 1, pp. 9–22.

Cunningham, R.E. & R.J.J., Williams. (1980). *Diffusion of Gases and Porous Media*, Plenum, New York.

Cussler, E. W. (1997). *Diffusion: mass transfer in fluid systems*, 2nd ed., Cambridge University Press, Cambridge.

Don Scott, H. (2000) *Soil physics: agricultural and environmental applications*. Iowa state university press, Iowa.

Evett, S.R. (2001). Water and energy balances as soil-plant-atmosphere interfaces, in: *Soil Physics Companion*, Warrick, A. W, (Ed.),127-188, CRC Press, Boca Raton, Florida.

Fang, C. & J. B. Moncrieff. (1999), A model for soil CO2 production and transport 1: Model development. *Agricultural and Forest Meteorology*. Vol. 95, No. 4, pp. 225-236, ISSN 0168-1923, DOI: 10.1016/S0168-1923(99)00036-2.

Fenton, R.K. & W.G., Whitford. (1978). The Burrow Environment of the Banner-Tailed Kangaroo Rat, Dipodomys spectabilis, in Southcentral New Mexico. *American Midland Naturalist*. Vol. 99, No. 2, pp. 270-279.

Fick, A. (1855). Ueber Diffusion. *Annals of Physics*, Vol. 170, No. 1, pp. 59- 86.

Ford, D. C., & P. W. Williams. (1989). *Karst Geomorphology and Hydrology*, Chapman and Hall, London.

Fukuda, H. (1955). Air and vapor movement in soil due to wind gustiness. *Soil Science*, Vol. 79, pp. 249-256.

Granger, R. A. (1995). *Fluid mechanics*. Rinehart & Winston, New York.

Hillel, D. (1998). *Environmental Soil Physics*, Academic, San Diego, California.

Hirst, W. & G.E., Harrison. (1939). The diffusion of radon gas mixtures, *Proceedings of the Royal Society A*, Vol. 169, pp.573–586.

Ho, C.K. (2006). Vapor transport processes, in: *Gas Transport in Porous Media*, C. K. Ho & S. W. Webb, (Ed.), pp. 27- 46, Springer, Dordrecht, the Netherlands.

Holford, D.J., S.D., Schery, J.L., Wilson & F.M., Phillips. (1993). Modeling radon transport in dry, cracked soil. *Journal of Geophysical Research*, Vol. 98, pp. 567–580.

Houghton, R.A. & G.M., Woodwell. (1989). Global warming change, *Scientific American*. Vol. 260, pp. 18-26.

Ishihara, Y., E., Shimojima & Harada, H. (1992). Water vapor transfer beneath bare soil where evaporation is influenced by a turbulent surface wind, *Journal of Hydrology*, Vol. 131, No. 1-4, pp. 63-104.

Jenkinson D.S., D.E., Adams & A. Wild. (1991). Model estimates of CO_2 emissions from soil in response to global warming. *Nature*, Vol. 351, pp. 304–306.

Kamai, T., N. Weisbrod & M.I., Dragila. (2009). Impact of ambient temperature on evaporation from surface-exposed fractures, *Water Resources Research*, Vol. 45, W02417, doi:10.1029/2008WR007354.

Kleineidam, C., R., Ernst & F., Roces. (2001). Wind induced ventilation of the giant nests of the leaf-cutting ant Atta vollenweideri. *Naturwissenschaften,* Vol. 88, No. 7, pp. 301-305, DOI:10.1007/s001140100235.

Kimball, B.A. & E.R., Lemon. (1971). Air turbulence effects upon soil gas exchange. *Soil Proceedings - Soil Science Society of America,* Vol. 35, pp.16-21.

Kimball, B.A. & E.R., Lemon. (1972). Theory of soil air movement due to pressure fluctuations. *Agricultural and Forest Meteorology,* Vol. 9, pp. 163-181.

Lambers, H., F.S., Chapin & T.L. Pons. (2008). *Plant physiological ecology,* 2nd ed., Springer.

Lanchao, L., & F. Amir. (1998). An analysis of two-phase fl ow stability in a thermosyphon with tube separator. *Applied Thermal Engineering,* Vol. 18, pp. 441–455.

Lapwood, E.R. (1948). Convection of a fluid in a porous medium. *Proceedings of the Cambridge Philological Society,* Vol. 44, pp. 508–521.

Lenhard, R.J., M. Oostrom, C.S. Simmons & M.D. White. (1995). Investigation of density-dependent gas advection of trichloroethylene: experiment and a model validation exercise, *Journal of Contaminant Hydrology,* Vol. 19, pp. 47–67.

Liu, Z. & J. Zhao. (2000). Contribution of carbonate rock weathering to the atmospheric CO_2 sink, *Environmental Geology,* Vol. 39, No. 9, pp. 1053-1058.

Marshall, T.J. (1958). A relation between permeability and size distribution of pores, *Journal of Soil Science,* Vol. 9, pp. 1–8.

Marshall, T. J. (1959). The diffusion of gases through porous media, *Soil Science Society of America Journal,* Vol. 10, No. 1, pp. 79- 82.

Massmann, J. & D.F. Farrier. (1992). Effects of atmospheric pressures on gas transport in the vadose zone. *Water Resources Research,* Vol. 28, No. 3, pp. 777- 791.

Millington, R.J. & J.P. Quirk. (1961). Permeability of porous solids, *Transactions of the Faraday Society,* Vol. 57, No. 7, pp.1200–1207.

Moldrup, P., T. Olesen, P. Schjonning, T. Yamaguchi & D.E. Rolston. (2000). Predicting the gas diffusion coefficient in undisturbed soil from soil water characteristics, *Soil Science Society of America Journal,* Vol. 64, No. 3, pp. 94- 100.

Morgan, R.P.C. (2005). *Soil erosion and conservation.* 3rd (Ed.) Blackwell publishing, MA, USA.

Nachshon, U., N. Weisbrod & M.I. Dragila. (2008). Quantifying air convection through surface-exposed fractures: A laboratory study, *Vadose Zone Journal,* Vol. 7, No. 4, pp. 948- 956.

Nachshon, U., N. Weisbrod, M.I. Dragila & A. Grader. (2011). Combined evaporation and salt precipitation in homogeneous and heterogeneous porous media, *Water Recourses Research,* Vol. 47, W03513, doi:10.1029/2010WR009677.

Nativ, R. & I. Nissim. (1992). Characterization of a Desert Aquitard— Hydrologic and Hydrochemical Considerations. *Ground Water,* Vol. 30, pp. 598–606.

Nazaroff, W.W. (1992). Radon transport from soil to air, *Reviews of Geophysics* , Vol. 30, pp. 137-160.

Nield, D.A. (1982). Onset of convection in a porous layer saturated by an ideal gas. *International Journal of Heat and Mass Transfer,* Vol. 25, pp. 1605–1606.

Nilson, R.H., W.B. McKinnis, P.L. Lagus, J.R. Hearst, N.R. Burkhard, & C.F. Smith. (1992). Field measurements of tracer gas transport induced by barometric pumping, in: *Proc. of the Third International Conference of High Level Radioactive Waste Management,* Am. Nucl. Soc.pp. 710–716.

NOAA. (2011). Trends in atmospheric carbon dioxide, in: Earth system research laboratory, Available from: http://www.esrl.noaa.gov/gmd/ccgg/trends/global.html.

Penman, H.L. (1940a). Gas and vapor movements in the soil. I. The diffusion of vapors through porous solids, *Journal of Agricultural Science*, Vol. 30, pp. 437–462.

Penman, H.L. (1940b). Gas and vapor movements in the soil. II. The diffusion of carbon dioxide through porous solids, *Journal of Agricultural Science*, Vol. 30, pp. 570–581.

Pirkle, R.J., D.E. Wyatt, V. Price & B.B. Looney. (1992). Barometric pumping: The connection between the vadose zone and the atmosphere. In: *National Ground Water Association, Proceedings of the Focus Conference on Eastern Regional Ground Water Issues*. 13-15 October, 1992. Boston. pp 427-442

Rastogi, M., S. Singh & H. Pathak. (2002). Emission of carbon dioxide from soil, *Current Science*, Vol. 82, pp. 510–517.

Reicosky, D.C., W.A. Dugas & H.A. Torbert. (1997). Tillage-induced soil carbon dioxide loss for different cropping systems. *Soil & Tillage Research*, Vol. 41, pp. 105–118.

Ronen, D., H. L. Wiener, E. R. Graber, O. Dahan & N. Weisbrod. (2010). Simultaneous counter-flow of chlorinated volatile organic compounds across the saturated-unsaturated interface region of an aquifer, *Water Research*, Vol. 44, No. 7, pp. 2107-2112, ISSN 0043-1354, DOI: 10.1016/j.watres.2009.12.016.

Rossabi, J. (2006). analyzing barometric pumping to characterize subsurface permeability, in: *Gas Transport in Porous Media*, C. K. Ho & S. W. Webb, pp. 279- 289, Springer, Dordrecht, the Netherlands.

Ryan, M. G. & B.E. Law. (2005). Interpreting, measuring, and modeling soil respiration, *Biogeochemistry*, Vol. 73, pp. 3- 27.

Scanlon, B.R., S.W. Tyler & P.J. Wierenga. (1997). Hydrologic issues in arid, unsaturated systems and implications for contaminant transport. *Reviews of Geophysics*, Vol. 35, pp. 461–490.

Scanlon, B.R., J.P. Nicot & J.W. Massmann. (2001). Soil Gas Movement in Unsaturated Systems, in: *Soil Physics Companion*, Warrick, A. W. pp. 297-341, CRC Press, Boca Raton, Florida.

Schaschke, C. (1998). *Fluid Mechanics: Worked Examples for Engineers*. PP. 300. Institution of Chemical Engineers, Rugby.

Shemin, G. (1997). A governing equation for fluid flow in rough fractures. *Water Resources Research*, Vol. 33, pp. 53–61.

Smith, J.A., A.K. Tisdale & H.J. Cho. (1996). Quantification of natural vapor fluxes of TCE in the unsaturated zone at Picatinny arsenal, New Jersey, *Environmental Science & Technology*, Vol. 30, No. 7, pp. 2243–2250.

Thorstenson, C. D. & D. W. Pollock. (1989). Gas transport in unsaturated zones: Multicomponent systems and the adequacy of Fick's laws,*Water resources research*, Vol. 25, No. 3, pp. 477-507, doi:10.1029/WR025i003p00477.

Vogel, S., C. Ellington & D. Kilgore. (1973). Wind-induced ventilation of the burrow of the prairie-dog Cynomys ludovicianus. *journal of Comparative Physiology A: Neuroethology, Sensory, Neural, and Behavioral Physiology*, Vol. 85, NO. 1, pp 1-14.

Weast, R.C. (1980). *Handbook of chemistry and physics*. 6th (Ed.), CRC Press, Cleveland, OH.

Weeks, E.P. (1993). Does the wind blow through Yucca Mountain, 45–53, in: *Proc. of Workshop V: Flow and Transport through Unsaturated Fractured Rock Related to High-*

Level Radioactive Waste Disposal, D.D. Evans and T.J. Nicholson (eds.), U.S. Nucl. Reg. Comm., NUREG CP-0040.

Weeks, E.P. (1994). Thermal and wind pumping as mechanisms for passive vapor extractions in hilly terrain. *Transactions - American Geophysical Union,* Vol. 75, No. 44, pp.263.

Weihermüller, L., M. Lamers & M. Reichstein. (2011). Introduction to production, transport and emission of trace gases from the Vadose zone to the Atmosphere. *Vadose Zone Journal,* Vo. 10, pp. 151-155.

Weisbrod, N., R. Nativ, E. Adar & D. Ronen. (2000). Salt accumulation and flushing in unsaturated fractures in an arid environment. *Ground Water,* Vol. 38, No. 3, pp. 452-461.

Weisbrod, N., & M.I. Dragila. (2006). Potential impact of convective fracture venting on salt-crust buildup and ground-water salinization in arid environments. *Journal of Arid Environments,* Vol. 65, pp. 386–399.

Weisbrod, N., M. Pillersdorf, M.I. Dragila, C. Graham & J. Cassidy. (2005). Evaporation from fractures exposed at the land surface: Impact of gas-phase convection on salt accumulation. pp. 151–164. In: *Dynamics of fluids and transport in fractured rock.* Vol. 162. Am. Geophysical Union, Washington, DC.

Weisbrod, N., M.I. Dragila, U. Nachshon & M. Pillersdorf. (2009). Falling through the cracks: The role of fractures in Earth-atmosphere gas exchange, *Geophysical Research Letters,* Vol. 36, L02401, doi:10.1029/2008GL036096.

Wyatt, D. E., D. M. Richers & R. J. Pirkle. (1995). Barometric pumping effects on soil gas studies for geological and environmental characterization, *Environmental Geology,* Vol. 25, No. 4, pp. 243-250. Url: http://dx.doi.org/10.1007/BF00766753.Doi: 10.1007/BF00766753

17

Reforming CO_2 into Fuel Using a TiO_2 Photocatalyst Membrane Reactor

Akira Nishimura[1] and Eric Hu[2]
[1]*Mie University*
[2]*The University of Adelaide*
[1]*Japan*
[2]*Australia*

1. Introduction

Due to mass consumption of fossil fuels, global warming and fossil fuels depletion have become the serious global environmental problems in the world. After the industrial revolution, the averaged concentration of CO_2 in the world has been increased from 280 ppmV to 387 ppmV. Therefore, it is necessary to develop new energy production technologies with less or no CO_2 emission. It is reported that CO_2 can be reformed into fuels eg. CO, CH_4, CH_3OH and H_2 etc. by using TiO_2 as the photocatalyst under ultraviolet (UV) light illumination (Adachi et al., 1994; Anpo & Chiba, 1992; Aurian-Blajeni et al., 1980; Dey et al., 2004; Henglein & Gutierrez, 1983; Hirano et al., 1992; Inoue et al., 1979; Ishitani et al., 1993; Kaneco et al., 1999; Ogura et al., 1992; Takeuchi et al., 2001). If this technique could be applied practically, a carbon circulation system would then be established: CO_2 from the combustion of fuel is reformed, using solar energy, to fuels again, and true zero emission can be achieved. Many R&D works on this technology have been carried out, using TiO_2 particles loaded with Cu, Pd, Pt to react with CO_2 dissolved in solution (Adachi et al., 1994; Goren et al., 1990; Halmann et al., 1984; Hirano et al., 1992; Ibusuki, 1993; Ishitani et al., 1993; Kawano et al., 1993; Lo et al., 2007; Tseng et al., 2002; Yamashita et al., 1994). Recently, nano-scaled TiO_2 (Pathak et al., 2004; Qu et al., 2005; Xia et al., 2007), porous TiO_2 (Cecchet et al., 2006), TiO_2 film combined with metal (Cueto et al., 2006; Wu & Lin, 2005), and dye sensitized TiO_2 (Ozcan et al., 2007), are developed for this process. However, the fuel concentration in the products achieved an all the attempts so far is still to low, ranging from 10 ppmV to 1000 ppmV, to be practically useful (Adachi et al., 1994; Dey et al., 2004; Goren et al., 1990; Halmann et al., 1984; Hirano et al., 1992; Ishitani et al., 1993; Kaneco et al., 1999; Lo et al., 2007; Pathak et al., 2004; Tseng et al., 2002; Xia et al., 2007). For the fuels to be practically useful, the concentration of produced fuels should be exceed the lowest combustible concentration of each fuel. For example, for CH_4 and CO, 5.3 vol.% and 12.5 vol.% is required, respectively. Therefore, the big breakthrough in increasing the concentration level is necessary to advance the CO_2 reforming technology.
According to the calculation by the author, the mass transfer time of 10^5 - 10^{-1} s is much slower than the photo reaction time of 10^{-9} - 10^{-15} s in this process. Therefore, the mass transfer is thought to be the main factor contributing to the slow photocatalytic reaction. Another reason courses the low reforming rate is the re-oxidization of the products.

Namely, due to the reaction surface covered by products, the further movement of the reactants to the reaction surface is prevented and the reverse reaction, i.e. re-oxidization, which produces CO_2 from CO and CH_4, occurs. Therefore, it is desirable that the products, i.e. CO and CH_4 are removed from the reaction surface as soon as they are produced. The reactants i.e. CO_2 and water vapour can then continue to react on the reaction surface, and the fuel production can be sustained under this non-equilibrium reaction condition. In other words, by removing the products away, the reaction is forced to head to one direction. The gas separation membrane is usually used in the gas separation processes like H_2 production from hydrocarbon, O_2 enrichment from the air, and CO_2 capture of the industrial power plants. Since the molecular diameters of reactants of CO_2 and water vapour are smaller than that of CO and CH_4 (CO_2: 0.33 nm, water vapour: 0.28 nm, CH_4: 0.38 nm, CO: 0.38 nm) (Nakagawa, 1988), the promotion of the reaction by gas separation is thought to be possible and was attempted in this study. This is a novel approach aiming to improve CO_2 reforming performance over the TiO_2. No similar attempts have been reported yet.
Since this research approach is very novel, the following subjects are set in this study:

i. Optimization of preparation condition, especially rising speed (*RS*) of gas separation membrane from the TiO_2 sol solution in dip-coating process in order to select the optimal TiO_2 film coating conditions to prepare the membrane for the reactor of gas circulation type,
ii. Verification of the concept of this study which is promotion of CO_2 reforming performance by gas separation and circulation,
iii. Proposal of TiO_2 photocatalyst power system with zero CO_2 emission for the future.

This chapter introduces the authors' approach to research and develop the TiO_2 photocatalyst membrane reactor consisting of TiO_2 photocatalyst and gas separation membrane. At first, the preparation procedure of TiO_2 film coated on gas separation membrane is introduced. To optimize the preparation condition, the surface structure and crystallization characteristics of TiO_2 film coated on gas separation membrane are analysed. In addition, the CO_2 reforming and permeation performance of TiO_2 film coated on porous gas separation membrane is evaluated by the batch type reactor. Finally, the CO_2 reforming performance of TiO_2 film coated on porous gas separation membrane is investigated by the gas circulation type reactor and the concept of this study which is promotion of CO_2 reforming performance by gas separation and circulation is verified.

2. Research and development on TiO_2 photocatalyst membrane reactor

Since TiO_2 photocatalyst membrane reactor is a novel approach to improve CO_2 reforming performance over the TiO_2, it is necessary to verify the effect of combination of gas separation membrane and TiO_2 on CO_2 reforming performance. Therefore, the authors investigated the preparation procedure of TiO_2 film coated on gas separation membrane by sol-gel and dip-coating method and the experimental operation conditions to promote the CO_2 reforming performance of the TiO_2 photocatalyst membrane reactor. The *RS* which influences the thickness and physical and chemical structure of TiO_2 film coated on gas separation membrane was investigated. The surface structure and crystallization characteristics of TiO_2 film coated on gas separation membrane, under the various *RS* conditions, were analysed by SEM (Scanning Electron Microscope), EPMA (Electron Probe Micro Analyzer) and XPS (X-ray Photoelectron Spectroscopy) to understand the impact of difference of *RS* on the surface structure and crystallization characteristics of TiO_2 film, as the first step. The CO_2 reforming and permeation performance of TiO_2 film coated on gas separation membrane was evaluated by the batch type reactor in order to select the optimal TiO_2 film coating conditions to prepare the membrane for the reactor of gas circulation type. In other words, the ideal TiO_2 film for this application should have large reaction surface areas and high crystallization characteristics but does not block the pores in gas separation membrane. After the suitable TiO_2 film coating conditions are known, the CO_2 reforming performance of TiO_2 film coated on porous gas separation membrane was investigated by the gas circulation type reactor. The effectiveness of gas separation and gas circulation using the gas separation membrane on CO_2 reforming performance was compared with the results obtained from the batch type reactor experiment.

2.1 Preparation of TiO_2 film coated on gas separation membrane

Sol-gel and dip-coating method was used for preparing TiO_2 film in this study. Figure 1 shows the flow chart of the sol-gel and dip-coating method. TiO_2 sol solution was made by mixing $[(CH_3)_2CHO]_4Ti$ (purity of 95 wt.%, Nacalai Tesque Co.), anhydrous C_2H_5OH (purity of 99.5 wt.%, Nacalai Tesque Co.), distilled water, and HCl (purity of 35 wt.%, Nacalai Tesque Co.).

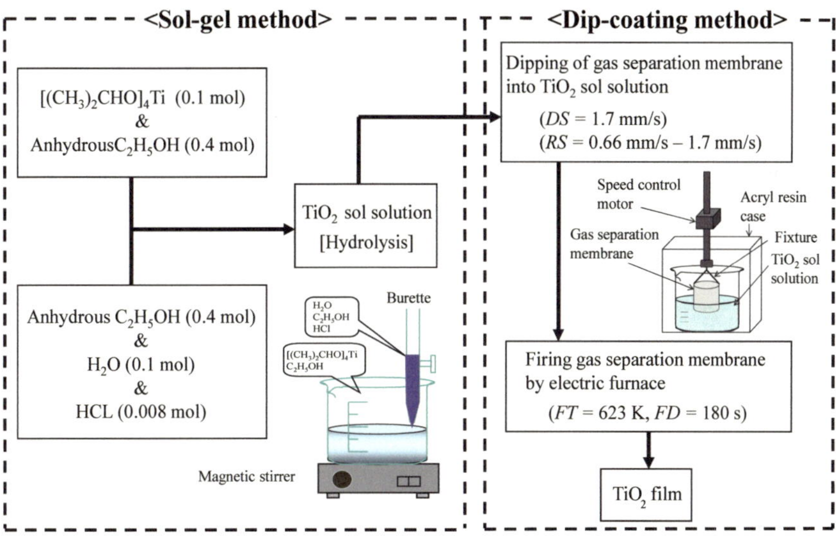

Fig. 1. Sol-gel and dip-coating method to prepare TiO_2 film in this study

The gas separation membrane (silica-alumina gas separation membrane, Noritake Co., Ltd.), which was the porous multilayer ceramic tube shown in Fig.2, was dipped into TiO_2 sol solution and pulled up at the fixed speed. Then, it was dried out and fired under the controlled firing temperature (*FT*) and firing duration time (*FD*), resulting that TiO_2 film was fastened on the surface of gas separation membrane. Coating number (*N*) was fixed at 1. *FT* and *FD* was set at 623 K and 180 s, respectively. *RS* varied from 0.66 mm/s to 1.7 mm/s. Downing speed (*DS*) of gas separation membrane into TiO_2 sol solution in dip-coating process was kept at the constant speed of 1.7 mm/s. Table 1 lists the physical properties of the gas separation membrane. It can be seen from Table 1, the mean pore size of silica layer is not ideal, as it is not between the molecular diameter of reactants and that of products, as required. It is difficult to find the gas separation membrane with the ideal pore size. However, the gas separation membrane selected is capable of separating gases through both molecular sieving diffusion and so called Knudsen diffusion mechanisms, therefore it can be used. The Knudsen diffusion can separate the gases whose molecular diameters are smaller than the pore size of silica layer. Since the molecular diameter of reactant and that of product is actually different as described above, we have decided to adopt this gas separation membrane.

Fig. 2. Gas separation membrane

	Thickness (μm)	Mean pore size (nm)	Void ratio (-)	Permeability (m^2)
1. Silica (SiO_2) layer	0.2	0.4	0.27	5.44×10^{-22}
2. γ-alumina (Al_2O_3) layer	2	4	0.44	8.88×10^{-20}
3. α-alumina (Al_2O_3) layer	100	60	0.39	1.76×10^{-17}
4. α-alumina (Al_2O_3) supporter	1000	700	0.40	2.45×10^{-15}
Silica layer is the top layer of this gas separation membrane.				
γ-alumina layer is the second layer. α-alumina layer is the third layer.				
α-alumina supporter is the bottom layer of gas separation membrane.				

Table 1. Physical properties of gas separation membrane

2.1.1 Analysis result of TiO_2 film coated on gas separation membrane by SEM

Figures 3 and 4 show SEM images of TiO_2 film prepared under various *RS* conditions. These SEM images were taken with 200 times and 1500 times magnification under the condition of acceleration voltage of 15 kV and current of 3.0×10^{-8} A. The silica layer covers one third of surface area of gas separation membrane used in this study at the center and the alumina layer is exposed except for the area covered by silica layer. Then, SEM images of TiO_2 film coated were taken for the silica covered area and the alumina area separately. From these figures, it can be seen that the number of clucks of TiO_2 film coated on alumina layer is less than that on silica layer, resulting that the amount of TiO_2 coated on alumina layer is larger than silica layer. Since the pore size of alumina layer is larger than that of silica layer as listed in Table 1, it can be thought that TiO_2 sol solution flows into the alumina layer more easily than the silica layer in dip-coating process. Therefore, it seems that TiO_2 film coated on alumina layer is fixed more strongly than that on silica layer.

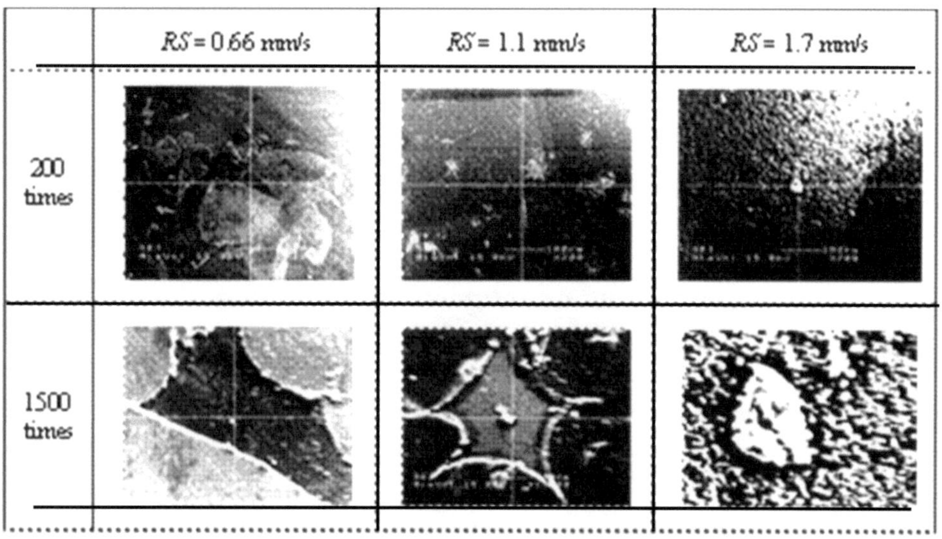

Fig. 3. SEM images of TiO_2 film coated on silica layer prepared under various *RS* conditions

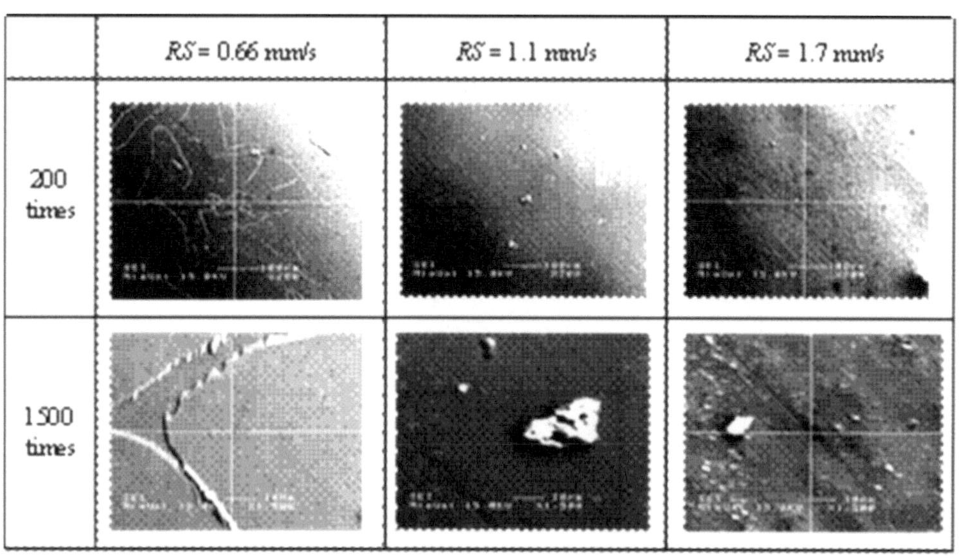

Fig. 4. SEM images of TiO_2 film coated on alumina layer prepared under various *RS* conditions

2.1.2 Analysis result of TiO_2 film coated on gas separation membrane by EPMA

Figure 5 demonstrates EPMA images of TiO_2 film prepared under various *RS* conditions. These EPMA images are taken by 1500 times magnification under the condition of acceleration voltage of 15 kV and current of 3.0×10^{-8} A. Tables 2 and 3 list the distribution of Ti concentration detected for silica layer and alumina layer, respectively.

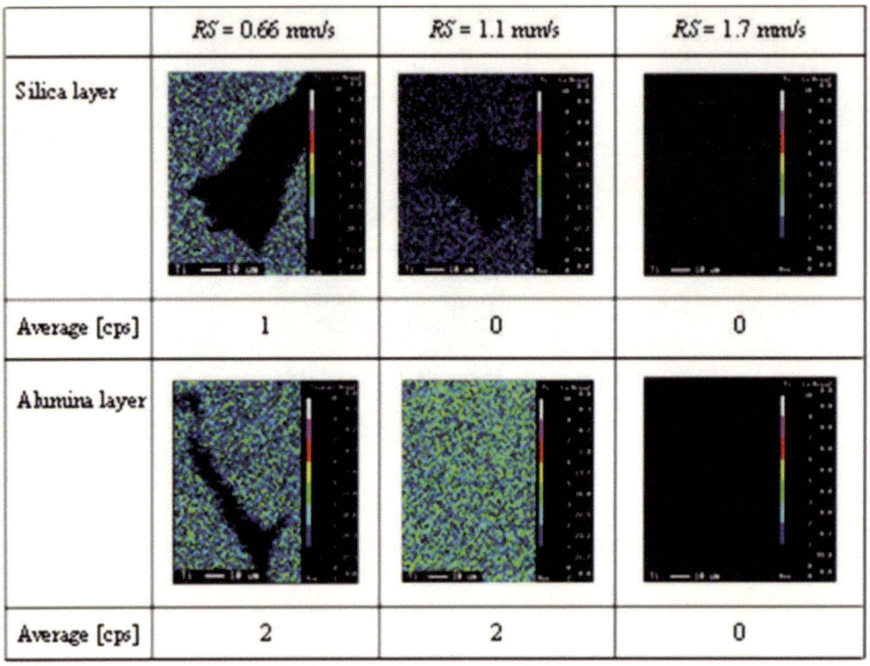

Fig. 5. EPMA images of TiO_2 film prepared under various *RS* conditions

	RS = 0.66 mm/s	*RS* = 1.1 mm/s	*RS* = 1.7 mm/s
Concentration of detected Ti (cps)			
10	0	0	0
10~8	0	0	0
8~7	0.1	0	0
7~6	0.5	0	0
6~5	5.0	0.5	0
5~3	8.3	1.8	0
3~2	14.5	6.1	0.3
2~1	20.1	17.2	2.8
1~0	51.4	74.4	96.9
0	0	0	0

Table 2. Concentration distribution of Ti detected by EPMA (silica layer)

	RS = 0.66 mm/s	*RS* = 1.1 mm/s	*RS* = 1.7 mm/s
Concentration of detected Ti (cps)			
10	0	0	0
10~8	0.1	0.3	0
8~7	0.2	0.7	0
7~6	0.7	1.8	0
6~5	7.5	13.7	0
5~3	11.4	16	0
3~2	18.6	22.5	0
2~1	24.3	23.2	0.2
1~0	37.1	21.7	99.8
0	0	0	0

Table 3. Concentration distribution of Ti detected by EPMA (alumina layer)

In Fig.5, the concentration distribution of Ti detected in observation area is indicated by the difference of colour. Light colours, e.g. white, pink and red mean that the amount of Ti is large, while dark colours like black, blue and green mean that the amount of Ti is small. EPMA detects the each element whose crystallization characteristic is memorized in advance. Therefore, if the large concentration of Ti is detected, it means that the amount of crystallized TiO_2 coated on gas separation membrane is large. The average concentration of Ti in the observation area is also shown in Fig.5. According to Fig.5, the average concentration of Ti detected in observation area for alumina layer is larger than that for silica layer. It can be said that TiO_2 film is coated in the pores of alumina layer mainly. From Fig.5, Tables 2 and 3, it reveals that the concentration of Ti is reduced with the increase in *RS*. Generally speaking, the thickness of TiO_2 film becomes thick and hubbly with the increase in *RS*. The thermal stress is acted on the interface between TiO_2 film and gas separation membrane in the firing process, resulting that formation of large clucks and detachment of TiO_2 film occur. Consequently, the concentration of Ti is reduced when *RS* is high.

2.1.3 Analysis result of TiO_2 film coated on gas separation membrane by XPS

Figures 6 and 7 show the intensity distributions of Ti detected in silica layer and alumina layer, respectively. These XPS data were obtained under the condition of ion acceleration voltage of 4 kV and pass energy of 112 eV. The samples were sputtered by Ar ion laser whose acceleration voltage of 2 kV. The sputtering speed was 15 nm/min, which was estimated by assuming the sample as SiO_2. The electron orbits of detected elements which were Ti, Si and Al were set at 2p. From these figures, it is known that the sputtering time of *RS* = 1.1 mm/s is the shortest among various *RS* conditions. However, the intensity of detected Ti is over 80000 cps for *RS* = 1.1 mm/s. It can be said that the amount of Ti is large with *RS* = 1.1 mm/s, resulting that fine TiO_2 film is prepared. Regarding *RS* = 0.66 mm/s, it is seen that the intensity of Ti over 60000 cps can be detected up to about sputtering time of 25 min for silica layer. Though the intensity of Ti detected in alumina layer is smaller than that in silica layer, the detecting period of Ti in alumina layer is almost equal to that in silica layer. According to Fig.5, the average concentration of Ti for *RS* = 0.66 mm/s is the largest

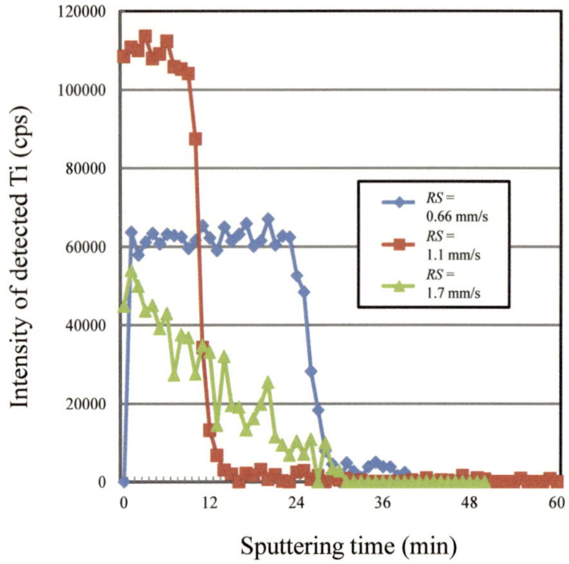

Fig. 6. Intensity distributions of detected Ti in silica layer

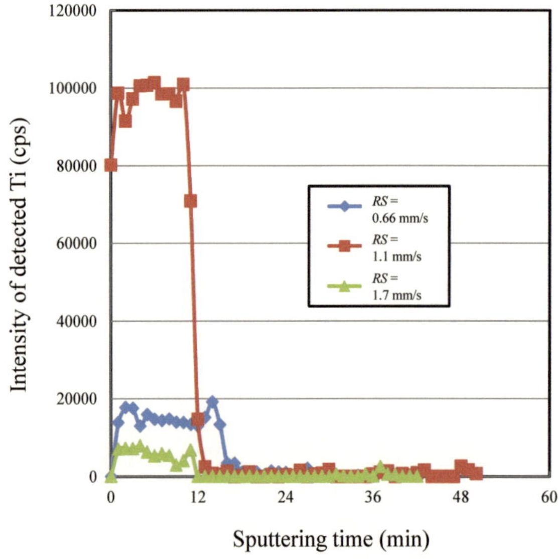

Fig. 7. Intensity distributions of detected Ti in alumina layer

among various *RS* conditions. From these results, one can conclude that the largest amount of TiO_2 film is coated on gas separation membrane with *RS* = 0.66 mm/s. Meanwhile, the intensity of detected Ti with *RS* = 1.7 mm/s is the lowest among various *RS* conditions. In addition, the average concentration of Ti is 0 cps as shown in Fig.5. Consequently, it is clear that the amount of TiO_2 film coated on gas separation membrane for *RS* = 1.7 mm/s is very small.

2.2 Performance of TiO_2 photocatalyst membrane reactor under batch type operation

Figure 8 illustrates the CO_2 reforming reactor that is termed as CO_2 reformer, with TiO_2 film coated on gas separation membrane. This reactor consists of one gas separation membrane with TiO_2 film (150 mm (*L.*)x6 mm (*O.D.*)x1 mm (*t.*), whose reaction surface is equal to the outer surface area: 2.26×10^{-3} m^2 and gas filling volume: 2.88×10^{-4} m^3) , one quartz glass tube (266 mm (*L.*)x42 mm (*O.D.*)x2 mm (*t.*)), and four UV lamps (FL16BL/400T16, Raytronics Corp., 400 mm (*L.*)x16 mm (*D.*)) located at 20 mm apart from the surface of gas separation membrane symmetrically. These parts are assembled with stainless plates by bolts and nuts. The center wave length and mean light intensity of UV light illuminated from energy UV lamp is 365 nm and 2.4 mW/cm², respectively. This is similar to the average light intensity level of UV ray in solar radiation in the daytime.

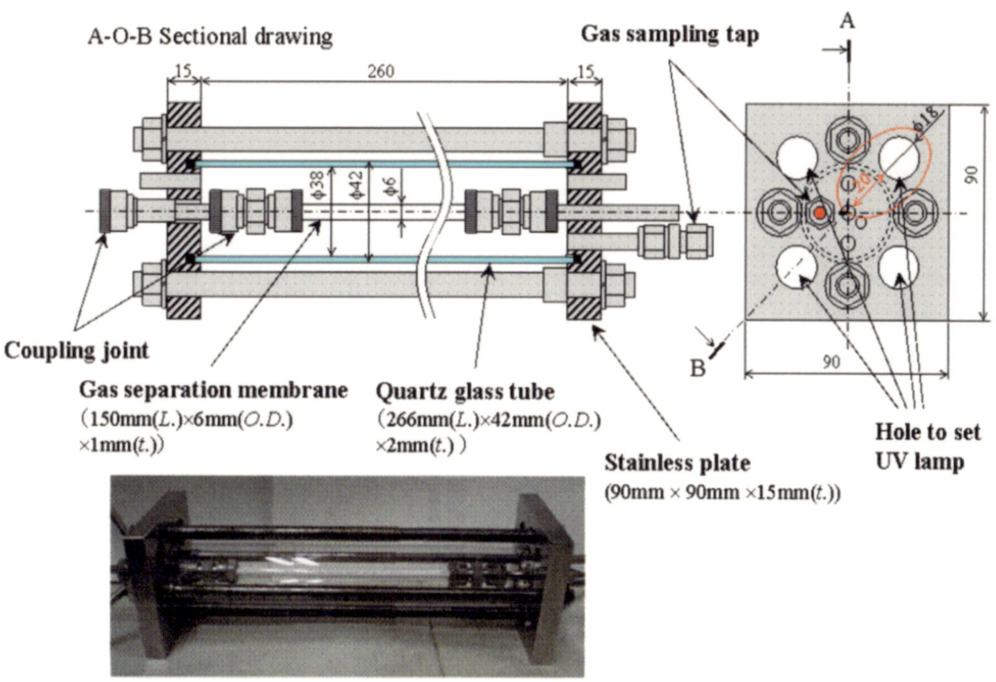

Fig. 8. CO_2 reformer composed of TiO_2 film coated on gas separation membrane (UV lamp is removed from this figure for understanding the inside of CO_2 reformer)

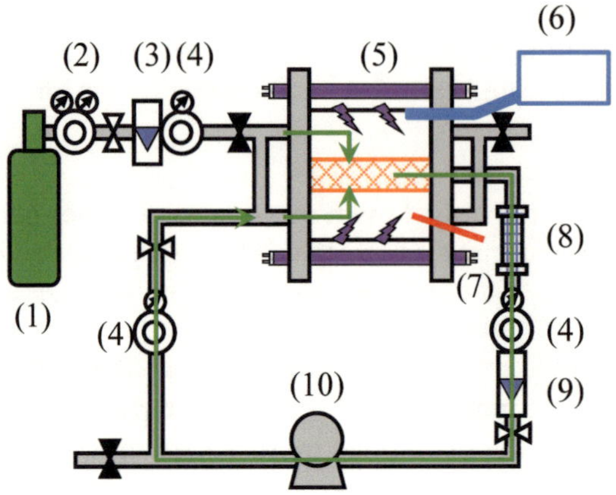

(1) CO_2 gas cylinder (99.995 vol%)
(2) Regulator
(3) Mass flow controller
(4) Pressure gauge
(5) CO_2 reformer
(6) Dew point meter
(7) Thermocouple
(8) Gas drier (Silica gel)
(9) Mass flow meter
(10) Tube pump

Fig. 9. CO_2 reforming and permeation experiment system

Figure 9 illustrates the whole experimental system set-up, which is termed as membrane reactor. With this membrane reactor, not only batch type but also gas circulation type experiment can be conducted. When it is used for batch type experiment, the valves located at inlet and outlet of CO_2 reformer are closed. The membrane reactor is composed of CO_2 reformer, CO_2 gas cylinder, mass flow controller (MODEL3660, KOFLOC), mass flow meter (CK-1A, KOFLOC), pressure gauge, gas drier and tube pump (WM-520S/R2, Iwaki Pumps). In the CO_2 reforming experiment by the batch type reactor, CO_2 gas whose purity was 99.995 vol.% was flowed through the CO_2 reformer as a purged gas for 15 min at first. After that, the valves located at inlet and outlet of CO_2 reformer were closed. After confirming the gas pressure and gas temperature in the reactor was at 0.1 MPa and 298 K, respectively, the distilled water of 1.00 mL (55.6 mmol) was injected into the CO_2 reformer and UV light illumination was started at the same time. This water was vaporized after injected into the reformer. Despite the heat of UV lamp, the temperature in CO_2 reformer was kept at about 343 K during the CO_2 reforming experiment. The amounts of the injected water and the CO_2 in the batch type reactor are 55.6 mmol and 13.0 mmol, respectively. The gas in CO_2 reformer was sampled every 24 h in CO_2 reforming experiment. The gas samples were analysed by FID gas chromatograph (GC353B, GL Science) and methanizer (MT221, GL Science). The concentration of water vapour and the temperature in CO_2 reformer was measured by dew point meter (VAISALA HUMICAP HMT330, VAISALA) and thermocouple, respectively. In this experiment, only CO was detected as the product. In the

CO_2 permeation experiment, the CO_2 reforming and permeation experimental system shown in Fig.9 was arranged.

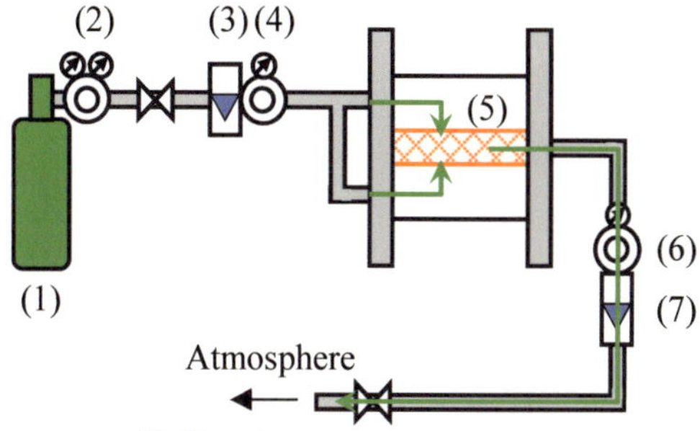

Fig. 10. Arranged CO_2 reforming and permeation experimental system for CO_2 permeation experiment

Figure 10 illustrates the CO_2 reforming and permeation experimental system for CO_2 permeation experiment by batch type. In the CO_2 permeation experiment, the CO_2 permeation flux was measured under the condition that the absolute pressure and temperature of supply gas to the apparatus was 0.10-0.40 MPa and 298 K, respectively. The flow rate of supply gas was set at 500 mL/min by mass flow controller. The flow rate of permeation gas was measured by mass flow meter. In the CO_2 reforming experiment carried out by the membrane reactor of gas circulation type, UV light was illuminated under the same condition of batch type reactor until the steady reaction state was confirmed. After that, the gas circulation by tube pump was started. The suction pressure and flow rate of permeation gas was controlled to evaluate the effect of gas separation and circulation on CO_2 reforming performance of this membrane reactor. The suction pressure and flow rate of permeation gas was set at 0.2 MPa and 0.39 mL/min, respectively. The produced CO would be removed from the CO_2 reformer to outside of the system by switching the outlet valve of CO_2 reformer on and off when needed. The distilled water of 1.00 mL (55.6 mmol) or 3.00 mL (166.8 mmol) was injected into CO_2 reformer when the CO_2 reforming experiment under the condition of batch type reactor was established. The gas samples taken every 24 h from CO_2 reformer were analysed by FID gas chromatograph and methanizer. The concentration of water vapour and the temperature in the CO_2 reformer were also measured.

2.2.1 CO_2 reforming by the membrane reactor of batch type

Figure 11 shows the CO concentration change in products with illumination time of UV light for several TiO_2 films prepared under various *RS* conditions. According to our previous studies, the reversal of superiority or inferiority on CO_2 reforming performance of TiO_2 film among selected parameters was confirmed until UV light illumination time of 48 h. However, this reversal was not confirmed and the superiority or inferiority among selected parameters was kept after UV light illumination time of 72 h. From this reason, in this study, the data is obtained only up to UV light illumination time of 72 h for the purpose of determining the best condition for promotion of CO_2 reforming performance. The distilled water of 1.00 mL (55.6 mmol) was injected into CO_2 reformer at the beginning of the CO_2 reforming experiment.

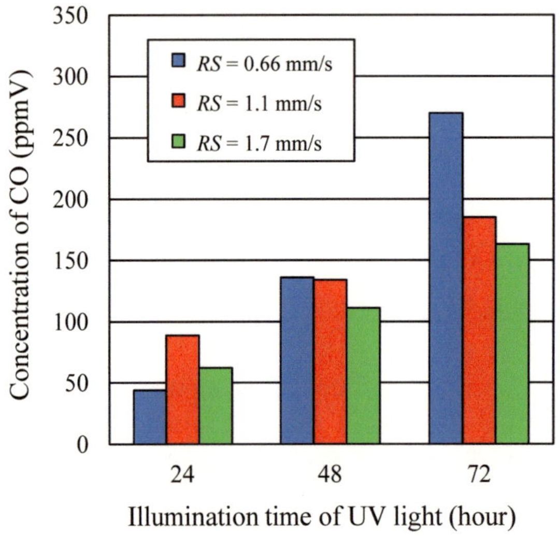

Fig. 11. Concentration change in produced CO with illumination time of UV light for several TiO_2 film prepared under conditions of different *RS*

From Fig.11, it is known that the concentration of CO is increased with decreasing *RS* values. Referring to the images of the SEM, EPMA and XPS shown in Figs.3, 4, 5, 6 and 7, the reason of this is thought to be that the amount of TiO_2 coated on gas separation membrane becomes larger when *RS* decreases within the range of 0.66 - 1.7 mm/s. Under slow *RS* condition, TiO_2 sol solution is easy to remain in the pore of silica and alumina layer in the dip-coating process, and TiO_2 film coated becomes thin and even. Consequently, the fine and strong TiO_2 film is prepared.

According to the reaction scheme shown in Fig.12, the number of electron and hydrogen ion (H^+) decides the type of product in the reaction. In this experiment, CH_4, C_2H_4 and the other hydrocarbons were not detected by gas chromatograph, due to less H^+ in the reaction. Therefore, the amount of water vapour injected into the reactor is an important parameter to be investigated since it is the source of H^+. From the reaction scheme, water vapour of 1 mol to CO_2 of 1 mol is necessary to produce CO of 1 mol. In this experiment, the amount of substance of injected water and CO_2 charged in the batch type reactor is 55.6 mmol and 13.0 mmol, respectively, resulting that the molar ratio of water vapour to

CO_2 is 4.28. Although the amount of water vapour injected seemed sufficient for this reaction, the change in temperature distribution and the concentration distribution of water vapour in the batch type reactor with time need to be checked to confirm what had happened.

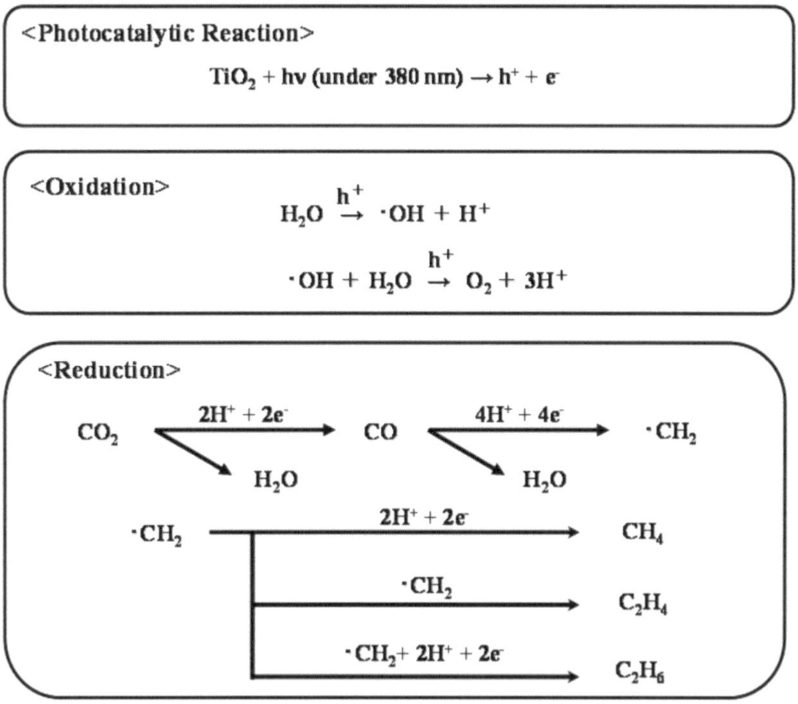

Fig. 12. Reaction scheme of CO_2 reforming into fuel by TiO_2 photocatalyst

Figure 13 shows the concentration change of water vapour in CO_2 reformer with illumination time of UV light during CO_2 reforming for several TiO_2 films prepared under various *RS* conditions. From this figure, it can be known that the concentration of water vapour for *RS* = 0.66 mm/s is the smallest among several *RS* conditions during UV light illumination from 0 h to 24 h. Under the same conditions, the temperatures in CO_2 reformer rose up to about 343 K during UV light illumination from 0 h to 24 h, and then remained steady irrespective of *RS*. The water injected had fully evaporated by UV light illumination at the time of 24 h. The reason why the concentration of water vapour for *RS* = 0.66 mm/s is the smallest can be explained by the results of XPS analysis. According to the results of XPS analysis, the largest amount of TiO_2 film is coated on gas separation membrane for *RS* = 0.66 mm/s, and the TiO_2 film is coated deeply to thickness direction in the pores of silica and alumina layer. The CO_2 reforming is carried out well and consumes more water, compared to the other *RS* conditions in the oxidization process of the reaction scheme. Therefore, the concentration of water vapour and CO for *RS* = 0.66 mm/s becomes low during the period from 0 h to 24 h, and the concentration of CO for

RS = 0.66 mm/s is the largest among the investigated *RS* conditions as shown in Fig.11. Although the water is produced in the reduction process of CO_2 reforming by TiO_2, this water vapour seems to be adsorbed by TiO_2 film or gas separation membrane. Therefore, the concentration of water vapour did not increase irrespective of *RS* as shown in Fig.13. According to saturated steam table, the saturation concentration of water vapour at 343 K is 307545 ppmV, while the 55.6 mmol of water injected, if all evaporates, just makes the concentration of 53040 ppmV in CO_2 reformer theoretically. Therefore, the all of water injected in CO_2 reformer is thought to be vaporized. On the contrary, the measured concentration of water vapour obtained in CO_2 reforming is just 25000 ppmV as shown in Fig.13. The gap between theoretical and experimental results might be caused by water vapour adsorption with TiO_2 film and gas separation membrane in the experiment of CO_2 reforming.

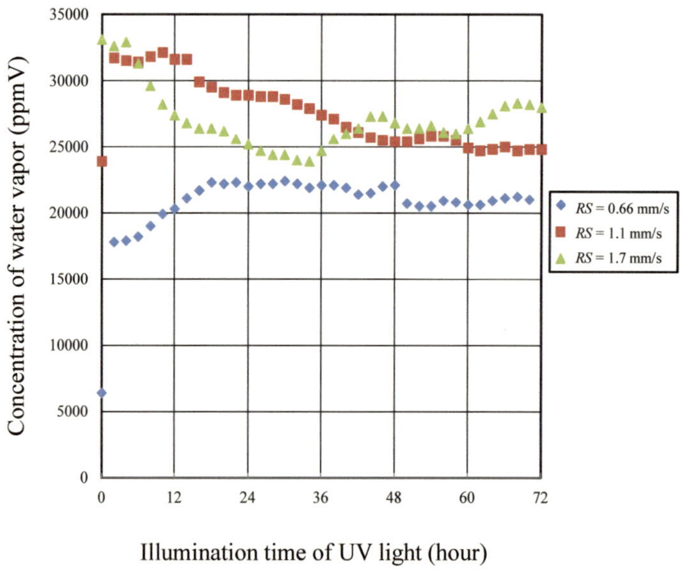

Fig. 13. Concentration change of water vapour in CO_2 reformer with illumination time of UV light during CO_2 reforming for several TiO_2 films prepared under various *RS* conditions

2.2.2 CO_2 permeation by the membrane reactor of batch type

Figure 14 shows the relationship between CO_2 permeation flux and pressure difference for several TiO_2 film prepared under various *RS* conditions. The pressure difference is known by subtracting the gas pressure after penetrating the gas separation membrane from the gas pressure before. The CO_2 permeation flux is calculated by the following equation:

$$F_{CO_2} = \frac{V_p}{A_p t_p} \tag{1}$$

where F_{CO_2}, V_p, A_p and t_p are CO_2 permeation flux (mol/(m²·s)), volume of permeated gas (mol), outer surface area of gas separation membrane (m²) and gas separation time (s), respectively.

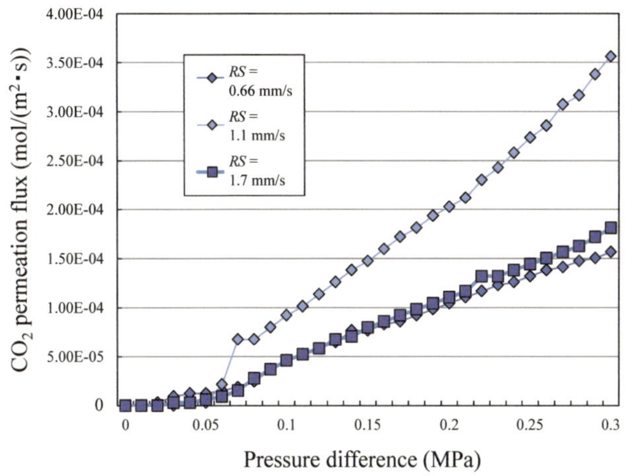

Fig. 14. Relationship between CO_2 permeation flux and pressure difference for several TiO_2 film prepared under various *RS* conditions

Comparing these results at pressure difference of 0.30 MPa, it is known that the CO_2 permeation flux for *RS* = 1.1 mm/s is the highest among these *RS* conditions. According to XPS analysis as shown in Figs. 6 and 7, the sputtering time of detecting Ti for *RS* = 1.1 mm/s is the shortest among various *RS* conditions, indicating that the depth of coated TiO_2 film diffused into the silica and alumina layers of gas separation membrane is the shallowest. Consequently, the highest CO_2 permeation flux is obtained at *RS* = 1.1 mm/s. On the other hand, regarding *RS* = 0.66 mm/s and 1.7 mm/s, the sputtering time of detecting Ti is longer though the intensity of Ti detected is lower as shown in Figs. 6 and 7, indicating the depth of coated. TiO_2 film diffused into the silica and alumina layers is deeper, thus the CO_2 permeation flux is lower.

2.2.3 Selection of the optimum coating condition

To select the optimum coating condition of TiO_2 film which would lead to the highest CO_2 reforming and permeation performance, the results by SEM and EPMA analysis and the results of CO_2 reforming and permeation experiment by the batch type reactor are compared and analysed. Figure 15 shows the comparison of the results between the concentration of produced CO and the CO_2 permeation flux for various *RS* conditions. In Fig.15, the concentration of CO at UV illumination of 72 h and CO_2 permeation flux at pressure difference of 0.30 MPa are shown. It can be seen that the concentration of CO is decreased with increasing *RS* gradually. On the other hand, the CO_2 permeation flux peaks at *RS* = 1.1 mm/s. Therefore, the optimum *RS* is different from the viewpoint of CO_2 reforming and permeation performance. Since the main goal of this study is to promote the CO_2 reforming performance, we have selected the *RS* = 0.66 mm/s as the optimum coating condition.

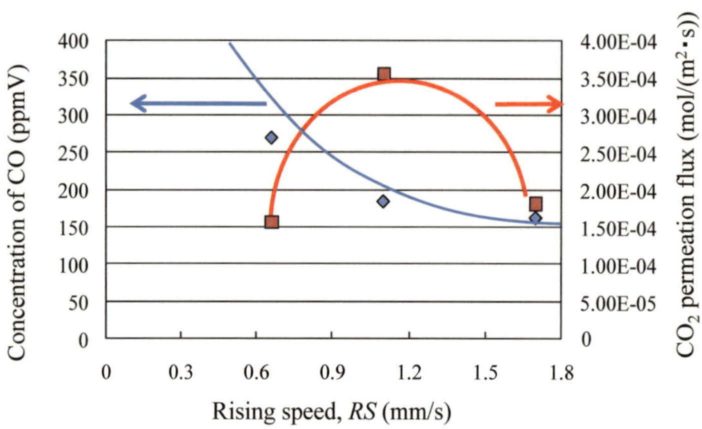

Fig. 15. Comparison of the results between concentration of produced CO and CO_2 permeation flux for each condition

2.3 Performance of TiO_2 photocatalyst membrane reactor under gas circulation type operation

According to the reaction scheme shown in Fig.12, CO is re-oxidized with the O_2 that is a by-product in this reaction. After attaining to the steady reaction state, the concentration of CO is decreased. This is the opposite reaction toward CO_2 reforming into fuel. Moreover, since the photocatalytic reaction occurred on the reaction surface, it is easy for the reaction surface to be covered by the products, which would stop the further reaction to happen. Therefore, removing the product of CO and CH_4 from the reaction surface as well as transporting the reactants, i.e. CO_2 and water vapour to the reaction surface quickly are necessary to promote further reaction and prevent the re-oxidization of CO. In this study, a tube pump and a gas separation membrane are used to realize this desirable measure for the promotion of CO_2 reforming performance.

2.3.1 CO_2 reforming by the membrane reactor of gas circulation type

Figure 16 shows the concentration change of CO produced with illumination time of UV light. The distilled water of 1.00 mL (55.6 mmol) was injected into CO_2 reformer when the CO_2 reforming experiment by batch type reactor started. To show the effect of gas separation and circulation on the CO_2 reforming performance, the gas circulation by tube pump only starts after the steady reaction state is reached. The steady reaction state was defined as the state at which the concentration of CO no longer increases along the time. Since the concentration of CO is diluted with the CO_2 in the pipe lines of the gas circulation type reactor after starting gas circulation, the concentration of CO is corrected by the following equation:

$$C_c = \frac{V_{total} C_d}{V_{batch}} \tag{2}$$

where C_c , V_{total} , C_d and V_{batch} means corrected concentration of CO (ppmV), total gas volume inside the experimental apparatus including the gas volume in the pipe lines (m^3), detected concentration of CO (ppmV), and total gas volume inside the experimental apparatus in the case of batch type reactor (m^3), respectively. The experiment by batch type and gas circulation type was carried out during the period from 0 h to 216 h and from 216 h to 480 h, respectively.

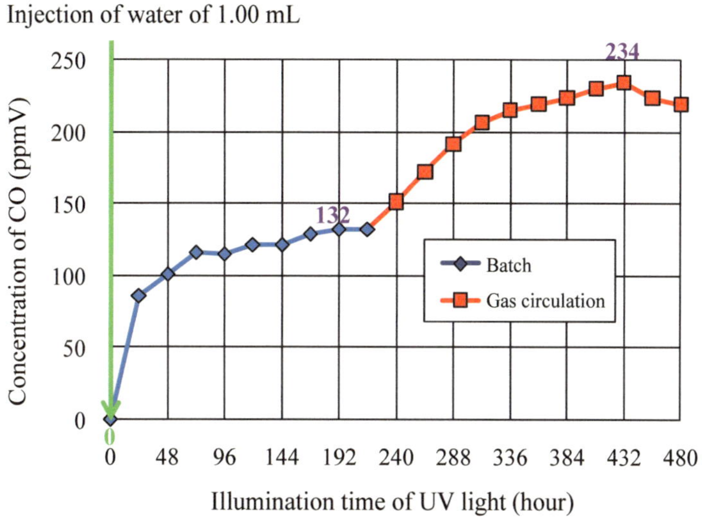

Fig. 16. Concentration change in produced CO with illumination time of UV light (Injection of water of 1.00 mL at the start of CO_2 reforming experiment)

It is observed that the concentration of CO in the reactor keeps increasing until 72 h and starts to be steady after 72 h in the experiment by batch type reactor. The highest concentration of CO which is 132 ppmV is obtained at 192 h after illuminating UV light. Since the concentration of CO is not increased over 192 h, it is determined that the experiment by batch type reactor reaches the steady reaction state at 192 h. After gas circulation, the concentration of CO starts to increase again, and peaks at 234 ppmV at UV light illumination of 432 h, which demonstrated the positive effect of gas separation and circulation on CO_2 reforming performance. To show that the steady reaction state and inverse reaction have occurred or not clearly, the change in production rate of CO with illumination time of UV light is shown in Fig.17. Production rate of CO can classify the reaction state into progressive, steady and inverse reaction state by positive, 0 and negative value, respectively. The production rate of CO, in Fig.17, which is calculated by Eq. (3):

$$R_{CO} = \frac{C_C}{t_{\text{int}}} \tag{3}$$

where R_{CO} and t_{int} means production rate of CO (ppmV) and gas sampling interval (h), respectively. The R_{CO} used for calculating is 24 h.

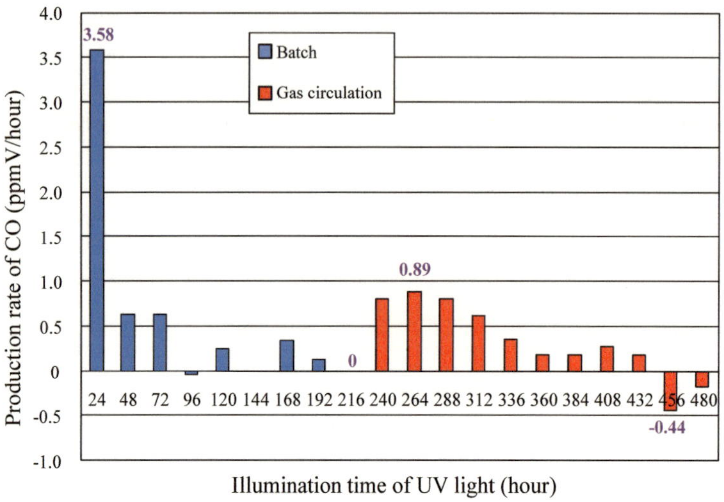

Fig. 17. Change in production rate of CO with illumination time of UV light (Injection of water of 1.00 mL at the start of CO_2 reforming experiment)

In the experiment by batch type reactor, it can be seen that the production rate of CO peaks at 3.58 ppmV/h at the UV light illumination time of 24 h and is decreased afterwards gradually. The production rate of CO which is 0 means the reaction steady state is reached. The negative value of the production rate means the inverse reaction, i.e. re-oxidization occurs. In the experiment by gas circulation type reactor, the production rate of CO after starting the gas circulation peaks at the highest value of 0.89 ppmV/h in the period from 240 h to 264 h. However, the production rate of CO became smaller after the passage time of 48 h, i.e. after the total illumination time of UV light of 264 h. Comparing the production rate of CO after starting the gas circulation and that at steady state of batch type reactor except for the period from 0 h to 24 h, the effect of gas separation and circulation on CO_2 reforming can be verified. However, the production rate of CO and the concentration of CO are still lower than the target value levels set for this study. Figure 18 shows the concentration of the water vapour during the CO_2 reforming experiment by batch type and gas circulation type reactor with illumination time of UV light. Since the concentration of water vapour is diluted with the gas in the pipe lines of the gas circulation type reactor after starting gas circulation, the concentration of water vapour is corrected by the following equation:

$$C_{c-H2O} = \frac{V_{total} C_{d-H2O}}{V_{batch}} \tag{4}$$

where C_{c-H2O}, V_{total}, C_{d-H2O} and V_{batch} means corrected concentration of water vapour (ppmV), total gas volume inside the experimental apparatus including the gas volume in the pipe lines (m^3), measured concentration of water vapour (ppmV), and total gas volume inside the experimental apparatus in the case of batch type reactor (m^3), respectively. From this figure, it is known that the highest concentration of water vapour in the experiment by batch type reactor and gas circulation reactor is 32400 ppmV and 65387 ppmV, respectively. As described above, if the water vapour is saturated in the reactor at 343 K, it should have the concentration of 307545 ppmV. The 1.00 mL, water injected into batch type reactor and gas circulation reactor, if all evaporated, could make the vapour concentration of 53040 ppmV and 112968 ppmV, theoretically. As not sure the lower water concentration measured in caused by water vapour was adsorbed by membrane or not all of water injected was vaporized, more water was injected in order to evaluate the effect of amount and timing of water injection.

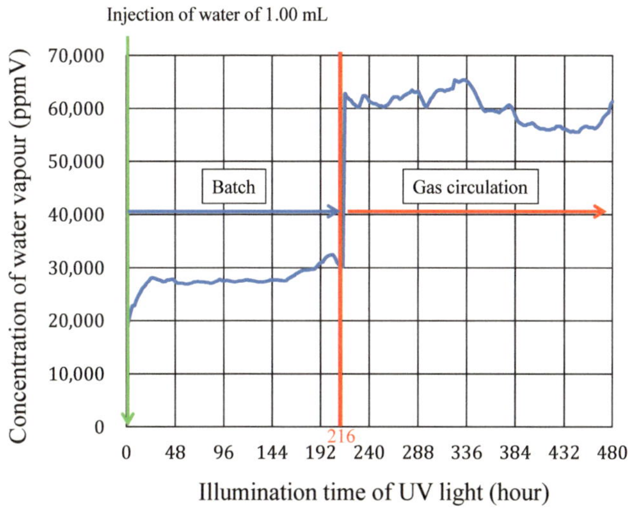

Fig. 18. Concentration change in water vapour during the CO_2 reforming experiment by batch type and gas circulation type reactor with illumination time of UV light (Injection of water of 1.00 mL at the start of CO_2 reforming experiment)

The further experiment plan was based on the assumption that at the steady and inverse reaction states, there was not sufficient water in the system. Therefore, the water was injected into CO_2 reformer when the steady and inverse reaction state in the CO_2 reforming was confirmed for not only batch type but also gas circulation type experiment. The amount of water injected was 1.00 mL at every time in this experiment. If the steady state was maintained after the injection of water in the batch type experiment by reactor, the gas circulation experiment then starts.

Figure 19 shows the concentration change in produced CO with illumination time of UV light. The initial distilled water of 1.00 mL (55.6 mmol) was injected into CO_2 reformer when the CO_2 reforming experiment by batch type reactor started. It had shown that the concentration of CO in the batch type reactor kept increasing until 168 h and the concentration of CO reached 136 ppmV. After the water of 1.00 mL was added into CO_2 reformer at 216 h when the steady state was confirmed, the concentration of CO increased again and attained to 186 ppmV at 480 h. Since the steady state was confirmed again at 504 h, another 1.00 mL of water was added into CO_2 reformer again. However, the concentration of CO did not increase any more, indicating the steady state maintained. After gas circulation started from 576 h, the concentration of CO started to increase again, and peaked at 179 ppmV at total UV light illumination of 624 h. Since the steady state in the experiment by gas circulation type reactor was confirmed at 648 h, the further 1.00 mL of water was injected into CO_2 reformer. However, the concentration of CO did not increase further, indicating the water inside system was sufficient and its effect was peaked.

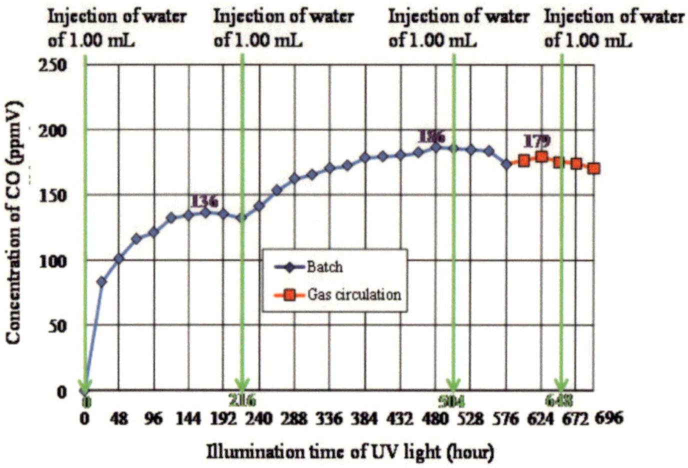

Fig. 19. Concentration change in produced CO with illumination time of UV light (Injection of water of 1.00 mL many times)

Figure 20 shows the concentration change in water vapour inside the system in the water adding experiment described above. From this figure, the measured concentration of water vapour obtained in CO_2 reforming experiment by batch type reactor is almost 25000 ppmV with total 3.00 mL water injected.

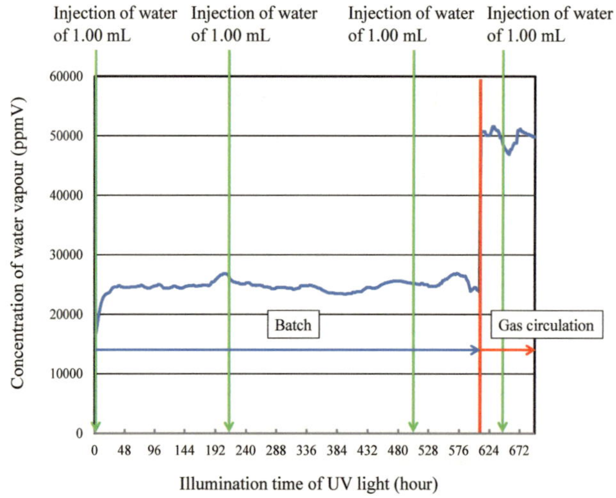

Fig. 20. Concentration change in water vapour during the CO_2 reforming experiment by batch type and gas circulation type reactor with illumination time of UV light (Injection of water of 1.00 mL many times)

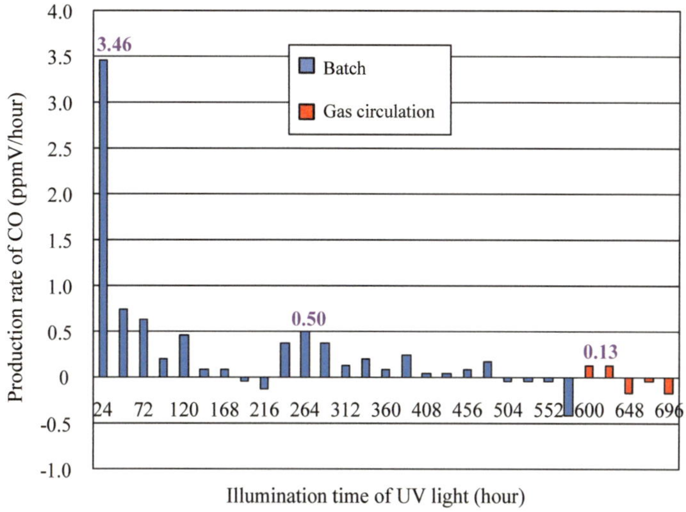

Fig. 21. Change in production rate of CO with illumination time of UV light (Injection of water of 1.00 mL many times)

Figure 21 shows the change in production rate of CO with illumination time of UV light. From this figure, it can be seen that there are two peaks of production rate of CO in CO_2 reforming experiment by batch type reactor, and there is one peak of production rate of CO in CO_2 reforming experiment by gas circulation type. The first peak was obtained at UV light illumination of 24 h with total of 1.00 mL water in the system, which was injected at the start of CO_2 reforming experiment. According to Fig.21, the production rate of CO decreases, after peaking at 24 h, gradually and reaches negative value at UV light illumination of 216 h. As mentioned above, since the steady reaction state was confirmed at UV light illumination of 216 h, the another 1.00 mL water was added into CO_2 reformer. As a result, the second peak was obtained at UV light illumination of 264 h. After that, the production rate of CO decreases again. Although further 1.00 mL water was added into CO_2 reformer again at UV light illumination of 504 h, the production rate of CO remains negative value. After gas circulation from total UV light illumination of 576 h, the production rate of CO increases and peaks at total UV light illumination of 600 h. From these results, the effect of water injection on the promotion of CO_2 reforming performance is verified. However, both the highest concentration and the highest production rate of CO in this CO_2 reforming experiment are lower than those in the case of only total 1.00 mL water injected. In addition, compared to the case of total amount of water injected of 1.00 mL, the effect of switching batch type reactor to gas circulation type reactor on the promotion of CO_2 reforming performance is not confirmed in this experiment.

Nevertheless the above described results seem reveals that the optimum timing of water injection is the very beginning of CO_2 reforming experiment. Therefore, one more experiment was conducted, i.e. the distilled water of 3.00 mL (166.8 mmol) was injected into CO_2 reformer at the very beginning of the experiment. Figure 22 shows the concentration change in produced CO with illumination time of UV light. The experiment by batch type and gas circulation type reactor was carried out during the period from 0 h to 168 h and from 168 h to 264 h, respectively. The highest concentration of CO which is 126 ppmV is obtained at UV light illumination of 144 h. Since the concentration of CO is not increased over 144 h, indicating that the experiment by batch type reactor attains to the steady reaction state, the gas circulation was started. After gas circulation, the concentration of CO starts to increase again, and peaks at 171 ppmV at UV light illumination of 240 h.

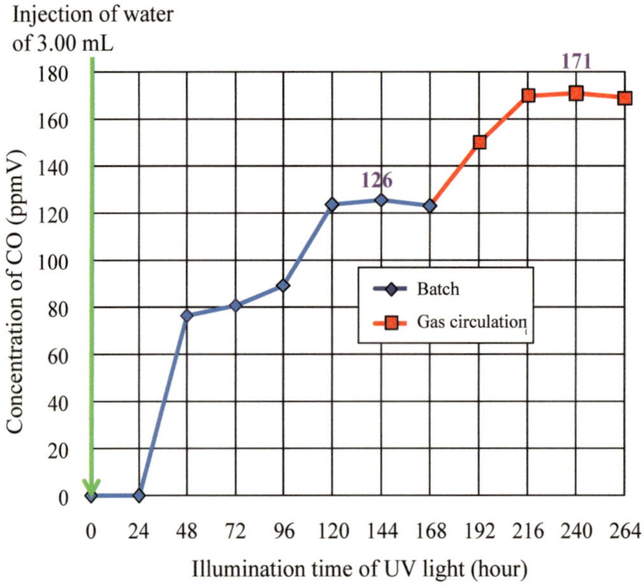

Fig. 22. Concentration change in produced CO with illumination time of UV light (Injection of water of 3.00 mL at the start of CO_2 reforming experiment)

Figures 23 and 24 show the comparison in production rate of CO and change rate of water vapour with illumination time of UV light for the amount of water injected of 1.00 mL and that of 3.00 mL, respectively. The change rate of water vapour which is calculated by Eq. (5):

$$R_{H_2O} = \frac{\Delta C_{H_2O}}{t_{\text{int}}} \tag{5}$$

where R_{H_2O}, ΔC_{H_2O} and t_{int} means change rate of water vapour (ppmV/hour), the amount of increase in concentration of water vapour (ppmV) and measurement interval of water vapour (h), respectively. The used for calculating R_{H_2O} is 24 h.

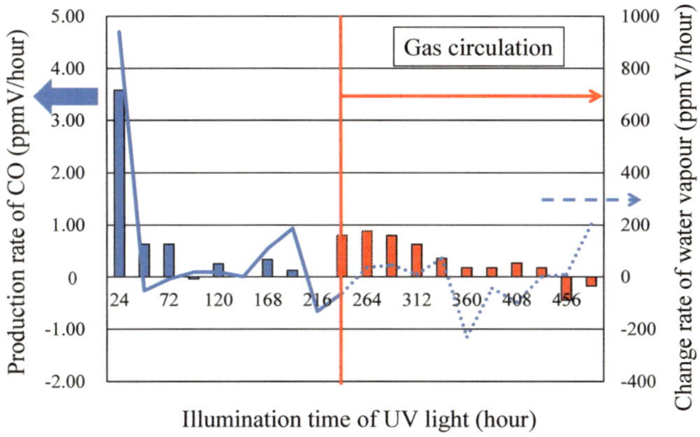

Fig. 23. Change in production rate of CO and change rate of water vapour with illumination time of UV light for the amount of water injected of 1.00 mL

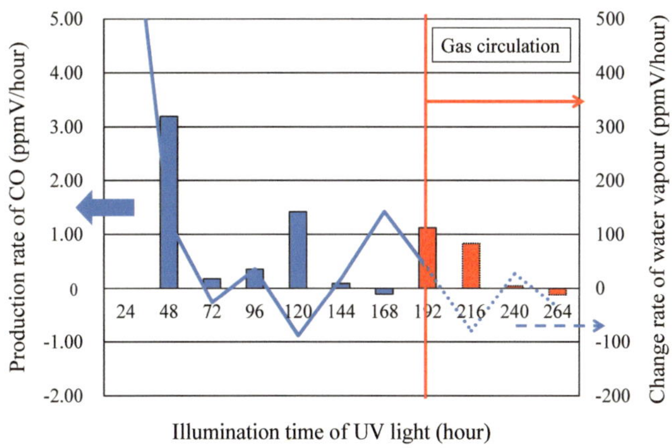

Fig. 24. Change in production rate of CO and change rate of water vapour with illumination time of UV light for the amount of water injected of 3.00 mL

From these figures, it can be seen that the concentration of water vapour in both cases, decreases rapidly from 0 h to 48 h. Since the highest production rates of CO for the amount of water injected of 1.00 mL and 3.00 mL are obtained from 0 h to 48 h, the CO_2 reforming is carried out well in the period. While the concentration of water vapour increases with illumination time of UV light from 0 h to 24 h due to temperature rise in CO_2 reformer, the change rate of water vapour closes to 0 in the period from 48 h to 96 h irrespective of the amount of water injected. Therefore, it can be thought that the amount of water consumed by photocatalytic reaction is balanced out by the amount of water vaporized due to the heat of UV lamp in the period from 48 h to 96 h. After starting gas circulation, the change rate of water vapour keeps at low level and decreases gradually irrespective of the amount of water injected, while the production rate of CO rises. Since the CO_2 reforming performance is promoted by gas separation and circulation operation, the water vapour is consumed by the CO_2 reforming reaction. In addition, the water vapour is also adsorbed by drier, which is installed to protect the mass flow meter, in pipe line of gas circulation type reactor during gas circulation. The water concentration increase due to temperature increase is balanced out by both consumption by CO_2 reforming reaction and adsorption by drier. Therefore, the change rate of water vapour keeps low and decreases gradually after starting gas circulation. Consequently, it reveals that CO_2 reforming performance of gas circulation type reactor is declined by increasing the amount of water injected due to decreasing the concentration of water vapour.

Therefore, it can be concluded, too much water in that system, no matter when it was added, would not help improving the CO_2 reforming performance.

2.4 Proposal to establish the carbon circulation system using TiO_2 photocatalyst membrane reactor

As described above, the CO_2 reforming performance of TiO_2 photocatalsyt membrane reactor is still low. To enrich the product i.e. CO further, a hybrid system combining TiO_2 photocatalsyt membrane reactor with fuel concentrator is proposed as illustrated in Fig. 25.

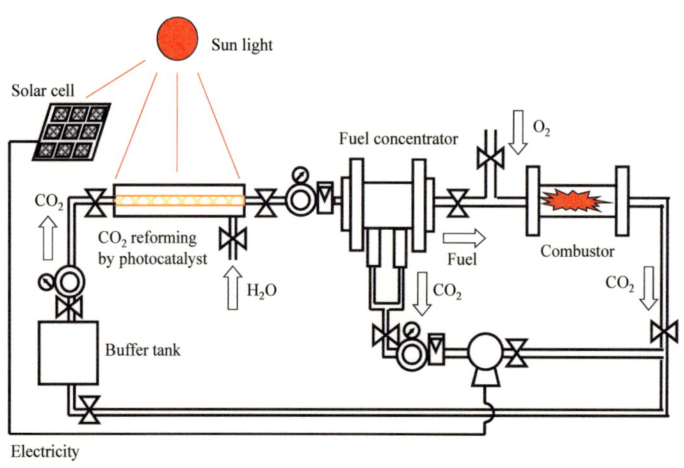

Fig. 25. Hybrid system combining TiO_2 photocatalyst membrane reactor with fuel concentrator

The fuel concentrator is a type of gas separation membrane. According to authors' previous study, the concentration of CO could be further enriched by the gas separation membrane which was composed of multiple hollow fibers. The concentration of CO of 3 vol.% in pre-mixed gas with CO_2, which simulated the maximum concentration of product by CO_2 reforming in the previous our studies, could be enriched by 6 times (Nishimura et al., 2007). If the pump necessary to enrich the fuel by the gas separation membrane as well as operating TiO_2 photocatalyst membrane reactor in the hybrid system can be powered by the electricity generated by solar cell, and the photocatalytical reaction is powered by the sun light, the proposed hybrid system is a true power system with zero CO_2 emission.

3. Conclusion

This chapter introduces the recent research and development of the TiO_2 photocatalyst membrane reactor consisting of TiO_2 photocatalyst and gas separation membrane. The following conclusions were obtained.

i. According to characterization by SEM, EPMA and XPS, the amount of TiO_2 film coated on gas separation membrane is reduced with increasing *RS*, and the largest amount of TiO_2 film is obtained for *RS* = 0.66 mm/s among various *RS* conditions investigated in this study.
ii. According to CO_2 reforming experiment by batch type reactor, the concentration of CO is decreased with increasing *RS* gradually. On the other hand, the CO_2 permeation flux peaks at *RS* = 1.1 mm/s. Since the main goal of this study is to promote the CO_2 reforming performance, the *RS* = 0.66 mm/s is selected as the optimum coating condition in this study.
iii. According to CO_2 reforming experiment by gas circulation type reactor, the positive effect of gas separation and circulation on CO_2 reforming performance is confirmed. However, too much water in that system which can not be consumed in CO_2 reforming process, no matter when it was added, would not help improving the CO_2 reforming performance.
iv. A concept power system with zero CO_2 emission, which consists of the TiO_2 photocatalsyt membrane reactor and fuel concentrator is proposed.

4. References

Adachi, K.; Ohta, K. & Mizuno, T. (1994). *Solar Energy*, Vol.53, No.2, pp.187-190

Anpo, M. & Chiba, K. (1992). *J. Mol. Catal.*, Vol.74, pp.207-212

Aurian-Blajeni, B.; Halmann, M. & Manassen, J. (1980). *Solar Energy*, Vol.25, pp.165-170

Cecchet, F.; Alebbi, M., Bignozzi, C. A. & Paolucci, F. (2006). *Inorg. Chim. Acta*, Vol.359, pp.3871-3874

Cueto, L. F.; Hirata, G. A. & Sanchez, E. M. (2006). *J. Sol-Gel Sci. Technol.*, Vol.37, pp.105-109

Dey, G. R.; Belapurkar, A. D. & Kishore, K. (2004). *J. Photochem. Photobiol. A : Chem.*, Vol.163, pp.503-508

Goren, Z.; Willner, I., Nelson, A. J. & Frank, A. J. (1990). *J. Phys. Chem.*, Vol.94, pp.3784-3790

Halmann, M.; Katzir, V., Borgarello, E. & Kiwi, J. (1984). *Solar Enregy Mater.*, Vol.10, pp.85-91

Henglein, A. & Gutierrez, M. (1983). *Ber. Bunsenges. Phys. Chem.*, Vol.87, pp.852-858

Hirano, K.; Inoue, K. & Yatsu T. (1992). *J. Photochem. Photobiol. A : Chem.*, Vol.64, pp.255-258

Ibusuki, T. (1993). *Syokubai*, Vol.35, pp.506-512

Inoue, T.; Fujishima, A., Konishi, S. & Honda, K. (1979). *Nature*, Vol.277, pp.637-638

Ishitani, O.; Inoue, C., Suzuki, Y. & Ibusuki, T. (1993). *J. Photochem. Photobiol. A : Chem.*, Vol.72, pp.269-271

Kaneco, S.; Kurimoto, H., Shimizu, Y., Ohta, K. & Mizuno, T. (1999). *Energy*, Vol.24, pp.21-30

Kawano, K.; Uehara, T., Kato, H. & Hirano, K. (1993). *Kagaku to Kyoiku*, Vol.41, pp.766-770

Lo, C. C.; Hung, C. H., Yuan, C. S. & Wu, J. F. (2007). *Solar Energy Mater. Solar Cells*, Vol.91, pp.1765-1774

Nakagawa, T. (1988). *Hyoumen*, Vol.26, No.7, pp.499-509

Nishimura, A.; Fujita, M. & Kato, S. (2007). *Proceedings of The 6th Asia Pacific Conference on Sustainable Energy and Environmental Technologies*, CD-ROM, Bangkok, Thailand, May 7-11, 2007

Ogura, K.; Kawano, M., Yano, J. & Sakata, Y. (1992). *J. Photochem. Photobiol. A. : Chem.*, Vol.66, pp.91-97

Ozcan, O.; Fukruk, F., Akkaya, E. U. & Uner, D. (2007). *Top. Catal.*, Vol.44, pp.523-528

Pathak, P.; Meziani, M. J., Li, Y., Cureton, L. T. & Sun, Y. P. (2004). *Chem. Commun.*, pp.1234-1235

Qu, J.; Zhang, X., Wang, Y. & Xie, C. (2005). *Elecrrochem. Acta 50*, pp.3576-3580

Takeuchi, K.; Murasawa, S. & Ibusuki, T. (2001). *World of Photocatalyst*, p.148, Kougyouchousakai, ISBN 4-7693-7063-6, Tokyo

Tseng, I. H.; Chang, W. C. & Wu, J. C. S. (2002). *Appl. Catal. B : Environ.*, Vol.37, pp.37-48

Wu, J. C. S. & Lin, H. M. (2005). *Int. J. Photoenergy*, Vol.7, pp.115-119

Xia, X. H.; Jia, Z. J., Yu, Y., Liang, Y., Wang, Z. & Ma, L. L. (2007). *Carbon*, Vol.45, pp.717-721

Yamashita, H.; Nishiguchi, H., Kamada, N. & Anpo, M. (1994). *Res. Chem. Intermed.*, Vol.20, pp.815-823

18

Power Generation Using Nonconventional Renewable Geothermal & Alternative Clean Energy Technologies

Basel I. Ismail
Lakehead University
Canada

1. Introduction

Advanced nonconventional renewable & alternative clean energy technologies which are used for generation of electricity have shown real promise and received renewed interest in recent years due to an increasing concern of environmental issues of greenhouse gas (GHG) emissions, being responsible for global warming & climate change, environmental pollution, and the limitations and conservation of natural energy resources. One of these innovative & emerging technologies is non-conventional, renewable and clean low-temperature geothermal energy (LTGE) technology. The vast low-temperature geothermal resources found widely in most continental regions have not received much attention for electricity generation. Continuous development of innovative drilling and ORC power generation technologies and other factors make this nonconventional and renewable energy source one of the best future viable, alternate and available source to meet the required future electricity demand worldwide, significantly reducing GHG emissions and mitigating global climate change. Section 2.1 of this chapter presents some novel applications of using LTGE resources and section 3.1 presents the fundamental concept of LTGE for power generation using ORC binary technology and discusses its limitations, environmental & economic considerations, and energy-conversion performance aspects. Another innovative alternative clean energy technology is thermoelectric (TE) power generation. Most of the recent research activities on applications of TE power generation have been directed towards utilisation of industrial waste heat (Riffat & Ma, 2003) where the cost of fuel input is cheap or free. In this large-scale application, TE power generators offer a potential alternative of green electricity generation powered by waste-heat energy that would contribute to solving the worldwide energy crisis, and the same time help reduce environmental global warming. The relatively low conversion efficiency of TE generators has been a major cause in restricting their use in electrical power generation. Recently, there has been a renewed research interest in TE technology due to emerging novel TE materials. Section 2.2 of this chapter presents innovative applications of using TE power generation technology and section 3.2 presents TE fundamental concept and discusses its limitations, energy-conversion performance, and material considerations for novel TE power generators. Thermophotovoltaic (TPV) power generation is another promising alternative clean energy source technology. There have been some causes for limiting the applications of TPV power generation technology. The improvement of TPV converters materials and total system optimization has recently attracted the attention of many research activities. Section 2.3 of this chapter presents novel applications of using TPV technology and section 3.3 presents its fundamental concept and discusses its limitations, energy performance aspects, and novel TPV system specifications.

2. Innovative features, developments & applications

2.1 Nonconventional low-temperature geothermal energy for power generation

Geothermal energy means the natural heat energy from the Earth. The source of geothermal energy is the continuous heat energy flux flowing from the interior of the Earth towards its surface. Unlike other conventional and renewable energy sources, geothermal energy has unique features, namely it is available, stable at all times throughout the year, independent of weather conditions, and has an inherent storage capability (Hammons, 2004). Distinct from fossil fuels, geothermal energy is also considered to be an environmentally friendly clean energy source which could significantly contribute to the reduction of GHG emissions by replacing fossil fuels for electrical power generation (Chandrasekharam & Bundschuh, 2008). The geothermal resources of the Earth are huge. The part of geothermal energy stored at a depth of 3 km is estimated to be 43,000,000 EJ (corresponding to 1,194,444,444 TWh) which is much larger compared to all fossil fuel resources, whose energy equivalent is 36,373 EJ, put together (Chandrasekharam & Bundschuh, 2008). The utilization of geothermal energy is usually divided into the part used for electricity generation and the part used directly for direct heating (non-electrical) applications. It was reported that geothermal energy provides approximately 0.4% of the world global power generation, with a stable long-term growth rate of approximately 5% (Ruggero Bertani, 2007) Recently, this form of renewable and green energy source has grown in 25 countries (Panea et al., 2010), with installed geothermal-electric capacity totalling up to 11 GW_e in 2010 (Ruggero Bertani, 2007), and is increasingly contributing to the electric power supply worldwide.

Geothermal energy resources vary broadly from one location to another, depending on the depth and temperature of the resource, the rock chemical composition and the abundance of ground water (Gupta & Roy, 2007). Geothermal energy sources differ in temperature from about 50 °C to 350 °C. The high-temperature geothermal resources (with temperature > 200 °C) are typically found in volcanic regions and island chains, whereas the medium-temperature (150-200 °C) and low-temperature geothermal resources (<150 °C) are usually found widely in most continental regions and by far the most commonly available geothermal resource (Chandrasekharam & Bundschuh, 2008; Gupta & Roy, 2007). The increase in temperature with depth in the Earth's crust can be expressed in terms of what is known as the geothermal temperature gradient. Down to the depths accessible by drilling with modern technology (i.e. over 10 km), the average geothermal gradient is about 2.5-3.0 °C/100 m (Dickson & Fanelli, 2005). For example, at depth around 3,000 m below ground level, the estimated temperature is 90 °C. There are, however, regions in which the geothermal temperature gradient is far from the average value. For example, in some geothermal areas the gradient is ten times the average value due to geothermal structure and composition of these areas (Dickson & Fanelli, 2005). It was estimated that the world net electricity demand is going to increase by approximately 85% from 2004 to 2030, rising from 16,424 TWh (in 2004) to 30,364 TWh in the year 2030. It was also reported that the emissions of GHG from geothermal power plants, in general, constitute less than 2% of the emission of these gases by fossil-fuelled power plants (Chandrasekharam & Bundschuh, 2008; Dickson & Fanelli, 2005). To comply with future energy demands, potential renewable & alternative energy sources should meet the following criteria: (1) the sources should be large enough to sustain a long-lasting energy supply to generate the required electricity for the country, (2) the sources should be technically and economically accessible, (3) the sources should have a wide geographic distribution, and (4) the sources should be environmentally friendly and thus should be low GHG emitters in order to make significant contribution to global warming mitigation (Chandrasekharam & Bundschuh, 2008). Low-temperature geothermal energy resources remarkably satisfy all of these criteria. This vast low-temperature/low-enthalpy geothermal energy resource has already been utilized for electric power generation by some countries, such as USA, Philippines, Mexico, Indonesia, Iceland, Austria, and Germany (Chandrasekharam & Bundschuh, 2008; Cui et al., 2009). The installations of several commercial low-temperature geothermal power systems in these countries have substantially proved the ability of low-temperature geothermal fluids to generate green electricity (Chandrasekharam & Bundschuh, 2008). In most developing countries, low-temperature geothermal resources have not received much attention for electricity generation. The main reason for not utilizing these resources by most developing countries (and several industrialized countries) for commercial exploitation is that they are not considered as economically feasible for generating electricity (Chandrasekharam & Bundschuh, 2008). Developing countries, in general, need to benefit from these new and continually improving technologies for using potential low-temperature geothermal resources for generating electricity (Chandrasekharam & Bundschuh, 2008; Galanis et al., 2009). It should be noted that for many developing countries, the use of LTGE resources is not new. Many of developing countries have been using these available resources for the past centuries for direct heating applications (Chandrasekharam & Bundschuh, 2008).

Recent increases in the cost and uncertainty of future conventional energy supplies for power generation are improving the attractiveness of low-temperature geothermal resources. Continuous development of innovative drilling and power generation technologies makes this nonconventional, renewable and clean energy source the best future viable, alternate and available source to meet the required future electricity demand worldwide, significantly reducing GHG emissions and mitigating global climate change (Chandrasekharam & Bundschuh, 2008). Generating electricity from low-temperature geothermal resources (water-dominated resources) can be effectively achieved using a Binary-Cycle technology which is also known as Organic Rankine Cycle (ORC) technology (Chandrasekharam & Bundschuh, 2008; Dickson & Fanelli, 2005; DiPippo, 2008). Low-temperature geothermal ORC technology has virtually no GHG emissions to the atmosphere (DiPippo, 2008; Hettiarachchi et al., 2007) and is an attractive energy-conversion technology due to its simplicity and its limited number of components, all of them being very common and commercially available. A number of successful & innovative ORC binary power plants were installed in different locations (e.g. remote and rural sites) worldwide which demonstrate the ability of this promising alternative technology to utilize renewable low-temperature geothermal energy sources for generating electricity. For example, two plants were installed in Nevada, USA in 1984 and 1987 with electric power generation capacity of 750 and 800 kW_e, respectively (Chandrasekharam & Bundschuh, 2008). The production wells supply geo-fluid (water) temperature at 104 ^oC with a flow rate of 60 l/s to these plants. The ORC binary fluid used was initially R-114 but due to non-availability of this working fluid the plant switched to iso-pentane in 1998. In another location near Empire, Nevada, approximately four 1 MW_e units were installed and commissioned in 1987. Two geothermal production wells with geo-fluids temperature of 137 ^oC were used (Chandrasekharam & Bundschuh, 2008). In

1998, a third well with geo-fluid temperature of 152 °C was drilled to maintain the capacity of the plant at approximately 4 MW_e. The modular approach was used so that high plant availability factors of 98% and more were achievable (Hammons, 2004). In 1987, another plant was installed and commissioned in Taiwan with an electric power generation of 300 kW_e. The plant draws geo-fluids from a 500 m deep well at a temperature of 130 °C. It was reported that the power generated from this facility was sold to the national power grid at 0.04 US$/kWh (Chandrasekharam & Bundschuh, 2008). In 1986, a low-temperature geothermal ORC unit (Mulka plant) with a power capacity of 15 kW_e was commissioned in Australia. The unit was coupled to a geothermal production well which was drilled down to a depth of 1,300 m, and supplying geo-fluid at 86 °C. The unit was operated non-stop for about three and a half years, showing frequency stability and response to load changes (Rosca et al., 2010). In 1992, a binary ORC power generation unit which utilized a low-temperature geothermal water resource with a temperature ranging from 90 to 115 °C was tested at a location near Larderello, Italy. The geothermal power plant generated between 800 and 1,300 kW_e of electricity (Rosca et al., 2010). In Germany, the first low-temperature geothermal power plant using ORC technology was installed at Neustadt-Glewe, with a power capacity of approximately 230 kW_e using a geo-fluid temperature of 98 °C (Ruggero Bertani, 2007). Another plant was commissioned in Thailand in 1989, with an installed capacity of 300 kW_e. The actual production was reported to vary from 150 to 250 kW_e and the geo-fluid temperature is 116 °C with a flow rate of approximately 8 l/s (Chandrasekharam & Bundschuh, 2008). Recently, in 2006, the first binary ORC plant which utilizes a low-temperature geothermal resource at a temperature of 74°C (reported by (Ruggero Bertani, 2007) to be the lowest low-temperature geothermal energy resource worldwide) was installed with a power generation capacity of 200 kW_e. In Japan, binary ORC technology was experimentally operated for 5 years starting in 1993 by NEDO (Yamada & Oyama, 2004). In Altheim, Austria, a geo-fluid of temperature 106 °C is utilized both for district heating and electric power generation using a binary plant technology as shown in Figure 1. The net electric output of this plant is 500 kW, selling to the electric grid 1.1 GWh in 2006 (Ruggero Bertani, 2007).

Fig. 1. Photograph of Altheim geothermal binary power plant in Austria (Ruggero Bertani, 2007).

2.2 Thermoelectric power generation as an alternative clean energy technology

A thermoelectric (TE) power generator is a solid-state device that provides direct energy conversion from thermal energy (heat) due to a temperature gradient into electrical energy based on "Seebeck effect". The thermoelectric power cycle, with charge carriers (electrons) serving as the working fluid, follows the fundamental laws of thermodynamics and intimately resembles the power cycle of a conventional heat engine. Thermoelectric power generators offer several distinct advantages over other power generation technologies (Riffat & Ma, 2003; Yadav et al., 2008): they are simple, compact and safe devices; they are environmentally friendly; they have very small size and virtually weightless; they are capable of operating at elevated temperatures; they are extremely reliable (typically exceed 100,000 hours of steady-state operation) and silent in operation since they have no mechanical moving parts and require considerably less maintenance; they are flexible power sources; they are suited for small-scale and remote applications typical of rural power supply, where there is limited or no electricity; and they are not position-dependent.

The major drawback of the thermoelectric power generator is its relatively low conversion efficiency (typically ~5% (Rowe & Min, 1998)). This has been a major cause in restricting their use in electrical power generation to specialized fields with extensive applications where reliability is a major consideration and cost is not. Applications over the past decade included industrial instruments, military, medical and aerospace (Riffat & Ma, 2003; Rowe & Min, 1998), and applications for portable or remote power generation (Stevens, 2001). However, in recent years, an increasing concern of environmental issues of GHG emissions, in particular global warming has resulted in extensive research into nonconventional technologies of generating electrical power and thermoelectric power generation has emerged as a promising alternative green energy technology. Enormous quantities of waste heat (low-grade) energy are discharged into the earth's environment much of it at temperatures which are typically too low to recover using conventional electrical power generators. Thermoelectric power generation (also known as thermoelectricity) offers a promising technology in the direct conversion of this low-grade thermal energy, such as waste-heat energy, into electrical power (Rowe, 2006). In this waste heat powered thermoelectric technology, it is unnecessary to consider the cost of the thermal energy input, and consequently thermoelectric power generators' low conversion efficiency is not a critical drawback (Riffat & Ma, 2003; Rowe, 1999). Thermoelectric generators have also been used to provide small amounts of electrical power to remote regions for example Northern Sweden, as an alternative to costly gasoline-powered motor generators (Rowe, 1999). In fact, more recently, they can be used in many cases, such as those used in cogeneration systems (Yodovard et al., 2001), to improve overall efficiencies of energy conversion systems by converting waste-heat into electricity (Yadav et al., 2008).

In general, the cost of a thermoelectric power generator essentially consists of the device cost and operating cost. The operating cost is governed by the generator's conversion efficiency, while the device cost is determined by the cost of its construction to produce the desired electrical power output (Riffat & Ma, 2003). Since the energy conversion efficiency of a module is comparatively low, thermoelectric generation using waste-heat energy is an ideal application. In this case, the operating cost is negligible compared to the module cost because the energy input (being the fuel) cost is cheap or free. Therefore, an important objective in thermoelectric power generation using waste-heat energy is to reduce the cost-per-watt of the devices. Moreover, cost-per-watt can be reduced by optimising the device geometry, improving the manufacture quality and simply by operating the device at a potentially larger temperature difference (Riffat & Ma, 2003). In addition, in designing high-performance thermoelectric power generators, the improvement of thermoelectric properties of materials and system optimization have attracted the attention of many research activities (Chen et al., 2005). Their performance and economic competiveness appear to depend on successful development of more advanced thermoelectric materials and thermoelectric power module designs. Vast quantities of waste heat generated from various sources are continuously discharged into the earth's environment much of it at temperatures which are too low to recover using conventional electrical power generators. Thermoelectric power generation, which presents itself as a promising alternative green technology, has been successfully used to produce electrical power in a range of scales directly from various sources of waste-heat energy. Enormous amounts of heat energy are typically rejected from industry, manufacturing plants and power utilities as gases or liquids at temperature which are too low (<450 K) to be used in conventional power generating units. In this large-scale application, thermoelectric power generators offer a potential alternative of electricity generation powered by waste-heat energy that would contribute to solving the worldwide energy crisis, and the same time help reduce environmental global warming. A photograph of a thermoelectric power generator used in natural gas field to directly produce power for cathodic protection of the well and gas line is shown in Figure 2. In this application, the thermoelectric device used the temperature difference between hot and cold legs of a glycol natural gas dehydrator cycle (Weiling & Shantung, 2004).

Fig. 2. Photograph of a thermoelectric power generator produced power for cathodic protection of the well and gas line (Weiling & Shantung, 2004).

Thermoelectric power generators have also been successfully applied in recovering waste-heat energy from steel manufacturing plants. In this application, large amounts of cooling water are typically discharged at constant temperatures of approximately 90 °C when used for cooling ingots in steel plants. When operating in its continuous steel casting mode, the furnace provides a steady-state source of convenient piped water which can be readily converted by thermoelectric power generators into electricity. It was reported that total electrical power of approximately 8 MW would be produced employing currently available modules fabricated using Bi_2Te_3 thermoelectric modules technology (Rowe, 2006). Another application where thermoelectric power generators using waste-heat energy have potential use is in industrial cogeneration systems (Riffat & Ma, 2003; Yodovard et al., 2001; Min & Rowe, 2002). For example, Yodovard et al. (2001) evaluated the potential of waste heat thermoelectric power generation for diesel cycle and gas turbine cogeneration in the manufacturing industrial sector in Thailand. It is reported that gas turbine and diesel cycle cogeneration systems produced electricity estimated at 33% and 40% of fuel input, respectively (Yodovard et al., 2001). The useful waste heat from stack exhaust of cogeneration systems was estimated at approximately 20% for a gas turbine and 10% for the diesel cycle. The corresponding net power generation was approximately 100 MW. Recently, the possibility of utilizing the low-grade heat energy generated from incinerated municipal solid waste has also been considered. For example, it was reported that an on-site experiment using a 60 W thermoelectric module installed near the boiler section of an incinerator plant, achieved an estimated conversion efficiency of approximately 4.4% (Rowe, 2006). A thermoelectric power generator produced by the Japanese Energy Conservation Center, which used waste heat as energy source to generate an electric power density of 100 kW/m^3 is shown in Figure 3. A waste heat-based thermoelectric power generator is also used in a domestic central heating system with the modules located between the heat source and the water jacket (Rowe, 2006). In this application, for example, the heat output provided by the gas/oil burner passes through the generator before reaching the central heating hot-water exchanger. The generator converts approximately 5% of the input heat energy to electrical power, the remainder of 95% transfers to the hot water heat exchanger for its intended use in heating the radiator system (Rowe, 2006).

Fig. 3. Photograph of a thermoelectric power generator produced by the Japanese Energy Conservation Center, which used waste heat as energy source to generate an electric power density of 100 kW/m^3 (Weiling & Shantung, 2004).

Waste-heat energy can also be utilized proportionally from 20-50 kW wood- or diesel-heated stoves (Nuwayhid et al., 2003), especially during the winter months in rural regions where electric power supply is unreliable or intermittent, to power thermoelectric generators. For example, a thermoelectric power generator to produce electricity from stove-top surface temperatures of 100-300 ^oC was designed and evaluated (Nuwayhid et al., 2003). In this application, two commercially available thermoelectric modules were considered and 100 W of electric power output was targeted for a minimum domestic use. A similar application is reported in (Rowe, 2006). In this application, thermoelectric power generators were used to generate small amounts of electrical power to remote regions for example Northern Sweden, as an alternative to costly gasoline powered engine generators. The generator uses heat from a wood-burning stove with the cold-side cooled with a 12 volt, 2.2 W fan. The generator produces approximately 10 watts of electric power. The utilization of waste-heat energy from exhaust gases in reciprocating internal combustion engines (e.g. automobiles) is another novel application of electricity generation using thermoelectric power generators. Although a reciprocating piston engine converts the chemical energy available in fossil fuels efficiently into mechanical work a substantial amount of thermal energy is dissipated to the environment through exhaust gas, radiation, cooling water and lubricating oils. For example, in a gasoline-powered engine, approximately 30% of the primary gasoline fuel energy is dissipated as waste-heat energy in the exhaust gases; waste-heat energy discharged in the exhaust gases from a typical passenger car travelling at a regular speed is 20-30 kW (Riffat & Ma, 2003). A comprehensive theoretical study concluded that a thermoelectric generator powered by exhaust heat could meet the electrical requirements of a medium sized automobile (Rowe, 1999). It was reported that among the established thermoelectric materials, those modules based on PbTe technology were the most suitable for converting waste-heat energy from automobiles into electrical power (Rowe, 1999). Wide-scale applications of thermoelectricity in the automobile industry would lead to some reductions in fuel consumption, and thus environmental global warming, but this technology is not yet widely proven (Riffat & Ma, 2003).

2.3 Thermophotovoltaic alternative energy technology for direct power generation

A thermophotovoltaic power generator is also a solid-state device which provides direct energy conversion from radiant energy (radiative heat) emitted from a heated object into electrical power. In photovoltaic (PV) energy-conversion systems, electric energy is directly generated out of photons that are absorbed by the PV cell. In conventional PV systems these photons (in the visible range wavelengths) originate from the sun of a temperature of approximately 6000 K and at a distance of approximately 150 x 10^6 km, but also other photon sources can be used. Another potential alternative is the use of photons having near- or infrared radiation wavelengths emitted by a heat source (at a distance of perhaps only a few centimeters) as used in combustion-driven systems in industry or in residential heating systems. Since these photons originate from a thermal source this type of photovoltaic energy conversion is called thermophotovoltaic (TPV) energy conversion. The electric power density of the TPV cells is typically much higher than that of solar energy driven PV cells since the radiation intensity of combustion-driven radiant sources can be made much greater than that of the sun. For example, Coutts (1999) reported that the power density of a typical solar PV device is of the order of 0.1 W/cm^2, whereas that for a TPV device is potentially to be 5-30 W/cm^2. In TPV devices or systems, an object, typically a selective emitter, is heated to temperatures typically higher than 1000 K; 1300-2000 K is considered a practical range. The resulting radiation (in form of photons) is absorbed by semiconductor photovoltaic (PV) cells which convert photons into electricity. Photons not absorbed by the TPV cells are reflected back by a reflector (a photon recirculating device) to the emitter (Coutts, 1999; Lal & Blakers, 2009). The main components of a TPV device and energy conversion processes involved in the operation of it will be discussed in more detail in later sections of this chapter.

The study of TPV energy conversion might date back to more than 40 years. Coutts (1999) and Nelson (2003) reported that the possibility of using infrared radiation and converting it to electricity using TPV technology was first appreciated by Professor Pierre Aigrain. In particular, Aigrain proposed this direct energy conversion concept during a series of lectures on a number of technical topics given while he was a visiting professor at MIT (Cambridge, MA) in late 1960 and early 1961. A substantial research effort followed the Aigrain disclosure in the 1960s. For example, the MIT faculty authored a series of conference papers (e.g. IEEE Photovoltaic Specialists Conference) and journal articles related to this technology (Coutts, 1999; Nelson, 2003). The US Army at Fort Monmouth played a significant role in advancing TPV technology. More particularly, the US Army needed portable power sources with a low noise, and TPV was an excellent candidate for this application (Nelson, 2003). Nelson (2003) also reported that General Motors (GM) was the most active industrial contributor in TPV technology development during this early period. Figure 4 shows an advanced silicon concentrator solar cell based TPV prototype system (Qiu & Hayden, 2006). In the mid 1970s, the pace of TPV development slowed significantly when the US Army chose another technology (thermoelectric) to satisfy the need for concealed power sources. According to the US Army, TPV development was not sufficiently advanced when compared with the older, more reliable thermoelectric approach. TPV technology development, benefited from the energy crises of the 1970s during which time worldwide interest was focused on renewable energy sources. TPV has experienced a revival since the early 1990s mainly because of the technology revolution and availability of high-performance TPV energy converters (Coutts, 1999) and the technological improvements in the area of selective emitter and low bandgap PV cells (Chia & Feng, 2007). More recently, there has been a renewed interest in the TPV clean power generation technology due to an increasing concern of environmental issues of GHG emissions, the limitations of energy resources, and, more particularly, the attractive features of this technology over other technologies as an alternative source of clean energy.

Fig. 4. Photograph of a silicon concentrator solar cell based TPV prototype system (Qiu & Hayden, 2006).

Like TE power technology, TPV power generators offer numerous potential distinct advantages over other technologies (Coutts, 1999; Chia & Feng, 2007; Badescu, 2005; Qiu & Hayden, 2006): they are silent, flexible, and reliable power sources; they are simple, compact, safe and transportable and have high power density; they are clean and environmentally friendly, if the heat source is fuelled by clean energy resource (e.g. solar, nuclear, biofuel, waste heat, etc.), or have low-emissions with low-emission burner-fuel system; they have fuel versatility and are characterized by a rapid start-up; they have no mechanical moving parts and require considerably less maintenance; they are suited for small-scale and remote applications typical of rural power supply, where there is limited or no electricity; and they are independent of the sun (if non-solar heat sources are used).

TPV energy conversion technology offers a potential application in direct generation of electricity in portable or remote power generation, stand-alone domestic gas-furnaces, silent electrical power supplies on recreational vehicles, hybrid electric vehicles, cogeneration of electricity and heat, and many others (Coutts, 1999; Nelson, 2003). Coutts (1999) reported that the greatest potential for TPV power generation is in its application to large-scale recovery of high-temperature waste heat from many industrial processes, such as glass, aluminum, steel, castings, etc. In this waste heat powered TPV technology, it is unnecessary to consider the cost of the thermal energy input. The application of this alternative clean technology in converting waste-heat energy directly into electrical power can also improve the overall efficiencies of energy conversion systems. There have been some causes for limiting the applications of TPV power generation technology. For example, Lal and Blakers (2009) reported that the main limiting factors of TPV systems are spectral filtering, TPV conversion efficiency, the optimization of individual components rather than the system as a whole, adequate systems modeling and thermal management. It was also reported that an important commercial limitation of TPV technology is the difficulty of accessing suitable cost-effective solar cells (Lal & Blakers, 2009). Therefore, in designing high-performance thermophotovoltaic power generators, the improvement of thermophotovoltaic converters materials and total system optimization have recently attracted the attention of many research activities. Their performance and economic competiveness appear to depend on successful development and demonstration of more advanced and innovative TPV converters materials and system designs.

3. Fundamental concepts and energy conversion performance aspects

In the following sections, the fundamental concepts, energy-conversion performance aspects and other considerations of the nonconventional renewable low-temperature geothermal power generation, thermoelectric clean power generation, and thermophotovoltaic power generation are presented and discussed.

3.1 Fundamental concept and energy performance aspects of low-temperature geothermal power generation using binary ORC technology

A schematic diagram showing a low-temperature geothermal ORC binary-fluid system used for electric power generation is shown in Figure 5. In this system, the first (primary) fluid being the geo-fluid (brine) is extracted from the low-temperature geothermal resource through the production well. The geo-fluid carries the heat from the liquid-dominated resource (thus called the geo-fluid heat carrier) and efficiently transfers this heat to the low-boiling point (BP) organic working fluid (the secondary/binary fluid) using an effective heat exchanger. Shell-and-tube heat exchangers are widely used (Chandrasekharam & Bundschuh, 2008). The Organic Rankine Cycle (ORC) is a thermodynamic Rankine cycle that uses the organic working fluid instead of steam (water). In this binary-fluid system, the low-boiling point organic liquid absorbs the heat which is transferred by the geothermal fluid and boils at a relatively much lower temperature (compared to water) and as a result develops significant vapor pressure sufficient to drive the axial flow or radial inflow turbine. The turbine is coupled to an electric generator which converts the turbine mechanical shaft power into electrical power. The organic working fluid expands across the turbine and then is cooled and condensed in the condenser before it is pumped back as a liquid to the heat exchanger using a condensate pump to be re-evaporated, and the power cycle repeats itself. One of the most important performance criteria in low-temperature geothermal ORC power generation technology requires the optimal selection of the ORC organic working fluid. Organic fluids used in binary ORC technology have inherent feature (compared to water) and that is they have low boiling temperature and high vapor pressure at relatively low temperatures, compared with steam (water) (Dickson & Fanelli, 2005).

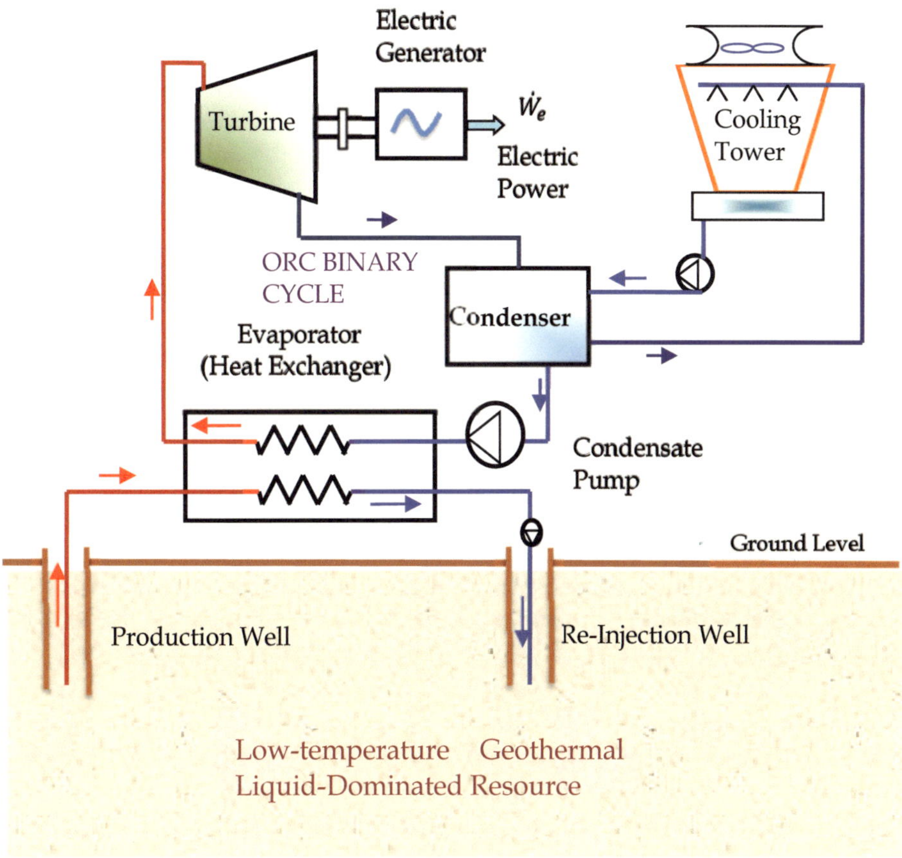

Fig. 5. A schematic diagram showing the basic concept of a low-temperature geothermal binary ORC system for electrical power generation.

Typical ORC organic fluids may include pure hydrocarbons (e.g. pentane, butane, propane, etc), refrigerants (e.g. R134a, R218, R123, R113, R125, etc), or organic mixtures (Panea et al., 2010; Saleh et al., 2007; Hung, 2001; Wei, 2007). The optimal energy conversion performance of a low-temperature geothermal ORC power generation system depends mainly on the type of organic fluid being used in the system. The selection of the type of organic fluid is typically based on the following criteria (Chandrasekharam & Bundschuh, 2008; Hettiarachchi et al.,2007; Saleh et al., 2007):

- The ORC organic fluid should result in high thermal efficiency by allowing maximum utilization of the available low-temperature geothermal heat source.
- It should be environmentally friendly; less in ozone depletion potential (ODP) and global warming potential (GWP).

- It should be safe (non-flammable and no-toxic) and non-corrosive.
- It should have a low-boiling temperature and should evaporate at atmospheric pressure.
- It should lead to optimum design and cost effectiveness of the ORC system.
- It should not react or disassociate at the pressures and temperatures at which it is used.
- It should have suitable thermal stability and high thermal conductivity.
- It should have appropriate low critical temperature and pressure.
- It should result in low maintenance.
- It should have small specific volume, low viscosity and surface tension.

It should be noted that many binary ORC fluids may not meet all these criteria (Chandrasekharam & Bundschuh, 2008) but the selection of the organic fluid should be optimized, in terms of the above requirements, while meeting the demanded power generation. In general, binary ORC systems exhibit great flexibility, high safety (installations are perfectly tight), and low maintenance (Wei, 2007). It was reported that the selection of suitable organic fluids for application in binary ORC systems for generating electricity still deserves extensive thermodynamic and technical studies (V. Maizza & A. Maizza, 2001).

3.1.1 Environmental & economic considerations of low-temperature geothermal binary ORC power technology

In general, geothermal energy is relatively pollution-free and considered to be a clean technology (Dickson & Fanelli, 2005) and it tends to have the largest technological potential compared to other renewable energy sources (Hammons, 2004). In particular, GHG emissions are typically zero when low-temperature geothermal energy reservoirs are utilized using ORC binary technology, since all of the produced geo-fluid is injected back into the reservoir (Hammons, 2004). One of the effective ways of getting rid of hazardous chemical constituents of geothermal water (e.g. trace metals) is re-injection. Low-temperature geothermal binary power generation systems are far less environmentally intrusive than alternative power generation systems in several respects, for example they are essentially zero-GHG emission systems and have low land usage per installed megawatt (DiPippo, 2008). As far as physical environmental effects, geothermal projects may cause some kind of disruption activities as other same size and complexity of civil engineering projects. Also, the locations of excavations and sitting of boreholes and roads will have to be taken into account, soil and vegetation erosion, which may cause changes in ecosystems, has to be watched. It should be noted that many geothermal installations are in remote areas where the natural level of noise is low and any additional noise is very noticeable (Dickson & Fanelli, 2005). There is a relatively larger production of waste-heat energy in geothermal systems, and this needs to be dissipated in an environmentally acceptable way. In low-temperature geothermal binary ORC power systems the thermal impact is much reduced by disposing of waste geothermal water using deep re-injection approach so that the thermal impact of the waste heat becomes insignificant (Dickson & Fanelli, 2005). Appropriate measures should be applied to prevent leakage of the binary working fluid from ORC power generation units to the environment (Yamada & Oyama, 2004); normally the installations of these units are made perfectly tight to meet high safety standards.

Generating electricity using geothermal ORC technology is very cost-effective and reliable (Chandrasekharam & Bundschuh, 2008; Dickson & Fanelli, 2005). Table 1 compares electrical energy costs produced by various renewable energy technologies. The cost of geothermal energy for generating electricity is favourable compared to other energy sources. The reported costs of low-temperature based small geothermal power plants vary from 0.05 to 0.07 US$/kWh for units generating < 5 MW_e (Chandrasekharam & Bundschuh, 2008).

Renewable Energy Source	Current Energy Cost (US cents/kWh)	Turnkey Investment Cost (US$/$kW_e$)	Potential Future Energy Cost (US cents/kWh)
Geothermal	2-10	800 - 3,000	1-8
Wind	5-13	1,100 - 1,700	3-10
Solar photovoltaic	25-125	5,000 - 10,000	5-25
Solar thermal	12-18	3,000 - 4,000	4-10
Biomass	5-15	900 - 3,000	4-10
Tidal	8-15	1,700 - 2,500	8-15
Hydro	2-10	1,000 - 3,000	NA

Table 1. Energy and investment costs for electric power production from different renewable energy sources (Hammons, 2004; Dickson & Fanelli, 2005).

The unit cost of electricity generated from low-temperature geothermal based small power plants is compared in Table 2. Moreover, the unit cost of electricity from small-scale geothermal plants (< 5 MW_e) is much lower than the average cost of 0.25 US$/kWh supplied through diesel generators (Chandrasekharam & Bundschuh, 2008). The total investment for a geothermal power plant mainly includes the following types of costs: (1) cost of exploitation, (2) cost of drilling, (3) cost of power plant (capital cost of design and construction), and (4) operating & maintenance costs (Chandrasekharam & Bundschuh, 2008). The first two types are referred to as subsurface costs whereas the other two are referred to surface costs. For small-scale geothermal power plants (< 5 MW_e) utilizing low-temperature resources, the subsurface cost typically accounts for approximately 30% of the total investment costs whereas the surface cost accounts for the remaining 70%.

Net Power (kW_e)	Capital Cost (US$/net kW_e) Geothermal Resource Temperature (°C)			O&M Cost (US$/year)
	100	120	140	
100	2,786	2,429	2,215	21,010
200	2,572	2,242	2,044	27,115
500	2,357	2,055	1,874	33,446
1000	2,143	1,868	1,704	48,400

Table 2. Unit cost of electricity generated from low-temperature based small power plants (Chandrasekharam & Bundschuh, 2008; DiPippo, 2008).

Generating electricity using low-temperature geothermal ORC technology is very reliable due to its advanced technological aspects. However, the maintenance costs and shutdowns could be reduced when the technical complexity of the plant is on a level that is accessible to local technical personnel or to experts who are readily available (Dickson & Fanelli, 2005). As mentioned before, geothermal ORC power generation plants are normally constructed and installed in small modular power generation units. These units can then be linked up to create power plants with larger power production rates. Their cost depends on a number of factors, but mainly on the temperature of the geothermal fluid produced, which influences the size of the ORC turbine, heat exchangers and cooling system. It was reported (Dickson & Fanelli, 2005) that the total size of the plant has little effect on the specific cost, as a series of standard modular units is linked together to obtain larger power capacities. It was also reported (Panea et al., 2010) that the modular units have a satisfying economic efficiency, because modular construction reduces installation time and costs. Ultimately, the economic viability of the geothermal power plant depends on its ability to generate revenue in the long-term.

3.1.2 Ideal and actual energy performance of low-temperature geothermal binary power generation systems

The theoretical overall performance of low-temperature geothermal binary systems can be evaluated using the fundamental thermal efficiency of a heat engine, given as (Cengel & Boles, 2008)

$$\eta = \frac{\dot{W}_{net,out}}{\dot{Q}_{geo,in}} \tag{1}$$

also,

$$\eta_{th} \equiv 1 - \frac{\dot{Q}_{cond}}{\dot{Q}_{geo,in}} \tag{2}$$

where, $\dot{W}_{net,out}$ is the net power output delivered by the geothermal power system (in kW); $\dot{Q}_{geo,in}$ is the thermal power supplied by the geo-fluid from the available geothermal resource (in kW); and $\dot{Q}_{cond}$ is the thermal energy rejected in the condenser (in kW). For quick estimate purposes, a correlation is proposed (Dickson & Fanelli, 2005) to calculate the actual net power output (with rough accuracy) as a function of the available thermal power from the geo-fluid flow and inlet temperature of the geo-fluid, given by

$$\dot{W}_{net,act} = \left(\frac{1}{278}\right)\left[\left(0.18T_{geo,in} - 10\right)\dot{Q}_{geo,in}\right] \tag{3}$$

Substituting Eq. (3) in Eq. (1), the estimated thermal efficiency of the low-temperature based geothermal power generation system, as a function of geo-fluid inlet temperature (in °C) at the production well, is given by

$$\eta_{th} \approx \left(\frac{1}{278}\right)\left(0.18T_{geo,in} - 10\right) \tag{4}$$

For example, using Eq. (4) it can be estimated that a thermal efficiency of approximately 5.5% could be achieved for power generation with a geo-fluid extracted from a low-temperature geothermal resource available at 140 °C. The thermal efficiency as a function of the low-temperature geothermal heat resource temperature, T_H (in K), and ambient temperature, T_a (in K) is given by (DiPippo, 2007)

$$\eta_{th} \cong \left(\frac{58}{100}\right)\left(\frac{T_H - T_a}{T_H + T_a}\right) \tag{5}$$

So for example, with a geothermal heat resource temperature of 140 °C and ambient temperature of 20 °C, the thermal efficiency is estimated to be 10%. It should be noted that Eq. (5) is valid for resource temperatures between 100 and 140 °C. The estimated net power output delivered by the geothermal power system can also be determined using (DiPippo, 2007)

$$\dot{W}_{net,out} \cong 2.47\dot{m}_{geo}\left(\frac{T_H - T_o}{T_H + T_o}\right)(T_H - T_C) \tag{6}$$

Where, $\dot{m}_{geo}$ is the geo-fluid mass flow rate. As can be noted from Eq. (5), for low-temperature geothermal resources, the power plant thermal efficiency is very dependent on the ambient temperature, which determines the heat sink temperature. It should be also noted that the above correlations given by Eqs. (3) through (6) provide quick estimate of the thermal efficiency and net power output, and therefore for more accurate calculations and system performance predictions, a detailed thermodynamic energy balance analysis should be performed to predict the net power, the available geothermal heat, and overall thermal efficiency using Eq. (1). Low-temperature geothermal binary power generation plants tend to have low thermal efficiencies: 10-13% reported by (DiPippo, 2008), 2.8-5.5% reported by (Gupta & Roy, 2007), and 5-9% reported by (Hettiarachchi et al., 2007) since the geothermal energy is produced at low enthalpy levels. Maximizing generating power capacity is normally sought from these power plants by maximizing the geo-fluid flow rate (depending on the capability of the production well) with a limited geo-fluid temperature available from the geothermal resource. It was reported (Chandrasekharam & Bundschuh, 2008) that low-temperature geothermal production wells with geo-fluid temperature < 150 °C and geo-fluid flow rate > 900 l/min could generate electric power ranging from 50 to 700 kW_e. When appropriate, multiple production wells could be installed using the same low-temperature geothermal energy reservoir so that a number of ORC power generation units could be cascaded to obtain larger power production rates from the plant (Gupta & Roy, 2007). Limited by the second-law of thermodynamics, the ideal (absolute maximum) efficiency of a thermoelectric power cycle, such as the low-temperature geothermal ORC power cycle, operating as a reversible heat engine between a heat source at a temperature T_H and a heat sink at a temperature T_L is Carnot efficiency, given as (Cengel & Boles, 2008)

$$\eta_c = \eta_{\max} = 1 - \frac{T_L}{T_H} \tag{7}$$

For example, for an ORC power system using a geo-fluid extracted from a low-temperature geothermal heat source at 140 °C (413.15 K) and a heat sink (condenser) at 40 °C (313.15 K), the maximum ideal Carnot efficiency can be calculated using Eq. (7) to be approximately 24.2%. For an actual (irreversible) geothermal ORC cycle operating between the same temperature limits would have lower efficiency. Another measure of the performance of the low-temperature geothermal ORC power plant can be obtained using the Second-Law of thermodynamics in the form of exergetic efficiency, η_{ex}, given as

$$\eta_{ex} \equiv \frac{\dot{W}_{net}}{\dot{E}x} \tag{8}$$

The exergetic efficiency in Eq. (8) is defined as the ratio of the actual net power output from the power generation system to the maximum theoretical power that could be extracted from the geo-fluid at the geothermal resource state. This involves determining the rate of exergy carried by the geo-fluid to the ORC power system. Typically, the design and operation of geothermal binary power generation systems should be optimized in order to increase their thermal and exergetic efficiencies guided by the Carnot ideal efficiency.

3.2 Fundamental concept and energy performance aspects of waste-heat (low-grade) direct energy conversion into electricity using thermoelectric clean technology

As introduced previously, thermoelectric power generation is based on a phenomenon called "Seebeck effect". When a temperature difference is established between the hot and cold junctions of two dissimilar materials (metals or semiconductors) a voltage is generated. In fact, this phenomenon is applied to thermocouples that are extensively used for temperature measurements. Based on this Seebeck effect, thermoelectric devices can act as electrical power generators. A schematic diagram of a simple thermoelectric power generator operating based on Seebeck effect is shown in Figure 6.

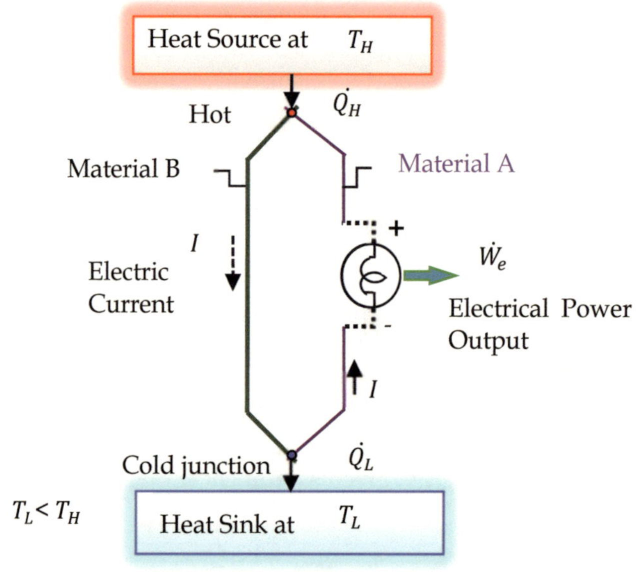

Fig. 6. A schematic diagram showing the basic principle of a simple thermoelectric power generator operating based on Seebeck effect.

As shown in Figure 6, heat is transferred at a rate of $\dot{Q}_H$ from a high-temperature heat source maintained at T_H to the hot junction, and it is rejected at a rate of $\dot{Q}_L$ to a low-temperature sink maintained at T_L from the cold junction. Based on Seebeck effect, the heat supplied at the hot junction causes an electric current to flow in the circuit and electrical power is produced. Using the first-law of thermodynamics (energy conservation principle) the difference between $\dot{Q}_H$ and $\dot{Q}_L$ is the electrical power output $\dot{W}_e$. It should be noted that this power cycle intimately resembles the power cycle of a heat engine (Carnot engine), thus in this respect a thermoelectric power generator can be considered as a unique heat engine (Cengel & Boles, 2008). Figure 7 shows a schematic diagram illustrating the main components and arrangement of a conventional single-stage thermoelectric power generator. As shown in Figure 7, the TE power generator is composed of two ceramic plates (substrates) that serve as a foundation, providing mechanical integrity, and electrical insulation for *n*-type (heavily doped to create excess electrons) and *p*-type (heavily doped to create excess holes) semiconductor thermoelements. In thermoelectric materials, electrons and holes operate as both charge carriers and energy carriers. There are very few modules without ceramic plates, which could eliminate the thermal resistance associated with the ceramic plates, but might lead to mechanical fragility of the module.

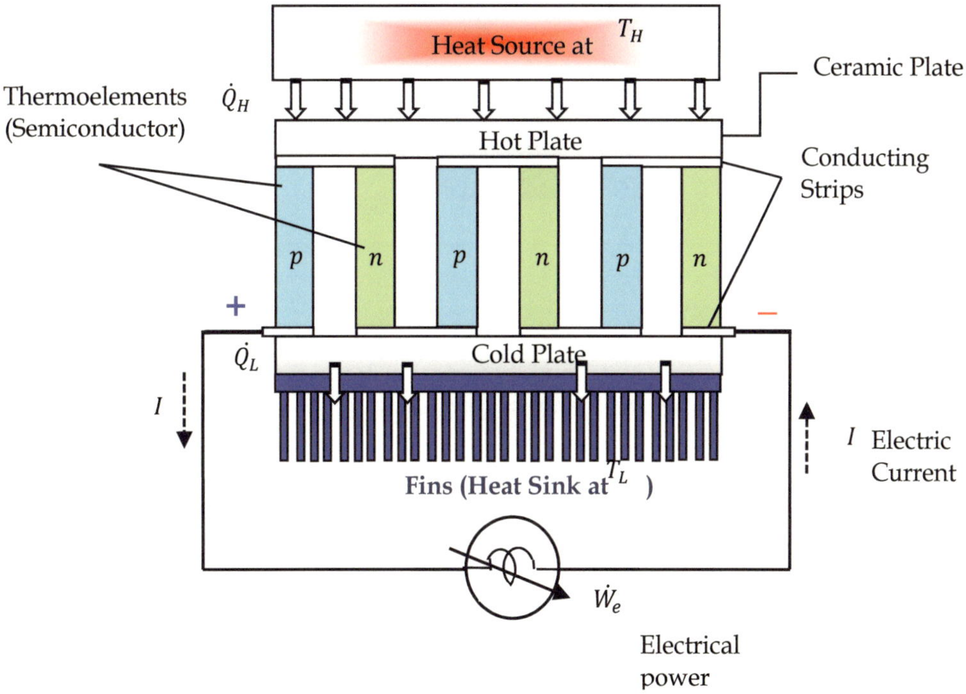

Fig. 7. A schematic diagram illustrating arrangement and main components of a typical TE power generator.

The ceramic plates are commonly made from alumina, but when large lateral heat transfer is required, materials with higher thermal conductivity (e.g. beryllia and aluminum nitride) are desired. The semiconductor thermoelements (e.g. silicon-germanium, lead-telluride based alloys) that are sandwiched between the ceramic plates are connected thermally in parallel and electrically in series to form a thermoelectric device (module). More than one pair of semiconductors are normally assembled together to form a thermoelectric module and within the module a pair of thermoelements is called a thermocouple (Riffat & Ma, 2003). The junctions connecting the thermoelements between the hot and cold plates are interconnected using highly conducting metal (e.g. copper) strips as shown in Figure 7. The sizes of conventional thermoelectric devices vary from 3 mm^2 by 4 mm thick to 75 mm^2 by 5 mm thick. Most of thermoelectric modules are not larger than 50 mm in length due to mechanical consideration. The height of single-stage thermoelectric modules ranges from 1 to 5 mm. The modules typically contain from 3 to 127 thermocouples (Riffat & Ma, 2003). There are multistage TE devices designed to meet requirements for large temperature differentials. Multi-stage thermoelectric modules can be up to 20 mm in height, depending on the number of stages. Photographs of single- and multi-stage thermoelectric modules are shown in Figure 8. The power output for most of the commercially-available thermoelectric power generators ranges from microwatts to multi-kilowatts (Riffat & Ma, 2003; Rowe, 1999). For example, a standard thermoelectric device consists of 71 thermocouples with the size of 75 mm^2 can deliver electrical power of approximately 19 W (Riffat & Ma, 2003).

(a) (b)

Fig. 8. Photographs of (a) single-stage, and (b) typical pyramid three-stage configuration TE modules (Sources: http://www.customthermoelectric.com, http://www.ferrotec.com).

3.2.1 Energy conversion & performance aspects of TE power generators

The performance of thermoelectric materials can be expressed using (Rowe, 2006)

$$Z = \frac{\alpha^2}{kR}, \tag{9}$$

where Z is the thermoelectric material figure-of-merit, α is the Seebeck coefficient given by

$$\alpha = -\frac{\Delta V}{\Delta T}. \tag{10}$$

R is the electric resistivity and k is the total thermal conductivity. This figure-of-merit may be made dimensionless by multiplying by $\bar{T}$ (average absolute temperature of hot and cold plates of the thermoelectric module, K), i.e.,

$$Z\bar{T} = \frac{\alpha^2 \bar{T}}{kR} \tag{11}$$

The term α^2/R is referred to as the electrical power factor. In general, a thermoelectric power generator exhibits low efficiency due to the relatively small dimensionless figure-of-merit ($Z\bar{T} \leq 1$) of currently available thermoelectric materials. The thermal efficiency of a thermoelectric power generator defined as the ratio of power delivered to the heat input at the hot junction of the thermoelectric device, is given by (Rowe, 2006)

$$\eta_{th} = \frac{\dot{W}_e}{\dot{Q}_H} \tag{12}$$

Limited by the second-law of thermodynamics, the ideal (absolute maximum) efficiency of a thermoelectric power generator operating as a reversible heat engine is Carnot efficiency, given by Eq. (7). The maximum actual conversion efficiency of an irreversible thermoelectric power generator can be estimated using (Min & Rowe, 2004)

$$\eta_{TE} = \eta_{Carnot}\left[\frac{\sqrt{1+Z\bar{\bar{T}}}-1}{\sqrt{1+Z\bar{\bar{T}}}+T_L / T_H}\right] \tag{13}$$

The value of the figure-of-merit is usually proportional to the conversion efficiency. The dimensionless term $Z\bar{T}$ is therefore a very convenient figure for comparing the potential conversion efficiency of modules using different thermoelectric materials.

3.2.2 Considerations of novel thermoelectric materials for TE power generators

A large amount of research in thermoelectric materials has focused on increasing the Seebeck coefficient and reducing the thermal conductivity, especially by manipulating the nanostructure of the thermoelectric materials. Because the thermal and electrical conductivity correlate with the charge carriers, new means must be introduced in order to conciliate the contradiction between high electrical conductivity and low thermal conductivity as indicated by Weiling and Shantung (Weiling & Shantung, 2004). Among the vast number of materials known to date, only a relatively few are identified as thermoelectric materials. Today's most thermoelectric materials, such as Bismuth Telluride (Bi_2Te_3)-based alloys and Lead Telluride (PbTe)-based alloys, have a $Z\bar{T}$ value of around unity (at room temperature for Bi_2Te_3 and 500-700K for PbTe). However, at a $Z\bar{T}$ of 2-3 range, thermoelectric power generators would become competitive with other power generation systems (Riffat & Ma, 2003; Weiling & Shantung, 2004). In general, effective thermoelectric materials should have a low thermal conductivity but a high electrical conductivity. Rowe (2006) reported that conventional thermoelectric materials (those which are employed in commercial applications) can be conveniently divided into three groupings based on the temperature range of operation. Alloys based on Bismuth (Bi) in combinations with Antimony (An), Tellurium (Te) or Selenium (Se) are referred to as low temperature materials and can be used at temperatures up to around 450 K. The intermediate temperature range – up to around 850 K is the regime of materials based on alloys of Lead (Pb) while thermoelements employed at the highest temperatures are fabricated from Silicon-Germanium (SiGe) alloys and operate up to 1300 K. Although these materials still remain the cornerstone for commercial and practical applications in thermoelectric power generation, significant advances have been made in synthesising new materials and fabricating material structures with improved thermoelectric performance. Efforts have focused primarily on improving the material's figure-of-merit, and hence the TE conversion efficiency, by reducing the lattice thermal conductivity (Rowe, 2006).

Attempts are also being made to improve the competitiveness of thermoelectrics in directions other than by improving the figure-of-merit. In particular, efforts have focused on increasing the electrical power factor, decreasing cost and developing environmentally friendly materials. For example, considering the electrical power factor as the dominant parameter, it has initiated a search for materials with high power factors rather than conversion efficiency. Considerable success has been made in synthesising materials, particularly attractive for waste-heat recovery. For example, it is reported in (Rowe et al., 2002) that the rare earth compounds $YbAl_3$, although possessing a relatively low figure-of-merit, has a power factor at least double that of any other reported in the literature, which operates over the temperature range of a waste heat source. MgSn has almost the same performance but costs less than 25% of the price (Rowe, 2006). Another recent direction to improve the competitiveness of thermoelectric materials, other than by improving the figure-of-merit, is by developing novel thermoelectric module shapes. As discussed previously, thermoelectric modules have typically plate-like shapes (see Fig. (8)) and fabricated from bulk semiconductors such as Bi_2Te_3 and PbTe, making them rigid and unsuitable for covering relatively large surfaces that are curved or non-flat (e.g. circular tubes) used in waste heat recovery applications. Also, this conventional configuration is suitable for applications where the flow of heat is perpendicular to the ceramic plates. In addition, in order to improve thermal contact to heat sources of arbitrary geometry, it is desirable to fabricate thermoelectric modules which can conform easily to a surface. Therefore, recent research has been focused on developing novel flexible- and cylindrical-based shapes of thermoelectric power generators. For example, Yadav et al. (2008) proposed and demonstrated the use of flexible and cost-effective thermoelectric power generator based on thin film thermoelectric on flexible fiber substrates. Min and Rowe (2007) have also recently developed a novel tube-shape thermoelectric module for power generation.

3.3 Fundamental concept and energy conversion aspects of direct generation of electricity using thermophotovoltaic alternative technology

The basic theory and operation of a thermophotovoltaic power generation system is shown in Figure 9. The TPV system mainly consists of the following components: (i) a high-temperature source of heat, (ii) a radiator (absorber/emitter component), (iii) a reflector (spectral controller), (iv) a semiconductor converter, (v) a low-temperature heat-sink, and (vi) an electrical load with a power conditioner, as shown in Figure 9. Typical fuel sources used in driving TPV systems includes (Coutts, 1999): (1) solar-fuelled TPV, bio-fuelled TPV, (3) nuclear-fuelled TPV, (4) liquid-hydrocarbon-fuelled TPV, (5) diesel-fuelled TPV, (6) propane-fuelled TPV, and (7) natural-gas-fuelled TPV. In a TPV device, the high-temperature heat source (e.g. a flame generated from a burner) produces thermal energy which is then absorbed by the emitter (radiator). The absorbed heat is then radiated (emitted) in form of high-energy photons by the emitter which is received and absorbed by the TPV semiconductor converter. There are basically two different types of emitters, namely broadband emitters and selective emitters. Blackbody is considered to be a typical broadband emitter. Selective emitters normally exhibit a high emittance in the spectral range utilizable for the TPV converter photocells (Chia & Feng, 2007). Some of the unabsorbed long-wavelength photons can be recirculated, using the reflector (also known as spectral filter), back to the emitter to be reabsorbed for maximum conversion efficiency. The TPV semiconductor converter (photovoltaic cell) then converts the photons into electrical power when connected to an electrical load. The optional heat sink is used to dissipate the low-grade heat to cool the converter.

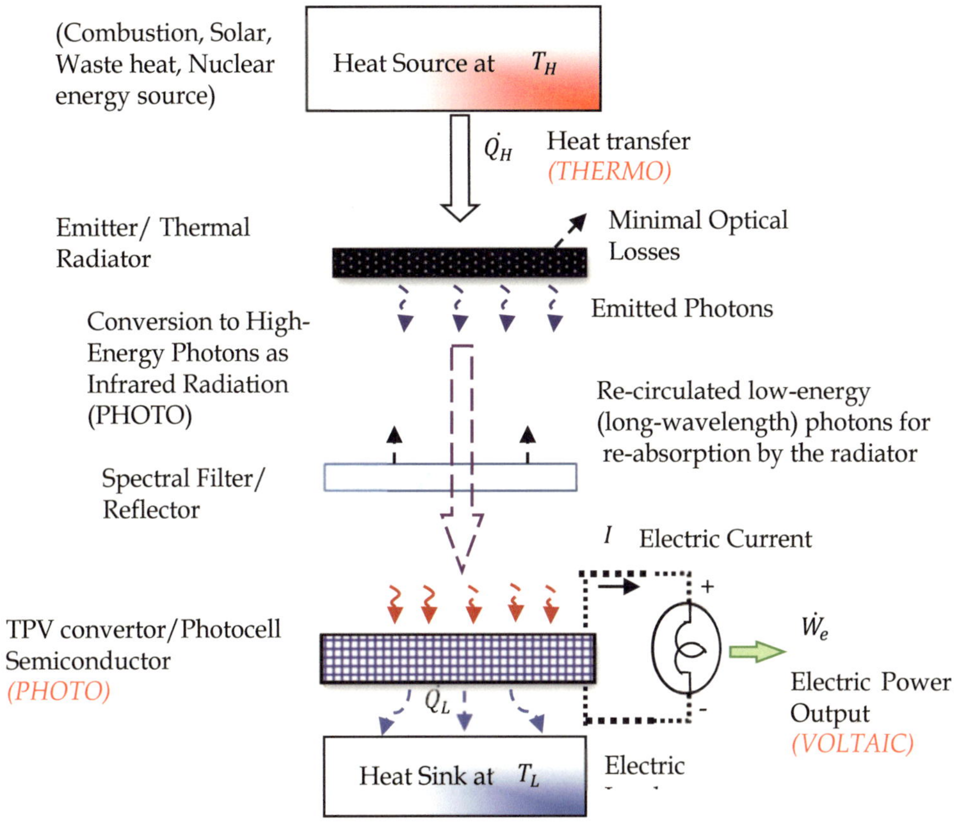

Fig. 9. A schematic diagram illustrating the basic concept of a simple thermophotovoltaic electric power generation system.

3.3.1 Energy-conversion aspects of TPV power generation systems

A thermophotovoltaic power generator is also a thermodynamic heat engine so that its thermal efficiency is given previously by Eq. (12). The output electric power $\dot{W}_e$ can be expressed in terms of DC electric current and voltage generated by a TPV system, as

$$W_e = IV \qquad Z = \frac{\alpha^2}{kR} \tag{14}$$

Lal and Blakers (2009) reported that the energy-conversion efficiency of the TPV system can be expressed as the ratio of electrical power output to the power required to keep the blackbody radiating emitter at a constant temperature, given by

$$\eta_{TPV} = \frac{I_{SC} V_{OC} FF}{\left(P_{radiated} - P_{returned}\right)} \tag{15}$$

Where, the power returned is the power recirculated by the reflector (spectral filter) and the Fill Factor (FF) of the photocell converter is given by

$$FF = \frac{I_{mp}V_{mp}}{I_{sc}V_{oc}} \tag{16}$$

Also, limited by the second-law of thermodynamics, the ideal efficiency of a thermal power cycle, which takes place as the TPV power system continuously performs, operating as a reversible heat engine is Carnot efficiency given previously in Eq. (7). In general, the higher the temperature of the heat source, the higher the ideal efficiency at a fixed heat-sink temperature. More specifically, Coutts (1999) reported Carnot efficiency in terms of the emitter (radiator) absolute temperature, T_{emit}, and the TPV photocell converter absolute temperature, T_{cell} , given by

$$\eta_{\max} = 1 - \frac{T_{cell}}{T_{emit}} \tag{17}$$

For example, for a TPV system performing between a radiator temperature at 500 °C (773.15 K) and photocell at 40 °C (313.15 K), the maximum efficiency of the TPV system can be calculated using Eq. (17) to be approximately 60%. However, for a real TPV power cycle operating between the same two temperature limits would have lower efficiency due to its thermal losses and inherent irreversibilities. In general, the ideal efficiency sets the upper limit for the thermodynamic performance of a TPV system and aids in optimizing designs and performances of these power generation systems. The overall energy conversion efficiency of a TPV system is the product of the discrete efficiencies mainly of the heat source, the emitter (radiator), the reflector, and the converter, given by (Chia & Feng, 2007)

$$\eta_{TPV} = \eta_{HS}\eta_{E}\eta_{R}\eta_{C} \tag{18}$$

Datas and Algora (2009) indicated that the overall efficiency of a TPV system cannot generally be formulated as the multiplication of the efficiency of each of the isolated components that makes up the system. It is because the relationship between these components is very complex and the efficiency of each component depends on all of them. They (Datas & Algora, 2009) recommended that the whole TPV system has to be considered simultaneously in determining its overall conversion efficiency. Badescu (2005) reported that experimental results showed that the overall TPV system efficiency is typically between 11 and 26% as far as broad-band (i.e. blackbody) emitters are considered.

3.3.2 Thermophotovoltaic system performance specifications

As discussed earlier, the TPV system mainly consists of four discrete components (excluding the load and optional heat sink), the possibly most complicated, in terms of technology, of which is the semiconductor photocell converter as reported by Coutts (1999). The incident radiation on the TPV photocell converter is typically provided by a radiating surface at a temperature in the approximate range of 1300-1800 °C (Coutts, 1999). During TPV energy conversion, the photon energy of the thermal radiation must be matched to the bandgap of the TPV converter photocells. It was reported (Qiu & Hayden, 2006) that most of the studies on TPV have focused on developing low-bandgap photocells (0.50–0.74 eV) such as GaSb and GaInAsSb photocells. These photocells are capable of converting a significant portion of infrared radiation from thermal radiators at a typical radiator temperature of 1473–1723 K, which is of interest to practical combustion systems. It was also reported (Qiu & Hayden, 2006) that the typical range of bandgaps for achieving high conversion efficiencies using emitter temperatures of 1000-2000 K is between 0.5 and 0.75 eV or slightly higher. Lal and Blakers (2009) reported that ideally, TPV photocells would have a bandgap of 0.6-0.9 eV to take into account of the temperature range of 1300-2000 K for a practical TPV emitter. The conventional silicon photocells have low cost (~ US$ 0.1/cm^2), high efficiency (> 20% under sunlight) (Lal & Blakers, 2009) and commercially available in large quantities (Qiu & Hayden, 2006), but the operating bandgap of 1.1 eV exceeds the optimum value. Lal and Blakers (2009) recommended the use of novel silicon silver photocells technology since it offers a method of fabricating highly efficient (>20%) thin silicon photocells in TPV systems. Silver-based photocells can be made up to a factor of 10 times thinner than conventional silicon photocells; current silver photocells have typical length 50-120 mm, width of 0.5-2 mm and thickness 20-100 μm (Lal & Blakers, 2009). The development of a low-cost low-bandgap TPV photocell which is described in (Heide et al., 2009). Chan et al. (2010) reported that in recent years, most of the research on TPV systems has been directed towards relatively lower-temperature (900-1100 ^{o}C) applications. Chan et al. (2010) also reported important benefits of using lower-temperature TPV as compared to its high-temperature counterpart; namely, (1) low-bandgap photocells convert a much larger portion of the infrared spectrum into electricity, thus enhancing output power density and maximizing efficiency, (2) the reduced temperature induced stresses, larger spectrum of available materials, and better material stability thus extending the possible design space. It was reported (Coutts, 1999) that the radiated spectrum extends across a wide wavelength range and, for a radiator temperature of 1500 K, the peak in the radiated spectrum occurs at about 2.5 μm. The emissivity is unity over the entire spectral range for a blackbody radiant surface. Badescu (2005) reported that using broadband emitters (radiators) is simpler than using selective emitters, it is easier to manufacture, may be more durable, and is less labor-intensive. Qiu and Hayden (2007) reported that an optical filter (reflector) reflects the non-convertible infrared energy back to the radiator and thus the use of an optical filter improves the spectral efficiency for optimal TPV system performance.

4. Conclusion

An increasing concern of environmental issues of greenhouse gas emissions, being responsible for global warming & climate change, environmental pollution, and the limitations and conservation of natural energy resources have recently resulted in extensive research into innovative nonconventional renewable & alternative clean energy technologies for generating electrical power. Some of these innovative & emerging technologies include: renewable low-temperature geothermal energy, thermoelectric and thermophotovoltaic green power generation technologies. This chapter introduced and presented fundamental concepts of these technologies. Innovative features, applications, limitations, specifications, and energy performance aspects of these nonconventional power generation technologies were also presented and discussed. A number of successful renewable and clean low-temperature geothermal energy binary ORC plants were installed in different locations worldwide which demonstrated the ability of this promising green technology for generating electricity. The vast low-temperature geothermal resources found widely in most continental regions have not received much attention for electricity generation. Continuous development of innovative drilling and ORC power generation technologies and other factors make this nonconventional, renewable and clean energy source one of the best future viable, alternate and available source to meet the required future electricity demand worldwide, significantly reducing GHG emissions and mitigating global climate change. Much of the recent research activities on applications of TE power generation have been directed towards utilisation of industrial waste heat. In this large-scale application, TE power generators offer a potential alternative of green electricity generation powered by waste-heat energy that would contribute to solving the worldwide energy crisis, and the same time help reduce environmental global warming. Future developments in this area seem to focus onto finding more innovative TE materials that could handle higher temperatures from various industrial heat sources at a feasible cost with more efficient energy performance. Future developments of more novel TE module geometries and configurations would make this technology more practical and attractive in applications where sources of waste heat have arbitrary shapes. There have been some causes for limiting the applications of TPV power generation technology. Some of the limiting factors of TPV systems applications and advancement are spectral filtering, low TPV conversion efficiency, the optimization of individual components rather than the system as a whole, adequate systems modeling and thermal management. One of the future TPV technology developments would be in designing high-performance TPV power generators, the improvement of TPV converters materials and total system energy-conversion optimization.

5. Acknowledgement

The author of this chapter wishes to acknowledge the funding contribution by Goldcorp Canada Ltd.-Musselwhite Mine that mainly supported the collaborative geothermal energy & heat pump (GHP) network technology, a contracted research project with Lakehead University, and for covering the cost of publishing this chapter with INTECH.

6. References

Badescu, V. (2005). Upper bounds for solar thermophotovoltaic efficiency. *Renewable Energy*, Vol. 30, pp. 211-225, ISSN: 0960-1481.

Cengel, Y. A., & Boles, M. A. (2008). *Thermodynamics: an engineering approach* (6th ed.), McGraw-Hill press, ISBN: 978-0073529214, New York.

Chan, W., Huang, R., Wang, C., Kassakian, J., Joannopoulos, Celanovic I. (2010). Modeling low-bandgap thermophotovoltaic diodes for high-efficiency portable power generators. *Solar Energy Materials & Solar Cells*, Vol. 94, pp. 509-514, ISSN: 0927-0248.

Chandrasekharam, D., & Bundschuh, J. (2008). *Low-enthalpy geothermal resources for power generation*, CRC Press Taylor & Francis Group, ISBN: 978-0-415-40168-5, New York.

Chen, L., Li, J., Sun, F., & Wu, C. (2005). Performance optimization of a two-stage semiconductor thermoelectric-generator. *Applied energy*, Vol. 82, pp. 300-312, ISSN: 0306-2619.

Chia, L. C., & Feng, B. (2007). The development of a micropower (micro-thermophotovoltaic) device. *Journal of Power Systems*, Vol. 165, pp. 455-480, ISSN: 0378-7753.

Coutts, T. J. (1999). A review of progress in thermophotovoltaic generation of electricity. *Renewable and Sustainable Energy Reviews*, Vol. 3, pp. 77-184, ISSN: 1364-0321.

Custom Thermoelectric. Available from: http://www.customthermoelectric.com

Cui, J., Zhao, J., Dai, C., & Yang, B. (2009). Exergetic performance investigation of medium-low enthalpy geothermal power generation. *IEEE Computer Society*, Vol. ,pp. 636-639, ISSN: 978-0-7695-3819-8.

Datas, A., & Algora, C. (2009). Analytical model of solar thermophotovoltaic systems with cylindrical symmetry: ray tracing approach. *Progress In Photovoltaics: Research and Applications*, Vol. 17, pp. 526-541.

Dickson, M. H., & Fanelli M. (2005). *Geothermal energy: utilization and technology,* Earthscan, an imprint of James & James (Science Publishers) Ltd. in Association with the International Institute for Environment and Development, ISBN: 1-84407-184-7, USA.

DiPippo, R. (2008). *Geothermal power plants: principles, applications, case studies and environmental impact* (2nd ed), Elsevier, ISBN: 978-0-7506-8620-4, New York.

DiPippo, R. (2007). Ideal thermal efficiency for geothermal binary plants. *Geothermics*, Vol. 36, pp. 276-285, ISSN: 0375-6505.

Ferrotec. Available from: http://www.ferrotec.com

Gupta, H., & Roy, S. (2007). *Geothermal energy: an alternative resource for the 21st century,* Elsevier, ISBN 978-0-444-52875-9, New York.

Galanis, N., Cayer, E., Roy, P., Denis, E. S., & Desilets, M. (2009). Electricity generation from low temperature sources. *Journal of Applied Fluid Mechanics*, Vol. 2, No. 2, pp. 55-67, ISSN: 1735-3645.

Hammons, T. J. (2004). Geothermal power generation worldwide: global perspective, technology, field experience, and research and development. *Electric Power Components and Systems*, Vol. 32, pp. 529-553, ISSN: 1532-5008.

Heide, J., Posthuma, N. E., Flamand, G., Geens, W., & Poortmans, J. (2009). Cost-efficient thermophotovoltaic cells based on germanium substrates. *Solar Energy Materials & Solar Cells*, Vol. 93, pp. 1810-1816, ISSN: 0927-0248.

Hettiarachchi, H. D. M., Golubovic, M., Worek, W. M., & Ikegami, Y. Optimum design criteria for an organic Rankine cycle using low-temperature geothermal heat sources. *Energy*, Vol. 32, pp. 1698-1706, ISSN: 0360-5442.

Hung, T.-C. (2001). Waste heat recovery of organic Rankine cycle using dry fluids. *Energy Conversion and Management*, Vol. 42, pp. 539-553, ISSN: 0196-8904.

Lal, N. N., & Blakers, A. W. (2009). Sliver cells in thermophotovoltaic systems. *Solar Energy Materials & Solar Cells*, Vol. 93, pp. 167-175, ISSN: 0927-0248.

Maizza, V., & Maizza, A. (2001). Unconventional working fluids in organic Rankine-cycles for waste energy recovery systems. *Applied Thermal Engineering*, Vol. 21, pp. 381-390, ISSN: 1359-4311.

Min, G., & Rowe, D. M. (2007). Ring-structured thermoelectric module. Semicond. *Sci. Technol.*, Vol. 22, pp. 880-883, ISSN: 0268-1242.

Min, G., Rowe, D. M., & Kontostavlakis, K. (2004). Thermoelectric figure-of-merit under large temperature differences. *Journal of Physics D: Applied Physics*, Vol. 37, pp. 1301-1304, ISSN: 0022-3727.

Min, G., & Rowe, D. M. (2002). "Symbiotic" application of thermoelectric conversion for fluid preheating/power generation. *Energy Conversion and Management*, Vol. 43, pp. 221-228, ISSN: 0196-8904.

Nelson, R. E. (2003). A brief history of thermophotovoltaic development. *Semicond. Sci. Technol.*, Vol. 18, pp. S141-S143, ISSN: 0268-1242.

Nuwayhid, R. Y., Rowe, D. M., & Min, G. (2003). Low cost stove-top thermoelectric generator for regions with unreliable electricity supply. *Renewable Energy*, Vol. 28, pp. 205-222, ISSN: 0960-1481.

Panea, C., Rosca, G. M., & Blaga, C. A. (2010). Power generation from low-enthalpy geothermal resources. *Journal of Sustainable Energy*, Vol. 1, No. 2, pp. 1-5, ISSN: 2067-5538.

Qiu, K., & Hayden, A. C. S. (2007). Thermophotovoltaic power generation systems using natural gas-fired radiant burners. *Solar Energy Materials & Solar Cells*, Vol. 91, pp. 588-596, ISSN: 0927-0248.

Qiu, K., & Hayden, A. C. S. (2006). Development of a silicon concentrator solar cell based TPV power system. *Energy Conversion and Management*, Vol. 47, pp. 365-376, ISSN: 0196-890.

Riffat, S. B., & Ma, X. (2003). Thermoelectrics: a review of present and potential applications. *Applied Thermal Engineering*, Vol. 23, pp. 913-935, ISSN: 1359-4311.

Rosca, M. G., Karytsas, K., & Mendrinos, D. (2010). Low enthalpy geothermal power generation in Romania. *Proceedings World Geothermal Congress*, Bali, Indonesia, April 2010.

Rowe, D. M. (2006). Thermoelectric waste heat recovery as a renewable energy source. *International Journal of innovations in Energy Systems and Power*, Vol. 1, pp. 13-23.

Rowe, D. M. (1999). Thermoelectrics, an environmentally-friendly source of electrical power. *Renewable Energy*, Vol. 16, pp. 1251-1265, ISSN: 0960-1481.

Rowe, D. M., Kuznetsov, V. L., Kuznetsova, L. A., & Min, G. (2002). Electrical and thermal transport properties of intermediate-valence $YbAl_3$. *J. Phys. D: Appl. Phys.*, Vol. 35, pp. 2183-2186, ISSN: 0022-3727.

Rowe, D. M., & Min, G. (1998). Evaluation of thermoelectric modules for power generation. *Journal of Power Sources*, Vol. 73, pp. 193-198, ISSN: 0378-7753.

Ruggero Bertani, E. (2007). World geothermal generation in 2007, *GHC Bulletin*, pp. 1-12.

Saleh, B., Koglbauer, G., Wendland, M., & Fischer, J. (2007). Working fluids for low-temperature organic Rankine cycles. *Energy*, Vol. 32, pp. 1210-1221, ISSN: 0360-5442.

Stevens, J. W. (2001). Optimal design of small ΔT thermoelectric generation systems. *Energy Conversion and Management*, Vol. 42, pp. 709-720, ISSN: 0196-8904.

Wei, D., Lu, X., Lu, Z., & Gu, J. (2007). Performance analysis and optimization of organic Rankine cycle (ORC) for waste heat recovery. *Energy Conversion and Management*, Vol. 48, pp. 1113-1119, ISSN: 0196-8904.

Weiling, L., & Shantung, T. U. (2004). Recent developments of thermoelectric power generation. *Chinese Science Bulletin*, Vol. 49, No. 12, pp. 1212-1219, ISSN: 1001-6538.

Yadav, A., Pipe, K. P., & Shtein, M. (2008). Fiber-based flexible thermoelectric power generator. *Journal of Power Sources*, Vol. 175, pp. 909-913, ISSN: 0378-7753.

Yamada, S., & Oyama, H .(2004). Small capacity geothermal binary power generation system. *Fuji Electric Review* , Vol. 51, pp. 86-89.

Yodovard, P., Khedari, J., & Hirunlabh, J. (2001). The potential of waste heat thermoelectric power generation from diesel cycle and gas turbine cogeneration plants. *Energy Sources*, Vol. 23, pp. 213-224, ISSN: 0090-8312.

19

Four Steps to the Hydrogen Car

Paul McGuiness[1,2] and Urša Pirnat[3]
[1]*Department of Nanostructured Materials, Jožef Stefan Institute*
[2]*Center of Excellence NAMASTE, Ljubljana*
[3]*University of Nova Gorica*
Slovenia

1. Introduction

Hydrogen has long been touted as the fuel of the future; however, it is becoming increasingly clear that the next generation of cars will be hybrid vehicles - combining a small, conventional petrol or diesel engine with a battery-powered electric motor - and that these hybrid vehicles are a stepping stone to the purely electric vehicle, which will be powered by a battery that can be quickly and easily recharged at home or on the road. From the point of view of hydrogen-powered vehicles, this leads us to four critical research questions that we will attempt to cover in this chapter. Does this mean that the hydrogen car has been squeezed out of the picture? Just why are hydrogen vehicles taking so long to realise as a viable form of transport? And what changes or improvements in terms of science and technology are required before our roads see hydrogen cars and, of course, the hydrogen filling stations required to fuel these vehicles? Here we have identified the four critical barriers that need to be overcome before the hydrogen car becomes a reality and have assessed each one in terms of the likelihood of success. Before there is a hydrogen future for the car we need to solve the problems related to on-board hydrogen storage, developing a hydrogen infrastructure, simple and clean hydrogen production and a cheaper fuel cell for converting the hydrogen into electricity.

Of these, storing enough hydrogen in the car is already a practical reality. Toyota is one of a number of companies that has successfully demonstrated a passenger vehicle with a range in excess of 600 km on a single tank of hydrogen. They have achieved this not with a heavily insulated storage vessel that contains liquid hydrogen at –253°C, nor by using some exotic hydrogen-storage material like a rare-earth transition-metal alloy or carbon nanotubes, but instead by developing an extremely strong, carbon-fibre tank that can safely store hydrogen in gaseous form at pressures of up to 680 bar. Of course compression is still an expensive process, but compared to liquefaction or the weight costs of metal-hydride stores, a pressurised-gas tank looks like the best solution. In Section 2 we look at the properties of hydrogen and point to the reasons why other forms of hydrogen storage will find it hard to compete with the pressurised tank.

A hydrogen infrastructure could be introduced quickly, provided there is sufficient financial investment. There is, of course, the obvious chicken-and-egg situation: who will build this infrastructure if there are no vehicles, and who will invest in a hydrogen car if there is nowhere to re-fuel it. But the position is changing. Germany plans to have more than 1000 refuelling stations by 2020. In the United States, efforts are underway to have 50+ hydrogen stations in the Los Angeles area. Estimates for the whole of the US suggest that the country will need tens of thousands of hydrogen stations to provide access for the majority of the population, with the costs expected to be a massive - but still realisable - $35 billion (a figure similar to that quoted for the Alaska North Slope gas pipeline). Carrying pure hydrogen down steel pipes also carries with it the substantial risk of hydrogen embrittlement. This is a well-know problem associated with many steels, whereby relatively small amounts of hydrogen are absorbed by the steel, but the expansion caused either by the presence of interstitial hydrogen atoms of the reformation of hydrogen molecules in the voids in the materials can lead to cracking and subsequent leaking from the pipeline. In this part of the chapter, Section 3, as well as looking at some of the regional-based efforts to construct an infrastructure in Germany, the UK and the US, we also devote some space to some of our recent work on dramatically reducing the amount of hydrogen permeation into the steel by using a very thin coating of conventionally applied TiAlN.

An appropriate and economically viable fuel cell requires technological improvements that will take time and money. Unlike conventional internal-combustion engines fuel cells are anything but robust and durable. In particular, fuel cells contain delicate membranes and expensive catalysts. However, the past 10 years have seen a series of improvements that mean fuel-cell vehicles can operate in low-temperature environments - something that was not possible until very recently. Furthermore, the introduction of nanotechnology in the production of platinum catalysts is driving down prices, making the materials costs similar to those associated with conventional catalytic converters. Section 4 looks at the problems of expensive catalyst materials and points to research which is currently underway to substantially reduce these costs.

However, hydrogen production remains as the most difficult problem to solve, requiring a scientific breakthrough to produce hydrogen sufficiently easily and cleanly. Today, the vast majority of the world's hydrogen is produced by the steam reforming of fossil fuels, but taking into account the depletion of fossil fuels and the associated harmful environmental effects, the production of hydrogen from renewable sources is the most critical problem to overcome. For this reason there are huge development projects focusing on renewable paths for hydrogen production, including some of the following: fermentation, biological water splitting, conversion of biomass and wastes, solar thermal water splitting, renewable electrolysis, photo-electrochemical water splitting - and all these options need to be explored, since it is unlikely that a single solution will be appropriate for every situation. Solar energy is, of course, a virtually inexhaustible and widely available energy source, which leads to the idea of renewable hydrogen in which the large-scale use of solar cells would be coupled to an electrolyzer. However, there is another option, which has been neglected for far too long, and is actually the most attractive and cleanest approach to solar hydrogen: the use of photo-electrochemical cells. The operating principle is simple, the photo-active semiconductor material is immersed in water and photo-generated electrons are directly used to reduce and oxidise water. Within this system the functions of the solar cell and the electrolyzer are combined in a single device. However, the material requirements and the complex mechanism of photo-electrochemical splitting with which a highly efficient solar-to-hydrogen conversion in the visible part of spectra would be achieved are far from simple. Any photocatalyst material has to meet several criteria, i.e., it has to absorb visible light, have a good level of conductivity for photo-exiton transfer, has to straddle the electrochemical potential for reduction and oxidation, has to be very stable in the various media and last - but not least -

should be economically acceptable. In Section 5 we look at the photo-electrochemical splitting of water and recent developments to increase the efficiency of the process.

2. How can we store enough hydrogen in a car?

Before looking more closely at the storage options for hydrogen, it will be useful to look first at the characteristics of the Universe's most common element. Hydrogen in the form of a gas was first recognised as a discrete substance by Henry Cavendish (1731-1810) in the 18th century; although others, like Robert Boyle (1627-1691), had probably prepared it earlier. Cavendish, like his French rival Antoine Lavoisier (1743-1794), also made the important discovery that hydrogen (which he referred to as "inflammable air") reacted with oxygen to produce water. He also made the observation that the reaction produced no chemical by-products. Although this turns out not to be completely true when hydrogen burns in air. In this case small quantities of nitrogen oxides are normally produced as well. An important characteristic of hydrogen as a potential fuel is that it is almost entirely non-polluting in terms of emissions associated with the on-board fuel and the vehicle - emissions that might be associated with the production of hydrogen at some other location are, of course, a completely different question.

Another very important characteristic of hydrogen is that it does not occur naturally. Despite being the most abundant element in the Universe, representing some 75% of matter in terms of mass, hydrogen gas is extremely rare on Earth, quite simply because its light weight means that it can escape the gravitational pull of the planet much more easily than heavier gases. However, this is not to say that the element hydrogen is in anyway rare on Earth: two of water's three atoms are hydrogen, hydrogen is a major constituent of all the flora and fauna on the planet, and hydrogen makes up 61 percent of all the atoms in the human body.

Hydrogen has physical and chemical properties that suggest it is a good candidate for a fuel: it is odourless, colourless, stable, and does not react with oxygen until sufficient energy is inputted to begin an exothermic reaction that results in the production of water.

Hydrogen is the lightest of the elements and stores a lot of energy in its chemical bond. In terms of being useful as a fuel, we need to note a few of its physical characteristics. The density of hydrogen gas is 0.0899 grams per litre, which is about two-thirds the density of air. In liquid form, hydrogen has a density of 70.99 g/l and it boils at -253°C. These properties are part of the reason why hydrogen has the highest energy-to-weight ratio of commonly available fuels. For example, 1 kg of hydrogen contains the same amount of energy as 2.1 kg of natural gas or 2.8 kg of petrol. Putting it in simple terms, if we think about energy per gram of fuel, hydrogen is an easy winner - hydrogen stores more than two and a half times more energy than petrol if we compare equal weights of the two substances. But in terms of volume, hydrogen looks a little less promising: hydrogen has an energy density of 9 MJ/litre, while petrol has an energy density of 31 MJ/litre. In other words, even if we think of hydrogen in the liquid state (and we will ignore all the low-temperature factors for now) a hydrogen fuel tank would need to be at least three times larger than a petrol tank if our vehicle is going to have the same range on the road.

So, storing enough hydrogen - and in an easily accessible form - is one of the four major challenges that need to be overcome. Detailed on-board hydrogen-storage-system requirements were drawn up by the FreedomCAR and Fuel Partnership (a collaboration among the DOE, the U.S. Council for Automotive Research, and the major energy and utility companies), with the most recent revision being in 2009. These requirements are very detailed; however, it is generally considered that a car needs a minimum of 5kg of hydrogen on board in order to provide a practical range of 500 km. Although this is much less than the 50 kg or so of petrol or diesel that a car currently carries around when the tank is full, hydrogen is not available to us in such a convenient form. Under conditions of standard temperature and pressure, hydrogen is a gas. This means that storing enough hydrogen gas at a pressure of around 1 bar would involve a huge balloon-like tank that is a totally impractical proposition.

The most obvious solution appears to be liquid hydrogen. In theory, only about 14.2 MJ/kg (and here we talk about a kg of liquid hydrogen) have to be removed to cool hydrogen gas from room temperature (298 K) to 20.3 K and then condense the hydrogen gas into the liquid form at atmospheric pressure. In practice, hydrogen is liquefied in a complex, multi-stage process involving compression and expansion that is combined with a counter-flow heat exchange and a certain amount of heat recovery using expansion turbines. The whole procedure makes use of other liquefied gases: first, a three-stage vapour-compression system is used for the initial cooling from room temperature; second, using a multi-stage nitrogen expansion the hydrogen gas is cooled to 77 K; and third, there is a multi-stage helium compression and expansion to obtain liquid hydrogen at 20.3 K.

However, the situation is not so simple or energy efficient as we have just described. A typical medium-sized liquefaction plant, one producing about 200 kg of liquid hydrogen per hour will consume about 50 MJ/kg, well over three times the theoretical figure. The reason for this is that the theoretical figure does not contain all the electrical, thermal and mechanical losses. A large plant, producing a 1000+ kg of hydrogen per hour can reduce this energy-consumption figure to about 35 MJ/kg, and feasibility studies for plants producing 10,000+ kg of hydrogen per hour have suggested that this figure could be reduced to about 30 MJ/kg.

A graph showing how the amount of energy required for the liquefaction of hydrogen for different liquid-hydrogen-plant facilities is shown below.

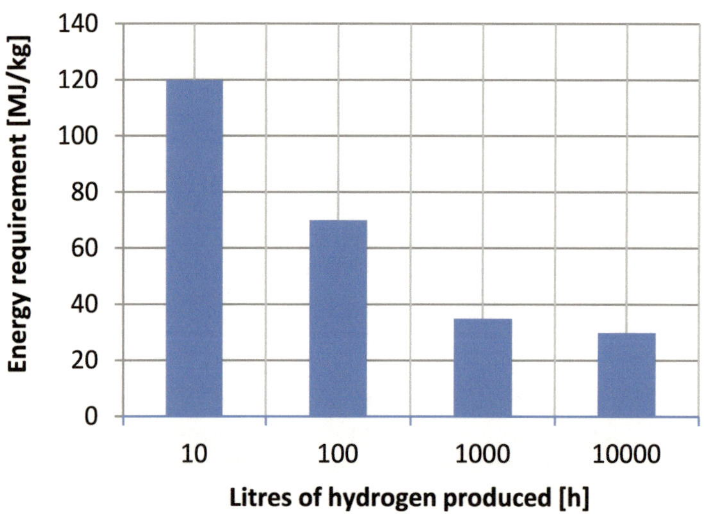

Fig. 1. Amount of energy required for the liquefaction of hydrogen for different liquid-hydrogen-plant facilities

Of course the absolute energy costs for producing the hydrogen are not the only consideration. It is important that we consider how much energy we use to liquefy the hydrogen compared to how much energy we can get out of the stored liquid hydrogen to drive our car.
For very small operations producing less than 10 kg per hour the energy needed to liquefy the hydrogen may exceed the energy available for use in the liquid hydrogen. Such a situation clearly makes no practical sense. However, even for a plant producing 10,000+ kg of hydrogen per hour (and no such plant currently exists) we would consume at least 25% of the available energy for the engine or the fuel cell of the car. The best practical figure for us to use would be 30%. In other words, 1.3 units of energy would have to be supplied during the hydrogen liquefaction process in order to have available 1 unit of energy.
A graph showing the percentage of available energy required to liquefy the hydrogen from room temperature for different liquid-hydrogen-plant facilities is shown below.

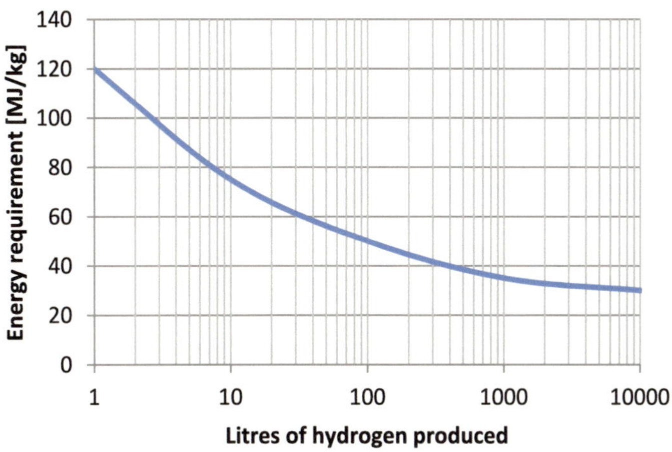

Fig. 2. Percentage of available energy required to liquefy the hydrogen from room temperature for different liquid-hydrogen-plant facilities

The best known problem, although not well understood, is that of boil-off. It occurs because the liquid-hydrogen tank cannot be perfectly insulated and pressure must not be allowed to build up inside the tank because the pressures would very quickly become impossible to contain safely. The extent of these losses depends on many factors, but a few percent per day is as low as we can realistically expect. This means that if our car was parked in a - well ventilated - garage it would lose about a quarter of a full tank in a week.
Perhaps the most elegant way of storing hydrogen is to form a reversible metal hydride. Some of the best researched examples include MgH_2, $NaAlH_4$, $LaNi_5H_6$, and $TiFeH_2$. These materials have an excellent energy density by volume - for example, the hydrogen atoms in $LaNi_5H_6$ are closer together than the hydrogen atoms in liquid hydrogen - however, their energy density by weight is often poor. The best metal-hydride stores are close to about 1.5 weight percent. That means to be able to store 5 kg of hydrogen we will need a metal-hydride storage tank weighing in excess of 330 kg - about the weight of two Ducati 848 Superbikes.
Most metal hydrides require high temperatures to desorb the hydrogen for use in the car. This energy cost is reduced with some of the more popular alloys, which require lower

desorption temperatures; however, if this interaction is too weak the re-hydriding pressure becomes too high. This problem appears to be the most difficult to overcome.
The most obvious, but for a long time overlooked, method of storing hydrogen is simply to compress the gas. The conventional gas cylinders found in any laboratory are filled to a maximum pressure of about 150 bar. To get a sufficient driving range and a reasonable size to fit within the confines of a car's structure, we require much higher pressures of 700-800 bar. Furthermore, in order to withstand everyday abuse and to facilitate handling these conventional cylinders are made from steel. This makes them far too heavy for automotive use. Recent advances in carbon-fibre technology have led to the practical realisation of such tanks. Quantum Technologies, a company based in California, is one of the leaders in this field. They recently demonstrated an all-composite 700-bar hydrogen tank, known as the QUANTUM TriShield, which has been bust tested to 1650 bar, making it comply with the European Integrated Hydrogen Project (EIHP) specifications. The EIHP is one of the main drivers of global regulatory standards.
The main problems associated with hydrogen gas tanks relate not just to the high pressures involved, but also to weight, volume, shape and cost. Weight is the least problematic. The use of carbon fibre and lightweight alloys puts compressed-hydrogen tanks streets ahead of solid-state storage and relatively close to the weights associated with conventional fuel tanks. In terms of weight, there are two directions currently being pursued. The first of these are the so-called cryo-compression tanks. These tanks use a liquid-nitrogen jacket to cool the tank to 77 K, so increasing its volumetric capacity by a factor of four. The second approach involves making the tank a more integral part of the car's structure, making use of the strength of other parts of the car to store the hydrogen, possibly by keeping the hydrogen within cellular load-bearing structures.
These storage tanks also require very safe and reliable pressure-management systems, able to allow the hydrogen to flow smoothly out of the tank to the fuel cell or internal combustion engine. Many companies have patented and proprietary systems for preventing any dangerous release of hydrogen during use. Quantum's patented 700-bar tank includes an in-tank regulator that provides additional safety by containing the high pressure in the tank and allowing a maximum delivery pressure of just 10 bar.
Such tanks are very sophisticated and, of course, expensive. This will be a major hurdle that needs to be overcome before these tanks could become widely available, but history shows that once technologies have been proved to work, if the demand is high, then the price will fall rapidly.

3. How will we introduce a safe and reliable hydrogen infrastructure?

At present there is no country anywhere in the world with any serious intention to introduce a hydrogen infrastructure for fuelling a transport system. In fact, a few years ago there were fewer than 4000 km of low-pressure hydrogen pipelines in the United States and Europe, combined. The pipelines that exist in the US are primarily for the transportation of hydrogen for use in the de-sulphuring of crude oil to produce diesel and petrol. The longest of these is the Texas-Louisiana pipeline that runs along the Gulf Coast.
A hydrogen pipeline infrastructure is basically the same as the pipeline infrastructure used for the transport of natural gas; it would have to involve similar components, such as transmission pipelines and compressor stations. Although there are several thousands of kilometres of hydrogen pipelines throughout the world, they are only designed to transport

the hydrogen over relatively short distances, with the production facility being located relatively close to the site of the end user. With the exception of a few longer stretches of pipeline, most of the world's several thousand kilometres of pipeline are made up of relatively short sections of a few hundred metres, carrying hydrogen at quite low pressures, typically less than 100 bar. However, the safety records associated with these pipelines are very good.

It is important to consider hydrogen's density when we make comparisons with natural gas pipelines. For any given pressure the density of hydrogen is about one-third that of natural gas. However, for an equivalent pipe diameter and gas pressure, the flow of hydrogen is approximately three times faster than that of natural gas. On this basis, and assuming large-scale hydrogen compressors could be made to operate as effectively as the natural-gas variants, we can envisage similar pipeline diameters of 1–2 m. The actual diameters would of course depend on other factors, particularly demand, achievable pressures and the regulations in force in any particular country or locality.

A lot of people have put forward the idea that we could make use of the existing natural gas infrastructure as the basis for a new hydrogen infrastructure. Natural-gas pipelines could be used in an unmodified form to transport 20% hydrogen in natural gas. However, modifying these same pipelines to carry pure hydrogen would require a number of issues to be addressed: many of the steel components in these pipelines could begin to suffer from embrittlement, and the sealing requirements for hydrogen are much stricter than for natural gas.

We have recently been involved in a major European project, called Hy-NANO-Im, funded by the EU as part of MNT.ERA-NET programme to look closely at the problem of preventing the permeation of hydrogen into steel structures in order to reduce the problems associated with the embrittlement of hydrogen pipelines based on steel. The starting points for the project were that the coating must be effective, cheap, stable and easy to apply. There have been many attempts to produce such anti-permeation coatings in the past, but all of them have failed in at least one aspect of the requirements for widespread application.

The experimental proof of a barrier's efficiency is the relative reduction of the steady permeation flux. Its definition is the ratio of the steady permeation rate through the uncoated membrane, i.e., the steel, versus the steady permeation rate through the coated membrane, termed "the permeation reduction factor" (PRF):

$$PRF = \frac{j_{uncoated}}{j_{coated}} . \tag{1}$$

Both permeation rates must be obtained under identical conditions in terms of the driving pressure and the temperature.

When the hydrogen migration through the planar, homogenous membrane, i.e., the steel, is limited by hydrogen atom diffusion in the material, the steady-state permeation rate j follows the Richardson equation

$$j = \frac{P}{d}\sqrt{p} , \tag{2}$$

where P is the permeability coefficient, p is the hydrogen driving pressure and d is the membrane thickness. The coefficient P is a material property and is a product of the hydrogen diffusivity and solubility. Applying the Richardson equation to a composite (coated) planar membrane enables the calculation of an effective permeation coefficient. The effective permeability $P_{eff.}$ is based on the sum of the permeation resistances for each layer, analogous to electrical resistors in series. For a two-layered membrane, i.e., a hydrogen-resistant coating on steel, it is then

$$\frac{d}{P_{eff.}} = \frac{d_1}{P_1} + \frac{d_2}{P_2}, \quad (3)$$

where the indices denote layer 1 and 2, while the membrane thickness is $d = d_1 + d_2$.

The permeability coefficient of a coating material can thus be obtained from the known layer thicknesses and the steady-state permeation rates through the coated and uncoated membranes.

The coatings reported prior to our Hy-NANO-Im project included dense alumina, only 1 μm thick, which produced a permeation reduction factor (PRF) of 1000 at 800°C, as reported by Levchuk et al. (2004). Erbium oxide, again 1 μm thick, has also been recognized as an efficient barrier, with a PRF of the order of 1000 (Levchuk et al. 2007). And an even more efficient barrier was made from Al–Cr–O, giving a PRF between 2000 and 3500 at 700°C (Levchuck et al. 2008). However, these coatings would not be particularly cheap or easy to apply.

In the Hy-NANO-Im project we decided to look at transition-metal-nitride hard coatings, which have been employed since the early 1980s for the protection of tools. Such coatings are most commonly prepared by physical vapour deposition (PVD) at a typical deposition temperature of around 450°C. Starting from the ubiquitous TiN, several ternary coatings have been developed, with the greatest usage probably achieved by TiAlN. This coating is distinguished by a high hardness (3300 HV), a good oxidation resistance (up to 850°C), a chemical inertness and a low thermal conductivity (PalDey et al. 2003). In its basic form, a TiAlN coating has an atomic ratio of approximately Ti:Al = 50:50, a columnar microstructure and a B1 cubic structure. Several TiAlN-based coatings have been developed over the past 10 years or so, including AlTiN (with an increased share of aluminium) and nanolayer TiAlN/TiN.

Among the diffusion-related properties, oxidation resistance (Inoue et al. 1997) and thermal stability (L Hultman, 2000) have been studied extensively. With the aim being to enhance the coating's performance in cutting operations, several options have been explored, such as multilayers or the addition of another element. However, in contrast to TiN, where there are many reports on its application as a diffusion barrier (M.Y. Kwak et al. 1999), the data on TiAlN is scarce. Of specific interest is the study by Man et al., (2004) who deposited TiAlN coatings on AISI 316L steel and measured the hydrogen permeation flux. They obtained the lowest permeation flux of 1.2 10^{-11} mol cm^{-2} s^{-1} for an aluminium share of x=0.4.

These permeation results combined with our experience in depositing dense films as hard coatings made us interested in testing the effectiveness of TiAlN barriers with respect to hydrogen when applied to steel for infrastructure and hydrogen-pipeline applications. In the Hy-NANO-Im project we looked at applying a TiAlN film of 5 μm, deposited on Eurofer steel samples using magnetron sputtering, which resulted in a PRF of 5800 to 20,000 at 400°C.

A cross-section TEM study provided valuable information on the microstructure and the phase composition of the coating. A qualitative EDS analysis confirmed the TiAlN composition of the film. Figure 3 shows the TiAlN/Eurofer interfacial region. The interface between the substrate and the film is smooth and abrupt. TiAlN crystallites extend from the interface in the form of elongated, columnar grains. They become clearly visible after 100 nm or so. The width of these parallel grown columnar grains is 20-50 nm and they are up to 500 nm long. Close to the thin edge of the film we observe, in weak contrast, wavy horizontal bands, *i.e.,* TiAlN growth zones. These bands are a consequence of the sample rotation. In the two-fold rotational geometry the sample travels along a complex trajectory within the deposition chamber, and thus periodically approaches towards and moves away from the target (Panjan et al. 2008). Although the targets are identical, a multilayer character becomes visible, as a consequence of the differences in the growth rate. The typical width of the bands is 10-15 nm, which corresponds to the amount of material deposited in each cycle.

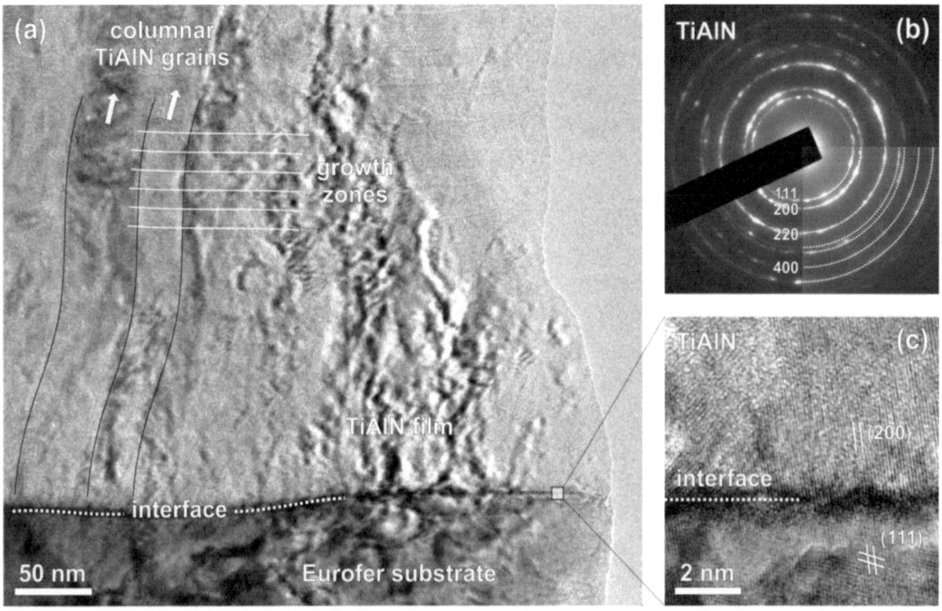

Fig. 3. Bottom part of the TiAlN film at the contact with the Eurofer substrate. (a) TiAlN grows in the form of 20-50-nm-wide columnar grains that extend from the TiAlN/Eurofer interface. Growth zones are parallel with the interface. (b) SAED pattern recorded from the TiAlN film shows diffraction rings stemming from randomly oriented TiAlN crystallites. Measured d-values correspond to the TiN-related cubic structure. (c) A high-resolution TEM image of the TiAlN/Eurofer interface shows that columnar TiAlN grains in the film grow on the top of the Fe grains of the substrate in a semi-epitaxial relation.

The SAED pattern recorded from a larger area of the TiAlN film shows the diffraction rings characteristic for a face-centred cubic TiN-related structure with slightly larger *d*-spacings. All of the diffraction rings can be attributed to the cubic phase; therefore, we could exclude the possible presence of any hexagonal AlN- or TiN-related phase. Along the interface we can regularly observe a certain degree of epitaxy between the Fe grains in the substrate and the TiAlN grains in the film. One such situation is shown in the close-up in Fig. 4c, where (111) lattice planes of iron continue into the (200) lattice of TiAlN. Because of the structural dissimilarity in the other directions a full-scale epitaxial growth cannot be expected in this system. The columnar texture of the TiAlN coating is retained throughout the film's cross-section all the way to the surface. Figure 4 shows the surface area of the TiAlN film. The surface itself is smooth with a roughness comparable to the roughness of the substrate (Fig. 4a). In the ion-milled parts of the TiAlN surface we can clearly recognize the columnar grains and the growth zones, which are characteristic for the film's method of deposition (Fig. 4b).

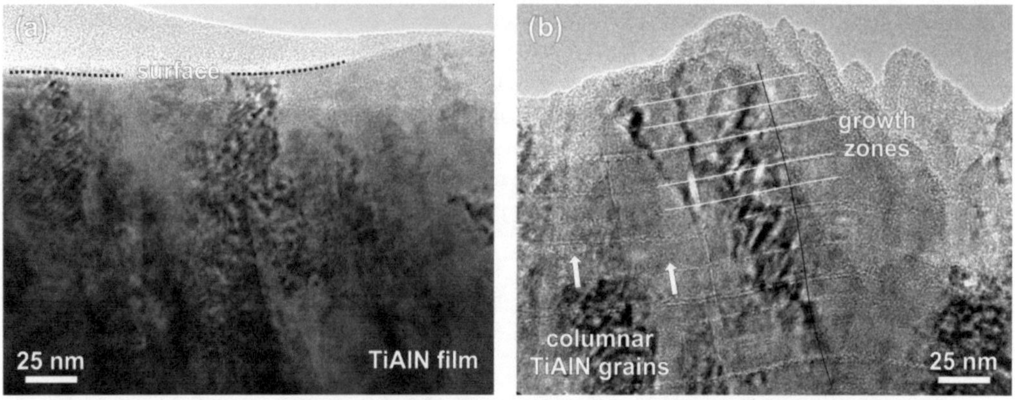

Fig. 4. Top part of the TiAlN film. (a) Smooth surface embedded in epoxy resin visible as amorphous layer above the surface line. (b) ion-milled surface region showing columnar TiAlN grains and distinct growth zones.

The measurements made on the uncoated membrane were performed to test the system's performance, reproducibility and to compare our data with previously published data. Since our system has a very high sensitivity, but the measurements with such a sensitivity last a long time, we decided to fix the temperature of the experiments at 400°C. The permeation flow of hydrogen through the uncoated membranes was observed immediately on a 5-Pa FS CM (capacitance manometer) gauge after the hydrogen was suddenly introduced to the upstream side. The steady flows at various values of the upstream pressure from p_{up} = 60–120 kPa are shown in Fig. 5 for the uncoated steel. The measured flow is proportional to the square root of the upstream pressure, indicating a diffusion-limited permeation regime. The small differences between the flows through the three uncoated samples can be related to slight variations in the surface composition or to impurities in the hydrogen gas. These impurities were kept at a low level by leading the gas through a trap chilled by liquid nitrogen. The average time-lag of 6.5 s was determined from several repeated cycles giving us a diffusivity constant of $D(400°C) = 6.4{\cdot}10^{-9}$ m^2/s. The permeability coefficient was determined to be P(400°C) = $1.7{\cdot}10^{-11}$ mol H_2/m s $\sqrt{Pa}$ and the hydrogen solubility was S(400°C) = $2.6{\cdot}10^{-3}$ mol H_2/m^3 $\sqrt{Pa}$. The obtained parameters are in very good agreement with the published transport parameters for hydrogen and deuterium in Eurofer steel (Esteban et al (2007) and Levchuck et al. (2004b)). When testing the TiAlN-coated sample at 1 bar of H_2 we observed that the permeation flux was both substantially lower and that the time to reach a maximum flow value was longer. The first exposure to 1 bar and the subsequent pressure change on the downstream side is given in Fig. 6.

Three repeated exposures of the steel to approximately 1 bar of hydrogen were conducted and the exposure lasted for some hours. Between those exposures the membrane was allowed to outgas in UHV for some tens of hours, which was sufficient to restore the low background. Two values for the PRF were recorded within the first 4 days: 13,300 and 20,000. During this period the membrane was kept at 400°C all the time.

However, an evaluation of a barrier's effectiveness is not simple and no unique criterion has been accepted (Perujo & Forcey 1995). Nevertheless, using the PRF as a means of assessing the capability of a given coating to act as a permeation barrier is still the most common way to enable such a comparison. It is generally determined in similar permeation devices, but at various temperatures and pressures. This makes any PRF comparison and prediction of its long-term operation even less reliable. In any case, we made a comparison of our PRF value measured at a planned 400 °C against the highest published values of PRF for different barrier-film materials deposited by various techniques.

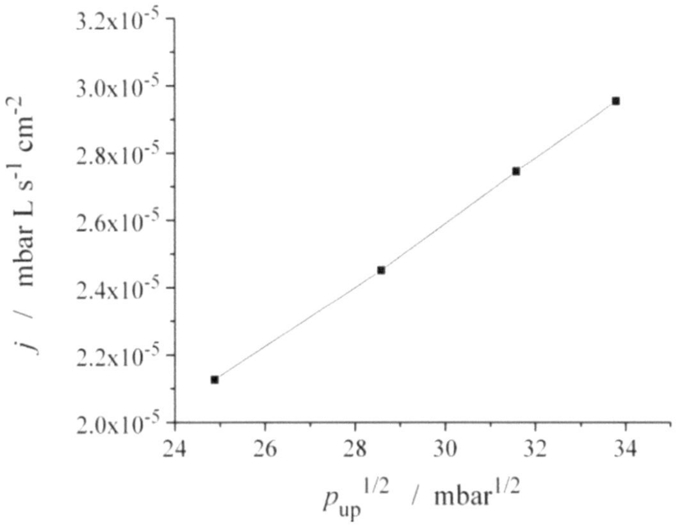

Fig. 5. Permeation rate versus square root of the upstream pressure from 62 000 Pa to 114 000 Pa

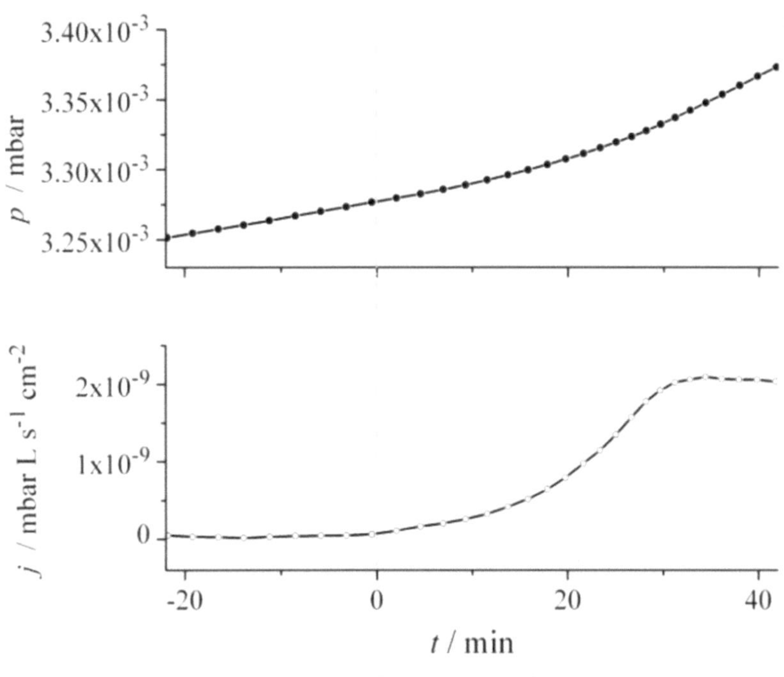

Fig. 6. Pressure evolution and permeation flux obtained from pressure derivative versus time before and after introducing 83000 Pa H_2 at the upstream side. Hydrogen was introduced at t = 0.

Table 1 summarizes the results on the barriers reported so far on the Eurofer substrate. Among the three published reports, all used arc evaporation for the deposition of the barrier film. One of the features of arc evaporation is the incorporation of droplets into the growing film, which may have a negative influence on the film's permeability. Nevertheless, the authors applied two varieties of arc evaporation, where this problem is significantly reduced. In magnetron sputtering, applied in our case, the droplet problem is largely irrelevant and the resulting compact microstructure contributes to the permeability of the film. Therefore, it is not surprising that our tested TiAlN film exhibited the highest PRF of any reported barrier materials applied so far to a Eurofer substrate.

Material	Deposition method	Thickness	Evaluation temperature	PRF	Ref
Al_2O_3	Filtered arc evaporation	1 μm	700, 750, 800°C	1000	Levchuck et al. (2004)
Er_2O_3	Filtered arc evaporation	1 μm	400-700°C	800	Levchuck et al. (2007)
Al-Cr-O	Pulse enhanced electron emission arc evaporation	2 μm	600-700°C	2000-3000	Levchuck et al. (2008)
TiAlN	Magnetron sputtering	5 μm	400°C	5800-20000	McGuiness et al. (2010)

Table 1. Comparison of the highest achieved PRF values for various barrier types applied on Eurofer.

Despite the common usage of PRF to assess a film's capability, it is not the best quantity for an intercomparison since it does not take into account the substrate's properties (material, thickness) and the film thickness. When the permeation through a coated membrane is assumed to be diffusion limited, then the permeability coefficient P of the film is the proper quantity that is independent of the substrate and dimensions. The permeation coefficient at 400°C for our measurements on the TiAlN film give $P_{TiAlN} = 2.9 \cdot 10^{-17}$ mol H_2/m s $\sqrt{Pa}$, $1.3 \cdot 10^{-17}$ mol H_2/m s $\sqrt{Pa}$ and $8 \cdot 10^{-18}$ mol H_2/m s $\sqrt{Pa}$ corresponding to the following PRF = 5800, 13300 and 20000. From ref. [9] (the only publication on permeation measurements through a TiAlN film deposited on AISI 316L steel) we calculated the $P_{TiAlN}(550°C) = 4.5 \cdot 10^{-16}$ mol H_2/m s $\sqrt{Pa}$ for the least permeable film investigated. The references [2-4] used in Table 2 unfortunately do not provide sufficient data to calculate P in the temperature range close to 400°C. The exception is ref. [3], where $P_{Er2O3}(600°C) = 4.1 \cdot 10^{-16}$ mol H_2/m s $\sqrt{Pa}$ could be calculated. From a rough extrapolation of $P_{eff.}$ for an Al_2O_3-coated membrane [2], the permeability coefficient P_{Al2O3} (600°C)$\approx 2 \cdot 10^{-16}$ mol H_2/m s $\sqrt{Pa}$ and P_{Al2O3} (400°C)$\approx 4 \cdot 10^{-17}$ mol H_2/m s $\sqrt{Pa}$ could be extracted. The latter value is in good agreement with the P_{TiAlN} obtained on our sample, while the P values at ≈600°C lie in the $(2\text{-}5) \cdot 10^{-16}$ mol H_2/m s $\sqrt{Pa}$ range for TiAlN [9], Er_2O_3 [3] and Al_2O_3 [2], suggesting that all three coating materials could have similar effectiveness as a hydrogen permeation barrier.

Our present approach seems a good choice, since TiAlN films are today deposited routinely as hard coatings. We also believe that the deposition parameters may be further optimized to fulfil the demands for barrier films as required for hydrogen pipelines as part of an international hydrogen infrastructure that could be used for automotive refuelling stations.

So far, hydrogen infrastructures for transport systems have been developed only at the regional level. The best-known and oldest pipeline is the Rhine-Ruhr pipeline that runs for a distance of more than 200 km. It was completed in 1938 and carries hydrogen through 30-cm diameter pipes at a pressure of approximately 15 bar. The original pipeline is still working, although the complete network is now much extended. In 2009 the organisation H2-Netzwerk-Ruhr set out a plan to become the "Silicon Valley" of hydrogen technology. The region of North Rhine Westphalia is the centre of many hydrogen-related activities. In Bottrop and Herten, the regenerative production of hydrogen takes place, industrial hydrogen is available in a pipeline from Marl, and Duisburg provides know-how through its centre for fuel-cell technology. In addition, the region operates a fleet of fuel-cell vehicles (Hychain) and in 2010 the city of Essen hosted the world hydrogen conference.

Other early starters in the regional pipelines were Isbergues in France (1973), Zeebrugge in Holland (1985), Rotterdam (1997). There are many other regional pipeline infrastructures in the planning stage or beginning construction. There are ideas to produce hydrogen in Iceland, with the plan to use a pipeline to transport it to Scotland, although at the moment this remains only as an idea, with the costs being described as prohibitive. In Germany, the National Organization Hydrogen and Fuel Cells Technology (NOW) is convinced that hydrogen will play a key role in the future and that hydrogen and battery technologies are in fact complementary rather than competitive. However, they admit that one of the primary difficulties of introducing to the market an alternative to gas-powered cars—whether powered by a battery or fuel cell—is building up the number of new cars on the road while creating a new fuelling infrastructure. They have said that hydrogen supplies could be trucked in initially, but eventually pipelines carrying the fuel to filling stations would be required. Daimler, who work closely with NOW, have admitted that fuelling infrastructure has not become real to the extent that is required, but they believe that the German government's stimulus plan is expected to fund new hydrogen filling stations to complement the four already located within the country. Four is not very many, but it is a start.

In the UK, Wales has taken a very forward-looking view on the subject of hydrogen and a hydrogen infrastructure. The organisation H2Wales has already identified that the transition to a hydrogen economy will require a number of intermediate steps, rather than a single-step change. These transition steps include the adoption of hydrogen as an additive to conventional fuels like diesel or CNG for internal combustion engines, electric vehicle drive development, hydrogen storage improvement and fuel-cell developments. In February 2010 the Welsh Secretary, Peter Hain, announced that the M4 motorway in south Wales is to become a "hydrogen highway". The scheme, to extend into south west England, is aimed at making hydrogen and electric-powered vehicles a viable alternative to petrol-driven machines. Under the plan, Wales will lead in developing alternative fuels, including hydrogen from renewable sources. The aim is to create an extensive renewable refuelling infrastructure.

In California, the California Hydrogen Infrastructure Project, led by Air Products and Chemicals, Inc. and the DOE, aims to conduct a comprehensive, California-based, multi-year hydrogen infrastructure demonstration project to develop and demonstrate strategically located fuelling stations. The infrastructure is expected to include flexible modes of supply to deliver liquid and gaseous hydrogen. A key feature of this project is that it will demonstrate "real-world" performance and generate data to help assess the feasibility of achieving the DOE's infrastructure goals. Honda, Nissan, and Toyota have agreed to provide a sufficient number of fuel-cell vehicles to the project at no expense, and BMW will also provide vehicles at no expense, operating with hydrogen internal combustion engines (H2-ICE) and onboard liquid-hydrogen storage to allow the project to demonstrate yet another important infrastructure component - liquid hydrogen refuelling. By encompassing a variety of vehicles, customer profiles, and fuelling experiences, this proposed infrastructure project will obtain a picture of real market needs.

The above examples are just a few of the many projects that are being carried out on relatively local scales by a wide range of different groups with the funding coming from a mix of public and private sources, which more or less input from car makers. It is clear that the technology is sufficiently mature, much can be taken from the existing natural-gas pipelines, and improvements like our own high PRF coatings to reduce permeation and embrittlement, means that all that is required to have a fully operating hydrogen infrastructure is the political initiative. It has been suggested that a country like China, which does not as yet have a massive "petrol/diesel" infrastructure is in the best place to leapfrog straight to a hydrogen-based fuelling infrastructure. However, the fact that we still have plenty of oil and that it is still very cheap means that for the moment this does not look economically attractive to the Chinese.

4. Is it possible to make the fuel cells cheaply enough?

The fuel cell is an electrochemical cell that converts chemical energy from a fuel into electricity. This electricity is generated from the reaction between hydrogen and an oxidizing agent, usually the oxygen in the air. The reactants flow into the cell, and the reaction products flow out of it, while the electrolyte remains within it. Fuel cells can operate continuously as long as the necessary reactant and oxidant flows are maintained. This makes fuel cells very different to conventional batteries, because they consume the reactant, in this case hydrogen, from an external point, rather than carrying all the fuel within the device itself. Figure 7 shows a schematic diagram of a fuel cell.

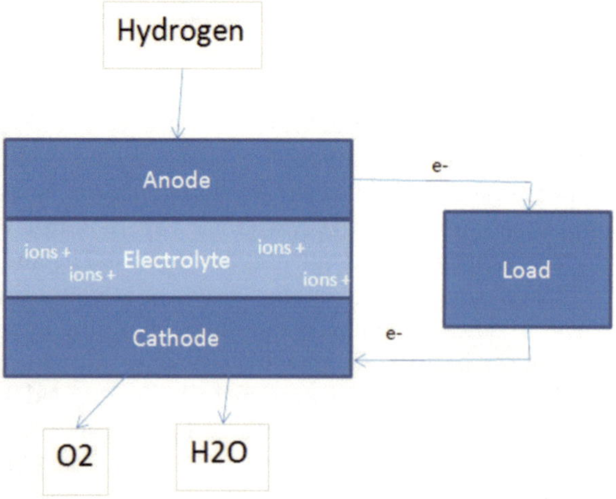

Fig. 7. Schematic diagram of a fuel cell

The first fuel cell was demonstrated by the Welsh scientist Sir William Robert Grove (1811-1896) in 1839. Interestingly, this makes the fuel cell an older technology than either the four-stroke or two-stroke internal combustion engine. Over the past 170 years many different types of fuel cells have been developed, with each type of fuel cell having a particular electrolyte, efficiency, working temperature and cost. Solid-oxide fuel cells (SOFCs), for example, have an electrolyte of O^{2-} - conducting ceramic oxide, and an efficiency of approximately 65%, a working temperature of up to 1100°C and are relatively cheap compared to other types of fuel cells. They also have the advantage that they can tolerate a lot of impurities in the hydrogen fuel, but the very high operating temperature makes them completely unsuitable for automotive applications. Molten carbonate fuel cells (MCFCs) operate in a similar manner, except that the electrolyte consists of molten carbonate, which is a negative ion and an oxidizing agent. Of the existing 10+ types of fuel cell, only the proton-exchange-membrane fuel cell has a realistic possibility of being used as the source of electricity to drive a car.

Proton-exchange-membrane fuel cells, sometimes referred to as polymer-electrolyte membrane fuel cells (PEMFCs) were developed in the 1960s for the NASA Gemini space missions, although were not - as is often said - used on the first flights to the moon on the Apollo missions. The key features of PEMFCs are their low weight, the low gas pressures required, and their relatively low operating temperatures. However, the critical material with respect to this type of technology is the special polymer electrolyte membrane (PEM). A PEM is a semipermeable ionomer membrance which is impermeable to gases, but is able to conduct protons - the basic function in a fuel cell. The most common PEM goes by the trade name Nafion and is made by the company DuPont. These membranes are not cheap, but the real cost problem associated with PEMFCs is the catalyst material - platinum. The positions of the various elements in the fuel cell are shown in the schematic diagram below (Figure 8).

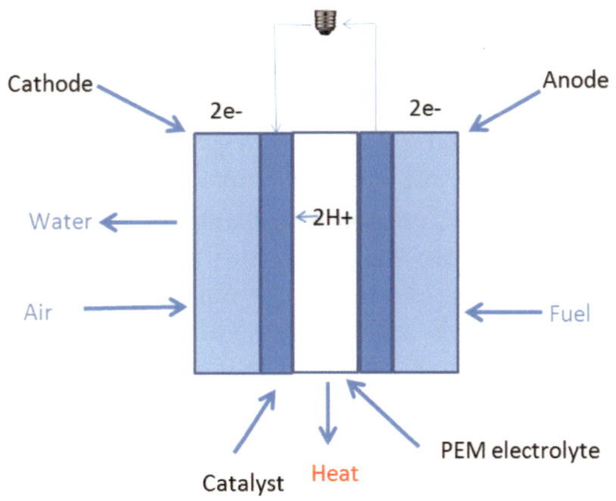

Fig. 8. Schematic diagram of a polymer-electrolyte membrane fuel cell

Platinum is prohibitively expensive, the current price is over $1,800 per ounce (about €50/gram). Even being able to reduce the amount of platinum in an automotive fuel cell to 10 grams will make it very hard to compete with the battery-based systems. However, recently, researchers at Alamos National Laboratory have developed a platinum-free fuel cell that employs carbon, iron and cobalt on the cathode of the PEM cell. During tests these fuel cells yielded high power, good efficiency and promising longevity. However, this is just one in a line of many such announcements that have been made over the years, there have been reports of replacing platinum with nickel and cobalt, titatnium dioxide nanocrystals and even carbon nanotubes, but all of these failed to live up to a lot of the early promise. Nevertheless, it is clear that progress is being made, and if scientists can also overcome the problems associated with robustness, then fuel cells could fulfil the promise that they have been showing for more than 100 years.

5. How can we produce the hydrogen easily and cleanly?

Currently, energy is cheap, but it is also environmentally polluting, and even if global warming remains a matter of debate, it is surely an illusion that the emission of carbon dioxide, which exceeded 30 billion metric tonnes in 2010, will have no effect on the planet. So the issue of solving humanity's number one problem during the next 50 years, the energy needed for an increasing population, is no longer just the depletion of fossil fuels but also the steam age alert. (Grimes et al. 2008) In addition to the fact that hydrocarbons have extraordinary value as the source of chemicals used to produce goods and other essentials for living, depletion is occurring rapidly. In a collective effort we succeeded in depositing vast amounts of CO_2 in the atmosphere, today reaching the highest concentrations in at least 500 000 years. If we keep burning fossil fuels to provide more than 40 Terawatts of energy we use every year the earth's atmospheric CO_2 can reach levels so that we end up cooking ourselves alive. Therefore, most of the energy scenarios predict, as the only solution, the use of renewable sources. The most fundamental of the sustainable energy sources is incident solar radiation and actually all others are simply a consequence of the sun's power: wave, wind, biomass and even fossil fuels, which is, by photosynthesis, solar energy stored over a geological time scale. At a power of 1000 W/m^2, the solar incident energy on the earth's surface by far exceeds all human energy needs. If 0.16% of the earth's surface would be covered with photovoltaic (PV) panels of 10% efficiency, this would meet the world's total energy demand. While 0.16 % appears to be a small number, it still corresponds to a huge surface area. (Gratzel 2005) It is a matter of scale and economics. And where would hydrogen be important in this scenario? Hydrogen already plays a vital role in our life, e.g., in fertilizer, chemicals and the petrol industry, but the real scenario is that today, 96% of production is based on the steam reforming of natural gas, and therefore it is dirty.

However, solar cells only generate electricity during daytime, so even mesoscopic solar cells require a practical way to store such large amounts of energy. Also, in the case of electricity from wind turbines, it is a fact that wind is inherently unreliable and so faces similar similar problems to PV electricity, i.e., that the grid cannot handle it if the wind blows at too many places at once. Therefore, in Hamburg they will install facilities with electrolysis to produce hydrogen and so keep the grid stable. Costs are still an issue for such solutions, but actually it is question of whether we can afford not to have hydrogen infrastructure if we want to be renewable.

Fuel from sunlight can be achieved by the reaction of splitting water $2H_2O \rightarrow H_2 + O_2$, which is a 4-electron reaction and therefore not so complicated and efficient as a 6- or 8-electron evolved reaction for producing methanol or methane. The route to solar hydrogen can therefore be coupled to PV and an electrolyser. Actually, if we connect an efficient solar cell with an efficient electrolyzser, the total efficiency for standard technology is estimated to be between 8 to 10 % and the estimated hydrogen cost would be €8 per kg. The main issue is again cost!

The Hydrogen Economy, which is a short name for obtaining fuel from sunlight and water, is actually here, we can do it using electrolysers powered by renewable energy. But the direct splitting of water/artificial photosynthesis is a much better way. So, here is the idea, which was neglected for far too long, the so-called Holy Grail of material science, a Direct Photoelectrochemical Water Spliting by sunlight with PEC cells. The photoelectrolsis of water is the direct splitting of water into hydrogen and oxygen using only sunlight as the input. The semiconductor material is placed into water and illuminated with sunlight, driving water splitting directly, and this configuration combines the solar cell and the electrolyzer into a single device. Although this approach is certainly a step into the right direction, there is still plenty of development research needed. We know it will work because nature has been doing it for billions of years, but we have different obstacles than nature, we do not have time, which means we must be concerned by efficiency and cost and that we need better light-harvesting semiconductors. (Turner 2008)

The splitting of water into hydrogen and oxygen, the sustainable and removable fuels of H_2-based economy, involves the following reactions:

$$\text{Reduction}: 2H_2O + 2e^- \rightarrow H_2 + 2OH^-,\ E°(H_2/H_2O) = -0.41/0.0\text{ V at pH} = 7/0\ (\text{vs NHE})$$

$$\text{Oxidation}: 2H_2O \rightarrow O_2 + 4H^+ + 4e^-,\ E°(H_2O/O_2) = -0.82/+1.23\text{ V at pH} = 7/0\ (\text{vs NHE})$$

$$E°_{\text{overall}} = E°(H_2O/O_2) - E°(H_2/H_2O) = 1.23\text{ V}$$

Thus, the overall reaction $H_2O \rightarrow H_2 + ½\ O_2$ is endothermic with an increase in Gibbs free energy ($\Delta G°$= 237 kJ mol). Water absorbs solar light in the IR region (λ = 1008nm) with a photon energy too low to drive the splitting. An advanced possibility to overcome thermodynamic potential is the use of a photo-electrochemical cell (PEC) with a photo-active catalyst (inorganic semiconductor), which is immersed in water. The basic principle of PEC is that when the semiconductor is illuminated with photons with an energy that is equal to or larger than the band gap, electrons are excited from the valance band (VB) into the conduction band (CB). These photo-generated electrons are directly or even better, through a cocatalyst, used to reduce water to hydrogen at the catalyst/water interface. The electron hole that remains in the valance band migrates to the surface, where it oxidizes water to form oxygen gas. The designed targets of water-splitting researchers in that field is rather ambitious. An estimated price of 3 \$/kg H_2 would be possible to achieve with photoelectrode materials that would cost round for 160 \$/m^2 if the solar-to-hydrogen efficiency would be 10 % and the life time would exceed 15 years.

This simple principle operates only if many material requirements are completely fulfilled simultaneously. Firstly, suitable band-edge positions of the photocatalyst are required to enable the reduction/oxidation of water. The photochemical water reduction occurs if the flat band potential (for doped semiconductors, this equals the bottom of the CB) exceeds the water reduction potential, $E°(H_2/H_2O)$. For water oxidation the VB edge should exceed the potential of water oxidation $E°(H_2O/O_2)$. The high chemical stability of the photocatalyst in different corrosive media and under illumination as well as long life-time is of crucial importance. Hence, chemical reactivity/stability depends on of the type of molecular orbitals that the CB and VB are composed of and is the definite characteristics of oxide semiconductors. These requirements limit the use of most of non-oxide semiconductors with a small band gap since they either dissolve or passivate in water. Furthermore, strong visible light absorption is required.

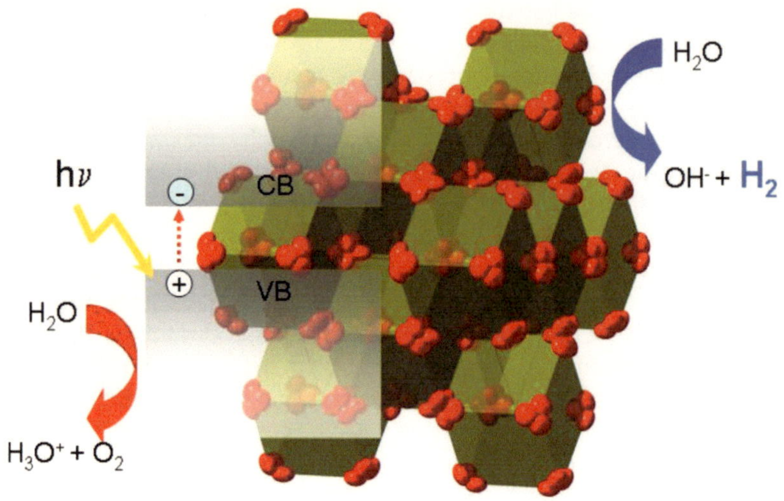

Fig. 9. Schematic presentation of water splitting by semiconductor photocatalysis

The first step in the photochemical solar-to-hydrogen conversion is the light absorption process and it has a definite energy limit, the value of which is the band gap. Also, the spectral region, in which the semiconductor absorbs light, is determined by the band gap. For this reason large band-gap materials only absorb at short wave lengths (UV region) and will not absorb as many photons from the solar spectrum as small-band-gap semiconductors. The optimum value of the band gap considering the energy required for splitting and the absorption within the visible range of solar light, should be somewhere around 1.9 and 3.1 eV. In addition, an efficient charge transport in the bulk is important to avoid any trapping of the photogenetared holes and the recombination of the exitons.

The problem is that as a consequence of the co-dependence of all the limiting material requirements, the energy efficiency of the solar-to-hydrogen conversion using suspended photocatalysts is low. Exclusive principles for the absorption of visible light versus the chemical stability of photocatalysts are the basic problem for low photoactivity. The fact is that stable materials that can split water with strong photoactivity have band gaps too large (> 3eV) and do not respond to visible light. In contrast, efficient light absorbers with small bad gaps are more chemically and photo reactive. Additional problems related to electronic band structure are unfavourable band-edge positions.

Photocatalytic splitting also requires an efficient charge separation and charge transport to the semiconductor surface, where at the semiconductor/water interface the reduction/oxidation takes place. Therefore, another problem that appears as a consequence of low conductivity is the slow kinetics of charge transfer. This can cause the electrons or electron holes not to reach the surface or the oxidation of the semiconductor itself. Finally, the performance of such a system could also be hindered due to either electrochemical resistance at the interface or/and backward reaction.

Most of the photocatalysts have so far been prepared by solid-state synthesis and the effects of variable material preparation conditions have not been fully investigated. From the chemical viewpoint there is a number of phase systems, which have never been subject of photocatalytic studies. A significant opportunity for improving water splitting is also within an investigation of the cocatalyst effect.

The pioneers of solar-hydrogen are Fujisima and Honda, who first reported photo-assisted water electrolysis using rutile (TiO_{2-x}) as photocatalyst of a Photo-Electrical Cell (PEC) (Fujisima 1972) An important additional characteristics of rutile is that unlike other photo-sensitive semiconductor materials such as non-oxide valance semiconductors (Si, GaAs, etc.) it is resistant to both corrosion and photo-corrosion in an aqueous environment. Consequently, extensive investigations of TiO_2 as a photo-electocatalyst for water-splitting have followed.

In the late 1970s after the general photocatalysis concepts were reported the titania catalysts were improved significantly. The enhanced evolution of H_2/O_2 took place on a TiO_2 catalyst to which a Pt cocatalyst had been added via photodeposition. Importantly it was shown that the spectral response of a TiO_2 single crystal can be extended into the visible range by doping. The presence of Cr^{3+} in the rutile lattice has been found to shift the optical absorption from 415 to 550 nm and a 10 % increase in sunlight conversion was measured. Although doping with Al^{3+} did not have a significant response in the visible range, it enhances the absorption in the UV region.

The Gratzel group published studies in the 1980s on the development of a multicomponent catalysts composed of colloidal TiO_2 particles doped with Nb_2O_5 and joined with Pt or RhO_2 particles as a cocatalyst. The next important improvement was achieved with the addition of $[Ru(bpy)3]^{2+}$ as a visible-light sensitizer - dye - and the system was able to split water under visible-light irradiation. The overall result of the Gratzel group showed that catalytic activity is strongly dependent on doping levels of the catalyst, the pH, temperature, etc.

Parallel to investigations of TiO_2 the same kinds of investigations were performed on other potential photocatalysts such as Fe_2O_3, WO_3, $SrTiO_3$. Also semiconductors based on tantalates, niobates, other mixed-oxide semiconductors and non-oxide semiconductors such us CsS, GaAs, etc, were the subject of the same interest, i.e., the direct splitting of water under visible-light irradiation. (Lewis)

The new millennium has brought progress in nanoscience with new, advanced techniques that can be applied to the fabrication of semiconductors (van de Krol et al. 2008). At the same time environmental concerns and awareness have continued to grow together with increasing energy demands. For that reason the interest in the solar generation of hydrogen and oxygen, the renewable and sustainable future energy carriers, has dramatically increased.

The new results in visible-light photocatalytic activity were published by a few of the world's top scientific groups. Even though the primary work of Gratzel is focused on photovoltaic cells, the principles that govern the visible-light photoactivity as well as the processing of semiconductor catalysts are the same as for solar hydrogen production. An important contribution from scientific and technological point of view is the tandem cell for water splitting by visible light. The low-cost tandem device is composed of a WO_3 or Fe_2O_3 nanocrystalline thin film on top (absorbs the blue part of vis-spectra and creates exciton and oxidize water) of the dye-sensitized TiO_2 cell above (absorbs the green and red parts of the visible spectra and generates hydrogen).

According to some recent studies the visible-light activity can be significantly increased using $BiVO_4$, $AgNbO_3$, $PbBi_2Nb_2O_9$ and $InMO_4$, M = Nb, Ta, due to a more appropriate band gap (Kudo et al. 1999). The narrower band gap of these compounds occurs due to the formation of a valance band consisting of theO_{2p} orbital hybridized with either Pb_{6s}, Bi_{6s} or

Ag_{4d}. The electronic excitation from this valence band to the Nb_{4d} conduction band is reported to facilitate the direct water splitting. $InTaO_4$ and $InNbO_4$ with band gaps in the range 2.8-2.4 eV were found to split water, too. (Osterloh 2008) The difference in unit-cell volumes for both TaO_6 and NbO_6 octahedral leads to a change in the lattice parameters, slightly affecting their photocatalytic activity, mainly due to a variation in the conductance-band levels formed by Ta_{5d} and Nb_{4d} orbitals.

Direct splitting under visible-light irradiation with $In_{(1-x)}Ni_xTaO_4(x = 0\text{-}0.2)$ was reported. The solid solution, prepared with a solid-state reaction, was impregnated with a Ni cocatalyst. The authors stressed that the Ni doping reduced the band gap from 2.6 eV (undoped) to 2.3 eV (doped) and that this is attributed to internal d-d transitions in partly filled d shells. A significant achievement was reported by Kato et al. for 0.2 wt % of co-catalyst NiO and 2 wt % of doped with La, the $NaTaO_3$ in comparison to undoped $NaTaO_3$ a nine times increase in H_2 and O_2 was found under the UV light illumination. The quantum yield was reported to be 56% at 270 nm - the best ever reported. The authors obtained 100-700-nm sized nanoparticles, which is smaller in contrast to undoped NiO/NaTaO3 (2-μm). They showed that with a controlled particle size, crystallization and surface topology an important improvement can be achieved. They created a nano-step morphology, which helped to reduce the unwanted electon-hole recombination. In addition, by loading the cocatalyst in the form of "ultra-fine" particles they created the separate active sites for H_2 and O_2 evolution, avoiding any back-reaction during water splitting.(18)

Recently, Osterloh published an overview on his research on inorganic materials as catalysts for photochemical water splitting. He stressed that many photocatalysts that work under UV light irradiation have been discovered, but only recently, in the past ten years or so, the investigation focused on enhancing photocatalytic performance in the visible range and photocatalytic materials interesting for solar hydrogen production. He pointed out that the effects of material preparation conditions, catalysts (nano)structures, nano-scale morphology, cocatalyts loading and similar are insufficiently investigated and, actually, that this topic is in its early stages of research and development. A new challenge is also in the application of quantum-size effects to tailor the electronic structure and photoactivity, since all these system were not so much fabricated using nanotechnology techniques.

The current record holder for the splitting of pure water in terms of quantum efficiency is NiO-modified La/$KTaO_3$. Unfortunately, this excellent 56 % efficiency is achieved under UV illumination.(18) For water splitting under visible light, to date the best solar-to-hydrogen efficiency of 2.5 % has been obtained with Cr/Rh-modified GaN/ZnO (Maeda 2006)

6. Conclusions

The situation is still far from certain; however, hydrogen vehicles still have a very strong chance, even though the competition, especially from pure electric vehicles, will be tough. Some of the difficulties have almost been overcome. The situation with storage suggests that the use of modern materials will provide the answer to high-pressure gaseous storage, as a better alternative to the liquid or solid-state options. In terms of infrastructure and distribution, this is largely a question of making the appropriate investments. Many well-funded projects around the world, which are at the regional rather than national level, show that this can be done. And although transporting hydrogen through steel pipes is not completely without problems, our own work has shown that we can reduce the problem of hydrogen permeation into steel to the point where it can be described as negligible. Fuel cells are an old technology, and one that works well, but the fragility of the device and the expense of some of the components are dampening interest, but once these problems are overcome with better materials and processes there is no reason to believe that fuel cells will always be too expensive. The big one is making the hydrogen cheaply and in a renewable form. The science is well understood, but so far we do not have the materials. Nevertheless, we believe that what it comes down to in the final analysis is whether direct photo-electrochemical water splitting can be sufficiently well developed over the next 5-10 years to produce cheap, plentiful and environmentally acceptable hydrogen that can be made readily available where it is required. If it can, then it will be hydrogen that fuels the car of the future.

7. References

G.A. Esteban, A. Pen˜a, I. Urra, F. Legarda, B. Riccardi, J. Nuc. Mater. 367–370 (2007) 473–477

Fujisima, Honda, Electrochemical Photolysis of Water at Semiconductor Electrode, Nature, 238, (1972) 37-38

Gratzel, Mesoscopic Solar Cell for Electricty and Hydrogen Production from Sunlight, Chemistry Letters, 34, 1, (2005) 8-13

Grimes, Varghese, Ranjan, Light, Water, Hydrogen. The solar Generation of Hydrogen by Water Photoelectrolysis, Springer (2008)

L. Hultman, Vacuum 57 (2000) 1–30

S. Inoue, H. Uchida, Y. Yoshinaga, K. Koterazawa, Thin Solid Films 300 (1997) 171–176

Kudo, K. Omri, H. Kato, A novel aqueous process for preparation of crystal form-controlled and highly crystalline BiVO4 powder from layered vanadates and its photocatlytic properties. J. Am. Chem. Soc. 121, (1999) 11459-11467

M.Y. Kwak, D.H. Shin, T.W. Kang, K.N. Kim, Thin Solid Films 339 (1999) 290–293

D. Levchuk, F.Koch, H.Maier, H.Bolt, J. Nuc. Mater. 328 (2-3) (2004) 103-106

D. Levchuk, F. Koch, H. Maier, H. Bolt, J. Nuc. Mater. 328 (2004b) 103–106

D. Levchuk, S. Levchuk, H. Maier, H. Bolt, A. Suzuki, J. Nuc. Mater. 367–370 B (2007) 1033-1037

D. Levchuk, H. Bolt, M. Doebli, S. Eggenberger, B. Widrig, J. Ramm, Surf. Coat. Technol. 202 (2008) 5043-5047

Lewis, Light work with water, Nature, 414,(2001) 589-590

Maeda, K. Teramura, D. Lu, T. Tanaka, N. Saito, Y, Inoue, K. Domen, Photocatalyst releasing hydrogen from water splitting in sunlight. Nature, 440, (2006) 295

P. J. McGuiness., M Čekada, V Nemanič, B Zajec, A Rečnik. Surf. coat. technol., 205 (2011) 2709-2713, doi: 10.1016/j.surfcoat.2010.08.133.

B.Y. Man. L Guzman, A Miotello, M Adami, Surf. Coat. Technol. 180-181 (9) (2004) 9-14

Osterloh, Inorganic Materials as Catalyst for Photochemical Splitting of Water, Chem. Mater., 20, (2008) 35-54

S. PalDey, S.C. Deevi, Mat. Sci. Eng. A 342 (2003) 58–79

M. Panjan, M. Čekada, P. Panjan, A. Zalar, T. Peterman, Vacuum 82 (2008) 158–161

A. Perujo, K.S. Forcey, Fus. Eng. Des. 218 (1995) 224-230
Turner, The other half of the equation, Nature materials,7, (2008) 770-771
R. van de Krol, Y. Liang, J. Schoonman, Solar hydrogen production with nanostructural metal oxides, J. Mater. Chem., 18, (2008) 2311-2320

20

Using Micro Cogeneration Technologies to Enhance the Sustainable Built Environment

Esam Elsarrag and Yousef Alhorr
Gulf Organisation for Research and Development
Qatar Science and Technology Park, Doha
Qatar

1. Introduction

The accompanying CO_2 and other greenhouse gas emissions have been identified as a primary cause of global warming and the demand for electricity is expected to rise sharply within two decades. New energy-efficient, environmentally friendly power systems are, therefore, urgently required to ensure a sustainable built environment and also to meet the current building regulation standards.Buildings, new and old account for 50% of global energy consumption and related emissions. As citizens of the world, we have to reduce our reliance on ever dwindling supplies of fossil fuel. The imperatives for this are energy security and climate change. Estimates of the cut needed within our lifetime range 60% to 80% of current demand. Although, energy resources are more depleted and fossil fuel cost is increasing in addition to the effect of burning these fuels on the environment. The accompanying CO_2 and other greenhouse gas emissions have been identified a primary cause of global warming and the demand for electricity is expected to rise sharply within two decades. Different green building certification schemes are deployed in different countries including LEED in USA, BREEAM in UK, ESTIDAMA in Abu Dhabi, QSAS in Qatar, etc. to encourage creating a sustainable built environment and encourage the adoption and application of green building practices.

New building regulations are enabling new regimes of distributed power generation, to which fuel cell systems are ideally suited.

Conventional electricity generation is inherently inefficient, using only about a third of the fuel's potential energy. In applications where heating or cooling is needed as well, the total efficiency of separate thermal and power systems is still only about 45%, despite the higher efficiencies of thermal conversion equipment.

The recent development of efficient thermal prime movers for distributed generation is changing the focus of the production of electricity from large centralized power plants to local generation units.

Micro Combined Heat and Power (micro-CHP) technologies use fuel, e.g. natural gas, but provide electricity as well as heat. The two main systems use either reciprocating engines or Stirling engines, see Fig. 1. The size of the European Union (EU) market of domestic micro-CHP (1 to 10 kW) exceeded the 90 millions of units deploying about 6.2 millions of new installations per year with a development plan for 2020 shows a quick evolution of micro-

CHP solutions to a higher efficiency comparable to condensing boilers. CHP was assessed in field tests in Germany, the UK and some other EC countries [1].

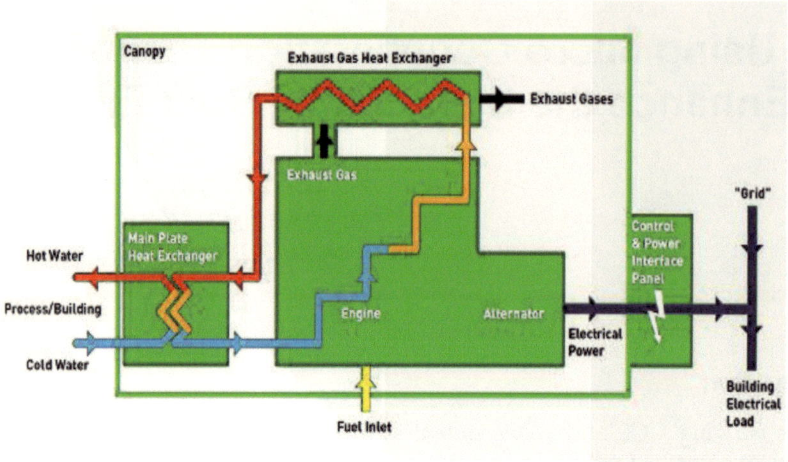

Fig. 1. Prime mover CHP diagram

Fuel cells, e.g. PEMFC, are also an alternative source of power; they provide a means of supplying electricity and heat and improving the built environment [2].

The current European policy pathway on the energy savings will enhance and accelerate this evolution with the implementation of highly efficient systems like the Proton Exchange Membrane Fuel Cells (PEMFC) and Solid Oxide Fuel Cells (SOFC) [3].

Fuel Cells are used primarily for transportation applications and some stationary applications due to their fast start-up time, low sensitivity to orientation, and favourable power-to-weight ratio. They have recently passed the test or demonstration phase and have partially reached the commercialization stage due to the impressive worldwide research effort [4]. They are different from batteries in that they consume reactant, which must be replenished, while batteries store electrical energy chemically in a closed system. Additionally, while the electrodes within a battery react and change as a battery is charged or discharged, a fuel cell's electrodes are catalytic and relatively stable.

Fuel cells also have an environmental advantage over batteries, since certain kinds of batteries require special disposal treatment. Fuel cells provide a much higher power density, packing more power in a smaller space.

Despite the currently promising achievements and the plausible prospects of PEMFCs, there are many challenges remaining that need to be overcome before they can successfully and economically substitute for the various traditional energy systems. With the many promising research efforts in overcoming these challenges, the most important tools for the commercialization will be the technical data and information from a real PEMFC application test [5].

The largest deployment of micro-CHP is in Japan at the year 2009, where over 90,000 units are in place where six Japanese energy companies launched the 300 W -1 kW PEMFC. In the UK, it is estimated that about 1,000 micro-CHP systems were in operation as of 2002. These are primarily Stirling and reciprocating engines [6]. Of the 24 million households in the UK, as many as 14 to 18 million are thought to be suitable for micro-CHP units.

Current building regulations enforce building designers to consider micro cogeneration as an alternative means of supplying energy especially. This chapter explores the importance of running a CHP plant to achieve maximum reduction in CO_2 emissions. A range of constraints, design and technical issues, is currently affecting the wide-scale deployment of microgeneration. Idle plant accrues no benefits, so it is important that the CHP plant operates for as many hours as possible. Such issue makes the design of a micro cogeneration technology is not quite as straightforward. The site heat and electricity demand must be properly assessed to prevent a CHP plant from being incorrectly sized.
Building designers are left with several questions; how to size a CHP plant? Does the integration of absorption chillers require over sizing the CHP unit? How much a CCHP plant contributes to carbon reduction? How to integrate a CHP or CCHP plant with other renewable sources such as biomass boilers to achieve a maximum reduction in CO_2 emissions in a development?
Section 2 of this chapter reviews the application of current micro cogeneration, the combination with district heating and incorporation of CHP into a trigeneration scheme. Section 3 assesses the use of a CHP unit when coupled to absorption/electric chillers, as well as the interactions with biomass boilers, to allow for setting up multi-generation systems for combined local production of different energy vectors. It investigates through a detailed study the maximum carbon reduction that could be achieved in a mixed use development. Section 4 examines the use of commercially available low-cost PEMFCs in dwellings to provide power for lighting in a dwelling and shows the advantages on both energy and carbon reduction when using the technology in dwellings.

2. Micro cogeneration review

Combined Heat and Power, CHP, have demonstrated superior efficiency for years in industrial plants, universities, hotels, hospitals and mixed use developments. They can be employed over a wide range of sizes, applications, fuels and technologies [7]. The resulting electricity can be used either wholly or partially on-site. It is particularly efficient when employed as a source for district heating to provide hot water, space heating and electricity for a number of linked buildings. Figure 2 shows the typical percentage of electricity and heat from a prime mover CHP.

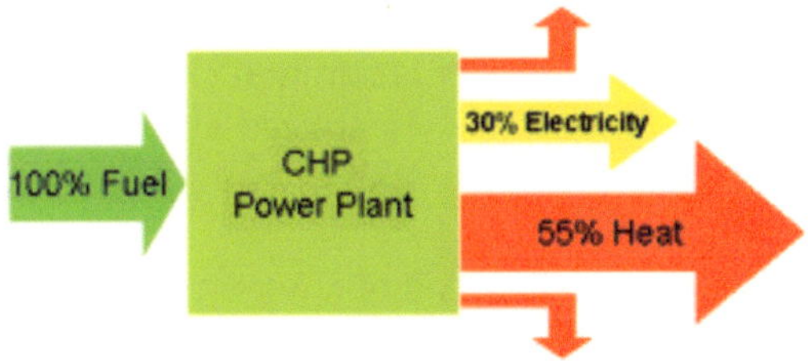

Fig. 2. Typical percentage of electricity and heat for a prime mover

The combination of cogeneration and district heating is very energy efficient. A thermal power station which generates only electricity typically converts 30 % of the fuel input into electricity. The major part of the energy is wasted in form of heat and dissipated to the environment. A cogeneration plant recovers that heat and can reach total energy efficiency beyond 85%. The heat is often obtained from a cogeneration plant burning fossil fuels but increasingly biomass, although heat-only boiler stations, geothermal heating and central solar heating are also used.

CHP can be incorporated into a trigeneration scheme to provide cooling alongside heat and power from the same energy source, see Fig. 3. Here excess heat produced is cooled by absorption chillers linked to the CHP system. This provides chilled water for cooling to be circulated around a building or community. This is particularly useful for schemes that require a large amount of air conditioning. This is also known as combined cooling, heat and power (CCHP) [8].

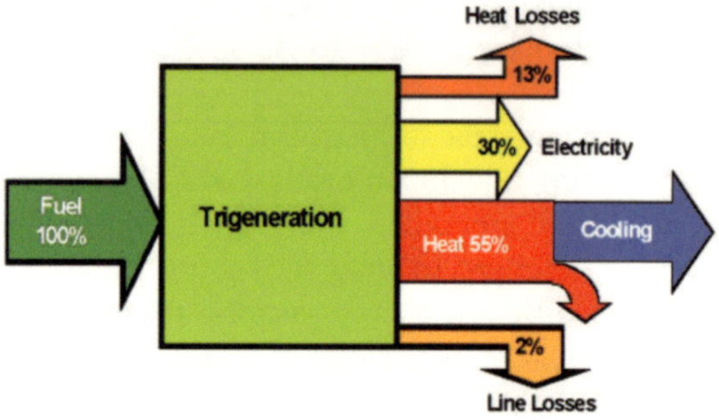

Fig. 3. Trigeneration diagram

CCHP could be combined with district or central cooling systems, as any system which provides building cooling through the distribution of chilled water, hot water or steam from a central plant. Thus, cooling achieved through distribution of district hot water or steam to drive absorption chillers located in buildings is also considered district cooling.

Usually space heating and hot water storage tanks serve as a heat sink for reasonable waste heat utilization. In summer, the heat demand is much lower but the heat of the electric generation process can be transformed into cooling energy by an absorption chillers. The absorption cycle, see Fig. 4 uses a heat driven concentration difference to move refrigerant vapours from the evaporator to the condenser.

The high concentration side of the cycle absorbs refrigerant vapours. Heat is then used to drive off these refrigerant vapours thereby increasing the concentration again.

Lithium bromide is the most common absorbent used in commercial cooling equipment, with water used as the refrigerant. Smaller absorption chillers sometimes use water as the absorbent and ammonia as the refrigerant. Heat fired absorption chillers can be provided from a central plant or by local heat fired absorption chillers connected to the district heating network.

The trigeneration is of concern and attraction of a lot of researchers in terms of application, economic and performance. In the application fields, an investigation in a typical supermarket for cooling, heating and power requirements was described and a number of CCHP options involving the use of different cooling and engine technologies were reviewed [9]. The investigation calculated and compared the energy savings of the different options against typical conventional supermarket technology. Typical energy demand profiles and economical proposals of trigeneration plants for an airport and typical results for a large airport was presented [10].

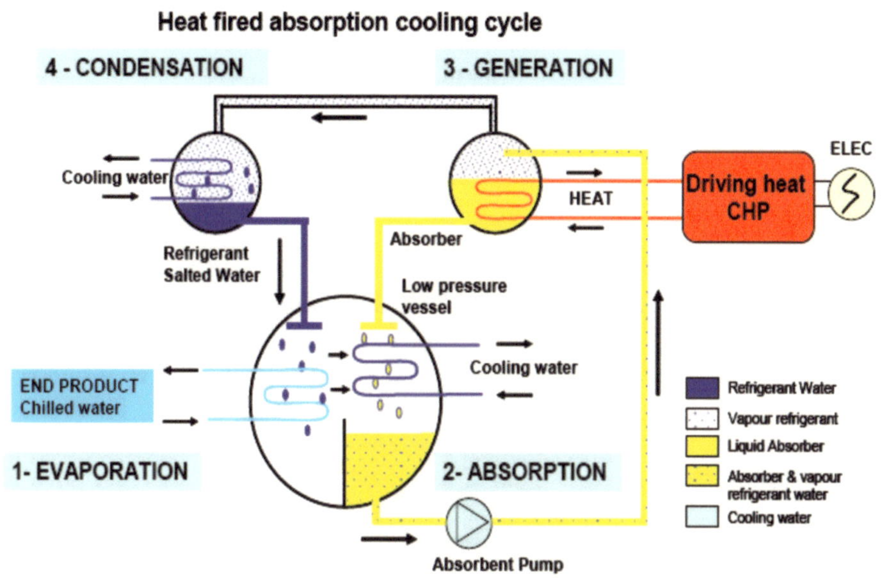

Fig. 4. Absorption cooling cycle

The evaluation of the energy efficiency and economic feasibility of a small scale trigeneration system for CCHP with a Stirling engine showed that a CCHP system saves fuel resources and has the assurance of economic benefits [11]. A study [12] proposed a modified exergo-economic optimisation overcoming the difficulties encountered when applying the traditional ethodologies to CCHP plants serving civil buildings, characterised by irregular demand-profiles.

The performance of CHP and CCHP systems was carried out by several researchers including static calculation methodology for evaluating the primary energy consumption for CCHP and separate productions and analysis of energy saving performances of various types and sizes systems in heating and cooling modes and improving energy use in cogeneration systems [13]. Investigations on the potential of CHP and CCHP systems for reducing the energy use and the emission of hazardous greenhouse gases were carried out [14-16]. In the experimental tests to evaluate the performance and emissions of a diesel engine generator and the performance of the trigeneration system the CO_2 emissions per unit kWh of useful energy output from trigeneration were reduced by 67.2–81.4% compared to those from single generation [15]. A study suggested the inclusion of an additional gas boiler and calculated the prices of the energies produced from such a trigeneration system [17].

Residential and small-scale commercial fuel cells are now becoming available to fulfill both electricity and heat demand from one system. Fuel cell technology in a compact system converts natural gas, propane, and eventually biofuels into both electricity and heat. Fig. 5 shows the chemical process for different fuel cells [18].

There is much interest in proton exchange membrane fuel cells (PEMFCs). There have been pre-commercial demonstrations for stationary power and CHP applications in Canada and Germany of small 3 kWe systems and larger 250 kWe systems. PEMFCs deliver high power density and offer the advantages of low weight and volume, compared to other fuel cells. PEMFCs, see Fig. 6, use a solid polymer as an electrolyte and porous carbon electrodes containing a platinum catalyst. Materials used include a fluorocarbon polymer backbone,

similar to Teflon, to which sulfonic acid groups are attached. The use of a solid polymer electrolyte eliminates the corrosion issues associated with liquid electrolyte fuel cells.
They need only hydrogen and oxygen from the air to operate and do not require corrosive fluids like some fuel cells. They are typically fuelled with pure hydrogen supplied from storage tanks or onboard reformers.

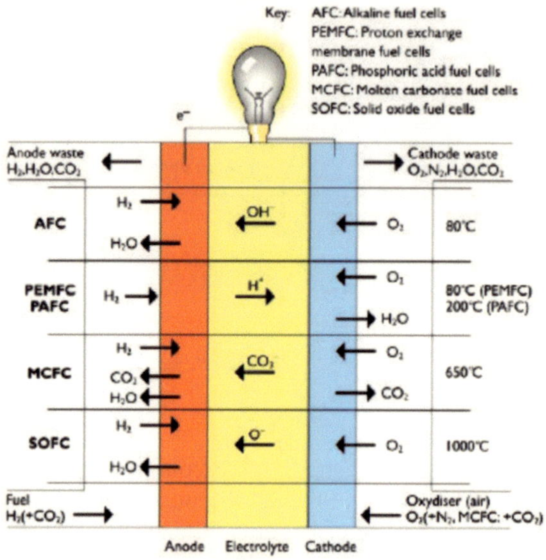

Fig. 5. Chemical process for different fuel cells

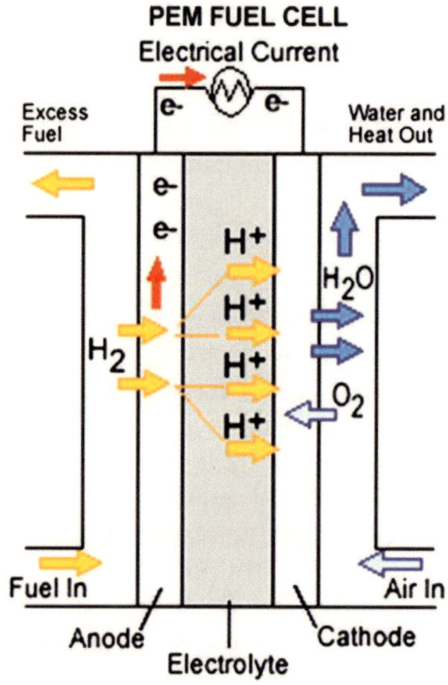

Fig. 6. The proton exchange fuel cell

Because the proton exchange membrane fuel cell uses a solid electrolyte, a significant pressure differential can be maintained across the electrolyte. This allows the operation of the fuel cell with low-pressure fuel and high-pressure air in order to optimise performance. In practice there is a trade-off between the energy and financial cost associated with compressing air and the improved performance. They operate at relatively low temperatures, around 80°C (176°F) which allows them to start quickly and results in less wear on system components, resulting in better durability. However, they require that a noble-metal catalyst (typically platinum) be used to separate the hydrogen's electrons and protons, adding to system cost. The platinum catalyst is also extremely sensitive to CO poisoning, making it necessary to employ an additional reactor to reduce CO in the fuel gas if the hydrogen is derived from an alcohol or hydrocarbon fuel. This also adds cost. Developers are currently exploring platinum /ruthenium catalysts that are more resistant to CO [18].

Fuel cell systems are presently the focus of intense research efforts among industrial, academic, and government organizations, but this work is primarily concentrated on technical hurdles to fuel cell commercialization.

A description, detailed design, energy savings and an economic analysis proving the technical and economical feasibility of the installation of the design of a laboratory-scale high-pressure electrolyser was investigated by researchers [19]. An examination of the performance of a PEM water electrolysis cell using $Ir_xRu_yTa_zO_2$ anode electro-catalysts showed that the best cell voltage they achieved was 1.567 V at 1 Acm^{-2} [20].

Fuel cells have a heat to power ratio of roughly 1:1 with overall efficiencies of around 80% when fired on hydrogen. Fuel cells powered with pure hydrogen have potential power efficiencies up to 45% i.e. 45% of the hydrogen is converted into electrical energy. However, when we add a reformer to convert other fuels to hydrogen, this can drop significantly. Fuel cell systems can maintain high efficiencies at loads as low as 50%, exhibiting characteristics that are ideal for use in buildings where much of the time is spent at low load [21].

3. Application of micro cogeneration in a mixed use development

Deploying CHP and CCHP systems in mixed use development or any application requires more attention and proper design by engineers. This section will provide analysis of deploying the proper size of CCHP and biomass for maximum carbon reduction in a mixed use development.

The development is residential flats and offices. Figure 7 below shows the annual energy demand by percentage of different mixed use development based on the percentage net internal floor area. It can be observed that the hot water demand decreases with the increase of the offices area but the cooling demand percentage remains approximately constant. The carbon emission produced from the thermal load decreases with the increase of the offices area, Fig. 8.

Atypical case study of 50% offices and 50% residential, by area, will be discussed and analysed. A district heating plant is proposed for the development as they can provide higher efficiencies, diversity of fuel type, and better pollution control than localised boilers. It will distribute the heat generated in a centralized location for the residential and commercial heating requirements such as space and water heating. The core element of a district heating system is usually a cogeneration plant, CHP, or a heat only boiler station. Both have in common that they are typically based on combustion of primary energy

carriers. The difference between the two systems is that, in a cogeneration plant, heat and electricity are generated simultaneously, whereas in heat only boiler stations - as the name suggests - only heat is generated.
In this study a CHP is coupled to absorption/electric chillers, as well as the interactions with renewable sources, to allow for setting up multi-generation systems for combined local production of different energy vectors such as electricity, heat, cooling power.
The main objective of adopting composite multi-generation systems as it may lead to significant benefits in terms of higher energy efficiency, reduced CO_2 emissions, and enhanced economy.

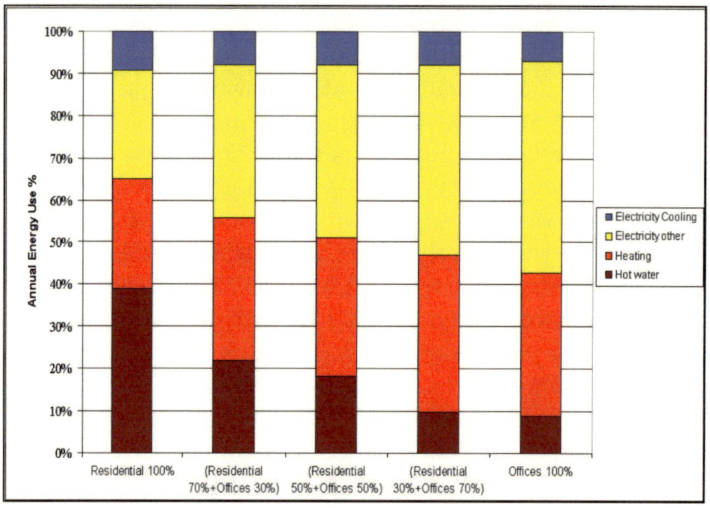

Fig. 7. Annual energy percentage for different mixed schemes

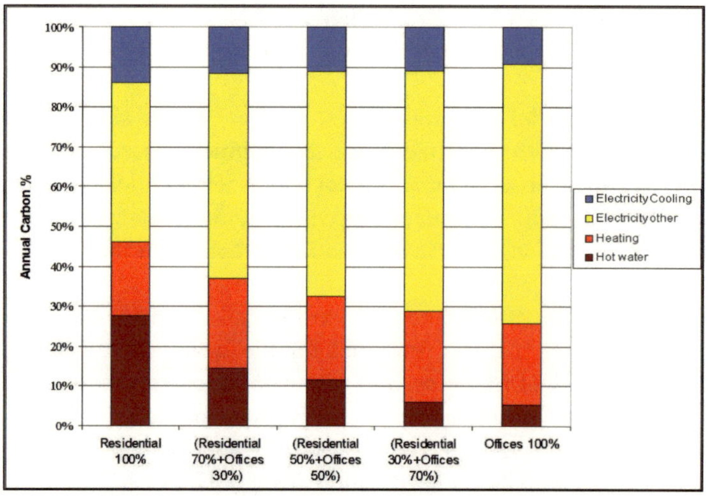

Fig. 8. Annual CO_2 emissions percentage for different mixed schemes

Biomass boilers are also proposed to be used in the development. They use a biological material derived from living, or recently living organisms, as the fuel. In the context of biomass for energy this is often used to mean plant based material, but biomass can equally apply to both animal and vegetable derived material. Energy produced from biomass residues displaces the production of an equivalent amount of energy from fossil fuels, leaving the fossil carbon in storage. Whereas fossil fuels are becoming increasingly expensive, some forms of biomass fuels are sustainable, renewable and almost carbon-neutral. This makes biomass a very sensible and economically viable option for many heating projects in relation to homes, communities, offices and industry. A typical biomass fuel is wood, in the form of logs or wood chips, but other energy crops such as straw can be used to fire biomass boilers. All of these biomass input fuels are renewable and almost carbon-neutral, in that the CO_2 which is released via the combustion process is cancelled out by the CO_2 which is absorbed by the plant when it is growing. Consequently, biomass could replace much of our current use of fossil fuels, in several different forms.
The proposed heating and cooling strategy for the proposed development is shown in the diagram below, Fig. 9. The main fuel is natural gas and the CHP is the lead boiler, top up by the biomass and then the gas boiler to meet the peak demand. The electricity produced by the CHP will be connected to the grid.
In summer; the heat produced by the CHP is used to meet the hot water demand and portion of the cooling need.

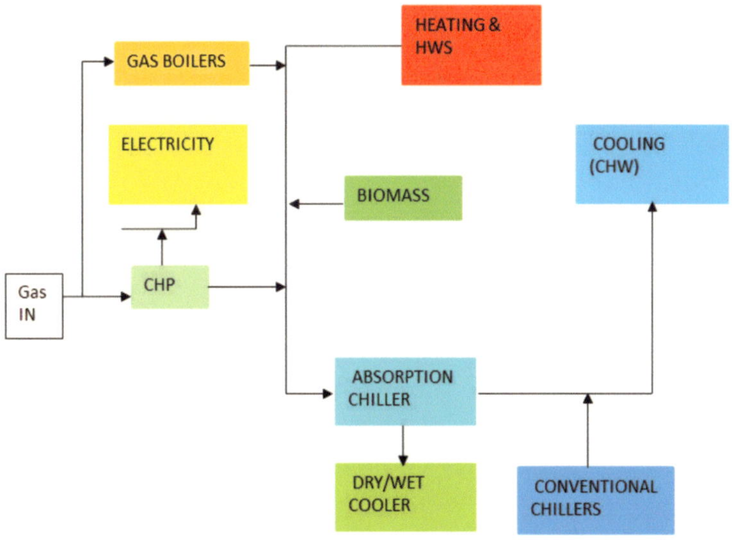

Fig. 9. Schematic diagram of the proposed district plant

The key to a successful building gas CCHP plant selection exists in that the detailed building heat, cool and power load status and appropriate unit capacity to ensure high hourly operation and waste heat recovery. Fig. 10 shows the combined hourly heating, hot water and power load for office building and residential building (50% of the floor area each) in a typical winter day in London.

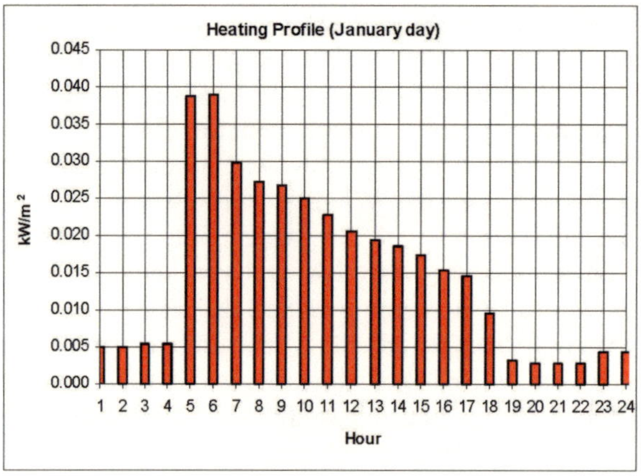

Heating

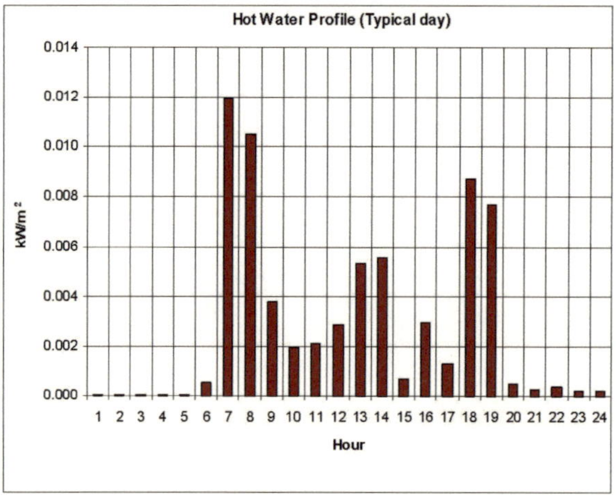

Hot water

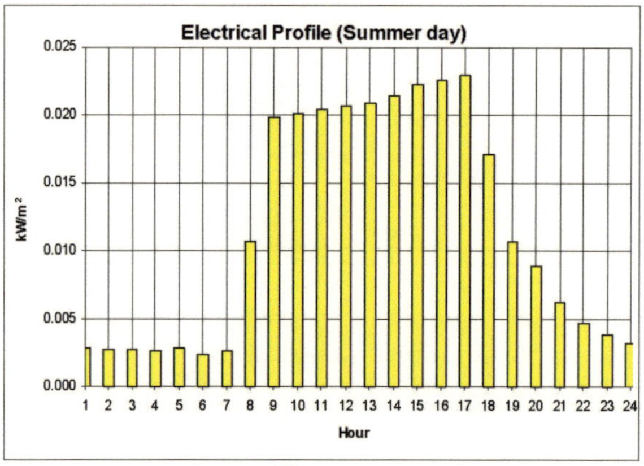

Power

Fig. 10. Hourly profile for a typical winter day

3.1 Carbon emission calculation

3.1.1 Carbon reduction from CCHP

The carbon reduction from a CHP plant can be calculated from the simplified formula below developed by the authors.

$$R_{CHP} = \frac{G \times CHP_{th}\%}{HPR} \times \left(CF_g \times HPR \times \left(1 - \frac{1}{\eta_b}\right) + \left(CF_e - CF_g \times \eta_b\right) \right) \quad (1)$$

Where R is the carbon reduction kg/year; G is the total gas energy in the base building, kWh/year; HPR is the heat to power ratio generated by the CHP; $CHPth\%$ is the percentage of heating demand met by CHP, CF_g is the carbon factor for gas; CF_e is the carbon factor for electricity and η_b is the boiler efficiency.
The percentage carbon reduction, $R_{CHP}\%$ can be defined as the ratio of the carbon reduction obtained by the CHP, $RCHP$, to the total carbon from the base building generated by natural gas, G, and electricity, E.

$$R_{CHP}\% = \frac{R_{CHP}}{CF_g \times G + CF_e \times E} \quad (2)$$

The absorption cooling percentage, $AC\%$, is calculated considering the domestic hot water demand in summer. The carbon reduction achieved from using the absorption chiller, R_{AC}, is calculated by:

$$R_{AC} = AC\% \times C \times (CF_e - COP_{cc} \times (\frac{ACCF}{COP_{AC}} + CF_e \times Par\%) \quad (3)$$

Where $ACCF$ is the carbon dioxide burden of the heat supply to the absorption chiller, $kgCO_2/kWh$; η_e is the CHP electrical efficiency; η_{th} is the CHP thermal efficiency; COP_{cc} is the seasonal energy efficiency ratio of the conventional chiller; COP_{AC} is the coefficient of performance of the absorption chiller; and $Par\%$ is the percentage of the parasitic power used by the absorption chiller.

$$ACCF = (\frac{CF_g}{CHPe\%} - CF_{re}) \times Ratio \quad (4)$$

The total CO_2 reduction, R_{CCHP}, is the sum of the carbon reduction due to the operation of the CHP and absorption chiller.

$$R_{CCHP} = R_{AC} + R_{CHP} \quad (5)$$

The percentage of total CO_2 reduction, $R_{CCHP}\%$, is given by:

$$R_{CCHP}\% = \frac{R_{CHP} + R_{AC}}{CF_g \times G + CF_e \times E} \quad (6)$$

3.1.2 Carbon reduction from biomass boiler

The percentage contribution of the annual thermal load by biomass, $Bio\%$, is the difference between the maximum percentage, $Max\% <100\%$, and the CHP percentage.

$$Bio\% = (Max\% - CHP\%) \tag{7}$$

$$R_{Bio} = (CF_g - CF_{bio}) \times Bio\% \times G \tag{8}$$

$$R_{Total}\% = \frac{R_{CHP} + R_{AC} + R_{Bio}}{CF_g \times G + CF_e \times E} \tag{9}$$

Where CF_{bio} is the carbon factor for biomass, R_{Bio} is the carbon reduction for biomass and R_{Total} is the total carbon reduction.

3.2 Results

Table 1 shows the case study development area assumptions used for the proposed development and substituted in equations 1 to 9. The results are depicted graphically in figures 11 to 15. Figure 11 shows the annual and monthly energy consumption. The heating and domestic hot water demand is about 50% of the annual demand.

	Residential	Offices
Area weight	50%	50%
Gas Benchmark kWh/m^2	75	82
Electricity Benchmark kWh/m^2	30	98
Cooling Benchmark kWh/m^2	10	14
DHW%	60%	11.4%
CHP η_e	30%	
CHP η_{th}	45%	
Heat to power Ratio HPR	1.5	
Conventional Chiller COP	3	
Absorption Chiller COP	0.68	
Carbon dioxide factor for conventional boiler $kgCO_2/kWh$	0.194	
Carbon dioxide factor for grid electricity $kgCO_2/kWh$	0.422	
Carbon dioxide factor for displaced grid electricity $kgCO_2/kWh$	0.568	
Maximum thermal percentage Max%	80%	

Table 1. The mixed use development

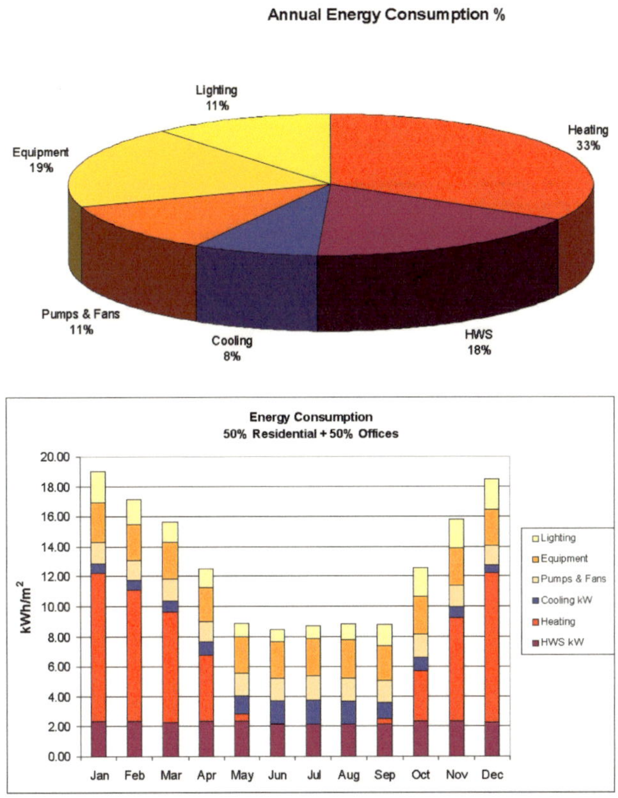

Fig. 11. Annual and monthly energy consumption

The relation between the peak load and annual thermal demand for the mixed use development is shown in Fig. 12. It can be observed that 80% of the annual thermal demand can be attained by 40% of the peak heating load.

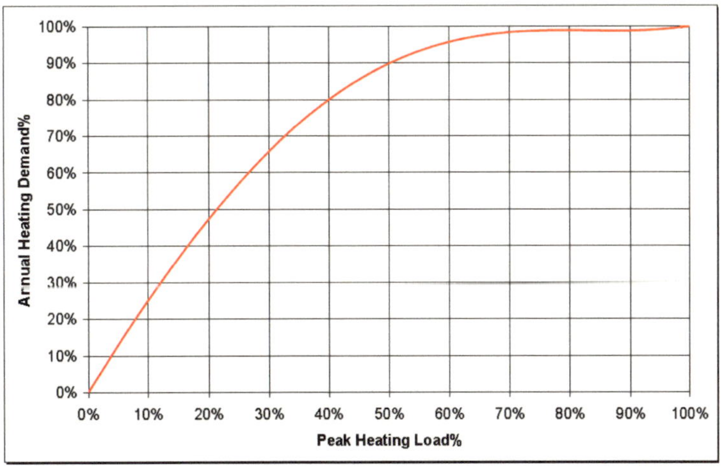

Fig. 12. Demand duration curve

For maximum carbon reduction the thermal load should be met by low carbon source. The operational condition of CCHP system with cooling and heating network interconnecting in one typical day in summer is illustrated in Fig. 13. It can be observed that by maximising the CHP more waste heat is generated in summer and using CCHP system will recover the waste heat in an absorption chiller.

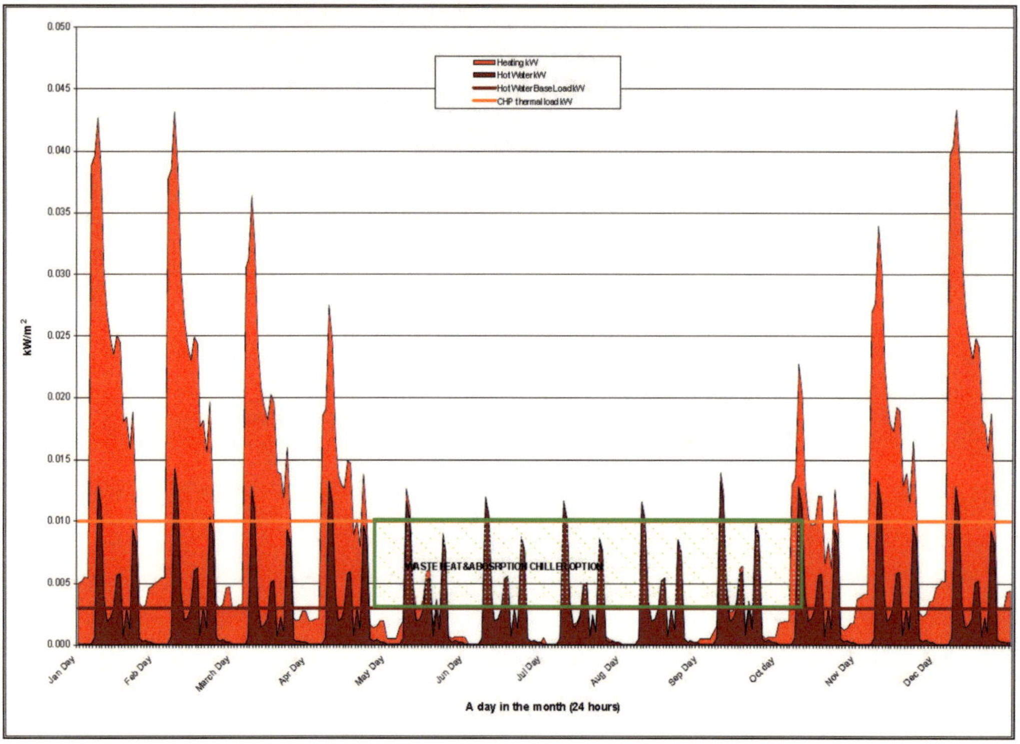

Fig. 13. Waste heat in summer

Equations (6) to (9) are depicted graphically in Fig. 14. As shown the maximum thermal load contribution by the CHP and/or biomass is limited by 80%. The maximum CO_2 reduction obtained by the CCHP is 26.0% however; the absorption cooling does not exceed 1.5% reduction attributable to the total emissions mainly because of the parasitic power used by the absorption chiller and the low coefficient of performance compared to modern vapour compression machines.

The use of a biomass boiler will reduce the CO_2 emissions from the development by 27.5% which supplies 80% of the annual thermal demand. However, unlike most gas fired boilers, wood boilers are limited in their ability to rapidly modulate heat output.

Therefore sizing should be considered to meet the base load and provide more responsive plant, such as gas-fired boilers, to meet peaks in demand. On another hand, the duty of the CHP plant is limited by the demand for heating so that normally only a small fraction of the electrical demand can be produced by the CHP without dumping heat. Therefore the point of intersection could be considered as the optimum operation of the CHP and biomass boiler. At this point 50% of the CO_2 emissions reduction is from the CCHP (42% of the annual thermal load) and the other 50% from the biomass boiler (38% from the annual

thermal load). The total CO_2 reduction is about 26%. However, at this point the contribution of absorption cooling in CO_2 reduction does not exceed the 0.5%.
The contribution of the CHP and biomass as a percentage of the annual thermal load is shown in Fig. 15.

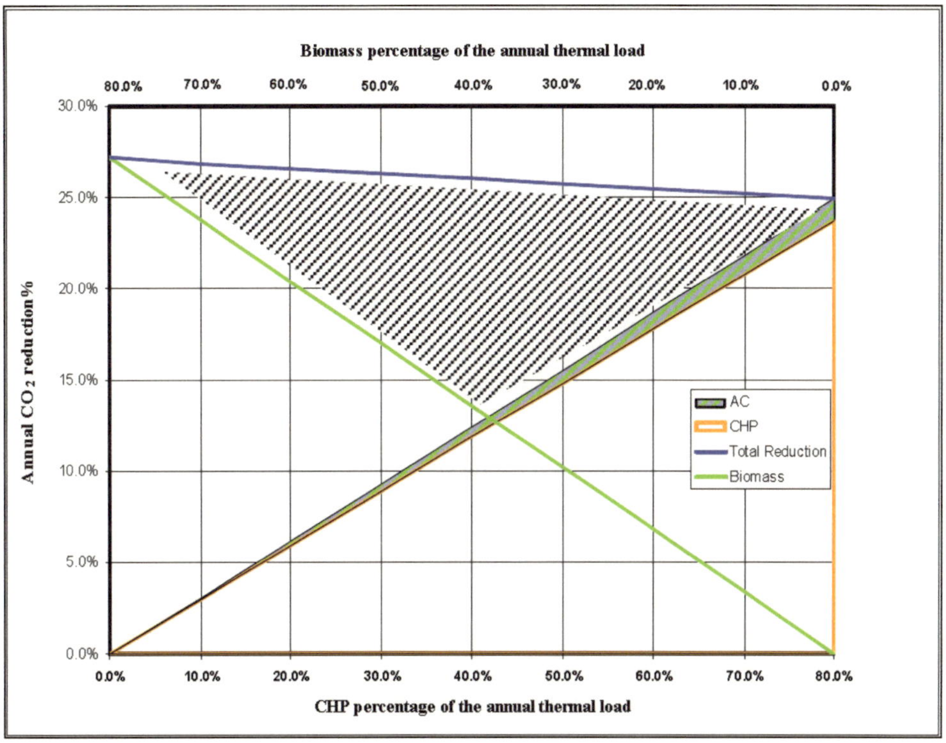

Fig. 14. Optimum CHP-biomass boiler

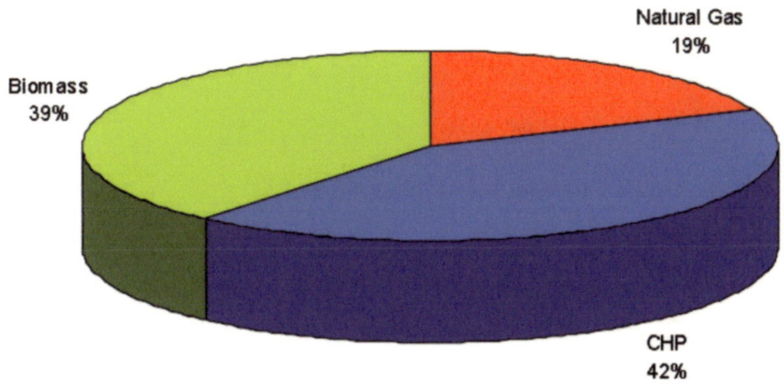

Fig. 15. Contribution of CHP and biomass as percentage of the annual thermal demand

4. Application of using fuel cells in dwellings

The main purpose of this study is to examine the use of commercially available low-cost PEMFCs in dwellings. The PEMFC will be used to provide power for lighting in a dwelling. The dwelling under study has an area of 150 m^2. The calculation of lighting use in a dwelling is based on the proportion of fixed low energy lighting outlets installed, and on the contribution of daylight.

4.1 Energy and CO_2 lighting calculations

Allowing for fixed low-energy outlets In UK houses, the average energy consumption for lighting is taken as 9.3 kWh/m^2 annually if no low-energy lighting is used [21].

$$E_B = 9.3 \text{ kWh} / \text{m}^2 \tag{10}$$

The Standard Assessment Procedure, SAP 2005calculation takes account of fixed lighting outlets with low energy lamps, by including a correction factor C_1 [22]:

$$C_1 = 1 - 0.5 \times \frac{N_{LE}}{N} \tag{11}$$

Where N_{LE} is the number of fixed low energy lighting outlets and N is the total number of fixed lighting outlets.

Analysis of typical house types gives the following approximate correction factor, C_2, for lighting energy use depending on the ratio of glass area to floor area, glass transmittance and light access factor.

$$C_2 = 52.2G^2 - 9.94G + 1.433 \text{ if } G \leq 0.095, \; C_2 = 0.96 \text{ if } G > 0.095 \tag{12}$$

$$G = \frac{\sum 0.9 \times A_w \times g_L \times FF \times Z_L}{TFA} \tag{13}$$

Where *FF* is the frame factor, A_w is the area of a window, m^2, *TFA* is the total floor area m^2, g_L is the light transmittance factor, Z_L is the light access factor. The annual energy used for lighting in the house, E_L kWh/year, is then:

$$E_L = E_B \times TFA \times C_1 \times C_2 \tag{14}$$

The reduction in lighting energy, kWh/year, use due to low energy lights is:

$$\Delta E_L = E_B \times TFA \times (1 - C_1) \times C_2 \tag{15}$$

The power and hydrogen consumed is calculated by [23]:

$$P = Vc \times I \times n \tag{16}$$

$$H_{used} = 1.05 \times 10^{-8} \times I \times \frac{T}{\rho_h} \tag{17}$$

Where *P* is the power in Watts, V_c is the cell voltage in Volts, *I* is the current in mA, *n* is the number of cells, H_{used} is the hydrogen used in Litre, T is the time in seconds and ρ_h is hydrogen density in kg/m^3.

4.2 Results

The energy reduction due to the use of 50% energy saving light fittings is about 25%, as shown in Fig. 16. The calculated energy and CO_2 emissions after using the low energy light fittings are shown in Fig. 17. The results show that the annual electrical energy demand is 4,500 kWh, the percentage of the lighting energy of the total electrical energy is about 26%.

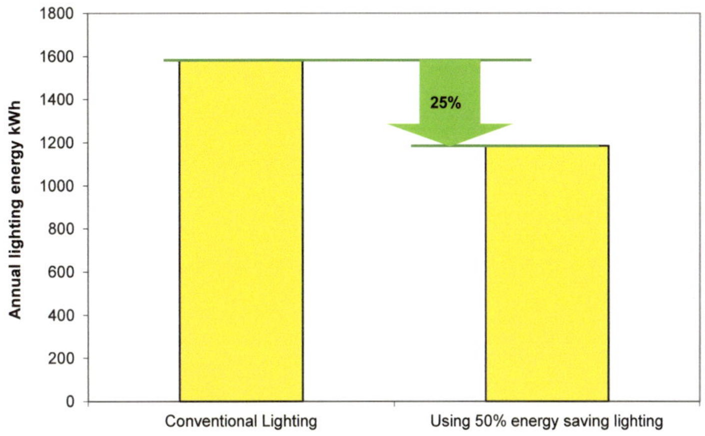

Fig. 16. Lighting energy consumption

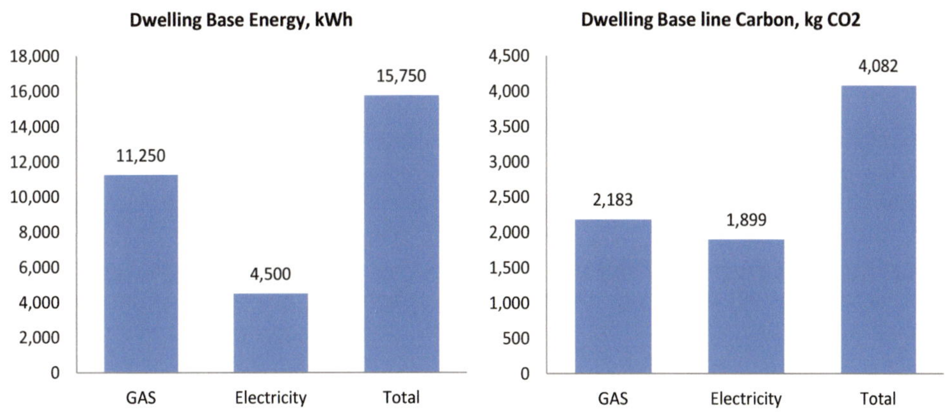

Fig. 17. Dwelling energy consumption and carbon emissions

The use of 100W PEMFC, as baseline, to generate electricity for the lighting will reduce the dwelling energy and CO_2 emissions. The calculations require finding out the annual demand that will be served by the PEMFC. The annual energy demand calculated from Eq. 14 is used to produce the hourly profile shown in Fig.18 [24]. The percentage of electrical demand met by the PEMFC is about 11% and reduces the annual CO_2 emissions by 8.7% as shown in Fig. 19. The results encourage the use of fuel cell in dwellings.

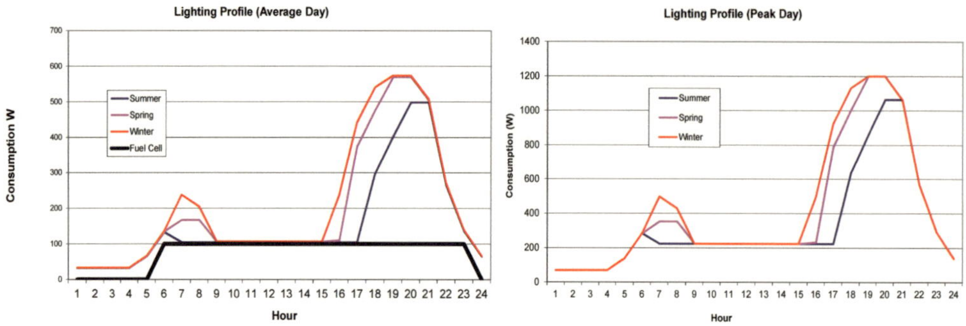

Fig. 18. Lighting energy consumption profile

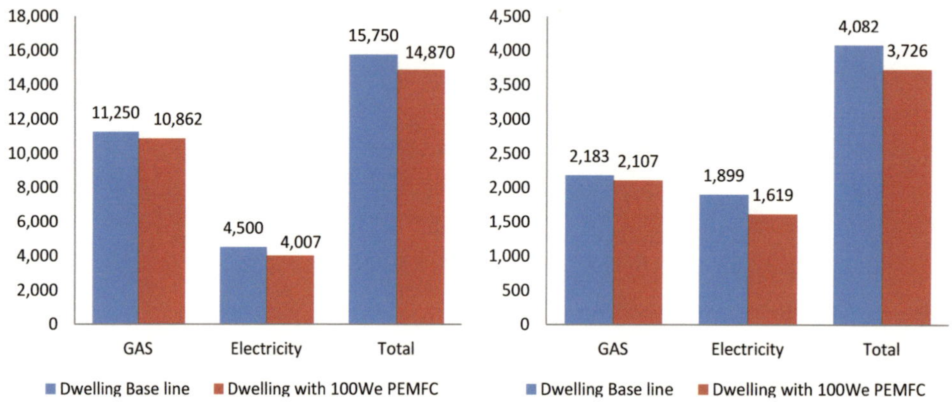

Fig. 19. Dwelling energy and carbon emissions reductions

5. Conclusions

Commercial building's heat and power demand is the most recently developed, both from the market perspective as well as in terms of the technology itself. The amount of power produced in a commercial building's cogeneration system can be less than, equal to, or greater than the local demand. The heat obtained from the system is generally used for heating large volumes of water destined for a variety of uses, including absorption cooling. CHP systems can be employed over a wide range of sizes, applications, fuels and technologies. The resulting electricity can be used either wholly or partially on-site. It is particularly efficient when employed as a source for district heating to provide hot water, space heating and electricity for a number of linked buildings. This chapter discussed the consideration of CHP as a leading option and alternative means of supplying energy. A full CHP feasibility study was carried out. However, before CHP assessment was done, all passive design and energy efficiency measures were conducted. The development heating and hot water demands were assessed to prevent the CHP from being incorrectly sized.

It is assumed that the electricity generated is utilised on site or exported back to the electricity grid and this can be worthwhile, particularly where on-site demand is low. The current electricity incentive, supplying electricity to the grid, means matching CHP capacity to heat load although the most cost effective solution often involves some modulating capability and/or heat storage. The CHP was sized using daily demand profiles in order to

accurately determine the actual amounts of heat that can be supplied to the building. Thermal store has been considered to smooth the demand profiles as it has a significant effect on the overall performance of the CHP system and its contrbution in reducing CO_2 emissions.
This chapter investigated the contribution of micro combined heat and power and renewables into CO_2 reduction in a mixed use development and dwellings and provided guidance to the designers of employing energy waste and low equipment usage in a traditional combined cooling, heating and deploying CCHP and biomass heating technologies for both energy efficiency and carbon reduction in different applications.
Here, the micro CHP operates as the lead boiler to maximise savings. The biomass boiler tops up the heating energy demand and the gas boiler will meet the spike and peak loads.
Detailed energy demand daily and monthly profiles for heat are established to accurately sizing CHP. In order to increase the size of the CHP further analyses have been developed to identify alternative conditions that would improve the viability considering the energy profile of the development and thermal store size in winter and the feasibility of using heat-driven absorption chilling plant, CCHP, to extend the base load heat demand into the summer months.
The mixed use development case study showed that the use of absorption chillers will result in additional small reduction in CO_2 and the maximum reduction did not exceed 1.5%. The maximum CO_2 reduction obtained by using the combination of CHP and biomass boilers was about 26%. The study suggests a careful consideration when sizing CCHP and biomass technologies and provides a methodology to estimate the contribution of each technology into carbon reduction. In this study the CHP unit size was considered to meet the base load and provide more responsive plant, such as gas-fired boilers, to meet peaks in demand.
The current study also encourages the use of fuel cells in dwellings and other application as it will have a significant impact on reducing carbon emissions. For instance, the study, showed that the use of 100W PEMFC in a dwelling for lighting power reduced the electrical demand by 11% and the dwelling CO_2 emissions by 8.7%. The results give important pointers to how micro cogeneration technologies can be used in residential, commercial and other applications in tandem with other energy-saving technologies.
Adopting composite multi-generation systems will lead to significant benefits in terms of higher energy efficiency, reduced CO_2 emissions, and enhanced economy considering that the duty of the CHP plant is limited by the demand for heating so that normally only a small fraction of the electrical demand can be produced by the CHP without dumping heat. The limitations of cogeneration technology in the commercial sector such as low power requirements have been overcome by the incentive programs which encourages deploying CHP and CCHP plants in such regions when applicable.

6. References

[1] Kuhn, V., Klemes, J., Bulatov, I. (2008). MicroCHP: Overview of selected technologies, products and field test results. *Applied Thermal Engineerin,* Vol.8, No.16, pp. 2039-2048

[2] Microgeneration Strategy (2006), Our Energy Challenge, Power from the People,

[3] Department of Trade and Industry available from www.dti.gov.uk

[4] John, V., Henry, W., Fuel Cells: A Technology Forecast, Texas State Technical College, ISBN 0978650389 available from http://publishing.tstc.edu

[5] Jung-Ho, W., (2007), Applications of proton exchange membrane fuel cell systems. *Renewable and Sustainable Energy Reviews* , Vol. 11, pp. 1720 -1738

[6] Takahiro K. (2008), Development Strategies toward Promotion and Expansion of Residential Fuel Cell Micro-CHP System in Japan, Osaka Gas Co., available from www.igu.org/html/wgc2009/papers/docs/wgcFinal00801.pdf
[7] http://en.wikipedia.org/wiki/Home_fuel_cell
[8] Micro Generation Heat and Power. Available from http:// business.ezinemark.com/ micro-combined-heat-and-power-319df04cb61.html
[9] CHPA (2011). Combined Cooling Heat & Power (CCHP), Combined Heat & Power Association, available from http://www.chpa.co.uk/combined-cooling-heat--power-cchp_188.html
[10] Maidment, G. & Tozer, R. (2002). Combined Cooling Heat and Power in Supermarkets. *Applied Thermal Engineering*, Vol.22, pp. 653–65
[11] Cardona, E., Piacentino, A., Cardona, F. (2006). Energy Saving in airports by Trigeneration. Part I: Assessing Economic and Technical Potential. *Applied Thermal Engineering,* Vol.26, pp. 1427-1436
[12] Kong, X., Wang, R., Huang, X. (2004). Energy Efficiency and Economic Feasibility of CCHP Driven by Sterling Engine. *Energy Conversion and Management*, Vol.45, No.9, pp. 1433–1442
[13] Cardona, E., Piacentino, A. (2007). Optimal Design of CHCP Plants in the Civil Sector byThermoeconomics. *Applied Energy,* Vol.84, pp. 729–748
[14] Hui, L., Lin, F., Kecheng, G., Jiang, Y. (2006). Energy Utilization Evaluation of CCHP Systems. *Energy and Buildings*, Vol.38, pp. 253- 257
[15] Joel, H. & Augusto, S. (2003). Trigeneration: An Alternative for Energy Savings. *Applied Energy, Vol.*76, pp. 219–227
[16] Lin, L., Yaodong, W., Tarik, S., Tom, R., Stuart, T., Shengchuo, Z., Jincheng, H., Yunxin, H., Xiaodong, H. (2007). An Experimental Investigation of a Household Size Trigeneration. *Applied Thermal Engineering,* Vol.27, pp. 576–585
[17] Gianfranco, C. & Pierluigi, M. (2008). A Unified Model for Energy and Environmental Performance Assessment of Natural Gas-fueled Poly-generation Systems. *Energy Conversion and Management*, Vol.49, No.8, pp. 2069-2077 BSRIA http://fuelcellsworks.com; Fuel Cell Works website.
[18] Degiorgis L., Santarelli M., Cal, M. (2007). Hydrogen from Renewable Energy: A pilot Plant for Thermal Production and Mobility. *Journal of Power Sources*, Vol.171, pp. 237–246
[19] Marshall, A.T., Sunde, S., Tsypkin, M., Tunold, R. (2007). Performance of a PEM water electrolysis cell using Ir*x*Ru*y*Ta*z*O2 electrocatalysts for the oxygen evolution electrode. *International Journal of Hydrogen Energy*, Vol.32, pp. 2320 - 2324.
[20] Application Manual AM12. (1999). Small Scale Combined Heat and Power for Buildings. The Chartered Institution of Building Services Engineers. ISBN 0900953926
[21] OPDM. (2006). The Building Regulations 2006: Conservation of Fuel and Power Part L1A. Office of the Deputy Prime Minister, UK
[22] BRE. (2005). The Government's Standard Assessment Procedure for Energy Rating of Dwelling, SAP 2005. Building Reserach Establishment, UK
[23] Larminie, J., Dicks, A. (2000). Fuel Cell Systems Explained. Chichester, England, Wiley, ISBN 047084857X
[24] Hendron, R., Anderson, R., Christensen, C., Eastment M. (2004). Development of an Energy Savings Benchmarks for All Residential End-Uses. National Renewable Energy Laboratory, Colorado, available online from www.nrel.gov/docs/fy04osti/35917.pdf

21

Nuclear Methodology for Non-Destructive Multi-Elemental Analysis of Large Volumes of Soil

Lucian Wielopolski
Environmental Sciences Department
Brookhaven National Laboratory
USA

1. Introduction

The extent of global warming is determined by the net difference between the sun's incoming energy and that reflected back into the space. Changes in the reflected energy that depend on the conditions of the land's surface and of the atmosphere alter the global energy-budget, stimulating global warming. The dominant factor forcing climate change is the increasing concentrations of atmospheric greenhouse gases (GHGs), in particular that of CO_2 [Solomon et al., 2007]. The rise in CO_2 level from the pre-industrial revolution value of about 260 ppm to present-day concentrations of about 380 ppm, viz., ~46% increase, is recognized widely, albeit some dispute its anthropogenic origin [IPCC, 2007; Tans, 2011]. Extensive farming, the ever-increasing consumption of fossil fuels, land-management practices, the cement industry, and deforestation are the main drivers disrupting the tenuous balance between natural carbon release by the soil's biota and plant respiration, and carbon uptake by the photosynthesis of aboveground plants [Denman, 2007]. This active soil-atmosphere carbon exchange is an inextricable part of the global carbon cycle: hence, it plays a pivotal role in possibly slowing down, stopping, or even reversing this perilous escalation of global warming. With the predicted increase in fossil-fuel consumption, combating global warming necessitates a multi-pronged approach, including improvements in energy efficiency and the use of alternative energy sources. However, by themselves, these are insufficient unless we modify the ways in which we use and produce energy, specifically, the way we manage carbon. Restoring the depleted organic-carbon pools in soil by terrestrial sequestration worldwide is critical for controlling global warming, and for restoring the soil's quality, productivity, and so assuring food security [Lal et al., 2004; Lal et al., 2005]. Central to improving carbon management is having a better understanding of the belowground carbon processes and the ability to quantify them. However, gaining this knowledge poses special challenges because these processes are invisible to us. Soil is a dynamic, living system, a mix of living and dead-plant matter, and a mélange of belowground biota embedded in a matrix of solids, liquids, and gases [Johnston et al., 2004]. The current state-of-the-art in soil analysis comprises taking soil samples and subsequently analyzing them chemically by dry combustion in a laboratory. This method, although very well established, is destructive, time-consuming, labor intensive, and provides only point information in time and space. The throughput of the method is highly inadequate for present demands for precise soil analyses over large areas in a variety of soil types and ecosystems.

The centrality of soil carbon in the global carbon cycle, in soil-quality management practices, in land restoration, in monitoring terrestrial sequestration, and in precision farming are only but few examples implicating soil carbon as vital in extensive research worldwide. The specific research objective in this chapter is to demonstrate the feasibility and the unique characteristics of a novel nuclear method for non-destructive multi-elemental analysis in large soil volumes over large areas. To this end Section 2 briefly reviews the current standard chemical-methods for analyzing soil and the emergence of new modalities that improve upon the existing shortcomings. Section 3 reviews the basic nuclear-physics processes necessary for understanding the unique characteristics of the promoted system. Section 4 describes the promoted system, its response function, and simulations of it using probabilistic Monte Carlo methodology. In addition, section 4 discusses the issues with spectral analysis, interferences, calibration, and its unique characteristics. Section 5 demonstrates the feasibility and the uniqueness of the system in numerous field studies and various soil types. Section 6 summarizes the research, and Section 7 contains the bibliography used in this chapter.

2. Methodologies for soil analysis

2.1 *Ex situ* chemical analysis

Chemical analysis of soil samples in a laboratory, occasionally referred to as an *ex situ* method, evolved over about the last 150 years. In the 19th century, Rogers and Rogers (1848) reported that dichromate-sulfuric-acid solution could oxidize organic substances. After unsuccessful attempts by Warrington and Peake (1880), and Cameron and Breazeale (1904), Ames and Gaither (1914) accomplished a high recovery of organic substances with this mixture. Schollenberger (1927) introduced the titrimetric determination of unused chromic acid in the oxidation reaction with ferrous ammonium sulfate using several indicators; diphenylamine, o-phenanthroline, or N-phenylanthranillic acid. This method generally is referred to as wet combustion.

Rather (1917) introduced the technique of estimating soil organic matter (SOM) from the weight loss of soils on ignition (LOI), and later, Tabatabai and Bremner (1970) offered an automated CO_2 analyzer based on thermal-conductivity measurements of the effluent gases. These analyzers, founded on Walkley-Black procedures (dry combustion) [Walkley, 1935] have widely recognized limitations and are both lengthy and labor-intensive; nevertheless, they become the standard method used for measuring SOM for many years Tabatabai and Bremner (1991) defined the principles for upgrading dry combustion via automated total C analyses; it still is considered as the method of choice for assessing carbon in soil. Extensive efforts over the years went into characterizing and testing this approach, thereby establishing it as the state-of-the-art method for analyzing carbon and nitrogen in soil. Numerous books document the protocols for implementing soil analysis by dry combustion [Allison, 1965; Tiessen, 1993; Tan 2005; Chatterjee, 2009]. However, again, this method is lengthy and labor intensive as it entails collecting core samples, occasionally by excavating them, and transporting them to a laboratory. Before analysis, the samples are weighed, dried, weighed again, ground, and sieved to below 2 mm; thereafter, subsamples again are ground, and then weighed aliquots are analyzed and percent carbon by weight determined. However, for assessing the total carbon stored in the field, knowing the soil's bulk density is essential, which is problematic since the natural variability in the field must be accounted for [Ellert, 2001]. Finally, the information gained from soil analysis by dry combustion represents a point measurement in space and time that poses a problem when extrapolating to fields at the landscape-, regional-, or a higher level. Accordingly, in the last decade new approaches for soil analysis were introduced and tested that circumvent some of the constraints imposed by the chemical methodologies.

2.2 *In situ* new modalities

The increasing need and interest in assessing, and reassessing carbon inventories in soil, and clarifying its dynamics on large spatial- and temporal-scales with reduced uncertainties, not possible at earlier, stimulated novel implementations of three independent, well-established techniques. These are near-infrared spectroscopy (NIR) [Sudduth and Hummel 1996; Christy et al., 2003], laser-induced breakdown spectroscopy (LIBS) [Ebinger et al., 2003], and inelastic neutron scattering (INS) [Wielopolski, 2006], the last one being the subject of the present chapter. These three techniques respectively entail spectroscopy on the molecular-, atomic-, and nuclear-levels.

2.2.1 LIBS

LIBS is a well-established atomic-emission technique extensively applied for surface analyses in research and industry (Balzer et al., 2005; Essington et al., 2009; Merdes et al., 2007). It involves focusing a pulsed laser-beam ~50 μm diameter on a sample and ablating a small amount, ~10^{-9} cm^3, of the material (Martin et al., 2010). When sufficient energy is focused on the irradiated target, matter is vaporized, and its atomic- and molecular-constituents are ionized. Thus, the micro-plasma so created is apparent as a spark that expands rapidly, reaching local thermodynamic equilibrium within 0.5- to 1.0-μs, depending upon the sample's conditions. Thereafter, as the micro- plasma starts to cool down, ionized species within the plasma are populated with concomitant emissions of light from atomic- and molecular-transitions (Buckley, 2006; De Lucia et al., 2009; Hahn, 2009). Recently, this method was applied successfully for soil analyses [Cremers et al., 2001; Ebinger et al., 2003; Martin et al., 2003]. Since the volumes analyzed by the LIBS are very small, a fraction of a cm^3, samples from heterogeneous materials, like soils, require averaging the signals from up to 100 laser shots per spot. The entire LIBS spectra, rather than the peaks alone, are analyzed using principal-component analysis. This approach supports investigations of the impact of experimental parameters on the systems' mode of data collection, such as the laser's irradiation-wavelength and power [Martin, 2010].

2.2.2 NIR

William Herschel in 1800 discovered the infrared (IR) region of the electromagnetic spectrum, spanning the red end of visible light and microwaves; for over a century, it has been used as a diagnostic tool. It is utilized intensively for surface analyses since IR can penetrate a few cm in the most advantageous cases. The IR region of the spectrum is divided arbitrarily into three regions: The near-, mid- and far-infrared, named for their distance relative to the visible spectrum. The far-infrared, approximately 1,000–30 μm (400–10 cm^{-1}), lying adjacent to the microwave region, has low energy and is employed in rotational spectroscopy. The mid-infrared, approximately 30–2.5 μm (4,000–400 cm^{-1}), is valuable in exploring fundamental vibrations and the associated rotational-vibrational structures. The higher energy near-IR, approximately 2.5–0.8 μm (4,000–4,000 cm^{-1}), can excite overtone- or harmonic- vibrations. Hence, infrared spectroscopy is a tool responding to material structures on the molecular level, in particular to the C-C, C-H, N-H, and O-H bonds. Diffuse reflectance spectroscopy relies upon detecting the scattered part of the focused incident beam on a sample where it undergoes reflection, scattering, and transmission through the material. Numerous publications discuss employing IR in agricultural research and soil analyses; this information appears in recent reviews [Reeves, 2010; Shepherd and Walsh 2007]. IR spectra are complex and require advanced chemometrics wherein measured reflectance spectra are acquired from each sample, and statistical procedures are used to correlate reflectance and the data obtained from soil analyses in laboratories. Awiti (2008) and Reeves (2010) discuss the intricacies of analyzing IR spectra. The IR method is destructive, requiring sampling of soils, or alternatively plowing through them with a sensor mounted on the tip of a shank [Christy et al., 2003].

2.2.3 INS

Inelastic neutron scattering, the last of the three newly emerging modalities for carbon analysis in soil in the field, differs fundamentally from LIBS and IR in that it does not require collecting soil samples. It is non-destructive and analyses large volumes, approximately 0.3 m^3, in static- and large fields in scanning-modes of operation. The INS approach is based on the spectroscopy of gamma rays induced by nuclear interactions of fast- and thermal-neutrons with the nuclei of the elements present in soil. The spectra, measured during and after the end of irradiation are recognized, respectively, as prompt- and delayed-gamma-ray spectra. The former consist of two separate, but concurrently acquired spectra from inelastic neutron- scattering reactions during the fast neutron pulse, about 25 μs long, and from thermal neutron-capture (TNC) reactions occurring in the 75 μs between the neutron pulses. The delayed gamma-ray spectra are measured at the end of the total period of irradiation. Since all the measured gamma-ray spectra result from the nuclear reactions of neutrons with atomic nuclei, they are insensitive to the chemical state of the element, and the intensity of the measured signal is proportional to the total number of atoms of a given element in the interrogated volume. A unique capability of the INS system is its ability for scanning arbitrarily large areas, so providing a single mean value for its entirety [Wielopolski et al., 2011]. A description, characterization, and the results from using the INS system are detailed hereafter. Table 1 briefly summarizes the key characteristics of the new techniques; comparative reviews of the various methodologies are given in the literature [Chatterjee, 2009; Gehl and Rice, 2007; Schumacher, 2002].

Method	Principle	Penetration in soil (cm)	Sampled volume (cm^3)	Features
Dry Combustion	Small subsample is fully oxidized and amount of CO_2 measured	Depends on core size	Arbitrary, ~0.1 g analyzed	Destructive, labor-intensive, and time-consuming
Laser induced breakdown spectroscopy (LIBS)	Laser-ablated sample forms micro plasma that, upon cooling, emits light from the ionized atomic- and molecular-species	0.1	$\sim 10^{-2}$	Destructive, with high spatial resolution, ~1 mm, minimal sample preparation, presence of roots and rocks increases variability in carbon signal necessitating its normalization
Mid- and Near-infrared reflectance spectroscopy (MIRS/NIRS)	NIRS (0.4-2.5 μm) and MIR (2.5-25 μm) region utilized to quantify soil C; based on the absorption of C-H, N-H and O-H groups found in organic constituents	0.2-1	~10	Destructive, isotopic and carbon specie sensitive, strong matrix interferences, minimal sample preparation
Inelastic neutron scattering (INS)	Based on spectroscopy of gamma rays induced by high, 14 MeV, and low energy neutrons	30	$\sim 10^5$	Non-destructive, no sample preparation, multi-elemental, analyzes large volumes with scanning capability, provides true sequential measurements, has analytical expression for the response function

Table 1. Key features of the *in situ* soil carbon determination techniques

3. Neutron- and gamma-ray transport in matter

3.1 Neutrons

3.1.1 Neutron interactions with matter

Neutrons interact with atoms' nuclei via nuclear reactions, and so are insensitive to the chemistry of the elements in the sample. These interactions are stochastic, and depend on the neutron's energy that loosely is categorized as fast above 1 MeV, epithermal between 1

MeV and 2.7 eV, and thermal below the latter. The neutrons interact with matter via numerous types of interaction that are divided roughly into scattering and absorption. Elastic scatterings occur when a fast neutron collides with a scattering atom, and transfers energy to it, after which the neutron continues to move at a somewhat degraded speed through the medium. In this process, no nuclear excitation occurs. Alternatively, a neutron may scatter inelastically (INS), wherein the nucleus is raised temporarily to an excited state that instantenously decays to a ground state with concomitant emission of a neutron with lower energy than the incident one, and of specific gamma rays. Inelastic reactions are typified by threshold energies below which this reaction cannot occur. There also are processes by which a thermalized neutron is absorbed resulting in prompt- and delayed gamma rays, whereas for fast neutrons, the absorption reactions are accompanied by the emission of a particle and gamma rays [Evans, 1955]. Fig.1 summarizes these reactions, and partially list elements that were analyzed *in situ* in soil.

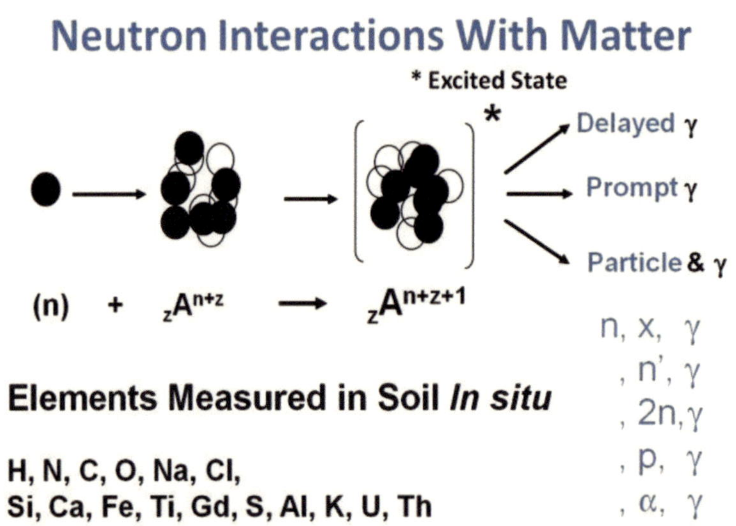

Fig. 1. The decay modes of excited nuclei following the absorption of a neutron, and a partial listing of the elements measured *in situ* in soil.

The path of an incident fast neutron intercepting the soil medium is random, and depends on the cross-sections (probabilities) for various reactions. Figure 2 depicts the conceptual temporal- and spatial-dependence of the excitations of the various elements in a large intact soil volume. Fast 14 MeV neutrons would penetrate the soil and, within 10^{-12} to 10^{-9} sec, induce gamma rays from inelastic reactions; alternatively, they may thermalize, slow down, and following more than 30 collisions, be absorbed, inducing prompt- and delayed-gamma rays.

3.1.2 Neutron attenuation in matter

Under the simplifying assumptions that neutrons are monoenergetic and non-changing in a perfectly narrow well-collimated beam, then the intensity of a neutron beam penetrating thickness x, I_x, is given by $I_0*exp(-\Sigma_t*x)$. Here, I_0 is the incident beam's intensity and Σ_t is the macroscopic cross-section, i.e., it is the probability per unit path-length that a neutron will undergo some kind of interaction as it moves about in the medium. For materials containing several elements, Σ_t is given as weighed summation of the individual cross-sections. However, in reality, we do not have a collimated beam, and neutrons change their energy as they slow down [Lamarsh, 1975]. Thus, in principle, the sampled soil is semi-infinite in size; however, practically the three mean free paths, $1/\Sigma_t$, in which 99% of the beam is attenuated are several tens of centimeters deep.

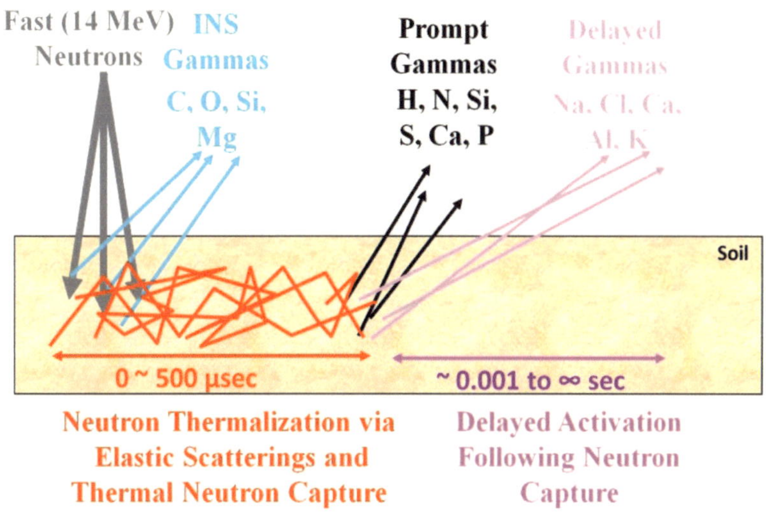

Fig. 2. Temporal- and spatial-neutron reactions with an incident 14-MeV neutron beam.

3.2 Neutron sources

Neutron sources are classified as small, medium, and large depending upon their size, operational complexity, energy, and intensity. Clearly, nuclear reactors are the largest, most prolific sources of neutrons; however, their industrial applications are extremely limited, except for generating electricity. Medium-sized neutron sources are based on plasma, light-ion accelerators, and high-energy photons devised for producing neutrons and for physics research. Small neutron source are the most suited for field applications. They include certain isotopes that emit neutrons following spontaneous fission, as for example, a ^{252}Cf source with a half life of 2.6 years, and an emission rate between 10^7 to 10^9 neutrons per second.

Another family of neutron sources is based on radioisotopes that decay via alpha particles, viz., radium, plutonium, and americium that impinge upon a low-Z elemental matrix, such as lithium, beryllium, carbon, or oxygen. Usually the source is prepared as a mixture of powders of the two materials. Typical emission rates for alpha-reaction

neutron sources range from 1×10^6 to 1×10^8 neutrons per second. The useful lifetime for such sources is highly variable, depending upon the half-life of the radioisotope that emits the alpha particles. Usual combinations of materials are plutonium-beryllium (PuBe), and americium-beryllium (AmBe).
Radioisotopes that decay by gamma emissions exceeding the neutron-binding energy co-located with beryllium or deuterium stimulate gamma-reactions. Two examples of this reaction with their decay products are 1) ^9Be + >1.7 Mev photon → 1 neutron + 2 ^4He; and, 2) ^2H + >2.26 MeV photon → 1 neutron + ^1H.
Sealed-tube neutron generators are the most popular neutron source in industry. They can be turned off at the end of use, so terminating radiation production, and they are operable either continuously or in a pulsed mode. A unique feature in more advanced neutron generators, referred to as associate particle neutron generators, is that the emitted neutrons can be tagged making their direction known. The sealed-tube neutron generators are designed as hermetic, sealed tube, compact accelerators that use deuterium–deuterium (D–D) and deuterium–tritium (D–T) reactions; D + ^2H → ^3He + n Q = 3.270 MeV, D + ^3H → ^4He + n Q = 17.590 MeV. These accelerators respectively generate neutrons of ~2.5 and ~14.1 MeV [Csikai, 1987]. Nominally, these sources produce between 10^7 to 10^{11} neutrons per second, and thousands of such small, relatively inexpensive systems were built over the past five decades. The number and variety of their applications are growing steadily [Chichester and Simpson, 2004].

3.3 Gama-ray interactions and attenuation

Gamma-rays interact with matter in several ways; however, we only need to take three processes into account. These are the photoelectric effect, significant for gamma-rays below 100 keV, Compton scattering that dominates above 100 keV and below 2MeV, and pair production above 2 MeV. These processes will attenuate and change the energy of the emitted gamma radiation that is recorded during gamma-ray spectroscopy. However, the elemental information is stored in the spectral peaks that represent the un-collided gamma-rays that traveled from the point of origin to the detector.
The attenuation of gamma-rays in a medium is similar to that of neutrons but it is governed by an attenuation coefficient, μ that equals $\mu_{pe} + \mu_C + \mu_{pp}$. The attenuation coefficient depends on the gamma-rays' energy and the soil's composition.

4. INS system

4.1 System description

INS system for quantitative elemental soil analysis is based on the spectroscopy of gamma rays induced by the fast neutrons inelastic scattering, and the prompt gamma-rays resulting from thermalized-neutron capture. An INS system comprises the following: A pulsed neutron generator operated at 10 kHz and a 25% duty cycle; NaI gamma-ray detectors; shadow-shielding material between the detectors and the neutron generator; nuclear spectroscopy electronics; and data acquisition on a laptop. The entire system is mounted on a cart about 30 cm above the ground and is powered by a 1 kW power generator, or by a bank of four 12 V batteries with a total capacity of 400 Ah feeding a 110 V inverter. During operation, when fully loaded, the system draws about 1.3 A. The INS system can be operated when it is positioned stationary in a fixed place, or in a scanning mode when towed across field. The INS system is shown in Fig. 3.

Fig. 3. Alpha prototype of an INS system with its main components marked.

The data typically are acquired for 30- to 60- minutes and displayed as INS and TNC spectra, as for example in Fig. 4; those particular spectra were acquired in a pine stand. A carbon peak is apparent in the INS spectrum, whereas nitrogen is shown in the TNC spectrum. The signal-to-noise ratio in these spectra can be improved using associated particle-neutron generators. The spectroscopy of gamma-rays induced by neutrons was discussed extensively for example [Alfasi and Chung, 1995; Nargolwalla and Przybylowicz, 1973].

4.2 System response function

The transport of neutrons and gamma-ray radiation in matter is very complicated and necessitates solving a five-dimensional integration over neutron energy, time, and space where there are two different spaces; the outside world denoted, x,y,z, and the inner world of the gamma-ray detector identified as x′, y′, and z′. Assuming steady-state conditions independent of time, the time integration becomes a simple time-multiplier. Under these conditions, the INS's response function is derived from basic principles and based on fundamental parameters can be written as Eq. 1;

$$C_N = k\int_T dt \int_E \iiint_V \varphi_n(x,y,z,E_n)\sigma(E)C_c(x,y,z)\rho_b(x,y,z)\Omega(x,y,z)At(E_\gamma x,y,z)dEdxdydz \int_D Det(E_\gamma,x',y',z')dx'dy'dz' \quad (1)$$

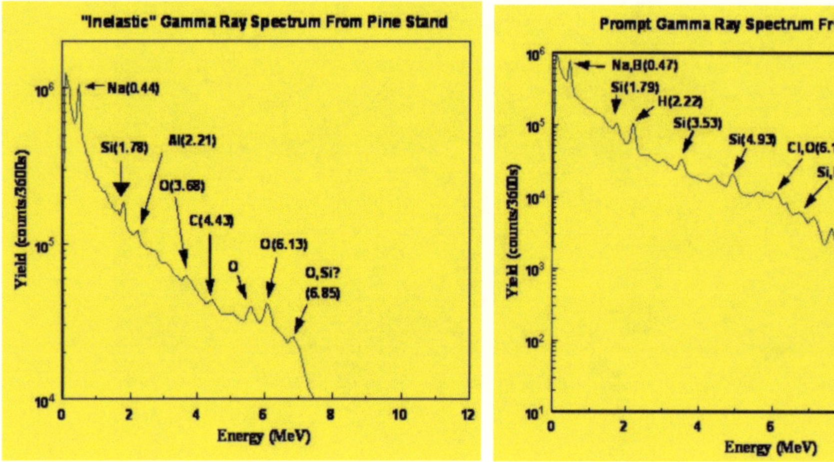

Fig. 4. A typical inelastic gamma-ray spectrum on the left, and prompt gamma-ray from neutron capture spectrum on the right. These spectra were acquired for 30 min in a pine stand.

where T is the time domain, E is neutron energy from incident energy of 14 MeV down to thermal energies 0.025eV, V the soil dimensions, which are semi-infinite in all directions. The simulation is limited to a parallelepiped volume 250 long by 200 cm wide and 50 cm deep, and D is the space occupied by the detectors. Inherent lateral homogeneity is assumed in this model. Thus, the number of counts in the carbon peak, C_N, given in Eq. 1 depends on the depth distributions of the various components in the equation; these components are shown in Fig. 5 in which

- $\varphi_n(x,y,z,E_n)$ (n/cm^2) is the calculated neutron-flux depth-distribution that decreases with depth;
- $\sigma(E)$ (cm^2) is the neutron cross-section that only depends on the neutron's energy;
- $C_c(x,y,z)$ (gC/cm3), the carbon-depth profile usually decreases, although it may assume different distributions;

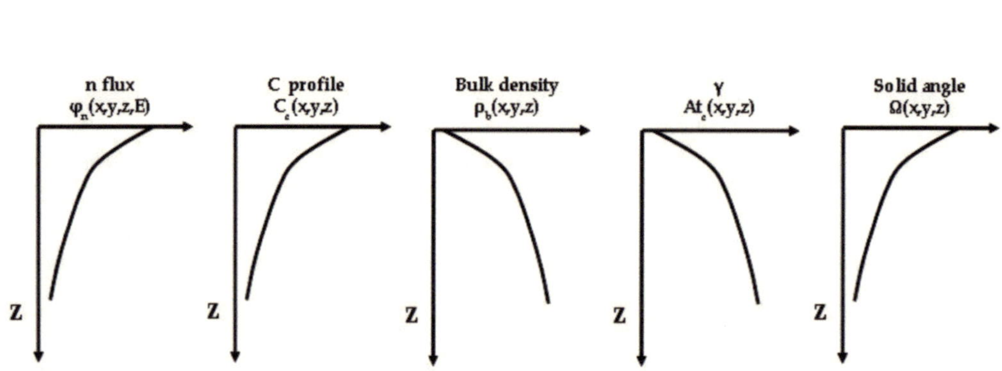

Fig. 5. Conceptual changes with depth in the key components in the INS system's response function.

- $\rho_b(x,y,z)$ (g/cm³), the soil's bulk density increases with depth;
- $\Omega(x,y,z)$ (fraction), the solid angle subtended by the detectors from the emission point of the gamma-rays in the soil;
- $At(E_\gamma x,y,z)$ (fraction), the attenuation of gamma-rays on their way to the detector, increases with depth; and,
- $Det(E_\gamma,x',y',z')$ (photopeak counts) complete the energy deposition in a detector and allocation in a spectrum.

There is no closed-form analytical solution to Eq. 1, regardless of the existence or lack of boundary conditions. Instead, the integral equation is estimated using the stochastic process of random-number generation and following the individual history paths for each particle. This approach is feasible since all the processes are well-defined. The mean probability of success from many histories is the estimate of the value of the integrals in Eq.1. This stochastic numeric approach is known as the Monte Carlo (MC) method [Lux and Koblinger, 1990]. In the following paragraph the MC methods is used for estimating various parameters of the INS system.

4.3 MCNP simulations

The INS system was simulated using a Monte Carlo neutron photon (MCNP) transport code [Breismeister, 1993; Pelowitz, 2005]. The simulation started with a neutron emission at the source, and terminated with a photon intercepting a detector. In some cases, the simulations was carried out for $10*10^9$ incident neutrons and soil composition was taken from mean values of the world soil's composition, Table 2 [Frank and Tolgyessy, 1993]; the soil bulk density was assumed as 1.4 g/cm³.

Element	O	Si	Al	Fe	C	Ca	K	Na	Mg	Ti	N	S	P
Weight (%)	49.0	33.0	7.10	3.80	2.00	1.37	1.36	0.63	0.63	0.46	0.10	0.09	0.08

Table 2. The median values of the distributions of the average top thirteen elements from the world soil series; soil bulk density is 1.4 g/cm³.

Because of the exponential attenuation of radiation in soil, the volume sampled by the INS is semi-infinite; however, a signal arriving from a large distance would be asymptotically vanishing. Instead, an effective sampling volume is defined as a volume from which 90-, 95-, or 99- percent of the detected signal arrives. Clearly, this volume would depend on the carbon's distribution in soil. For example, in the extreme case where the entire carbon content is located on the soil's surface the sampled volume would be zero according to this definition. Thus, these sampling volumes are calculated for the conservative condition when carbon is homogenously distributed throughout the soil's volume. Figure 6 illustrates the results of the latter, derived from 2.5 10^6 1 cm³ voxels; Table 3 summarizes the calculated maximum depth, defined as the height of the vortex, the footprint, the volumes, and sampled mass using a soil bulk density of 1.4 g/cm³.

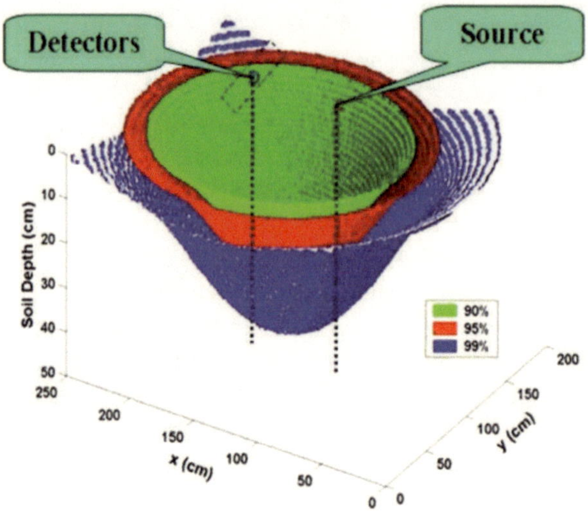

Fig. 6. Monte Carlo-calculated surfaces ascribing the volumes from which 90% (green), 95% (red), and 99% (blue) of the total detected carbon-signal emanated.

Carbon Response %	Depth (cm)	Footprint (m^2)	Volume (m^3)	Mass (kg)
90	25	2.4	0.23	326
95	31	3.3	0.37	522
99	44	7.1	0.79	1105

Table 3. Results from MCNP simulations derived from a test box of 200x250x50 cm^3 containing 2,500,000 one cm^3 voxels, and soil mass with bulk density 1.4 g/cm^3.

Figure 7 depicts the impact of changing the carbons depth profile from a homogeneous to a linear one, starting at 10% by weight at the surface and decreasing to 2% every 5 cm, it shifted the calculated height of the vortex from about 25 cm to 20 cm. The very large volumes obtained in these simulations are unique and counterintuitive to conventional wisdom in standard practice that use fixed very small volumes. These very different volumes relate to the system's calibration and are addressed in the next section.

4.4 System correlation versus calibration

Comparing the INS readings with those obtained by chemical analysis of soil samples is referred to as the calibration of the INS system. However, the term calibration in this context is misused; instead, the process should be referred to simply a correlation between them. To clarify these two concepts and provide consistency in their use, they are briefly reviewed here.

Calibration is the validation of specific measurement techniques and equipment against another measurement acquired, in way as similar as possible, from a second device with known- or assigned- correctness and termed the primary standard. Thus, standards and certified reference materials have designated value based on fundamental parameters or by direct comparison with a reference base. A primary standard usually is under the jurisdiction of a national standards body, such as the National Institute of Standards and Technology (NIST) in the United States. Therefore, at its simplest, two basic requirements constitute a calibration: (i) Using a device with known or assigned correctness in an absolute sense, or a primary- or secondary- device certified or traceable to the NIST's standard materials; and, (ii) measuring a secondary device in way as similar as possible to that of the primary device.

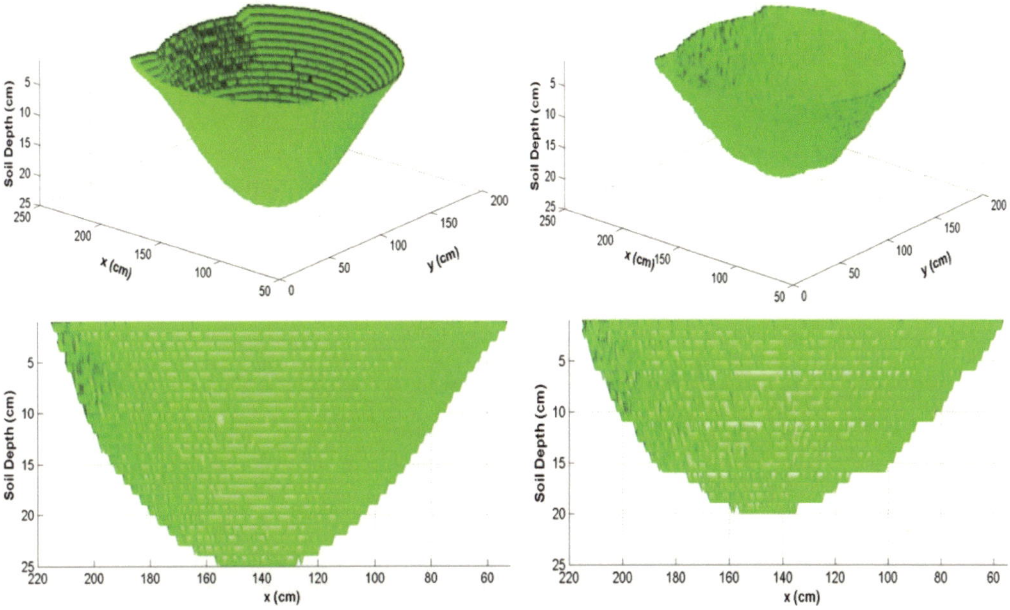

Fig. 7. Changes in the carbon depth profiles from (left) a homogeneous one to (right) a linear one affect the depth of the apex. The ruggedness of the surface on the right is due to lower statistics.

Certainly, soil analysis by the DC method does not qualify as a primary calibration device or as one with a calibrated reading. The standard materials used for initially calibrating a DC analyzer differ substantially from the various soil matrices being analyzed. In addition, as discussed above, the very large differences in the volumes sampled by each method, combined with the high heterogeneity in the field, represent different domains necessitating differing underlying assumptions. These conditions negate the possibility of calibrating any of the two methods discussed. Instead, two independent variables are compared using correlation, indicating the strength and direction of a linear relationship between these two random variables. In other words, correlation tests the hypothesis whether two independent variables are co-related, and measures the strength (departure) of two random variables from independence. In this broad sense, there are several correlation coefficients (Pyzdek, 2003) measuring the degree of correlation, adapted to the nature of the data; these are not discussed here.

In principle, each of the soil analytical methods could be calibrated in a strict sense against a known amount of carbon, albeit with a great deal of difficulty in preparing a proper standard material with a matrix identical to that of the soil. The slopes of true calibration lines, sensitivities of the methods, i.e., readings per gram carbon, would differ because of the diverse interactions involved and the varying number of carbon atoms in the dissimilar volumes sampled by each method. However, since all the methods attempt to assess the same entity, carbon, they would retain, for different but fixed volumes some proportionality. Therefore, it is logical that without any abrupt changes in the carbon profile, and within the instruments' linear domains, there will be a linear correlation between any two of the methods. In fact, under the ideal condition of no sampling errors, the slope of the correlation line would be the ratio of the sensitivities of each method.

We tested the proportionality concept by comparing INS readings against chemical analysis by dry combustion of soil samples drawn from an excavation of 40 cm by 40 cm and 40 cm deep. The soil was removed in layers 0-5, 5-10, 10-20, 20-30, and 30-40 cm thick, and prepared for analysis. Subsequently, three samples from each layer were pulverized further, and two aliquots from each sample analyzed for carbon. The INS reading was plotted against total surface carbon derived from each depth, Fig. 8 [Wielopolski et al., 2010]. The regression improves with increasing depth and beyond some depth the proportionality changes very little; there is little carbon in deeper layers.

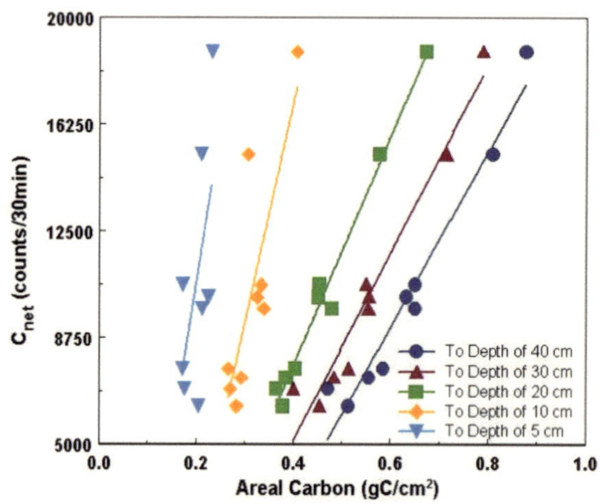

Fig. 8. Correlation of INS readings versus total surface carbon, soil plus roots, derived by chemical analysis of soil samples from various depths and summed-up.

4.5 Data analysis

Data analysis of experimental results is concerned with properly identifying the spectral peaks and with minimizing error propagation in their quantitative analysis. The latter focuses on possible spectral interferences.

4.5.1 Error propagation

The statistical nature of radioactive decay in nuclear counting was recognized soon after its discovery. Hence, in any sample containing a large number of radioactive atoms or excited nuclei, some average number, N, is counted per unit time fluctuating around that average. The statistics of nuclear counting follows a binomial distribution, which for a large number of counts, about $N > 12$, can be approximated by a normal distribution with a mean value, N, and standard deviation (SD) that equals the square-root of N (sqrt(N)) [Evans, 1955].

By extension, in nuclear spectroscopy, the gamma-ray events in the detector are represented by the number of counts falling into contiguous energy intervals (channels) surrounding a peak in a spectrum (Fig. 4). The total number of counts in a given energy-interval T_t following counting time of T minutes is due to the incident-signal counting rate, S_r times T, and the background-counting rate, B_r times T. Thus, $T_t = S_rT + B_rT$ are linear with time T, and, conversely, the net number of counts associated with an element (E) of interest, S_rT, is given by the difference $T_t - B_rT$. The INS's net counts are converted to conventional units of areal density (g E/m^2) by dividing the net signal by the system's sensitivity, s, defined as the number of counts acquired during a counting period T, S_rT, per gram element per unit area; k is a proportionality constant with matching units of g E/m^2. Thus $s = S_rT/k$, also is the slope of the regression line that correlates INS yield *versus* the soil's carbon concentration. The experimentally determined quantities B_r, S_r and s represent the key performance parameters of an INS system from which other parameters are derived. Using the general uncertainty estimator of a function f(x,y,z...) given, to a first approximation, by Eq. 2 [Bevington, 1969], it can be shown that the SD of S_rT, σ_S, equals sqrt($T_{tot} + B_rT$) or to sqrt(($S_r + 2B_r$)T). Thus, the signal is proportional to the counting time and the error is proportional to the square-root of the time. In Eq. 2, the partial derivatives of the function with regard to the parameters x_i that are being search are added in quadrature,

$$\sigma_f^2 = \sum_i [\sigma_i^2 (\frac{\partial f}{\partial x})_i^2] \tag{2}$$

Extending the counting time improves the signal, and the error increases correspondingly but at a slower pace [Wielopolski, 2011]. The application of statistics to nuclear counting data is mandatory to assess the precision with which measurements are made. It should be emphasized that only the uncertainty due to the randomness of the counting statistics are addressed here; any other significant sources of uncertainty must be added in quadrature to the overall estimate of the precision.

4.5.2 Methods for spectra analysis

There are several analytical approaches and commercial software for analyzing spectral peaks. The simplest one assumes a linear background under the peak shown in Fig. 9 as a straight line "c""d".

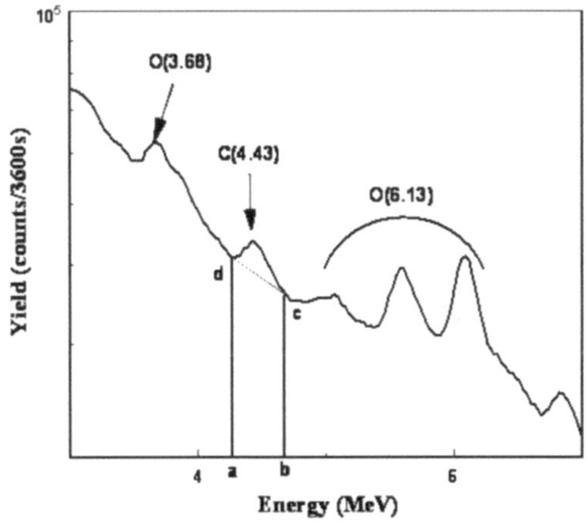

Fig. 9. Expanded INS spectrum in Fig. 4 shows oxygen- and carbon-peaks. The trapezoidal background under the carbon peak is bound between 'a" and "b".

Thus, the net number of counts in the peak is defined as the total counts minus the background delimited by the area enclosed by the trapezoid "abcd". This approach is valid provided that the peak is clearly defined and there are no overlapping peaks distorting the area with extraneous counts. The least-squares method offers a more advances analysis wherein the peak is fitted with one or more Gaussian functions, thus partly resolving the problem of overlapping peaks. However, with complete overlap, as might occur when an identical gamma-ray is produced by an interfering element, it is more difficult to resolve. However, since an excited element generally produces more than a single gamma-ray with fixed ratios among them, it is possible to resolve interfering peaks by fitting an entire spectrum instead of the peak alone; this is the library-least-squares method (LLS) [Arinc at al., 1976]. The measured spectra from pure elements, referred to as elemental standard library, are least-squares-fitted to an unknown measured spectrum revealing any discrepancies between the synthetically modeled spectrum and the measured one. The fundamental assertion in the LLS method is that a measured unknown spectrum is a linear superposition of standard reference libraries plus an error term. The multipliers of the standard libraries are found by minimizing the error term, or of the reduced χ^2 given by equation Eq. 3 [Gardner et al., 1975; Wielopolski, 1981; Wielopolski and Cohn].

$$\chi^2 = \sum_{i=1}^{n} \left[b_i - \sum_{j=1}^{m} x_j a_{ij} \right]^2 / \sigma_i^2 (n-m), \tag{3}$$

where:

b_i - counting rate in channel i for the composite spectrum;
a_{ij} - counting rate of pure element j in channel i per unit amount of component j;
x_j - amount of component j in the unknown;
m - number of components;
n - number of channels;
σ_i^2 - variance of the random error in channel i.

The LLS method requires an extra effort in deriving good elemental libraries with good statistics; in turn, this approach reduces the reported error for the analyzed peak intensities.

4.5.3 Peak interferences

During the INS and TNC processes, most elements emit a multiplicity of gamma-rays, some of which may overlap with those emitted from other elements. The gamma-rays emitted by a given element bear fixed relationships among themselves; thus, analyzing an undistorted peak enables the correction of a distorted one with interference. Alternatively, using the LLS method with a good set of elemental libraries, the interfering peaks can be assessed and corrected values obtained. For example, thermal neutron capture in a ^{28}Si yields 4.94-MeV gamma-rays that generate a single escape peak, C_{SEP}, at 4.94 − 0.511 = 4.43 MeV that overlaps the carbon peak. This peak is generated in the TNC spectrum that, depending on the neutron-generating pulsing regimen, must be adjusted by the ratio of the live times, LT_{INS}/LT_{TNC} (Mitra and Wielopolski, 2005). Thus, the contribution of the escape peak to the carbon peak in the INS spectrum is $C_{SEP} = Si_{(4.43)}\ LT_{INS}/LT_{TNC}$. A second process affecting the carbon peak is the decay of excited silicon (^{28}Si) to the ground state via several cascades feeding the 1.78-MeV level that, in itself, decays to the ground state. A fraction of these cascades originates at the 6.23-MeV level generating a 6.23 − 1.78 = 4.45 MeV gamma-rays (C_{CAS}) that overlap the carbon peak. The ratio between the 4.45 MeV and 1.78 MeV gamma-rays is constant, and was calculated theoretically as 0.0547 [Herman et al., 2007]. Furthermore, the 1.78-MeV Si peak itself must be corrected for $^{28}Si(n,p)^{28}Al$ reactions that, with a decay time of 2.25-min half-life, contribute to the 1.78-MeV line. Since all the interferences are associated with peaks acquired concurrently with the carbon peak, the corrections are made for each individually measured spectrum. Thus, $C_{CAS} = [Si_{(1.78)} - Al_{(1.78)}\ LT_{INS}/LT_{TNC}]0.0547$. Estimates suggest that the errors associated with these theoretically- and experimentally- determined correction factors are about 10%. Thus, the corrected number of counts, N_c, in the carbon peak with N_p is

$$N_c = N_p - C_{SEP} - C_{CAS} \tag{4}$$

Figure 10 shows carbon calibration in a sand pit using a synthetic soil of sand mixed with known different amounts of carbon before and after correcting for silicon interference. The intercept was reduced from 5029 to 348±327, which includes the 0 value.

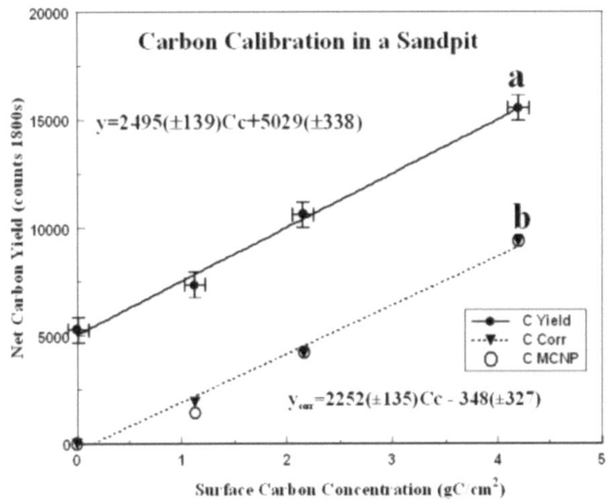

Fig. 10. Calibration of an INS system over a sand pit filled with synthetic soils; (a) Calibration before correction for Si interference, and, (b) after correcting using Eq. 4, for the silicone interference with the carbon peak. The circles indicate simulation of calibration using Monte Carlo calculations normalized at point b.

Differences may occur in a correction factor when operating the system in a scanning mode due to differences in delayed activation detected by the detectors when they operate in the stationary mode versus scanning when they are moved away from an activated area. Large changes in the soil's composition, particularly the presence of aluminium, may affect the correction [Wielopolski et al., 2008].

4.6 Mean field value (scanning)

The INS and TNC processes are very fast, ~10^{-9} s, and the INS's footprint is very large, about 2.5 m^2. Thus, in one millisecond, a very long time for nuclear processes to occur, the INS system towed at a speed of 10 km/h (or about 2800 mm/s) would move about 3 mm, a distance that is negligible in relation to the footprint; hence, the system can be perceived as stationary. Thus, continuous scanning of a large field at reasonably constant speed yields a single value that represents a physically averaged value for the field. The assessed mean from continuous scan of the entire field is analogous to a simple mean or a geo-statistically average of a set of discrete samples measured by conventional chemical methods. This scanning capability introduces significant changes in conventional wisdom in soil sampling and analysis. This consideration was tested by comparing INS stationary measurements with conventional analysis of core samples taken from the same sites, and then by comparing the calculated mean value with that determined by scanning a transect containing the sampled sites. There was a good agreement in organic soil; however, further clarification is required in a pasture [Wielopolski et al., 2011].

5. Results

Using an INS alpha prototype system, numerous field measurements were carried out across Montana, Arizona, Alabama, North Carolina, Maryland, New Hampshire, Pennsylvania, and Ohio. They were taken on federal-, state-, and privately owned lands; each required a different permit to operate a neutron generator, i.e., a radiation-producing device (RPG). The soil types tested included pure organic soils, histosols, various types of agricultural soils, abandoned surface mine fields, and forests.

5.1 Correlation between INS and chemical analysis

One study was carried out in the Blackwood Division of the Duke Forest near Durham, NC (35°58′41.430″ N, 79°5′39.087″ W). This is a complex site with woody and rocky components in the soil matrix; it included grassland, a pine forest, and a hardwood forest. The grassland, dominated by the C3 grass shrub, *Festuca arundinacea,* was mowed at least once annually for hay. The pine site was planted in 1983 with loblolly pine (*Pinus taeda* L.) in the overstory, and has been unmanaged since. The hardwood site is an unevenly aged, 80- to 100-yr-old oak (*Quercus* sp.)–hickory (*Carya* sp.) forest, also never managed [Wielopolski, et al., 2008, 2010]. Three INS static measurements were taken on each site; nine excavations, 40x40x40 cm^3 at the same INS locations gave soil samples for chemical analyses by dry combustion. The results from the three sites were combined to correlate the INS measurements with chemical assays; these were corrected for the solid fraction of the soil.

Similar results were combined from three sites in Ohio. The first site, designated pastureland, is located at the USDA's North Appalachian Experimental Watersheds (NAEW) in Coshocton County OH (40°16′19″N and 81°51′35″W). The soil is derived from a fine-loamy, mixed, mesic Typic Hapludultsoess-like silty deposit overlying a loamy textured material weathered from sandy shale and siltstone bedrock. Orchard grass (Dactylis glomerata L.) is the dominant species in the watershed under pasture, and the watershed's slope is about 10% [Kelley et al., 1975].

The second location sited at the Ohio Agricultural Research and Development Center (OARDC) Muck Crops Branch in Willard, OH, (41°0′37.80″N and 82°44′0.39″W). The soils are classified as the Linwood muck, a loamy mixed euic, mesic, Terric Halosaprist (Elder and Lal, 2008). This soil contains almost 80% organic matter; the main crops grown there are lettuce (*Lactuca sativa*), radish (*Raphanus sativus*), and other leafy species. The third location is an unmanaged natural forest adjacent to the cultivated organic soil, dominated by white oak (*Quercus alba L.*) and red oak (*Quercus rubra L.*), also in Willard, Ohio. The measured soil bulk densities varied from 0.37 to 1.7 g/cm^3 [Wielopolski et al., 2011]. Fig.11 illustrates the correlations for Duke Forest and the Ohio sites; the former also shows the 95% confidence limits.

5.2 Scanning

The capability of the INS system to non-destructively scan fields continuously and provide a mean value for the scanned field was highlighted previously and evaluated by comparing a transect scan with mathematical averaging of discrete points on the transect. Two scans performed in a wheat field, 100 m by 100 m, in Montana, and in a 30 m by 30 m corn field in Maryland demonstrated the feasibility for scanning arbitrarily large areas. Figure 12 shows the global positioning system (GPS) traces of the scans in the wheat- and corn-fields. The large wheat field allowed well-structured scans whereas random scanning was more suitable for the small corn field

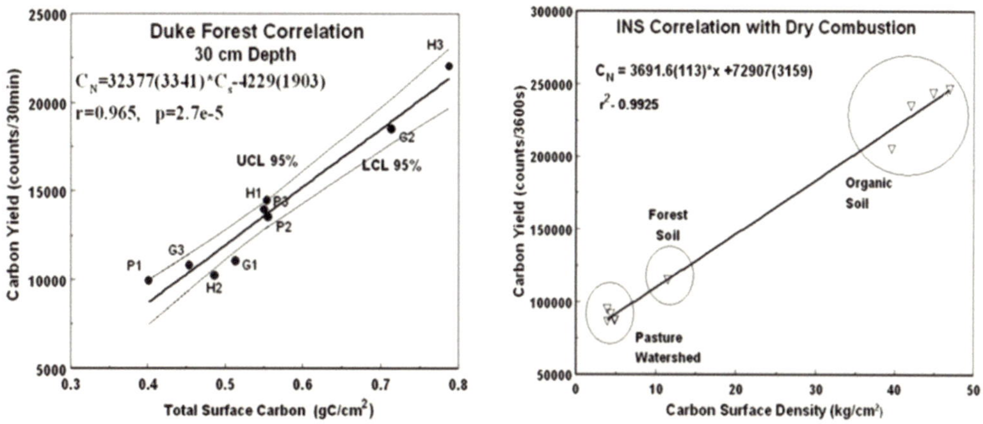

Fig. 11. Regression lines between INS yield and chemical analysis of soil samples from Duke Forest on the left, and from the Ohio sites on the right. The symbols G, P, and H, respectively, stand for grass, pine forest, and hardwood forest, respectively.

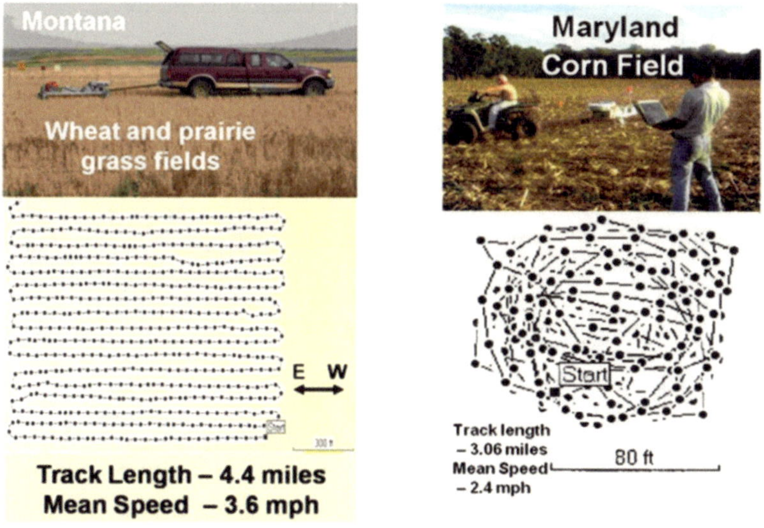

Fig. 12. GPS traces of the scans obtained from of a wheat- and a corn-field by towing the INS system over them. The figures show the total length of the scans and the mean speeds. In the corn field (right), the scan was monitored in real time on a slave computer to assure better coverage of the field.

5.3 Forest and mine fields

It is difficult to estimate the effect on the soils carbon stores of forest-management practices because of the long time-scale of the response and the heterogeneity of forest soils. With traditional sampling methods, changes in forest soils have been impossible to detect, even following clear-cutting; repeated sampling rarely detected changes of less than 20% [Johnson et al., 1992; Yanai et al. 2003]. Furthermore, the logistics of sampling in forests are uniquely complex not only in bringing the equipment to the site, but in dealing with the presence of boulders and fallen woody logs; accordingly, the sampling and analysis protocols must be reformulated. Testing the INS system in a forest environment [Wielopolski, Yanai et al., 2010] led us to the realization that the system's weight must be reduced before it was usable on regular basis. In comparison, conventional sampling of forest soil is especially troublesome due to the frequency of boulders; a better approach is to excavate quantitative pits, but this is a very time-consuming procedure and its scope limited. Figure 13 illustrates using the INS system in Bartlett forest, NH, and sampling there with a motorized coring machine. The difference in the footprints of both methods is apparent, making the use of the INS less error-prone.

Similarly, sampling abandoned minefields under restoration poses challenges because of the rocks therein. There are nearly 3.2 Mha of permitted mine lands in the United States, almost 75% of which are located in the Northeast Appalachian coal-mining region. Estimates of potential carbon sequestration in reclaimed mined land soils in the northeastern US are widely divergent: -0.02- [Amichev et al., 2004], 0.25- [Intergovernmental Panel on Climate Change, 1998], and 1.9-Mg C ha^{-1} yr^{-1} (Akala and Lal, 2000). Mine land soils typically are highly heterogeneous, even at very small scales; their drainage, texture, mineralogy, rock fragments, bulk density, and pH differ greatly, all of which influence the soils carbon dynamics and sequestration. The INS's unique capabilities will augment tremendously the monitoring of the restored mine lands. Figure 14 shows an example of minefield soil riddled with stony material; the value of using the INS instrument is contrasted with the effort of conventional soil sampling when it is possible, if at all.

Fig. 13. On the left an INS set up measures the forest soil. On the right, a mechanized coring tool is used to sample soil from the forest floor. Many cores were abandoned because of the belowground rocks.

Fig. 14. Comparison of sampling abandoned minefield restored with its original overburden using the INS system on the left, and an attempt using conventional coring on the right, many cores could not be sampled deeper than 10 cm.

5.4 Geological sequestration

We demonstrated the value of employing the INS to monitor belowground changes in carbon content caused by CO_2 leakage from a zero-emission research and technology (ZERT) facility that tests instrumentation for assurance-monitoring of geological sequestration. The explored hypothesis was as follows: An increase of CO_2 in the soil's atmosphere would slow and/or kill the growth of roots that would decrease the belowground level of carbon. In two consecutive tests over two years, a decrease in carbon content of 15% and 7% was registered [Wielopolski and Mitra, 2010; Wielopolski, 2011]. The correlation between the CO_2 leak and belowground carbon level is yet to be determined. All surface monitoring devices used for monitoring geological sequestration offer point-in time and in-space measurement that may prove insensitive to the expected very low leakages from the storing formations. Alternatively, using INS offers the possibility of monitoring a time-integrated signal that might be more sensitive. Moreover, the INS provides an area-integrated signal, invaluable where prior knowledge of the leakage site might be unknown.

6. Conclusions

Inelastic neutron scattering is a newly emerging modality for soil analysis that differs fundamentally from all other modalities, new and past. However, for better understanding of the INS's unique characteristics, since not all soil researchers may be familiar with the nuclear spectroscopy features, some basic principles involved with it were introduced. The INS is based on well-understood fundamental principles of nuclear physics and demonstrates very unique capabilities that extend beyond those of the current state-of-the-art in soil sampling paradigms while at the same time revolutionize the existing ones. For example, the INS's larger footprint than in conventional coring is apparent in Figs. 13 and 14. Using an advanced AP neutron- generator, it is possible to extend the measurements non-destructively to gain information on a depth profile of an element, or to monitor repeatedly changes in a region at a specified depth in soil. These data, in turn, enable the development of new models for assessing carbon stocks to a depth of 100 cm. The key outstanding features of INS comprise its non-destructive, i.e., remote, analysis of large samples of soil in static- and scanning-modes of operation and multi-elemental analysis in true sequential order, i.e., measurements taken at the same spot. A brief review of other modalities highlighted the uniqueness of the INS system. Because of the lateral variability in carbon distribution, the large footprint of the INS provides a better-averaged value, hence changing sampling requirements. This is particularly true when scanning large areas where a single scan provides a mean value of a field that otherwise would require tens or more sampling spots. Although not shown, economic advantages of the INS system can be demonstrated.

Using INS alpha prototype demonstrated the correlation between INS signal and that measured by the standard chemical method by dry combustion; the difference between correlation and calibration when comparing signals from different domains was emphasized. It is important to understand that when comparing vastly different modalities proportionality is important.

Thus, we demonstrated the feasibility of the INS system as a viable new modality for carbon analysis in soil. Its unique characteristics improved and expanded the scope of possibilities to observe in depth the belowground processes that otherwise are not visible to naked eye without disturbing the environment. The INS capabilities to monitor soil carbon may prove invaluable in promoting trading with carbon credits. Overall, it is an important addition to the arsenal of instrumentation improving our understanding and monitoring of carbon and other elements in soil.

7. References

Akala, V.A., and R. Lal. 2000. Potential of mine land reclamation for soil organic carbon sequestration in Ohio. Land Degrad. Dev. 11: 289–297.

Alfassi Z.B. and Chung C. (1995). *Prompt gamma activation analysis.* CRC Press, Boca Raton, London, Tokyo. ISBN: 0-8493-5149-9.

Allison L.E., (1965) Organic Carbon. In: *Methods of soil analysis*, C.A. Black C.A. D.d. Evans, J.L. White, L.E. Ensminger, and P.E. Clark (eds.), pp1367-1378. American Society of Agronomy, Madison Wisconsin, USA.

Ames, W.J. and Gaither, E.W. 1914. Determination of carbon in soils and soil extracts. J. Ind. Eng. Chem. 6:561.

Amichev, B., J.A. Burger, and J.A. Rodrigue. (2004). Carbon sequestration by forests and soils on mind land in the Midwestern and Appalachian coalfields: Preliminary results. p. 20–46. In Proc. of the 2004 Natl. Meeting of the American Society of Mining and Reclamation and the 25th West Virginia Surface Mine Drainage Task Force, Lexington, KY. 18–24 Apr. 2004. ASMR, Lexington.

Arinc F., Gardner R.P., Wielopolski L. and Stiles, A.R., (1976). Application of the last-squares method to the analysis of XRF spectral intensities from atmospheric particulates collected on filters, Advances in X-Ray Analysis, 19:367-380.

Awiti A.O., M.G. Walsh, K.D. Shepherd, and J. Kinyamario. (2008). Soil condition classification using infrared spectroscopy: A proposition for assessment of soil condition along a tropicalforest-cropland chronosequence. Geoderma, 143:73-84.

Balzer H., M. Hoehne, V. Sturm, and R. Noll. (2005). Online coating thickness measurement and depth profiling of zinc coated sheet steel by laser-induced breakdown spectroscopy. Spectrochemica Acta Part B, 60:1172-1178.

Bevington, F.P.; Robinson, D.K. (1969). *Data Reduction and Error Analysis for the Physical Sciences*; McGraw-Hill Book Company: New York, NY, USA.

Breismeister, J.F. (Ed.), (1993). MCNP-A general purpose Monte Carlo n-particle transport code version 4A. Los Alamos National Laboratory, NM, Report; LA-12625-M.

Buckley S.R. (2006). LIBS comes on strong. Laser Focus World. 44:95-98.

Cameron, F.K., and J.F. Breazeale (1904). The organic matter in soils and sub soils. J. Am. Chem. Soc. 26:29-45.

Chatterjee A., R. Lal, L. Wielopolski, M.Z. Martin, and M.H. Ebinger, (2009). Evaluation of Different Soil Carbon Determination Methods, Crit. Rev. Plant Sci., 28(3):164-178.

Chichester D.L. and Simpson J.D. (2004). Compact accelerator neutron generators. American Institute of Physics, Technology DECEMBER 2003/JANUARY 2004, p22-25.

Christy CD, P. Drummond, D.A. Laird (2003). An on-the-go spectral reflectance sensor for soil. ASAE Paper No 031044. 2003 ASAE Annual Meeting, Las Vegas, NV.

Csikai J. (1987). *Handbook of fast neutron generators.* CRC Press Inc., Boca Raton, Florida. ISBN: 0-8493-2967-1.

Cremers, D.A., M.H. Ebinger, D.D. Breshears, P.J. Unkefer, S.A. Kammerdiener, M.J. Ferris, K.M. Catlett, and J.R. Brown. (2001). Measuring total soil carbon with laser-induced breakdown spectroscopy (LIBS). J. Environ. Qual. 30:2202–2206.

De Lucia Jr. F.C., J.L. Gottfried, C.A. Munson, and A.W. Muiziolek. (2009). Current status of standoff LIBS security applications at the United States Army Research Laboratory. Spectroscopy. 24:32-38.

Denman, K.L., G. Brasseur, A. Chidthaisong, P. Ciais, P.M. Cox, R.E. Dickinson, D. Hauglustaine, C. Heinze, E. Holland, D. Jacob, U. Lohmann, S Ramachandran, P.L. da Silva Dias, S.C. Wofsy and X. Zhang (2007). Couplings Between Changes in the Climate System and Biogeochemistry. In: *Climate Change 2007: The Physical Science Basis. Contribution of Working Group I to the Fourth Assessment Report of the Intergovernmental Panel on Climate Change* [Solomon, S., D. Qin, M. Manning, Z. Chen, M. Marquis, K.B. Averyt, M.Tignor and H.L. Miller (eds.)]. Cambridge University Press, Cambridge, United Kingdom and New York, NY, USA.

Ebinger MH, M.L. Norfleet, D.D. Breshears, D.A. Cremers, M.J. Ferris, P.J. Unkefer, M.S. Lamb, K.L. Goddard, C.W. Meyer, (2003). Extending the applicability of laser-induced breakdown spectroscopy for total soil carbon measurement. Soil Sci Soc Am J 67:1616–1619.

Elder, J.W., Lal, R., 2008. Tillage effects on gaseous emissions from an intensively farmed organic soil in Northern Central Ohio. Soil and Tillage Research 98, 45–55.

Ellert B.H, H.H. Janzen, B.G, McConkey, (2001). *Measuring and comparing soil carbon storage.* In: Lal R, et al. (ed) Soil processes and the carbon cycle. CRC Press, Boca Raton, FL, pp 131–146

Essington M.E., G.V. Melnichenko, M.A. Stewart, and R.A. Hull. 2009. Soil metals analysis using Laser-induced breakdown spectroscopy (LIBS). Soil Sci. Soc. Am. J., 73:1469:1476.

Evans R.D. (1955). *The atomic nucleus.* McGraw-Hill Book Company, New York, Toronto, London.

Frank, V., and Tolgyessy, J. (1993). The chemistry of soil, In: Tolgyeddy (Ed.) *Chemistry and biology of water, air and soil environmental aspects.* Elsevier, ISBN 13: 978-0-444-98798-3

Gardner R.P., Wielopolski L. and Verghese K. (1978). Mathematical techniques for quantitative elemental analysis by dispersive X-ray fluorescence. J. Radio. Chemistry, 43:611-643.

Gehl R.J. and C.W. Rice, (2007). Emerging technologies for *in situ* measurement of soil carbon. Climatic Change (2007) 80:43–54

Hahn D.W. (2009). Laser-induced breakdown spectroscopy for analysis of aerosol particles: the path toward quantitative analysis. Spectroscopy, 24:26-33.

Herman, M., R. Capote, B.V. Carlson, P. Obložinský, M. Sin, A. Trkov, H. Wienko, and V. Zerkin. 2007. EMPIRE: Nuclear reaction model code system for data evaluation. Nucl. Data Sheets 108:2655–2715.

International Panel on Climate Change, IPCC, (2007). Summary for Policymakers. In: *Climate Change 2007: The Physical Science Basis. Contribution of Working Group I to the Fourth Assessment Report of the Intergovernmental Panel on Climate Change* [Solomon, S., D. Qin, M. Manning, Z. Chen, M. Marquis, K.B. Averyt, M.Tignor and H.L. Miller (eds.)]. Cambridge University Press, Cambridge, United Kingdom and New York, NY, USA; http://www.esrl.noaa.gov/gmd/ccgg/trends/ (accessed on 9 March 2011).

Intergovernmental Panel on Climate Change. (1998). Special report on land-use, land-use change, and forestry. Chapter 4. Additional human-induced activities. Available at http://www.grida.no/climate/ipcc/land_use/ (verified 5 Oct., 2008). UNEP/GRID-Arendal, Arendal, Norway.

Johnson, C. E., C. T. Driscoll, T. J. Fahey, T. G. Siccama and J. W. Hughes. (1995). Carbon dynamics following clear-cutting of a northern hardwood forest ecosystem. p. 463-488. In J. M. Kelly and W. W. McFee (eds.). *Carbon: Forms and Functions in Forest Soils.* American Soc. of Agronomy, Madison, WI.

Johnston CA., P. Groffman, D.D. Breshears, Z.C. Cardon, W. Currie, W. Emanuel, J. Gaudinski, R.B. Jackson, K. Lajtha, K. Nadellhoffer, D. Nelson Jr., W.M. Post, G. Tetallack, and L. Wielopolski (2004). Carbon cycling in soil. Front. Ecol. Environ., 2(10):522-528.

Kelley, G.E., Edwards, W.E., Harrold, L. (1975). Soils of the North Appalachian Experimental Watersheds. U.S. Government. Printing Office, Washington, DC. (Miscellaneous, publication no. 1296).

Lal R., T.M. Sobecki, T. Iivari, and J.M. Kimble (2004). *Egradation in the United States: extend, severity, and trends,* Lewis Publishers, Boca Raton, London, New York, and Washington, D.C., ISCN: 1-56670-534-7.

Lal R., N. Uphoff, B.A. Stewart, and D.O. Hansen (Eds.) (2005). *Climate change and global food security.* Taylor & Francis Boca Raton, London, New York, Singapore, ISBN: 0-8247-2536-0.

Lamarsh J.R., (1975). *Introduction to Nuclear Engineering.* Addison-Wesley Publishing Company, Menlo Park, London, Amsterdam, Don Mills, Sydney. ISBN# 0-201-04160-x.

Lux I. and Koblingrt L. (1990). *Monte Carlo particle transport Methods: neutron and photon calculations.* CRC Press, Boca Raton, Ann Arbor, Boston. ISBN: 0-8493-6074-9.

Martin M.Z., N. Labbe, N. Andre, S.D. Wullschleger, R.D. Harris, and M.H. Ebinger. 2010. Novel multivariate analysis for soil carbon measurements using laser-induced breakdown spectroscopy Soil Sci. Soc. Am. J. 74:87-93.

Martin M.Z., S.D. Wullschleger, C.T. Garten Jr., and A.V. Palumbo, (2003). Laser-induced breakdown spectroscopy for the environmental determination of total carbon and nitrogen in soils. Applied Optics 42:2072-2077.

Merdes D.W., J.M. Suhan, J.M. Keay, D.M. Hadka, and W.R. Bradley. 2007. The investigation of laser-induced breakdown spectroscopy for detection of biological contaminants on surfaces. Spectroscopy, 28:38.

Mitra, S., and L. Wielopolski. (2005). Optimizing the gate-pulse width for fast neutron induced gamma-ray Spectroscopy. Proc. SPIE 5923:592308, doi:10.1117/12.614569.

Nargolwalla S.S. and Przybylowicz E.P. (1973). *Activation analysis with neutron generators.* John Wiley & Sons, New York, London, Sydney, Toronto. ISBN: 0=471-63031-4.

Pelowitz D.B. (Ed.) (2005). MCNP^{TX} user's manual version 2.5.0.Report LA-CP 05-0389.

Pyzdek, T. 2003. Quality engineering handbook. Marcel Dekker, New York.

Rather, J.B. 1917. *An accurate loss on ignition method for determination of organic matter in soils.* Arkanasas Agric. Exp. Stn. Bull 140.

Reeves III J.B. 2010. Near- versus mid-infrared diffuse reflectance spectroscopy for soil analysis emphasizing carbon and laboratory versus on-site analysis: Where are we and what needs to be done. Geoderma, 158:3-14.

Rogers, R.E., and W.R. Rogers (1848). New method of determining the carbon in native and artificial graphite, etc. Am. J. Sci. 2:352.

Schollenberger, C.J, (1927). A rapid approximate method for determining soil organic matter. Soil Sci. 24: 65-68.

Schumacher B.A, (2002). Methods for determination of total organic carbon (TOC) in soil and sediments. United States Environmental Protection Agency (USEPA), Environmental Sciences Divisions National Exposure Research Laboratory, Report# NCEA-C-1282, EMASC-001, April 2002.

Shepherd K.D and M.G. Walsh. 2007. Infrared spectroscopy - enabling an evidence-based diagnostic surveillance approach to agriculture and environmental management in developing countries. J. Infrared Spectrosc. 15:1-19.

Solomon, S., D. Qin, M. Manning, R.B. Alley, T. Berntsen, N.L. Bindoff, Z. Chen, A. Chidthaisong, J.M. Gregory, G.C. Hegerl, M. Heimann, B. Hewitson, B.J. Hoskins, F. Joos, J. Jouzel, V. Kattsov, U. Lohmann, T. Matsuno, M. Molina, N. Nicholls, J. Overpeck, G. Raga, V. Ramaswamy, J. Ren, M. Rusticucci, R. Somerville, T.F. Stocker, P. Whetton, R.A. Wood and D. Wratt, (2007). Technical Summary. In: *Climate Change 2007: The Physical Science Basis. Contribution of Working Group I to the Fourth Assessment Report of the Intergovernmental Panel on Climate Change* [Solomon, S., D. Qin, M. Manning, Z. Chen, M. Marquis, K.B. Averyt, M. Tignor and H.L. Miller (eds.)]. Cambridge University Press, Cambridge, United Kingdom and New York, NY, USA, p-19.

Sudduth KA, Hummel JW (1996) Geographic operating range evaluation of a NIR soil sensor. Trans ASAE. 39:1599–1604

Tabatabai, M.A., and J.M. Bremner (1970). Use of the Leco automatic 70-second carbon analyzer for total carbon analysis in soils. Soil Sci. Soc. Am. Proc. 34:608-610.

Tabatabai, M.A., and J.M. Bremner (1991). Automated instruments for determination of total carbon, nitrogen, and sulfur in soils by combustion techniques. In: *Soil Analysis,* p. 261-286. Smith, K.A. Ed., Marcel Dekker, New York.

Tan K.H., (2005). *Soil sampling, preparation, and analysis,* pp339-364.Taylor and Francis, Boca Raton, London, New York, Singapore. ISBN: 0-8493-3499-3.

Tans, P. *Trends in Atmospheric Carbon Dioxide – Mauna Loa;* US Department of Commerce/ National Oceanic and Atmospheric Administration: Boulder, CO, USA. Available online: http://www.esrl.noaa.gov/gmd/ccgg/trends/ (accessed on 9 March 2011).

Tiessen H., J.O. Moir (1993). Total and organic carbon. In: *Soil sampling and methods of analysis.* M.R. Carter (ed.), pp187-199. Lewis Publishers, Boca Raton, Ann Arbor, London, Tokyo. ISBN: 0-87371-861-5.

Warrington, R., and W.A. Peake (1880). On the determination of carbon in soils. J. Chem. Soc. (London) 37:617-625.

Walkley A. (1935). An examination of methods for determining organic carbon and nitrogen in soils. J Agr Sci., 25:598–609.

Wielopolski L. (2011). Geological Carbon Sequestration: A New Approach for Near-Surface Assurance Monitoring. Int. J. Environ. Res. Public Health, 8:818-829.

Wielopolski L., A. Chatterjee, S. Mitra, and R. Lal. (2011). *In Situ* Determination of Soil Carbon Pool by Inelastic Neutron Scattering: Comparison with Dry Combustion. Geoderma, 160:394-399.

Wielopolski L. and S. Mitra. 2010. Near-surface soil carbon detection for monitoring CO2 seepage from a geological reservoir. Environ Earth Sci. 60:307–312

Wielopolski L., K. Johnsen, and Y. Zhang. 2010. Comparison of soil analysis methods based on samples withdrawn from different volumes: Correlations versus Calibrations. Soil Sci. Soc. Am. J. 74:812-819.

Wielopolski L. R.D. Yanai, C.R. Levine, S. Mitra, and M.A. Vadeboncoeur. (2010). Non-destructive Carbon Analysis of Forest Soils Using Neutron-induced Gamma-Ray Spectroscopy. Forest Ecol. Manag., 260:1132-1137.

Wielopolski L., G. Hendrey, K. Johnsen, S. Mitra, S.A. Prior, H.H. Rogers, H.A. Torbert. (2008). Non-destructive system for analyzing carbon in the soil. Soil Sci. Soc. Am. J. 72(5):1269-1277, 2008.

Wielopolski, L. (2006). Inelastic neutron scattering for assessing soil carbon. In: R *Encyclopedia of Soil Science* (Ed), Second Edition, Taylor and Francis, New York, 1:863-866.

Wielopolski L. and Cohn S.H. (1984). Application of the library least-squares analysis to a whole body counter spectra derived from an array of detectors. Med. Phys., 11:528-533.

Wielopolski L. (1981). Application of Si(Li) detector response function for the analysis of a tungsten library standard, Nuclear Instr. and Methods, 190:177-180.

Yanai, R.D., W.S. Currie, and C.L. Goodale. 2003. Soil carbon dynamics after forest harvest: an ecosystem paradigm reconsidered. Ecosystems 6:197-212.

22

The Role of Plant Breeding and Biotechnology in Meeting the Challenge of Global Warming

Nigel G. Halford
Rothamsted Research
United Kingdom

1. Introduction

The availability of water is already a major limitation for plant production in many parts of the world and shortages of water are recognised as major threats to food security. Drought is often, although not always, accompanied by high temperatures and although these two stresses provoke different responses in plants the dividing line is often blurred as one stress exacerbates the effects of the other. While both drought and heat stress are already problems, the range over which they impact seriously on crop yields and the frequency with which they do it are both predicted to increase as a result of global warming. Plants will also have to cope with a steeply rising atmospheric CO_2 concentration.

If the predictions of climate change are correct, global warming will cause changes in temperature at a rate unmatched by any temperature change over the last 50 million years. For example, the temperature changes that happened between ice ages and warm interglacial periods as a result of the glacial cycles of the past million years were of 4 to 7 °C (Solomon et al., 2007). Although these changes in temperature were large, they occurred relatively gradually, with the global warming at the end of each ice age taking approximately 5,000 years (Solomon et al., 2007). The current rate of global climate change is much more rapid. The upper end of the range predicted by Global Climate Models in the International Panel for Climate Change (IPCC) 4th Assessment Report is a 5°C increase in global mean temperature by the end of the century. As well as the mean increase in temperature, there is predicted to be an increase in the frequency and severity of extreme weather events.

The potential consequences of such events for food production can be seen in the effects of the severe Australian drought of the last decade, which reached crisis point in 2007-2008, and the Russian drought of 2010. Both Australia and Russia are major wheat exporters in normal years, but in 2007-2008 exports of wheat from Australia fell to 1.7 million tonnes, compared with 7.5 million tonnes in 2005-2006, while in July 2010 the Russian government banned wheat exports altogether to protect domestic supply. The price of wheat grain on the London International Financial Futures & Options Exchange (LIFFE) rose in 2008 to 198 pounds (GBP) per tonne. It fell back to below 100 GBP per tonne in 2009 but rose to a new record of 200 GBP per tonne in December 2010. At the time of writing it is at 198 GBP per tonne for July 2011.

In both the Australian and Russian droughts, lack of rain was accompanied by high temperature and both stresses would likely have affected grain yield. Drought stress can be devastating at any time during wheat development, while increased temperature shortens the growing period, reducing yield. Heat stress at flowering is especially damaging, resulting in much lower grain number and substantial yield losses. During the hot summer in northern Europe in 2003, for example, with maximum temperatures up to 38 °C in the UK and 40 °C in France, wheat production fell by about 20 %.

There is another angle to the issue of global warming for agriculture in that agriculture is a major contributor to greenhouse gas emissions and will be expected to contribute to efforts to cut emissions in order to avoid the worst case scenario predictions of global temperature increases. In the UK, for example, farming is responsible for about 7.4 % of total greenhouse gas emissions and the Climate Change Act (2008) commits the country to an 80 % reduction in emissions by 2050 across all sectors of the economy. Much of the greenhouse gas emissions from agriculture come from the methane that is produced from microbial breakdown of animal waste. The technology to capture this methane through anaerobic fermentation is being developed and there is an embryonic industry growing to provide it to farmers. It is not yet cost-effective for most farms but cost may not be the deciding factor if governments interfere in the market as they have done with other biofuels. As far as crops are concerned, the major 'carbon footprint' comes from the use of fossil fuels to produce fertilisers, herbicides and pesticides. Intensive agriculture is and will be required in order to meet the demand for food and fuel, but its dependency on fossil fuels is not sustainable and will have to be addressed.

Concerns about climate change are also driving the use of crops to produce biofuels. Peak oil, the point when the maximum rate of global petroleum extraction is reached, after which the rate of production enters terminal decline, may be imminent, but bioethanol from sugar and starch crops, biodiesel from oil crops and biomass for electricity production would still struggle to compete with fossil fuels in open competition. However, governments in many countries have intervened in the market to promote the use of biofuels. In Europe, for example, the Directive on the Promotion of the Use of Biofuels and Other Renewable Fuels for Transport, 2003/30/EC (also known as the Biofuels Directive), stipulated that measures had to be taken across the EU to replace 5.75 % of all transport fossil fuels with biofuels by 2010. The Renewable Energy Directive/Fuel Quality Directive of 2009 then imposed targets of 20 % of all energy and 10 % of transport fuel to come from renewable sources by 2020. These directives effectively created a protected market for biofuels in Europe.

The problems brought on by or in response to climate change add to a long list of issues facing agriculture. World population passed six billion in 1999. In 2011 it will pass seven billion and it is predicted to peak at nine billion around the middle of this century before starting to decline slowly. In addition, greater prosperity in highly populated, economically rapidly emerging countries such as China and India is leading, understandably, to a demand for a better diet, in particular a demand for more meat. There are also increasing problems of fresh water supply, competition for land use, the need for wildlife conservation, soil erosion, salination and pollution. Arguably, the era of crop surplusses and cheap food is over and we have entered one where food security is at the top of political and scientific agendas.

Plant breeders face the challenge of developing new crop varieties that will meet the increasing demand for food and biofuels in a rapidly changing and unpredictable environment. New varieties will be required with a set of physiological traits suited for the conditions that are likely to prevail ten to twenty years ahead. Here, modelling will play an important part.

This chapter reviews the role that plant breeding and agriculture will play in meeting the exceptional challenges of climate change, and the progress that has been made in developing the genetic tools that plant breeders will require. It describes the role of modelling and systems biology, current knowledge on plant responses to drought, heat and high CO_2, the often ignored issue of the effect of stress on crop quality and food safety, and the use of established and novel crops for biofuel production.

2. The role of modelling

The fact that plant breeders will have to develop new varieties for climatic conditions that do not currently prevail makes crop simulation modelling particularly important (reviewed by Semenov and Halford, 2009). Crop simulation modelling, sometimes referred to as an application of systems biology, has been used increasingly as a guide to experimentation. The aim is to enable a broad understanding of complex processes that would not be possible by studying single or small groups of genes responding, for example, to one particular environmental stimulus. Mathematical equations are developed from the analysis of large datasets from transcriptomic, proteomic and metabolomic analyses of plants subjected to a range of environmental conditions. Experiments are then carried out to test hypotheses derived from the model and the results of these experiments are used to refine the model. In the case of climate change, crop simulation modelling is used in conjunction with modelling of future climates.

There are several examples of the use of crop simulation models in assessing the impact of climate change. Carbone and co-workers (Carbone et al., 2003), for example, used modelling to predict the response of soybean and sorghum to varying climate change scenarios in the south-eastern USA. The study predicted yield decreases of up to 69 % in soybean and 51 % in sorghum in response to higher temperatures and reduced rainfall, although these decreases could be mitigated considerably by changing sowing date and variety, and if increased atmospheric CO_2 concentrations were considered.

Ewert and co-workers (2002) used crop simulation models to predict the effects of elevated CO_2 concentration on wheat yield, using data from free-air CO_2 enrichment (FACE) in Arizona, USA, and open-top chamber experiments in Germany. The authors concluded that crop simulation models could be used to predict wheat growth and yield for different CO_2 and drought treatments in a field environment, but that there was still uncertainty about the combined effects of CO_2 concentration and drought that required further testing and enhancement of the model.

Jamieson and co-workers (2000) also tested three crop simulation models against data from FACE wheat experiments in Arizona, but in this case included results of experiments in which the amount of applied nitrogen as well as atmospheric CO_2 concentration was varied. The models all predicted yield trends in terms of green area index, biomass accumulation and yield. The study showed that changes in CO_2 concentration affected light use efficiency, whereas nitrogen application affected the green area index.

Oleson and co-workers (2007) predicted some positive consequences of climate change for European agriculture when the impact of increased CO_2 concentration was included, but with considerable uncertainty arising from emission scenarios, the climate and crop simulation models that were used, and local soil and climatic conditions. The study predicted that the area where temperature was sufficient for maize cultivation in Europe would expand 30–50 % by the end of the century, and that there would be increases of 35-54

% in net primary productivity in northern Europe as a result of a longer growing season and higher CO_2 concentrations. It showed large increases in yield of winter wheat for northern Europe, but smaller increases or even decreases in southern Europe.

Richter and Semenov (2005) used the *Sirius* crop simulation model to evaluate the impact of climate change on drought indicators and yield of winter wheat in the United Kingdom. Climate scenarios were constructed using a stochastic weather generator for the 2020s and 2050s and compared with a 1960–1990 baseline. The study predicted that soil moisture deficit and potential yield loss due to drought would increase in the future, especially on shallow soils. However, this would be offset by a CO_2-related increase in radiation use efficiency, so that average wheat yields would be likely to increase by 15–23 % by the 2050s.

Semenov et al. (2009) used *Sirius* to predict the effects of temperature and drought on UK wheat yield. Some researchers had suggested that the impact of drought on UK wheat yield would increase with climate change (Foulkes et al., 2007; Witcombe et al., 2008) and emphasised the importance of breeding for drought-tolerant varieties. The potential effects of an increase in temperature had been considered to be of secondary importance. However, Mitchell and co-workers (1993) had reported that a temperature of 27 °C or higher half-way through anthesis resulted in a high number of sterile grains and potentially high yield losses and this was followed by a series of studies predicting severe effects of temperature increases on wheat yield (Wheeler et al., 1996; 2000; Ferris et al., 1998).

Climate scenarios constructed by using the LARS-WG stochastic weather generator (Semenov, 2007) predict warmer, drier summers for the UK by 2050, but wetter winters, compared with the baseline of 1960-90. The predicted increases in maximum temperature for the 2050 high emissions scenario are between 2 and 4 °C and the probability of maximum temperature exceeding 27 °C around flowering would be affected significantly by such large increases in temperature mean. However, the crop simulation model predicts that this would be offset somewhat because of wheat phenology; that is the fact that wheat development is driven by thermal time (in other words the rate of wheat growth and development is temperature-dependent), so flowering would occur earlier in the season, before the summer temperature peak. This would also mitigate the effect of summer drought because the crop would avoid the most severe summer drought stress by maturing earlier. Indeed, the model actually predicted that the impact of drought stress on wheat in the UK would decrease with climate change, despite the drier and warmer summers, and that an increase in the frequency of heat stress around flowering represented a greater risk for sustainable wheat production (Semenov et al., 2009; Semenov and Halford, 2009). Challinor et al. (2010), on the other hand, used crop simulations to show wheat crop failure rates increasing under climate change in Northeast China due to increasing extremes of both heat and water stress. The predictions in all of these studies are based on a number of assumptions and in my view it would be premature, for example, to discount the potential yield losses from drought for wheat in the UK in the mid to late part of the century. However, breeding for heat stress tolerance is clearly an important target.

In conclusion, not all of the predicted effects of climate change on yield are negative. Effects vary from one location to another and are also variety-dependent. However, the predicted yield losses for some crops in some areas are spectacular, demonstrating clearly that the impact of climate change on agriculture needs to be addressed urgently. The other main conclusion is that crop simulation modelling can sometimes predict surprising and counter-intuitive effects on crops, and identify traits that had been considered to be of secondary importance but which should be given top priority.

3. Drought and heat tolerance

Plants adopt several strategies to avoid the effects of drought. For example, if drought is most likely to occur in late summer (typical of the UK and northern Europe) they may avoid it by growing, flowering and setting seed before this time. Other 'avoidance' strategies include the development of deeper and more extensive root systems, allowing the plant to obtain more water from the soil to survive dry periods. Plants have also evolved responses that enable them to survive even if they do become short of water. These are referred to as tolerance traits and they differ from species to species and between different varieties, developmental stages, organs and tissue types. The plant hormone, abscisic acid (ABA), plays a key role, initiating a network of signalling pathways involving multiple protein kinases (enzymes that attach a phosphate group to another protein, affecting its activity) and transcription factors (proteins that regulate the expression of genes).

Transcription factors involved in drought stress responses include ABA response element binding proteins (AREBPs), members of the zinc finger homeodomain (ZFHD)-1, myeloblastosis (MYB) and myelocytomatosis (MYC) families, dehydration-responsive element binding protein (DREB)-1 and -2, and the NAC family (reviewed by Semenov and Halford, 2009). Over-expression of another transcription factor, plant nuclear factor-Y (NF-Y), has been shown to improve drought tolerance in maize.

Some of the signalling pathways that respond to drought are also activated when heat stress is applied. In some cases this may be misleading because elevated temperatures will cause water stress unless ambient humidity is adjusted to prevent it, and this sometimes obscures the fact that heat stress presents a plant with its own specific problems. As described in the previous section, high temperature causes wheat and other cereals to develop and mature more quickly; it also brings about an increase in respiration and an inhibition of photosynthesis. This is caused by a reduction in the activity of ribulose 1,5-bisphosphate carboxylase/oxygenase (Rubisco) and the efficiency of photosystem II. The reduction in Rubisco activity occurs because the enzyme responsible for maintaining its activity, Rubisco activase, is unstable at even moderately high temperatures. This makes the genetic manipulation of Rubisco activase to improve its stability at high temperatures a potentially important target.

Transcription factors that are specifically associated with heat stress include heat shock factors (HSFs) (reviewed by Semenov and Halford, 2009) and over-expression experiments with HSFs have resulted in increased thermo-tolerance in transgenic plants. Heat stress can cause RNA, proteins and other molecules to fold incorrectly, affecting their assembly, translocation, turnover and activity. So-called molecular ‚chaperones', such as heat-shock proteins (HSPs), that keep proteins and RNA in their correct conformation, are expressed to mitigate this problem and these have therefore also attracted much attention.

It is not clear yet whether the manipulation of any of these genes would provide a consistent improvement in drought and/or heat tolerance under field as opposed to laboratory conditions. At present there are no transgenic crop varieties being marketed on the basis of improved drought or heat tolerance, but all of the major plant biotechnology companies have such varieties in development.

4. Plant responses to high CO_2

Atmospheric CO_2 concentration 500 million years ago was approximately twenty times what it is today. Its long-term decline reflects the loss of carbon from the biosphere as it is incorporated into rocks occurring at a faster rate than the release of carbon into the atmosphere through volcanic activity. In the very long term this loss of carbon from the biosphere is a threat to life on Earth because CO_2 concentration will eventually fall below the point where plants can photosynthesise efficiently. There have been blips in this decline in the past, probably as a result of periods of unusual volcanic activity, such as during the so-called 'mid-Cretaceous superplume'. Currently we are on the up-curve of another blip, this time resulting from human activity, and it is the effect of this blip over the coming decades and centuries that is the more immediate threat. The atmospheric CO_2 concentration has risen from a pre-industrial level of 270 parts per million (ppm) to 390 ppm and is rising at 1-2 ppm per year, taking it to levels not seen for 20 million years.

How crop plants respond to rapidly rising CO_2 levels will be key to food security over the coming decades; in particular, will elevated CO_2 levels have beneficial effects on plant growth and crop yield that will offset the negative effects of increased temperature and drought stress? A substantial amount of research has been carried out over the last twenty years or so on the long-term acclimation of crop plants to elevated CO_2 using free-air CO_2 enrichment (FACE). In these experiments, CO_2-enriched air is passed over the plants in an experimental plot from a system of pipes, with the flow controlled in response to the data fed back from sensors in the plot. The reader is referred to the catalogue of publications arising from these experiments from Steve Long, Don Ort, Andrew Leakey, Elizabeth Ainsworth and colleagues at the University of Illinois, and to a detailed review by Leakey et al (2009), which lists the following major conclusions: Elevated CO_2 stimulates photosynthetic carbon gain and net primary production, but Rubisco activity is down-regulated; nitrogen use efficiency is improved; water use is decreased; dark respiration is stimulated through transcriptional reprogramming of metabolism; carbon gain in C4 plants is stimulated under drought stress, although photosynthesis is unaffected; yield is increased but by less than might be expected.

A key finding of the FACE experiments and a significant potential problem in meeting the challenges of climate change is that photosynthesis of crop plants does not increase as much as predicted in response to elevated CO_2 concentrations (Long et al., 2006); in addition, a decreased proportion of photosynthate is partitioned to the harvested organs. Understanding why this is and how it can be remedied is an important challenge. Part of the explanation may be that exposure to elevated CO_2 causes an increase in sugar levels in leaves and a decrease in expression of Rubisco genes (Cheng et al., 1998). Feedback regulation of photosynthetic gene expression in response to sugars was first demonstrated by Jen Sheen in 1990 (Sheen, 1990), using photosynthetic gene promoter/reporter gene fusions to show repression by glucose or sucrose in a maize protoplast system. The expression of the Rubisco small subunit (*RbcS*) and other genes encoding proteins involved in photosynthesis have also been shown to be reduced by glucose in cell suspensions of *Chenopodium rubrum* and by sucrose in oilseed rape cell cultures.

Sugars also affect the expression of genes encoding isocitrate lyase and malate synthase, the two key enzymes involved in the glyoxylate cycle, as well as α- and β-amylase. These experiments, which were performed by several research groups, were reviewed by Sheen (1994). Subsequently, micro-array experiments showed that 444 Arabidopsis genes are up-regulated by glucose (Price *et al.*, 2004), including those involved in biotic and abiotic responses, carbohydrate metabolism, N metabolism, lipid metabolism, inositol metabolism, secondary metabolism, nucleic acid related activities, protein synthesis and degradation, transport, signal transduction, hormone synthesis and cell growth or structure. A similar number (534) are down-regulated. Clearly, changes in carbohydrate metabolism and sugar concentration as a result of increases in CO_2 levels could have profound effects on gene expression and metabolism. However, a systems analysis of photosynthetic metabolism under elevated CO_2 suggested that 'dynamic reprogramming' and 'co-operativity' of the chloroplast network made the system more robust in the face of abnormal conditions, minimising large and possibly damaging fluctuations in metabolites (Luo et al., 2009).

Increased atmospheric CO_2 concentration has other effects on plant leaves. For example, it causes changes in stomatal density (Woodward, 1987) in a response that involves detection by mature leaves and the initiation of a systemic signal that brings about changes in developing leaves even if those leaves are not experiencing elevated CO_2 levels themselves (Lake et al., 2001; Coupe et al., 2006). Teng et al. (2006) found that elevated CO_2 not only reduced the stomatal density of Arabidopsis leaves, but also stomatal conductance and transpiration rate. It also increased chloroplast number and size, the latter possibly due to an increase in starch grain size and number. In another study, Bloom et al. (2010) showed that elevated CO_2 inhibited the assimilation of nitrate into organic nitrogen compounds and suggested that this, rather than feedback inhibition by sugars, was largely responsible for lower than expected photosynthetic rates in plants exposed to high CO_2 concentrations.

To date, while many studies have investigated the effects of high CO_2 concentration on plants, few have reported attempts to improve plant performance at high CO_2. Clearly, the mechanisms of feedback inhibition of photosynthesis and the inhibition of nitrogen assimilation by high CO_2 are potential targets for elucidation and manipulation.

5. Potential effects on crop quality and food safety

An often ignored aspect of global warming is the effect it will have on crop quality and food safety, as opposed to yield. For example, temperatures greater than 35 °C during wheat grain development cause changes in the expression of different groups of seed storage proteins, with consequent effects on dough quality (Blumenthal et al. 1993; Irmak et al., 2008). Heat stress is also likely to affect the concentrations of free amino acids and sugars, with potentially profound effects on processing properties. Free asparagine and proline, for example, accumulate in many plant species in response to stress (Lea et al., 2007; Lea and Azevedo, 2007). On the other hand, moderate increases in summer temperature in the UK may actually reduce the levels of free asparagine and other amino acids in wheat (Curtis et al., 2009).

Temperature during cultivation is a major factor in determining sugar concentrations in potatoes because the processes of photosynthesis, transpiration, translocation of carbohydrates and respiration are all temperature-dependent. The optimum temperature range for most varieties is quite narrow, between 15 °C and 20 °C (Kumar *et al.*, 2004). High temperatures during grain filling also cause an increase in sucrose, reducing sugars and sugar phosphates, and a reduction in starch in wheat (Jenner, 1991; Gooding et al., 2003).

Drought causes an increase in sugar content in wheat and its close relatives, barley and rye, although rye is generally more tolerant of stresses than wheat or barley, and there are considerable inter-varietal differences, with tolerant varieties accumulating higher concentrations of sugars than intolerant varieties (reviewed by Halford et al., 2011). Drought also causes an accumulation of sugars in the leaf sheath of rice (Cabuslay *et al.*, 2002) and affects enzymes involved in the conversion of sucrose to starch, with a concomitant effect on grain filling (Yang *et al.*, 2003). Sucrose synthase activity, for example, has been shown to be substantially enhanced by mild drought stress and to correlate with starch accumulation in the grain (Yang *et al.*, 2003). High temperature and drought stress also cause sugar accumulation in maize.

Drought imposes an osmotic stress on plants and plants respond by changing the partitioning between soluble and insoluble carbohydrates to maintain osmoregulation. Fructan accumulation, for example, plays a role in dehydration tolerance in cereals, and has been shown to improve rye liposome stability during drying and rehydration (Hincha *et al.*, 2007). Simple sugars and long chain carbohydrate polymers, including fructan, also improve heat shock tolerance (Fu *et al.*, 1998). Interestingly, the capacity for osmotic adjustment in plants is enhanced by elevated CO_2, possibly because of the energy requirement of the process (Pérez-López *et al.*, 2010).

Changes in sugar and amino acid concentrations in potato tubers, cereal grain and other crop products are important because they have a major impact on processing properties through the participation of these metabolites in the Maillard reaction. This reaction, which was named after the French chemist, Louis Camille Maillard, who first described it in 1912, comprises a series of non-enzymatic reactions between reducing sugars, such as glucose, fructose or maltose (although sucrose can participate after high temperature breakdown) and amino groups, principally those of amino acids. The Maillard reaction only occurs during high temperature cooking and processing, mainly in foods prepared by frying, baking and roasting. Its products include melanoidin pigments, which are complex polymers that are responsible for the brown colour in fried, baked and roasted foods. The reaction also provides complex mixtures of compounds that impart flavour and aroma. (Mottram, 2007; Halford et al., 2011). The particular compounds formed give different cooked foods their characteristic aroma, and depend on the amino acid and sugar composition of the food as well as the processing conditions. Unfortunately, some of the products of the Maillard reaction are undesirable. One of these is acrylamide, which is neurotoxic, carcinogenic and genotoxic in rodents and has been classified as a probable human carcinogen by the World Health Organisation. Acrylamide forms when the amino acid that participates in the later stages of the Maillard reaction is asparagine. The reduction of acrylamide levels in food is now recognized as a difficult and important problem for the agricultural and food industries and anything that increased the acrylamide forming potential of crop products would be a serious problem.

6. Energy crops

A potentially major contribution that agriculture can make in the mitigation of climate change is through the development of energy crops; that is crops from which biomass is obtained for burning to produce electricity, or crops from which bioethanol or biodiesel is produced for use in liquid fuels. Simplistically, the benefit in terms of reducing CO_2 emissions is that the CO_2 that is released when the biomass or fuel is burnt is first acquired from the atmosphere as the crop grows. In reality, many crops require a considerable input of fossil fuels for their cultivation, for example in the production of agrochemicals and fertilisers, the oil and diesel that is used to power farm machinery and vehicles, transport, storage and processing. The carbon equation for different crops and production systems varies, and how it is calculated is a controversial issue. Some scientists question the ethics and usefulness of biofuels altogether (Walker, 2010).

The use of plant material for fuel is nothing new, of course. Humans are believed to have started using fire in a controlled way for warmth and cooking some three quarters of a million years ago. The dependence on plant products for fuel, however, has declined as human civilisations have developed and alternative sources of energy have been discovered. One of the consequences of this is that fuel and food production from agriculture have never previously been in conflict. That is not the case now, however, with agriculture expected to meet the challenge of providing food security for an expanding world population while also helping to meet the demand for fuel as the approach of peak oil and the need to reduce CO_2 emissions forces the development of alternatives to fossil fuels.

The prospect of using biomass crops for electricity production, either for co-firing with fossil fuels or for burning on their own, has led to considerable research on the possible development of non-food crops specifically for this market (Halford and Karp, 2011). These include fast-growing trees such as willow and poplar, 'giant' grasses, such as Miscanthus, as well as reeds switchgrass and algae. The potential market is obviously huge, but so far these novel crops have not been adopted on a large scale, possibly because the burning of plant biomass is one of many possible 'renewable' alternatives to fossil fuels for heat and power generation, such as wind, hydro-electric and solar energy. In contrast, there are few alternatives to replace transport fuels, unless radical changes occur, such as the widespread adoption of battery or hydrogen-powered vehicles. The number of vehicles on the roads is continually rising and unless CO_2 emissions from the transport sector can be curbed, they will counter any reductions achieved elsewhere. There is also a political aspect to this, with some countries wishing to reduce their reliance on a few major producers for their fuel needs. As a result, the last decade has seen a massive increase in the production and use of bioethanol and biodiesel from plants.

Bioethanol is produced from sugars and the major sugar crops, sugar cane and sugar beet, are potential sources of feedstock. Indeed, production of ethanol from sugar cane has been an established industry in Brazil for several decades. Sugars can also be produced from starch through enzymatic digestion, which means that cereal grain is another potential feedstock. This industry is more recent but it too is now well-established, particularly in the USA, where the annual growth rate in bioethanol production, almost entirely from maize starch, was 25 % in the second half of the last decade, resulting in bioethanol production taking a third of the US maize crop in 2010. In Europe, the major starch crop is wheat, and a plant designed to produce bioethanol, operated by Ensus on Teeside in the UK, is expected to take 1.2 million tonnes of grain per year (about 8 % of the total UK wheat crop in a good year) when fully operational, and produce 400 million litres of ethanol, 350 thousand tonnes of animal feed and 300 thousand tonnes of carbon dioxide for use in the manufacture of soft drinks. A similarly-sized plant will come on line in the UK in 2011, operated by Vivergo, a joint venture between British Sugar, BP and DuPont. With a number of smaller plants in the pipeline, one fifth of the UK's wheat harvest could be used for fuel rather than food production by 2015. Such a massive change in grain use will undoubtedly affect the market and there is likely to be some controversy and conflict if food prices rise as a result. However, production at the Ensus plant has stalled due to a fall in demand arising from slow implementation of EU Biofuel Directives and competition from imported bioethanol. While production is expected to re-commence later in 2011, there are clearly uncertainties in the market.

Sugars from starchy grains are, of course, used in the production of alcoholic drinks by malting, fermenting and distilling. However, the traditional malting process has been considered too complex for bioethanol production for fuel and a 'dry-grind' process has been favoured in which the entire kernel is ground into a coarse flour, then slurried with water. The resulting mash is then cooked, treated with enzymes, fermented and distilled. The first enzyme to be added is α-amylase, an endoamylase that acts at random locations along the starch chain to yield shorter glucan chains. Ultimately its products are maltotriose, maltose and limit dextrin (a mixture of branched and unbranched glucans) and the process is known as gelatinisation and liquefaction. Gluco-amylases are then added to produce smaller sugars that can be fermented; this process is known as saccharification.

The drivers for plant breeding and biotechnology in the biofuel market are starch content and fermentability, and the quality and value of the co-product of fuel production, which is a protein-rich animal feed. Syngenta, for example, have produced a GM maize variety with a highly thermostable α-amylase gene, *amy797E*, from the thermophilic bacterium, *Thermococcales*. This variety is claimed to give a better yield of ethanol in the dry-grind process. Monsanto, on the other hand, have targeted the nutritional value of the animal feed co-product and, together with Cargill in a joint venture called Renessen, has developed a variety of maize (Mavera) that has been genetically modified to increase its lysine content. Lysine is an essential amino acid and cereal grains are generally too low in it to provide a balanced diet on their own. This means that animal feed must contain other sources of lysine, for example soybean or oilseed rape meal. The high lysine trait is imparted by a bacterial gene encoding a lysine-insensitive dihydrodipicolinate synthase. Dihydrodipicolinate synthase is an enzyme in the pathway for lysine synthesis which is feedback-inhibited by lysine, thereby providing the major regulatory control for flux through the pathway. The bacterial enzyme is not affected by lysine, allowing the amino acid to accumulate to levels beyond what it would in an unmodified plant. Mavera maize is being grown entirely for US domestic bioethanol production with high-lysine animal feed as a valuable co-product.

Biodiesel is derived from plant oils, usually after the fatty acids have been esterified with methanol to create fatty acid methyl esters (FAMEs). Biodiesel production is now an important and rapidly-growing industry in Europe. All of the major oil crops are potential sources, including oilseed rape (canola), soybean, sunflower and palm oil (Halford and Karp, 2011, and chapters therein). Less well-known, non-food crops, may also enter this market, including Jatropha, Pongamia and algae (Halford and Karp, 2011). Jatropha is the name given to a large group of succulent trees and shrubs, sometimes given the common name physic nut, within the Euphorbiaceae family. Jatropha oil is toxic (it is commonly called vomit or purge oil), but that does not affect its use for biofuel production, and Jatropha could be cultivated in sub-tropical and tropical countries on land that is currently considered too poor for food production. Pongamia is another medium-sized tree, suitable for tropical and sub-tropical cultivation, which produces large, oil-rich seeds. The attraction of algae as a source of oil for biofuel is its enormous potential yield (ten times that of oil palm, per hectare) and very low cost (less than a hundredth that of crude oil). Despite the potential of these novel, non-food crops, however, it is oil from well-established crops that is currently being diverted into biodiesel production.

As with the use of grains for ethanol, the impact of the use of plant oils for fuel on food prices is likely to provoke conflict, and the economic and ethical pros and cons are already controversial issues. This is exemplified in oilseed rape. Ironically, oilseed rape was first grown on a large scale in Europe during World War II to produce oil for industrial uses. At that time, erucic acid made up about half of oilseed rape oil. This particular fatty acid can be used to make transmission oils, oil paints, emulsions for photographic film and paper, healthcare products and plastics (in the form of its derivative, erucamide), as well as biodiesel. Erucic acid is still used for these purposes, but erucic acid is toxic, so oilseed rape varieties being grown for industrial oil have to be separated from varieties grown for food and feed. Edible varieties were produced through intensive plant breeding in the thirty years following World War II to reduce the levels of erucic acid and other toxins to the point considered acceptable for human consumption. The first low erucic acid varieties were grown in Canada in 1968. Nevertheless, oilseed rape did not get its seal of approval for human consumption (Generally Recognized as Safe) from the Food and Drug Administration of the USA until 1985. Canadian producers then came up with the name Canola (**Can**adian **o**il **l**ow erucic **a**cid) for edible oilseed rape oil. This name was adopted all over North America not only for the edible oil but also for the crop itself. The oil of these varieties is made up mainly of oleic acid (60 %), linoleic acid (20 %) and α-linolenic acid (10 %). The prevalence of these varieties means that they are the main source of oil for the new market in biodiesel. However, the rapidly increasing demand for biodiesel may lead to more cultivation of high erucic acid varieties and conflicting demands on plant breeders and growers for varieties with different oil profiles suitable for different end uses.

The debate on the use of plants to produce energy and fuel is a complex one, but energy crops do have the potential to provide a source of renewable energy which can reduce CO_2 and other greenhouse gas emissions and mitigate global warming. Competition for land, water, fertiliser and other inputs, and the conflicting demands of food and fuel security, make the issue a difficult one, with many shades of grey. Biomass/biofuel production was originally envisaged as being centred on dedicated non-food crops with the potential to produce high biomass yields with relatively low fertiliser inputs, such as willow, poplar and some perennial grasses. However, it is the use of sugar cane and starchy grain crops such as wheat and maize to produce bioethanol and oil crops for biodiesel that has taken off. These crops are, obviously, also used for food, and they require high fertiliser and other chemical inputs, which means that the carbon saving that they offer is much reduced. Nevertheless, as long as the market remains, the technology will move on. One possibility is that cell wall polysaccharides in the straw from crops such as wheat and maize will provide fuel feedstock in the future, complementing rather than competing with the production of food. These polysaccharides are currently much less amenable than starch to digestion and breakdown to simple sugars for fermentation, but solutions to that will be found through either modification of the plants or the discovery and development of microbial enzymes that will break these polysaccharides down.

7. Conclusion

I have reviewed the current state of kowledge on plant responses to drought, heat and high CO_2, the effect of stress on crop quality and food safety, and the use of crops for biofuel production, all in the context of global warming and climate change. The review emphasises the important role of modelling and systems biology in enabling plant scientists and breeders to produce varieties that will cope with extremes of drought, temperature and CO_2 concentrations in future climates.

Food prices have risen sharply in recent years as a result of increasing demand from a growing world population, increased prosperity in countries such as India, China and Brazil, all with huge populations aspiring to eat a better diet, the rapidly expanding use of biofuels from food crops, competition for land and water, and a series of extreme weather events, notably in Australia and Russia. These weather events may or may not be linked to global warming, but climate models predict that global warming will cause not only a rise in temperature at a rate not seen in the last 50 million years, but also an increase in the frequency and severity of extreme weather events. These events will impact severely on crop yield. The charitable organisation, Oxfam, released a report in May 2011 predicting that food prices will continue to increase and that the cost of staple foods will double in 20 years unless world leaders take action to reform the global food system, with climate change accounting for half of the increase. Minimising food price increases and ensuring food security for all will be a major challenge in the coming decades and will require improved varieties and the wide adoption of agronomic best practice. Plant breeding and agriculture are, therefore, two of the key areas that will decide how we cope with climate change and how mankind emerges from the 21st century. There is an urgent need for investment in crop science, as well as knowledge transfer to ensure that advances in our understanding of plant responses to heat, drought and CO_2 feed through into improved varieties.

8. Acknowledgment

Rothamsted Research receives grant-aided support from the Biotechnology and Biological Sciences Research Council (BBSRC) of the United Kingdom.

9. References

Bloom, A.J., Burger, M., Asensio, J.S.R. & Cousins, A.B. (2010). Carbon dioxide enrichment inhibits nitrate assimilation in wheat and Arabidopsis. *Science* 328, 899-903.

Blumenthal, C. S., Barlow, E. W. R. & Wrigley, C. W. (1993). Growth, Environment and wheat quality - the effect of heat stress on dough properties and wheat gluten proteins. *Journal of Cereal Science* 18, 3-21.

Cabuslay, G.S., Ito, O. & Alejar, A.A. (2002). Physiological evaluation of responses of rice (Oryza sativa L.) to water deficit. *Plant Science* 163, 815-827.

Carbone, G. J., Kiechle, W., Locke, C., Mearns, L. O., McDaniel, L. & Downton, M. W. (2003). Response of soybean and sorghum to varying spatial scales of climate change scenarios in the southeastern United States. *Climatic Change* 60, 73-98.

Challinor, A.J., Simelton, E.S., Fraser, E.D.G., Hemming, D. & Collins, M. (2010). Increased crop failure due to climate change: assessing adaptation options using models and socio-economic data for wheat in China. *Environmental Research Letters* 5, 034012.

Cheng, S.H., Moore, B. & Seeman, J.R. (1998). Effects of short- and long-term elevated CO_2 on the expression of ribulose-1,5-bisphosphate carboxylase/oxygenase genes and carbohydrate accumulation in leaves of *Arabidopsis thaliana* (L.) Heynh. *Plant Physiology* 116, 715-723.

Coupe, S.A., Palmer, B.G., Lake, J.A., Overy, S.A., Oxborough, K., Woodward, F.I., Gray, J.E. & Quick, W.P. (2006). Systemic signalling of environmental cues in Arabidopsis leaves. *Journal of Experimental Botany* 57, 329-341.

Curtis T.Y., Muttucumaru N., Shewry P.R., Parry M.A.J., Powers S.J., Elmore J.S., Mottram D.S., Hook S. & Halford N.G. (2009) Effects of genotype and environment on free amino acid levels in wheat grain: Implications for acrylamide formation during processing. *Journal of Agricultural and Food Chemistry* 57, 1013-1021.

Ewert, F., Rodriguez, D., Jamieson, P., Semenov, M.A., Mitchell, R.A.C., Goudriaan, J., Porter, J.R., Kimball, B.A., Pinter, P.J., Manderscheid, R., Weigel, H.J., Fangmeier, A., Fereres, E. & Villalobos, F. (2002). Effects of elevated CO2 and drought on wheat: testing crop simulation models for different experimental and climatic conditions. *Agriculture Ecosystems & Environment* 93, 249-266.

Ferris, R., Ellis, R.H., Wheeler, T.R. & Hadley, P. (1998). Effect of high temperature stress at anthesis on grain yield and biomass of field-grown crops of wheat. *Annals of Botany* 82, 631-639.

Foulkes, M.J., Sylvester-Bradley, R., Weightman, R. & Snape, J.W. (2007). Identifying physiological traits associated with improved drought resistance in winter wheat. *Field Crops Research* 103, 11-24.

Fu P., Wilen R.W., Robertson A.J., Low N.H., Tyler R.T., & Gusta L.V. (1998). Heat tolerance of cold acclimated Puma winter rye seedlings and the effect of a heat shock on freezing tolerance. *Plant and Cell Physiology* 39, 942-949.

Gooding, M.J., Ellis, R.H., Shewry, P.R. & Schofield J.D. (2003) Effects of restricted water availability and increased temperature on the grain filling, drying and quality of winter wheat. *Journal of Cereal Science* 37, 295-309.

Halford, N.G. & Karp, A. (Eds) (2011). Energy Crops. Royal Society of Chemistry, Cambridge. ISBN: 978-1-84973-032-7.

Halford, N.G., Curtis, T.Y., Muttucumaru, N., Postles, J. & Mottram, D.S. (2011). Sugars in crop plants. *Annals of Applied Biology* 158, 1-25.

Hincha, D.K., Livingston, D.P., Premakumar, R., Zuther, E., Obel, N., Cacela, C. & Heyer, A.G. (2007) Fructans from oat and rye: Composition and effects on membrane stability during drying. *Biochimicae et Biophysica Acta - Biomembranes* 1768, 1611-1619.

Irmak, S., Naeem, H.A., Lookhart, G.L. & MacRitchle, F. (2008). Effect of heat stress on wheat proteins during kernel development in wheat near-isogenic lines differing at Glu-D1. *Journal of Cereal Science* 48, 513-516.

Jamieson, P.D., Berntsen, J., Ewert, F., Kimball, B.A., Olesen, J.E., Pinter, P.J.J., Porter, J.R. & Semenov, M.A. (2000). Modelling CO_2 effects on wheat with varying nitrogen supplies. *Agriculture, Ecosystems and Environment* 82, 27-37.

Jenner C.F. (1991). Effects of exposure of wheat ears to high temperature on dry matter accumulation and carbohydrate metabolism in the grain of 2 cultivars. 1. Immediate responses. *Australian Journal of Plant Physiology* 18, 165-177.

Kumar, D., Singh, B.P. & Kumar, P. (2004). An overview of the factors affecting sugar content of potatoes. *Annals of Applied Biology* 145, 247-256.

Lake, J.A., Quick, W.P., Beerling, D.J. & Woodward, F.I. (2001). Plant development: Signals from mature to new leaves. *Nature* 411, 154.

Lea, P.J., Sodek, L., Parry, M.A.J., Shewry, P.R. & Halford, N.G. (2007). Asparagine in plants. *Annals of Applied Biology* 150, 1-26.

Lea, P.J. & Azevedo, R.A. (2007). Nitrogen use efficiency. 2. Amino acid metabolism. *Annals of Applied Biology* 151, 269-275.

Leakey, A.D.B., Ainsworth, E.A., Bernacchi, C.J., Rogers, A., Long, S.P. & Ort, D.R. (2009). Elevated CO_2 effects on plant carbon, nitrogen, and water relations: six important lessons from FACE. *Journal of Experimental Botany* 60, 2859-2876.

Long, S.P., Ainsworth, E.A., Leakey, A.D.B., Nösberger, J. & Ort, D.R. (2006). Food for thought: Lower-than-expected crop yield stimulation with rising CO_2 concentrations. *Science* 312, 1918-1921.

Luo, R., Wei, H., Ye, L., Wang, K., Chen, F., Luo, L., Liu, L., Li, Y., Crabbe, M.J.C., Jin, L., Li, Y. & Zhong, Y. (2009). Photosynthetic metabolism of C3 plants shows highly cooperative regulation under changing environments: a systems biological analysis. *Proceedings of the National Academy of Sciences USA* 106, 847-852.

Mitchell, R.A.C., Mitchell, V., Driscoll, S.P., Franklin, J. & Lawlor, D.W. (1993). Effects of increased CO_2 concentration and temperature on growth and yield of winter wheat at two levels of nitrogen application. *Plant, Cell and Environment* 16, 521-529.

Mottram D.S. (2007). The Maillard reaction: source of flavour in thermally processed foods. In *Flavours and Fragrances: Chemistry, Bioprocessing and Sustainability*, pp 269-284. Ed R. G. Berger. Berlin, Springer-Verlag.

Olesen, J.E., Carter, T.R., Diaz-Ambrona, C.H., Fronzek, S., Heidmann, T., Hickler, T., Holt, T., Quemada, M., Ruiz-Ramos, M., Rubaek, G.H., Sau, F., Smith, B. & Sykes, M.T. (2007). Uncertainties in projected impacts of climate change on European agriculture and terrestrial ecosystems based on scenarios from regional climate models. *Climatic Change* 81, 123-143.

Pérez-López, U., Robredo, A., Lacuesta, M., Muñoz-Rueda, A. & Mena-Petite, A. (2010). Atmospheric CO2 concentration influences the contributions of osmolyte accumulation and cell wall elasticity to salt tolerance in barley cultivars. *Journal of Plant Physiology* 167, 15-22.

Price, J., Laxmi, A., St Martin, S.K. & Jang, J.C. (2004). Global transcription profiling reveals multiple sugar signal transduction mechanisms in Arabidopsis. *Plant Cell* 16, 2128-50.

Richter, G.M. & Semenov, M.A. (2005). Modelling impacts of climate change on wheat yields in England and Wales - assessing drought risks. *Agricultural Systems* 84, 77-97.

Semenov, M.A. (2007). Development of high resolution UKCIP02-based climate change scenarios in the UK. *Agricultural and Forest Meteorology* 144, 127–138.

Semenov, M.A. & Halford, N.G. (2009). Identifying target traits and molecular mechanisms for wheat breeding under a changing climate. *Journal of Experimental Botany* 60, 2791–2804.

Semenov, M.A., Martre, P. & Jamieson, P.D. (2009). Quantifying effects of simple wheat traits on yield in water-limited environments using a modelling approach. *Agricultural and Forest Meteorology* 149, 1095-1104.

Sheen, J. (1990). Metabolic repression of transcription in higher plants. *Plant Cell* 2, 1027-1038.

Sheen, J. (1994). Feedback control of gene expression. *Photosynthesis Research* 39, 427-438.

Solomon, S., Qin, D., Manning, M., Marquis, M., Averyt, K., Tignor, M.M.B., LeRoy Miller, H. & Chen, Z. (2007). Climate Change 2007: The Physical Science Basis. Contribution of Working Group I to the Fourth Assessment Report of the Intergovermental Panel on Climate Change., pp. 996. NY: Cambridge University Press.

Teng, N., Wang, J., Chen, T., Wu, X., Wang, Y. & Lin, J. (2006). Elevated CO2 induces physiological, biochemical and structural changes in leaves of Arabidopsis thaliana. *New Phytologist* 172, 92-103.

Walker, D. (2010). Biofuels – for better or worse? *Annals of Applied Biology* 156, 319-327.

Wheeler, T.R., Batts, G.R., Ellis, R.H., Hadley, P. & Morison, J.I.L. (1996). Growth and yield of winter wheat (*Triticum aestivum*) crops in response to CO_2 and temperature. *Journal of Agricultural Science* 127, 37-48.

Wheeler, T.R., Craufurd, P.Q., Ellis, R.H., Porter, J.R. & Prasad, P.V.V. (2000). Temperature variability and the yield of annual crops. *Agriculture Ecosystems & Environment* 82, 159-167.

Witcombe, J.R., Hollington, P.A., Howarth, C.J., Reader, S. & Steele, K.A. (2008). Breeding for abiotic stresses for sustainable agriculture. *Philosophical Transactions of the Royal Society B-Biological Sciences* 363, 703-716.

Woodward, F.I. (1987). Stomatal numbers are sensitive to increases in CO_2 from preindustrial levels. *Nature* 327: 617-618.

Yang J., Zhang J., Wang Z., Zhu Q. & Liu L. (2003). Activities of enzymes involved in sucrose-to-starch metabolism in rice grains subjected to water stress during filling. *Field Crops Research* 81, 69-81.

23

Modeling Renewable Energy Consumption for a Greener Global Economy

Perry Sadorsky
Schulich School of Business
York University
Canada

1. Introduction

Climate change, resource depletion, and energy security are powerful forces helping to stimulate interest in renewable energy consumption. Renewable energy (defined broadly as energy generated from geothermal, wind, solar, tide and wave, wood, waste, and biomass) can help to address carbon dioxide related climate change, is not depletable, and is not susceptible to energy security issues the way that oil is. In response to these forces, it is widely expected that renewable energy will play an increasingly greater role in the overall share of energy consumed. This has helped to generate an increased interest into the relationship between renewable energy consumption and economic growth.

Climate change is now recognized as one of the biggest threats to society (Stern 2007, 2009). Climate changes such as rising temperatures, rising sea levels, acidification of the world's oceans, altered rain fall patterns, and greater variability in weather patterns, have the potential to seriously disrupt business activity. In particular, climate change can affect business activity through industry specific risks like regulatory and physical risks and company specific risks like reputation, litigation and competitive risks (Labatt & White, 2007). According to the World Resources Institute, 61.4% of global GHG emissions come from the energy sector[1] which means that any serious attempt to control greenhouse gas emissions will have to reduce fossil fuel consumption and increase renewable energy consumption. Effectively reducing fossil fuel consumption will require putting a price on carbon (either through taxes on carbon intensive products or a cap and trade emissions trading system) with the result that carbon based energy products like oil will increase in price. Pricing carbon will raise the price of carbon intensive energy products and help to encourage a substitution to low carbon energy sources.

Resource depletion and in particular, peak oil concerns stem from Hubbert (1956) who, in one of the scenarios he considered, forecast that US domestic oil production would peak in 1970. According to data provided by BP (2010), US domestic oil production did peak in 1970 at 11 million barrels of oil per day. Hubbert (1971) went on to predict that global oil production would peak in 2000. To date, global oil production has not yet peaked, but the general consensus is that it will soon (Verbruggen & Marchohi, 2010; Almeida & Silva, 2009). While there are lots of different authors predicting the year for peak global oil

---

[1] http://cait.wri.org/figures.php?page=World-FlowChart&view=100

production, the consensus seems to be that peak oil will occur before 2015 (Almeida & Silva, 2009). If proponents of peak oil are correct and global oil production is soon going to peak, then what happens after the peak? The most likely scenario would be a lengthy adaptation phase during which energy efficiency is increased, short-term transitional energy sources are developed, and long-term sustainable energy sources are developed (Almeida & Silva, 2009). Peak oil brings higher oil prices and the revenues from higher oil prices can be used to support financing of renewable energy.

Energy security issues refer to the fact that many of the largest oil consuming countries are net oil importers and that much of the world's proven oil reserves are located in just a few countries. In 2009, four countries, Saudi Arabia, Iran, Iraq, and Kuwait accounted for 46.3% of the world's 1.333 trillion barrels of proven oil reserves (BP, 2010)[2] [3].In addition to having to pay a hefty bill for imported oil, this makes oil importing countries vulnerable to oil supply disruptions (oil embargos, terrorist attacks, political unrest) and oil price shocks. In 2009, the four largest oil consuming countries (US (21.7%), China (10.4%), Japan (5.1%), and India (3.8%)) were all oil net importers[4]. The United States imports approximately 66% of the oil that it consumes and in 2009, imported 3.25 billion barrels of oil (BP, 2010). At an average price of $61.92 per barrel for West Texas Intermediate crude oil, the total cost of US oil imports in 2009 was $201 billion dollars or $16.8 billion per month. Now suppose that at sometime in the near future, the US imports the same amount of oil, but the average price rises to $100 per barrel. In this case, the bill for oil imports would be $325 billion dollars or $27 billion per month. This is a lot of money to pay for imported oil, especially for a country that has a large trade deficit and an enormous federal budget deficit[5] [6]. Looking to the future, the 2009 US bill for imported oil is likely a minimum value because for the first six months of 2009 the US was in the worst economic downturn since the Great Depression of the 1930s. As the US economy recovers from the 2008-2009 recession oil demand and oil imports will surely rise.

As a result of these developments on climate change, resource depletion, and energy security it would be useful for policy makers to have an idea of how renewable energy consumption, income, oil consumption, and oil prices interact at the global level. The purpose of this paper is to develop and estimate a vector autoregression (VAR) model to investigate the dynamic interactions between these variables for the global economy. The model is used to analyze short-term dynamics and to make forecasts into the future until the year 2030. One of the advantages of conducting the analysis at the global economy level is that the results of this paper can then be compared with some of the results from international organizations like the International Energy Agency who makes forecasts of oil consumption and energy usage to 2030.

2. Literature review

To date there have been several published papers looking at the relationship between renewable energy consumption and various macroeconomic variables (like income, oil

---

[2] This does not include 143 billion barrels of oil from the Alberta tar sands.

[3] In 2009, Saudi Arabia, Iran, Iraq, and Kuwait accounted for 19.8%, 10.3%, 8.6% and 7.6% respectively of the world's proven oil reserves.

[4] The numbers in parentheses represent each country's respective share of global oil consumption.

[5] In January 2011, the US imported close to 9 million barrels of oil per day. At an average price of $89 per barrel, the monthly bill for imported oil was $24.8 billion dollars.

[6] In early 2011, the US federal budget deficit was estimated at $1.6 trillion dollars.

prices, capital, labour) at the country or regional level (see for example, Apergis & Payne, 2010a, 2010b, 2011; Chang et al., Lee, 2009; Chien & Hu, 2007,2008; Marques et al., 2010; Menyah & Wolde-Rufael, 2010; Sadorsky, 2009a, 2009b)[7]. The consensus from most of this research is that increases in income are a major driver behind increased renewable energy consumption.

Chien and Hu (2007) use data envelope analysis to analyze the effects of renewable energy on the technical efficiency of 45 economies over the period 2001-2002. They find that increasing the use of renewable energy improves an economy's technical efficiency while increasing the use of traditional energy (fossil fuel) decreases technical efficiency.

Chien and Hu (2008) use structural equation modeling techniques to analyze the effects of renewable energy on GDP for 116 countries in 2003. Their results show that renewable energy correlates positively with capital formation and that increasing the use of renewable energy increases GDP through the impact the impact on capital formation. Increasing the use of renewable energy has little impact on the trade balance.

Sadorsky (2009a) uses vector autoregression techniques to analyze the relationship between renewable energy consumption, income, oil prices and CO2 emissions in the Group of 7 (G7) economies over the period 1980-2005. In the long-term, increases in real GDP per capita and carbon dioxide emissions per capita are found to be major drivers behind increases in G7 renewable energy consumption per capita.

Sadorsky (2009b) uses panel cointegration techniques to investigate the relationship between renewable energy consumption and GDP for a sample of 18 emerging economies followed over the period 1994-2003. In the long-term, increases in real GDP per capita are found to be a major driver behind increases in renewable energy consumption per capita for emerging economies. No short-term evidence of statistically significant Granger causality is found between GDP and renewable energy consumption.

Chang et al. (2009) use threshold econometric techniques to investigate the relationship between economic growth and renewable energy supply for OECD member countries over the period 1997 to 2006. The threshold approach allows for a distinction on how renewable energy supply depends upon different economic growth regimes (like high or low). They find that countries characterized by high economic growth are able to respond to high energy prices with increases in renewable energy. Countries with low economic growth tend to be relatively non-responsive to energy price changes.

Apergis and Payne (2010a) use panel cointegration techniques to examine the relationship between renewable energy consumption and economic growth for a panel of 20 OECD countries over the period 1985-2005. The theoretical framework uses an aggregate production function relating output to labour, capital, and renewable energy. They find evidence of bidirectional causality between renewable energy consumption and economic growth in both the short-run and the long-run.

Apergis and Payne (2010b) use panel cointegration techniques to examine the relationship between renewable energy consumption and economic growth for a panel of 13 Eurasia countries over the period 1992-2007. As in Apergis and Payne (2010a) the theoretical model is based upon an aggregate production function which relates capital, labour, and renewable energy to output. They find empirical support for a feedback relationship between renewable energy consumption and output.

---

[7] In the macroeconomic-energy literature, renewable energy is usually defined as non-hydro electricity generated from geothermal, wind, solar, tide and wave, wood, waste, and biomass.

Marques et al. (2010) uses panel regression techniques to investigate the relationship between renewable energy consumption, political factors, socioeconomic factors, and country specific factors for a panel of 24 European counties covering the period 1990-2006. They find that lobby efforts from the fossil fuel sector, and CO2 emissions reduce renewable energy consumption, while reducing energy self sufficiency promotes renewable energy consumption.

Menyah and Wolde-Rufael (2010) use vector autoregression techniques to study the relationship between carbon dioxide emissions, renewable energy consumption, nuclear consumption and real GDP for the US over the period 1960-2007. They find causality running from nuclear energy consumption to CO2 emissions but no causality running from renewable energy consumption to CO2 emissions. There is evidence of causality running from GDP to renewable energy.

Apergis and Payne (2011) use panel cointegration techniques to examine the relationship between renewable energy consumption and economic growth for a panel of 6 Central American countries over the period 1980-2006. Results from a panel error correction model indicate bidirectional causality between renewable energy consumption and economic growth in both the short- and long-run.

3. Empirical model

A vector autoregression (VAR) is used to empirically investigate the relationship between renewable energy consumption, income, oil consumption, and oil prices. One of the advantages of using a VAR is that the researcher does not need to provide prior assumptions about which variables are response variables and which variables are explanatory variables because in a VAR all variables are treated as endogenous. This means that in a VAR, each variable depends upon the lagged values of all the variables in the system. This allows for a much richer data structure that can capture complex dynamic properties in the data (Brooks, 2002).

A vector autoregression can be written in the following way.

$$y_t = B_0 + \sum_{j=1}^{p} B_j y_{t-j} + u_t, \Sigma(u_t u_t^{'}) = \Sigma \tag{1}$$

In the above equation, the variable y is a n vector of endogenous variables, B_0 is a n vector of constants, and B_j is a nxn matrix of regression coefficients to be estimated. The error term, u, is assumed to be independent and identically distributed.

The VAR is estimated using annual data on global renewable energy consumption, income (GDP), oil consumption and oil prices from 1980 to 2008. The time frame is determined by data availability on renewable energy consumption. In estimating the VAR, all variables are expressed in natural logarithms in order to reduce heteroskedasticity. A linear time trend is included in the estimation of the VAR. The estimated coefficients from a VAR are of little use by themselves and it is usual to use impulse response functions to investigate the dynamic interaction between variables in a VAR. Analysis proceeds by estimating the model and using impulse response functions to investigate the dynamic interaction between the variables. The outcome of this analysis can then be used to provide forecasts and make policy recommendations.

4. Data

For this study, annual data over the period 1980-2008 is collected on renewable energy generation (REN), income (RGDP), oil consumption (OILCONSUMPTION), and real oil prices (ROIL). It is difficult to obtain data on renewable energy consumption and so following other authors, data on renewable energy generation is used as a proxy for renewable energy consumption. Since electricity cannot be stored, the difference between renewable energy consumption and renewable energy generation is not likely to be large. World renewable energy generation (defined as non-hydro electricity generated from geothermal, wind, solar, tide and wave, wood, waste, and biomass) data is available from the US EIA[8]. Data on world GDP is available from the World Bank online data base[9]. Data on world oil consumption and West Texas Intermediate crude oil prices are available from

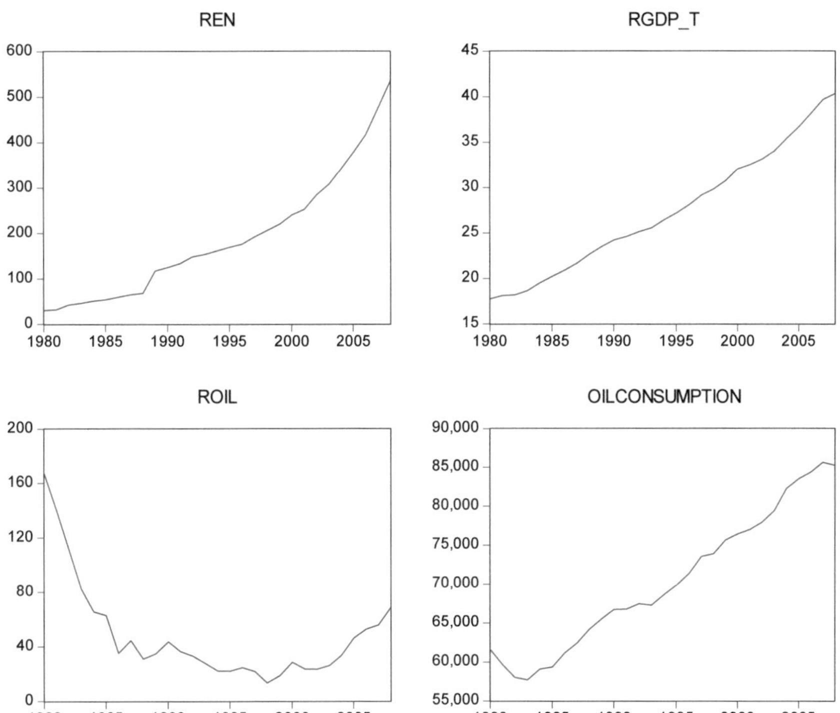

Fig. 1. Data plots

BP (2010). Oil prices are average oil prices for the year. Renewable energy is measured in billions of kilowatt hours (KW-H), real GDP is measured in 2000 US dollars, oil prices are measured in US dollars per barrel, and oil consumption is measured in thousands of barrels per day. Real oil prices are constructed by deflating nominal oil prices by the US CPI (available from the Federal Reserve Economic Database)[10].

[8] http://www.eia.doe.gov/cfapps/ipdbproject/IEDIndex3.cfm?tid=6&pid=29&aid=12
[9] http://databank.worldbank.org/ddp/home.do?Step=12&id=4&CNO=2
[10] http://research.stlouisfed.org/fred2/

Data plots are shown in Figure 1. Real world GDP, measured in trillions of US 2000 dollars (RGDP_T), has been growing along a linear trend line. Since 2000, global renewable energy consumption has been growing faster than in the period before 2000. Notice how real GDP and oil consumption tend to move upwards over time with similar slopes. Real oil prices display a U shaped pattern with the maximum value obtained in 1980. Real oil prices in 2008 are half of their value in 1980.

	REN	RGDP	ROIL	OILCONSUMPTION
Mean	10.198	2.926	-3.173	1.162
Median	8.266	3.207	-2.494	1.492
Maximum	54.345	4.563	42.619	3.511
Minimum	3.384	0.331	-58.167	-3.142
Std. Dev.	9.761	1.044	25.351	1.577
Skewness	3.626	-0.493	-0.183	-1.163
Kurtosis	16.470	2.579	2.337	4.306
t stat	5.528	14.830	-0.662	3.899
Jarque-Bera	273.032	1.342	0.670	8.299
Probability	0.000	0.511	0.715	0.016
Observations	28	28	28	28

Variables are calculated as log returns

Table 1. Summary statistics over 1980-2008 for variables in % growth rates

Table1 shows summary statistics for the variables measured in percentage growth rates. The average annual growth rate for renewable energy is 10.198% per year and this value is only slightly larger than its standard deviation. Real GDP has a higher average annual growth rate (2.926%) than oil consumption (1.162%). Real oil prices have a negative average annual growth rate and real oil prices also have the largest standard deviation (about 7 times the size of the absolute mean value). The t statistics indicate that the average annual growth rates are statistically significant for each of renewable energy, GDP, and oil consumption.

	REN	RGDP	ROIL	OILCONSUMPTION
REN	1.000	-0.016	0.123	-0.051
RGDP	-0.016	1.000	0.250	0.734
ROIL	0.123	0.250	1.000	0.184
OILCONSUMPTION	-0.051	0.734	0.184	1.000

Table 2. Correlations for variables measured in growth rates

Table 2 reports ordinary correlation coefficients. Renewable energy consumption correlates positively with real oil prices and negatively with each of the other two data series. The highest correlation is between real GDP and oil consumption (0.734).

5. Empirical results

In estimating the VAR, all variables are expressed in natural logarithms. A linear time trend is included in the estimation of the VAR. The lag length, p, of the VAR is chosen using the approach of Toda and Yamamoto (1995). Their approach to modeling is to estimate the VAR in levels and use a conventional lag length selection criteria like AIC or SIC to select the lag length and then to add an extra lag for the possibility that some or all of the variables contain unit roots. The Toda and Yamamoto (1995) approach does not require pre-testing for cointegration between the variables and can accommodate VAR models where some variables have unit roots and some variables are stationary. For the renewable energy consumption, income, oil consumption, and oil price VAR model, the SIC selects a lag length of one. Adding one extra lag length for the possibility that some or all of the variables contain a unit root establishes a VAR with p=2[11]. Since impulse response functions are used to interpret the results, some care must be taken to ensure that the VAR is stable because, strictly, speaking, impulse response functions are only valid if the VAR is stable (Lutkepohl, 2005). If the modulus of each eigenvalue of the matrix B is strictly less than one the estimated VAR is stable. Figure 2 shows the inverse roots of the AR characteristic polynomial are each less than unity, which satisfies the stability condition.

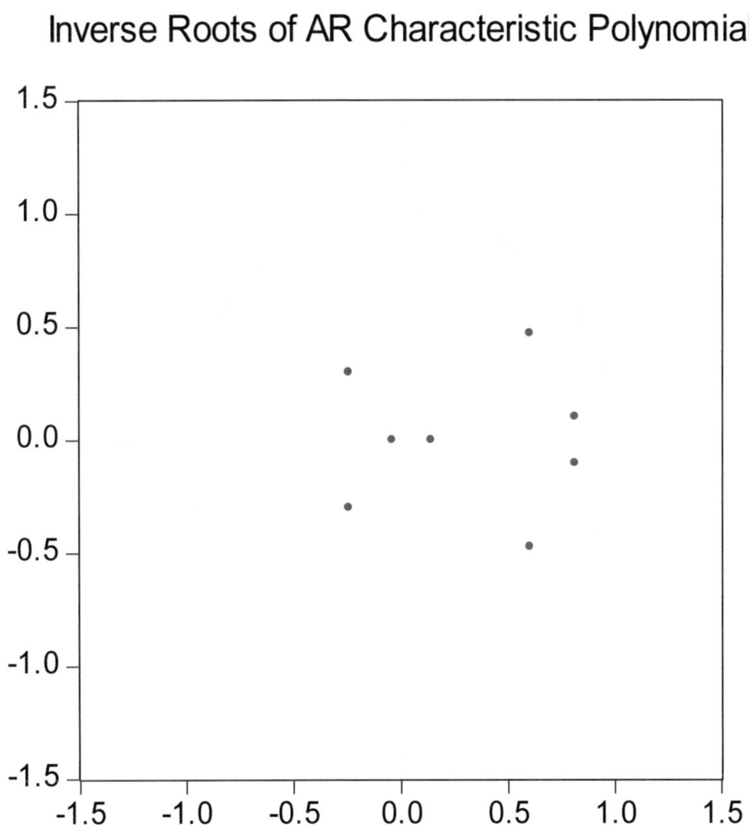

Fig. 2. Stability for VAR with 2 lags

[11] All estimation was carried out using Eviews 7.0.

Table 3 reports LM tests for no autocorrelation of the residuals. All p values are larger than 1% (and only one p value is smaller than 5%) indicating that the residuals do not exhibit any strong evidence of autocorrelation. Normality tests confirm that at the 5% level, the residuals from the VAR are normally distributed (Table 4). The results of Figure 2, and Tables 3 and 4 indicate the VAR is adequately specified and can be used for constructing useful impulse response functions.

Lags	LM-Stat	Prob
1	21.62	0.16
2	26.88	0.04
3	16.42	0.42
4	14.55	0.56
5	9.81	0.88
6	9.23	0.90
7	11.82	0.76
8	8.39	0.94
9	18.81	0.28
10	16.93	0.39
11	23.09	0.11
12	21.11	0.17

Probs from chi-square with 16 df.

Table 3. LM tests on residuals

Component	Jarque-Bera	df	Prob.
1	2.218373	2	0.3298
2	4.695628	2	0.0956
3	3.448962	2	0.1783
4	3.670008	2	0.1596
Joint	14.03297	8	0.0809

Table 4. Normality tests on residuals

Figure 3 reports the generalized impulse response functions of each variable in response to a one standard deviation shock to each of the other variables[12]. The impulse responses are drawn with plus and minus 2 standard error bands to form 95% confidence intervals. As expected, each variable responds positively and statistically significantly to its own shock. The more interesting results are how a variable responds to shocks from other variables.

[12] Generalized impulse response functions, which are not sensitive to the ordering of the variables in the VAR, are reported.

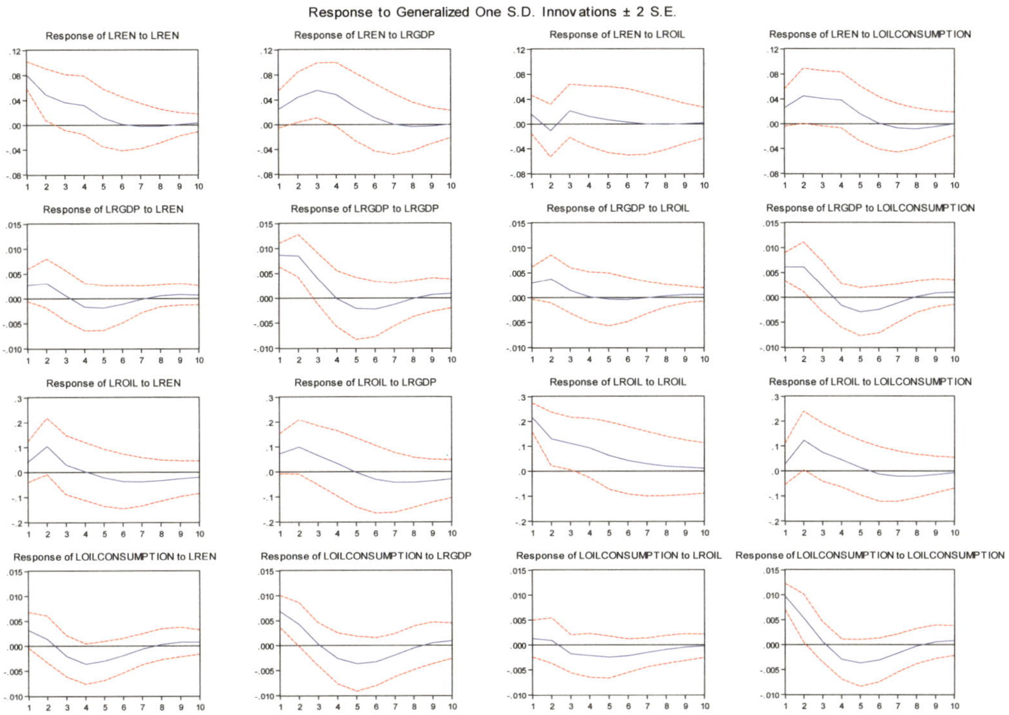

Fig. 3. Impulse response functions

A one standard deviation shock to income has a positive and statistically significant impact on renewable energy consumption for four years. This result is important in establishing that global consumption of renewable energy increases with increases in global income. A one standard deviation shock to oil consumption has a positive and statistically significant impact on renewable energy consumption for three years. This result is consistent with the pattern of global energy consumption over the period 1980-2008. As global energy consumption has risen specific sectors like oil and renewable have grown and so we expect to see a relationship between oil consumption and renewable energy consumption (Table 1). These results are important in establishing that increases in income and oil consumption are each important drivers of increased renewable energy consumption.

A one standard deviation shock to real oil prices has a positive and statistically significant impact on income in the first year of the shock. A one standard deviation shock to oil consumption has a positive and statistically significant impact on income for two and a half years. This result is important in establishing that increases in oil consumption leads to increase economic growth. This makes sense when one thinks about the production side of the economy. Energy is an essential input into the production of most goods and services and increasing the factor s of production like capital, labour and energy increases output.

A one standard deviation shock to income increases real oil prices over the first two years and this result is marginally statistically significant. In year two, real oil prices respond in a positive and statistically significant way to a one standard deviation shock to oil consumption.

A one standard deviation shock to income has a positive and statistically significant impact on oil consumption for years one and two. This result is important in establishing that higher economic growth leads to greater oil consumption. Oil consumption does not respond to shocks to real oil prices in a statistically significant way.

Figure 4 reports accumulated impulse response function. Whereas Figure 3 traces out the effects of a shock across time, Figure 4 reports the accumulated effect of a shock across time. The accumulated response of renewable energy consumption to a shock to income is positive and statistically significant for 6 years. The accumulated response of renewable energy consumption to a shock to oil consumption is positive and statistically significant for 5 years. The accumulated response of real oil prices to a shock to oil consumption is positive and statistically significant for 3 years. The accumulated impulse response functions in Figure 4 also more clearly show the feedback relationship between income and oil consumption. The accumulated response of income to a shock to oil consumption is positive and statistically significant for 4 years. The accumulated response of oil consumption to a shock to income is positive and statistically significant for 4 years.

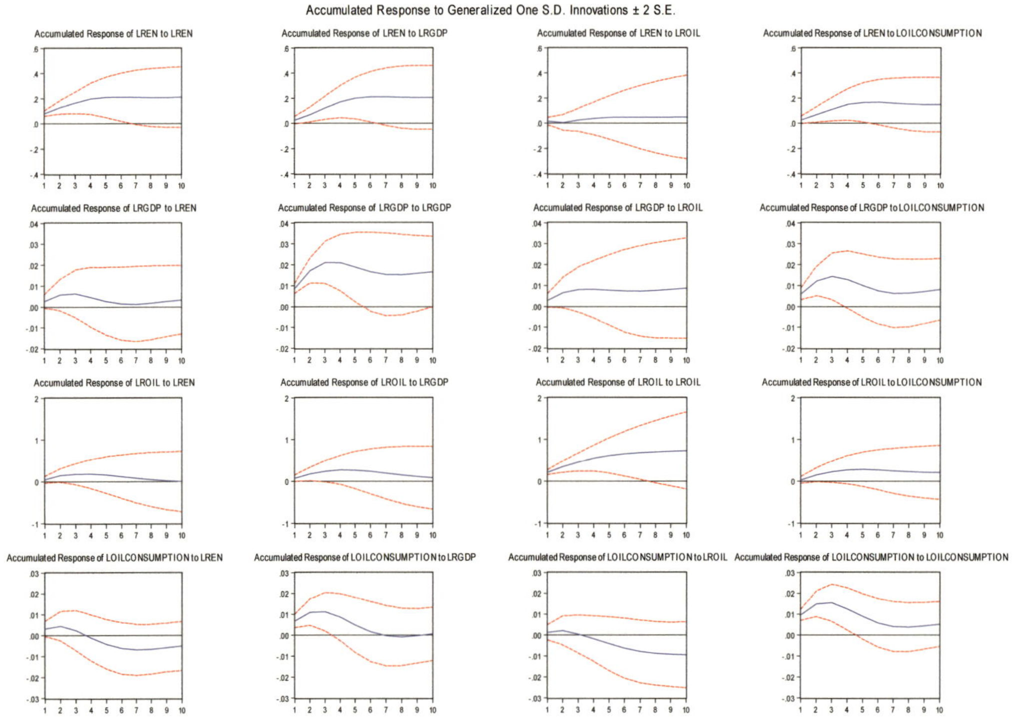

Fig. 4. Accumulated impulse response functions

6. Forecasts

The VAR model can also be used to construct out-of-sample dynamic forecasts over the period 2009-2030. Two sets of forecasts are presented, dynamic and stochastic. In constructing dynamic forecasts, historical data for lagged endogenous variables are used if they are dated prior to the first period of forecasting. After words, forecasts from previous periods are used as lagged endogenous variables. For the dynamic forecasts, the error terms in equation (1) are ignored. For stochastic forecasts, the error terms in equation (1) are modeled using Monte Carlo simulation techniques. The model predicts a distribution of outcomes for each variable at each observation which facilitates the computation of standard error bands. Stochastic forecasts are produced using Monte Carlo simulation techniques assuming each endogenous variable is distributed normally. The number of replications is set at 10,000[13].

The VAR model is estimated over the period 1980-2008 and dynamic forecasts are made over the period 2009-2030. The dynamic forecasts show that renewable energy consumption, GDP, and oil consumption are each forecast to rise along a fairly straight trend line. Real oil prices dip in 2010 before resuming an upward trend.

Table 5 summarizes the dynamic forecasts for the years 2015, 2020, 2025, and 2030 in a table, making it easier to read the actual forecasted values. In the year 2030, global oil consumption is forecast to be 117 million barrels of oil per day, and renewable energy consumption is forecast to be 3.4 trillion KW-H. Between 2009 and 2030, oil consumption is predicted to grow by 1.5% per year and renewable energy consumption is predicted to growth by 8.5% per year. Real global GDP (in 2000 US dollars) is forecasted to be $77 trillion in 2030. The VAR model forecasts real (2000 US dollar) oil prices. In order to get a forecast of nominal oil prices, the US CPI must be forecast. The US CPI is modelled over the period 1980-2008, using a linear trend (intercept is equal to -10.66 and slope coefficient is equal to 4.09). The R square from this regression is 98%. Real oil prices are forecasted to be $247 per barrel in 2030 and nominal oil prices are forecasted to be $216 per barrel in 2020 and a whopping $550 per barrel in 2030.

These forecasts for oil consumption and renewable energy consumption are fairly close to those provided by the International Energy Agency (IEA, 2007). The International Energy Agency (2007, 80), for example predicts global oil consumption of 116 million barrels per day in 2030. Over the period 2005-2030 oil is expected to have an average annual growth rate of 1.3% while renewable energy consumption has an average annual growth rate of 6.7%[14].

Stochastic dynamic forecasts are shown in Figure 6 along with plus and minus two standard error bands. Oil consumption and real GDP are each estimated fairly precisely as shown by the tightness of the error bands. In 2030 global oil consumption is forecast at 117 million barrels per day and the 95% confidence interval ranges from 114 million barrels per day to 120 million barrels per day. Renewable energy consumption is reasonably well estimated. In 2030, renewable energy consumption is forecast at 3.4 trillion KW-H with a 95% confidence interval ranging from 2.6 trillion KW-H to 4.4 trillion KW-H. The greatest uncertainty is associated with the forecasts of real oil prices. In 2030 real oil prices are forecast at $247 per barrel with a 95% confidence interval that ranges from $118 per barrel to $516 per barrel. This wide confidence interval puts nominal oil prices in 2030 between $263 per barrel and $1149 per barrel.

---

[13] A higher number of replications increases precision but for this particular model there is not much difference in the forecasts calculated from 10,000 replications compared to those calculated from 5,000 replications.

[14] The IEA forecasts do change from year to year depending upon global economic conditions. IEA (2009), for example, in response to the global recession of 2008-2009 forecasts oil consumption in 2030 to be 105 million barrels per day.

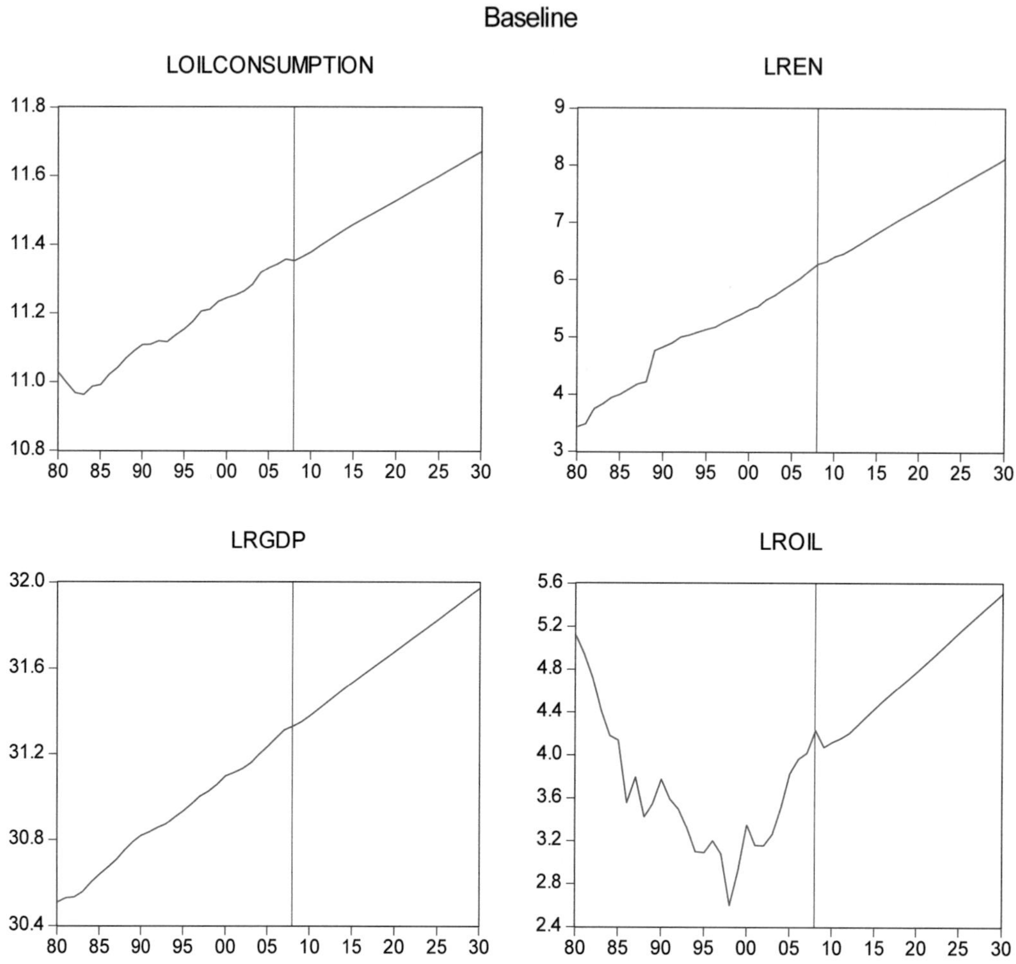

Fig. 5. Dynamic forecasts

Year	Oil consumption	Renewable	RGDP	Real Oil	Nominal Oil
2015	94753	920	49455	84	136
2020	101601	1414	57209	119	216
2025	109148	2180	66352	172	347
2030	117175	3359	76914	247	550

Oil consumption measured in 1,000s of barrels per day, renewable energy consumption in billions of KW-H, RGDP in billions of 2000 US dollars, and real oil prices in 2000 US dollars. Nominal oil prices measured in current US dollars per barrel.

Table 5. Dynamic forecasts

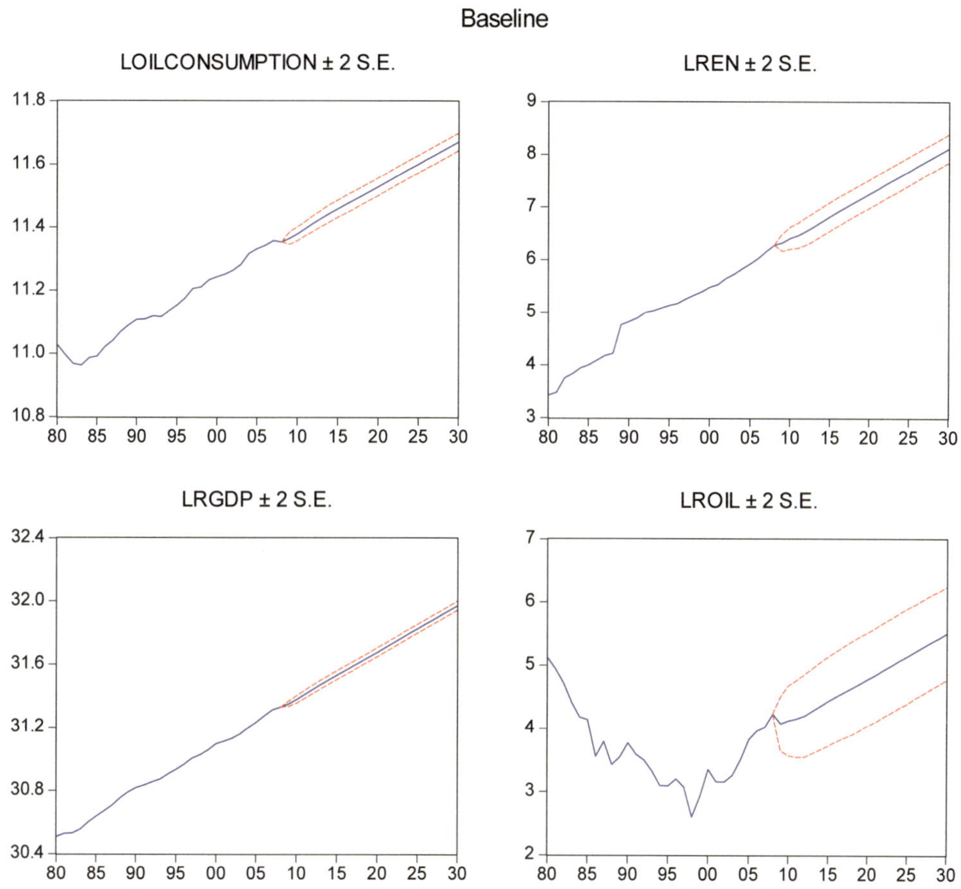

Fig. 6. Stochastic dynamic forecasts

A few comments are in order regarding the assumptions under which the forecasts were made. The forecasts are made under the assumptions of no major energy policy or environmental policy changes that would dramatically reduce the consumption of oil and fossil fuel sources in favour of more renewable energy. In other words, the assumption is made that no major global climate change treaty, like a successor to Kyoto, is signed and ratified that creates a big reduction in the usage of fossil fuels. The forecasts also assume that peak oil will not result in severe shortages of oil such that the oil consumption forecasts cannot be met. By comparison, the International Energy Agency (2010) predicts that a combination of factors including slowing global economic growth, rising oil prices, increased usage of renewable, and concerns about peak oil (mostly realized through rising oil prices) will limit global oil consumption to 100 million barrels of oil per day in 2035.

7. Policy implications

The results from impulse response functions show that positive shocks to income increase renewable energy consumption. This is an important result showing that increasing global GDP will increase the consumption of renewable energy. This means that economic policies that speed economic growth and development will lead to increases in renewable energy consumption. Rising incomes lead to more disposable income and individuals with higher incomes have more concern for the environment which increases the usage of renewable energy. From a policy perspective, government policy at the national and international level aimed at increasing income is good for increasing renewable energy consumption. Government policies that can increase income and wealth generation include stable and well designed monetary and fiscal policy, well functioning labour markets, policies that focus on increasing innovation and productivity, and an economic landscape free of corruption and rent seeking. While these policies need to be enacted at the national level, international organizations like the IMF, World Bank, and United Nations need to promote the benefits of sound economic and environmental policies to their member countries.

The results from impulse response functions show that positive shocks to oil consumption increase renewable energy consumption. The relationship between renewable energy consumption and oil consumption is of less interest because as global energy consumption grows so too do the individual energy sector components like oil and renewables.

The results from impulse response functions show that positive shocks to real oil prices have very little impact on renewable energy consumption. This result is somewhat counter intuitive to what is often presented in the popular media. Media stories often point to higher oil prices as a primary driver behind the expansion of the renewable energy business and increases in renewable energy consumption. This view is also supported by economic theory. Rising oil price should create incentives to substitute away from expensive fossil fuels towards alternatives like renewable. Over the period which the VAR was estimated (1980-2008) real oil prices were not continuously rising but instead resembled more of a U shape. In fact, over the estimation period, the highest recorded real oil prices were in 1980. The rather weak relationship between oil prices and renewable energy has been noted by others. In a somewhat different context, Henriques and Sadorsky (2008) and Sadorsky (2010) have looked at the impact of oil prices on the stock prices of clean energy companies. Henriques and Sadorsky (2008) find that the stock prices of clean energy companies are more responsive to changes in technology then to changes in oil prices. In a slightly different context, investigating the relationship between risk factors and clean energy stock prices, Sadorsky (2010) finds that company sales growth has a negative impact on company risk (CAPM beta) while oil price increases have a positive impact on company risk.

Looking into the future, it is expected that if real oil prices rise by enough, then a substitution effect from oil to renewable energy will follow. The uncertainty is that it is not clear just how high oil prices will have to rise in real terms to create a significant substitution effect. As Figure 1 shows, real oil prices in 2008 were just half of their value in 1980. In a world of rapidly rising oil prices, government policy can play a role in speeding up the substitution from oil to renewable energy.

What if issues like climate change, resource security, or resource depletion become major constraining factors, and that it is not possible to secure 117 million barrels of oil per day in 2030 or major climate change initiatives are implemented to drastically reduce the amount of CO_2 being emitted? In these cases, additional government policy is called for to further increase the usage of renewable energy. One approach to increasing the usage of renewable energy consumption at the global economy level is to use the stabilization wedges approach (http://cmi.princeton.edu/wedges/) advocated by the Carbon Mitigation Initiative at Princeton University (http://cmi.princeton.edu/).

The Carbon Mitigation Initiative (CMI) identifies that currently, 8 billion metric tons of carbon dioxide are being released globally into the atmosphere from the burning of fossil fuels. In the next 50 years this amount will double. The lead researchers on the project, Robert Socolow and Stephen Pacala, have identified 15 stabilization wedges with each wedge representing a strategy to reduce annual carbon emissions by 1 billion metric tons[15]. The technology currently exists to make each one of these wedges operational. The efficiency category contains four wedges (raise the fuel economy of 2 billion cars from 30 mpg to 60 mpg, reduce the average number of miles travelled from 10,000 to 5,000, increase efficiency in heating, cooling, lighting, and appliances by 25%, increase coal-fired power plant efficiency from 40% to 60%). The fuel switching category has one wedge (replace 1400 coal fired electric plants with natural gas-powered facilities). The carbon capture and storage category has three wedges (introduce carbon capture and storage at 800 large coal-fired plants or 1,600 natural gas-fired plants, use carbon capture in carbon derived hydrogen plants, use carbon capture in carbon derived synthetic fuels plants). The nuclear category has one wedge (double the current global nuclear energy capacity). The wind category has one wedge (increase wind electricity capacity by 15 times relative to today for a total of 2 million windmills). The solar category has two wedges (install 350 times the current capacity of solar electricity, use 40,000 square kilometers of solar panels to produce hydrogen for fuel cell powered cars). The biomass fuels category has one wedge (increase ethanol production by 15 times by creating biomass plantations with area equal to 1/6 of world cropland). The natural sinks category has two wedges (eliminate tropical deforestation, adopt conservation tillage practices in all agricultural worldwide).

According to the CMI calculations, global carbon emissions from burning fossil fuels are currently around 8 billion metric tons a year. They identify three possible paths for carbon emissions. Path one is to continue along the same current business as usual trend line for carbon emissions. Following this path results in a projected 16 billion metric tons of carbon dioxide being released into the atmosphere from burning fossil fuels in 2057. In 2057, carbon dioxide emissions would reach 800 ppm and average global temperatures would rise 9°F. Path two is designed to hold carbon emissions from fossil fuels at today's rate of 8 billion metric tons per year. This means that 8 wedges have to be cut by 2057. Further cuts are required after 2057. In 2057, this path leads to a carbon concentration of 525 ppm and an average global temperature increase of 5.4°F. The third option calls for a reduction in emissions below the present value. A total of 12 wedges are cut. In 2057, this path leads to a carbon concentration of 450 ppm and an average global temperature increase of 3.6°F. One of the real advantages of the carbon wedge stabilization concept is that each wedge represents a reduction in carbon dioxide emissions of one billion metric tons and that the technology currently exists to make each wedge fully operational.

Many approaches to increasing the usage of renewable energy and reducing the usage of fossil fuels advocate taxing fossil fuels and subsidizing renewables and in doing so create incentives to shift consumption form fossil fuels to renewables. Another approach is to focus on energy security and create industrial policy to reduce the dependence on oil.

---

[15] http://cmi.princeton.edu/wedges/intro.php

Some, like Lovins et al (2004) have already laid out a viable road map for how, by 2050, the United States economy can be flourishing with no oil at all. They advocate investing $180 billion over the next decade to eliminate U.S. oil dependence. The U.S. current imports approximately 66% of the oil consumed on a daily basis. Rather than spending money on importing oil, investments can be made to lessen and gradually eliminate the dependence on imported oil. In their calculations, this investment will result in a gross savings of $130 billion. They advocate using a four step approach that relies on creative destruction to revitalize the energy sector. First, oil efficiency must be doubled by using advanced but proven technologies to design and build ultralight vehicles. Second, business and public policies must be developed to accelerate the design and manufacturing of light weight materials, like carbon-fiber composites, that can be used in buildings, vehicles, heavy trucks and airplanes. Third, petroleum products must be replaced with biofuels. Fourth, increase efficiency in the natural gas sector to save half the projected 2025 use of natural gas. These savings will mean more natural gas can be used to make hydrogen and this will provide a convenient secure path to the hydrogen economy. While this four step plan is specifically designed for the United States, there is no reason why parts or all of this strategy could not be adopted in other countries.

8. Conclusions

This paper develops and estimates a VAR model of renewable energy consumption, income, oil consumption, and real oil prices in order to investigate the dynamic interaction between these variables. The model is estimated using global data over the period 1980-2008. The VAR model fits the data well. The results from impulse response functions show that positive shocks to income or oil consumption increase renewable energy consumption. Shocks to real oil prices have little impact on renewable energy consumption.

One of the important policy implications of these results is that government policy at the national and international level aimed at increasing income is good for increasing renewable energy consumption. Examples of good government policies that can increase income and wealth generation include stable and well designed monetary and fiscal policy, well functioning labour markets, policies that focus on increasing innovation and productivity, and an economic landscape free of corruption and rent seeking. While these policies need to be enacted at the national level, international organizations like the IMF, World Bank, and United Nations need to promote the benefits of sound economic and environmental policies to their member countries.

If other issues like climate change, resource security, or resource depletion become major constraining factors, then additional government policy is called for to further increase the usage of renewable energy. Government can tax fossil fuels and subsidize renewable energy. In particular government can provide subsidies in the form of R&D credits, low cost loans, and production tax credits to speed the development and roll out of renewable energy products. Government can provide renewable portfolio standards and subsidies to consumers and early adopters of new renewable energy products. Government can also create new energy efficiency and carbon-fiber composites industrial policy designed to lessen the dependence on oil. This new industrial policy would increase energy efficiency by using new existing technologies to design and build ultralight vehicles and to accelerate the design and manufacturing of light weight materials, like carbon-fiber composites, that can be used in buildings, vehicles, heavy trucks and airplanes.

9. References

Almeida, P. & Silva,P. (2009). The Peak Of Oil Production-Timings And Market Recognition. *Energy Policy*, Vol.37, No.4, (April 2009), pp. 1267-1276, ISSN: 0301-4215

Apergis, N. & Payne, J. (2010a). Renewable Energy Consumption And Economic Growth: Evidence From A Panel Of OECD Countries. *Energy Policy*, Vol.38, No1, (January), pp. 656-660, ISSN: 0301-4215

Apergis, N. & Payne, J. (2010b). Renewable Energy Consumption And Growth In Eurasia. *Energy Economics*, Vol.32, No.6, (November 2010), pp. 1392-1397, ISSN: 0140-9883

Apergis, N. & Payne, J. (2011). The Renewable Energy Consumption-Growth Nexus in Central America. *Applied Energy*. Vol.88, No.1, (January 2011), pp.343-347, ISSN: 0306-2619

Brooks, C. (2002). *Introductory Econometrics for Finance*, Cambridge University Press, Cambridge, United Kingdom. ISBN 0-521-79367-X

BP *Statistical Review of World Energy*, June 2010 (www.BP.com)

Chang, T-H.; Huang, C-M & Lee, M-C (2009). Threshold Effect Of The Economic Growth Rate On The Renewable Energy Development From A Change In Energy Price: Evidence From OECD Countries. *Energy Policy*, Vol.37, No.12, (December 2009), pp. 5796-5802, ISSN: 0301-4215

Chien, T. & Hu, J-L (2007). Renewable Energy And Macroeconomic Efficiency Of OECD And Non-OECD Economies. *Energy Policy*, Vol.35, No.7, (July 2007), pp. 3606-3615, ISSN: 0301-4215

Chien, T. & Hu, J-L (2008). Renewable Energy: An Efficient Mechanism To Improve GDP. *Energy Policy*, Vol.36, No.8, (August 2008), pp. 3045-3052, ISSN: 0301-4215

Henriques, I. & Sadorsky, P. (2008). Oil Prices And The Stock Prices Of Alternative Energy Companies. *Energy Economics*, Vol.30, No.3, (May 2008), pp. 998-1010, ISSN: 0140-9883

Hubbert, M.K. (1956). Nuclear Energy And Fossil Fuels. Publication Number 95, Shell Development Company, Houston, Texas, 1956. Presented on the Spring Meeting on Drilling and Production Practice, March 7–9, *American Petroleum Institute*, San Antonio, Texas

Hubbert, MK. (1971). The Energy Resources On The Earth. In: *Energy and Power, a Scientific American Book*. W. H. Freeman & Co., San Francisco, pp. 31–40, ISBN: 0716709384

International Energy Agency. (2007). World Energy Outlook, Paris, ISBN: 978-92-64-02730-5

International Energy Agency. (2010). World Energy Outlook, Paris, ISBN: 978-92-64-08624-1

Labatt, S. & White, R. (2007). *Carbon Finance*, John Wiley & Sons, Hoboken, New Jersey. ISBN-13 978-0-471-79467-7

Lovins, A.; Datta, E.K.& Bustnes, O-E, (2004). *Winning the Oil Endgame*, Rocky Mountain Institute, Colorado, ISBN 10: 1881071103

Lutkepohl, H. (2005). *New Introduction to Multiple Time Series Analysis*, New York, Springer, ISBN 978-3-540-26239-8

Marques, A.C.; Fuinhas, J.A. & Pires Manso, J.R. (2010). Motivations Driving Renewable Energy In European Countries: A Panel Data Approach. *Energy Policy*, Vol.38, No.11,(November 2010), pp. 6877-6885, ISSN: 0301-4215

Menyah, K. & Wolde-Rufael, Y. (2010). CO2 Emissions, Nuclear Energy, Renewable Energy And Economic Growth In The US. *Energy Policy*, Vol.38, No.6, (June 2010), pp. 2911-2915, ISSN: 0301-4215

Sadorsky, P. (2009a). Renewable Energy Consumption, CO2 Emissions And Oil Prices In The G7 Countries. *Energy Economics*, Vol.31, No.3, (May 2009), pp. 456-462, ISSN: 0140-9883

Sadorsky, P. (2009b). Renewable Energy Consumption And Income In Emerging Economies. *Energy Policy*, Vol.37, No.10, (October 2009), pp. 4021-4028, ISSN: 0301-4215

Sadorsky, P. (2010). Modeling Renewable Energy Company Risk. *Energy Policy*, in press, doi:10.1016/j.enpol.2010.06.064, ISSN: 0301-4215

Stern, N. (2007). *The Economics of Climate Change: The Stern Review*, Cambridge: Cambridge University Press, ISBN: 9780521700801

Stern, N. (2009). *The Global Deal: Climate Change and the Creation of a New Era of Progress and Prosperity*, Public Affairs, New York, ISBN 978-1-58648-669-3

Toda, H.Y.& Yamamoto, T. (1995). Statistical Inference In Vector Autoregressions With Possibly Integrated Process. *Journal of Econometrics* Vol.66, No.1-2, (March-April 1995), pp. 225–250, ISSN: 0304-4076

Verbruggen, A. & Al Marchohi, M. (2010). Views On Peak Oil And Its Relation To Climate Change Policy. *Energy Policy*, Vol.38, No.10, (October 2010), pp. 5572-5581, ISSN: 0301-4215

24

Influence of Global Warming on the RC Structures and Durability Monitoring in Civil Engineering

Guofu Qiao et al.*
School of Civil Engineering
Harbin Institute of Technology, Harbin
China

1. Introduction

Global warming has or will affect all aspects of human life on Earth. We will focus on the influence of global warming in civil engineering and the structural health monitoring (SHM) technologies here. Obviously, reinforcing concrete (RC) structures are the most important structural style in civil engineering. With the development of global warming and further deterioration of the environment, the conditions of RC structures now becomes more atrocious than that of before. In the past few decades, many tremendous bridges, skyscrapers, super dams and other huge harbor works have been built. These structures are expected to serve safely for tens of years to hundreds of years. Unfortunately, the corrosion of the reinforcing steel, called as "cancer of the steel", has been a world-wide problem which deteriorates the durability of RC structures and degrades the serviceability severely. The corrosion of RC structures has resulted in very high repair costs, which sometimes are even much greater than the initial construction cost, and in some extreme situations, leaded to collapse of the structure. The corrosion of the reinforcing steel in concrete will then be a fatal attack to the durability of RC structures. Fortunately, structural health monitoring systems enable scientists and engineers to identify the status of structures, and then provide the foundation for safety assessment, service-life prediction, maintenance & reinforcement and full-life design of structures.

2. Mechanism of the reinforcing steel's corrosion

Generally, the hydrated reactions of the cement can be shown as Eq. (1) to Eq. (7). As the hydration reaction of cement is completed, the cement should been changed into a complex multi-phase and heterogeneous body[1]. Hydration product normally provides a high

*Tiejun Liu[2], Guodong Sun[3], Yi Hong[4], Baoguo Han[1], Huigang Xiao[1], Zhichun Zhang[4] and Jinping Ou[1,5]
[1]*School of Civil Engineering, Harbin Institute of Technology, Harbin, China*
[2]*Shenzhen Graduate School, Harbin Institute of Technology, Shenzhen, China*
[3]*Department of Computer Science and Technology, Tsinghua University, Beijing, China*
[4]*Center for Composite Material and Structure, Harbin Institute of Technology, Harbin, China*
[5]*School of Civil & Hydraulic Engineering, Dalian University of Technology, Dalian, China*

$$2C_3S + 11H \rightarrow C_3S_2H_8 + 3CH \quad (1)$$

$$2C_2S + 9H \rightarrow C_3S_2H_8 + CH \quad (2)$$

$$C_3A + 3C\overline{S}H_2 + 26H \rightarrow C_6A\overline{S}_3H_{32} \quad (3)$$

$$2C_3A + C_6A\overline{S}_3H_{32} + 4H \rightarrow 3C_4A\overline{S}H_{12} \quad (4)$$

$$C_4A\overline{S}H_{12} + 2C\overline{S}H_2 + 16H \rightarrow C_6A\overline{S}_3H_{32} \quad (5)$$

$$C_4AF + 3C\overline{S}H_2 + 21H \rightarrow C_6(A,F)\overline{S_3}H_{32} + (F,A)H_3 \quad (6)$$

$$C_4AF + C_6(A,F)\overline{S_3}H_{32} + 7H \rightarrow 3C_4(A,F)\overline{S}H_{12} + (F,A)H_3 \quad (7)$$

degree of protection to the reinforcing steel against corrosion, by virtue of the high alkalinity (CH, PH > 13.5) of the pore solution. Under high alkalinity, steel remains passivated. The basic factors and fundamental corrosion processes of the reinforcing steel's corrosion are illustrated as Fig.1. When the pH value of the pore solution drops to low values due to the carbonation reaction $Ca(OH)_2 + CO_2 + H_2O \rightarrow Ca(HCO_3)_2 + CaCO_3$ or as sufficient chloride ions penetrate into the reinforcement, the protective film is destroyed and the reinforcing steel is depassivated. Furthermore, corrosion in the form of rust formation and loss in cross-section of the rebar occur in the presence of oxygen and water (humidity).

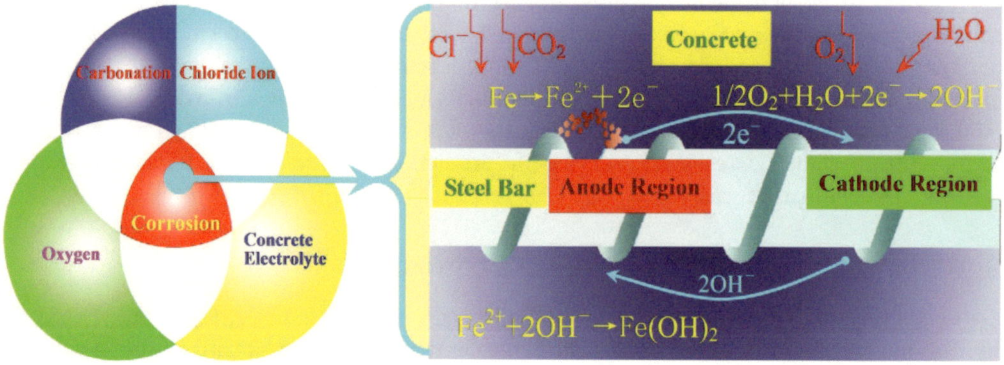

Fig. 1. Schematic plan of the reinforcing steel's corrosion process in concrete

Essentially, the corrosion process is an electrochemical reaction process. CO_2 and Cl^- often lead to general corrosion and pitting corrosion, respectively. During the general corrosion process, Fe atom is oxidized to $Fe(OH)_2$, Fe_3O_4, $Fe(OH)_3$ or $Fe_2O_3 \cdot nH_2O$ at the anode region.

The oxidation state depends on the content of O2 in the concrete. On the other hand, O atom at the cathode region is reduced to O^{2-}.The micro anode and cathode on the surface of the reinforcing steel form the microcell. Compared with that of the general corrosion, the pitting corrosion process is much more complicated. It is generally agreed that the barrier oxide layer of the metal is a highly defective structure, with the point defects being metal and oxygen vacancies and metal interstitials. Macdonald has developed a point model (PDM) to descript the growth and breakdown of passive films[2]. The fundamental basis of the general PDM is illustrated in Fig.2, in which the basic reactions that lead to the generation and annihilation of point defects at the interfaces (Reactions (1)-(6)), together with film dissolution (Reaction (7)), are displayed.

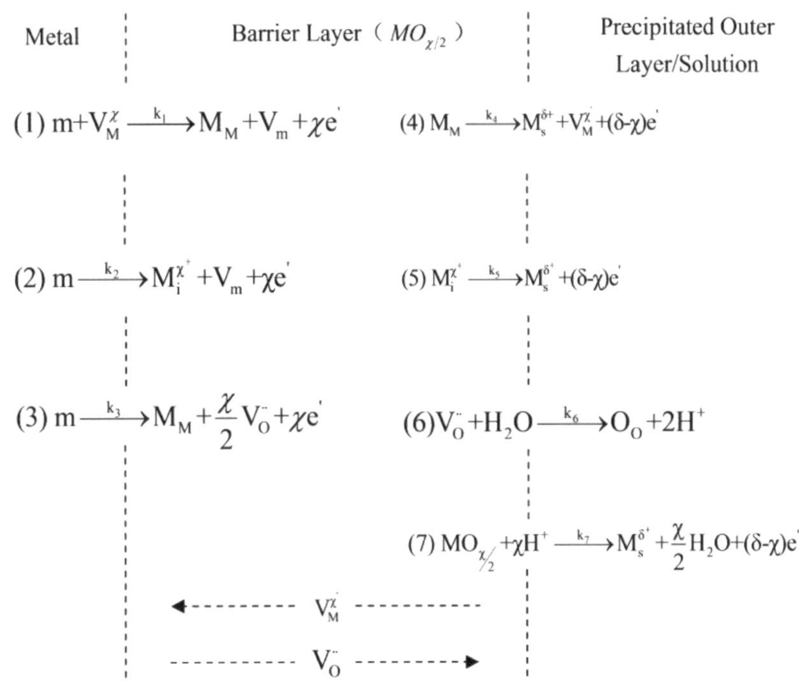

Fig. 2. Schematic of physico-chemical processes that occur within a passive film according to the Point Defect Model

According to D.Macdonald's PDM, the anion Cl^- is able to enhance the flux of cation vacancies through the barrier layer, such that under favorable conditions (voltage, PH, [Cl^-]) vacancy condensation will occur at the metal/barrier layer interface and hence passivity breakdown will ensue. The initial reaction occurring at the film/solution interface is the absorption of an aggressive anion (Cl^-) into a surface oxygen vacancy followed by a Schottky-pair-type reaction. Anion absorption leads to the generation of cation vacancies at the film/solution interface and, hence, to the flux of cation vacancies across the passive film to the film/metal interface. If the cation vacancy diffusivity is so high that the flux of cation vacancies can not be accommodated by that of cations generated at the metal/film interface, a cation vacancy condensate will be formed, causing the local thinning of the film or detachment from the metal. Once the condensate has grown to a critical size, the film ruptures and hence a rapid local attack occurs, either due to complete dissolution or else to the stress in the film that induces a mechanical or structural instability. The postulated reactions of the aggressive anions Cl- in this process are illustrated in Fig.3.

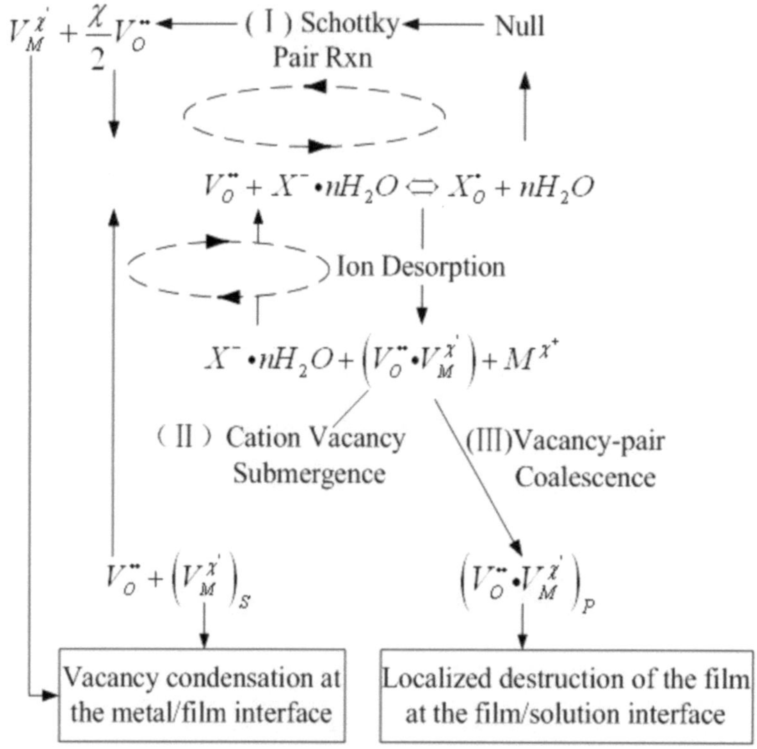

Fig. 3. Postulated reactions for cation vacancy generation at the barrier layer/solution interface according to the Point Defect Model

Although the general corrosion and the pitting corrosion process are very complicate, it must keep in mind that the electron is presented and transferred inside the barrier layer or on the surface of the reinforcing steel. The energy from the steel-smelting process is delivered by the electrochemical reactions and physicochemical processes described above. Furthermore, this delivered energy will be applied as the power source of the self-powered wireless corrosion monitoring sensor in future.

3. Characterize the corrosion by the EIS and EN techniques

3.1 Review of the electrochemical measurement techniques

Equivalent circuit (EC) is the transfer function of the electrochemical system. The EC in Fig.4 is the universal transfer function which can be used to simulate and characterize the electrochemical characteristics of the steel-concrete system[3,4].

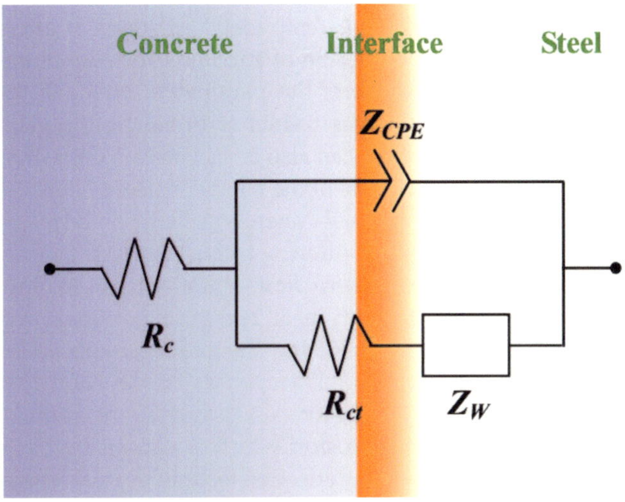

Fig. 4. EC of steel-concrete system. R_c, R_{ct}, Z_W and Z_{CPE} are the resistance of concrete, Faraday resistance, impedance of the diffusion process and the impedance of CPE, respectively.

R_c and R_{ct} are the resistance of the concrete and reactive resistance of the corrosion, respectively. Z_W and Z_{CPE} represent the diffusion impedance of oxygen and dispersion impedance of the concrete-steel interface, respectively. The diffusion of the oxygen results in the Warburg impedance. Z_w can be calculated as follows:

$$Z_w = \frac{1}{Y_{0W}}(j\omega)^{-\frac{1}{2}} = \frac{\sigma}{\sqrt{\omega}} - j\frac{\sigma}{\sqrt{\omega}} \tag{8}$$

Where σ is the Warburg coefficient. Y_{0W} is the admittance as follows,

$$Y_{0W} = \frac{nFC_s\sqrt{D}}{\gamma Z_F^0 |I_F|} \tag{9}$$

Where Z_F^0 is Faraday impedance without diffusing effect, γ is the reaction order of the reactant; I_F is the faradic current; n, F are the constants, C_s is the concentration at electrode's surface. The coarse interface of the steel-concrete, isolated reaction region and the heterogeneous nature of concrete are the most important factors causing the dispersion effect. This phenomenon is depicted by the constant phase element (CPE). The CPE is expressed as Eq. (10) by IUPAC.

$$Z_{CPE} = \frac{1}{Y_{0dl}}(j\omega)^{-n} = \frac{1}{Y_{0dl}}\omega^{-n}\left(\cos\frac{n\pi}{2} - j\sin\frac{n\pi}{2}\right) \tag{10}$$

Where Y_{0dl} is basic admittance and n is the constant in the region (0, 1],

$$Y_{0dl} = \frac{\omega^{-n}}{|Z_{CPE}|} \tag{11}$$

The dispersion effect and the diffusion of the oxygen can prolong the balance time of the electrochemical system, or even lead the system never to reach the balance state. According to the results of V.Feliu, J.A.Gonzalez and C.Andrade[3,4], dispersion effect and diffusion effect can influence the response of the excitation of galvanostatic step at initial time region and long term region, respectively. Therefore, the parameters of R_c, R_{ct} and Z_{CPE} can be extract accurately by fitting the experimental results at initial time region. Applying the experimental results at long term region, Z_W can also be calculated. Once the elements in the EC are calculated, the corrosion characteristics of the concrete-steel system can be clarified.
The electrochemical corrosion status of the steel-concrete system can be recognized by active monitoring technique (AMT) and passive monitoring technique (PMT). As the EC in Fig.4 is the universal transfer function for general corrosion, abundantly actuating signals can be applied to excite the concrete-steel system in linear region or nonlinear region. Therefore, the electrochemical elements which include R_c, R_{ct}, Z_W and Z_{CPE} can be extracted in time domain or frequency domain by analyzing the responses of the concrete-steel system to these actuating signals. Table 1 listed the major AMTs which are applied extensively in corrosion measurement. For the pitting corrosion which is caused by Cl^-, electrochemical emission spectrum (EES) in PMTs is the most effective technique to recognize the corrosion status of the reinforcing steel. Essentially, electrochemical noise (EN) reflects the intrinsic information of the initial, metastable, repassive and stable stages during the pitting corrosion process. Listening passively to the reinforcing steel confides how the corrosive mediums intrude the RC structures, the pitting corrosion status can be identified. Many methods have been developed to analyze the EN data in time domain[5], frequency domain[6] or in chaos domain[7]. Wavelet transform (WT) has been proposed as an alternative tool to overcome the limitations of FFT in the analysis of non-stationary signals of the EN data. It can provide the information on transients in time domain and the possibility of working with non-stationary signals. The previous research has shown that It is much better to decompose the EN data during the pitting process of the reinforcing steel to wavelet space, and extract the characteristics of the EN based on WT much more effectively[8].

AMT	Actuating Signal	Response Signal
Electrochemical Impedance Spectrum(EIS)	Current/Voltage	Voltage/Current
Electrochemical Frequency Modulation(EFM)	Current/Voltage	Voltage/Current
Harmonic Analysis(HA)	Current/Voltage	Voltage/Current
Transient Galvanostatic Decay	Current	Voltage
Transient Potentiostatic Decay	Voltage	Current
Potentialdynamic Scan	Voltage	Current
Linear Polarization	Current/Voltage	Voltage/Current
Coulostatic Method	Electric Charge	Voltage/Current

Table 1. AMTs in corrosion measurement

Electrochemical Impedance Spectrum(EIS) technique has been widely used as a corrosion measurement tool. A number of publications have reviewed both the theory and many areas of application [9-12]. These reviews suggest that AC impedance is a useful technique for determining mechanistic and kinetic information about the corrosion processes occurring at corroding interfaces. The experimental impedance data $Z(j\omega)_{exp}$ may be well approximated by the impedance $Z(j\omega)_{equiv}$ of an equivalent circuit made up of ideal resistors, capacitors, inductors and various distributed elements. In such a circuit, elements represent the various processes involved in the transport of charge and mass in the system being investigated. Various equivalent circuits have been proposed to explain the impedance response of film-covered surfaces, which involve cathodic and anodic reactions in the system. In the last few decades, many researchers [13, 14] have done lots of work about steel corrosion. By extracting the information from the high, medium and low frequency region of EIS, many qualitative and quantitative parameters about electrochemical process of carbon steel in cement mortar can be gotten. The corrosion process of carbon steel in cement mortar can be reflected by these parameters.

Electrochemical noise (EN) technique has been widely used to study the corrosion process of various metallic materials, such as metals [15, 16] and organic coated metals [17, 18], and it can provide useful information about the corrosion mechanism by measuring the fluctuations of the current and potential simultaneously, which are generated during the corrosion process. One of the advantages of electrochemical noise technique is that it does not conduct any disturbing signal. Therefore it is able to avoid the artificial disturbances to the system during the measurement [19]. The sensitivity of EN measurement is much higher than that of the other traditional techniques to the localized corrosion process [20]. Many methods have been developed to analyze the EN data [21], including statistical analysis [22, 23] and spectral analysis [24-26]. However, the most commonly-used EN analysis methods (statistical and spectral) mentioned above are devised for stationary signals that do not show distinctive transients .The main disadvantage of those methods is that they analyze signals by averaging the features across the whole time record [27]. DC trend removal method is another problem which obstacles the application of ENA in practical engineering.

Wavelet transform (WT) has been proposed as an alternative tool to overcome the limitations of FFT in the analysis of non-stationary signals of EN data [28, 29]. It can provide the information on transients in time and the possibility of working with non-stationary signals, so it has been used to differentiate corrosion type and to study corrosion mechanism of process [30, 31]. Only some work of EN for studying the corrosion of reinforcing steel in concrete has been reported [32, 33]. A. Legat[33] found that the electrochemical noise technique was able to follow the high corrosion dynamics in concrete, and the measured signals containing certain fluctuations indicated that the process of corrosion initiation consisted of a sequence of several events. Their results of the measurements also revealed that the distribution of anodic and cathodic sites could alternate during the wetting and drying of concrete. Hu et al. [34] studied the corrosion behavior of reinforced steel in the simulated concrete pore solution by using electrochemical noise, and they analyzed the EN data by discrete wavelet and provided a criterion to determine the threshold value controlling the turning of corrosion state of steel in concrete. The aim of this work is to study the corrosion process of reinforcing steel in cement mortar and illustrate the fact that the EDP is an effective indicator of the pitting corrosion in corrosion monitoring

3.2 EIS Performance of the RC system

Three-electrode system has been used in the experiment. The working electrode is Q235 carbon steel which is widely used in civil engineering. The specimens are embedded in epoxy resin. Working electrode's surface is polished with 400,800, 1200 and 2000 grit silicon carbide paper and cleaned with ethanol and acetone. Its working area is 1.13cm^2. When steel corrosion in concrete is studied by EN, the carbon steel specimen is embedded in cement mortar (Fig.5).When EIS is measured, one of the two identical working electrodes is substituted for graphite electrode. The mass ratio of sand, cement and water is 3:1:0.6. SCE reference electrode is used to measure potential.

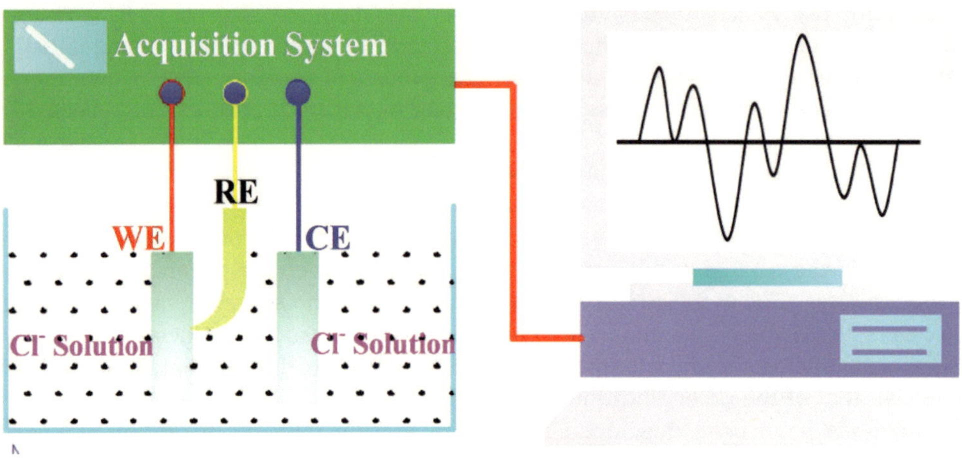

Fig. 5. Schematic plan of the three-electrode electrochemical measurement system

Two kinds of influencing factors which are temperature and concentration of NaCl solution are considered during the experiment. EN is measured under the condition that the temperature is 50°C and the concentration of NaCl solution is 3.0%. The sampling frequency is 2Hz. As temperature and concentration of NaCl solution is changed from 26°C to 50°C and 0.5% to 5.0%, respectively, EIS is measured. Princeton Applied Research Potentiostat/Galvanostat Model 263A is used. The frequency range of EIS measurement is 10^{5}Hz to 10^{-3}Hz. In order to accelerate the seepage velocity of NaCl solution to the steel bar surface in concrete, all the cement mortar specimens are put into the pressure pan. We have kept the pressure of the span at 7MPa for 60mins by the pressure pump before unloading it. Then all the specimens are immersed in different concentration of NaCl solution for 70days.

The Nyquist plots at 26°C and 50°C are shown in Fig.6., Fig.7 and Fig.8 show the Bode plotsat 26°C and 50°C, respectively[8]. Concentration of sodium chloride solution changes from 0.5% to 1.0%, 3.0% and 5.0%. From the Nyquist and Bode plots we can see that there are two pieces of capacitive impedance arc. The capacitive impedance arcs at high frequency (from 10^{5}Hz to 10^{3}Hz) are generated by the sodium chloride solution outside the cement mortar layer. The resistance of the solution is no more than a few hundred Ohms. The other capacitive impedance arcs are caused by the cement layer. The equivalent circuit (EC) of cement layer is composed of resistance element R_c and capacitive element C_c. Obviously, the capacitive impedance arc has been flattened in the medium frequency region (between KHz to 100mHz). This phenomenon is often caused by CPE. Smearing phenomenon has arisen at the low frequency region.

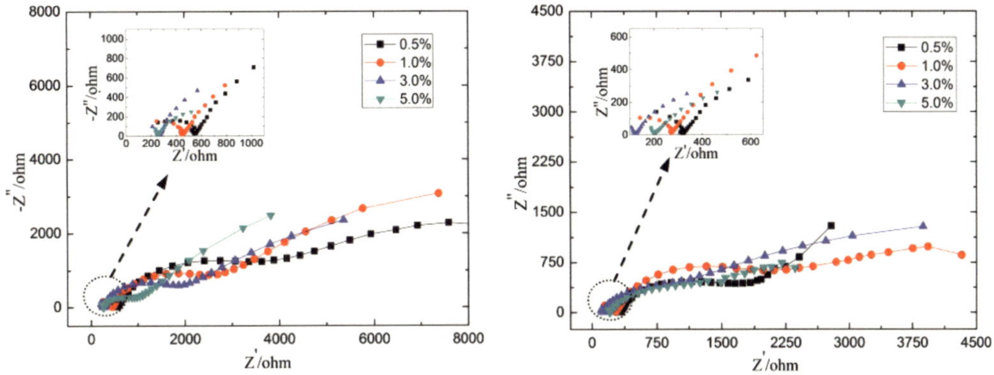

(a) 26°C (b) 50°C

Fig. 6. Nyquist plots of Q235 carbon steel in cement mortar with different NaCl solution

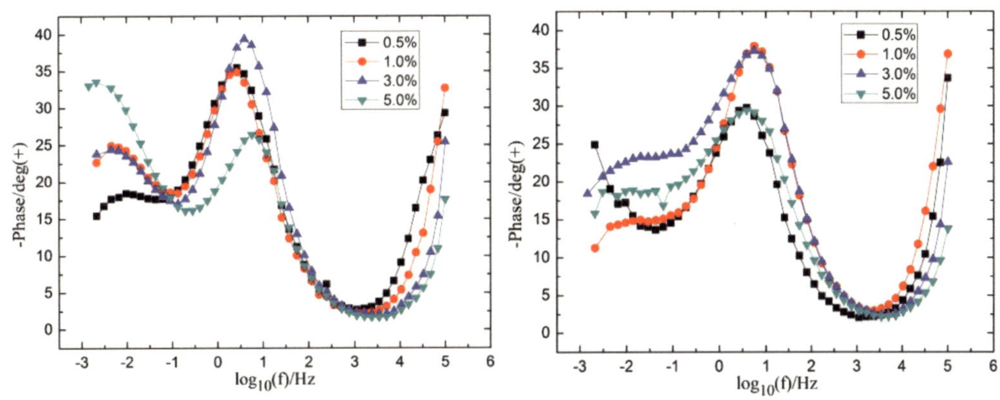

(a) 26°C (b) 50°C

Fig. 7. Bode plots of Q235 carbon steel in cement mortar with different NaCl solution

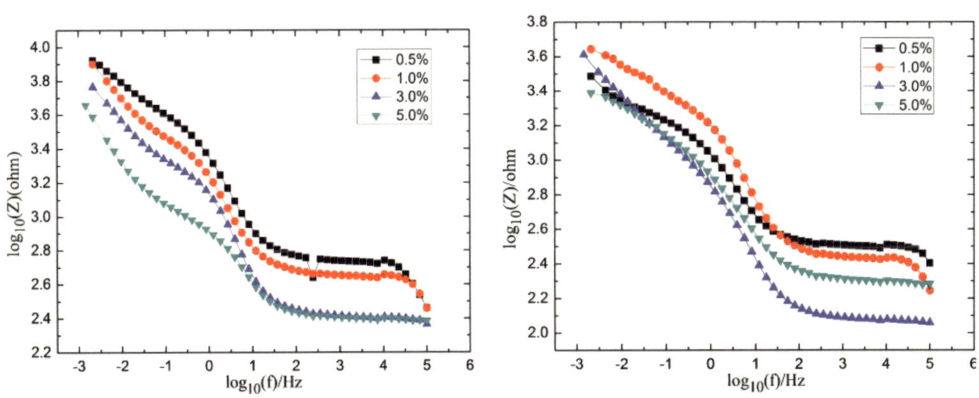

(a) 26°C (b) 50°C

Fig. 8. Impedance-frequency plots of Q235 carbon steel immersed in different NaCl solution

According to the Nyquist and Bode plots feature, the much more detailed EC in Fig.9 is used to simulate the Q235 carbon steel corrosion in the cement mortar. R_{sol} is the resistance of chloride sodium solution. R_c and C_c are the resistance and capacity of cement layer respectively. Because there exists smearing phenomenon in the Nyquist plots, Warburg impedance exists in the electrochemical process. W represents the diffusing effect of Oxygen. Q represents the dispersion effect. By fitting the experiment results with the EC in Fig. 7 by ZSimpWin3.20 software, we can obtain the value of the parameters in the EC. The fitting results of 26°C and 50°C are listed in Table 2 and Table 3, respectively.

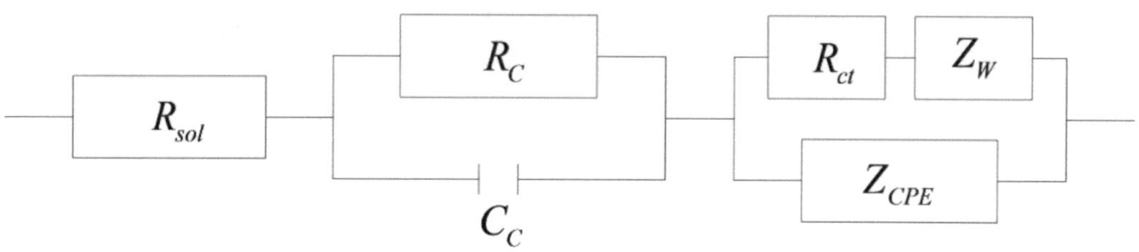

Fig. 9. Equivalent circuit for carbon steel in cement mortar

26℃	R_s $\Omega\,cm^2$	R_c $\Omega\,cm^2$	10^9C_c F cm^2	$10^{-3}R_{ct}$ $\Omega\,cm^2$	10^3Y_{0W} $\Omega^{-1}\,cm^{-2}\,S^{1/2}$	10^4Y_{0dl} $\Omega^{-1}\,cm^{-2}\,S^n$	n
0.5%	191.6	342.6	9.604	3.998	1.777	1.074	0.7155
1.0%	132.8	316.5	6.254	2.340	1.489	1.162	0.7723
3.0%	123.2	247.8	80.80	2.606	1.692	1.374	0.9339
5.0%	109.6	219.3	25.54	0.6511	2.783	1.482	0.7936

Table 2. Fitting results of $R_s(R_cC_c)((R_{ct}Z_W)Z_{CPE})$ at 26℃

50℃	R_s $\Omega\,cm^2$	R_c $\Omega\,cm^2$	10^9C_c F cm^2	$10^{-3}R_{ct}$ $\Omega\,cm^2$	10^3Y_{0W} $\Omega^{-1}\,cm^{-2}\,S^{1/2}$	10^4Y_{0dl} $\Omega^{-1}\,cm^{-2}\,S^n$	n
0.5%	127.66	289.4	3.897	1.349	4.942	1.900	0.7329
1.0%	100.2	257.3	46.74	0.2906	2.169	2.720	0.8054
3.0%	108.1	218.2	21.15	0.1172	0.994	3.596	0.7625
5.0%	93.18	195.0	2.005	0.1507	6.326	3.469	0.6060

Table 3. Fitting results of $R_s(R_cC_c)((R_{ct}Z_W)Z_{CPE})$ at 50℃

Either at 26°C or 50°C, there are two semi-circles in the Nyquist plots. This indicates that there are two time constants. One is caused by the cement mortar around the carbon steel WE (at the high frequency region), the other is caused by the electric double layer (at the medium frequency region).According to Fig.6 a), with the increasing of the NaCl

concentration, the diameter of the semi-circle in Nyquist plot decreases gradually and the shape of the EIS spectra shows noticeable change. The phase angles in Fig.7 a) increase gradually as the concentration of NaCl solution increases from 0.5% to 5.0%. Also, R_{ct} values decrease from 3998Ω·cm² to 0.6511Ω·cm² and Y_{0dl} values increase from 10740Ω^{-1}·cm^{-2}·S^n to 14820Ω^{-1}·cm^{-2}·S^n at same time. Although there is a fluctuation of R_{ct} which is 2606Ω·cm² in value as the concentration of NaCl solution is 3.0%, the R_{ct} will decrease with the increasing of the NaCl solution concentration. During this process, the passive film has been destroyed by the chloride ion. The chloride ion which is from the environment penetrates into the mortar and accumulates in the steel/mortar interface. This can cause the initiation and propagation of pitting corrosion. The destroying process of the chloride ion is accelerated by increasing the concentration of NaCl solution. The results show that the high concentration of NaCl solution will accelerate the influence of diffusion effect.
Smearing phenomenon has appeared in the low frequency region. This indicates that the diffusion process is the controlling process. Warburg impedance Z_w in Eq. (8) is often used to describe the semi-infinite plane diffusion's characteristic of oxygen. The Warburg diffusion process results from the diffusion of oxygen through the mortar to the steel/mortar interface. During this period, the system is limited by the diffusion process of oxygen. The Y_{0w} shows a fluctuant tendency in Table 2 and Table 3, which indicates that the concentration of NaCl solution has no significant influence on the diffusion process of the oxygen. But as the temperature is increased to 50℃, the value of Y0w is greater than its value with the same concentration of NaCl solution at 26℃.So the diffusion process proceeds more easily with the increasing of the temperature.

3.3 EN Feature of the RC system

3.3.1 Wavelet backgrounds

The wavelet analysis is a relatively new way of signal processing. It can overcome the problems of traditional Fourier transform. The continuous wavelet transform was proposed by Grossman and Morlet in 1984 [35]. Daubechies [36] constructed families of compactly supported wavelets. Mallat introduced the multi-resolution signal decomposition (MRSD) algorithm [37]. Since then the research in this field has progressed rapidly. The detailed description about wavelet transform can be found in the correlative literatures [38-40]. A wavelet is an oscillatory、real or complex function of zero average and finite length. The wavelets approach consists essentially in representing the time record $x_n(t), n = 1,2,...,N$ by special functions. By scaling and translating the father wavelets φ and the mother wavelets ψ as equation (12) and equation (13), the special functions which is a linear combination of basis functions $\varphi_{j,k}$ and $\psi_{j,k}$ can be generated.

$$\varphi_{j,k}(t) = 2^{-j/2}\varphi\left(2^{-j}t - k\right) = 2^{-j/2}\varphi\left(\frac{t - 2^j k}{2^j}\right) \tag{12}$$

$$\psi_{j,k} = 2^{-j/2}\psi\left(2^{-j}t - k\right) = 2^{-j/2}\psi\left(\frac{t - 2^j k}{2^j}\right) \tag{13}$$

Where $k = 1,2,\cdots,N/2^J$; and $J = 1,2,\cdots J$; J is often a small natural number which depends mainly on N and the basis function.

The time signal $x(t)$ can be represented as follows:

$$x(t) \approx \sum_{k} s_{J,k}\varphi_{J,k}(t) + \sum_{j=1}^{J}\sum_{k} d_{j,k}\psi_{j,k}(t) \tag{14}$$

Where $s_{J,k}$ and $d_{j,k}$ are so-called wavelet coefficients.

Unlike conventional techniques, wavelet decomposition produces a family of hierarchically organized decompositions. At each level j, the j-level approximation $A_j(t)$, and a deviation signal called the j-level detail $D_j(t()$ can be calculated according to the following equations.

$$A_j(t) = \sum_{k} s_{J,k}\varphi_{J,k}(t) \tag{15}$$

The detail $D_j(t)$ is defined as,

$$D_j(t) = \sum_{k\in Z} d_{j,k}\psi_{j,k}(t), j,k \in Z \tag{16}$$

$$d_{j,k} = \int_{-\infty}^{+\infty} x(t)\psi_{j,k}(t)dt \tag{17}$$

Decomposed process of the signal by wavelet transform can be illustrated as Fig.10.

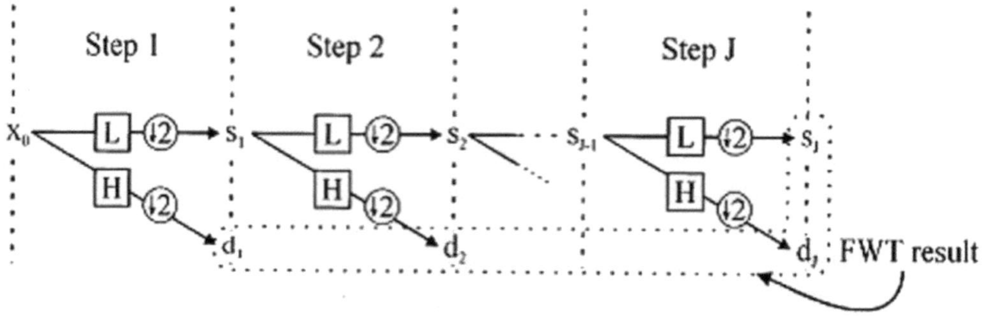

Fig. 10. General scheme of the FWT algorithm

The contribution of each crystal to the overall signal can be estimated by energy distribution plot (EDP), which is the plot of the relative energy accumulated by each crystal versus the crystal name. The term "crystal" is used because the wavelet coefficients in a crystal correspond to a set of translated wavelet functions arranged in a regular lattice. The energy E of the whole signal is calculated as follows:

$$E = \sum_{n=1}^{N} x_n^2, \quad n = 1,2,\ldots,N \tag{18}$$

The relative energy corresponding to each crystal can be calculated as follows:

$$E_j^d = \frac{1}{E}\sum_{n=1}^{N/2^j} d_{j,n}^2, \; j = 1,2,\ldots,J \tag{19}$$

$$E_j^s = \frac{1}{E}\sum_{n=1}^{N/2^j} s_{J,n}^2 \tag{20}$$

Because the Symlet wavelet is orthogonal, the following equation is satisfied:

$$E = E_J^s + \sum_{j=1}^{J} E_j^d \tag{21}$$

According to Eq. (19), (20) and (21), the EDP can be plotted[8].

3.3.2 Wavelet analysis of EN

The electrochemical potential noise (EPN) and electrochemical current noise (ECN) are listed in Fig.11 and Fig.12 respectively[8]. The NaCl solution is not injected at this time. From Fig.11 we can see there is a DC trend from -0.61315V to 0.61689V in the EPN data. The amplitude of the EPN in Fig.11 is very low (10^{-5}V degree).Fig.12 shows the ECN before NaCl solution is injected. The current amplitude is 10^{-7}A degree which almost remains unchanged in all the time-record. Also, there is a DC trend in the ECN data. The frequency of the signal in ECN and EPN is very high. From Fig.16 and Fig.12 we can see that the signals are caused by the measuring apparatus and the concrete-steel system. But the amplitude of EPN and ECN is very low and the accuracy is satisfied.

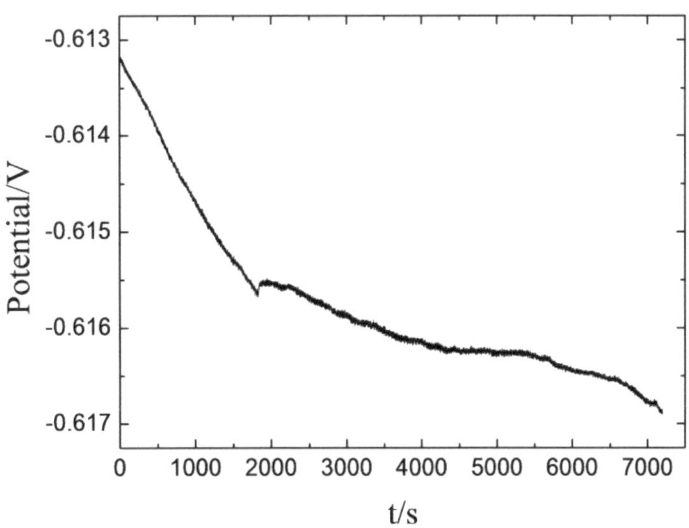

Fig. 11. EPN before NaCl solution injected

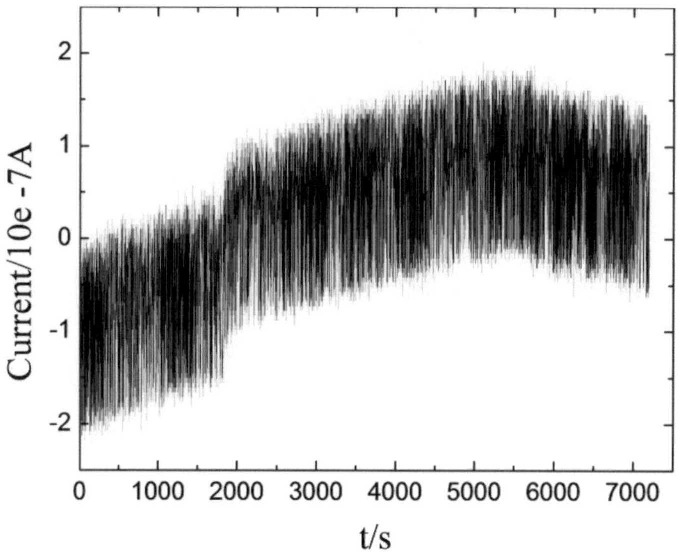

Fig. 12. ECN before NaCl solution injected

Fig.13 and Fig.14 show the EPN and ECN which are sampled after 3.0% NaCl solution is injected into the specimen respectively. The sampling frequency is 2Hz. Also, there is a DC trend in Fig.13 and Fig.14. The DC trend of EPN in Fig.13 is much greater than the DC trend of ECN in Fig.14. There is characteristic of pitting corrosion which is shown as EPN transients in Fig.13. The smaller Fig. in Fig.13 shows a characteristic transient which is chosen stochastically. From the smaller Fig. we can see that the amplitude of the EPN transient which changes from -0.534V to -0.538V is very large (The degree of EPN in Fig.11 is

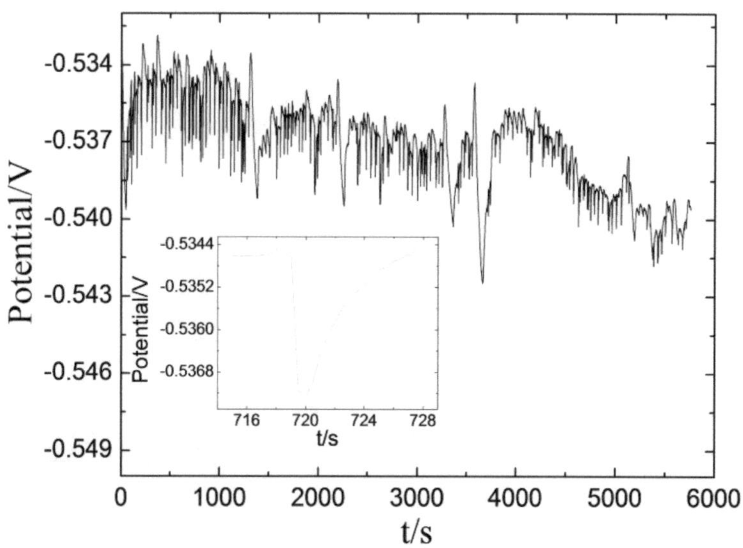

Fig. 13. EPN after 3.0% NaCl solution injected

only 10^{-5}V). The duration of the transient is several tens of seconds (It is 20 seconds in the smaller Fig.). The characteristic of the transient is that it decreases suddenly and recovers slowly. There is a much gentler DC trend in Fig.14. The amplitude of the ECN is 10^{-6}degree (It is 10^{-7}degree in Fig.12). The minus sign of the ECN means that the current is flows from identical CE to the WE. The smaller Fig. in Fig.14 shows the characteristic transient of the ECN. It increases sharply and recovers slowly. It must be kept in mind that here the current 'increases' is from plus sign to minus sign. Also its duration is several tens of seconds. According to D.Macdonald's PDM, the aggressive anion Cl^- is able to enhance the flux of cation vacancies through the barrier layer. Under favorable conditions (voltage, PH, [Cl^-]) vacancy condensation will occur at the metal/barrier layer interface and hence passivity breakdown will ensue. So, the fresh surface of the metal will generates EPN and ECN transients.

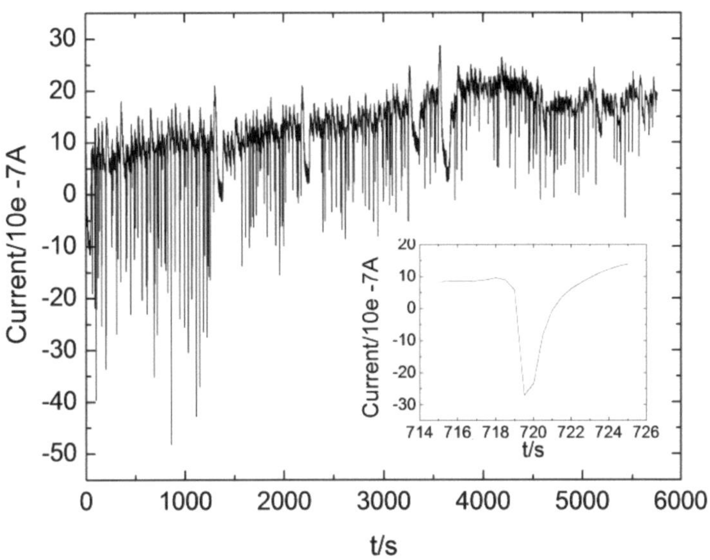

Fig. 14. ECN after 3.0% NaCl solution injected

Sym4 wavelet is used to decompose the EPN and ECN signal from j=1 scale to j=8 scale. The crystals from d_1 to d_8 are the details of the signal. a_8 is the approximation part of the signal. From the decomposed results we can see that the signal energy is stored in approximation crystal a_8 and detail crystals which are from d_1 to d_8.The approximation a_8 is the principal part of the DC trend. It contains most of the energy of the signal. In order to eliminate the DC trend influence, the energy of a_8 is not be considered in the calculation. In this way, we can remove the DC trend.

Fig.15 and Fig.16 show the EDP of EPN in Fig.11 and Fig.13, respectively[8]. According to Fig.15, the detail signal energy accumulates in crystal d_1 to d_4. They are the high frequency information of the signal in Fig.11. The energy in crystal d1 to d4 almost occupies 80% of the whole detail energy. But in Fig.16, most of the energy accumulates in the low frequency crystal, such as d_4, d_5, d_6 and d_7. According to Fig.13 and Fig.14, we can see that the low frequency parts in Fig.16 correspond to the transient of the pitting corrosion. The fact that the energy distribution changes from high frequency parts in Fig.15 to low frequency parts in Fig.16 shows that the EPN transients emerge and become the dominant parts gradually. From Fig.17 we can see that most of the energy of the detail information in Fig.12 accumulates in crystal d_1, d_2, d_3 and d_8. There is a little difference between Fig.17 and Fig.15. In Fig.15 the energy in crystal d8 is very small(less than 0.025), but in Fig.17 the energy in crystal d_8 is large than 0.2. This can be caused by the frequency component in ECN. Although we have removed most of the DC trend in Fig.14 by eliminating the crystal a_8, the very low frequency which is very less will exist in crystal d_8.So we can see the higher energy ratio of crystal a_8 in Fig.17 and Fig.18. When 3% NaCl solution is injected into the specimen by high pressure pump, we can see that the energy in d_4、d_5、d_6 and d_7 becomes more significant. In Fig.17, the energy in crystal d_4、d_5、d_6 and d_7 is less than 0.2 of the whole detail information, but in Fig.18, it increases to no less than 0.6. This also means that the current transient becomes the main component in the ECN. The current transient is caused by the pitting corrosion, so the change of EDP can show the corrosion mechanism.

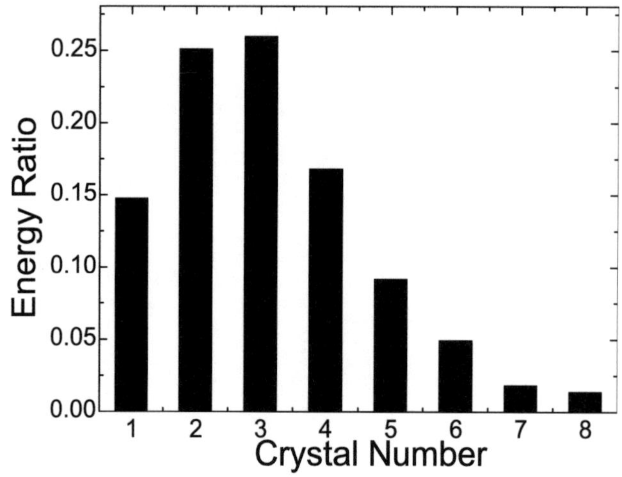

Fig. 15. EDP of EPN before NaCl solution injected

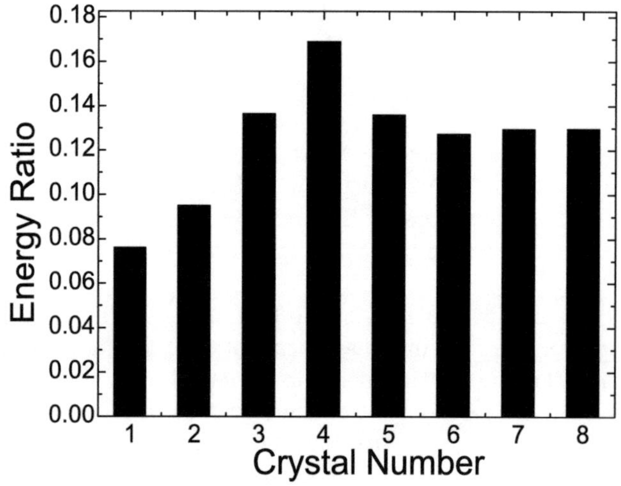

Fig. 16. EDP of EPN after NaCl solution injected

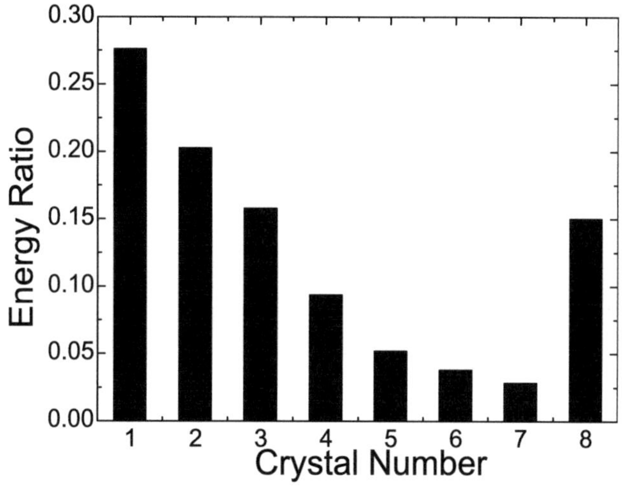

Fig. 17. EDP of ECN before NaCl solution injected

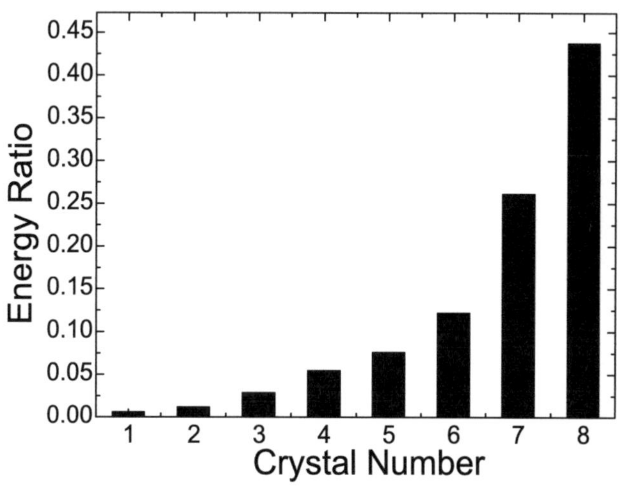

Fig. 18. EDP of ECN after NaCl solution injected

4. Corrosion sensors and networks

4.1 Solid-state reference electrode

Monitor and control the corrosion of reinforcing steel in concrete require reliable measurement of stable potentials. Therefore, embeddable reference electrodes(RE) is the key element in corrosion monitoring of concrete structures for long-term monitoring and potentiostatically controlled cathodic protection of reinforcement in concrete[41]. A practical RE have to obey the following conditions: it must be stable, invariant to chemical change in concrete, tolerant to climatic conditions and have the ability to pass small currents with a minimum of polarization and hysteresis effects, display long term performance, be cost effective and result from an environmentally safe manufacturing process [42]. Traditional liquid references, such as the standard hydrogen electrode (SHE), the saturated calomel electrode (SCE) and the copper/copper sulfate reference electrode (CSE), etc., are conventionally used for the most laboratory experiments in most conditions due to its reliable performance. Unfortunately, the short service life and fragile structure determine these reference electrodes can not be used in the field. Recently, several embeddable REs, such as MnO_2 [43-47], Metal-Metal Oxide (MMO) [48, 49] and $NiFe_2O_4$ [50], have been prepared and attempted to apply in RC structures. European Federation of Corrosion (EFC) has also reviewed the electrochemistry and characteristics of embeddable reference electrodes for concrete, and reported the service status of the six kinds of embeddable REs which have been implanted in Gimsøystraumen Bridge (in Norway) few years ago[51]. It can be seen clearly from the foregoing that there is an utmost need to develop and evaluated the performance of reliable and maintenance-free REs for use in concrete [52, 53]. Compared with that of the other REs, the RE based on Ag/AgCl is stable, easily prepared and industrialized conveniently, and have been used in electrochemistry for a long period[54]. Although the RE based on Ag/AgCl can not be applied directly in reinforcing concrete structures currently, its performance can be improved greatly by the elaborate design of the materials and the structure. PTFE, Al_2O_3 and AgCl powder are applied as the carrier material and enhanced substance to improve the behavior of the RE, respectively.

A novel Ag/AgCl based solid-state RE has been prepared and characterized to provide the key element for the potential control of the corrosion monitoring sensors(see Fig.19)[55]. Long-term stability, temperature response, anti-polarizability and anti-disturbance of ions have been investigated. The results indicate that the balance potential of the reference electrode is 3.64 mv±1mv (vs.SCE) at 25°C. The temperature response coefficient of the RE is -0.51 mv/°C±0.03mv/°C in [10°C, 70°C]. Also, the anti-polarizability of the solid-state RE meets the requirements of the electrochemical measurement in the field. Additionally, the reference electrode is insensitive to the concrete admixtures which include Na_2SO_4, NaCl, $NaNO_3$ and $NaNO_2$.

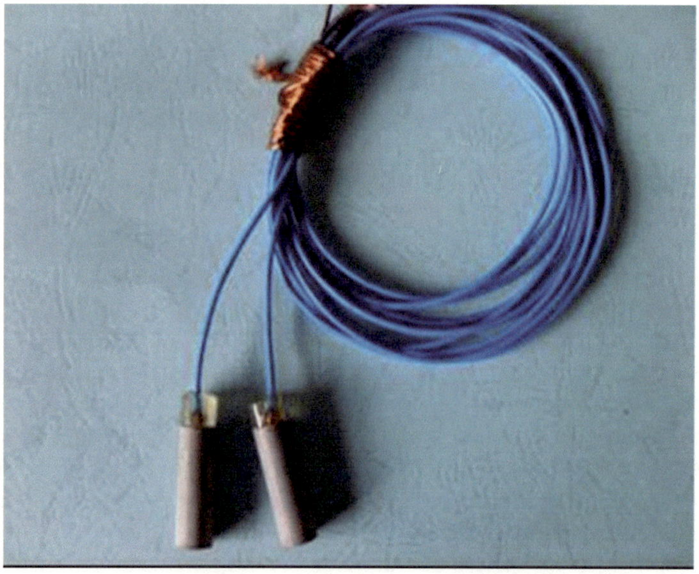

Fig. 19. Solid-state reference electrode

4.2 Wired corrosion monitoring sensors

The five-electrode corrosion sensors have been applied to monitor the corrosion state of the reinforcing steel in the beams[55]. There are four kinds of elements which include one reference electrode(RE), one counter electrode(CE) and three identical working electrodes(WEs) in the corrosion sensor. Considering the field applicability, a home-made all solid-state reference electrode (ASSRE) has been used to setup the five-electrode corrosion sensor. The materials of the counter electrode and working electrodes are the graphite rod and the Q235 carbon steel, respectively. One of the three working electrodes, CE and ASSRE are integrated as classical three-electrode system to execute the galvanostatic step measurement. The other two identical working electrodes and ASSRE are applied to obtain the EN data. The sampling frequency of EN is 2Hz. Fig.20 shows the photograph of the corrosion sensors.

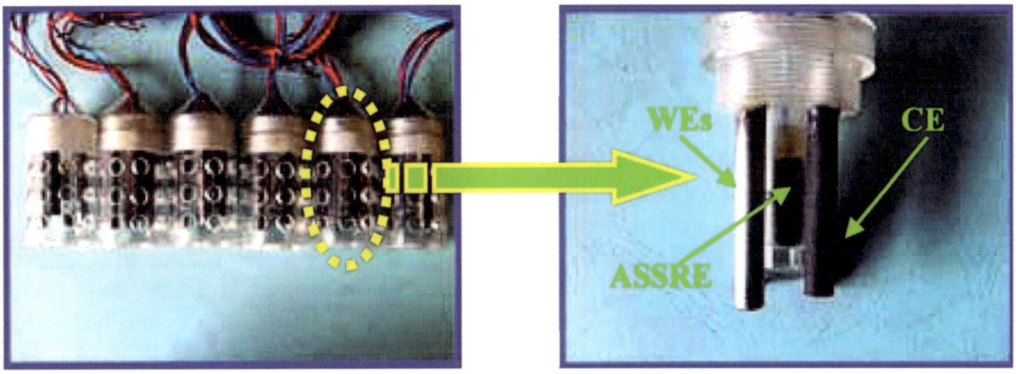

Fig. 20. Five-electrode corrosion sensors

The current confining sensor has been developed for RC structures[56]. Fig 21 illustrates the geometric configuration of the sensor. The red ring is the counter electrode(CE). The other two yellow rings beside the CE are the current confining rings. Two all solid-state reference electrodes (ASSREs) are applied to sense the potential of the concrete between the counter

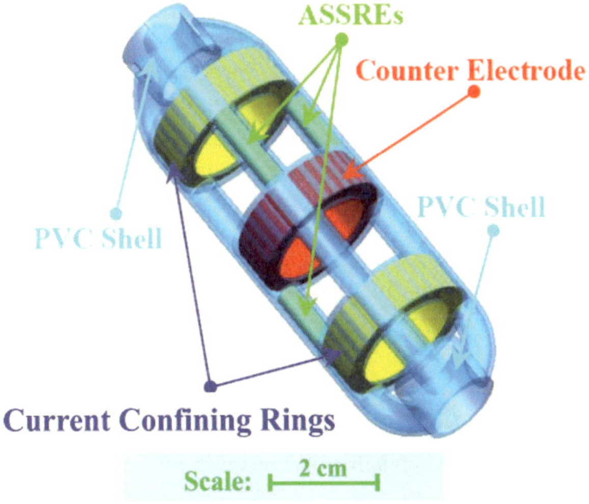

Fig. 21. Solid state and current confining corrosion sensor

electrode and the current confining rings. The third ASSRE is used as the normal reference electrode(RE) of the traditional three-electrode system in electrochemical techniques to measure the potential of the working electrode(WE). Here, WE is the reinforcing steel in RC structures. The potential of the counter electrode and the current confining rings is controlled by the embedded microcircuit of the data acquisition and processing system. The geometric configuration of the sensor is optimized by the Finite Element Method (FEM). the electric fluxline can be effectively confined in a fixed area on the surface of the steel bar between the CE and the other two rings. Therefore, the corrosion rate of the reinforcing steel in RC structures can be measured accurately by the optimized corrosion sensor.

4.3 Self-powered wireless corrosion monitoring sensors & networks

For an airtight civil engineering structure(for example, concrete beam), while a wire collection system is used for monitoring corrosion, the wire put inside concrete beam needs to prolong outside concrete beam to send the corrosion sensor's information, which will accelerate the corrosion process because the wire's laying possibly makes reinforcing steel inside the beam contact with air. Wireless sensor networks with wide application especially in structural health monitoring[57-60], matured in recent years represent an innovative solution to current corrosion monitoring systems, due to their low-cost, small size, low-power and wireless communication capability.

The most interesting and fascinating fact is that the corrosion process is an energy release process. This vital energy provides the possibilities to realize in-situ, real-time and on-line corrosion monitoring of RC structures by wireless energy-harvesting sensors and networks. We investigate deeply the generation of corrosion energy in RC structure and its availability as a power source. Powered by the energy harvested from corrosion environment, a new smart sensor platform is then designed to detect the corrosion status and report the samples via wireless links to the end user. Developing energy harvesting sensors is of critical importance for in-situ, real-time and on-line corrosion monitoring of RC structures because of its intrinsic capability of providing unbounded lifetime. This represents a tremendous step towards realization of practical corrosion monitoring.

For a wireless sensor network, energy is usually provided from either solar power[61], structure vibration[62], chemical batteries, or lithium batteries. The wireless corrosion monitoring sensors need to be put inside concrete beam in an airtight way, which require that wireless sensor can work for a long time without outside energy supply as corrosion is a slow process(about tens years). Thus it is necessary to solve the problem of energy supply for the sake of monitoring corrosion with wireless sensor networks.

In the proposed self-powered wireless corrosion monitoring sensor network system shown in Fig.22[63], we use the electrochemical noise produced by corrosion not only as sensing signal but also as energy source for wireless sensor, the process of monitoring corrosion is the following: unit of gathering energy collect electrochemical noise, while energy reach the working requirement of wireless sensor, the other units of wireless corrosion sensor are in activation. Then micro-processing unit gathers and process the sensor signal--electrochemical noise, the sensor signal processed is little in volume for wireless transmission because the energy needed by wireless transceiver unit is much more largely than the one of micro-processing unit. All the above units are put together inside airtight concrete structure to integrate into a self-powered wireless corrosion monitoring sensor, lots of such sensors forming self-organizing networks send their signal information to wireless base station. Wireless base station may receive the information and control the sensors, remote users can visited wireless base station to know the sensors' status and corrosion information of monitored objective. The designed self-powered wireless corrosion sensor is exhibited in Fig.23[64].

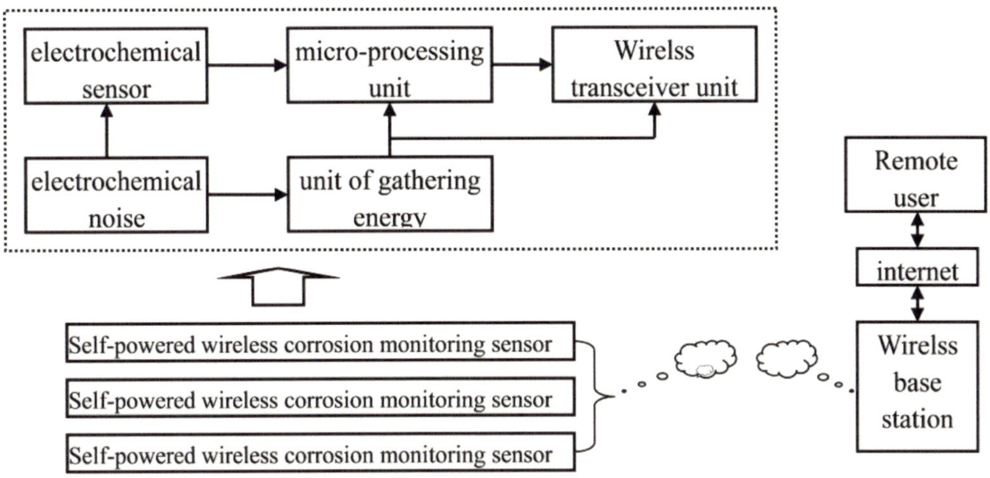

Fig. 22. Self-powered wireless corrosion monitoring networks

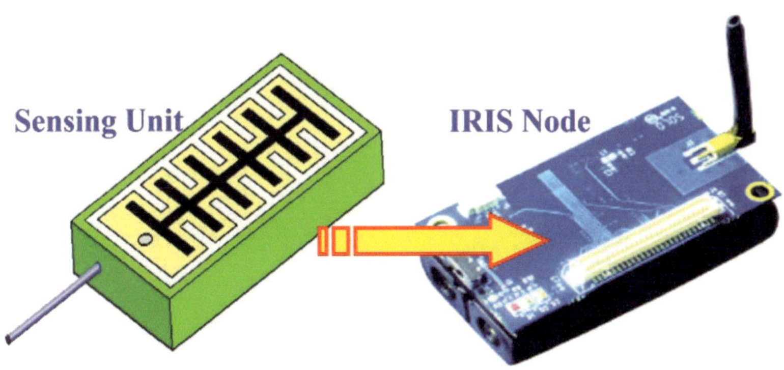

Fig. 23. Self-powered wireless corrosion monitoring sensor

5. Application of the corrosion monitoring system in civil engineering

5.1 Corrosion monitoring of the RC beams

The corrosion sensors presented in the previous sections have been applied to monitor the corrosion status of the RC beams (see Fig.24). The electrochemical measuring techniques which are seemed as the software for corrosion monitoring based on galvonastatic step excitation and electrochemical noise have been developed to extract the characteristics of general corrosion and pitting corrosion, respectively[65].

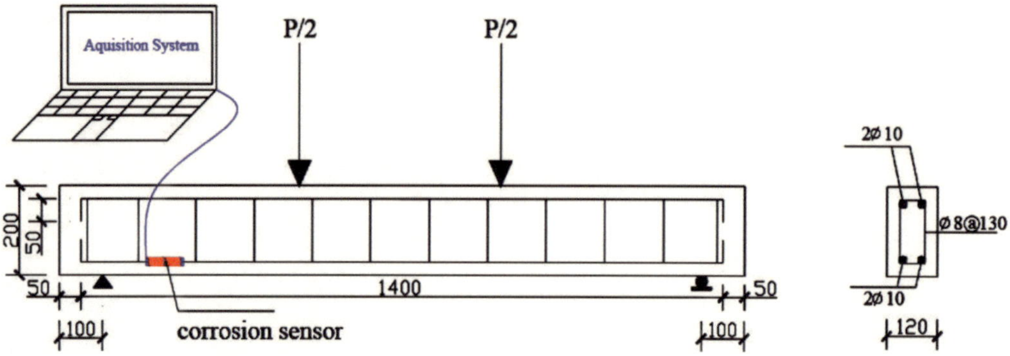

Fig. 24. Distributed bars, sectional dimension and the position of the corrosion sensors.

5.1.1 Half-cell potential map

Fig.25 shows the half-cell potential results of the beams. The half-cell potential of all the bars are higher than -200mv (Vs. CSE) before corrosion. According to ASTM C876-91, the corrosion probability of the steel bar is no more than 5% and the deterioration effect of the corrosion can be neglected. However, the potential shifts towards negative severely after corrosion. The potential in most measuring points is less than -400mv (Vs. CSE) after corrosion. Some of the measuring results are even less than -1000mv. This can be explained by the non-homogeneous corrosion of the steel bar. Actually, there are many defects in the

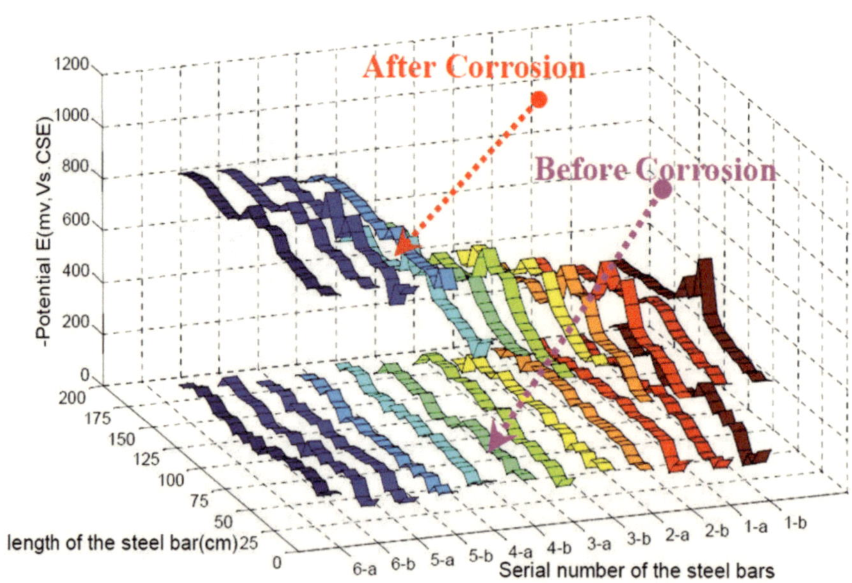

Fig. 25. Half-cell potential of the RC beams

concrete. The corrosion of the steel bar in these defects is much more severely than that of the other sections. The corrosion probability of the beam is more than 95% after corrosion. Comparing the results of the beam1# to that of the beam 6#, the potential becomes much more negative when the corrosion is more severe. But it must be mentioned that the corrosion products and the water saturated degree of the concrete can influence the accuracy of the potential results. There are minimums in each bar's half-cell potential. Such as the steel bar 1-b, the minimum of the potential is -725.2mv (Vs.CSE) in the middle of the beam. This phenomenon is due to the local corrosion of the steel bar. The reddish-brow corrosion product perspires from the concrete crevice during the accelerating process.

5.1.2 Transient potential response

Five pieces of RC beams which is from the beam 2# to the beam 6# have been tested in the experiments. Considering the length of the paper, we only show the typical results of the beam 2# in Fig.26. According to the plots of the potential response, we can see clearly that the potential response can not reach the steady state even as the current has been applied for 1000s. This is due to the diffusion process of the oxygen. The accelerating corrosion process has consumed too much oxygen in the pore of the concrete to meet the requirements of the cathode reaction. Thus, the electrochemical process is controlled by the diffusion process. The method which has been derived in the previous section is used to analyze the data of the potential response. The fitting results are listed in Table 4. According to the results, the resistance of the concrete decreases gradually. The mean value of the concrete resistance decreases from 546.89Ω·cm² to 275.79Ω·cm² as the accelerating degree increases gradually from the beam 2# to the beam 6#. There are much more Cl^- concentrated in the pore of the concrete in the beam 6# than that of the beam 2#. The corrosion products of the steel bar have precipitated in the pore. Therefore, these two aspects can lead to the decrease of concrete resistance.

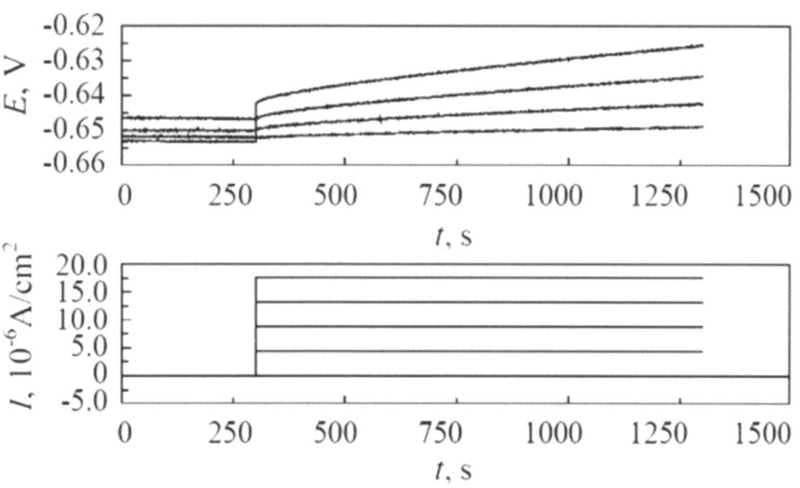

Fig. 26. The potential response of the beam 2#

Parameters		R_c $\Omega \cdot cm^2$	R_p $\Omega \cdot cm^2$	C_{NI} $10^{-4}F/cm^2$	τ_c s	K	σ $\Omega \cdot cm^2 \cdot s^{-1/2}$
2#	5uA/cm2	537.64	422.84	3.8355	0.16218	0.81683	38.457
	10 uA/cm2	551.48	415.65	3.1425	0.13062	0.80325	32.309
	15uA/cm2	545.57	437.50	3.2663	0.14290	0.84104	30.391
	20 uA/cm2	552.86	444.21	3.3560	0.14908	0.80152	30.971
3#	5uA/cm2	285.12	404.26	6.1530	0.24874	0.74654	13.694
	10 uA/cm2	251.19	415.95	3.9900	0.16596	0.76499	12.344
	15uA/cm2	271.53	416.49	4.2485	0.17695	0.70294	12.251
	20 uA/cm2	297.32	440.24	6.2499	0.27515	0.55664	12.353
4#	5uA/cm2	419.49	376.16	4.0576	0.15263	0.74447	15.297
	10 uA/cm2	435.69	391.05	3.4091	0.13331	0.90131	17.379
	15uA/cm2	448.40	387.00	3.1530	0.12202	0.88753	18.756
	20 uA/cm2	444.86	397.21	3.3036	0.13122	0.87004	19.112
5#	5uA/cm2	323.99	422.92	4.7930	0.20271	0.95230	17.243
	10 uA/cm2	310.63	432.48	4.1350	0.17883	0.84943	18.119
	15uA/cm2	305.13	428.50	4.1777	0.17901	1.00000	18.314
	20 uA/cm2	307.18	431.41	4.5445	0.19605	0.84993	18.934
6#	5uA/cm2	267.63	350.31	7.1249	0.24959	0.74114	19.628
	10 uA/cm2	278.01	373.64	6.3498	0.23725	0.74691	22.026
	15uA/cm2	280.34	372.49	5.9271	0.22078	0.90379	24.376
	20 uA/cm2	277.16	373.59	6.3188	0.23606	0.84188	25.230

Table 4. Fitting results of the response to galvanostatic step excitation

Also, the polarization resistance decreases gradually from the beam 2# to the beam 6#. This means that the content of the chloride ion can influence the corrosion rate of the steel bar. Much more Cl^- can destroy the passivity film much more severely. The magnitude of the non-ideal capacity C_{NI} is 10^{-4} F/cm2. The magnitude of the time constant $\tau_c = C_{NI}R_P$ is 10^{-1} second. R_P and C_{NI} are the essential characteristics related to the electrochemical process and the double layer. They do not depend on the WE's area, the resistance of the concrete and other factors. Therefore, τ_c which has many advantages than that of Half-cell potential can be used to determine qualitatively the corrosion status of the structure. According to the results of τ_c, the steel bar in the concrete has been corroded severely. Most of the K values are less than 1.This means that the CPE presents in the EC. The magnitude of σ and σ/R_p are $10\Omega \cdot cm^2 \cdot s^{-1/2}$ and more than 10^{-2}, respectively. These results show that the diffusion process controls the electrochemical reactions.

5.1.3 Electrochemical noise results

The EN of the beams from 1# to 6# has been obtained before and after accelerating corrosion. Also, we only show the results of the beam 2# here in Fig.27 and Fig.28, respectively. According to Fig.27 a) and Fig.28 a), we can see there is an obvious trend in the electrochemical potential noise (EPN) before and after corrosion. The electrochemical current noise (ECN) is much more stationary than that of the EPN in Fig.27 and Fig.28. The magnitudes of the EPN and ECN before corrosion are 2mv and 10^{-7}A, respectively. The high frequency components (≤2 Hz) exhibit as white noise in the EN before corrosion. The distribution of the EN takes on the band form before corrosion. However, a large number of transients present in the EN record after corrosion. Compared to the EN before corrosion, the magnitude of the transients is very enormous after corrosion. The magnitudes of EPN and ECN are 10mv and 10^{-6}A, respectively. The transient increases sharply and recovers slowly, and its duration is tens of seconds. According to D.Macdonald's PDM, the aggressive anion Cl^- is able to enhance the flux of cation vacancies through the barrier layer. Under favorable conditions (voltage, PH, [Cl^-]), vacancy condensation will occur at the metal/barrier layer interface, and hence passivity breakdown will is confirmed. So, the fresh surface of the metal will generate EPN and ECN transients.

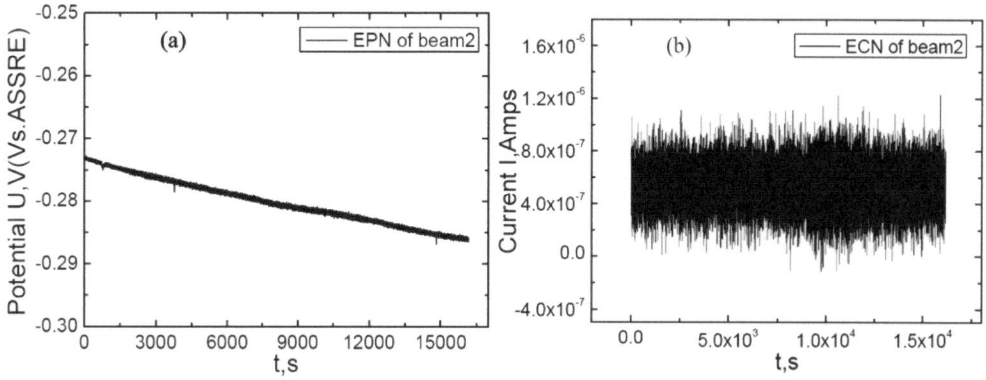

Fig. 27. EN of the beam 2# before corrosion. (a) Electrochemical potential noise. (b) Electrochemical current noise.

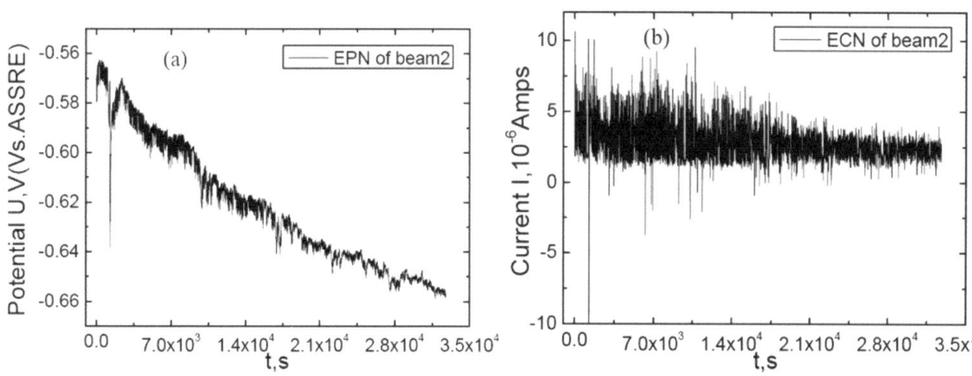

Fig. 28. EN of the beam 2# after corrosion. (a) Electrochemical potential noise. (b) Electrochemical current noise.

Sym4 wavelet has been applied to analyze the EN and the depth of the decomposing scale is 8 levels. The energy distribution plots of the EN before and after corrosion have been shown in Fig.29 and Fig.30, respectively. There are 9 parts of the signal after decomposition of the EN. The approximation of the EN which includes the trend of the EN is the lowest frequency part (≤0.004Hz). So, the signal of this part has been removed as the trend of the signal. The energy distribution plots in Fig.29 and Fig.30 are the Energy ratio Vs. Crystal number which is from 1 to 8. According to Fig.29a), the energy of the EPN is concentrated on the crystal 1, 2 and 3 before corrosion. The frequency of these parts which is between 1Hz and 2-3Hz is much higher than that of other parts. It is much more obvious in Fig.29 b) that the energy is concentrated greatly on the crystals from 1 to 3. The energy of crystal 1 to 3 is no less than 70 percent of the signal in Fig.29 b). The signal in this much higher frequency range can be caused by the white noise of the system and the general corrosion of the steel-concrete system.

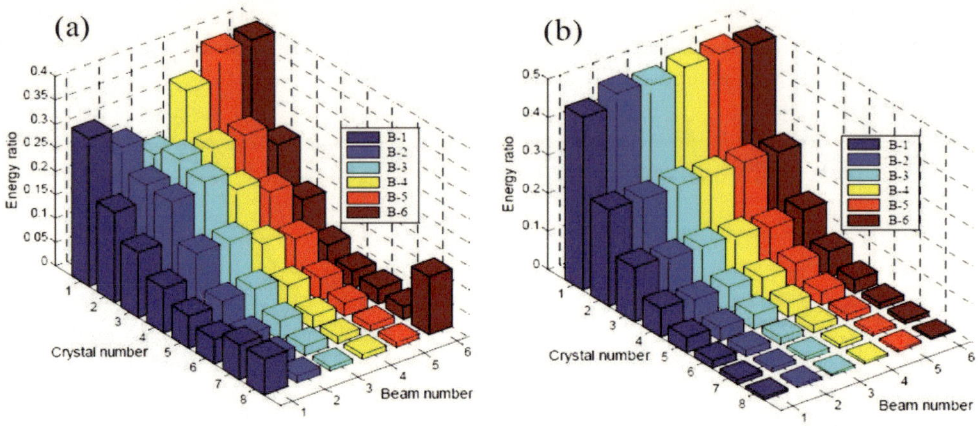

Fig. 29. Energy distribution plot of the EN before corrosion. (a) Electrochemical potential noise. (b) Electrochemical current noise. The histograms whose colors are from dark blue to dark red are the energy distribution plots of the beams which are from the beam1# to the beam 6#. The crystals which are from No.1 to No.8 are the energy concentrated on the different scales. The frequency range of the crystals which are from No.1 to No.8 decrease gradually. So the EN on No.1 level is the most detail part of the EN signal, but the EN on No.8 is much coarser than that of No.1.

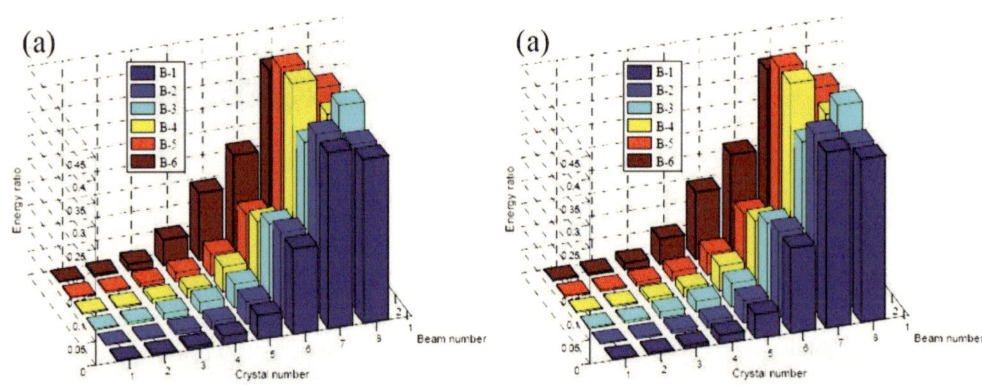

Fig. 30. Energy distribution plot of the EN after corrosion. (a) Electrochemical potential noise. (b) Electrochemical current noise. The means of the histograms in the Figures are same as that of Fig.29.

Compared with the results before corrosion in Fig.29, the energy distribution plot changes greatly after corrosion in Fig.30. The sum of the energy of the EPN on crystal 5, 6, 7 and 8 is no less than 80 percent of the overall signal's energy which the approximation has been removed. The frequency of the crystals from 5 to 8 is no more than 2-5 Hz. The condition of ECN in Fig.12 b) is similar with that of the EPN in Fig.12 a). The energy on crystal 5, 6, 7 and 8 is more than 70 percent of the signal. The obvious difference of the EDP before and after corrosion is due to the different components in the EN. There are a large number of transients in the EN after corrosion. The transient is generated by the breakdown and recovery of the passivity film during the pitting corrosion process. The attenuation life of the transient is from tens of seconds to a few hundred seconds. Much more energy is concentrated on the crystals which are from No.5 to No.8. The change of the EDP indicates the change of the corrosion type. According to the change of the energy distribution on the different crystal, the pitting corrosion is verified. The energy which is distributed on the crystal 5, 6,7and 8 is the intrinsic characteristics of pitting corrosion of the steel bar in concrete. The EDP of the EN can be applied as the benchmark to identify qualitatively the presence of the pitting corrosion.

5.2 Corrosion monitoring of the bridge pier

The corrosion monitoring techniques and sensors presented in the previous sections have been integrated as corrosion monitoring system in Fig.31, and the system has been applied in the corrosion monitoring of the bridge pier (see Fig.32). The location of the corrosion sensors have been optimized according to the hot spot of the force and the deterioration caused by the ambience. The system has serviced for 1.5 year, and no corrosion sites have been founded till now.

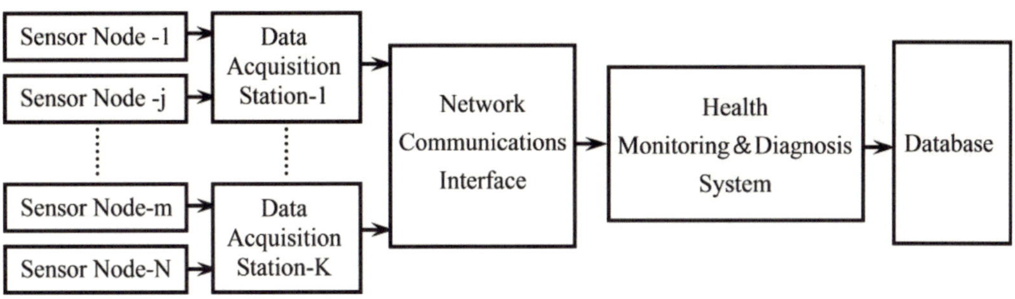

Fig. 31. Corrosion Monitoring System

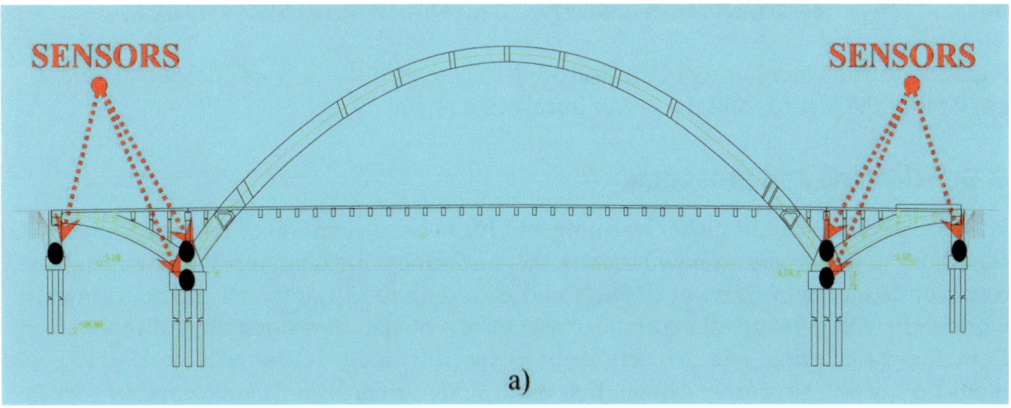

Fig. 32. Corrosion monitoring of the bridge pier. Scheme of the sensors' location a), Bird's eye view of the pier b) , and the sensor installation c)

6. Conclusions and prospects

With the development of global warming and further deterioration of the environment, the conditions of RC structures now becomes more atrocious than that of before. Reinforcement corrosion is the main cause of damage and early failure of reinforced concrete structures worldwide with subsequent enormous costs for maintenance, restoration and replacement. Corrosion monitoring will be very helpful for the safety assessment and service life prediction of RC structure. Although several kinds of corrosion monitoring sensors and systems have been applied in petroleum industry, chemical industry and industrial factory buildings, there are many factors which handicap the development of corrosion monitoring in civil engineering structures. In the past few years, The Government of China has provided special found to do the research in the field of corrosion monitoring in concrete. The project includes four parts of work which are Research of Corrosion Monitoring Method, Development of Corrosion Sensor Platform, System Integration Technique and Safety Assessment & Life Prediction of Structure. The electrochemical essence of the corrosion process determinates that the corrosion monitoring sensor based on the electrochemical theory can be used effectively to realize the object directly. This provide the chance and challenge simultaneously to the researchers in the field of electrochemistry, material, computer, electronics, signal processing, mechanics, etc. A series of originality innovations should be developed in future.

The realization of the wired and self-powered wireless corrosion monitoring sensors and networks can be considered as a milestone in the field of SHM. Besides it can be used in civil engineering, the advantages of power harvesting, wireless, long-life and advanced EN technique, etc, determine the system can be applied widely in most formidable conditions, such as nuclear industry, petrochemical engineering and many other corrosion monitoring needed conditions.

7. Acknowledgment

This research is supported by grants from NSFC (Project No.: 51008098 and 50538020), National Science Fund for Distinguished Young Scholars (Project No.:50925829), Research Innovation Found of HIT (Project No.: HIT.NSRIF.2009 LPQQ57500095) and SZDCCE 10-09. Also, the financial support from Jilin Provincial Communications Department through the transport projects (Project No.: 2010-1-7) is grateful acknowledged.

8. References

[1] Sidney Mindess; J.Francis Young; David Darwin. (2002). *Concrete*. Prentice-Hall Inc., Canada

[2] D.Macdonald; Brian M.Marx. Development of Advanced Electrochemical Emission Spectroscopy for Monitoring Corrosion in Simulated DOE Liquid Waste. (2004). *Annual report of the project DE-FG07-97ER62515*

[3] V.Feliu; J.A.Gonzalez; C.Andrade. (1998). Equivalent Circuit for Modelling the Steel-Concrete Interface.I.Experimental Evidence and Theoretical Predictions. *Corros. Sci.*, Vol. 39, pp. 864-882

[4] V.Feliu; J.A.Gonzalez; C.Andrade. (1998) Equivalent Circuit for Modelling the Steel-Concrete Interface. II. Complications in Applying the Stern-Geary Equation to Corrosion Rate Determinations. *Corros. Sci.*, Vol. 40, pp. 995-1006

[5] Z.Sun; F.Mansfeld. (1999). Localization Index Obtained from Electrochemical Noise Analysis. *Corrosion*, Vol. 55, Issue 10, pp. 915-918

[6] Y.Y.Shi; Z.Zhang; J.Q.Zhang. (2008). Dimensional Analysis Applied to Pitting Corrosion Measurements. *Electrochimica Acta.*, Vol. 53, pp. 2688-2698

[7] X.F. Liu; H.G. Wang; H.C. Gu. (2006). Fractal Characteristic Analysis of Electrochemical Noise with Wavelet Transform. *Corros. Sci.*, Vol. 48, pp. 1337-1367

[8] Guofu Qiao, Jinping Ou. (2007). Corrosion Monitoring of Reinforcing Steel in Cement Mortar by EIS and ENA. *Electrochimica Acta.*, Vol. 52, pp. 8008-8019
[9] S. Turgoose; R.A. Cottis. (1991). The Impedance Response of Film Covered Metals, *Conference Proceedings: Electrochemical Impedance Analysis and Interpretation*, San Diego, CA
[10] Li, T.C. Tan and J.Y. Lee. (1996). *Corros. Sci.*, Vol. 38, pp. 1935-1939
[11] D.D. Macdonald; M.C.H. McKubre. (1981). Electrochemical Impedance Techniques in Corrosion Science. *Electrochemical Corrosion Testing, ASTM STP 727*, pp. 110
[12] D.D.Macdonald. (2006). Reflections on the History of Electrochemical Impedance Spectroscopy. *Electrochimica Acta*, Vol. 51, pp. 1376–1388.
[13] Y. Chen; W.P. Jepson. (1999). EIS Measurement for Corrosion Monitoring under Multiphase-flow Conditions. *Electrochimica Acta*, Vol. 44, pp. 4453-4464
[14] Schiessl P. (1988). *Corrosion of Steel in Concrete*. RILEM Technical Committee 60-CSC. New York: Chapman & Hall
[15] A. Aballe; M. Bethencourt; F.J. Botana; M. Marcos, J.M. Sanchez-Amaya. (2004). *Corros. Sci.*, Vol. 46, pp. 1909
[16] A. Nagiub; F. Mansfeld. (2001). *Corros. Sci.*, Vol. 43, pp. 2147
[17] R.J.K. Wood; J.A. Wharton; A.J. Speyer; K.S. Tan. (2002). *Tribol. Int.*, Vol. 35, pp. 631
[18] F. Mansfeld; L.T. Han; C.C. Lee; C. Chen; G. Zhang; H. Xiao. (1997). *Corros. Sci.*, pp. 255
[19] J.G. Yu; J.L. Luo; P.R. Norton. (2002). *Electrochim. Acta.*, Vol. 47, pp. 4019
[20] F. Mansfeld; Z. Sun; C.H. Hsu. (2001). *Electrochim. Acta.*, Vol. 46, pp. 3651
[21] A. Nagiub; F. Mansfeld. (2002). *Electrochim. Acta.*, Vol. 47, pp. 2319
[22] A. Aballe; M. Bethencourt; F.J. Botana; M. Marcos. (1999). *Electrochim. Acta.*, Vol. 44, pp. 4805
[23] R.J.K. Wood; J.A. Wharton; A.J. Speyer; K.S. Tan. (2002). *Tribol. Int.*, Vol. 35, pp. 631
[24] R.A. Cottis. (2001). *Corrosion*, Vol. 57, pp. 265
[25] J.F. Chen; W.F. Bogaerts. (1995). *Corros. Sci.*, vol. 37, pp. 1839
[26] Y.F. Cheng; J.L. Luo; M. Wilmott. (2000), *Electrochim. Acta.*, Vol. 45, pp. 1763.
[27] U. Bertocci; J. Frydman; C.Gabriclli; F.Huet; M.Keddam. (1998). *Electrochem. Soc.*, Vol. 145, pp. 2780
[28] A. Aballe; M. Bethencourt; F.J. Botana; M. Marcos. (1999). *Electrochim. Acta.*, Vol. 44, pp. 4805
[29] A. Aballe; M. Bethencourt; F.J. Botana; M. Marcos. (1999). *Electrochem. Commun.*, Vol. 1, pp. 266
[30] F.H. Cao; Z. Zhang; J.X. Su; Y.Y. Shi; J.Q. Zhang. (2006). *Electrochim. Acta*, Vol. 51, pp. 1359
[31] C. Cai; Z. Zhang; F.H. Cao; Z.N. Gao; J.Q. Zhang; C.N. Cao. (2005). J. *Electroanal.Chem.*, Vol. 578, pp. 143
[32] L. Mariaca; A. Bautista; P. Rodriguez; J.A. Gonzalez. (1997). *Mater. Struct.*, Vol. 30, pp. 613
[33] A. Legat; M. Leban; Z. Bajt. (2004). *Electrochim. Acta*, Vol. 49, pp. 2741
[34] R.G. Hu; C.J. Lin; B. Zhao; R.G. Du. (2005). *Proceedings of 16th International Corrosion Congress*, Paper 258, Beijing
[35] A. Grossman; J. Morlet. (1984). *SIAM J. Math*, Vo. 15, pp. 723
[36] I. Daubechies. (1988). *Commun. Pure Appl. Math*, Vol. 41, pp. 909
[37] S. Mallat. (1989). *IEEE Trans. PAMI*, Vol. 11, pp. 674
[38] I. Daubechies. (1992). *Ten Lectures on Wavelets*, SIAM Press, Philadelphia, USA

[39] G. Strang; T. Nguyen. (1996). *Wavelets and Filter Banks*, Wellesley-Cambridge Press, Wellesley, MA

[40] S. Mallat. (1998). *A Wavelet Tour of Signal Processing*, Academic Press, Oval Road, London

[41] Frank J. Ansuini; James R. Dimond. (2001). Long-term Field Tests of Reference Electrodes for Concrete- Ten Year Results. *Corrosion*, Paper No. 01296, pp. 1-13

[42] Leigh Ann Pawlick; Glenn E. Stoner; Gerardo G. Clemena. (1998). Development of an Embeddable Reference Electrode for Reinforced Concrete Structures. *Final Contract Research Sponsored by Virginia Transportation Research Council (A Cooperative Organization Sponsored Jointly by the Virginia Department of Transportation and the University of Virginia)*, No. VTRC 99-CR1, pp.1-93

[43] Arup, H.; Sørensen, B. (1992). A New Embeddable Reference Electrode for Use in Concrete. *Corrosion*, Paper No. 208, NACE, Houston, TX, USA

[44] Arup, H.; Klinghoffer, O.; Mietz, J. (1997). Long Term Performance of MnO_2-Reference Electrodes in Concrete. *Corrosion*, Paper No. 243, NACE, Houston, TX, USA

[45] Muralidharan S.; Tae-Hyun, H.; Jeong-Hyo, B., et al. (2006). Electrochemical Studies on the Solid Embeddable Reference Sensors for Corrosion Monitoring in Concrete Structure. *Materials Letters*, Vol. 60, pp.651–655

[46] Muralidharan, S.; Tae Hyun Ha; Jeong Hyo Bae; Yoon Cheol Ha; Hyun Goo Lee; Dae Kyeong Kim. (2007). A Promising Potential Embeddable Sensor for Corrosion Monitoring Application in Concrete Structures. *Measurement*, Vol. 40, pp.600-606

[47] Muralidharan, S.; Saraswathy, V.; Thangavel, K., et al. (2008). Electrochemical Studies on the Performance Characteristics of Alkaline Solid Embeddable Sensor for Concrete Environments. *Sensors and Actuators B. Chemical*, Vol. 130, pp. 864–870

[48] Duffó, G.S., Farinab, S.B., Giordano, C.M. (2009). Characterization of Solid Embeddable Reference Electrodes for Corrosion Monitoring in Reinforced Concrete Structures. *Electrochimica Acta*, Vol. 54, pp.1010–1020

[49] Castro, P.; Sagüés, A. A.; Moreno, E. I., et al. (1996). Characterization of Activated Titanium Solid Reference Electrodes for Corrosion Testing of Steel in Concrete. *Corrosion*, Vol. 52, No. 8, pp. 609–617

[50] Muralidharan, S.; Saraswathy, V.; John Berchmans, L., et al. (2010). Nickel Ferrite ($NiFe_2O_4$): A Possible Candidate Material as Reference Electrode for Corrosion Monitoring of Steel in Concrete Environments. *Sensors and Actuators B. Chemical*, Vo.145, pp. 225–231

[51] Myrdal, R. (2007). The Electrochemistry and Characteristics of Embeddable Reference Electrodes for Concrete. *European Federation of Corrosion Publications*, No. 43, Woodhead Publishing Limited and CRC Press LLC. Institute of Materials, Minerals & Mining

[52] John P. Broomfield; Kevin Davies; Karel Hladky. (2002). The Use of Permanent Corrosion Monitoring in New and Existing Reinforced Concrete Structures. *Cement & Concrete Composites*, Vol. 24, pp. 27–34

[53] Gustavo S. Duffó; Silvia B. Farina. (2009). Development of an Embeddable Sensor to Monitor the Corrosion Process of New and Existing Reinforced Concrete Structures. *Construction and Building Materials*, Vol. 23, pp.2746–2751

[54] C. P. Atkins; M. A. Carter; J. D. Scantlebury. (2001). Sources of Error in Using Silver/Silver Chloride Electrodes to Monitor Chloride Activity in Concrete. *Cement and Concrete Research*, Vol. 31, pp.1207-1211

[55] Guofu Qiao; Huigang Xiao; Yi Hong; Yuelan Qiu. (2011). Preparation and Characterization of the Solid-State Ag/AgCl Reference Electrode for RC Structures. *Sensor Review*. (Revised)

[56] Guofu Qiao; Tiejun Liu; Yi Hong; Jinping Ou. (2011) Optimization Design of a Corrosion Monitoring Sensor by FEM for RC Structures. *IEEE Sensor Journal*. DOI:10.1109/JSEN.2011.2112644

[57] J. P. Ou; H. W. Li. (2003). *SPIE*, Vol.5099, pp. 356-362

[58] Y. Yu; H. W. Li; J. P. Ou. (2004). *The 3rd International Symposium on Instrumentation Science and Technology*, Vol.1, pp.741-748, Xi an

[59] Y. Yu; J. P. Ou. (2006). *High Technique Letter*, Vol.12, pp.358-362

[60] S. Glaser; H. Li; M. Wang, et al. (2007). *Smart Structures & Systems*, Vol. 3, pp. 221-244

[61] Kohvakka Mikko; Hannikainen Marko; Hamalainen, et al. (2003). *IECON Proceedings (Industrial Electronics Conference)*. pp. 1499-1504

[62] Scott Meninger; Jose Oscar Mur-Miranda; Rajeevan Amirtharajah, et al. (2001). *IEEE Transaction on very large scale integration (VLSI) systems*. Vol.9, pp.64-76

[63] Yan Yu; Guofu Qiao; Jinping Ou. (2010). Self-powered Wireless Corrosion Monitoring Sensors and Networks. *IEEE Sensor Journal*, Vol. 10, pp. 1901-1902

[64] Guodong Sun; Guofu Qiao. (2011). Corrosion Monitoring Sensor Networks with Energy Harvesting. *IEEE Sensor Journal*. DOI: 10.1109/JSEN.2010.2100041.

[65] Guofu Qiao; Huigang Xiao; Jinping Ou. (2011). Identification of the Reinforcing Steel's Corrosion State in RC beams Based on Electrochemical Sensor. *Sensor Review*. (In Press)

25

Tuvalu Visualization Project - Net Art on Digital Globe: Telling the Realities of Remote Places

Makiko Suzuki Harada, Hidenori Watanave[1] and Shuuichi Endou[2]
[1]*Faculty of System Design, Tokyo Metropolitan University*
[2]*NPO Tuvalu Overview Industrial Art Course; Faculty of System Design, Tokyo*
Japan

1. Introduction

1.1 About this study

Tuvalu Visualization Project is a net-art-work on digital globe that tells the realities of remote places. Existent Internet archive contents were able to have only the collection of landscape photos, but were not able to show reality of the people and life. And also not able to develop communication between users and remote places based on archives.Then this study propose the method of at the point of design using the digital globe. Our art work based on the proposal technique was able to be experienced by many users at real exhibitions, especially through the internet ,the Web site access is 13270 times. At real exhibition, the work enabled users to inspect of the situation of the local area three-dimensionally, as well as the creation of solidarity through the visualization of pseudo-synchronous communication. This archive work was developed by only open sourse software and anyone can download and distribute free-of-charge from the internet. Furthermore, the authors attempt to use the result of this study in helping support the international philanthropic activities by individual and non-profit organization.

As stated above, this chapter state the Background and purpose, the chapter2 will state the Related Studies, the chapter3 will state the composition of this art work in detail.

The chapter4 will state the cosideration and the chapter5 will state the Conclusion and Outlook.

1.2 Background and purpose of this study

Through the popularization of search engines and social network services, it has become possible to easily obtain via internet the information relating to remote places, not available before. However, because the information search solely takes place based on keywords and meta tags, the more sensational and one-sided information-such as major accidents and disasters-tend to be communicated firstly, and more multifaceted and unaffected truth is not easily communicated. Such tendency is especially strong in developing countries and remote islands, due to the low saturation level of the internet use, making it difficult to webcast individually. This situation prevents the understanding between cultures and people living in remote places, with the presence of the language barrier adding to the problem. Therefore, the phenomenon happening in distant and far places is likely to be taken as things of no concern, and is difficult to be considered in relation to one's daily life.

This study has the purpose of analyzing different devices appropriate to create an opportunity to reconsider globally the phenomenon in far and distant places in relations to people's daily lives. Furthermore, the authors attempt to use the result of this study in helping support the international philanthropic activities by individual and non-profit organizations. As a starter, the authors have visualized on the digital globe the "realities"of the remote place, which is many-sided in its state. Furthermore, we have developed an internet content which includes the pseudo-synchronous communication function relating to it. The "Tuvalu Visualization Project" (Fig.1) presented in this thesis is an net art work with both the functions of a photographic archive of the people and their daily life scenes of the remote islands in the South Pacific, the Tuvalu Islands, broadcasted via the digital globe on the internet, as well as a communication platform.

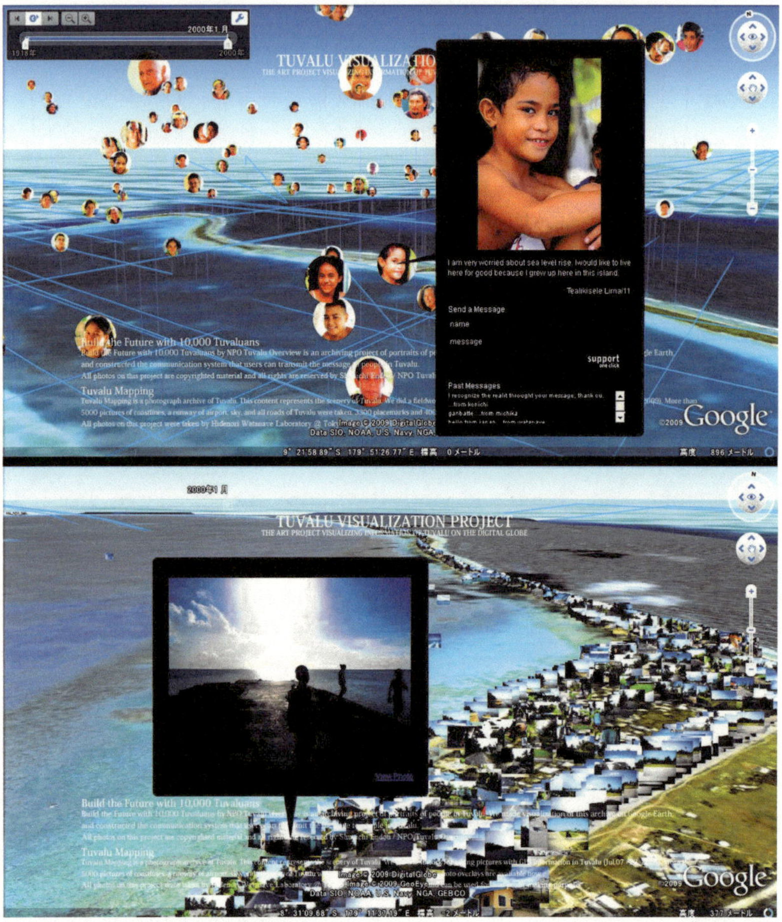

Fig. 1. Tuvalu visualization project

1.3 The definition of "Reality" in this study

The word "Reality" has the meaning of "the real state of affairs". The origin of the word comes from the Buddhist term "tattrasyalakSaNam", referring to the truth unable to be measured by the heart, and separating from the human's superficial words. In the case of sociology, this word is used to mean the general image which emerged from the collection of abundant data concerning the phenomenon, as in "The Collaborate Study: the Realities of the Damages due to the Atomic Bomb in Hiroshima-Nagasaki", in which researchers in various fields analyzed the complicated aspects of the atomic bombing. Furthermore, in the case of field of arts, especially in regards to the activity of photographers, the word has been used to mean the "real state of affairs" of events and accidents, expressed through photographs of people and scenes, not communicated by the mass media. Based on this, this study defines the word "reality" as the state of the townspeople living locally as well as their daily life scenes, which is not able to be communicated through the one-sided report relating to remote places. Also, the art work used as two of its components the "Record of Portraits and Interviews" as well as "Landscape Photographs" collected in the Tuvalu Islands. The choice of the subject matter and material will be mentioned in more details in 3.2.

2. Related studies

Needless to say, the internet fraught with problems mentioned in 1.1, is also a favorable place for those who expresses and sends information individually. The "Net Art" with the internet being the medium, accessible by people all over the world, has been utilized as a means to send the author's message throughout the world. In recent years, there are cases, such as the "Niko Niko (Smile) Animation", where the connection itself is expressed which was formed between the people through the art work, intended to be a communication platform. Such works are beginning to replace the existent media, i.e. the television. Out of the preceding art works, this study has especially referred to those that connect stories and places complementarily, and which has the intension of reinforcing each area as well. The digital map rapidly spreading in recent years, has been considered suitable for the purpose of "knowing and extracting the connection between our real location and the world". This is in accordance with the purpose of this study mentioned in 1.1. As examples of preceding cases of net art works utilizing the digital map, there exist the "Sensorium" which re-expresses the various data existent in the world by visualizing it upon the digital map; the "CCTV" which is a part of the global panopticon where the user watches a local area using surveillance cameras; and the "Field-Works" which made it possible to relive the past memories of another by the placement of photographs and images in the virtual space based on GPS information. These are net art works broadcasted one-way via the internet. However, in recent years, works of two-way communication have been released, in which the general user contribute data, such as the "Life Slice Camera", the "Jikukan-Poemer+Kakiko map", and the "Sakura Mapping". The "Sakura Mapping" especially is an excellent example of individually oriented daily life communicated publicly on the digital globe. The general user visualized photographs of cherry blossoms taken by mobile phone cameras with GPS function by mapping them on to the digital globe, the "Google Earth". Moreover, there also exist the feature of being able to be viewed easily by anyone using a personal computer at home, where it is broadcasted via the already widespread internet service. However, these works have only the collection of landscape photos as its theme, and is not able to show the state of the people. Therefore, it is insufficient to "tell the realities of the remote places", as is the purpose of this study. Also, it has become more difficult today to transmit images of people, due to the increased feeling of resistance to publicly exhibiting images of people on the internet, in view of the protection of personal information. This problem can be solved by developing the social activity of obtaining the permission to publicly broadcast via the internet. Through diligently explaining the purpose in producing the work, the understanding of people, who are the objects of photography, may be gained. There exist an effort to communicate the images of the people living in remote places, at the same time joining the social movement, which is called the "Crisis in Darfur", with the theme of the Darfur Crisis. In this piece of work, it is possible to view the state of remote places in detail on the digital globe, as well as to inspect the images of the people who live there and their experiences. Even though "the stories and the places tie closely together, and the reality of the tragedy happening in the remote place is communicated, with the possibility of increasingly evoking the viewer's action", there still remains the misgiving of it "taken as things of no concern" as mentioned beforehand. On the other hand, in the precedented case of the net art work with the intention to be a communication platform, in relation to the uploaded contents, the "Niko Niko (Smile) Animation", through adding the communication function of pseudo-synchronously replaying the comments of the user, the sense of solidarity between the users is created. It is also successful in producing the sense of the everyday life, as if talking over a coffee table. This point is the focus of this study, and the authors have chosen to create a new net art work combining both the contents on the digital globe, showing the realities of remote places through covering the images of people and the scenes of their daily life thoroughly, as well as the pseudo-synchronous communication function related to it. The next chapter will state the composition of this art work in detail.

3. Composition of the art work

3.1 Purpose of the art work

The authors, with the above discussion in mind, have decided to create a net art work, combining the contents on the digital globe that covers the images of the people and the scenes of their daily life extensively - telling the realities of remote places - and the pseudo-synchronous communication function related to it. Through this art work, by communicating not only sensational and one-sided information, but also providing many-sided realities of the remote place, it is possible to create a sense of solidarity using the pseudo-synchronous communication. It is also possible to create an opportunity to recognize anew the local problems globally, by viewers overlapping the phenomenon-thought of as things of no concern up til now-in their own lives.

3.2 Choice of theme and collection of source materials

From the olden times, photographs were used as a means to bring out the realities of the object taken. Many photographic art work exist which has high artistic value, stepping into the interior of a person. Also, in comparison to the yet unified sound/video format, the imaging format, i.e. jpeg, is possible to be displayed on almost all platforms, and the size of the data is relatively small. For these reasons, photographs are used as the subject matter in this art work. Moreover, as stated in chapter 2, in the course of publicly exhibiting images of people via the internet, it is essential to have solidarity with the social activities developed at the local level. With this point in mind, Tuvalu was taken up as the theme of this art work. It will be stated in detail hereafter.

Tuvalu is an island nation in the South Pacific, known for having the danger of its land going under water, due to the rise of the sea level originating from global warming. It is also known as the country for which domain name, .tv, was allotted. While articles tying together Tuvalu and submerging of its land, or relating to the .tv domain, rank high in the search engine, it is difficult to know the real state of the people living in Tuvalu and scenes of their daily life on the other hand. Moreover, because the internet access in Tuvalu is dependent on the communication circuit of the artificial satellite, with it being narrow band and unstable (as of July, 2009), it is difficult to webcast individually. The Non-Profit Organization Tuvalu Overview(TOV) has started a social action program since 1998 with a project called, "Build the Future with 10000 Tuvaluans". It is part of the movement to communicate the lifestyle and culture of Tuvalu and the people living there, as well as the damage they face. TOV undertook to interview all 10,000 Tuvaluan citizens throughout the nine islands within the Tuvalu nation in order to draw out the realities of the people of Tuvalu. The real voices, such as their values, dreams, and wishes for the developed countries, are archived as interview records, and at the same time the portrait of each citizen is photographed. This project has proceeded mainly as photo exhibitions at actual places. However, the webcasting on the internet has been limited to the conventional Web site. From these conditions, Tuvalu was chosen as the theme of the net art work, and the portraits and interview records of 400 citizens of the Island of Nukulaelae, in the Southern part of Tuvalu - fruit of the "Build the Future with 10000 Tuvaluans" project- were utilized in creating the art work.

In the "Build the Future with 10000 Tuvaluans" project, the photographs of scenes of daily life are not covered enough, and it is one of several elements in communicating the realities of a remote place. In order to collect the landscape photographs of Tuvalu, the authors did a fieldwork from July 5th through the 24th in the year 2009. In communicating the reality of Tuvalu, it is desirable to gather photographs based on the point of view of the local people, not affected by the preconceptions of Tuvalu held by a foreigner as the photographer. Thereupon, in the attempt to minimize the inclusion of the will of the photographer, it was worked out that the photographs were taken every 10 (seconds) at a set angle, without looking through the finder. In the event of taking the photographs, the GPS unit, "GPS-CS3K" and a digital camera were used together, and 5017 images with GPS information were taken. After excluding those blurred or with low light intensity, 3602 images were mapped out. In addition to that, high resolution landscape photographs taken in various places in Tuvalu - through the "Build the Future with 10000 Tuvaluans" project - were also used.

3.3 Selecting the platform

The digital globe, "Google Earth", can be downloaded free-of-charge from the internet, and it is a 3-Dimensional digital globe software, able to be used on Windows/Mac/Linux. The end-user can zoom in on the preferred place on Earth, and view the detailed satellite images mapped out on the three-dimensional land form. Also, the viewer is able to grasp in global scale the distance and the position of the place in relation to the place of one's living. Furthermore, a variety of expression is possible through kml (KeyholeMarkupLanguage) in "Google Earth", where one can document the contents by using functions such as the place mark, overlay, polygon, and time slider. Also, by applying the JavascriptAPI, an excellent feature is seen where the communication over the internet of contents including the geographic and time information is made simple. These merits satisfy one of the purpose stated in chapter 3.1, which is to "communicate the many-sided realities of remote place by connecting it to a place on Earth". Added to this, it has been downloaded a total of over one hundred million times, and it is speculated that many numbers of users are employing this software. This point satisfies the purpose of "creating the solidarity through pseudo-synchronous communication". With these points in mind, the "GoogleEarth" was chosen as the platform. Still, the state of Tuvalu was made easy to grasp, for the entire region of

Tuvalu was the object of high resolution satellite images (aerial photographs) mapped out on "GoogleEarth" (as of July, 2009).

3.4 Portraits and interview records

The portraits and interview records from the "Build the Future with 10000 Tuvaluans" are placed as a content, designed as a net art where the viewers are able to browse all the portraits related to a place, or equivalently in a 3D space, taking advantage of the feature of "Google Earth" in which contents are able to be placed in 3-D space.

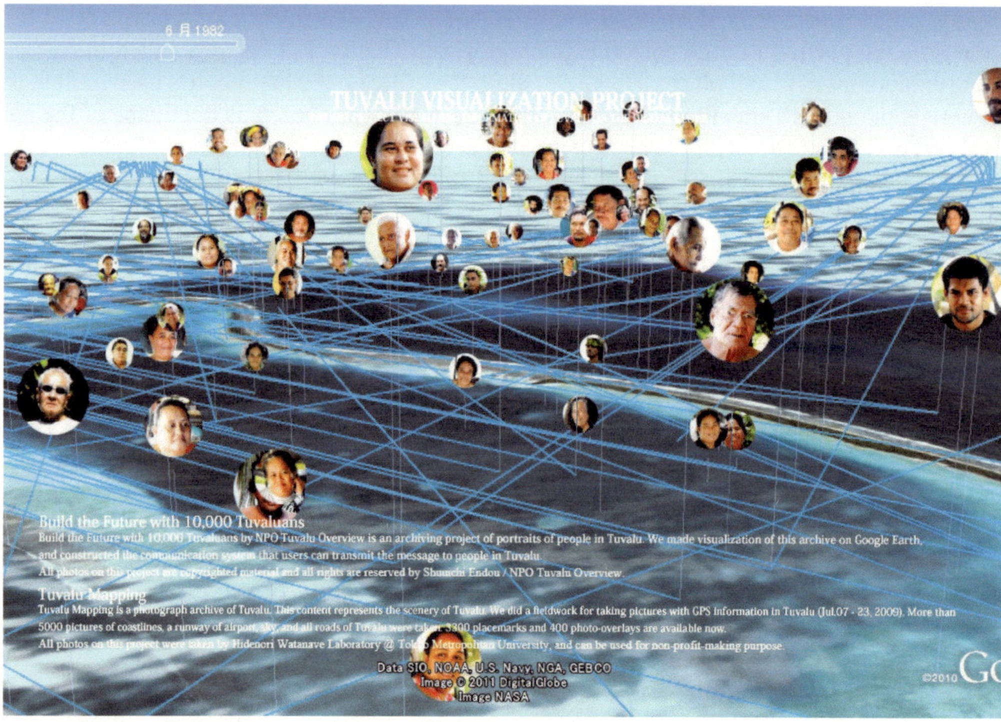

Fig. 2. Portraits mapping

As in the precedented cases stated in chapter 2, if the group of portraits are mapped out based on the point of photography, there exists the merit of being able to grasp the information of the people connected to a place. However, many icons will be crowded within a small area, and the visibility will be lacking. Also, by the habit of each viewer, it is expected to produce a difference in the opportunity for inspection of each portrait. Therefore, it has been designed to switch between two modes of display, one based on the positional information, and the other randomly positioned. In the case of the random positioning, the group of portraits are displayed randomly as icons in the air above the whole region of Nukulaelae Island, the place of photography (Fig. 2). In addition to the latitude and longitude, the altitude is also made random, and the user is able to tilt the camera in order to see the photographs equivalently. The zoom-in/out and pan function can be used to freely fly between the icons, enabling the user to find the preferred person. The two modes of display can be switched around by the viewer rereading the kml at an arbitrary timing. All icons has the timespan tag according to the age of the citizen photographed, and it is possible to narrow down the display based on the age using the Google Earth time slide (Fig.3).

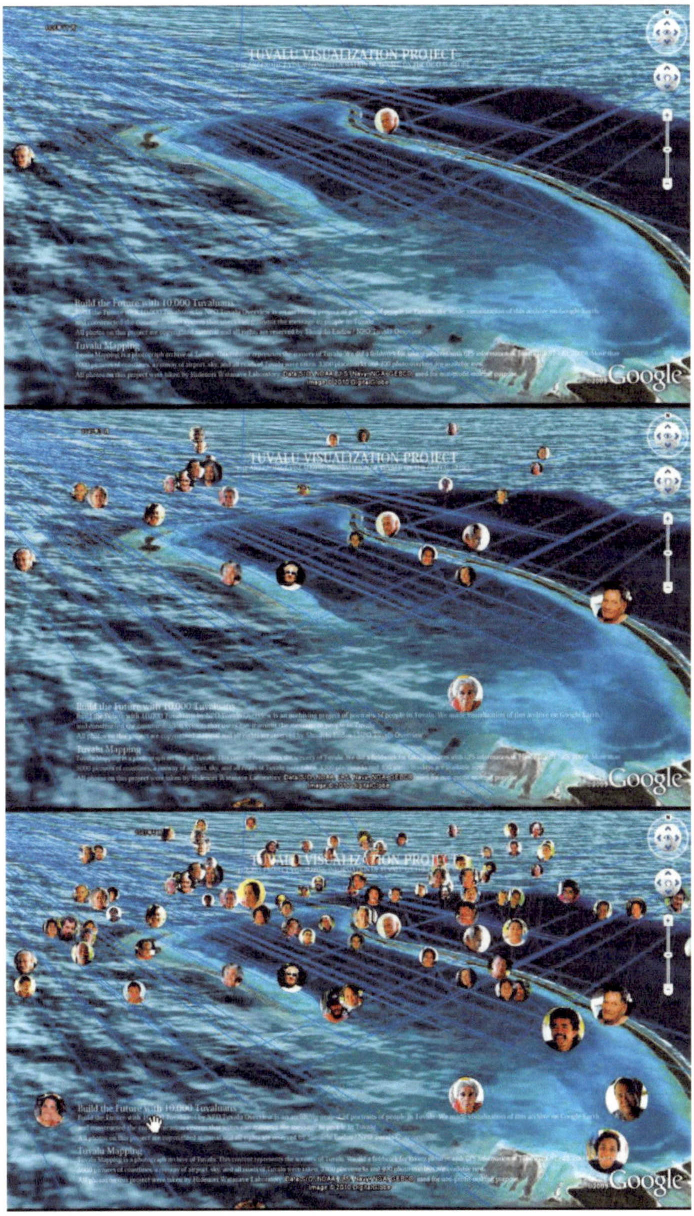

Fig. 3. Narrowing display by timespan

Clicking on the icon will display a balloon (Fig.4). Within the balloon, a large-sized photograph (Maximum width of 320 (pixel)), name, age, and a comment entry form are displayed.

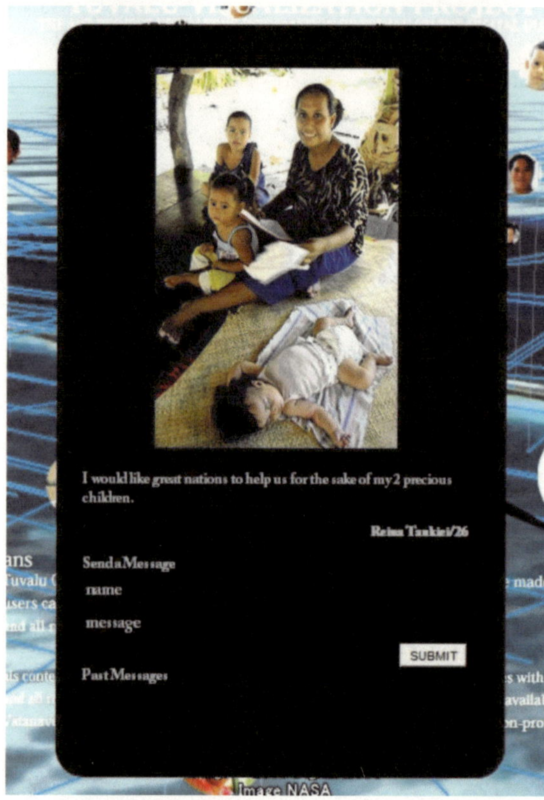

Fig. 4. Balloon display

The mapping of landscape photographs, the other contents enabling the visualization of the "realities of the daily life", will be stated later.

3.5 Sending the comment and visualization of the communication

In order to realize the "pseudo-synchronous communication function" mentioned in 3.1, a system has been implemented to visualize pseudo-synchronously the communication exchanged between the Tuvaluans and the users through lines on the digital globe. After the user enters the name and comment into the comment entry form in the balloon explained in 3.4, a line is drawn between the location of the user on the Earth and the portrait icon of the person whom the comment was sent to. The location of the user is estimated by the IP geolocation method. As the comment is sent, the global IP address of the user is verified with the MaxMind GeoIP database (MaxMind) of the server simultaneously, estimating the location information. A line is drawn on the Google Earth through the following system. When the network link kml of the client side accesses the PHP for line output of the server side, the kml file, including each location information of both the user and the portrait icon, is downloaded. All the comments contributed so far will be displayed within the balloon. Through this system, the record of the communication is expressed as the group of lines drawn from various places on Earth.

This allows the pseudo-synchronous visualization of the expanse of global communication including Tuvalu (Fig.5).

Fig. 5. Displaying lines

3.6 Mapping of the landscape photographs

As another content to tell the realities of a remote place, mapping of photographs with GPS information, as well as super high resolution photo overlay were created. All photographs were mapped out at the points of shooting, and displayed as icons on the Google Earth. Because the visibility is decreased when the icons are mapped out on the ground surface, sinking into the satellite images, all the icons are set at an altitude of +15[m] above ground surface. Through this design, the photographs and the ground surface are displayed separately by tilting the camera. It is possible to search the landscape photograph of liking, while flying around freely between the icons by using the zoom-in/out and panning functions (Fig. 6)

Moreover, twenty landscape photographs taken all over Tuvalu have been implemented as super high resolution photo overlay. Each photo is divided into 256 pieces to be uploaded into the server. They are displayed in multi-layers on the 3D land form in the Google Earth using the photooverlay tag (Fig.7). Also, images with resolution dependent on the degree of zooming-in are automatically displayed in the streaming format.

Fig. 6. GPS photograph mapping

Fig. 7. UHD photo overlay

3.7 Overlay changing according to its altitude

In order to naturally deepen the understanding of Tuvalu by the user carrying out the operation, it is designed so the detailed information on the nation of Tuvalu and this project will be displayed gradually (Fig. 8) as the user zoom-in to Tuvalu.

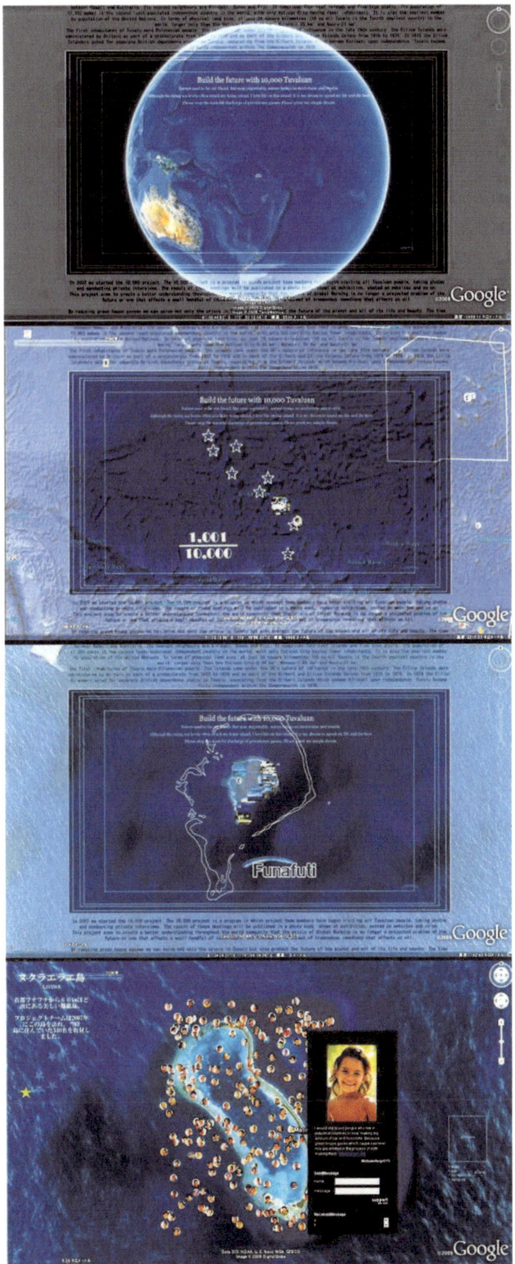

Fig. 8. Overlay changing according to its altitude

3.8 System structure

The system structure of the "Tuvalu Visualization Project" is described below:
3.4 Portrait and interview record: it is stored in the database (MySQL) after being correlated to each other, and it is retrieved through the PHP script generating the kml dynamically. The location oriented display and random display mode is switchable by the use of the network link.
3.5 Sending the comment and visualization of communication: the comments to the database is stored by retrieving the PHP script of the server from the interface of Google Earth. When the comment record and the line display is updated, the newest data is loaded to the client side when the view is refleshed on Google Earth.
3.6 Landscape photographs (with GPS information): after granting the creative-commons license, they are uploaded to the photograph sharing service, and are retrieved from the static kml file.
3.6 Landscape photographs (for Ultra High Definision photograph overlay): after implementing the tile separation function, they are stored in the directory in the Web, and are retrieved from the static kml file.
3.7 Overlay: it is stored in the directory on the Web, and is retrieved from the static kml file.
All of these kml files are set to be read via the single network link kml file which the client will download. Still, the actual transmission uses the kmz file which is compression of kml. Also, all of the kml file mentioned above is made public on the website. It is recommended to be consulted together with the kml reference.

4. Considerations

The "Tuvalu Visualization Project" was able to be experienced by many user audience through the internet, as well as at real exhibitions.

4.1 Internet release and response

Out of the contents of this art work, 3.6 mapping of landscape photographs was released on the internet on August 3rd, 2009, and 3.4 mapping of portraits and interview record and contents of 3.5 visualization of communication were both released on August 30th, 2009. The actual results of the access at the time of writing this thesis (March 25th, 2010) is as below.
Web site access: 13270 times (PV)
kmz file downloads: 5120 times
YouTube replays: 6377 times
Comments contributed: 158, from 6 countries
It is mentioned many times in places such as the internet news sites, blogs, and twitter. The comment, "I can understand very well where Tuvalu is, what kind of faces the people living there have, and what they feel towards the rise of the sea level" is unattainable by just having the photographs and the interview record on the Web page. It is speculated that this art work has achieved its purpose of telling the multifaceted realities of remote place via the digital globe. Many comments were made which display the common awareness of the issue in the global scale, as well as an intimacy with the people of Tuvalu,as the following show: "You are very courageous we support you and are spreading your message in the U.S. Thank you for the insight and for speaking up for social justice." "Dear Valoa-I passionately want the same thing. You did not create the problem-we did! Greetings to you

and your family." It is conceivable that this fact exhibits the purpose of the art work, mentioned in 1.1, was achieved, which is to create a sense of solidarity through the visualization of the pseudo-synchronous communication - the feature of this work - through the webcasting via the digital globe.

4.2 Real exhibitions and the response of the audience

So far, this art work has been exhibited at conferences and festivals within and outside of the country, and had the opportunity to be experienced by a large number of audience. In many occasions, the audience viewed the screen from a distance at first. After having interest in the land of Tuvalu, portraits and landscape photographs expressed three-dimensionally on the digital globe, as well as the spread of communication visualized on the global scale, moved onto the actual operation, which involved inspecting the land of Tuvalu as well as the portraits and landscape photos, and contributing comments. Some examples of the feedbacks of the audience are as follow:

Those related to the land, such as "It is around here on Earth", "It is so far from Japan", "It is so close to sea level", "It is already sinking"

Those related to the photographs of the people of Tuvalu and their interview records, such as "The people's smiles are wonderful", "This person has such a dream", "I can see that they are normal people just like us"

Those related to the daily scenes, such as "It is interesting how the pigs are walking the public road", "It is unexpected to see many cars running", "It is a problem that a dumping ground for trash is created"

Those related to the communication function, such as "Comments from so many nations are contributed", "I want to write a comment, too"

These opinions were uttered sequentially as the audience followed the operation procedure. Looking at this, it is presumed that the communication of the reality of remote place using a digital globe - enabling inspection of the situation of the local area three-dimensionally, as well as the creation of solidarity through the visualization of pseudo-synchronous communication - appeal to the psychology of the audience, and achieve the purpose of this art work stated in chapter 1.1.

4.3 Problem areas and points of improvement

At the first time of this exivision, the client software of the Google Earth needs to be downloaded and installed and this has become a hurdle to view the art work. But now, the users can use this contents on Internet Browser without download and install. Also, in the event of internet publication, it is inevitable that most likely a user familiar with the operation of Google Earth will experience the art work. However, in the exhibits at real venues, it is not always the case. For the audience not familiar with the operation of Google Earth, they may only view the art work from a distance, or even if they did operate it, some were zooming onto a point not related to the art work. Even though such audience were able to operate more smoothly after receiving instruction from a staff, it took more time to understand the purpose of the art work. It is conceivable that improvement such as implementing operation interface or auto-pilot function, other than the keyboard and mouse is needed in regards to this point.

5. Conclusion and outlook

In this thesis, the net art work using the digital globe on the internet, the "Tuvalu Visualization Project" is presented, having the function of both the photography archive telling the realities of the people living in "Tuvalu" and their daily life, as well as the communication platform.
In conclusion, it is considerble that this fact exhibits the purpose of the study as stated in introduction.
The authors intend to keep creating opportunities to recognize anew the local problems globally, through a group of net art works starting with the above mentioned.
At the internet publication and real exhibits of this art work, there were opinions desiring projects with the same purpose in other parts of the world.
And also, it is presumed that this work can create a users' community through the visualization of pseudo-synchronous communication between remote place and users using a digital globe.
The curious instance, in the response to the internet as stated in 4.1,a user who moved to America from Tuvalu found an acquaintance residing in Tuvalu through the art work, and contributed a comment. This case went beyond the authors' supposition, and indicate the creation of a social network connecting the residents of Tuvalu and the audience with each other. But the users become enrolled in the archive only as senders of messages.
In the future, it is needed that to consider the way of work which not only owners of information can release the archive and collect comments, using the merit of Internet, but all users can provide their own information to archive and enjoy contributing their intelligence to community.
And the turning of the scape of the internet community is so fast, the analyses of access log and coordinating with the popular web contents and services at all times are necessary to continue inducing users to this work.
In the future, we plan to make improvements as stated above, as well as continue creating new archive work which is able to use in many fields.

6. Acknowledgements

The authors would like to show our deep appreciation to Photon Inc., DIGITAL CONTENTS EXPO/ASIAGRAPH2009, and each person involved in the Japan Media Art Festival held by the Agency for Cultural Affairs, for their utmost cooperation in creating this art work as well as in its exhibition.

7. References

INTERNET watch (2009). "The Protest Activity Against Iran on Twitter, etc., in Relations with the Democratization Strategy of the United States."In: Internet watch, 22.06.2009, Available from http://internet.watch.impress.co.jp/docs/news/20090622 295583.html

Ikuo. Ohash. (2009) Towards Construction of the Global Information Society - Movement and Issues of the International Society Surrounding the Digital Divide. (Feature Article, ICT Industry as a Turning Point)). *Creation of Intellectual Property, Knowledge cre-ation and integration*, Public Relations Department, Nomura Research Institute, 17(3),pp.6-35 (2009.03)

International Telecommunication Union. Internetindicators: Hosts, Users and Number of PCs http://www.itu.int/ITU-D/ict/statistics/

Kougaku. Takeda.(2006). *The Realities of All Existing Things and the Ultimate Nature of All Things (tathata)*. All Existing Things as Realities", "Origin of Things"; Journal of Indian and Buddhist Studies, 55(1), pp.92-96 (2006.12)

Shoji. Sagawa.(1999): *Joint Studies: Realities of Damage by the Atomic Bomb on Hiroshima and Nagasaki*, Shin-Nihon Publishing Company (1999.07)

Mao. Ishikawa.(1995). *Okinawa and Self-Defense Force Captured by a Female Photographer*, Koubunken (1995.05)

Nobuko. Oyabu. Stand - Choosing to Stand Up, Word of Press Ministries (2007.10)

STALLABRASS Julian: Internet art: the onlineclash of culture and commerce; Tate Publishing,United Kingdom (2003.04)

Niwango Inc.: "Nico Nico Douga"; Winner of Honorary Mention of Digital Communities category, Prix ARSElectronica 2008(2008.09)

Colin Ellard: YOU ARE HERE Why We Can Find Our Way to the Moon, but Get Lost in the Mall; Doubleday Publishing, United Kingdom(2009.07)

Sensorium: "Sensorium"; Golden Nica of .net cat-egory, Prix ARS Electronica 1987(1987.09).

NTT Intercommunication Center: ICCONLINE "Art Meets Media: Adventure of the Perceptions"
http://www.ntticc.or.jp/Archive/2005/art meets media/ Mediaonline/netartj.html

Masaki Fujihata: "Field-Works" http://www.?eld-works.net/

Life Slice Laboratory: "Life Slice Calendar and Life Slice World Map"; Exhibition of the 6th Japan Media Arts Festival Award-Winning Works (2003.02)

Masashige Motoe, Yasuto Nakanishi, Shohei Matsukawa: JIKUKAN-POEMER: Experimentation of Web Geographic Information System Using a Mobile Phone equipped with a Camera and a GPS at Tama Center and its Evaluation; Architectural Institute of Japan Academic Lecture Series (A-2, Fire Proof, Oceanography, Information System Technology) 2004,pp.531-532 (2004.07)

Photon: Sakura Mapping http://mapping.jp/

International Crisis Group: Crisis in Darfur
http://www.crisisgroup.org/en/key-issues/preventing-implosion-in-sudan.aspx

Henri Cartier-Bresson: "Henri Cartier-Bresson(Masters of Photography Series)"; Aperture,France (1997.09)

NPO Tuvalu Overview: "Build the Future with10000 Tuvaluans" http://10000.tv/

MaxMind,Inc.: MaxMind - GeoIP-IP Address Lo-cation Technology
http://www.maxmind.com/app/ip-location

Hidenori Watanave Laboratory: "Tuvalu Visual-ization Project" http://tv.mapping.jp/

Google,Inc.: KML Reference - KML - Google Code
http://code.google.com/intl/ ja/apis/kml/documentation/kmlreference.html

Mainichi Newspaper: Mainichi.jp"Living in Tuvalu: Recreating the Island Nation with a 3D Contents" http://mainichi.jp/life/ecology/graph/20090806/

GIGAZINE: "'Tuvalu Visualization Project' Mapping the Voices of the People Living in Tuvaluon Google Earth"
http://gigazine.net/index.PHP?/news/comments/20100204jmaf2009tuvaluvisuali zation project/

Hidenori Watanave, Makiko Suzuki, YuichiWatanabe and Shuichi Endo: "Tuvalu Visuaiization Project " ASIAGRAPH 2009 in Tokyo;Artech (Art Works And Technolo-Gies Exhibits), Miraikan, Japan (2009.10)

Shuichi Endo and Hidenori Watanave, "Tuvalu Visualization Project " Exhibition in Tu-,svalu Overview ' exhibit space at United NationsClimate Change Conference 2009 in Copenhagen(COP15), Copenhagen, Denmark (2009.11)

Hidenori Watanave/ Shuichi Endo: "Tuvalu Visualization Project;" Exhibition of the 13th Japan Media Arts Festival Award Winning Works, The National Art Center（2010.02）

Hidenori Watanave, Makiko Suzuki Harada andShuichi Endo: "Tuvalu Visualization Project (4KUHD Version) " 12th International Conference on;Virtual Reality - Laval Virtual, Laval Virtual Rev-olution 2010, Theatre de Herce - Salle Polyvalente(Multifunction Hall), Laval, France（2010.04）

26

The Role of Methane Emissions on Ancient and Present Climatic Changes

Guido Barone[1], Elena Chianese[2] and Angelo Riccio[3]
[1]*Department of Chemistry, "Federico II" University of Naples*
Complesso Monte S. Angelo Via Cintia, Naples
[2]*Department of Environmental Sciences "Parthenope" University of Naples*
Centro Direzionale – Isola C4, Naples
[3]*Department of Applied Sciences, "Parthenope" University of Naples*
Centro Direzionale, Isola C4, Naples
Italy

1. Introduction

In this Chapter we have reviewed the available literature to discuss how the changes in the atmospheric concentration of methane can affect the future of climate. There is an increasing interest in the exploitation of natural gas for the energy policy by all Countries. In fact its physical properties and its about constant chemical composition (as largely constituted by methane) recommend it to restrict pollutants during combustion processes and to give the best ratio between energetic content and CO_2 emissions, with respect to all other oil derivatives and fossil sources. Moreover natural gas seems to promise more conspicuous reserves than oil, then assuring a "methane peak" more delayed in time than "oil peak".

However, during the last century, we were in the presence of an increasing accumulation of methane in the atmosphere, as never was occurred in the last 800,000 years. Note that methane is much more effective (25 times) as Green House Gas (GHG) than CO_2. The main suspicion about the causes of this accumulation is charged on the leaks of gas that occurs during the extraction and management activities.

On the other hands, the increasing accumulation of methane in the atmosphere could influence the future of the climate because of two dangerous effects: the first is the fast achievement of a no-return threshold, beyond which the Global Warming (GW) become out of human control. This is made easier by a non linear positive feedback, promoted by the increasing of the same GW on the dynamics of emission processes from the natural sources (destabilization of hydrates, stimulation of metabolic production by plants and animals, etc.). For these complex reasons it will need to know better and, if possible, to control the emission mechanisms by natural sources and, besides, to study how ameliorate the present extraction technologies or invent novel ones that will guarantee an about null gas leaks. That needs to prevent the threat for the future of next human generations.

On these purposes we organized the presentation of this Chapter according to the following Sections.

In Section 2 we discuss briefly the biological and thermogenic origins of methane and its biological dynamics in the sediments. In Section 3 are discussed the general characteristics of all the methane sources and the details concerning the natural sources. In Section 4 the anthropogenic sources are presented in details. In Section 5 all data on fluxes are summarized and shortly discussed. In Section 6 the chemical and photochemical dynamics of methane in the atmosphere is discussed and finally its impact on the climatic changes is analyzed in the Conclusions.

Methane (CH_4) is an important Greenhouse gas (GHG) and its concentration in the atmosphere fluctuated during the last million of years between 320 ppbv (nmol/mol of dry air) (during glacial ages) and 780 (warm periods) up to the end of XIX Century, as deduced from the analysis of the composition of deep ice core drilled in the Antarctica (O'Connor et al., 2010; Loulergue et al., 2008). Before the end of the last glaciation, methane concentration diminished up to 380 ppbv (Monnin et al., 2001). It must be remembered that the ice cores extracted by the Greenland glacier show an abrupt change (200 ppbv in a century or 16 ppbv per 1 °C) in concomitance with the fast warming years at the end of last glaciations (Chappellaz et al., 1993; Huber et al., 2006; Wolff et al., 2010).

In the historical age the methane concentration in the atmosphere increased gradually up to 715 ppbv in 1750 (Etheridge et al., 1998) and 780 ppbv at the end of 1800. Finally a very large methane content in the atmosphere has been registered in the last century: the concentration really underwent an acceleration achieving in 2005 the value of 1,774.62 ± 1.22 ppbv (more than double its pre-industrial value) as measured by NOAA in both hemispheres (Forster, 2007) with an increase of about 1,000 ppbv since the beginning of the industrial era in the late 1700's, representing the fastest changes in this gas over at least the last 800,000 years. In 2008 CH_4 achieved 1787 ppbv (Dlugokencky & Bruhwiler, 2009) with an increase of 140 % respect to the start of industrial revolution. This is much higher than the increase of carbon dioxide (CO_2) in the same period (35 %).

Recently, conspicuous emissions of methane were observed from the circum arctic territories and were preliminarily attributed to the anaerobic decomposition of organic sediments on the deep of recent marshes formed by the fusion of permafrost due to the global warming (Walter et al., 2007). Also unexpected methane flows were observed during the 2010 year from the Central and Eastern Siberian Seas (Shakhova et al., 2010). The global warming indeed caused in the last years an abnormous increase in the summer mean temperature (about 4°C) of Siberian rivers and coastal seas.

Other phenomena were observed around the Svalbard Islands An alternative hypothesis concerning the decomposition of the methane hydrates seems more convincing (Barone & Chianese, 2009) in this case.

Moreover other dramatic episodes were registered some time ago in very different environments, as the abrupt emission of a great amount of natural gases near the Namibia coasts, were an animal slaughter occurred (Ward, 2007). This event, however, can be better attributed to a gravitational instability of the deep sea layers, oversaturated by gases produced anaerobically in the marine sediments.

The origin of the present increase, as so as the large changes of concentration of atmospheric methane in the past is discussed in this Chapter. Methane is a long-lived greenhouse gas because it is chemically stable and persists in the atmosphere. The chemical lifetime of CH_4 from removal through reactions with the hydroxyl radical (OH) is estimated at 9.6 years (Folland et al., 2001). Once emitted, however, CH_4 actually remains in the atmosphere for

what is known as a "perturbation lifetime" of approximately 12 years before removal and ultimate conversion to carbon dioxide (CO_2), mainly by chemical oxidation in the lower atmosphere, or troposphere (Solomon et al., 2007). It must be noted that methane is about 25 times more active as greenhouse gas than the CO_2 (O'Connor et al., 2010; Barone & Chianese, 2009; United States Environmental Protection Agency [U.S. EPA], 2010a).
The longer perturbation lifetime of CH_4 is primarily a result of feedbacks between CH_4, OH, and its by-product CO which is also removed by reactions with OH. Minor removal processes include reaction with chlorine in the marine boundary layer, adsorption on soil sinks, and stratospheric reactions. Increasing emissions of CH_4 reduce the concentration of OH, a feedback that may increase the atmospheric lifetime of CH_4 (Solomon, 2007a).
As OH reacts also with other short-lived pollutants, including Volatile Organic Compounds (VOCs), and tropospheric ozone, it interferes with the same VOCs and ozone precursor species and affects the CH_4 lifetime (Wuebbles, 2010).
For this reason, emissions of this gas into the air have a long-term influence on climate. Methane is a light gas and it diffuses and mixes very well throughout the atmosphere, much faster than it is removed: its global concentration can be accurately estimated from data at a few locations (Solomon, 2007b).
In order to understand how the different residence times of different Greenhouse gases, with largely different Radiation Forcing (RF) powers, can affect Global Warming it is important to evaluate their accumulation velocities in the atmosphere. In fact the methane concentration in the atmosphere (amongst those of all other GH gases) is regulated by the dynamical interplay between sources and sinks. Over a period of 100 years, each molecule of CH_4 has 25 times the direct global warming potential of a molecule of CO_2 (Solomon, 2007a).
That implies not only to define the values of the present emissions from the different sources, but also to understand how these values changed along the time in the last century and in the past and how will be possible to forecast next variations. It is likely to discuss two aspects, i.e. how the emissions can change in the next decades, due to new assessments and landuse changes (urbanization, deforestation, infrastructure cementifications, new agriculture and husbandry implementations, spontaneous increase of deserts or wetlands, flooding of coastal lands due to increase of sea level) and how these changes will impact on the climate.
In turn the same kind and intensity of the emission as so as the transformation of land availability and uses will be regulated by Global Warming.

2. Origins of methane

2.1 Recent biological origin of methane

A large portion of CH_4 emissions can be linked to the biological process of anaerobic decomposition, during which some bacteria, the **methanigen**, break organic matter down in the absence of oxygen. They are a specialized group of microbes that produce CH_4 through the process of methanogenesis. Some of the methane can be partly or completely oxidized by another group of bacteria, called methanotrophs, while the remainder can ultimately enter the atmosphere. Microbial production of methane occurs in a variety of settings, some deemed natural and others attributed to anthropogenic activities. For example, methanogenesis can occur both in natural wetlands (a natural source) and in human-influenced wetlands such as rice paddies (an anthropogenic source). Methanogenesis also takes place in other water bodies, submerged sediments, landfills and waste treatment facilities, and the digestive systems of animals arthropods and ruminants, some of the last domesticated (classified as an anthropogenic source) and some wild (classified as a natural source). Other natural sources of CH_4 include wildfires and geologic processes. Other anthropogenic sources include natural gas handling, biomass burning, and fossil fuel production.

The total flux of CH_4 into the atmosphere from all sources is currently estimated to be 566 Teragrams of CH_4 per year (Tg CH_4/y), which is more than double the pre-industrial value (Solomon, 2007a; U.S. EPA, 2010a). Based on analysis of known methane sources, observed isotopic abundances, and budget modelling, the Intergovernmental Panel on Climate Change (IPCC) estimates that the significant increase in atmospheric CH_4 levels observed over the last two centuries is primarily due to increasing anthropogenic emissions of CH_4, which are currently approximately 1.5 to 2.5 times the magnitude of natural emissions (Forster, 2007). In the same report, a similar estimate is reached by adding together the best estimates for the main natural source categories, which suggest that natural sources contribute 208 Tg CH_4/y to the atmosphere, or 37% of total global emissions. The remainder comes from anthropogenic sources, the largest of which are fossil fuels and natural gas systems exploitation, livestock, waste management and biomass burning (Solomon, 2007b; U.S. EPA, 2007).

2.1.1 Metabolic ways of methanogenesis in the wetland soil and marsh sediments

From complex natural polymers (cellulose, amylase, other polysaccharides and different vegetation proteins) several hydrolytic species of bacteria produce monomers (simple sugars or amino acids. Different fermentative bacteria begin to operate, by means of different fermentation ways, on these materials:

1. Acetate way: is the more direct one, giving acetate a helpful substrate for the *Methanigen bacteria* that produce methane;
2. Long chain fatty acid way (propionate, succinate, butyrate, lactate acids) that in turn can be transformed in acetate; analogously is the ethanol way;
3. Other species of bacteria (*syntrophs bacteria)* can oxidize fatty acids up to CO_2 releasing at mean time H_2 and before that the gas will released, there is the intervention of Homo-*acetogen bacteria* that transform the two molecular species, by an acetogenesis process, in the acetate, thus reconnecting to the first way.

2.2 Ancient (fossil) origins of methane as natural gas and formation of its hydrates

Different mechanisms have been invoked to explain the formation of methane deposits and natural sinks. Methane can originate by both thermo chemical and biochemical processes. In the first case, as other light hydrocarbons, it has the same origins of petroleum through a transformation of organic materials (dead microorganisms, zoo- and phytoplankton, algae, marine animals, and all kinds of life degradation products including dispersed terrestrial pollens). These materials fall on the ocean soundings, continental escarpment seafloor and marine trenches or lacustrine stagnant basins, together with a mixture of clays. This sedimentary mud can be covered by layers of sands and can transform in rock during slow repeated processes under high pressure and temperature.

As consequence of long period geological processes (in Holocene or much more ancient ages), these rocks subside by 2 - 4 or even 6 km depth. Moreover, due to the collision between tectonic plates, the rocks can curl, for the strong side compression forces, and also move from the marine trenches below continental platforms (Figure 1).

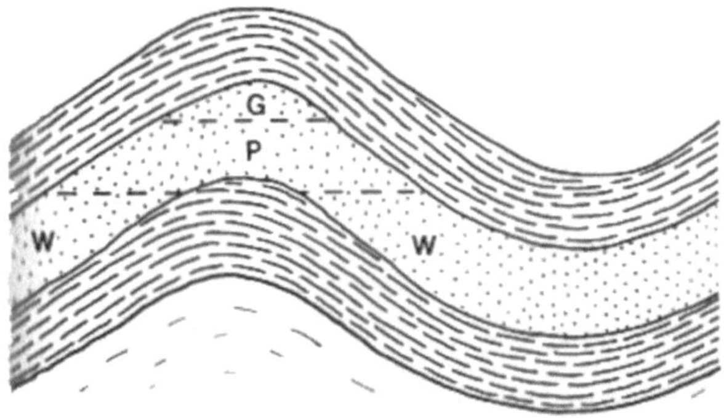

Fig. 1. Curling of layers of rocks containing petroleum reservoirs sealed up and down by other impervious clay rocks: P and G indicate the rock layers sodden (mainly sandstones) of petroleum and gas respectively and W the rock layers in whose interstices predominates water.

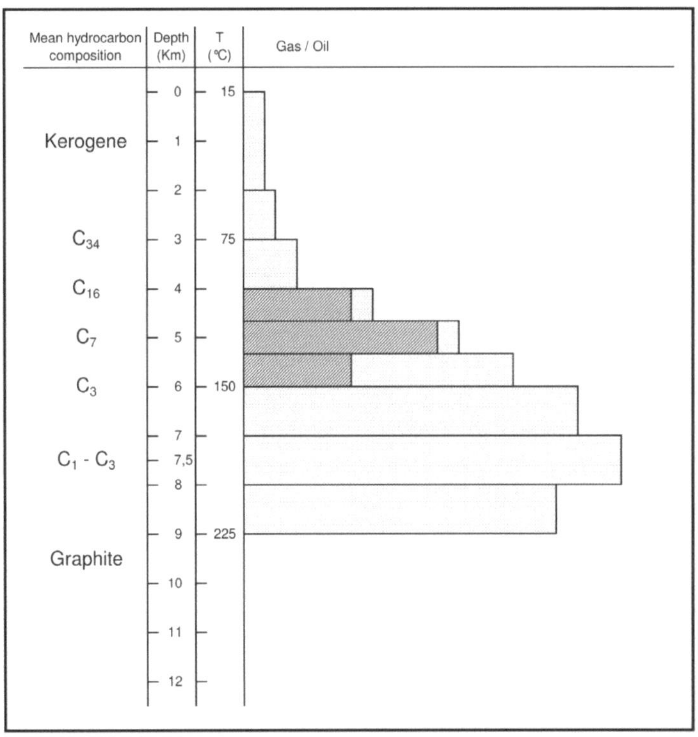

Fig. 2. Natural gas thermogenesis in the deep marine and terrestrial Earth layers. The two central columns report the achieved deep by the organic sediments and the temperatures they experienced. On the first column on the left is reported the mean hydrocarbon composition in which the kerogen blend is transformed, passing from a mixture of solid paraffin's, to crude oil, light oil, gases and finally to a graphite status. The histogram on the right gives a qualitative ratio amongst the gases and oil fractions produced at different temperatures and depths (i.e. pressures).

In both cases the subsided rocks experience temperatures between 80 and 160 °C or even higher. The organic matter in the interstices of these rocks (kerogen) undergoes a thermal degradation, in absence of oxygen, giving rise to a complex mixture constituted mainly by hydrocarbons: so originates a petroleum reservoir. The high temperatures of the deeper layers would promote the thermochemical production of hydrocarbons with small molecular weights (Figure 2).
However, when the rocks are oversaturated, expulsion and migration processes through pores or fractures occur until the oils and gases, blocked on the top and sides by impervious clay layers, encounter geological traps. These traps can be drilled and exploited for the oil extraction and refinement. The gases can remain free on the top of the trap, or solubilized in the oil or finally migrate through other thin geological fractures into other reservoirs, often located very far from the petroleum field.
Here, in the presence of sea water and at low temperatures, about 0 C, and on the high hydrostatic pressure exerted by the column of some kilometers of ocean and by thick layers of overhanging rocks, methane and other gases can be fixed as hydrate crystals (Figure 3).

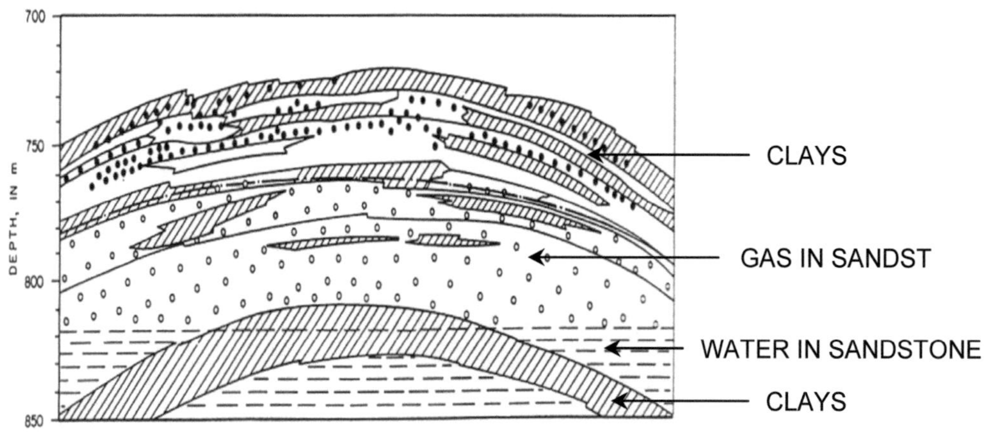

Fig. 3. Structure of a geological trap of natural gas hydrates, originated both by migration from petroleum reservoirs, or by local biological processes. (Adapted and redrawn from Barone & Chianese, 2009; Claypool & Kvenvolden, 1983 and Sloan & Koh, 2008). The filled spots in the upper layers represent gas bubbles eventually seeping trough fractures in the clays or overhanging rocks.

Different is the formation of hydrates under the Arctic permafrost: here the low temperature and high pressure forced the preexisting accumulation of methane and natural gas to transform partially into hydrates. In these areas the extensions of the two fields of hydrates and free gases are geographically partially overwhelmed. In all the less ancient, humid environments (marshes, peat-mosses, anoxic depths of crater-like lakes, and also under permafrost and marine sediments) the biochemical origin of methane seems largely preferred. In these cases the activity of methanigen bacteria is mainly involved in the biochemical transformation of organic sediments accumulated in previous years.
In the case of marine sediments these bacteria use the reduction of sulfate and nitrate as metabolic source of oxygen (Sloan & Koh, 2008). When the biochemical activity occurs on very recent sediments, methane and other gases (frequently H_2S) can gently and slowly seep continuously versus the surface, unless entrapment phenomena create, under high pressure,

instable mixtures of free gases and saturated solutions. The entrapment occurs very often in the upper layers of marine and oceanic sediments (400 - 500 m under the sea floor) and of permafrost (200 - 1000 m under the surface).

When the origins are more ancient, isotopic carbon analysis is not sufficiently accurate for settling between the different hypotheses, even if the presence of ^{13}C can be assumed as a proof for recent biochemical origins. The ratio R = methane versus (ethane plus propane) can better demonstrate the thermo-geochemical ($R < 10^2$) (Diaconescu e al., 2001; Mazzini et al., 2004; Sassen & Macdonald, 1994; Woodside et al., 2003) or the biochemical ($R = 10^3 - 10^5$) origin of gases (Claypool & Kvenvolden, 1983).

There are no sufficient proofs to support the hypothesis that origins of methane and its hydrate is a surviving memory of Earth primordial reducing atmosphere. The presence of N_2 together with CH_4 in natural double hydrates could be a weak indication for this hypothesis, as well as the presence of He in the evolved gas; but the last one could originate from the decay of radioactive elements in the neighbor of the trap.

2.3 Biological dynamics of methane in the organic sediments

Besides the methanigen productions, other biochemical transformations occur in the organic sediments that leave to the oxidation of methane. During the natural or human-promoted landfill processes, several aerobic bacteria colonies and associations of species develop and growth. These microorganisms oxidize methane, producing CO_2 that is dispersed in the atmosphere. Chemical transformations also occur in the upland soils and riparian areas, so the dynamic methane balance is negative and these layers act more as a sink, subtracting at all 30 Tg/yr of CH_4 from the atmosphere.

Much more efficient are the biochemical activity of the marine aerobic bacteria colonies and associations of species that develop and growth on the upper layers of the oceanic sediments. In these cases the need of oxygen for the oxidative processes is given by the reduction of nitrate and sulfate ions present in the seawater. It is the same oxygen source utilized by these microorganisms for their own metabolism. The trophic sites of both the anaerobic and aerobic colonies, however, do not have often well defined boundaries.

3. Sources of methane

Natural Sources of methane (about 30%):

- Wetlands;
- Upland soils and riparian zones;
- Vegetation
- Rivers and estuaries;
- Lakes;
- Hydrates from terrestrial and continental shield permafrost
- Oceans (from biota and organic dissolved or suspended products);
- Terrestrial and marine geological sources;
- Wild fires;
- Wild ruminants;
- Termites and other arthropods;
- Hydrates from submarine fields;
- Natural gas spontaneous emissions from fossil deposits (gas, carbon, petroleum, shale rocks, etc.).

Anthropogenic sources of methane (about 70%):

- Animals husbandry;
- Rice Paddies;
- Landfills
- Biomass combustion;
- Methane from biogenic origins: Waste disposal and waste treatment.
- Gas, petroleum and exploitations and production;
- Coal mining
- Hydrates exploitations;
- Shale rocks.

It is important to note that sometimes there are superposition of source sites (location of permafrost in continental shelves, gas hydrates, and geologic sources), or ambiguity in the definitions (e.g., there are small ponds in natural wetlands that are difficult to distinguish). Field studies may include flux measurements that have contributions from more than one of the sources listed here (O'Connor et al. 2010; U.S. EPA, 2010a).

This Chapter is organized into paragraphs covering natural sources and anthropogenic sources by type. However, the earth is a mosaic of these different source types, the boundaries between these source types are sometimes inexact (e.g., between a wetland and the emergent vegetation of a lake margin), and system changes that affect one source can also affect one or more other sources (O'Connor et al. 2010; U.S. EPA, 2010a).

The issue of methane in permafrost regions exemplifies this interconnectedness. The predicted thawing of permafrost with climate warming may contribute to increased natural source methane emissions to the atmosphere. However, this is a complex system response. There is not a lot of methane frozen into permafrost (unless it is a gas hydrate formation), so permafrost thaw will not release much methane directly. The released methane is probably being oxidized as it diffuses through 1 to 100 meters of thawed soil before reaching the atmosphere. However permafrost thawing implies the formation of marshes or can be associated with thermokarst erosion, which can form (or drain) lakes: in the first two cases the anoxic fermentation of organic sediments can occur giving methane emissions (Walter at al., 2007). Moreover, if gas hydrates are associated with permafrost, then permafrost thawing promotes destabilization of these hydrates, with methane release, potentially in large quantities (Barone & Chianese, 2009).

Another consideration is changes in wetland vegetation and moisture status associated with permafrost thawing; this would be an issue for some, but not all, permafrost landscapes. If the landscape gets wetter and the vegetation composition becomes more dominated by sedges, this could lead to increased methane emissions from a wet landscape - at least for years to decades; however, if the landscape gets drier (or stays relatively dry), then methane emissions would probably stay low (U.S. EPA, 2010a).

3.1 Detailed description of natural sources

3.1.1 Wetlands

Wetlands are ecosystems in which saturation with water is the dominant factor controlling soil development, as well as the species of plants and animals that are present. They are transitional areas, at the interface between upland environments and aquatic systems, and they are concentrated in the high latitudes, where frozen soils can inhibit water drainage, and in the tropics, where precipitation rates are high.

Because saturated soils create anoxic conditions, wetlands are an important natural source of CH_4, which is produced by anaerobic microbial processes that decompose dead plant. They cover about 5 percent of the Earth's surface. Emissions of CH_4 are a minor part of the larger amounts produced and consumed in wetland soils. The different types of bacteria, both producing and consuming it are affected differently by environmental factors (e.g., temperature, water level, and organic matter supply and characteristics). Therefore, a relatively small environmental change can result in a large change in flux by changing the balance between production and consumption (U.S. EPA, 2010a).

In recent years, sophisticated models have been applied for estimating emissions from wetlands. However there are substantial uncertainties affecting the emissions estimates, even if it is possible to estimate fluxes under changing environmental or climate conditions. The 2010 U.S. EPA Report on emission estimates were derived by taking a simple average of the many estimates that have been made since 2004. Flux estimates for high latitude wetlands range from 24 to 72 Tg CH_4/yr, with an average of 43 Tg CH_4/yr. Reported emissions from tropical wetlands range from 81 to 206

Tg CH_4/yr, with an average of 128 Tg CH_4/yr (Table 1). The 2010 U.S. EPA Report estimates a total mean emission value of 170 Tg CH_4/yr with an uncertain range of 108-278 Tg CH_4/yr (compare this estimate with that of: 100 - 230 Tg CH_4/yr (Denman et al., 2007) and the estimate of 225 Tg CH_4/yr (Keppler at al, 2003) that perhaps also encloses emissions from rice paddies. In this chapter the last ones are intrusted to the anthropogenic emissions (see paragraph 4.2).

The overall estimates reflect an increase over the figures from the U.S. EPA 1993 Report (U.S. EPA, 1993), suggesting that emissions from the tropical latitudes made the greatest contribution to global fluxes, achieving about the 75 percent of the total CH_4 emissions from wetlands worldwide.

Changes in land use and climate will affect CH_4 emissions from wetlands, with the potential for both large increases and large decreases. At high latitudes, changes in climate are thought to be the major factors driving changes in CH_4 emissions. For example, models using altered temperatures and rainfall suggest that emissions from northern wetlands could double by the end of the 21st century. In the tropics, changes in land use, such as draining or filling wetlands for other uses, are believed to be the major driver of change (U.S. EPA, 2010a).

3.1.2 Upland soils

Upland soils are well-aerated, not water-saturated, and generally oxidizing sites, with dry soil conditions. These conditions favour microbial processes that make these soils a sink for CH_4. Natural sink include upland soils associated with forests and grasslands under natural vegetation, but not agricultural lands.

Riparian zones, located at the interface of terrestrial and aquatic environments, are often permanently wet and rich in organic matter, with water saturated soil conditions and microbial available carbon that contribute to high rate of CH_4 production.

CH_4 sink capability of soils under natural vegetation (including upland and riparian soils) is estimated at 30 Tg CH_4/yr that are subtracted from the atmosphere (Table 1). Methods of estimating CH_4 budgets (i.e., the balance of sources and sinks) however contain significant uncertainties due to the large spatial and time variability (U.S. EPA, 2010a).

Many interrelated factors determine the sink strength for CH_4. Recent studies have found that the strength depends on oxidation by methanotrophic microbes in the soil, and therefore is influenced by environmental factors that control this oxidation rate. The primary factor is soil diffusivity, which controls the amount of CH_4 transferred into the soil and, therefore, its availability to methanotrophs. Other factors are soil organic carbon content, vegetation type, pH, bulk density, and drainage. In the future enlargement of these sink areas will depend on the human use of territory and climatic changes.

3.1.3 Water bodies

In all water bodies, microbial processes that occur both in the water column and in sediments produce CH_4. Different aquatic sources, including the deep waters of the open ocean, shallower coastal waters on the continental shelves, freshwater rivers and estuaries, where fresh and salt waters mix are considered.

The **oceans** are believed to be a minor natural source of CH_4. Continental shelf areas, estuaries, and rivers contribute to emissions of CH_4, but they have higher organic inputs and nutrient levels than the open ocean; they are relatively shallow, mixing is active and the gas produced in the sediments is easily transported as surface water and released to the atmosphere.

In the open ocean also, enhanced CH_4 production and flux can be found in upwelling areas, which are areas where the prevailing winds and currents bring nutrient-rich deep water to the sea surface. Upwelling areas tend to have higher rates of primary productivity, which in turn leads to more organic material falling down, depleting oxygen in the depth levels and creating here favourable conditions for methane-producing bacteria.

Open ocean emissions are low and dispersed over large areas, and thus they are difficult to resolve with techniques such as inverse modelling that use changes in atmospheric concentrations to estimate flux. The most recent emissions for the open ocean are estimated as 1.8 Tg CH_4/yr, a weighted average among less of 1 Tg CH_4/yr and about 4 Tg CH_4/yr.

Emissions from the **continental shelves** are somewhat higher than those from the open ocean, even though these environments cover a much smaller area. This difference likely reflects greater organic inputs and an increase in sedimentary contributions. An average of recent estimates gives a total flux of 5.5 Tg CH_4/yr.

Estuaries and rivers cover limited areas, yet they contain much biological activity. They are also sites of active mixing, enabling CH_4 produced in adjacent wetlands and shallow-water environments to be rapidly released to the atmosphere. Averaging the available estimates for estuaries and adding the only available estimate for rivers results in a total flux of 1.85 Tg CH_4/yr.

On overall, natural CH_4 emissions from oceans, estuaries, and rivers are estimated at 9.1 Tg CH_4/yr (U.S. EPA, 2010a) (Table 1). Natural emissions from oceans, estuaries, and rivers represent only about 2 percent of the total global CH_4 emissions to the atmosphere every year. They are expected to remain largely unchanged in the future.

3.1.4 Permafrost

Permafrost is soil, sediment, or detritus that is permanently frozen (temperature $< 0°C$) for at least two consecutive years. It is widespread and nearly continuous in the arctic, but also exists intermittently in the sub-arctic and boreal regions, and at high elevation. Permafrost is a natural source of CH_4, which could be released as it melts.

Emission estimates are based on measured concentrations of CH_4 in permafrost and estimates of contemporary permafrost degradation rates. Current CH_4 releases from permafrost are estimated to be 0 to 1 Tg CH_4/yr. There is now strong evidence that permafrost is melting, and that a substantial fraction of permafrost existing now will be melted within the next 100 years due to global climate change. However, it now also seems clear that the permafrost melting will be only a small direct source of CH_4. Indirectly, however, thawing permafrost is already impacting CH_4 emissions from other natural sources, particularly lakes and wetlands. Permafrost can contain ice wedges, which are lenses of frozen water that can be up to several meters thick. As permafrost thaws and ice wedges melt, this water can sometimes drain away, leading to ground subsidence or collapse, which in turn can alter drainage patterns. In this process, known as thermokarst erosion, lakes and wetlands can form or can dry up. This process affects CH_4 emission rates from high latitude lakes and wetlands.

3.1.5 Gas hydrates

Gas hydrates are ice-like crystals formed between water and gas molecules such as CH_4 under high pressure and ambient temperatures. Large quantities of CH_4 are currently trapped in hydrate form, occurring mainly on continental shelves and to a lesser extent below permafrost regions. Gas hydrates can store large amounts of the gases that they trap, and are stable within a specific range of temperature and pressure known as the hydrate stability zone. In Figure 4 thermodynamic stability ranges of methane and lighter hydrocarbons hydrates are reported together the domain of coexistence with liquid water, ice and natural gases.

Under current conditions, the CH_4 emissions from gas hydrates are small (estimated at 2 to 9 Tg CH_4/yr); however, the potential for significant CH_4 release from gas hydrates warrants close examination of this source (U.S. EPA, 2010a). A significant fraction, if not all, of these emissions are expected to be oxidized in the ocean water column (see paragraph 2.3).

Since 1993, there has been limited discussion of the current flux of CH_4 from gas hydrate reservoirs. Oceanic and onshore continental reserves are believed to be stable at present, which means that they are not currently emitting CH_4. Offshore continental shelf reserves are currently unstable, however, and are believed to emit CH_4 (Shakhova, 2010; Barone & Chianese, 2009). Estimates assume that the CH_4 being liberated from the gas hydrate form is released into the atmosphere. It is possible, however, that some or all of this gas is not actually emitted and, instead, it is oxidized or absorbed within the sediment or dissolved into the water column.

Due to their proximity to the Earth's surface (< 2,000 meters) and the sensitivity of the hydrate stability zone, gas hydrates will likely be affected by climate change. Pressure on hydrates is expected to change as a result of sea level rise and the melting of polar ice caps; however, temperature changes are likely to have a far more significant effect. CH_4 emissions from this source are likely to increase as temperature rises. Based on recent research, it is estimated that the increase in methane emissions due to an increase in ocean temperatures may be sufficient to overcome oxidation in the water column and result in significant atmospheric methane emissions. The magnitude of the methane emissions expected to reach the atmosphere due to release from destabilized hydrates upon ocean warming is, however, uncertain. The methane hydrates and their stability will be moreover discussed on paragraph 4.9, referring to the offshore extraction from the deep ocean sediments.

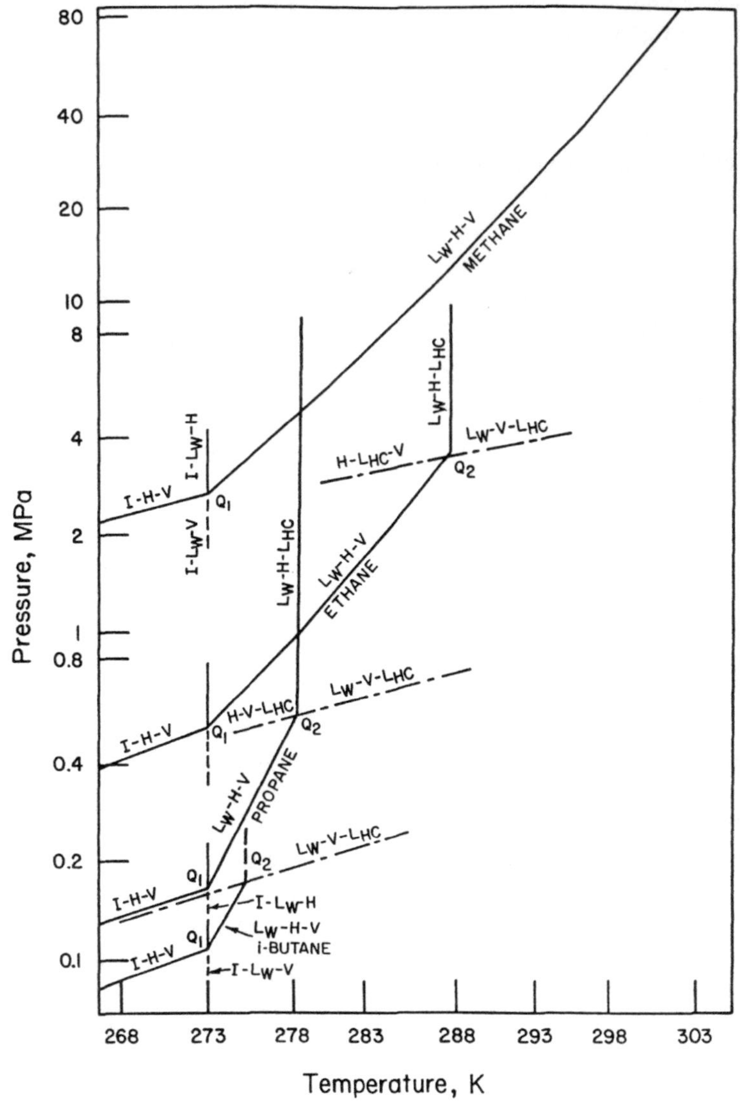

Fig. 4. Phase diagrams of simple natural hydrocarbons forming hydrates (from Barone & Chianese, 2009; adapted and redrawn from Sloan & Koh, 2008). H indicates hydrate, I ice, L_w and liquid water and vapour, L_{HC} liquid light hydrocarbons. Methane hydrate phase diagram presents a quadruple point (Q_1), the other gas hydrates two (Q_1 and Q_2).

3.1.6 Lakes and ponds

Lakes and ponds are naturally formed permanent water bodies dispersed on the land. This source category includes natural freshwater lakes but excludes impoundments and reservoirs (water bodies formed by dams), as greenhouse gas emissions from impoundments, reservoirs, and other engineering works are considered to be anthropogenic (paragraph 4.7).

CH_4 production rates depend on temperature, organic matter availability (food for the bacteria), and isolation from oxygen; these factors are influenced by climate, lake size and depth, and productivity of microscopic and macroscopic plants and animals, which create organic matter for CH_4 production when they die and sink to the bottom. There are four pathways for CH_4 emissions from lakes: bubbling, diffusion, plant-mediated transport, and seasonal overturning. Bubbling has been determined to be the dominant pathway for CH_4 flux, accounting for more than 90 percent of CH_4 emissions from lakes.

Based on recent estimates, lakes emit approximately 30 Tg CH_4/yr to the atmosphere. The number and total area of large lakes is well known, but some uncertainty involves the total surface area of small lakes and ponds. Lakes smaller than

1 km^2 constitute about 40 percent of the total global lake surface area. Because small lakes and ponds generally emit more CH_4 per unit area than large lakes, uncertainties about total surface area are a major factor in the overall uncertainty of the estimate (U.S. EPA, 2010a).

Climate warming impacts on permafrost and the development of thermokarst lakes could substantially affect future CH_4 emissions from lakes. It is estimated that emissions from lakes north of 45°N will eventually decrease, due to lake area loss and permafrost thaw. Before this long-term decline, though, a period of increased CH_4 emissions, associated with thermokarst lake development in the zone of continuous permafrost, would come. CH_4 emission rates from northern lakes could rise as high as 50 to 100 Tg CH_4/yr during this transitional period, which would last hundreds of years.

3.1.7 Geologic sources

CH_4 and other hydrocarbons can seep naturally from **geologic sources** on the deep Earth's crust. Some geologic CH_4 emissions are produced via what is known as macroseepage, which includes relatively large localized emissions from identified geologic features and events such as mud volcanoes and localized vents. Emissions can also result from diffuse soil exhalation or degassing in volcanically active or other geothermal regions, and from microseepage in petroliferous or hydrocarbon-containing sedimentary basins. Sources include both marine (underwater) and terrestrial (land-based) faults.

Surprisingly white crystal efflorescence formed by the same hydrates found in the oil pipelines were discovered in 1986 on the floor of Caribbean-Mexican Gulf, at 1000-3000 meters depth under the sea level, near the crust fractures from which petroleum emissions drip out (Sassen & Macdonald, 1994). In the following years the same discover was made on other marine floors at different depths. In these conditions high pressure favors the formation of the hydrates from the seeps and endothermic inclusion reactions, that cause a temperature decrease with respect to the deep ocean local temperature (2÷4 °C, almost constant). So, the methane and part of the other gases are selectively trapped in white or even dirty crystals, while the bigger molecules float up to the sea surface (Barone & Chianese, 2009).

Previous estimates of natural sources have either ignored this source or only evaluated "traditional" but actually minor sources such as high temperature magma-producing volcanoes. More recent estimates include emissions from mud volcanoes, other macroseepage locations, terrestrial microseepage, and submarine seeps.

Submarine estimates are extremely uncertain, particularly when estimating the proportion of emissions that are absorbed by ocean water before reaching the surface. In contrast, estimates of onshore emissions can be based on direct measurements and standard emission factor concepts applied to point sources (for individual features such as mud volcanoes) and more diffuse area sources (for microseepage). "Bottom-up" emissions estimates for both marine and terrestrial sources generally lie in the range of 32 to 74 Tg CH_4/yr. This range largely reflects uncertainty in estimating both the global number of sources and the proportion of emissions that actually reach the atmosphere, rather than being absorbed by ocean water. Recent isotopic constraints on the budget suggest a narrower range of 42 to 64 Tg CH_4/yr, based on the total budget of "fossil" (radiocarbon-free) methane in the atmosphere.

Relatively few climate- or human-related factors are believed to be capable of influencing CH_4 emissions from geologic sources. Some reports suggest decreased emissions associated with large-scale extraction of oil and gas, and increased emissions following deglaciation events and the corresponding increase in seismic activity (i.e., post-glacial rebound). While geologic CH_4 emissions have almost certainly changed in the past and are likely to continue to change in the future, these mechanisms are too speculative to be used as a basis to estimate even the potential direction of future changes in geologic CH_4 emissions.

3.1.8 Wildfires

Wildfires are fires in forests, grasslands, savannas, and shrublands. They can either be ignited by lightning strikes or started accidentally by humans, but do not include deliberate controlled burns for land-clearing activities. As they burn, wildfires release a number of greenhouse gases, particulates, and other air pollutants. When combustion is complete—for example, in dominantly flaming fires—the carbon in biomass is generally converted to CO_2.

When combustion is incomplete, particularly in smouldering fires, some carbon is released in the form of carbon monoxide (CO) and CH_4: smouldering of biomass, consisting of both living and dead organic matter, is the primary source of emissions of CH_4 from wildfires.

Wildfire emissions of CH_4 are estimated to range from 2 to 5 Tg per year. This range depends on the frequency and strength of wildfires, which in turn are determined by a number of factors, including type of vegetation burned, influences from weather (e.g., wind, humidity, temperature) and climate.

Future climate change is likely to lead to enhanced frequency of weather conditions associated with high wildfire risks in many regions of the world. Climate change could affect multiple elements of wildfires, including fire behaviour, ignition, fire management, and vegetation fuels. The recent episodes in Russia and Australia are symptomatic. Currently, no scenarios for future CH_4 emissions from global wildfires exist in the literature.

3.1.9 Plants

Plants, from the tropical forests and other unflooded ecosystems, have long been recognized as important conduits for CH_4 emissions, transporting CH_4 from anaerobic soils and sediments to the atmosphere. However, it is only recently that plants themselves have been considered a possible source of CH_4 production. In 2006, it was reported that plants emit CH_4 through an unidentified process under aerobic conditions, and that this previously unrecognized source could add up to as much as 62 to 236 Tg CH_4/yr, or 10 to 40 percent of global CH_4 emissions. A significant plant CH_4 source could help explain a number of gaps in the understanding of current and past global CH_4 budgets, including an apparent large unidentified CH_4 source in the tropics.

The initial estimate of the size of a possible plant CH_4 source has since been revised downward in a number of analyses that have either scaled emission rates measured in the laboratory by estimates of global plant production ("bottom-up" estimates) or worked backward from global budgets to determine how large a plant CH_4 source could be reconciled with current estimates of other CH_4 sources ("top-down" estimates). The range of greatest agreement among these estimates is about 20 to 60 Tg CH_4/yr (U:S: EPA, 2010a).

The recently proposed aerobic plant CH_4 source has not yet been incorporated into simulations of future CH_4 emissions. However, future plant emissions would likely depend on changes in the distribution of different vegetation types, as well as changes in environmental factors that might control emission rates. Current estimates attribute 35 to 50 percent of global plant emissions to tropical forests, with the second largest source, tropical savannah and grasslands, contributing about 20 percent. These estimates suggest that future plant emissions will depend largely on changes in climate and land use in the tropics.

3.1.10 Wild animals

CH_4 is also produced by enteric fermentation, a normal digestive process that occurs in ruminant animals such as bison, gnus, deer, elk, mountain goats, and sheep, as well as in some smaller rodent species. Note that IPCC reports do not consider zoogenic emissions. Note also that enteric fermentation occurs in cattle and other domesticated ruminants; however, emissions from livestock are considered anthropogenic, so only estimated emissions from wild animals are reported here. Estimates of emissions from wild animals range from 2 to 15 Tg CH_4/yr (U.S. EPA, 2010a).

Changes to land use, which alter the type of plants available for wild ruminants, could affect the diets of these animals and subsequently their rate of enteric fermentation. As human activity encroaches on wildlife ecosystems, wild animal populations will likely decrease due to habitat unavailability.

3.1.11 Terrestrial arthropods

Termites and other terrestrial arthropods produce CH_4 as a result of microbial degradation of ingested organic matter. In the years some investigation of CH_4 emissions from termites has resulted in more refined estimates of emissions from the various termite species, and has suggested that some CH_4 may be oxidized in termite mounds prior to atmospheric release. The high rates of CH_4 produced by arthropods are due to methanigen (methane-producing) bacteria, which are found in many species of termites. Based on recent research, termites and other terrestrial arthropods continue to be a small but not insignificant contributor to global CH_4 emissions, producing an estimated 2 to 22 Tg CH_4 per year.

The habitats for terrestrial arthropods are also linked to climate effects, resulting in shifting or reduction of dry soil ecosystems (e.g., in more northern environments) or drought, which are likely to decrease populations. Currently, no scenario for future CH_4 emissions from this source exists in the literature.

4. Anthropogenic sources (about 70%)

Anthropogenic sources of methane consist in:

- Rice Cultivation;
- Animals husbandry;
- Landfills;
- Waste disposal and waste treatment;
- Methane from biogenic origins;
- Biomass burning;
- Gas and petroleum refining and production;
- Coal mining
- Exploitation of Gas Hydrates;
- Exploitation of Shale rocks.

4.1 Rice cultivation

Methane is produced during flooded rice paddies by the anaerobic decomposition of organic matter in soils. Flooded soils are ideal environments for methane production because of their high levels of organic substrates, oxygen-depleted conditions, and moisture. The level of emissions varies with soil conditions and production practices as well as climate. The rice paddy global emissions of methane can estimate to be about 100 Tg/yr. The main part of the rice production is concentrated among the 45°N and 10°S in the Asian and Mediterranean countries; the U.S. production is a small part of the total (U.S. EPA, 2010b). From 1980 the production of rice increased of more than 40%, but the emission of methane become less large because of the advanced cultivation technologies. Several cultivation practices have shown promising for reducing methane emissions from rice cultivation: minor use of organic fertilizers, addition of urea and ammonium salts, reducing the ph of the paddies, alternate short drainage periods among them.

4.2 Animal husbandry: Livestock enteric fermentation and manure management

Among domesticated livestock, ruminant animals (cattle's, buffaloes, sheep's, goats, and camels) produce significant amounts of methane as part of their normal digestive processes. In the rumen, a large pre-stomach of these animals, symbiotic bacteria convert vegetation feed (at high cellulose content) by means of microbial fermentation, into products that can be digested and utilized by the animal. This microbial fermentation process, referred to as enteric fermentation, produces methane as a by-product, which can be exhaled by the animal. Methane is also produced in smaller quantities by the digestive processes of other animals, but emissions from these sources are insignificant. EPA and other Agencies have studied options for reducing methane emissions from enteric fermentation and have developed resources and tools to assist in estimating emissions and evaluating mitigation options (U.S. EPA, 2010b).

Methane is produced also during the anaerobic decomposition of organic material in livestock manure management systems. Liquid manure management systems, such as lagoons and holding tanks, can cause significant methane production and these systems are commonly used at larger swine and dairy operations. Manure deposited on fields and pastures, or otherwise handled in a dry form, produces insignificant amounts of methane as they are easily oxidized at air. On the overall the husbandry of animals introduce about 85 - 115 Tg/yr of methane into the atmosphere (Keppler at al., 2003; U.S. EPA, 2010b).

4.3 Landfills

Landfills are the second-largest human-related source of methane in the U.S., accounting for 23 percent of all methane emissions in 2007. Methane is generated in landfills and large open dumps as waste decomposes under anaerobic conditions. In the last case the favorable conditions occur if the wastes are piled up in high and huge hills without adequate aeration apparatus. The amount of methane created depends on the quantity and moisture content of the waste and the design and management practices at the site.

4.3.1 Waste disposal

Untreated deposits of urban solid wastes can produce methane, besides amines and hydrogen sulphide, if they stay for a long period without sufficient aeration. They can produce up to 25 Tg/yr of methane.

4.3.2 Wastewater treatment

Wastewater from domestic (municipal sewage) and industrial sources is treated to remove soluble organic matter, suspended solids, pathogenic organisms, and chemical contaminants. These treatment processes can produce methane emissions if organic constituents in the wastewater are treated anaerobic ally and if the methane produced is released to the atmosphere. In addition, the sludge produced from some treatment processes may be further biodegraded under anaerobic conditions, resulting in methane emissions. These emissions can be avoided, however, by treating the wastewater and the associated sludge under aerobic conditions or by capturing methane released under anaerobic conditions.

Waste disposal and management can produce up to (25 + 40) Tg/yr on the overall (Keppler et al., 2003).

4.4 Methane from biogenic origins

This is a promising energy source, with the utilization of both biomasses from natural origin and by the gasification or induced fermentation of wastes. The biogenic origins of methane, however, opened a lot of controversies about the real causes of the methane increase in the atmosphere in the last century.

4.5 Biomass burning

About 40 Tg/yr of methane are produced during the incomplete burning of vegetation and organic residues. The quantity of emissions depends on the burning technology, on the temperature of the flames, on the water content and on the carbon content of the biomass (Keppler at al., 2003).

4.6 Fossil fuels ascertained reservoirs

Before to treat the methane emissions from the management of fossil fuels, it is better to remember the estimate of their reservoirs.

The estimates for the Organic Carbon in Earth reservoirs are reported in Table 1.

Reservoir of Organic Carbon in the Earth	Gtons of carbon
Fossil fuels	5000
Gas hydrates	10000
Lands (including soil, biota, peat, detritus)	2790
Ocean (including biota, dissolved or dispersed organics)	983
Atmosphere	3.6

Table 1. Gtons of Organic Carbon in Earth reservoirs.

The estimate of the organic carbon dispersed in rocks and sediments could be equal nearly 1000 times the total known reservoirs.

4.6.1 Natural gas and petroleum systems

Methane is the primary component of natural gas. Methane losses occur during the production, processing, storage, transmission, and distribution of natural gas. Because gas is often found in conjunction with oil, the production, refinement, transportation, and storage of crude oil is also a source of methane emissions. The U.S. inventory report provides a detailed description on methane emissions from natural gas and petroleum systems and how they are estimated for what concern the U.S. reservoirs (U.S. EPA, 2010b).

4.6.2 Coal mining

Methane trapped in coal deposits and in the surrounding strata is released during normal mining operations in both underground and surface mines. In addition, handling of the coal after mining, results in methane emissions.

Overall emissions of methane in the atmosphere from the fossil fuel exploitation, extraction and transport are estimated to be about 110 Tg/yr (Keppler et al., 2003).

4.7 Hydrate exploitation

In this paragraph we discuss with much more details the technology of extraction of methane from the gas hydrates (clathrates) for both the enormous reservoirs, recently discovered, of this source and for the possible peril for the climate if the production will be carried out with not sufficient cautions, especially in the offshore and deep ocean exploitations.

From the middle of '30 years, the engineers, managing petroleum pipelines in the cold regions of Siberia, discovered the formation of some "dirty crystals" incrusting the pipes and often blocking the transmission of oil at temperatures near or below 0°C.

The analysis of these crystals showed that they are an incoherent mixture of crystals of inclusion compounds of methane and other small hydrocarbons in solid water. Actually the blend of crystals is mixed with the mud used for drilling and extraction processes. Similar situations were also observed in Canadian and Alaskan pipelines (Barone & Chianese, 2009; Sloan & Koh 2008). These inclusion compounds of gases in solid water were named gas hydrates and they take part of the large family of clathrate. In particular the methane hydrate (of definite stoichiometric composition: $8CH_4.46H_2O$) is thermodynamically stable around 0°C under a pressure range of few atmospheres (Barone & Chianese, 2009) (see paragraph 3.1.5).

Note that decomposition of 1 m^3 of solid methane hydrate (density 912 kg/m^3) produces 0.790 m^3 of liquid water and 156 m^3 at STP of gas (172 m^3 at 25 °C).

Extended deposits of gas hydrates, mainly methane hydrates, were discovered in several areas of West Siberia, besides the methane gas reservoirs, located in the depth, under the permafrost layers. The methane hydrate field found at the confluence of Mussayakha and Jennissei/Ob river mouths, offered a large methane source, actually exploited from 1971 up to 1980 (Makogon & Sloan, 1994; Makogon et al., 1996; Sloan & Koh, 2008).

Today the technological interest is devoted to ascertain the consistency and to study the possibility of exploiting the enormous reservoirs of natural gas hydrates present also under the permafrost of circumarctica territories of Alaska and Northwest Territories of Canada, and under the soundings of continental escarpments and oceans. These reservoirs occur near natural gas fields or strictly associated with them (Collett, 1993; Collett, 2007, Collett et al., 1994; Council of Canadian Academies, 2008; Dallimore & Collett, 2005; Ripmeester at al., 2005).

Although the permafrost reservoirs (found or estimated and including those extending below the proximate continental margins) represent only the 1% of the offshore fields, the highest accessibility and the possibility to use the existing transport infrastructures, for the natural gas transference, gives much more promising the exploitation of these fields. Intense international collaborative project was developed, involving Institutions and Industries from Canada, Japan, U.S., India and Germany (Collett, 1993; Collett et al., 1994; Collett, 2007; Council of Canadian Academies, 2008; Dallimore & Collett, 2005; Ripmeester at al., 2005).

Extended geophysical researches in marine environments were devoted to the petroleum offshore and deep ocean drilling technologies. The Deep Sea Drilling Project (DSDP) and the successive Oceanic Drilling Program (ODP), ROPOS Project and IODP Project for the petroleum research led to discover the presence of extended deposits of pure hydrocarbon hydrates, in equilibrium with free gases under the wide but tick layers of oceanic sediments. Large deposits of hydrates, mainly methane hydrates, were found out of the east and western coasts of U.S., as so as the western coasts of Mexico, Middle America and Peru and Chile Trenches. In particular on the Blake Bahamas Ridge and on Barklay Canyon out of Vancouver Island, at the Bush Hill in the Green Canyon, at the Tigershark field in Alaminos Canyon and in Mississippi Canyon of Mexican Gulf, on the southern summit of the Hydrate Ridge at the Oregon Margin, on Håkon Mosby site in the Norwegian Sea and around the Shetland islands (Dickens et al., 1997; Lu et al., 2005; Milkov & Sassen, 2002; Milkov & Sassen, 2003a; Milkov & Sassen, 2003b; Milkov et al., 2003; Kvenvolden, 1994).

An intensive effort was undertaken by Japanese Drilling Projects on the Nankai Trough (Okhotsk Sea): this project seems very promising for a possible exploitation and commercialization (Kurihara et al., 2005; Matsumoto et al., 2004; Takahashi et al., 2001; Tsuji et al., 2004; Uchida et al., 2004).

India-U.S. cooperation leads to drilling projects for the location of hydrate accumulation on the Indian continental margin, in particular on the shale fractured area occurring at the Krishna-Godowari basin. Chinese projects were focused on the South China Sea with increasing success. Other Countries of the Asian South East area, on the basis of this success, were encouraged to undertake their own exploration in the same zone. South Korea also tried to explore the Ulleung basin on the East Korean Sea (Fan et al, 2005; Park, 2006).

Recent reviews calculated a total methane volume of 1.2×10^{17} m^3 at STP or 74'400 Gtons in the oceanic hydrates, of which 4.4×10^{16} m^3 under the continental shelves. These quantities are two or three order of magnitude higher than the known natural free gas (Barone & Chianese, 2009; Klauda & Sandler, 2005; Milkov, 2004; O'Connor et al., 2010).

Note that the clathrate reserves present under the Antarctica permafrost or the North Pole Ocean soundings have not yet been evaluated.

The possibility of exploiting, as new energy sources, the enormous reservoirs of natural gas clathrates present under the permafrost of circumarctica territories and under the soundings of continental escarpments and oceans, has stimulated the interest for many exploitation programs. However, the deep drilling technologies now at disposal could destabilise the hydrate reservoirs and this, besides the drilling and extraction costs, is an actual restraint to the exploitation of this energy sources.
Many efforts also must be made by the mine engineers for improving or developing new available technologies for a safe utilization of these natural reserves without any threat of introducing in the atmosphere relevant and persistent quantities of CH_4. Very interesting is the advanced drilling and extracting technology experimented by Japanese project in the Nankai Trough. In order to not destabilize, with a vertical wall, the hydrate layers during the operation, a 90° bent exploring and extraction wells were successful employed, achieving the deposit by the side (Moridis et al., 2004; Moridis & Reagan, 2007a; Moridis & Reagan, 2007b; Moridis & Sloan, 2007; Moridis, 2008; Takahashi & Tsuji, 2005). This procedure moreover protects the ship bearing the extractive derrick, respect to an abrupt large accidental emission of gas that can cause the sinking of the navy or its fire (remember the Mexico Gulf disaster).

4.8 Shale rocks

Shale are sedimentary rocks enclosing any kind of fine-grained and laminated sheets, consisting at least of 30% of silt- and clay-sized-particles. They also contain substantial amounts of quartz small quantities of carbonates, feldspars, iron oxides, fossils and organic matter. Some organic-rich shales, called oil shales, contain kerogen (see paragraph 2.2) that, when present in large enough quantities, can yields oil if subjected to intense heat.
Shale is the must abundant of the sedimentary rocks, accounting for roughly 70% of this rock type in the Earth crust. They typically are formed in environments where muds, silts and other sediments were deposited by gentle transporting currents and became compacted, as , for example, the deep ocean floor, basins and shallow seas, river floodplains and playas. Most shales occur in extensive sheets several meters thick, though some develop in lenticular formations.
Shales are often found with alternate layers of sandstones or limestone. The laminated structure gives the property defined fissility, id est. they exhibits a tendency to split into thin layers that are usually parallel to the bedding-plane surface. Such physical properties as permeability and plasticity are largely dependent on the grain sizes of constituent minerals. Shales color is determined primarily by the composition: higher the organic content darker is the color; ferric oxides tone to red or purple, ferrous salts impart green or blue; lime turn to gray or yellow.
Shales are commercially important, having applications in the ceramic industry. They are valuable material for tile, brick and pottery manufactures and for Portland cement industry. In addition they were exploited for the extraction of bitumen and liquid petroleum.
In the past the gas extraction from the shale source was utilized by small enterprises and, later on, abandoned for economic reasons. The reservoirs of natural gas embedding the shale rocks are however very promising and can become an important energy source especially for the Countries that do not have direct access to other reservoirs (Howarth et al., 2001).
New technology, analogous to that experienced for the extraction of methane from the hydrates (Takahashi & Tsuji, 2005) was recently proposed for the exploitation and intensive extraction of natural gas from shale rocks as a new important source. In fact very large field

extensions of oil shales are present in the North of United States and Canada, sites that there are not subject to unstable conditions of foreign Countries (U.S. EPA, 2010c). Moreover it is easier to achieve the core of the shale field by means of an oblique coaxial well, without bents, due to the frequent presence of rock formations as hills.

Actually the exploitation of these resources imply the use of pressurized warm water or fluids (hydraulic fracking) to break the shale rocks and made free the gas (or oil). Moreover it is recommended to build up a pipeline, before to bore a new well, in order to avoid gas leaks. This can be not economically favorable and it will result in a discouraging option at the present energy market status (Council of Scientific Society Presidents, 2010). However the U.S. Government decided (2011) to finance research and new technology developments in order to exploit the enormous shale gas reservoirs existing in the North America, with the double goals: renew the U.S. gas reserves and export the gas in the foreign markets with a competitive price.

Moreover the permeability of this kind of shale rocks and the delay in connecting wells with pipelines can favor leaking emissions of gas into the atmosphere up to 200 times respect to the emissions of traditional well exploitation. So that, an intensive exploitation of oil shale for the gas extraction, at the present technological state of art, will cause a true disaster for the environment, especially in the first decades of activity, with a further increase of greenhouse gas concentration and consequent unforeseen perils for the Global Warming (Howarth et al., 2011).

5. Summary of the methane emissions

The literature data discussed in the Paragraphs 3 and 4, are finally collected in the Table 2, where are selected and compared with the results from Keppler et al., 2003; U.S. EPA, 2010a and Whiticar & Schaefer, 2007. The emissions from shales are not considered by EPA and other references quoted here, and not evaluated. Also the fugitive emissions from industries, transport, aviation and other activities are not here considered explicitly and grouped into the total anthropogenic sources (Whiticar & Schaefer, 2007).

	(U.S. EPA, 2010a)			(Keppler et al., 2003)
Source	Emissions estimate (Tg CH_4/year)	Uncertainty Range [(a)]	δ^{13} C(‰) [(b)]	Emissions estimate (Tg CH_4/year)
Wetlands (Total) - Northern/bogs - Tropical swamps	170.3 (42.7) (127.6)	 (27-72) (81-206)	 (-62) (-58.9)	225 (including rice cultivation)
Upland soils and riparian areas	-30[(c)]	Not available	Not available	
Oceans, rivers and estuaries	9.1	2.3-15.6	-58	15
Permafrost	0.5	0-1		
Lakes	30	10-50	-53.8	
Gas Hydrates		2-9 [(d)]	-62.5	10
Terrestrial and marine		42-64	-41.8	

	(U.S. EPA, 2010a)			(Keppler et al., 2003)
Source	Emissions estimate (Tg CH_4/year)	Uncertainty Range [a]	δ^{13} C(‰) [b]	Emissions estimate (Tg CH_4/year)
geological sources				
Wildfires		2-5	-25	
Vegetation		20-60	Not available	
Termites/arthropods	20 [e]	2-22	-63	20
Wild animals	8	2-15	-60.5	30
Total of natural sources	**208**		**-57[f]**	**300 (including rice cultivations)**
All sources (natural plus anthropogenic)	**566 [g] 586[i]**	**503-610[h, i]**	**-54.5[b]**	**610-625**
Rice cultivations				100[j]
Husbandry				85; 88.5 [k]
Coal mining, gas and petroleum exploitation				110; 95 [k]
Biomass burning				40; 21 [k]
Waste disposal + Waste treatment				40+25; 58 [k]
Total of anthropogenic sources				**310-325; 263-281[k] (including other minor sources)**
Natural sources as percent of the total	**37%**			

[a] Ranges presented here may reflect a compilation of several different estimates. Published estimates vary due in part to uncertainty in estimating the global number of point and diffuse sources and the average annual emissions from each individual source or source area.
[b] Mean value from Ref 66 and references therein.
[c] The balance for these sites is in favour of sinks more than sources.
[d] Spontaneous emissions.
[e] Estimates for terrestrial arthropods include termites. It is estimated that other arthropods could contribute up to 100 Tg CH4/year.
[f] From Lassey et al., 2007.
[g] Mean value for anthropogenic emissions (Wuebbles & Hayhoe, 2002).
[h] Because the relative contributions of emissions from each source to the total budget are not independent of each other (i.e., if one source is at the lower end of its estimated range, another may be at the higher), the ranges cannot be summed.
[i] From Denman et al., 2007.
[j] From Stern & Kaufmann, 2000.
[k] From Ungher et al., 2010.

Table 2. Current methane emissions from natural and anthropogenic sources

Several methods have been invoked to separate and determine the contributions of different sources to the atmosphere. One of the methods analyzes the relative abundance of the three principal isotopes of carbon (the stable isotopes ^{12}C and ^{13}C, and the unstable ^{14}C) that can suggest which processes are producing CH_4. This is because microbes producing CH_4

discriminate against the heavier isotopic form, ^{13}C, giving a gas depleted of this isotope. In contrast, combustion processes (wildfires, biomass burning) do not discriminate. That gives significance to the numbers reported in the fourth column of Table 1 and note (b).
Moreover the ancient sources of CH_4 (coal, oil, natural gas hydrates, geological sources) release gas that lacks radioactive ^{14}C, contrarily to the more modern sources. (This radioactive isotope originates from the action of cosmic rays on the atmospheric CO_2 that in turn can be metabolized by the photosynthetic organisms and enter in the feeding chain of living organisms, but it is not accumulated and rapidly decay at their death). The isotopic composition of each sink accounts for the gas fluxes in atmosphere, giving the isotopic signature of each source. However the signatures of different analogous sources overlap and can change seasonally. The addition of isotopic data however, can be a powerful constraint on possible source budgets.
By analyzing ice cores, it is possible to compare present-day methane concentrations and sources with historical (i.e., pre-industrial) records. This has been helpful in estimating how natural sources respond to changes in the environment before there was a significant anthropogenic input (Chappellaz et al., 1993; Keppler et al., 2003). For instance the pre-industrial total methane emission was estimated 233 Tg/yr mainly from Wetlands and unchanged sources from wild animals, plants, wildfires and oceans (Keppler et al., 2003).
Depletion of ^{14}C isotope in ice cores, as measured by the δ^{14} C(‰) analogous to that reported for ^{13}C in Table 2, can also help to evaluate the fossil CH_4 origin from geological sources.

6. Impact of methane on the climate changes

6.1 Methane as greenhouse gas

Radiative Forcing (RF) is a measure of how the energy balance of the Earth-atmosphere system is influenced when factors that affect climate are changed (solar radiation, Earth albedo, greenhouse gas concentration, aerosol particles). Radiative forcing is usually quantified as the "rate of energy change per unit area of the globe as measured at the top of the atmosphere", and is expressed in units of watts per square meter (W/m^2). When radiative forcing from a factor or group of factors is evaluated as positive, the energy of the Earth-atmosphere system will ultimately increase, leading to a warming of the system. In contrast, for a negative radiative forcing, the energy will ultimately decrease, leading to a cooling of the system. As of 2005, atmospheric CH_4 and N_2O are the second- and third-largest contributors to radiative forcing among greenhouse gases, after CO_2 (Forster et al., 2007).
The conspicuous water vapour contribution to the greenhouse effect is not considered in the RF calculations, as its concentration is quite constant, as confined between the defined limits represented by the mean vapour pressure of the ocean and the cold trap at troposphere-stratosphere boundaries. Other minor contributions to the RF became by the halogenated compounds as CFC (negative after Montreal Protocol) and other chlorine or bromine containing compounds and the long living HFC and FC (positive). The increase of RF from 1990 to 2010 was estimated about 0.47 W/m^2 for CO_2 and about 0.08 W/m^2 for CH_4 (Butler, 2010).

6.2 The last glaciations

At the end of the last glaciation, the rate of increasing of methane concentration in atmosphere achieved about 200 ppm per year, as proved by the analysis of Greenland ice carrots (Chappellaz et al., 1993; Huber et al., 2006; Wolff et al., 2010). This could be one of the training effects on the ancient and quite abrupt global warming. The main emission source, at this time, was the decomposition of hydrates from the permafrost on the continental shelves and escarpments, as they remained mainly uncovered by the ocean water, retired for their accumulation into the ice continental shields. The hydrate instability was due to diminished hydrostatic pressure of 6 -10 atmospheres, corresponding to the lowering of 60 – 100 meters of the sea level. Also, it must be considered that the stability zone of methane hydrates is defined both by the thermodynamic phase diagram, as reported in the Figure 4 of this chapter, and the combination of it with the hydrothermal and geothermal gradients, since the stability zone is confined in well defined values of depth (pressure) and temperature of the permafrost layers (Barone & Chianese, 2009; O'Connor et al., 2010). The emission of gas from the destroyed structure of hydrates can be considered a modulating effect of methane, with respect to the astronomical factors, on the bistability of climate in the last million of years (O'Connor et al., 2010). Methane indeed ruled the climate in very ancient eras (hundreds millions or billion of year ago, in very different atmospheric and geologic environments (Pavlov et al., 2000).

6.3 Recent concentration trends in atmosphere

Today the methane effects on climate warming occur in a totally different condition, i.e. at the summit of a warm period. In previous paragraphs and in Table 1 the methane emissions were mainly considered. The unique sinks reprinted from EPA, and reported here, are those of biogeochemical nature, as the upland soils (and riparian areas) that are estimated to subtract from the atmosphere about 30 Tg/y of methane (U.S. EPA, 2010a).
During the last Century the methane concentration in the atmosphere increased rapidly up to decade 1970 - 80; after that the increase was less clear and the concentration is stabilized up to the 2006 (Butler, 2010; Dlugokencky, 2003). Recent evidences, however, indicate that the concentration starts again to increase (Dlugokencky & Bruhwiler, 2009) achieving the value of about 1810 ppm at 2010 (Butler, 2010).
The rapid increase occurred in the last Century can be attributed in large part to the enormous development of mining and oil and gas extraction activities. The increase almost corresponds to the total anthropogenic emissions not balanced by the depletion of methane for the chemical processes in atmosphere. In the years 1980 – 2000 this balance was probably attained, both for the petroleum crisis and the Soviet Union economic collapse, occurred in the previous decades (remember that the estimated life time of CH_4 in atmosphere is about 8-10 years: Folland 2001, Shindell 2006) , and then a delay must occur between the restarting of increased emission and a clear signal of accumulation). In the first years of the new Century, the exploitation of oil and gas gives a new acceleration to the unbalanced accumulation of methane in the troposphere. It is possible that the new increased rate of accumulation could also be due to high precipitation regime on the tropics (increased emissions from wetlands) and by the warm temperatures on the Arctic (emissions from permafrost), both stimulated by the Global Warming (Dlugokencky & Bruhwiler, 2009; Butler, 2010).
In Table 1 the leaking methane emissions from coal mining and gas and petroleum extraction are in fact estimated to be 95 – 110 Tg/y (Keppler et al., 2003; Ungher et al., 2010). Those quantities are due to the incoercible losses from coal mines (especially in those working underground) because they must be continuously aerated in order to avoid dangerous concentrations of CH_4 and/or CO into the galleries, otherwise their moisture can explode spontaneously even in dilute compositions.

A conspicuous quantity of gas is dispersed into the air at the exit of wells for oil extraction (it is too much expensive to trap the gas and it is dispersed in the atmosphere). The loss is high in the case of oil management with high content of light fractions, as for those from Mexican Gulf. In the case of refinery plants for the fractionation of the crude oil, gas is preferentially burned.

6.4 Methane chemistry and the dynamics of atmospheric processes

However, in order to properly take into account the short- and long-term effects of this greenhouse gas on the Global Warming, it needs to consider the trend of the accumulation of CH_4 into the atmosphere in a quantitive manner, and not only qualitative as in the previous paragraphs. Actually, the understanding of what happened in the last centuries is object of several controversies. In this paragraph we consider only some simple questions. Any kind of prediction is based on the attempt to delineate the dynamics between the net emission of methane into the atmosphere and all the depletion processes. In this manner it is possible to try some perspective on climate change, based on the evaluation of the increased Radiative Forcing promoted by methane according to different future horizons.
These predictions suffer of the high uncertainty on the emissions factors and of the fact that it is not possible to simply scale down the emission factors: in fact the annual averages cannot be transformed in instant averages, as they results from a combination of fluxes emitted discontinuously during the year and site by site.
An estimate of methane depletion from the troposphere was attempted at 2004 (O'Connor et al., 2010; Lilieveld et al. 2004): the value is about 560 Tg/yr, the 90% (504 Tg/yr), due to fotooxidation initiated by the main radical reaction:

$$CH_4+OH^\bullet \rightarrow CH^\bullet{}_3+H_2O \qquad (1)$$

followed by a cascade of reactions involving the radical $CH^\bullet{}_3$ that give CO_2 at the end:

$$CH^\bullet{}_3+O_2 \rightarrow CH_3OO^\bullet \rightarrow HCO+H_2O \qquad (2)$$

$$HCO+OH^\bullet+h\nu \rightarrow H_2+CO_2 \qquad (3)$$

The $OH^\bullet$ concentration increases proportionally to the low troposphere (Planetary Boundary Layer, PBL) pollution. However the $OH^\bullet$ production mainly happens in urban areas, industrial districts and high traffic motorways. Then $OH^\bullet$ is partially destroyed before it can be mix with CH_4. Other oxidizing species (ozone, nitrogen oxides, their radical derivatives and the atomic oxygen in its own ground state) react very slowly with methane.
On the other hands CH_4 is a light molecule that can diffuse up to the stratosphere, where very fast reactions occur:

$$CH_4+O(^1D) \rightarrow CH^\bullet{}_3+OH^\bullet \qquad (4)$$

$O(^1D)$ indicates an oxygen atom in its singlet excited state. This species is fast consumed by reaction with water vapour in the troposphere, giving $OH^\bullet$ radicals in turn. In the stratosphere the presence of $O(^1D)$ is due to the photo dissociation of molecular oxygen caused by the solar UV radiation:

$$O_2+h\nu \rightarrow O(^1D)+O(^3P) \qquad (5)$$

where O(^{3}P) represent another electronic excited state of oxygen atom (triplet state) that cannot recombine with O(^{1}D). The last one can be consumed by reactions with water vapour, giving two OH$^\bullet$, or with molecular oxygen (O_2), giving in turn ozone (O_3).
CH_4 then can also react with O(^{3}P) and moreover with Cl atoms originated by the photo dissociation of anhydrous HCl (from volcanic activity) or chlorinated species of anthropogenic origins (CFC). In that manner about 25 Tg/yr of CH_4 are consumed into the stratosphere (Ridgwell et al., 1999). Finally about 30 Tg/y are absorbed by dry soils (Table 1).

7. Conclusions

The present accumulation trend of methane in the atmosphere is very alarming for the future of the climate. However, it needs still more accurate research and models for formulating scenarios at all convincing. As we declared in the Introduction Section the trends for the future decades must be referred to three interfering feedbacks and self-accelerating processes: the increased leaks of gas by the extraction activities, the increased emissions by the perturbed natural reservoirs and the depletion reactions in the atmosphere. Really a very rough estimate of the recent unbalanced accumulation in the troposphere, based on emissions rates of 566 – 586 Tg/yr (Table 1) (Denman et al., 2007; Wuebbles & Hayhoe, 2002) gives a mean value of 16±10 Tg of methane per year. Other references (Bousquet et al., 2006; Keppler et al, 2003), give higher value of methane emissions from which higher accumulation rates can be estimated during the next decades (up to 40 Tg/yr). Then, assuming for the next decade an increasing accumulation of 16 Tg/yr of methane, id est. 0.9% of the present concentration budget, the last will reach 1960 ppm in 2020; if we consider an increase of 40 Tg/yr it will reach 2200 ppm in 2020. Considering a 20 year Time Horizon (TH) from now, i.e. 2030, the methane concentration will reach 2120 or 2600 ppm. At 50 year TH from now, id est. at 2060 and at same increasing trend, methane concentration could achieve 2600 or 3800 ppm, respectively, that will promote a catastrophic climate changes. Other estimates are much more prudent, considering an increase up to 1790 – 1900 ppm only at the end of the century (Ferretti et al., 2005 and other reported in O'Connor et al., 2010).
However for the natural gas extraction wells, there are accepted losses from 1.7% to 6%, in dependence of the employed technology. It has estimated that, during the extraction of gas from the shale rocks, utilizing the fracking process of the layers by means of pressurized water, 3.6 - 7.9% of the gas (respect to the total gas extracted) is dispersed in the atmosphere (in other words from 30% to 400% more than the losses in the case of extraction from the gas fields) (Howarth et al., 2011; Wood et al., 2010). Then, the exploitation of shale rocks as new gas or oil source can open a serious threat, due to the high leaks of gas from these activities.
Very recently (Isaksen et al. 2011) it has been presented a detailed analysis of the feedback contributions to the Global Warming, mainly due to the stimulated methane emissions from the Arctic. These Authors depicted some emission scenarios considering the permafrost thaw, hydrate destabilization and submarine landslides. They use the global-scaled Oslo CTM2 (Chemical Transport model 2), that combines the meteorological inputs from the Reading ECMWF (European Center for Medium Range Weather Forecast) and the Second-Order Moments Scheme for the advecting transport (Prather 1986). The data base for anthropogenic and natural emissions (Oliver and Berdovsky 2001; Muller et al. 1992) were also used. The methane concentrations were calculated as suggested by the ESRL network (Dlugokencky et al. 2010) that has a horizontal resolution of 2.8 x 2.8 degrees and gives 60 vertical layers up to 50 km in the stratosphere. This approach have been validated by several

inter comparisons (Shindell et al. 2006, Isaksen et al. 2005, 2011 and References quoted therein). The strong reciprocal feedbacks of the spontaneous Arctic increasing emissions and Global Warming lead to an accumulation rate of methane from 2.5 to 5.2 (or more) times the present value. This methane burden involves an overall Radiative Forcing 250-400 % times higher than the present one. Moreover those results involve also a strong increase of water vapour and hydroxyl radicals in the stratosphere and ozone in the troposphere.

The conclusions are impressive as it is forecast an increase of the RF of 0.6 Wm^{-2} on a Time Horizon of 50 years for the methane increase alone, that become 0.7 Wm^{-2} considering the life time increase of methane promoted by its own increased concentration and 0.9 Wm^{-2} for the accumulation of CO_2 as final product of the CH_4 oxidation processes. Other contributions must be due to the accumulation of the other GHG.

These scenarios are much more alarming than those based on simple linear extrapolation of methane accumulation trends registered in recent years. They prelude to a "world without ice" (Pasken, 2009). That will be, paradoxically, a world with few fresh waters, furrowed not by rivers, but rather by discontinuous wades, because of the failure of perennial glaciers and springs.

8. Acknowledgements

Research supported by the Project ITM474529 from the Campania Regional Board.

9. References

Barone, G. & Chianese, E. (2009). Hydrates of Natural Gases and Small Molecules: Structures, Properties, and Exploitation Perspectives. *ChemSusChem*, 2 (11), 992-1008.

Bousquet, P.; Hauglustaine, D. A., Peylin, P., Carouge, C. & Ciais, P. (2006). Contribution of anthropogenic and natural sources to atmospheric methane variability. *Nature*, 443, 439-443.

Butler, J. H. (2010). The NOAA Annual Greenhouse Gas Index (AGGI). http://www.tyndall.ac.uk/sites/default/files/tyndallcoop_shale_gas_report_final.pdf

Chappellaz, J.; Blunier, T., Raynaud, D., Barnola, J. M., Schwander, J. & Stauffer, B. (1993). Synchronous changes in atmospheric CH_4 and Greenland climate between 40 and 8 Kyr BP. *Nature*, 366, 443-445.

Claypool, G. E. & Kvenvolden, K. A. (1983). Methane and other hydrocarbon gases in marine sediment. *Ann. Rev. Earth Planet Sci.*, 11, 299-327.

Collett, T.S. (1993). Natural gas hydrates of the Prudhoe Bay and Kuparuk River area, North Slope, Alaska. *American Association of Petroleum Geologists Bulletin*, 77 (5), 793-812.

Collett, T.S.; Godebol, S.P. & Ehlig-Economomides, C. (1994). *Proc.35thAnnual Meeting Canadian Institute Mining*, p. 1.

Collett, T.S. (2007). Artic gas hydrate energy assessment studies. *The Arctic Energy Summit*, (Anchorage, Alaska).

Council of Canadian Academies (2008). *Energy from gas hydrates: assessing the opportunities and challenges for Canada*, Report of the Expert Panel on Gas Hydrates.

Council of Scientific Society Presidents (2010) Letter from the council to President Obama and senior administration officials, dated May 4, 2010. Council of Scientific Society

Presidents, 1155 16th Avenue NW, Washington, DC 20036. Available at http://www.eeb.cornell.edu/howarth/CCSP%20letter%20on%20energy%20&%20 environment.pdf.

Dallimore, S.R. & Collett, T.S. (2005). Scientific Results from the Mallik 2002 Gas Hydrate Production Research Well Program, Mackenzie Delta, Northwest Territories, Canada, *Geological Survey of Canada Bulletin*, 585.

Denman, K.L.; Brasseur, G., Chidthaisong, A., Ciais, P., Cox, P.M., Dickinson, R.E., Hauglustaine, D., Heinze, C., Holland, E., Jacob, D., Lohman, U., Ramachandran, S., da Silva Dias, P.L., Wofsy, S.C. & Zhang, X. (2007). Couplings between changes in the climate system and biogeochemistry. In: *Climate Change 2007: The Physical Basis. Contribution of Working Group I to the Fourth Assessment Report of the Intergovernmental Panel on Climate Change.* Solomon, S., Qin, D., Manning, M., Chen, Z., Marquis, M., Averyt, K.B., Tignor, M. & Miller, H.L. (2007). Cambridge, UK, and New York, NY: Cambridge University Press.

Diaconescu, C. C.; Kieckhefer, R. M., & Knapp, J. H. (2001). Geophysical evidence for gas hydrates in the deep water of the South Caspian Basin, Azerbaijan. *Mar. Petrol. Geol.*, 18, 209-211.

Dickens, G. R.; Paull, C. K., Wallace, P. & the ODP Leg 164 Scientific Party (1997). Direct measurements of in situ methane quantities in a large gas-hydrate reservoir. *Nature*, 385, 426-428.

Dlugokencky, E. J.; Houweling, S., Bruhwiler, L., Masarie, K. A., Lang, P. M., Miller, J. B. & Tans, P. P. (2003). Atmospheric methane levels off: temporary pause or a new steady-state? *Geophys. Res. Lett.*, 30 (19), 1992.

Dlugokencky, E. J. & Bruhwiler, L. (2009). Observational constraints on recent increases in the atmospheric CH_4 burden. *Geophys. Res. Lett.*, 36.

Dlugokencky, E. J., Lang P. M.. Masarie K.A., (2010). Atmospheric methane dry air mole fraction from the NOAA ESRL Carbon Cycle Cooperative Global Air Sampling Network, 1983-2009, Version 2010-08-12, *NOAA*, Boulder, Colorado.

Etheridge, D. M.; Steele, L. P., Francey, R. J. & Langenfelds, R. L. (1998). Atmospheric methane between 1000 A.D. an d present: Evidence of anthropogenic emissions and climatic variability. *J. Geophys. Res.*, 103 (D13), 15979-15993.

Fan, S.; Zhang, J. & Wang, J. (2005). Progress of gas hydrate studies in China. *Paper 3047, proceeding*, ICGH2005, v. 3, pp. 1012-1021.

Ferretti, D.F., Miller, J.B., White, J.W.C., Etheridge, D.M., Lassey, K.R., Lowe, D.C., Meure, C.M.M., Dreier, M.F., Trudinger, C.M., van Ommen, T.D. & Langenfelds, R.L. (2005). Unexpected changes to the global methane budget over the past 2000 years. *Science* 309 (5741), 1714-1717.

Folland, C.K.; Karl, T.R., Christy, J.R., Clarke, R.A., Gruza, G.V., Jouzel, J., Mann, M.E., Oerlemans, J., Salinger, M.J. & Wang, S.W. (2001). Observed Climate Variability and Change. In: Climate Change 2001: The Scientific Basis. Contribution of Working Group I to the Third Assessment Report of the Intergovernmental Panel on Climate Change, Houghton, J.T., Ding, Y., Griggs, D.J., Noguer, M., van der Linden, P.J., Dai, X., Maskell, K. & Johnson C.A. (eds.) Cambridge, UK, and New York, NY: Cambridge University Press.

Forster, P.; Ramaswamy, V., Artaxo, P., Berntsen, T., Betts, R., Fahey, D.W., Haywood, J., Lean, J., Lowe, D.C., Myhre, G., Nganga, J., Prinn, R., Raga, G., Schulz, M. & Van

Dorland, R. (2007). Changes in Atmospheric Constituents and in Radiative Forcing. In: Climate Change 2007: The Physical Science Basis. Contribution of Working Group I to the Fourth Assessment Report of the Intergovernmental Panel on Climate Change, S. Solomon, D. Qin, M. Manning, Z. Chen, M. Marquis, K.B. Averyt, M.Tignor, and H.L. Miller (eds.). Cambridge, UK, and New York, NY: Cambridge University Press.

Hesse, R. (1986). Early diagenetic pore water/sediment interaction: modern offshore basins. *Geoscience Canada*, 13, 165-196.

Howarth, R. W.; Santoro, R. & Ingraffea, A. (2011). Methane and the greenhouse-gas footprint of natural gas from shale formations. Letter for "Climatic Change". DOI 10.1007/s10584-011-0061-5.

Huber, C.; Leuenberger, M., Spahni, R., Flückiger, J., Schwander, J., Stocjer, T. F., Johnsen, S., Landais, A. & Jouzel, J. (2006). Isotope calibrated Greenland temperature record over Marine Isotope Stage 3 and its relation to CH_4. *Earth Planet. Sci. Lett.*, 243, 504-519.

Isaksen, I.S.A., Zerefos C., Kourtidis K., Meleti C., Dalsoren S.B., Sundet J.K., Grini A., Zanis P., Balis D. (2005). Tropospheric ozone changes at unpolluted and semipolluted regions induced by stratospheric ozone changes. *J. Goephys.Res.*.110, D03202.

Isaksen, I.S.A., Gauss M., Myhre G., Walter Antony K.M., Ruppel C. (2011). Strong atmospheric chemistry feedback to climate warming from Arctic metrhane emissions. *Global Biogeochem. Cycles*, 25, GB 2002, 1-11.

Keppler, F.; Hamilton, J.T.G., Brass, M. & Röckmann, T. J. (2003). Methane Emissions from Terrestrial Plants under Anaerobic Conditions. *Nature*, 439, 187 - 191.

Klauda, J.B. & Sandler, S.I. (2005). Global distribution of methane hydrate in ocean sediment. *Energy Fuels*, 19 (2), 459-470.

Kurihara, M.; Funatsu, K., Kusaka, K., Yasuda, M., Dallimore, S.R., Collett, T.S., Hancock, S.H. (2005). In: Scientific Results from the Mallik 2002 Gas Hydrate Production Research Well Program, Mackenzie Delta, Northwest Territories, Canad, 2005, *Geological Survey of Canada Bulletin* 585, (Eds: S.R. Dallimore and T.S. Collett).

Kvenvolden, K.A. (1994). Natural Gas Hydrate Occurrence and Issues. *Annals New York Acad. Science*, 715, 232-246.

Lassey, K.R.; Etheridge, D.M., Lowe, D.C., Smith, A.M. & Ferretti, D.F. (2007). Centennial evolution of the atmospheric methane budget: What do the carbon isotopes tell us? *Atmos. Chem. Phys*, 7, 2119-2139.

Lilieveld, J.; Dentener, F. J., Peters, W. & Krol M. C. (2004). On the role of hydroxyl radicals in the self-cleansing capacity of the atmosphere. *Atmos.Chem Phys.*, 4, 2337-2344.

Loulergue, L.; Schilt A., Spahni, R., Masson-Delmotte, V., Blunier, T., Lemieux, B., Barnola, J. M., Raynaud, D., Stoker, T. F. & Chappellaz, J. E. (2008). Orbital and millennial - scale features of atmospheric CH_4 over than past 800000 years. *Nature*, 453, 383-386.

Lu, H.; Raupach, M.R. & Richards, K.S. (2005). Modelling entrainment of sedimentary particles by wind and water: A generalized approach. *J. Geophys. Res*, 110.

Makogon, Y.T. & Sloan, E.D. (1994). Phase Equilibrium for Methane Hydrate from 190 to 262 K. *J. Chem. Eng. Data*, 39, 351-353.

Makogon, Y.T.; Mehta, A.P. & Sloan, E.D. (1996). Structure H and Structure I Hydrate Equilibrium Data for 2,2-Dimethylbutane with Methane and Xenon. *J. Chem. Eng. Data*, 41, 315-318.

Matsumoto, R.; Tomaru, H. & Lu, H. (2004). Detection and evaluation of gas hydrates in the eastern Nankai Trough by integrated geochemical and geophysical methods. *Resource Geology,* 54, 53-68.

Mazzini, A.; Ivanov, M. K., Parnell, J., Stadnitskaya, A., Cronin, B. T., Poludetkina, E., Mazurenko, L. & Van Weering, T. C. E. (2004). Methane related antigenic carbonates from the Black Sea: geochemical characterisation and relation to seeping fluids. *Mar. Geol.*, 212, 153-181.

Milkov A.V. & Sassen, R. (2002). Economic geology of offshore gas hydrate accumulations and provinces. *Marine and Petroleum Geology*, 19, 1-11.

Milkov A.V. & Sassen, R. (2003a). Two-dimensional modelling of gas hydrate decomposition in the north-western Gulf of Mexico: Significance to global change assessment. *Global Planet. Change*, 36, 31- 46.

Milkov A.V. & Sassen, R. (2003b). Preliminary assessment of resources and economic potential of individual gas hydrate accumulations in the Gulf of Mexico continental slope. *Marine and Petroleum Geology*, 20, 111-128.

Milkov, A.V.; Claypool, G.E., Lee, Y.J., Dickens, G.R., Xu, W., Borowsky, W.S. & the Ocean Drilling Program Leg 204 Scientific Party (2003c). In situ methane concentrations at Hydrate Ridge, offshore Oregon: New constraints on the global gas hydrate inventory from active margins. *Geology*, 31, 833-836.

Milkov, A.V. (2004). Global estimates of hydrate-bound gas in marine sediments: How much is really out there? *Earth. Sci. Rev.*, 66, 183-197.

Monnin, E.; Indermühle, A., Dällenbach, A., Flückiger, J., Stauffer, B., Stocker, T. F., Raynaud, D. & Barnola, J. M. (2001). Atmospheric CO_2 concentrations over the last glacial termination. *Science*, 291, 112-114.

Moridis, G.J.; Collett, T.S., Dallimore, S., Satoh, T., Hancock, S. & Weatherhill, B. (2004). *Numerical Studies Of Gas Production From Several Methane Hydrate Zones At The Mallik Site,* Mackenzie Delta, Canada, JPSE 43, 219.

Moridis, G.J. & Reagan, M.T. (2007a). Strategies for Gas Production From Oceanic Class 3 Hydrate Accumulations. *Offshore Technology Conference* (Houston, Texas, U.S.A.), *OTC-18865.*

Moridis G.J. & Reagan, M.T. (2007b). Gas Production From Oceanic Class 2 Hydrate Accumulations. *Offshore Technology Conference* (Houston, Texas, U.S.A.), *OTC 18866.*

Moridis, G.J. & Sloan, E.D. (2007). Gas Production Potential of Disperse Low-Saturation Hydrate Accumulations in Oceanic Sediments. *J. Energy Conversion and Management*, 48 (6), 1834-1849.

Moridis, G.J. (2008). Paper LBNL University of California, 161E.

Muller J., (1992). Geographical distribution and seasonal variation of surface emissions and deposition velocities of atmospheric trace gases. *J. Geophys.Res.,* .97(D4), 3787-3804.

O'Connor, F. M.; Boucher, O., Gedney, N., Jones, C. D., Folberth, G. A., Coppell, R., Friedlingstein, P., Collins, W.J., Chappellaz, J., Ridley, J. & Johnos, C.E. (2010). Possible role of wetlands, permafrost, and methane hydrates in the methane cycle under future climate change: A review. *Rev. Geophys.*, 48, RG4005.

Oliver J. G.J., Berdowski J.J.M., (2001). Global emissions sources and sinks. in *"The climate System"* 33-78. Berdowski J.J.M., Guicherit R., Heij B.J., EDS: Balkema Publ., Lisse , Netherlands.

Park, K.P. (2006). Gas Hydrate Exploration in Korea. *Proc. of the 2nd International Symposium on Gas Hydrate Technology* (Daejeon, Korea).

Pasken, U. (2009). The reality worse then the prognosis. Latest IPCC report. NowPublic News coverage. Available at http://www.nowpublic.com/environment/reallity-worse-then-prognosis-latest-ipcc-report.

Pavlov, A.; Kasting, J. F., Brown, L. L., Rages, K. A. & freedman R. (2000). Greenhouse worming by CH_4 in the atmosphere of early Earth. *Journal of Geophysical research-Planets*, 105, 11981-11990.

Prather M.J. (1986)Numerical advection by conservation of 2nd-order moments, *J. Geophys.Res.*, .91(D6), 6671-6681.

Ridgwell, A. J.; Marshall, S. J. & Gregson, K. (1999). Consumption of atmospheric methane by soils: A process-based model. *Global Biogeochem. Cycles*, 13, 59-70.

Ripmeester, J.A.; Lu, H., Mondrakovski, J., Dutrisac, R., Wright, F., Dollimore, S. (2005). Structure and composition of hydrate in sediment recovered from Mallik 5L-38, Mackenzie Delta, NWT, Canada: X-ray diffraction, Raman and Solid-state NMR spectroscopy. *Geol. Surv. Can. Bull*, 585, 106 (89?).

Sassen, R. & Macdonald, I. R. (1994). Evidence of structure H hydrate, Gulf of Mexico continental slope. *Org. Geochem.*, 22, 1029-1032.

Shakhova, N.; Semiletov, I., Salyuk, A., Yusupov, V., Kosmach, D. & Gustafsson, O. (2010). Extensive methane venting to the atmosphere from sediments of the East Siberian Arctic Shelf, *Science*, 327, 1246-1250.

Shindell D.T. et al. (2006). Multimodel simulation of carbon monoxide: Comparison with observations and projected near-future changes. *J. Geophys.Res.*, 111, D19306.

Sloan E.D. & Koh C.A. (2008). *Clathrate Hydrates of Natural Gases (third ed.)*, Taylor & Francis Group, CRC Press, Boca Raton.

Solomon, S.; Qin, D., Manning, M., Chen, Z., Marquis, M., Averyt, K.B., Tignor, M. & Miller H.L. (2007a). *Climate Change 2007: The Physical Science Basis. Contribution of Working Group I to the Fourth Assessment Report of the Intergovernmental Panel on Climate Change*. Cambridge, UK, and New York, NY: Cambridge University Press.

Solomon, S.; Qin, D., Manning, M., Alley, R.B., Berntsen, T., Bindoff, N.L., Chen, Z., Chidthaisong, A., Gregory, J.M., Hegerl, G.C., Heimann, M., Hewitson, B., Hoskins, B.J., Joos, F., Jouzel, J., Kattsov, V., Lohmann, U., Matsuno, T., Molina, M., Nicholls, N., Overpeck, J., Raga, G., Ramaswamy, V., Ren, J., Rusticucci, M., Somerville, R., Stocker, T.F., Whetton, P., Wood, R.A. & Wratt. D. (2007b). Technical summary. In: *Climate Change 2007: The Physical Science Basis. Contribution of Working Group I to the Fourth Assessment Report of the Intergovernmental Panel on Climate Change.* Solomon, S., Qin, D., Manning, M., Chen, Z., Marquis, M., Averyt, K.B., Tignor, M. & Miller H.L. Cambridge, UK, and New York, NY: Cambridge University Press.

Stern, D.I. & Kaufmann, R.K. (2000). Is there a global warming signal in hemispheric temperature series: a structural time series approach. *Climatic Change*, 47, 411-438.

Takahashi, H.; Yonezawa, T. & Takedomi, Y. (2001). Exploration for natural hydrate in Nankai trough wells offshore Japan. *Offshore Technology Conference*, (Houston, Texas), *OTC 13040.*

Takahashi, H. & Tsuji, Y. (2005). Multi-Well Exploration Program in 2004 for Natural Hydrate in the Nankai-Trough Offshore Japan. *Offshore Technology Conference* (Houston, TX, U.S.A.) *OTC 17162*.

Tsuji, Y.; Ishida, H., Nakamizu, M., Matsumoto, R. & Shimizu, S. (2004). Overview of the MITI Nankai Trough Wells: A Milestone in the Evaluation of Methane Hydrate Resources. *Resource Geology*, 54, 3-10.

Uchida, T.; Lu, H., Tomaru, H. & The MITI Nankai Trough Shipboard Scientists (2004). Subsurface occurrence of in the Nankai Trough area: Implication for gas hydrate concentration. *Resource Geolog*, 54, 35-44.

Ungher, N.; Bond, T. C., Wang, J. S., Koch, D. M., Menon, S., Shindell, D. T. & Bayer, S. (2010). Attribution of climate forcing to economic sector. *PNAS*, 107 (8), 3382-3387.

U.S. EPA (United States Environmental Protection Agency) (1993). *Current and Future Methane Emissions From Natural Sources*. EPA-430-R-93-011. Washington: U.S. Environmental Protection Agency.

U.S. EPA (United States Environmental Protection Agency) (2007). *Inventory of U.S. Greenhouse Gas Emissions and Sinks: 1990-2005*. EPA-430-R-07-002.

U.S. EPA (United States Environmental Protection Agency) (2010a). *Methane and Nitrous Oxide Emissions From Natural Sources*. EPA-430-R-10-001.

U.S. EPA (United States Environmental Protection Agency) (2010b). *Inventory of U.S. Greenhouse Gas Emissions and Sinks: 1990-2008*. EPA-430-R-10-006.

U.S. EPA (United States Environmental Protection Agency) (2010c). *Greenhouse gas emissions reporting from the petroleum and natural gas industry*. Background Technical Support Document. Available at http://www.epa.gov/climatechange/emissions/ downloads10 Subpart-W_TSD.pdf. Accessed 3 January 2011.

Walter, K. M.; Smith, L. C. & Chapin III, F. S. (2007). Methane bubbling from northern lakes: present and future contributions to the global methane budget. *Phil. Trans. R. Soc.*, 365, 1657-1676.

Ward P. D., (January 2007). Impatto dal profondo. *Le Science* (Italian Edition of Scientific American), 1, 72-79.

Whiticar, M. & Schaefer, H. (2007). Constraining past global tropospheric methane budgets with carbon and hydrogen isotope ratios in ice. *Phil. Trans. R. Soc. A*, 365, 1793-1828.

Wolff, E. W.; Chappellaz, J., Blunier, T., Rasmussen, S. O. & Svensson, A. (2010). Millennial-scale variability during the last glacial: The ice core record. *Quat. Sci. Rev.*, 29 (21-22), 2828-2838.

Wood, R; Gilbert, P, Sharmina, M, Anderson, K, Fottitt, A, Glynn, S & Nicholls, F (2011) Shale gas: a provisional assessment of climate change and environmental impacts. Tyndall Center, University of Manchester, Manchester, England.

Woodside, J. M.; Modin, D. I. & Ivanov, M. K. (2003). An enigmatic strong reflector on sub bottom profiler records from the Black Sea - the top of shallow gas hydrate deposits. *Geo-Mar Lett.*, 23, 269-277.

Wuebbles, D. J. & Hayhoe, K. (2002). Atmospheric methane and global change. *Earth Sci. Rev.*, 177-210.

27

Enteric Methane Emission from Pigs

Henry Jørgensen, Peter K. Theil and Knud Erik Bach Knudsen
Aarhus University, Faculty of Science and Technology
Denmark

1. Introduction

The modern pig industry rely on relatively few feedstuffs mostly from cereals (corn, wheat, barley, oats, rye and rice), cereal co-products (different milling fractions, residues from biofuel and alcohol industries, etc.), cereal substitutes (tapioca, maniocca), legumes (peas, beans, lupins), protein concentrates (meal or cakes of soybean, rape, sunflower, cotton) and co-products from the sugar and starch industries to produce compounds feeds. The classical pig diet can also be characterised as relatively concentrated but an increased demand of high energy cereals for direct human use and increased availability of fibre rich ingredients from, for instance, the feed milling or starch extraction/fermentation industries have promoted an increased utilisation of fibre rich co-products in the pig feeds (Serena et al. 2007). Especially pregnant sows may be supplied with fibre rich diets without compromising their reproductive performance. The direct use of forage crop is also developing although at a rather limited scale and primarily in organic farming. Other benefits, such as increased well being of animals, improvement of the gut transit or reduction of stomach ulcers also favour an increased utilisation of fibre rich ingredients in pig feeds. An increased dietary fibre concentration is on the other hand associated with reduced available energy content of the diet if not combined with high energy ingredients such as animal fat or vegetable oil. The consequence is that the amount of feed required per kg meat produced is increased (Fernández et al. 1983; Lekule et al. 1990).
The present chapter will summarise our current knowledge concerning dietary and enteric fermentation that may influence the methane (CH_4) emission in pigs. Enteric fermentation is the digestive process by which carbohydrates are broken down by microorganisms into simple molecules for absorption into the blood stream of the animal. The data compilation is based on all experiments carried out in the respiration chambers at Research Centre Foulum during the last 20 years. It should, however, be stressed that none of the studies were specifically designed to investigate dietary or enteric pig factors that may influence the CH_4 emission.
The data used in the chapter is based on Danish pigs with reference to other European studies. In a global context Denmark's 12 million pigs (FAOSTAT, 2009) contribute with 1.3 % of the worlds pig population. The main number of pigs is in Asia (59.6 %) where the main pig population stay in China (47.8 % of the worlds pig population).
The objective of the chapter is therefore:
To obtain a general overview of the pigs' contribution to methane emission.

Where is the pigs' enteric gas produced and how is it measured.
The variation in methane emission and factors affecting the emission.
Possibility for reducing the enteric methane emission and the consequences.

2. Sites for enteric gas production

The majority of carbohydrates are digested by endogenous enzymes in the small intestine but the dietary fibre fraction will be fermented primarily in the lower gut resulting in production of short chain fatty acids (acetic-, propionic- and butyric acid), gases (carbon dioxide (CO_2), hydrogen (H_2) and methane (CH_4)), urea and heat. However, the stomach and small intestine is also colonised by microorganism, which lead to production of primarily H_2. This is illustrated in Figure 1A&B that shows the results from a study in which the gas concentration has been measured at 12 sites of the gastrointestinal tracts (Figure 1A) after feeding a low and high fibre diet with 6 and 27 % dietary fibre per kg DM, respectively (Jensen & Jørgensen, 1994).

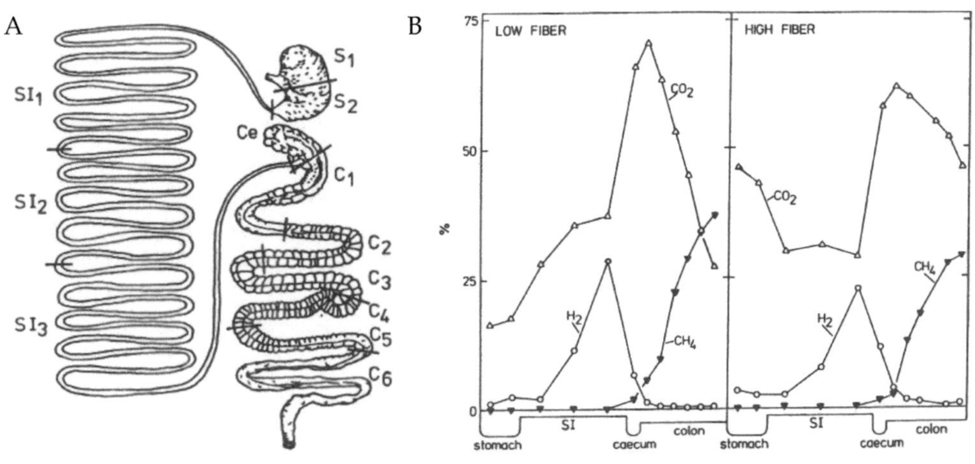

Fig. 1. Sampling sites in the gastrointestinal tract. SI, small intestine; S, stomach, Ce, caecum; C, colon. (A). Composition of gases from various regions of the gastrointestinal tract of pigs fed the low- and high-fiber diets (B). Data from Jensen & Jørgensen (1994).

The gas composition in the various segments shows similar trends for the two groups of pigs (Figure 1B). In both groups, low levels of H_2 were detected in the gas from the stomach, followed by a steady increase along the small intestine reaching a maximum (21 to 28 %) in the last third of the small intestine. Gas from the caecum and the first segment from the large intestine also contained substantial amounts of H_2, while the amounts of gas from the other segments of the large intestine were small in spite of the fact that hydrogen production is an obligate part of anaerobic fermentation (Miller & Wolin, 1974; Jensen, 1996). The produced short chain fatty acids can be utilised as an energy source and contributes with a significant part of energy for maintenance (Jørgensen et al. 1997). From an energy point of view, only CH_4 and H_2 are important as they correspond to combustible gases and represent a loss of energy. However, information on how dietary composition and intrinsic animal

factors influence gas production in pigs is rather limited. From an environmental point of view CH_4 is of great interest as it is the major contributor to the greenhouse gas emission. Carbon dioxide emission from livestock is assumed to be zero. The CO2 from the atmosphere used by plant photosynthesis is returned to the atmosphere as expired CO2 (IPCC, 2006). Therefore, the enteric CO2 production by animals can be regarded as neutral. However, a part of the C is returned as CH_4 and therefore requires separate considerations. Methane is a very potent greenhouse gas that contributes to the global warming and CH_4 is 21 times more potent than CO_2.
A recent Danish inventory of CH_4 emission from livestock enteric fermentation and manure management show that cattle contribute with 86 % of the enteric emission and the pig production with 10 % and horses with 3 % (Mikkelsen et al. 2011). The manure handling contributes with 30 % of the total CH_4 emission from livestock in Denmark and the distribution between pig and cattle manure is 45 and 50 %, respectively. The remaining part is attributed to the other animal species.
Carbon dioxide (CO_2) is produced in all living cells and can be regarded as a waste product when nutrients (carbohydrates, fat and proteins) are oxidised. As an example, glucose is oxidised completely, and quantitatively the stochiometry for glucose oxidation is (Stryer, 1988):

$$C_6H_{12}O_6 + 6\,O_2 \rightarrow 6\,CO_2 + 6\,H_2O \tag{1}$$

The produced CO_2 is then transported from the cells (organs) via blood either as dissolved CO_2, bound to haemoglobin or as bicarbonate to the lungs. I the lungs, CO_2 is being transferred to alveolar air and then expired during exhalation. The amount of CO_2 produced by an animal depends to a large extent on the live weight, physiological state, feed intake and physical activity of the animal, whereas the composition of the feed supplied plays a minor role (Just et al. 1983, Noblet and Shi 1994, Jørgensen et al. 1996, Jørgensen 1997, Schrama et al. 1998, Jørgensen et al. 2001, Theil et al. 2002, Theil et al. 2004, Pedersen et al., 2008, Jentsch et al. 2009).

3. Measurement of enteric methane and data description

3.1 Measurement of methane and carbon dioxide production

The respiration chambers consist of climatic controlled airtight rooms where the animals' energy metabolism can be measured as illustrated in Figure 2. The measured parameters are heat production and the animals' consumption of oxygen and production of carbon dioxide, CH_4 and H_2. The technique applied is the so called indirect principle which means that the atmospheric air is ventilated through the chambers and the amount of air is measured together with the concentration of oxygen, carbon dioxide, methane and hydrogen in both in-going and out-going air (Jørgensen, 2001; Jørgensen et al. 1996). When the consumed amount of oxygen and produced amount of carbon dioxide and methane is known, the heat production can be calculated (Christensen & Thorbek, 1987; Jakobsen et al. 2005).
In most cases both nitrogen (N) and energy balances have been performed. In brief, a typical balance experiment comprises a total period of 12 days, including 5-7 days for adaptation to the feed, metabolic cage and environmental conditions. Daily faeces and urine are collected quantitatively during the last 5-7 days as described by Just et al. (1983). During the collection

period, the metabolic cage with the pigs is placed in the respiration chambers and the amount of CH_4, CO_2 and H_2 produced and oxygen (O_2) consumed is measured and heat production is calculated during 2 x 24 h periods.

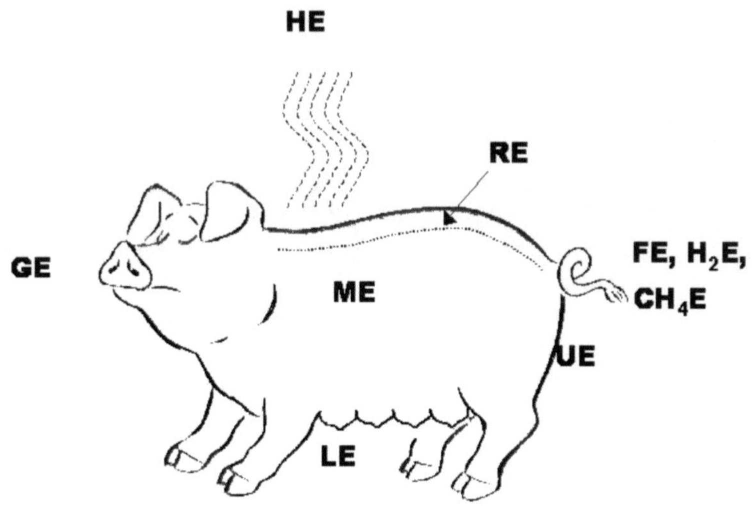

Fig. 2. Illustration of energy flow in a pig: GE, total gross energy intake; ME, metabolizable energy (energy available for metabolism); HE, heat energy - energy lost as heat; RE, retained energy, LE, lactation energy; UE, urine energy; FE, faecal energy; H_2E, hydrogen energy; CH_4E, CH_4 energy.

3.2 Overview of data

The respiration chamber was established at Research Centre Foulum in 1990 and numerous experiments have been carried out on growing - finishing pigs from 30 - 150 kg live weight. Other experiments with sows at different physiological stages (dry, pregnant and lactating sows including suckling piglets) have been carried out. In order to measure suckling piglets without contribution of the lactating sow, experiments have been carried out where milk replacer has been fed to the piglets.

The whole dataset (Table 1) comprises one experiment with piglets fed with milk replacer and a total of 16 experiments with growing pigs covering the weight range 25–150 kg live weight and 9 experiments with adult sows either dry, pregnant or lactating sows (including the litter). The main purpose of the various experiments has been influence/effect on the energy utilization of different diets or treatments. In Table 2 is shown the mean of 140 diets/treatments, which are based on 783 measurements of energy balance.

3.3 Presentation of results – expression of methane emission

Methane emission is in literature expressed in different ways. It is not only an animal's daily production that is of interest but also the emission relative to amount of feed ingested (in kg or energy equivalent or amount of fermentation substrates).

Study / Reference	Diets, g/kg DM	Balance, n	LW kg	CH_4 L/d	CO_2 L/d
Piglets					
Theil et al. 2007	Milk	12	4.9	0.13	94
Growing pigs					
Jørgensen et al. 1996	28-266 DF	24	81	6.7	991
Bornholt et al. unpublished	60-318 DF	28	67	3.2	637
Jørgensen et al. 1997	Infusion SCFA	86	92	1.0	825
Jørgensen et al 1996	36-207 Fat	48	52	3.2	621
Jørgensen 1997	55-143 DF	30	60	2.6	708
Jørgensen unpublished [1]	197-204 DF	24	68	3.4	834
Jørgensen et al. unpublished	123-156 DF	24	65	1.0	796
Jørgensen et al. 2001	192-311 DF	27	63	4.1	742
Carlson et al. 1999	247-303 DF	48	64	4.6	710
Theil et al. 2001 [2]	Breed*Temp	42	77	3.1	737
Wang et al. 2002	76-167 DF	48	51	3.6	824
Jørgensen unpublished	31-166 Fat	30	50	1.4	654
Jørgensen et al. unpublished	155-184 DF	56	83	2.8	829
Jørgensen et al. unpublished [3]	197-197 DF	16	77	4.4	897
Jørgensen et al. unpublished	165-290 DF	16	90	6.4	856
Serena et al. 2007	223-536 DF	37	61	3.7	989
Adult sows					
Olesen et al. 2001	193-409 DF	16	290	13.6	1406
Theil et al. 2004	199-339 DF	36	230	7.5	1226
Jørgensen unpublished	186-374 DF	9	231	8.2	1050
Jørgensen et al. 2010	164-294 DF	24	210	5.7	1198
Theil et al. 2002	173-229 DF	9	209	5.4	1287
Theil et al. 2004	31-113 Fat	25	206	14.4	1762
Jørgensen unpublished	308-536 DF	18	212	9.9	1108
Serena et al. 2009	249-536 DF	32	214	12.2	1084
Jørgensen unpublished	189-447 DF	18	208	12.0	1077

Table 1. Description of data. 1) Environmental temperature x dietary protein. 2) Breed of pigs x environmental temperature. 3) Feed particle sixe. DF, dietary fibre. SCFA, short chain fatty acids; Fat, dietary fat.

Item	n	Mean	Range of values	Standard deviation
Chemical composition				
Crude protein, g/kg DM	140	177	98 - 239	27
Crude fat, g/kg DM	140	69	26 – 257	42
Starch, g/kg DM	140	435	92 - 706	117
Total sugars, g/kg DM	103	45	13 - 220	33
Crude fibre, g/kg DM	54	85	13 - 266	54
Total dietary fibre, g/kg DM	140	215	0 - 536	113
Main results, piglets				
Live weight, kg	3	4.9	3.9 – 6.0	1.1
DM intake, kg/d	3	0.18	0.14 – 0.21	0.03
CO_2, L/d	3	94	74 - 115	21
CH_4, L/d	3	0.13	0.11 – 0.15	0.02
Total fibre fermented, %	3	0	0 – 0	0
Total fermentable fibre, g/d	3	0	0 – 0	0
DC energy, %	3	98	98 - 98	0.3
CH_4-energy, % GE	3	0.12	0.11 – 0.14	0.02
CH_4-energy, % DE	3	0.13	0.11 – 0.14	0.02
Main results, growing pigs				
Live weight, kg	101	67	25 - 152	24
DM intake, kg/d	101	1.77	0.87 – 3.06	0.45
CO_2, L/d	101	767	368 - 1350	192
CH_4, L/d	101	3.4	0.28 – 12.1	2.2
Total fibre fermented, %	101	56	32 - 86	12
Total fermentable fibre, g/d	101	182	26 - 455	93.4
DC energy, %	101	84	60 - 99	6.9
CH_4-energy, % GE	101	0.39	0.03 – 1.07	0.22
CH_4-energy, % DE	101	0.47	0.03 – 1.29	0.28
Main results, adult sows				
Live weight, kg	36	224	192 - 298	27
DM intake, kg/d	36	2.20	1.48 – 4.87	0.80
CO_2, L/d	36	1213	976 - 1837	220
CH_4, L/d	36	9.8	3.2 – 28.7	5.6
Total fibre fermented, %	36	62	30 - 87	13
Total fermentable fibre, g/d	36	417	150 - 999	191.4
DC energy, %	36	78	49 - 85	7.9
CH_4-energy, % GE	36	1.01	0.29 – 2.69	0.62
CH_4-energy, % DE	36	1.31	0.40 – 3.25	0.78

Table 2. Chemical composition of diets (mean of 140) and main results for the three classes of pigs: piglets, growing pigs and adult sows.

Total dietary fibre, calculated as the residual fraction after subtraction of the analysed content of sugars, starch, crude protein, crude fat and ash from the dry matter content. DC, digestibility coefficient; GE, gross energy; DE, digestible energy.

Thus to aid in presentation of the presented results in this chapter and also in comparing to other literature sources the following expression of methane emission is used: 1. Litre per day, which is the results measured. 2. Litre per kg dry matter (DM) or gross energy (GE) intake as not only the live weight (kg) of an animal is important but also the amount of substrate for fermentation. 3. Methane emission expressed per g fermented fibre. 4. Finally methane is also expressed per unit of digestible energy (DE) because it may be more relevant to relate the methane production to the amount of energy needed for animal production.

In the ICPP (2006) guidelines for national greenhouse gas inventories and the Danish emission inventory for agriculture (Mikkelsen et al. 2011), the CH_4 conversion rate expressed as Y_m = CH_4-energy in % of gross energy (GE) for calculating the total amount of CH_4 in Gg is used.

4. Dietary and intrinsic factors influencing methane emission

4.1 The variation in the chemical composition of feed and main results

Chemical composition of diets and main results are presented in Table 2. In total, CH_4 excretion amounted from 0.1 L/day with piglets reared artificially to 28.7 L/day corresponding to nearly zero at the lowest value to 3.3 % of digested energy (DE). This is of similar magnitude as reported by Jentsch et al. (2007) for female calves (50-75 kg body weight) but much lower than reported for ruminants where the loss as fermentation of CH_4 can be as high as 9-13 % of digestible energy (DE) (Johnson & Johnson, 1995).

4.2 Correlation between variables

The correlation between dietary characteristic and production of CH_4 and CO_2 is presented in Table 3. As expected there was a negative correlation between CH_4 and dietary protein, dietary fat and dietary starch and positive correlations to the fibre fractions. The highest correlation to the daily CH_4 excretion is found for fermented fibre per kg DM intake (r=0.86) as illustrated in Figure 3.

	g/kg DM					g/d	
	Protein	Fat	Starch	NSP	Total fibre	Total fibre	Total ferm. fibre
CH_4, L/d	-0.42	-0.23	-0.32	0.71	0.63	0.75	0.86
CO_2, L/d	-0.61	-0.47	0.18	0.33	0.34	0.65	0.62

Table 3. Pearson correlations of chemical compositions (g/kg DM, g/d) and excretion of methane and carbon dioxide (L/d) of 140 diets/treatments. P<0.05.

NSP, non starch polysaccharides. Total fibre, calculated as the residual fraction after subtraction of the analysed content of sugars, starch, crude protein, crude fat and ash from the dry matter content. Total fermentable fibre, the amount being fermented in the hind-gut of the intestine.

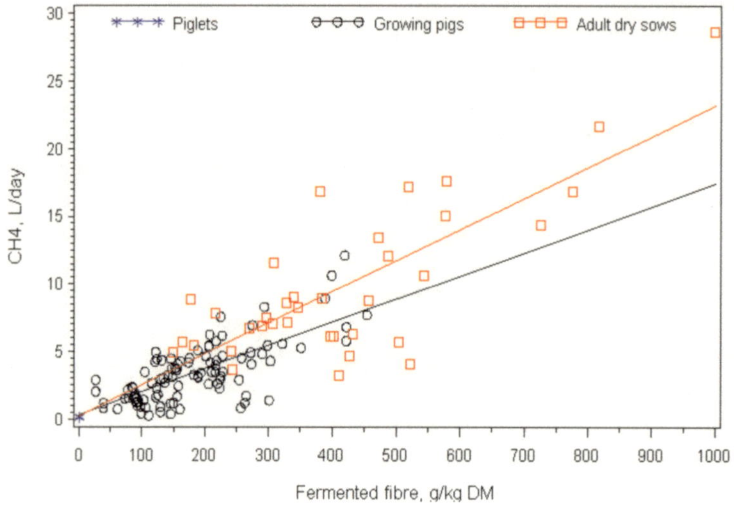

Fig. 3. Plot of CH_4 excretion (L/day) against total fermented fibre (g/kg Dry Matter intake) on the total dataset (n=140).

There is a strong relationship between CO_2 production and DM intake (r=0.87) and between CO_2 production and body weight (r=0.83). This is a reflection of the growth of the animals as the metabolic pathways involved in maintenance and growth require energy, which will lead to CO_2 production (Figure 4, black circles from growing pigs at maintenance + retention). In a comprehensive review of the metabolic studies from the Oscar-Kellner Institute it was shown that increasing the daily weight gain of animals results in an overall lower CO_2 emission (Jentsch et al. 2009) as higher daily weight gain results in relatively less amount of energy required for maintenance. From Figure 4 it can also be seen that the CO_2 production from the adult sows (red squares) up to a DM intake of approximately 3.5 kg/day is higher because the sows at the low feed intake are fed at maintenance where energy intake is equivalent to the amount of energy oxidised.

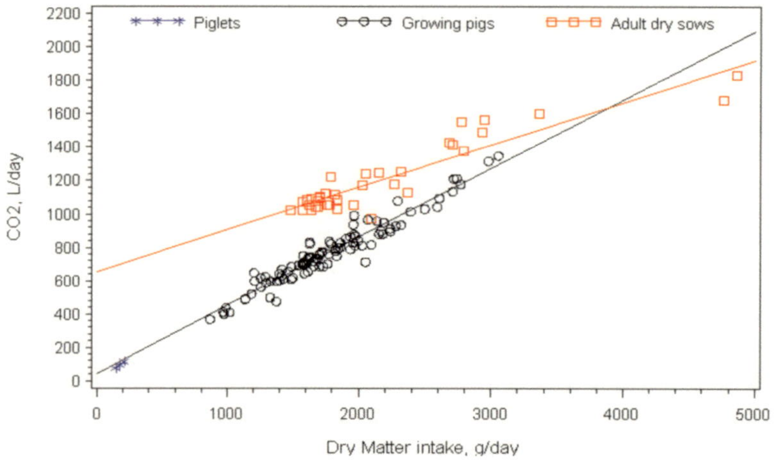

Fig. 4. Plot of CO_2 excretion (L/day) against total Dry Matter intake (g/day) on the total dataset (n=140).

4.3 Diurnal variation in methane and carbon dioxide

Table 2 shows the average excretion of CH_4 and CO_2 of the three classes of pigs. However, close inspection of the results revealed large diurnal variation depending on feeding and physical activity, and the digestion and fermentation in the intestine. Figure 5 show an example of diurnal variation in a growing pig fed a diet containing sugar beet pulp (a type of fibre easily fermented). The variation in CO_2 is mainly a reflection of feeding a meal and the subsequent activity (Le Goff et. al. 2002; Schrama et al. 1998 and review by Pedersen et al. 2008). The animals here are fed twice daily at 08.00 and 15.00 h. The lowest CO_2 concentration is found during night-time from around 20.00 in the evening to 07.00 the next morning. This is the period with the lowest heat production during the day and can be defined as basal metabolic rate or maintenance. The sudden increase in CH_4 concentration shown occasionally is related to the activity where the animals stand up and release some intestinal gas (flatulence).

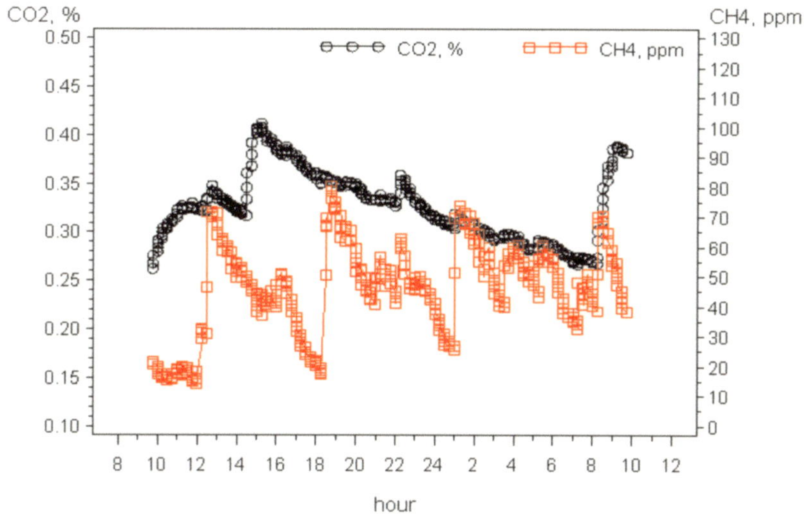

Fig. 5. Typical diurnal variation in concentration of carbon dioxide (CO_2) and methane (CH_4) in the outgoing air from the respiration chamber with a growing pig fed a diet containing sugar beet pulp. The gas from the respiration chamber is measured every 4 minutes.

4.4 Influence of dietary fat on enteric methane emission

Dietary fat can potentially influence the CH_4 production, as have been shown in ruminants (Beaucemin & McBurn, 2006), due to biohydrogenation of unsaturated fatty acids and enhanced propionate production. Fat sources with medium chain fatty acids i.e. coconut oil and palm oil is also shown to depress CH_4 production in ruminants (Machmüller et al, 2003). In two of our studies, growing pigs have been fed increasing amount of either rapeseed oil or fish oil (Jørgensen et al. 1996; Jørgensen unpublished) and one study with adult sows using animal fat (Theil et al. 2004). The concentration of dietary fat varied from 3 to 21 % but had only a small and insignificant influence on the CH_4 production. However, neither of these experiments was specifically designed to investigate the effect on CH_4 production and all diets had a relatively low level of dietary fibre (12 - 18 % dietary fibre), which is the main substrate for fermentation. Similar observation was done by Noblet & Shi

(1994) who did not find evidence of a digestive interaction between dietary fat and fibre. Pigs, as non-ruminants, digest the majority of various fat sources up to 90 %, (Jørgensen and Fernández, 2000; Jørgensen et al. 2003) and when feeding fish oil or coconut oil, as much as 94 % of the fat is digested at the terminal ileum (Jørgensen et al. 1996; 2000). It is therefore only limited amount of the dietary fatty acids that reach the large intestine with a potential for reducing the CH_4 production.

4.5 Influence of environmental temperature on enteric methane emission

In ruminants it is generally assumed that CH_4 production decreases with increasing passage rates associated with cold adaptation (McAllister et al. 1996). In order to evaluate if low ambient temperature decreases CH_4 production in pigs, results from three experiments where ambient temperature have varied in the range 12-29 °C were evaluated (Theil et al. 2001). A tendency was seen for a reduced CH_4 production at the low temperature independent of whether CH_4 was corrected for differences in feed intake or not.

4.6 Influence of body weight on methane emission

The pigs' gastrointestinal tract develops with age (and weight) towards increased capacity and ability to ferment the dietary fibre fraction of the diet. This is illustrated in Figure 6 for growing pigs.

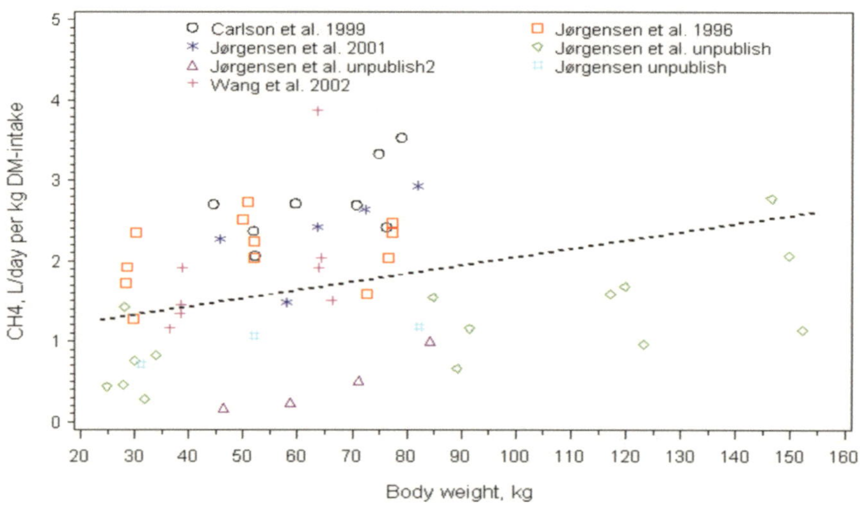

Fig. 6. Influence of body weight on fermentation/excretion of CH_4 when the excretion is corrected for feed intake (L/day per kg Dry Matter intake).

The results are from experiments where equal types of diet are fed through several periods (i.e. Latin square design). The results show some variation from period to period, but the general trend is increasing CH_4 emission in response to increasing body weight. However, as the pigs grow, daily feed intake increases as well and in order to test the influence of only body weight, the CH_4 production is corrected for differences in feed intake (DM intake) as shown in the Figure. Even with this correction there is a significant effect of body weight on CH_4 production, and the derived equations are shown below:

$$CH_4,\ L/day\ per\ kg\ DM\ intake = 1.01 + 0.0107 \times BW,\ kg,\ n = 55,\ R^2 = 0.71 \quad (2)$$

The effect of body weight was minor, however, significant in agreement with the studies by Christensen & Thorbek (1987) and Noblet & Shi (1994). However, Kirchgessner and co-workers concluded in a summary of several of their experiments that methanogenesis in pigs is fairly independent of live weight if feeding condition is kept constant.

4.7 Comparison between growing pigs and adults sows fed similar types of fiber

A comparison of the ability of growing pigs and sows to digest and utilise various fibre rich feedstuffs is shown in Figure 7. The data in Figure 7 is limited to experiments where the animals are fed the same type of fibre. The results demonstrate that both growing pigs and sows produce equal amounts of CH_4 per g fermented fibre. However, the sows had a much greater fermentative capacity, i.e. they can ferment a much larger amount of fibre per day.

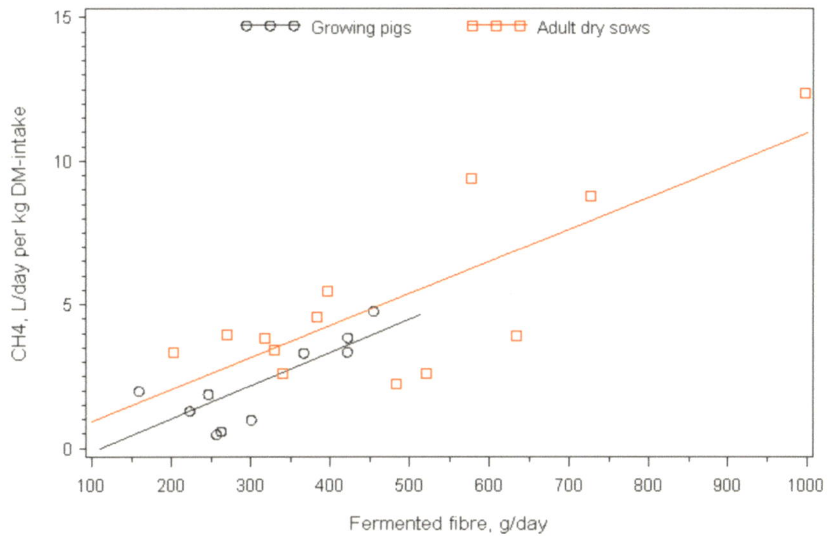

Fig. 7. Influence of body weight (growing pigs *vs* adult sows) on fermentation/excretion of CH_4 in experiments where the animals are fed equal types of fibers and the excretion is corrected for feed intake (L/day per kg Dry Matter intake).

In the actual experiments neither the sows nor the growing pigs were able to consume more of the most voluminous fibre feedstuffs. No significant difference between the slopes from the two classes of animals was found. Therefore, emission of methane in pigs per unit of feed ingested can be expressed as:

$$CH_4,\ L/day\ per\ kg\ DM\ intake = -7.57 + 0.0112 \times Fermented\ Fibre,\ g/day,\ n = 23,\ R^2 = 0.71 \quad (3)$$

A reduced feed intake allow more time for fermentation and as a result more enteric CH_4 is produced as found by Christensen & Thorbek (1987) in a study with reduced feed intake to growing pigs. This could also be the case when adult sows in the dry period and during pregnancy are fed relative restricted (2-2.5 kg DM per day) in order not to gain excessive weight because this may cause health problems around farrowing and lactation. In the last 3-4 weeks of the pregnancy period, the feeding level is in general increased to 3.5 kg DM per day. This was tested in three of the experiments with sows; however there was a non-significant decrease in CH_4 emission expressed per kg dry matter intake.

4.8 Comparison between growing pigs and adult sows–data from the whole database

When the whole database was used to compare the CH_4 emission in growing pigs and adult sows the results are shown in Figure 8.

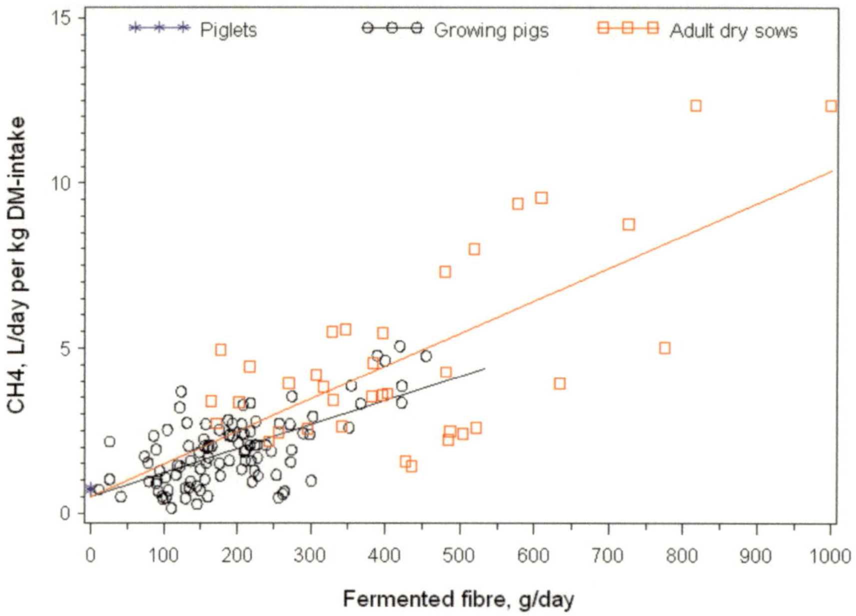

Fig. 8. Influence of body weight (growing pigs *vs* adult sows) on fermentation/excretion of CH_4 when plotting the whole dataset (n=140) the excretion is corrected for feed intake (L/day per kg Dry Matter intake).

The results show large variation in the fermentability of the different fibres depending on fibre type but it was also found that some animals had low CH_4 production independent of amount and type of fibre. When comparing the contribution of the different fibre rich feedstuffs to the CH_4 emission it should be kept in mind that the lower energy digestibility in fibre rich feedstuffs compared to concentrated feeds implies that the pigs must consume a larger quantity of feed to obtain the same production. When the CH_4 production was expressed as either L/day or L/day per kg DM intake, there was no significant difference between growing pigs and adult sows:

$$CH_4,\ L/day = 0.440 + 0.0206 \text{ x Fermented Fibre}, \quad g/day,\ n = 137,\ R^2 = 0.74 \tag{4}$$

$$CH_4,\ L / day\ per\ kg\ DM\ intake = 0.626 + 0.00894 \times Fermented\ Fibre,\ g / day,\ n = 137,\ R^2 = 0.64 \quad (5)$$

However, when the CH_4 emission was expressed relative to either GE or DE the slopes for growing pigs and sows were significant different and expressed as:

$$\begin{aligned} CH_4 - energy,\ \%\ GE &= 0.0628 + 0.00277\ (growing\ pigs) \times Fermented\ Fibre,\ g / kg\ DM \\ CH_4 - energy,\ \%\ GE &= 0.0628 + 0.00488\ (adult\ sows) \times Fermented\ Fibre,\ g / kg\ DM \\ & n = 137,\ R^2 = 0.72 \end{aligned} \quad (6)$$

$$\begin{aligned} CH_4 - energy,\ \%\ DE &= 0.0838 + 0.00376\ (growing\ pigs) \times Fermented\ Fibre,\ g / kg\ DM \\ CH_4 - energy,\ \%\ DE &= 0.0838 + 0.00606\ (adult\ sows) \times Fermented\ Fibre,\ g / kg\ DM \end{aligned} \quad (7)$$

On average, energy from CH_4 relative to DE amounted to 0.68% (Table 2). However, energy loss of CH_4 in adult sows was approximately three times higher (1.31% of DE) than in growing pigs (0.47% of DE). The reason for the higher loss of CH_4 in proportion of DE per g of fermented fibre in sows than in growing pigs is, as discussed by Jørgensen (2007), the sow's relatively greater capacity for fermentation. A factor of importance is also that non-pregnant sows were fed relatively less relative to the body size. The values estimated in current compilation are comparable to results obtained by Schrama et al. (1998) for growing pigs and Kirchgessner et al. (1991) and Le Goff et al. (2002) for adult sows or used by Vermorel et al. (2008) in an assessment of enteric methane emission by farm animals in France. As indicated in Table 2, the CH_4 production in sows could be as high as 3.25% of DE.

The average CH_4 production by growing pigs was estimated 0.39% of GE or 0.47% of DE, which is lower than the value for all classes of pigs (0.6% of GE) assumed in the report on emission of greenhouse gases from Danish agriculture (Mikkelsen et al., 2006; 2011). If the value of CH_4-energy (1.0% of GE or 1.3% of DE) is representative for diets for sows (lactating and non-lactating), then the value in the latter report is underestimated for this class of pigs.

4.9 Case study – methane emission when feeding low and high fiber diets to growing pigs and adult sows

In Table 4 are shown examples of the typical variations in diet composition that may influence the CH_4 emission from pigs (Sørensen and Fernandez, 2003). All the diets were formulated to satisfy the nutritive recommendations and they resemble diets used in practice.

Over the last two decades diets have been formulated closer to pigs' requirement with regard to amino acid composition to reduce N-pollution to the environment. This is illustrated with diet no. 5 and 6 (Table 4), where diet no. 6 is supplemented with free amino acids. Both diets having the same growth potential but diet no. 6 had reduced protein content and a reduced content of especially soybean meal. Because less fibre is fermented a lower CH_4 production can be expected. Sugar beet pulp was used to increase the concentration of dietary fibre in diet 7, which resulted in higher fermentation of fibre and, consequently, emission of CH_4 that was 60 % higher compared to diet 6 at the same daily gain.

Animal	Growing pigs				Adult sows		
Diet[1]	5	6	7		9	11	12
Composition, %							
Wheat	33.3	40.1	22.0		45.2	-	40.7
Barley	25.1	33.9	30.4		38.5	81.2	4.6
Sunflower cake	-	-	-		5.7	-	-
Rapeseed cake	-	-	-		3.0	-	-
Barley straw	-	-	-		5.0	-	-
Soybean meal	15.4	10.0	19.5		0.5	3.7	-
Peas	11.6	4.3	-		-	-	-
Sweet lupine	10.	6.6	10.9		-	0.5	10.0
Sugar beet pulp	2.0	2.0	15.0		-	12.5	42.8
DL-Methionine 40	0.14	0.20	0.13		-	0.05	0.13
L-Threonine 50	-	0.30	-		-	0.05	0.19
L-Lysine 50	-	0.70	-		-	-	-
Min.+Vit. Mix	2.46	1.90	2.07		2.10	2.00	1.48
Chemical composition							
Protein. % in DM	23.3	20.0	23.2		14.6	13.0	15.1
Dietary fibre, % DM	22.5	21.0	29.8		22.6	27.9	41.5
DM-intake. kg/d	1.68	1.70	1.80		1.85	1.83	1.81
DC energy, %	83	82	78		82	83	84
Ferm fibre, g/kg DM	139	123	194		106	180	335
Expected CH_4, L/d[2]	5.5	5.1	8.0		4.9	7.6	13.3
CH_4-energy, % GE[3]	0.45	0.40	0.60		0.58	0.94	1.70
CH_4-energy, % DE[4]	0.61	0.55	0.81		0.73	1.17	2.11

[1] The diet number refers to the diet in the paper of Sørensen & Fernandéz (2003). DC, digestibility coefficient; GE, gross energy; DE, digestible energy.
[2] Estimated using equation no 4.
[3] Estimated using equation no 6.
[4] Estimated using equation no 7.

Table 4. Examples on trend in diets for growing pigs and dry sows *data from Sørensen & Fernandéz (2003).

Restricted feeding is generally used for dry and gestating sows to avoid overweight, which can lead to farrowing and locomotion problems. An undesired effect of that are animal welfare problems such as stereotypic behaviour, which, however, can be reduced by feeding high fibre diets. The consequences of increasing the fibre content are illustrated by diet no. 11 and 12 relative to diet 9. Diet no. 9 contains fibre with relative high content of insoluble fibre from barley straw, which illustrates a situation where the sows eat some of the bedding material, which is common. Insoluble fibre is more resistant to fermentation and subsequently the amount of fermented fibre is the lowest of the three diets and the calculated CH_4 emission is low too. In diets no. 11 and no. 12, the amount of fermentable

fibre is increased by adding sugar beet pulp, which resulted in almost a tripling of the CH_4 emission when comparing diet no. 12 with diet 9.
Formulating diets closer to pigs' requirement and using a variety of feedstuffs it is thus possibly to reduce the CH_4 emission without compromising the animal performance. In ruminants there is in a meta-analysis shown strong indication that dietary tannins is related to CH_4 emission (Jayanegara et al. 2011), however, it remains to be shown if dietary tannins have any influence on the enteric CH_4 emission from pigs.

5. Implications on alterations in pig feeding

The enteric CH_4 production may be reduced by altering the feed composition of the diet. Formulation of diets rich in fat, starch and protein and low in fiber will be optimal if the major interest is to reduce the methane emission. The most prominent effect will be to reduce the dietary fiber content. However, feeding pigs is a complex issue, and feed composition and feeding strategies should satisfy a number of various aspects.

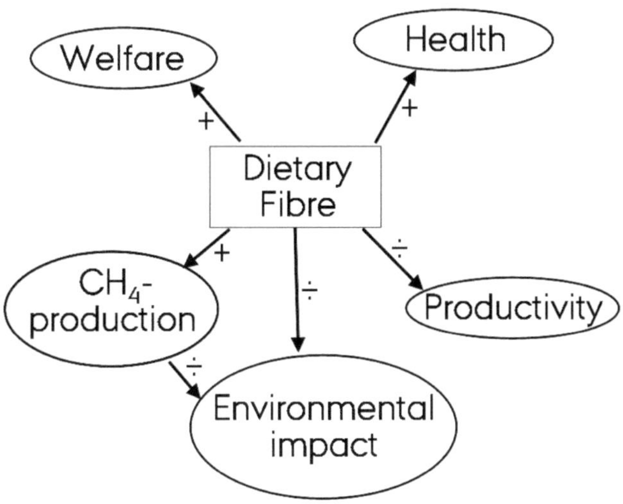

Fig. 9. Relationship between dietary fibre intake (the main source of enteric methane emission) and traits related to the environment and pig production. (÷ indicate negative effect and + indicate positive effect).

If low fiber diets are generally applied in pig feeds, it will improve their productivity, and the pollution load caused by excreted nutrients will be reduced. However, concomitantly low fiber diets will be detrimental for the animal health, fiber is important for the health of the intestine and consumption of fiber prevents e.g. gastric ulcers. Moreover, dietary fiber improve the welfare of e.g. sows by reducing gastric emptying rate (Jørgensen et al., 2010), hunger and stereotypic behaviour when they are fed restrictedly throughout gestation (Figure 9).

6. Conclusion

The main dietary contributor to the CH_4 enteric emission at all live weight of pigs is the dietary concentration of fibre, whereas neither dietary fat nor protein had any significant

impact on the CH_4 emission. The lack of effect of dietary fat on the CH_4 emission is possibly due to a very high digestibility of the fat in the small intestine.
Production of CH_4 by piglets is low and amounts to only 0.13 L/day or 0.1 % of GE. For growing pigs' feed, a low fibre diet or a standard diet, the CH_4 emission can be estimated from 0.2 to 0.5 % of GE corresponding to 3.4 L/day for an average slaughter pig and thus lower than the standard 0.6 % of GE recommended by ICPP (2006) or used by Mikkelsen et al. (2011). When feeding diets with higher fibre content, the CH_4 emission depends on type and might contribute up to 1 % of GE. Dry and gestating sows fed at maintenance have a CH_4 emission varying from 0.6 to 2.7 % of GE depending on feeding level and fibre type, whereas the CH_4 emission of lactating sows is estimated to be approximately 0.6 % of GE. In all the experiments the CH_4 emission of sows has been found to exceed the standard 0.6 % of GE recommended by the ICPP.
The Danish pig production is rather intensive meaning that the standard pig feed contain less dietary fermentable fibre than when pigs are raised under more extensive production e.g. countries in Asia. This implies that the methane emission from the global pig production could be higher than the average values found under Danish conditions.

7. References

Beauchemin, K.A. & McGinn, S.M. 2006. Methane emission from beef cattle: Effects of fumaric acid, essential oil, and canola oil. J. Anim. Sci. 84:1489-1496.

Christensen, K. & Thorbek, G. 1987. Methane excretion in the growing pig. Br. J. Nutr. 57:355-361.

FAOSTAT. 2009. FAO Statistical Database, Food and Agriculture Organization of the United Nations, Rome, Italy. Accessed 22.06.2011 at http://faostat.fao.org/site/573/default.aspx.

Fernández, J.A., Jørgensen, H. & Just, A. 1986. Comparative digestibility experiments with growing pigs and adult sows. Anim. Prod. 43:127-132.

IPCC. 2006. Guidelines for National Greenhouse Gas Inventories. Agriculture, forestry and other land use; Emission from livestock and manure management, volume 4, Chap. 10, 87pp.

Jakobsen, K., Theil, P.K. & Jørgensen, H. 2005. Methodological considerations as to quantify nutrients and energy metabolism in lactating sows. J. Anim. Feed Sci. 14:31-47.

Jayanegara, F., Leiber, F. & Kreuzer, M. 2011. Meta-analysis of the relationship between dietary tannin level and methane formation in ruminants from *in vivo* and *in vitro* experiments. J. Anim. Physiol- Anim. Nutr. In press doi:10.1111/j.1439-0396.2011.01172.x.

Jensen, B.B. & Jørgensen, H. 1994. Effect of dietary fiber on microbial activity and microbial gas production in various regions of the gastrointestinal tract of pigs. Appl. environ. Microbiol. 60:1897-1904.

Jensen, B.B. 1996. Methanogenesis in monogastric animals. Environ. Monit. Assess. 42:99-112.

Jentsch, W., Schweigel, M., Weissbach, H.S., Pitroff, W. & Derno, M. 2007. Methane production in cattle calculated by the nutrient composition of the diet. Arch. Anim. Nutr. 61:10-19.

Jentsch, W., Piatkowski, B. & Derno, M. 2009. Relationship between carbon dioxide production and performance in cattle and pigs. Arch. Tierzucht. 52:485-496.

Johnson, K.A. and Johnson, D.E. 1995. Methane emissions from cattle. J. Anim. Sci. 73:2483-2492.

Just, A., Jørgensen, H., Fernández, J.A., Bech-Andersen, S. & Enggaard Hansen, N. 1983 Forskellige foderstoffers kemiske sammensætning, fordøjelighed, energi- og proteinværdi til svin. 556. Report from the National Institute of Animal Science, Denmark , 99 pp.

Jørgensen, H. 1997. Energy utilization of diets with different sources of dietary fibre in growing pigs. In: K. J. McCracken, E. F. Unsworth, and A. R. G. Wylie (Eds.) Energy Metabolism of Farm Animals. pp. 367-370. CAB International, University Press, Cambridge.

Jørgensen, H. 2001. Energimålinger ved hjælp af respirationskamre. JordbrugsForskning 5:6-7.

Jørgensen, H. 2007. Methane emission by growing pigs and adult sows as influenced by fermentation. Livest. Sci. 109:216-219.

Jørgensen, H. & Fernández, J.A. 2000. Chemical composition and energy value of different fat sources for growing pigs. Acta Agric. Scand. Sect. A-Anim. Sci. 50:129-136.

Jørgensen, H., Jensen, S.K. & Eggum, B.O. 1996. The influence of rapeseed oil on digestibility, energy metabolism and tissue fatty acid composition in pigs. Acta Agric. Scand., Sect. A, Anim. Sci. 46:65-75.

Jørgensen, H., Zhao, X.Q. & Eggum, B.O. 1996. The influence of dietary fibre and environmental temperature on the development of the gastrointestinal tract, digestibility, degree of fermentation in the hind-gut and energy metabolism in pigs. Br. J. Nutr. 75:365-378.

Jørgensen, H., Larsen, T., Zhao, Z.Q. & Eggum, B.O. 1997. The energy value of short-chain fatty acids infused into the caecum of pigs. Br. J. Nutr. 77:745-756.

Jørgensen, H., Gabert, M.V., Hedemann, M.S. & Jensen, S.K. 2000. Digestion of fat does not differ in growing pigs fed diets containing fish oil, rapeseed oil or coconut oil. J. Nutr. 130:852-857.

Jørgensen, H., Bach Knudsen, K.E. & Theil, P.K. 2001. Effect of dietary fibre on energy metabolism of growing pigs and pregnant sows. In: A. Chwalibog and K. Jakobsen (Eds.) Energy Metabolism in Animals. pp. 105-108. Wageningen Pers, Wageningen.

Jørgensen, H., Fernández, J.A. Krogh Jensen, S. & Danielsen, V. 2003. Effect of fat sources and rapeseed lecithin on fat digestibility and utilization in young pigs. Pages 1254-1256 in Proceedings 11th International Rapeseed Congree. H. Sørensen ed. Samfundslitteratur Grafik, Gylling.

Jørgensen, H., Serena, A., Hedemann, M.S. & Bach Knudsen, K.E. 2007. The fermentative capacity of growing pigs and adult sows fed diets with contrasting type and level of dietary fibre. Livest. Sci. 109:111-114.

Jørgensen, H., Strathe, A., Theil, P.K. & Bach Knudsen, K.E. 2010. Evaluation of a simple non-invasive 13C breath test to evaluate diet effects on gastric emptying in pigs. Livest. Sci. 133:64-66.

Kirchgessner, M., Kreuzer, M., Müller, H.L. & Windisch, W. 1991. Release of methane and of carbon dioxide by the pig. Agribiol. Res. 44:103-113.

Le Goff, G., Le Groumellec, L., van Milgen, J., Dubois, S. & Noblet, J. 2002. Digestibility and metabolic utilisation of dietary energy in adult sows: influence of addition and origin of dietary fibre. Br. J. Nutr. 87:325-335.

Lekule, F.P., Jørgensen, H., Fernández, J.A. & just, A. 1990. Nutritive value of some tropical feedstuffs for pigs. Chemical composition, digestibility and metabolizable energy content. Anim. Feed Sci. Technol. 28:91-101.

McAllister, T.A., Okine, E.K., Mathison, G.W. & Cheng, K.-J. 1996. Dietary, environmental and microbiological aspects of methane production in ruminants. Can. J. Anim. Sci. 76:231-243.

Machmüller, A., Soliva, C.R. & Kreuzer, M. 2003. Methane-suppressing effect of myristic acid in sheep as affected by dietary calcium and forage proportion. Br J. Nutr. 90:529-540.

Mikkelsen, M.H., Gyldenkærne, S., Poulsen, H.D., Olesen, J.E. & Sommer, S.G. 2006. Emission of ammonia, nitrous and methane from Danish agriculture 1985 - 2002. Methodology and estimates. Research notes from NERI, no. 231. National Enviromental Research Institute. Ministry of the Environment, Denmark. 88 pp.

Mikkelsen, M.H., Albrektsen, R. & Gyldenkærne, S. 2011. Danish emission inventories for agriculture. Inventories 1985 - 2009. National Research Institute, Aarhus University. - NERI Technical Report No. 810. 136 pp.

Miller, T.L. & Wolin, M.J. 1974. Fermentations by saccharolytic intestinal bacteria. Am J. Clin. Nutr. 32:164-172.

Noblet, J & Shi, X.S. 1994. Effect of body weight on digestive utilization of energy and nutrients of ingredients and diets in pigs. Livest. Prod. Sci. 37:323-338.

Olesen, C.S. & Jørgensen, H. 2001. Effect of dietary fibre on digestibility and energy metabolism in pregnant sows. Acta Agric. Scand. Sect. A-Animal Sci. 51:200-207.

Pedersen, S., Blanes-Vidal, V., Jørgensen, H., Chwalibog, A., Haeussermann, A., Heetkamp, M.J.W. & Aarmink, A.J.A. 2008. Carbon dioxide production in animal houses: A literature review. Agric. Eng. Int. GICR Ejournal X:1-19.

Schrama, J.W., Bosch, M.W., Verstegen, M.V.A., Vorselaars, A.H.P.M., Haaksma, J. & Heetkamp, M.J.W. 1998. The energetic value of nonstarch polysaccharides in relation to physical activity in group-housed, growing pigs. J. Anim. Sci. 76:3016-3023.

Serena, A., Jørgensen, H. & Bach Knudsen, K.E. 2007. Nutritional value of co-products from vegetable food industry. Pages 473-491 in Paradigms in pig science. J. Wiseman, M. A. Varley, S. McOrist, and B. Kemp eds. Nottingham University Press, Nottingham.

Serena, A., Jørgensen, H. & Bach Knudsen, K.E. 2008. Digestion of carbohydrates and utilization of energy in sows fed diets with contrasting levels and physicochemical properties of dietary fiber. J. Anim Sci. 86:2208-2216.

Serena, A., Jørgensen, H. & Bach Knudsen, K.E. 2009. Absorption of carbohydrate-derived nutrients in sows as influenced by types and contents of dietary fiber. J. Anim Sci. 87:136-147.

Stryer, L. 1988. Biochemistry. 1089 pp. W.H. Freeman and Company. New York.

Sørensen, P. & Fernández, J.A. 2003. Dietary effects on the composition of pig slurry and on the plant utilization of pig slurry nitrogen. J. Agric. Sci. 140:343-355.

Theil, P.K., Jørgensen, H. & Jakobsen, K. 2002. Energy and protein metabolism in pregnant sows fed two levels of dietary protein. J. Anim. Physiol. Anim. Nutr. 86:399-413.

Theil, P.K., Jørgensen, H. & Jakobsen, K. 2004. Energy and protein metabolism in lactating sows fed two levels of dietary fat. Livest. Prod. Sci. 89:265-276.

Theil, P.K., Kristensen, N.B., Jørgensen, H., Labouriau, R. & Jakobsen, K. 2007. Milk intake and carbon dioxide production of piglets determined with the doubly labelled water technique. Animal 1:881-888.

Vermorel, M., Jouany, J.-P., Eugéne, M., Sauvant, D., Noblet, J. & Dourmad, J.-Y. 2008. Evaluation quantitative des émissions de méthane entérique par les animaux d´élevage en 2007 en France. INRA Prod. Anim. 21:403-418.

Wang, J.F., Zhu, Y.H., Li, D.F., Jørgensen, H. & Jensen, B.B. 2004. The influence of different fibre and starch types on nutrient balance and energy metabolism in growing pigs. Asian-Australian J. Anim. Sci. 17:263-270.

28

Human Security and Global Warming: Challenges Before Christianity and Islam in Nigeria

Dokun Oyeshola
Obafemi Awolowo University
Nigeria

1. Introduction

A discourse on the concept of global warming in this paper needs to be contextually placed within the ecological systems of the planet earth and human security. The planet earth is made up of ecological systems that manifest in the variety of biodiversity with its added values of goods and services they provide which are usually referred to as 'gifts of nature'. For instance, four major areas of these values may be identified. The first is the provision of services by which the harvestable products of foods, medicines, fresh water, wood and fibres, fuel wood and other non-wood products, including the genetic resources for improving these products, flow from ecosystems to support humans. In addition, biodiversity boosts ecosystem productivity where each species no matter how small, all have an important role to play. The second is in its regulating services for local and global climate, including carbon sequestration, regulation of water quality, diseases, pollution and flood. For example, a large number of plant species means a greater variety of crops, greater species diversity ensures natural sustainability for all life forms; and healthy ecosystems can better withstand and recover from a variety of disasters. Provision of cultural services is the third. This includes the aesthetic, spiritual, educational, scientific and recreational values. The last is in its supporting services it provides through its role in making the above three functions to happen. For instance, it mediates nutrient cycling, soil formation, pollination and primary production.

The ecosystem services are intrinsically linked to the constituents of human well-being of security, availability and access to the basic needs of livelihood of food, shelter and other materials. Other areas are those of health, recreation, eco-tourism, spiritual well-being and culture. Maintenance of biodiversity unites human generations of the past, present and future. It is therefore mandatory that natural resources which are an intrinsic part of biodiversity be protected. The major classes of these natural resources consist of agricultural land, forest land and its multiple products and services, natural land areas preserved for aesthetic, recreational and scientific purposes, fresh and salt water fisheries, mineral resources that include the mineral fuels and non-fuels, the renewable non-mineral energy sources of solar, tidal, wind and geothermal systems, water resources, the waste-assimilative capacities of all parts of the environment. Some natural resource stock is renewable by natural or artificial processes while others are non-renewable. Those that are renewable often depend on appropriate non-destructive methods of management as with farmlands, fisheries and waste disposal since some changes in natural resource systems are irreversible. In the assessment of natural resource stocks, it is important that interactions with other systems and potentially irreversible changes be taken into account.

Primarily, as a result of human activities the ecological make-up and harmonious relationships among the environmental constituent beings have been degraded and altered leading to global warming and human insecurity. Global warming leading to climate change lends to extreme climate events which implies an immense social cost (Oyeshola, 1995; 1998; Adinna, 2001). For instance, Hurricane Gilbert in 1988 caused damage estimated at over US$870 million in Jamaica (Foley, 1991: 47). Similarly, the memory is still fresh of recent hurricanes, landslides, tsunami, floods, earthquakes and fires. It is this dynamic, overwhelming, destructive capacity associated with climate change that makes it a different global, socio-economic, developmental and security problem. Unlike other problems like nuclear warfare, refuges and poverty which may be localized and whose solutions may be delayed, global warming is a problem that involves every nation and the solution to it cannot be delayed. Any delay in addressing the problem promptly and radically too may mean that the entire humanity may perish sooner than later! This is a deadly serious implication and Nigeria with its great population and abundant resources is a part of the planet earth which is being threatened by global warming.

Nigeria with a population of over 140 million is home to approximately 250 different ethnic groups, each with its language. The three largest, the Hausa-Fulani, the Yoruba and Igbo, together are less than two-thirds of all Nigerians. The country is almost evenly divided between Christianity predominantly in the South and Islam in the North. There are traditionalists in the land. The North is among the poorest predominantly Muslim regions in the world. In terms of income, most Nigerians are very poor, with oil wealth concentrated in the hands of a few elite. Militancy in the Niger Delta must be reckoned with if there will be peace in the land. Religious crisis is equally seriously at home in the country.

In addition to this landscape, water and electricity supply are notoriously erratic, if not non-existent in many places. Many of the country's roads and highways are in deplorable condition, many public buildings and facilities are decrepit and the inability of the Administration to tackle key infrastructural challenges continued to make the cost of production in Nigeria one of the highest in the world and by so doing putting operators in the sector at a disadvantage compared to their counterparts in other countries. In the face of this some corporations moved their headquarters away from Nigeria.

Corruption and glaring ineptitude of both the leadership and ranks and file of the Nigerian Police Force have seen crime rate spiralling to unprecedented levels. Armed robberies, political assassinations, kidnapping and brazen thudgery are of common occurrence. Specifically, the problem of poverty which is the bane of development in Nigeria, has not been adequately addressed in spite of National Policy on Poverty Eradication (2000) Programme.

In the context of environmental degradation, soil erosion is a serious problem in Nigeria. About three years ago, it led to gullies that threatened communities. For instance, deep gullies in Kebbi threatened communities in Jeda, Zuru, Shanga, Yahuri, Argungu, Arewa, Aleiro, Bagudo and Birnin Kebbi. In Awka over time, more than 1,000 persons and over 4,000 families have lost their ancestral lands to the menace of gully erosion in the three towns of Nanka, Ekwulobia and Nnewi. Flooding, another environmental problem occurs throughout the country in three main forms. The first is the urban flooding which mainly affects human livelihood and it occurs in towns located on flat or low lying terrain usually with poor drainage systems. Such towns include Lagos, Maiduguri, Aba, Warri, Benin and Ibadan. Deforestation has become a major environmental problem in Nigeria. This is due to hardwood logging and tremendous population growth being experienced in the country. Construction of developmental projects, their implementation and related activities leading to industries, airports, recreational places, park, mechanized farming, housing estates and so on also contribute to deforestation because they take space.

The problem of industrial pollution is enormous; it is a serious problem in Nigeria. For instance, Nigeria has about 5,000 registered industrial facilities and 10,000 small scale industries operating illegally within residential premises. Disposal of domestic and industrial wastes is a problem in Nigeria. Disposing of industrial waste is leading, in some cases, to nightmare menacing the physical environment. Some of the industrial wastes are often dumped on the ground, thrown into the stream/lakes or disposed of in a deep hole in the ground. The ultimate sufferer of these methods of disposal is the physical environment. It takes a healthy planet, to sustain healthy people. All the problems enumerated above constitute the problem of human security in Nigeria which the two dominant religions of Christianity and Islam must address.

In the face of the crises enumerated above, both late Yar 'Adua and Goodluck Jonathan Administrations expect something radical from both Christianity and Islam. They expected from the two statistically dominant religions that they make their contributions in helping to investigate problems related to the organization of community life (ethnicity, family life, youth restlessness and joblessness, corruption, justice, violence, development, traditional values, etc) which is at the heart of human security and to seek out solutions and perhaps most importantly of all, to initiate the process of reorganizing already fractured relations among human groups in the country. At various times, the government has called on religious leaders to team-up for the progress and peace of the nation not only by asking for prayers but also by inviting them to conscientise their constituencies in order to promote indigenous values of love, community relationship and profound spirituality. Vices of intolerance, hatred, cutthroat competition, disharmony, pride, arrogance, theft, covetousness and cheating arguably introduced to the land by default through western civilisation (Omolewa in the *Nation*, August 8, 2007: 41) must be confronted. By its tenets, Christianity is expected to take an active part in reducing inequality, poverty and injustice (Is. 11: 1-9; Acts 2: 40-47, Mt.5: 1-8; 2 Cor.4: 16-18) and in exploring steps toward a more satisfactory organization of societal life. Islam up-holds similar orientation. Poverty is real in Nigeria and it is one of the major concerns of the government as expressed in its Poverty Alleviation Programme and religions are to lend their helping hands. Furthermore, Christianity and Islam in Nigeria are expected to make clear what exactly is at stake as the world confronts itself in need of transformation if global warming and climate change for instance, will not put an end to the existence of humanity and the earth planet as a living planet (Simonis, 2007: 8 ff).

In Nigeria religion pervades all aspects of life of human persons to the point that the dividing line between sanity and madness in the behaviour and activities among some practitioners of religions is often blurred. In other words, in the experience of Nigeria religion has created religious crisis and promoted socio-economic development at the same time. Since the influence of religion is so powerful in Nigeria, it can then be taken that it is a commanding force that can bring about a desirable effect if properly utilised.

In the light of the above contextual background, the aim of this chapter therefore is to challenge the two dominant religions in Nigeria in the face of human insecurity which has become the national project especially from the point of view of the 7 point agenda of the ruling government. This will take place in the context of global warming. This is necessary because the devastating consequences of global warming will negate all efforts at securing and promoting human security, and this will be discussed not only as a religionist and practitioner and a Nigerian but also as an academic. The former categories may often be sentimental and apologetic as has been demonstrated in the context of Nigeria commissions set up at various times to find solutions to Nigerian religious crisis. A truly academic wants to be as objective as possible within the limit his value system imposes. In this way, radical but beneficial conclusions will be achieved.

The first part of this chapter takes up the two major concepts of human security and global warming. The second part begins with a discourse on some underling Christian and Islam theological principles in order to situate some of the assumptions. The second section of the second part takes up the explication of some internal sources of religious conflict like dogmatism in the interpretation of the sacred scriptures, passive acceptance of faith, gospel of prosperity and so on because the orientations they lead to influence adherents' actions and pronouncements while the third part discusses challenges before Christianity and Islam in relation to human security and global warming in Nigeria.

2. Human security and global warming: An understanding

2.1 Human security

After the end of World War 11, proponents of the classical tradition of international relations saw security narrowly as national security ('the protection or preservation of the state's territorial integrity and external sovereignty'). For them security denotes the absence of a military threat which generally means the protection of a state from external military attack (Makinda, 1996: 152). Willy Brandt understood security more broadly as he stated that:

> An important task of constructive international policy will have to consist in providing a new, more comprehensive understanding of 'security' which would be less restricted to the purely military aspects. In the global context true security cannot be achieved by a mounting up of weapons-defence in the narrow sense – but only by providing basic conditions for peaceful relations between nations, and solving not only the military but also the non-military problems which threaten them (Brandt, 1980: 214).

Similarly, the Palme Commission and its subsequent reports of 1982 and 1989 argued that the abolition or large reduction in weapons of mass destruction and conventional disarmament are prerequisites to providing momentum for progress on economic and social development and environmental conservation (The Palme Commission, 1989).

Furthermore, Gro Harlem Brundtland in *Our Common Future*, was more explicit about the inadequacy of security defined primarily in military terms. She demonstrated the tension between security understood primarily in military terms and environmental security. She equally criticized global militarism and the vested interests that profited from it and called on nations to 'turn away from the destructive logic of an 'arms culture' (World Commission on Environment and Development, 1987: 297-304).

All the above definitions and descriptions are security of nations and their relationships centred. And so if one thinks of security only in terms of nation-states only, one risks confining analysis of security to a threat to the survival of the state of which war and conquest are the most extreme but not the sole threats. Therefore the concept of security can be expanded to include natural forces that threaten security. Some of these are earthquakes, hurricanes, volcanoes, floods, mud slides or forest fires. Human agencies that can threaten security include criminals, lunatics, organised crime, civil war, revolution, local and regional wars and conflict of authority.

From the perspective of feminists, security must be defined in multidimensional terms which include freedom from both physical and structure violence. This perspective is further reflected in the National Organisation for Women which in its 1990 resolution on Women in Combat affirmed that:

> 80-90 per cent of casualties due to conflict since the Second World War have been civilians, the majority of them women and children. The strategy of rape in the war in Bosnia has alerted the world to an atrocity that has always existed in war time, although it is usually unreported. Women and children constitute 80 per cent of the global refugee population, a phenomenon usually attributed to military conflict: women are also the most frequent victims of domestic violence in all societies, a crime which is always under-reported, but one which ranges across regions, cultures and classes (United Nations, 1991). Violence against women is higher in militarized societies and in military families. Evidence such as this suggests that women are particularly vulnerable to militarism and war (Booth and Smith, 1995: 191).

Extending the definition of security to economic and environmental dimensions also highlights vulnerabilities of women as a 1981 Report on the UN Committee on the States of Women claimed:

> While women represent half the global population and one-third of the paid labour force and are responsible for two-thirds of all working hours, they receive only a tenth of world income and own less than 1 per cent of world property (United Nations, 1991:81-114 citing Jaggar, 1983: 138).

And within the structural adjustment policies in the south, women have been impacted heavily upon. They have to:

> Assume additional care-giving tasks when states are forced to cut back on social spending. Women are also particularly vulnerable to environmental degradation: in areas of Africa and Asia they are walking up to 10 kilometres a day to search for shrinking supplies of fuel-wood and water. Women's reproductive systems are particularly susceptible to the hazards of toxic wastes and industrial pollution and accidents (Oyeshola, 1995: 211; Booth and Smith, 1995:192).

In the postcolony era of many African countries particularly Nigeria, the term security, carries with it, in addition to what has been stated above a nuance which is worth noting. This can be described as security not only as control over territories and subjugated peoples, but also as a particular mode of masculinity. In patriarchal societies, masculine conceptions of security require control over women. In military states this translates into command over women which involves the formal or informal appointment of a 'commander of women' who is recognized by the military hierarchy, and who is arrogated the right to articulate and address the concerns of women.

It is from the above background that the Earth Summit of Rio in 1992 was coming. At the Earth Summit, one of the fundamental emerging principles in its declarations was that of principle 1 which stipulated that human person is at the heart of socio-political activities and that it is no longer economic growth per se as was the situation before. This new orientation has necessitated a paradigm shift in respect of the concept of security. Flowing from this the United Nations Development Programme's (UNDP) 1994 *Human Development*

Report argued that security should be expanded to include threats in seven areas namely, economic, food, health, environment, personal, community and political. This it refers to as human security.

Economic security requires an assured basic income for individuals, usually from productive and remunerative work or, as a last resort, from a publicly financed safety net. In this sense, only about a quarter of the world's people are presently economically secure. Food security which is closely related to economic security requires that all people at all times have both physical and economic access to basic food. According to the United Nations, the overall availability of food is not a problem, rather the problem often is the poor distribution of food and a lack of purchasing power. In the past, food security problems have been dealt with at both national and global levels. However, their impacts are limited. According to UN, the key is to tackle the problems that are related to access to assets, work and assured income.

Health Security aims to guarantee a minimum protection from diseases and unhealthy lifestyles. In developing countries, the major causes of death traditionally were infectious and parasitic diseases, whereas in industrialized countries, the major killers were diseases of the circulatory system. Today, lifestyle-related chronic diseases are leading killers worldwide, with 80 percent of deaths from chronic diseases occurring in low- and middle-income countries. According to the United Nations, in both developing and industrial countries, threats to health security are usually greater for poor people in rural areas, particularly children. This is due to malnutrition and insufficient access to health services, clean water and other basic necessities.

Environmental security aims to protect people from the short- and long-term ravages of nature, man-made threats in nature, and deterioration of the natural environment. In developing countries, lack of access to clean water resources is one of the greatest environmental threats. In industrial countries, one of the major threats is air pollution. Global warming, caused by the emission of greenhouse gases, is another environmental security issue.

Personal security aims to protect people from physical violence, whether from the state or external states, from violent individuals and sub-state actors, from domestic abuse, or from predatory adults. For many people, the greatest source of anxiety is crime, particularly violent crime. Community security aims to protect people from the loss of traditional relationships and values and from sectarian and ethnic violence. Traditional communities, particularly minority ethnic groups are often threatened. About half of the world's states have experienced some inter-ethnic strife. The United Nations declared 1993 the Year of Indigenous People to highlight the continuing vulnerability of the 300 million aboriginal people in 70 countries as they face a widening spiral of violence.

Political security is concerned with whether people live in a society that honours their basic human rights. According to a survey conducted by Amnesty International, political repression, systematic torture, ill treatment or disappearance was still practised in 110 countries. Human rights violations are most frequent during periods of political unrest. Along with repressing individuals and groups, governments may try to exercise control over ideas and information.

Since all these facets surround the human person therefore human security must necessarily embrace them and much more. For instance and in addition human security must embrace religion, human rights and all aspects of human well-being. It is within the context of the above that the two dominant religions in Nigeria are challenged to protect and promote human security and find solution to global warming which is a serious threat to humanity.

2.2 Global warming

Global warming may be referred to as the gradual increase of the temperature of the planet thus leading to its climate change. This accentuates the rise in sea level which culminates in flooding and melt of ice caps from polar and mountain glaciers. The impacts of rising sea level include loss of coastal ecosystems, flooding of cities, displacement of coastal inhabitants, and increased vulnerability to storm surges. Bangladesh, one of the poorest nations in the world is projected to lose 17.5 percent of its land if sea level rises about 40 inches (I metre). Tens of thousands of people are likely to be displaced, and the country's agricultural system will be adversely affected. Coastal flooding will threaten animals, plants, and fresh water supplies. The current danger posed by storm surges when cyclones hit Bangladesh is likely to increase .The Majuro Atoll in the Pacific Marshall Islands is projected to lose 80% of its land with a 20-inch (0.5m) rise in sea level. Many of the islands will simply disappear under the rising seas. A similar fate awaits other islands throughout the South Pacific and Indian Oceans, including many in the Maldives and French Polynesia. Coral reefs, which protect many of these islands, will be submerged, subjecting the local peoples to heightened storm surges and disrupted coastal ecosystems. Tourism and local agriculture will be severely challenged (Wiki Encyclopedia, 2010).

With sea level changes (Rodda, 1991: 13) due to global warming many of the world's richest and most heavily populated agricultural zones in low-lying lands will be flooded. Many beaches will be eroded and fresh water reserves can be in danger of pollution with salt water. The low-lying lands and heavily populated agricultural zones like the deltas of great rivers like the Ganges (Bangladesh), Nile (Egypt), the Mekong (Indo-China), the Indus (Pakistan), Limpopo river basin (Mozambique) (Houghton, 1994:93-96) and the Niger delta (Nigeria) are vulnerable. Other low-lying areas include Guyana, Papua New Guinea, India and Indonesia. Much of the Netherlands is 5 metres below sea level while Lagos is only 3 metres above sea level and many of the world's largest cities including Calcutta, Shanghai, Bangkok, Jarkata, Tokyo, London, new York, Miami and Venice (Foley, 1991:41).

Other impacts will include incidence of the outbreak of diseases such as yellow fever, dengue malaria and natural disasters. Cold winter weather reduces the spread of infectious diseases by killing infectious organisms and carrier species, such as mosquitoes. Warmer, wetter weather, on the other hand could increase the spread of malaria, dengue fever, and yellow fever. The possible increase in flooding and damage to water and sewage infrastructure can further encourage the spread of disease. Rising sea level requires many different local responses. For instance, urban areas on the U.S. coastline could be surrounded by rising sea water. Cities may require extensive infrastructure development to assure fresh water supplies, secure transportation, and protect people from flooding and storm surge. Sea walls can be built to protect cities and roads from rising seas. More robust building construction may also be required to withstand the increasingly intense storms that are likely to result from global warming. Fresh water supply is a concern as sea water penetrates ground water aquifers, which become brackish and less usable further inland.

At the end of the day, the cost of 'fight' back of sea level rise due to global warming will be usually huge as the experience of Nigeria in respect of the bar beach, in Lagos demonstrated. The low-lying lands and heavily populated agricultural zones are vulnerable to rising sea levels. Lagos is only three metres above sea level and shares the fate of similar cities like Calcutta, Shanghai, Bangkok, Jakarta, Tokyo, London, New York, Miami and Venice.

Nigeria cannot be indifferent to the effect of global warming as especially and effectively predicted by the 1988 Toronto Conference. Succinctly, these possible effects of global warming are:

i. Imperial human health and welfare;
ii. Diminish global food security, through increased soil erosion and greater shifts and uncertainties in agricultural production, particularly for many vulnerable regions;
iii. Change the distribution and seasonal availability of fresh water resources;
iv. Increased political instability and the potential for international conflict;
v. Jeopardise prospects for sustainable development and reduction of poverty;
vi. Accelerate extinction of animal and plant species upon which human survival depends and
vii. Alter yield, productivity and biological diversity of natural and managed ecosystems, particularly forests.

Already, Nigeria is experiencing coastal land sea water rise leading to serious fight of Fashola Administration of Lagos state to keep the sea in check by variety of strategies like sand filling and making concrete barriers across the beaches among others. Weather is becoming unpredictable making farming problematic. Already farming produce is not as buoyant as one would have wished to see. Lake Chad is typical of water masses that are drying up as a result of global warming and fresh water is not easily available as was the case some decades ago due to the drying up of sources of springs in some high lands. Besides, heat has become a common source of complaint by many citizens even if the country is situated in the tropics because the weather is becoming unusually hotter. Consequently, Nigeria cannot but team-up with international community to seek for solutions for global warming. Anything less than that threatens the very existence of the country and its people.

3. Some underling Christian and Islam theological principles

Here, it is necessary to establish some of the principles underlying Christian and Islamic theologies. Basic principles underlying these theologies consist of many elements. Some of these are first, that theology is the systematic study of God in himself and his relationships with all he has created both the living and non-living beings, and human response to the relationship is part of theology. Second, the tenets of faith affirm among other things that the entire universe including humanity is under the dominant domain of God, the creator of heaven and earth in spite of the seemingly powerful manifestation of the devil who will be humiliated ultimately at the end of time; truth and goodness that are attributes of God will finally triumph over evil. Third, and for Christianity human beings have distorted the pristine harmonious and beautiful state of affairs at creation (Gn. 1). By their actions violence, destruction and taking of life have become the hallmark of the landscape of human relations and environmental degradation as part of the consequences of human actions and activities that now endanger the very essence that sustains the ecosystem. In other words, human beings have introduced 'sin' into the world and by so doing damaged the natural delicate but harmonious relationships among all that inhabit it.

Fourth, Christian theology affirms that salvation of every person will primarily depend on his/her positive participation in the affairs of the world. But for both religions, there is reward and punishment after death for every person depending on how he or she lived his/her life. Lastly, Christian theology does not make appeal to a utopia government, not even a messianic one but only to a practice to be taken up over and over again; the practice being the only criterion of what is unacceptable namely that someone lacks something necessary (Matt. 25). In order that there will be no system that allows someone to lack basic necessities of life, the political restructuring of society becomes a task in which good theology must make an impact. In the final analysis, underdevelopment, violence, corruption and bad governance belong to the culture of death whereas God raises people to life. And in spite of the 'sins' of the world God is still at work in his world. Therefore, it is all those who undertake to create the conditions needed so that all may have life, life in abundance (Jn. 10: 10) that are the true sons and daughters of God. It is they that are proclaiming a living faith that builds peace and good governance, and promote and protect human security, and fight against global warming.

The basic assumption within Christian theologies is that the entire universe including humanity is under the domain of God, 'the creator of heaven and earth' and theology being the systematic and scientific study of God in himself and His relationship with all He has created. Good theology in this context means the authentic understanding of God and His relationship with His creation. Nigeria is a populous and religious country and its socio-political dominance in the continent is in no doubt. It enjoys natural resources, human and social capital. In spite of that, religious violence, ethnic clashes, corruption and almost collapse extended family and community values are a part of the national landscape. Religious fundamentalism of many practitioners of religion is primarily responsible for this state of affairs in the country, proliferation of Pentecostal and Charismatic Movements notwithstanding. It is on this premise that implication of fundamentalism is based with a view to moving Nigeria forward. There cannot be peace, good governance, progress, human security and wholesome environment without good theology.

Some religionists believe that religion is above criticism. Two myths seem to support this and keep faith beyond the fray of rational criticism. Ironically these myths seem to foster religious extremism and religious intolerance. The first one is that most religionists believe that there are good things that people get from religious faith for example, strong communities, ethical behaviour, spiritual experience that cannot be had elsewhere. The second is that many also believe that the terrible things that are sometimes done in the name of religion are the products not of faith per se but of human 'unredeemed' nature (base nature) forces like greed, hatred and fear for which religious beliefs are themselves the best (or even the only) remedy. The conflicts in Palestine (Jews v. Muslims), the Balkans (Orthodox Serbians v. Catholic Croatians; Orthodox Serbians v. Bosnia and Albanian Muslims), Northern Ireland (Protestants v. Catholics), Kashmir (Muslims v. Hindus), Sudan (Muslims v. Christians and Traditionalists), Nigeria (Muslims v. Christians), Sri Lanka (Sinhalese Buddhists v. Tamil Hindus), Indonesia (Muslims v. Timorese Christians) etc. arguably, are religious in character. Taken together, these myths seem to have granted perfect immunity to outbreaks of reasonableness in public religious discourse in Nigeria.

Many religious moderates have taken the apparent high road of pluralism, asserting the equally validity of all faiths, but in doing so they neglect to notice the unreconciliable sectarian truth claims of each. For instance, Mathews Ojo points this out among Christian Pentecostals and Charismatics when he stated that Each (Pentecostal and Charismatic Movements) has its uniqueness and often appeals to different classes of people. In fact, it is rather strange to group William Kumuyi' s strict holiness ethos with David Oyedepo's this-worldly concern, or associate Daniel Olukoya's MFM (Mountain of Fire and Miracles Ministries) steeped in the African cosmology of evil with the politically conscious Tunde Bakare's Latter Rain Assembly (Ojo, 2010: 25)

As long as many Christians believe that only their baptized will be saved on the Day of Judgement, they cannot possibly 'respect' the beliefs of others. Many Muslims and Jews generally take the same arrogant view of their own enterprises and have spent millennia passionately reiterating the errors of other faiths. In a place like Nigeria, a secular state where citizens could be stoned or beheaded for doubting the veracity of the Koran may never know peace and progress. There is clearly a sacred dimension to human existence and coming to terms with it could be the highest purpose of human life. However, the way the divine and secular is understood and lived makes the deference between 'hell' and 'heaven' (corruption, violence, assassination etc and human security and sane planet) in Nigerian context.

4. Some internal sources of religious conflict that threaten human security

4.1 Christianity

In the Christian context there are some inherent sources of conflict. One of them has to do with Dogmatism in the interpretation of the Christian Scriptures

4.1.1 Dogmatism in the interpretation of the Christian scriptures

One of the basic principles in the interpretation of the word of God in the contemporary world is to see every part of the sacred writings in its totality and the intention of God for all He has made. This principle is necessary because any person can interpret the tenets of Christian faith in order to produce a different picture of Christianity. There is no limit to what people can read out of the Scriptures if they are really determined. This has led to different interpretations of the Bible culminating in schisms, heresies, fundamentalism, religious ideologies, religious crises and violence. Nigeria has witnessed too many religious crises with their terrible consequences. One of the fundamental sources of religious crises in Christianity in Nigeria is 'fundamentalism' where the Scriptures could easily be manipulated to a definite purpose without a central authority to authenticate the 'right' or 'wrong' or 'bad' or 'good' of such an interpretation. Such an interpretation may lead to violence, passive acceptance of the faith (do not get involved in socio-political and economic life of the society i.e. governance), the gospel of prosperity that is, Christianity without the carrying of the Cross and health gospel (a Christian does not need a doctor since Christ has redeemed his followers from sin, poverty and sickness. What a Christian needs to do is to believe and claim his/her health by expressing that belief). This type of fundamentalism creates crisis for the society. For instance and as Matthews Ojo observes and he can be quoted extensively.

> Charismatic movements emerged in Nigeria on a reformist agenda in the early 1970s, to champion a new spirituality that is redemptive, prophetic, hopeful, and progressive, and which is anchored on the premise that a vibrant religion which was originally the capacity of re-ordering the society on a pietistic orientation. However by the late 1990s, this religion which was originally sectarian, promoting a strict holiness ethos, and on the fringe of the society, with membership numbering a few thousands, had moved to the centre stage, increased its membership to several millions, and had consequently dissipated its pietistic energy. Moreover, it has now accepted the values of larger society as it shifted to prosperity: mechanism for upward social mobility and materials aggrandizement. Hence, it is not strange to find the 'SU' of the 1970s who were then derided, becoming the founders of prosperity- type churches, riding the SUV vehicles,

and being portrayed as celebrities. Besides, men and women shifted their membership from Deeper Life Bible Church with its asceticism to Winners' Chapel with its theology that supports accumulation of wealth. In fact, many evangelists prefer to be known and addressed as bishops and archbishops with all the paraphernalia of power.

It is a paradox of religiosity and corruption in the society, of religious fundamentalism within a context of moral latitude and of inept governance, of a religion piety that revolves around self, while the collective is ignored, and of high religiosity within a failed state that is in constant motion but transition to nowhere in particular. William Kumuyi, Enock Adeboye, Mathew Ashimoliwo, Sunday Adelaja, Chris Oyakhilome and Daniel Odukoya - all Pentecostal pastors - could be exporting the Nigerian brand of Pentecostalism around the world, while their own country still needs the same panacea being offered freely to others. Indeed, both the Pentecostal religion and the personality steering it are important in any critical review of the dynamism of Pentecostalism in Africa. Whereas fourth century monastic movements in the Egyptian desert produced a crop of mystics and saints that distanced themselves from the larger society, and consequently kept to their saintliness, contemporary Pentecostalism reflects every kind of ambiguity that has characterized a nation in need of a saviour (Ojo, 2010: 5 and 6).

The following anomalies may now be treated to explicate some of the points raised above.

4.1.2 Passive acceptance of the faith

Some Christian denominations place great stretch on the end of the world (end-time). These proponents are keen to see 'biblical prophecies' being fulfilled everywhere. They dwell extensively on the apocalyptic (from the Greek word for 'vision') parts of the Bible namely the books of Daniel, Revelation, Ezekiel, Joel and Zephaniah. They give prominence to the appearance of plagues, famines and disasters of every kind. For them, all kinds of hardship and deprivation as foretold are ordained by God; they are by definition unavoidable.

This way of thinking encourages passivity and resignation especially among the poor and marginalised. Because the end is supposedly near, nothing else matters except preaching the Gospel, or fulfilling the great Commission (Go out to the whole world to proclaim the Good News (Mt. 28: 19)). Evangelism and winning souls for the Lord are all that matters; anything else is a distraction. The implication of this orientation is that it diverts attention from the socio-economic and political realities of the country by teaching that there is not time to do anything about them. They forget that Saint Paul 'changed' his mind when he said to his audience that any person who does not work must not eat. This was in the context of the early Christians waiting for the second coming of Christ which did not come. They were doing nothing while waiting for the second coming of Christ.

4.1.3 The gospel of prosperity

According to the adherents of this orientation, God has met all the needs of human beings in the suffering and death of Christ, and every Christian should now share the victory of Christ over sin, sickness and poverty. Therefore a believer has a right to the blessings of health and wealth won by Christ. And s/he can obtain these blessings merely by a positive confession of faith. S/he can 'claim' the benefits. The prosperity gospel phenomenon, in general, is arising in situations where the majority of people experience acute-poverty and even desperation while a minority enjoy the benefits of the consumer society. Some other characterises are well put by Shorter and Njiru:

- There is also a widely felt need for a cathartic experience, enabling individuals to rid themselves of guilt-feelings and other moral burdens, often related to the experience of poverty. It is a desire for social healing, as much as for inner healing. Salvation is basically a subjective assurance of deliverance. God is a 'problem-solving God', offering immediate relief from present affliction.
- Paradoxically, members of new religious movements are ready to surrender material possessions. This is a calculated sacrifice which sows the seed of future prosperity. It is a 'cargo-cult' faith.
- Often there is an Adventist or millenarian orientation in the new religious movements .. There is also a literal expectation of miracles, cures and other marvels. Finally, and most dangerous of all, there is a mystique of leadership, of accepting the unquestioned authority of an evangelist/prophet, who possesses special gifts and who is the vehicle for a divine message. Many leaders of new religious movements also seem to regard their operation as a business, as much as a religious avocation (Shorter and Njiru, 2001: 13).

This phenomenon makes wealth or prosperity a simple matter of faith and generous giving to a preacher. The more you give to God the better your benefit is the axiom. The orientation preaches personal salvation, a deliverance from ill health and poverty. The rich are not blamed for oppressing the poor. On the contrary they are evidence of God's blessing and are to be emulated. Poverty is said to be due to sin and lack of faith. It can be overcome by prayer, fasting and generous donations to the evangelist. Prosperity Gospel simply ignores the political and economic reasons for so much poverty that is around in the country. Dependent economics, fluctuating prices, overgrazing, global warming, environmental degradation, overpopulation, corruption, mismanagement, destabilisation and mistaken priorities and so on are empty phrases without relevance. Prosperity gospel like the passive acceptance of faith diverts attention from political and economic issues and working for peace and reconciliation. It leads to no social activity whatsoever. According to this gospel, only Christians matter. It is the Christian that will have job, food, education and good life. Only Christians are to be successful. So the plight of other members of the nation or the state or society or the situation of the general population is of no consequence to Christians. The charitable thing the Christian can do to the deprived unbelievers is merely to convert them so that they can prosper as well.

The gospel of prosperity tells people that material prosperity will be provided by a 'miracle-working God' or through the' law of sowing' - the more you give to the preacher the more God will give you. Thus the gospel of prosperity functions by diverting all attention from the economic and social situation. As far as the adherents of this gospel are concerned daily critiquing the reasons for why they are poor is not a virtue. The key biblical texts they often quote are:

- Mark 11: 23-24 ‘Whatever you ask for in prayer, believe that you receive it, and you will”
- Deuteronomy 5: 28-30 'All these blessings shall come upon you ... if you obey the voice of the Lord your God'.
- Malachi 3: 8-11 'Bring the full tithes into the storehouse ... and thereby put me to the test if I will not open the windows of heaven for you and pour down for you an overflowing blessing'.
- Mark 10: 19-30 'There is no one. who will not receive a hundredfold here in this life'.
- Philippians 4: 19 'My God will supply every need of yours according to his riches in glory in Christ Jesus'.

4.1.4 Health gospel

A corollary to the prosperity gospel is the health gospel. The basic principle of the health gospel is that Jesus redeemed us from sin, poverty and sickness. He bore our sickness (Isaiah 53: 4-5). All a Christian has to do is to believe, and claim his or her health by expressing that belief. For the proponent of this gospel, all sickness comes from Satan and that no Christian should be sick - to be sick is to lack faith or to have sinned. For them a Christian has no need of doctors. The difficulty about this orientation is its exclusive dependence on God's miraculous intervention at healing people. Yes, God is able to effect any healing but sometimes He does not heal for whatever reason.

The implication of this gospel on socio-political life is profound. It does not promote the necessary commitment to the eradication of diseases, child immortality and increase life expectancy of citizens of any nations. It is not its concern that any community is mobilised for child inoculation and provision of clean water. As far as the gospel is concerned it claims that it is faith that will turn around the state of crisis that our health institutions are experiencing. This type of Christianity leaves everything up to God and focuses only on the health of an individual. It is not capable of analysing the socio-political causes of deteriorating health services due to economic mismanagement, corruption, diverting available resources to military spending or to prestige projects. The gospel does not challenge oppressive and dehumanising structures that promote culture of death. Their followers are told that they are in the world but not of the world (Jn 15: 19). That may be true but this passage of the bible has been misinterpreted. Our world is God's world. All God created was good (Genesis 1) and so Christians must so believe. Arguably, the manifestation of evil forces is visible in the world but still it is God's world. Therefore forces of evil (sickness and diseases) must be radically confronted.

Thank goodness, in recent years many Christians have come to understand how 'the world' or 'human society' operates particularly through the influence of Social Sciences. Authentic Christians have come to realise that a lot of human suffering is caused by political, cultural and economic systems. These systems are not created by God, nor are they part of the nature of things even though they may appear that way. Political and economic systems are created by human beings or rather by particular group of human beings largely for their own benefits. Consequently, Christians realise that their duty to help the suffering extends beyond offering relief to suffering individuals; it extends to improving the system that inflicts so much hurt and damage in the first instance. To love one's neighbour means more than providing relief to those who are suffering, although this is important; it means confronting dehumanizing systems and working towards changing them. This implies entering 'the world' and using one's influence positively in society generally. Commerce, politics, religion, education, entertainment industry, technology, computer are not 'Satan's world' or systems or creation as some Christians would want people to believe. They are 'instruments' in our hands to be used in order to improve the world. A part of a Christian's task is to be at whatever human needs are to be met (Matthew 25: 31-36) and to cooperate readily with all those who are already meeting these needs. Establishing Churches or sects or fellowships or ministries as alternative society of human beings is misleading and unchristian. Rather Churches and sects or fellowships are expected to be the leaven of the society (You are the light of the world, you are the salt of the earth Mathew 5:13-16). It is the Lord who created the world and what God created is good!

4.1.5 Preaching 'bad theology' (fundamentalism)

Unfortunately, the former President Olusegun Obasanjo in the Nigerian context saw his presidency as God ordained and his battle of administration and policies as God's. The Nigerian Tribune of 11/1/2004, on the starting of a four-day warning strike called by the Nigeria Labour Congress (NLC) to protest increase of prices of petroleum products beautifully captured the divine mission syndrome when it stated of the president as:

- My presidency, God-ordained - Obasanjo (caption)
- President Olusegun Obasanjo has described his presidency as God-ordained and divinely directed, affirming that his regular prayer request is that 'Nigeria should be great.'
- The President said he had faith in the corporate survival and greatness of Nigeria, stating that of the two prayer requests he made to God when a pastor called him on his mobile phone, 'God has answered my prayer that Nigeria should be great. The second one is that I asked God to make me handsome, but God immediately answered no,' the President said while thanking God for making his son very handsome.
- President Obasanjo was speaking at the Aso Rock chapel after he was given an award of a great leader by the Lagos State chapter of the Pentecostal Association of Nigeria. The President, who was presented a plaque and two giant paintings, said he was touched that at a time like this in the life of Nigeria, he was recognised by men of God for his activities as a leader.
- 'I am greatly touched by this honour and by the timing. Actions like this encourage me. I thank you for recognising our good intention as a leader,' the President said.
- Presenting the award, a representative of Lagos PAN, Pastor Okonkwo and two of his other colleagues described the President as a tool of God for the reformation of Nigeria. Lauding the President for not bending in the face of crisis, Pastor Okonkwo said 'God sent the President to reform Nigeria and right the wrong of the past, stating that the President's reforms have divine inspiration' (The *Nigerian Tribune*, 11/10/04: 3)

President Olusegun Obasanjo is not alone. His counterpart, the formal President George W. Bush was equally fundamentalistly disposed as shown in the following passage.

> Many prominent Republicans belong to the Council for National Policy, a secretive Christian pressure group founded by the fundamentalist Tim LaHaye (co-author of the apocalyptic "Left Behind" series of novels). This organisation meets quarterly to discuss who knows what. George W. Bush gave a closed-door speech to the council in 1999, after which the Christian Right endorsed his candidacy. Indeed, 40 percent of those who eventually voted for Bush were white evangelicals. Beginning with his appointment of John Ashcroft as his attorney general, President Bush found no lack of occasions on which to return the favour. The departments of justice, Housing and Urban Development, Health and Human Services and Education now regularly issue directives that blur the separation between church and state. In his "faith-based initiative" Bush has managed to funnel tens of millions of taxpayer dollars directly to church groups, to be used more or less however they see fit (Harris, 2006: 155).

In Christian theology, it is not nation-states that get rid the world of evil- they are too often caught up in complicated webs of political power, economic interests, cultural clashes and nationalist dreams. The confrontation with evil is a role reserved for God and for the people of God when they faithfully exercise moral conscience. But God has not given the responsibility for overcoming evil to any nation-state. To confuse the role of God with that of President Obasanjo as a politician is a serious theological error that some might say borders on idolatry or blasphemy (*South Letter*, Issue 39: 51).

4.2 Islam

There is a problem of fundamentalism with Islam precisely because of the vision of life that it prescribes to all Muslims in the Quran and further elaborated in the literature of the hadith which recounts the sayings and actions of the Prophet.

In an examination of Islam and its relations with non-Muslims, Quixotic Media, the producers of the documentary film of Islam: What the West Needs to Know (2007), the following issues in six parts based on the canonical texts of Islam, its primary resource material are paramount and revealing.

1. **There is no God but Allah and Muhammad is his Prophet**: In the first part, it is argued by various commentators that Islamic violence stems from the teachings and examples of Muhammad and that the Qur'an prescribes and sanctions violence against non-Muslims.
2. **The Struggle**: In the second part, Walid Shoebat defines the word "Jihad" to mean the struggle to impose Allah's will over the earth, resulting in holy war against the non-Muslim world to bring it under the rule of Islam.
3. **Expansion:** In the third part, Bat Ye'or describes the expansion of Islam through conquest and presents historical evidence of enslavement and massacres of Christians, Jews, Zoroastrians and Hindus by Muslim invaders.
4. **War is Deceit**: In the fourth part, Robert Spencer and Serge Trifkovic discuss the Islamic principle of Taqiyya, Islamic dissimulation, which, the lecturers argue, enjoins Muslims to deceive non-Muslims to advance the cause of Islam.
5. **More than a Religion**: In the fifth part, the documentary asserts that "Islamic law governs every aspect of religious, political, and personal action, which amounts to a form of totalitarianism that is divinely enjoined to dominate the world, analogous in many ways to Communism".
6. **The House of War**: The final part covers the division of the world into Dar al-Islam ("the house of Islam" or "the house of peace"), the land governed by the Islamic law, and Dar al-Harb ("the house of war"), the land of non-Muslims. According to the film, Muslims are enjoined to bring the Dar al-Harb under the control of Islam. The film argues, "Muslims in Western nations are called to subvert the secular regimes in which they now live in accordance with Allah's command." (Wikipedia, the free encyclopaedia)

In the Quran, the following are equally revealing.

1. "They wished that you reject Faith, as they have rejected (Faith), and thus that you all become equal (like one another). So take not Auliya (protectors or friends) of them, till they emigrate in the way of Allah (to Muhammad). But if they turn back (from Islam), take (hold of) them and kill them wherever you find them, and take neither Auliya (protectors or friends) nor helpers from them."(Surah 4:89)
2. "(Remember) when your Lord revealed to the angels, "Verily, I am with you, so keep firm those who have believed, I will cast terror into the hearts of those who have disbelieved, so strike them over the necks, and smite over their fingers and toes. "(Surah 8: 12).
3. "Jihad is ordained for you (Muslims) though you dislike it, and it may be that you dislike a thing whicgood for you and that you like a thing which is bad for you. Allah knows but you know not AI-jihad (holy fighting) in Allah's cause (with full force of numbers and weaponry) is given utmost importance in Islam and is one of its pillars which it stands. By jihad, Islam is established, Allah's word is made superior, (His word being La ilaha which means none has the right to be worshipped but Allah), and His

Religion (Islam) is propagated. By abandoning Jihad (may Allah protect us from that) Islam is destroyed and Muslims fall into an inferior position; their honour is lost; their lands are stolen, their rule and authority vanish. Jihad is an obligatory duty in Islam on every Muslim, and he who tries to escape from this duty, or does not in his innermost heart wish to fulfil this duty, dies with one of the qualities of a hypocrite."(Surah 2: 216)

4. Fight against those who believe not in Allah, nor the last day, nor forbid that which has been forbidden by Allah and his messenger (Muhammad) and those who acknowledge not the religion of the scripture (Jews and Christians) , until they pay the Jizyah [a tax levied from the people of the scripture (Jews and Christians who are under the protection of Muslim government) with willing submission, and feel themselves subdued."(Surah 9: 29) "0 you believe! Fight those of the disbelievers who are close to you, and let them find harshness in you; and know that Allah is with those who are Al-Muttaqun (the pious- see V 2:2) believers of Islamic monotheism who fear Allah much (abstain from all kinds of sins and evil deeds which he has forbidden) and love Allah much (perform all kinds of good deeds which he has ordained)" (9: 123) those who believe, fight in the cause of Allah, and those who disbelieve, fight in the cause of Taghut (Satan). So fight you against the friends of Shaitan (Satan). Ever feeble indeed is the plot of Shaitan (Satan)."(Surah 4: 76).
5. "Let those (believers) who sell the life of this world for the Hereafter fight in the cause of Allah, and is killed or gets victory, we shall bestow on him a great reward" Surah 4: 74). "That you believe in Allah and his Messenger (Muhammad) and that you strive hard and fight in the cause of Allah with your wealth and your lives; that will be better for you, if you but know!" (if you do so) He will forgive you your sins and admit you into Gardens under which rivers flow, and pleasant dwellings in And (Eden) paradise; that is indeed the great success"(Surah 61 : 1112). "Have you seen those who were told to hold back their hands (from fighting) and perform As-Salat (lqamat-as-Salat), and give Zakat, but when the fighting was ordained for them, behold! A section of them fear men as they fear Allah or even more. They say: "Our Lord! Why have you ordained for us fighting? Would you have granted us respite for a short period? Say: "short is the enjoyment of this world. The Hereafter is (far) for him who fears Allah, and you shall not be dealt with unjustly even equal to scalish thread in the long slit of a date-stone." (Surah 4: 77).
6. "So, when you meet (in fight-jihad in Allah's cause) those who disbelieve, smite (their) necks till when you have killed and wounded many of them, then bind a bond firmly (on them, i.e. take them as captives). Thereafter (is the time) either for generosity (i.e. free them without ransom), or ransom (according to what benefits Islam), until the war lays down its burden. Thus (you are ordered by Allah to continue in carrying out jihad against the disbelievers till they embrace Islam and are saved from the punishment in the hell-fire or at least come under your protection), but ifit had been Allah's will, He will never let their deeds be 10st."(Surah 47: 4) "Then when the sacred Months (the 1st, i\ 11 the and lih months of the Islamic calendar) have passed, then kill the Mushrikun (see V2: 1 05 neither those who disbelieve among the people of the scripture (Jews and Christians) nor Al-Mushrikun (the idolaters, polytheist, disbelievers in the oneness of Allah, Pagans, etc'). whenever you find them, and capture them and besiege them, and lie in wait for them in each and every ambush. But if they repent and perform As-Salat (lqamat-as-Salat, and give Zakat, then leave their way free. Verily, Allah is oft-forgiving, most Merciful." (Surah 9: 5)

The roots of Islamic fundamentalism go back to the first centuries of Islam. But Islamic fundamentalism in its current context, theory, and power emerged after Ruhollah Khomeini came to power in Iran in 1979. The Khomeini regime transformed the idea of creating a global Islamic rule from an unachievable ideal to an achievable goal by many fundamentalist groups, and it also gave these groups global backing. For instance, the foreword of the regime's constitution reads, in part:

> Given the context of Iran's Islamic Revolution, which was a movement for the victory of all the oppressed over the oppressors, it provides the ground for continuation of the revolution inside and outside the country, specifically in spreading international links to other Islamic and people's movements, tries to pave the way for the creation of unique global ummah so the continuation of the struggle for the salvation of deprived and suffering nations can be settled.

Another part of the foreword, under the headline "Ideological Army," reads:

> The Army of the Islamic Republic and the Revolutionary Guards Corps carry not only the duty of protecting the borders but also ideological duty, i.e., Jihad for God and struggle to spread the rule of God's law in the world.

The Eleventh Act of the constitution reads:

> The government of the Islamic Republic of Iran is obligated to base its general policy on the coalition and unity of the Islamic nations and to try to fulfil the political, economic, and cultural unity of the Islamic world.

From the point of view of Islam, the world is divided into the 'House of Islam' and the 'House of War', and the latter designation indicates how many Muslims believe their differences with those who do not share their faith will be ultimately resolved. This implies that the only future devout Muslims can envisage is one in which all infidels have been converted to Islam, subjugated or killed. The tenets of Islam simply do not admit of anything but a temporary sharing of power with the 'enemies of God'.

The feature of Islam that is most troubling to non-Muslims especially in Nigeria and which apologists for Islam do much to obfuscate is the principle of jihad. Literally, the term can be translated as 'struggle' or 'striving' but it is generally rendered in English as 'holy war' and this is no accident. While Muslims are quick to observe that there is an inner (or 'greater') jihad, which involves waging war on one's own sinfulness, no amount of casuistry can disguise the fact that the outer (or 'lesser') jihad that is war against infidels and apostates is a central feature of the faith. Armed conflict 'in defence of Islam' is a religious obligation for every Muslim man. The duty of jihad is an unambiguous call to world conquest.

The presumption is that the duty of jihad will continue, interrupted only by truces, until all the world either adopts the Muslim faith or submits to Muslim rule'. There is just no denying that Muslims expect victory in this world, as well as in the next (Harris, 2006: 111).

The same idea is witnessed to and elaborated upon by the literature of the Hadith for instance,

i. Jihad is your duty under and ruler, be he godly or wicked.
ii. A single endeavour (of fighting) in Allah's Cause in the forenoon or in the afternoon is better than the world and whatever is in it.
iii. Nobody who dies and finds good from Allah (in the Hereafter) would wish to come back to this world even if he were given the whole world and whatever is in it, except the martyr who on seeing the superior of martyrdom, would like to come back to the world and get killed again (in Allah's Cause).

iv. He who dies without having taken part in a campaign dies in a kind of unbelief. Paradise is in the shadow of swords.

In the Muslim perception, conversion to Islam is a benefit to the convert and a merit in those who convert him. However, as Lewis rightly observed, in Islamic law, conversion from Islam is apostasy that can carry a capital offence for both the one who is misled and the one who misleads him. Furthermore, if a Muslim renounces Islam, even if a new convert reverts to his previous faith, the penalty is death (Lewis, 2003: 55). Another area of concern has to do with why should Islam cannot be critiqued or 'criticized'. Criticising Islam is currently a taboo; its religious beliefs are simply beyond the scope of rational discourse. Criticising a Muslim's ideas about God and the afterlife is thought to be impolite in a way that criticising his ideas about physics or geography is not. For instance, when a Muslim suicide bomber obliterates himself along with a score of innocents on a Jerusalem or Iraqi street, the role that faith played in his actions is invariably discounted; his motives must have been political, economic or entirely personal. Without faith, desperate people would still do terrible things but when do whatever terrible things they do in the name of faith, they are exonerated!

5. Human security and global warming: Challenges before the two religions in Nigeria

With the magnitude of poverty steering the nation in the face, the unacceptable high level of corruption, an unpardonable mismanagement in the high places and lack of commitment to sustainable development, fundamentalist interpretation of sacred scriptures ('bad' theology) has no place in a secular society like the one of Nigeria. In the Nigerian situation, what is paramount is good governance, the bed rock of progress and development. But it will not just happen; it has to be worked for, nurtured and developed. It calls for hard work, political engineering, courage and eternal vigilance. A theology that does not promote peoples active and progressive participation in socio-cultural, political and economic life of a nation is not worth the name. It is an aberration. Human person is central to good governance, sustainable development and religious fundamentalism does not seem to recognize or acknowledge this fact. Its understanding of human person in its relation to the world is superficial lacking authenticity and yet the consequences of the activities of fundamentalists on their society can be devastating. The entire nation may be engulfed by the flames of hatred, destruction of life and property and terrorism. Ultimately, religious fundamentalism, one of the major basis of religious conflict in Nigeria is a danger both to the wellbeing of people, progress and development of nations at large whereas good theology promotes life, affirms the supremacy of God in all things and establishes a harmonious and sustainable relationship among the living and non-living beings that inhabit the planet earth. Therefore, religion has a major role to play in the development and promotion of human security of the country bearing in mind that Nigeria is a 'religious' country.

Within the context of the above, the first challenge facing the two dominant religions in Nigeria in respect of human security and global warming is how to confront religious fanaticism in such a way that radical understanding of true religion will be enthroned in the praxis of all the adherents. Second, is how to support the government to pass laws and increase funding for counter religious crisis programmes. One of such legislative mechanisms will provide specific measures that certain government departments or agencies will need to take to prepare the country for the consequences of a fight against religious crisis. The purpose of this is to reduce the religious bodies' vulnerabilities to threats, preventing and deterring them responding to them and managing the consequences of religious crisis attack

Generally good governance of any state calls for eternal vigilance of the stake holders because their existence and security which includes territorial defence of the country, the physical, social and psychological quality of the lives of the citizens and society are at stake. The citizens need the necessary social, economic and political conditions conducive to happiness and relative prosperity that ensure their tranquillity and well-being. The government in the country cannot promote good governance with all its good intension because the system that it is running is systemically corrupt. The foundation of political institutions, security instruments and the economy of the country must be strengthened and internal injustices including ethnic social inequality challenged. It is good governance that can make all these happen.

The two religions can always call attention of the government to advance effective democracies as the long-term antidote to religious fundamentalism, to prevent attacks by religious 'terrorists', to deny terrorists the support and sanctuary of rogue political and religious rogue individuals, to deny terrorists control of any nation or group or individual as a base and launching pad for terror, and to lay the foundations and build the institutions and structures needed to carry the fight forward against religious crisis (terror) and help ensure ultimate success. Also the two religions in collaboration with the government can commit its vast resources of human beings and material towards achieving the dissolution of pervasive religious fanatical support that the religious fanatics currently receive from abroad and at home, helping them to rebound rapidly and consistently from the onslaught of the military forces. This can be done by freezing the assets of fanatically-affiliated persons and organisations and denying them safe haven. Furthermore, apprehended suspects are to be tried and punished according to the law if found guilty.

What is good for the human security and growth of Nigeria is a competent president with integrity and moral standing. A leader that should have the will and moral stamina to lead with courage and right the wrong issues that are stagnating the development of the country. Nigeria has all it takes to compete favourably with any developed nation on the surface of this earth. It has human and natural resources and what remains is a group of competent and sound leaders at all levels of governance. This is where the two dominant religions in the country come in. To start with, they are to lead by the good examples of the exemplary lives of both adherents and religious leaders. They must preach both by word of mouth and by their activities. They must always challenge constructively the government at all times along the lines of positive value of hard work, sincerity of purpose, corporate responsibility for the community and the nation at large, love for peace and provision of resources for and empowerment of the 'wretched of the earth'. When necessary, the government may be made to subsidise for food of the very poor and elderly because of the effect of extreme weather conditions on agricultural products and promote transportation and housing for the same category on the account of the devastating consequences of global warming.

Furthermore, Nigeria acclaims to be a secular state and in a civil society at the minimum, ideas can be criticized without the risk of physical violence. Unfortunately in the country certain things cannot be said about a book and a religion because such utterances carry the penalty of death, torture or imprisonment. In a country where rule of law is discriminately used, there cannot be human security. The two religions have a fundamental role to play in this context. Specifically, the two religions must contribute to human security in Nigeria by undergoing a radical transformation of religious fanaticism that often manifests in acts of terrorism. Unless the fanatical aspect of practice of religions can be reshaped to accommodate the rights of other Nigerians it will be difficult to experience human security and have time to fight meaningfully forces that are promoting global warming, the enemy of the planet entity. On the part of Christians, a Concordat with the state could be entered in by which the secular character of the country will be protected and if broken international community could intervene to restore the status quo.

Education prepares an individual to make a contribution to society (Bozimo, 2000; Idowu, 2001; Esere, 2004). Also, as Robert Bogan rightly stated, education is for defence of rights (Pontifical Commission Justitia et Pax, 1975: 38). Education may have forms namely formal, informal and non-formal (Fafunwa, 1974) but its primary aim is not just only to prepare a person for a particular kind of job or profession as important as that may be, but also the development of personality which involves the development of both the intellect and the will. In other words, education is the total process of human learning by which knowledge is imparted, faculties trained, will empowered and skills developed. Schooling is only one form in which education is provided. In the context of human security and global warming, the two religions have a role to play. By taking advantage of their sermons, public declarations, social ministries of schools, publications, hospitals, agriculture, banking and election monitoring, the two religions can exercise tremendous influence on the citizenry. Workshops on themes like 'One Planet one future' and 'Contextual Theology', if properly undertaken can change peoples' perspectives of their reality. In addition, the country has endorsed many international environmental treaties and protocols and has also established National Environmental Policy which has human security at its core. The goal of the policy is to achieve sustainable development in Nigeria and in particular to:

a. Secure for all Nigerians a quality of environment adequate for their health and well-being;
b. Conserve and use the environment and natural resources for the benefit of present and future generations;
c. Restore, maintain and enhance the ecosystems and ecological processes essential for the functioning of the biosphere to preserve biological diversity and the principle of optimum sustainable yield in the use of living natural resources and ecosystems;
d. Raise public awareness and promote understanding of essential linkages between environmental and development and to encourage individual and community participation in environmental improvement efforts; and
e. Cooperate in good faith with other countries, international organisations/agencies to achieve optimal use of transboundary natural resources and effective prevention or abatement of transboundary environmental pollution.

The strategies for implementation are human population; land use and soil conservation; water resources management; forestry, wildlife and protected natural areas; marine and coastal area resources; sanitation and waste management; toxic and hazardous substances, mining and mineral resources; agricultural chemicals; energy production; air pollution; noise; working environment; settlements, recreational space, greenbelts, monument and cultural property. These strategies are to lead to:

a. The establishment of adequate environmental standards as well as the monitoring and evaluation of changes in the environment;
b. The publication and dissemination of relevant environmental data;
c. Prior environmental assessment of proposed activities which may affect the environment or the use of a natural resource (*National Policy on the Environment*, 1989: 3-6).

In the context of environment and global warming the two dominant religions through their adherents and leaders could collaborate with the government and internationalise and practise the environmental policy of the federation. By so doing they will be equally be fighting the scourge of global warming and promoting human security. In order that this may take place all Nigerians are called to conversion so that truncated religious attitude of many citizens could be more embracing. All Nigerian religious people must care for creation because it is good, because its goodness is embedded in its wholeness and because humankind can only live well and meaningfully within this wholeness. They are dependent on the rest of nature and of the interdependencies which exist among different forms of life, between systems and the physical and chemical environment which encircles life on the Earth and between themselves and the rest of the universe.

The expression of God's love and beauty is meant to be shared by all life, human and non-human. Religious Nigerians must cherish the whole of creation, not for their sake alone, but for its sake and for God's sake, for God made and loves it all. Christians share in the cross of Jesus, dying to old and self-centred selves. They rise with him to newness of life. The whole creation will share in "the glorious liberty of the children of God" (Rom. 8: 21)

From the experience of Japan, Nigeria can learn. Japan experienced an extreme scarcity of the commodities for daily life especially during and immediately after the Second World War. There were hardly any noticeable deeds of 'kindness' seen among the people and blatant egoism was rampant everywhere. It was this reality that gave birth to Small Kindness Movement in the Japanese society. The objectives of the movement include:

a. Let people never fail to greet each other in the morning and in the evening.
b. When spoken to, people are to give a reply with a clear voice.
c. People are to accept kind deeds of others with gratitude from the bottom of their hearts and say 'Thank you'.
d. When someone says 'Thank you', the reply should be 'You are welcome'.
e. People should try not to throw away litter in the streets.
f. In trams and buses, people are to give seats to old people and mothers holding babies.
g. When noticing someone in trouble, people are to try to help.
h. People are to refrain from causing nuisance to others.

The contribution of the movement was invaluable. Among others, it transformed the relationship between individual identity and social harmony giving way to the expression 'You go your way, I go mine, but let's keep friendly' (Seiji Kaya, 1986: 387-390).

In Nigerian situation, social movements could do similar things to bring about the development and stability of the country, help Nigeria fight insecurity and global warming. But then the challenges must be adequately articulated and addressed with sincerity, passion and vigour within the context of the national history. The two religions may be instrumental.

6. Conclusion

At independence in 1960, Nigeria was enthusiastic to join the civilized world where democratic principles are the order of the day. It needed the necessary social, economic and political conditions conducive to promote human security of its citizens and wholeness of the planet. However, today at fifty, the journey to the promise land is still very far. Apart from poor governance, religious violence, academic and non academic workers' strikes, professional workers' strikes, unemployment, militancy, environmental degradation and corruption dominate the Nigerian landscape.

For the two religions of Islam and Christianity in Nigeria the planet earth is God's creation and a gift to humanity and must be related to accordingly. Therefore artificial distinction between the world of religion and the secular world must not exist. Affirmation of faith in the life after affirms this since what is done here on earth will determine the type of reward (heaven of hell) that will be meted to every individual. In this wise, working for human security and fighting global warming is mandatory on every adherent of the two religions. Failure to comply implies not giving glory to God the author of all. Unfortunately, religious fanaticism (fundamentalism) for instance has derailed the purposeful road to the promotion of human security and the fight against global warming giving way to hate, discrimination, destruction of life and property and environmental degradation culminating in global warming, a serious threat to the very existence of all lives and the planet.

The challenges before the two religions must include among others repositioning the radical tenets of the religions that will overcome their fanaticism that leads to destruction of life and property and endangers human security. Others will include insistence on the secular character of the Nigerian state, rediscovering the potent of radical education as opposed to 'banking system of education' (Freire, 1970, 1970b) in order to empower its adherents and promotion of good governance of the country. In relation to environmental degradation, the Ministry of Environment is established to protect and nurture the environment by upholding the precautionary principles of Rio Conference of 1992 for instance, the mandatory Environmental Impact Assessment report before the execution of any major developmental project, last Saturday environmental day of some States and ban on forest fire burning especially in preparation for land farming. Nigeria still has a long way to go in this respect and it is in this context that good leadership becomes paramount.

Good governance/leadership, the bed rock of progress and development will not just happen; it has to be worked for, nurtured and developed. It calls for hard work, political engineering, courage and eternal vigilance of the stake holders because their existence and security which includes territorial defence of the country, the physical, social and psychological quality of the lives of the citizens and society are at stake. The two religions and their leaders must collaborate with the government in addressing them seriously.

Finally, it may be affirmed that unless the government does remedy most if not all the anomalies (inadequate infrastructure, insecurity, violation of human rights, unemployment of the youth and so on) of the land on the one hand, and establish an enabling environment where conversion of heart of every citizen takes place on the other hand, and encourage collaboration of the two dominant religions, it is difficult to see how Nigeria can promote human security and effectively fight global warming locally and internationally.

7. References

Adinna, E.N. (2001), *Environmental Hazards and Management*, Snaap Press, ISBN 978-049-335-2, Enugu.

Booth, K and Smith, S. (eds.), (1995) *International Relations Theory Today*, Polity Press, Cambridge.

Bozimo, G. (2000) 'Economic Participation Decision-making and Women in Education: A Conceptual Analysis' *Journal of Women in Academic, 1* (1), 176-182.

Brandt, W., (1980) Independent Commission on International Development Issues, *North-South: A Programme for Survival*, Pan Books, ISBN 0 330 26140 1, London.

Bretherton, C. and Ponton. G (eds.), *Global Politics: An Introduction,* Blackwell, ISBM 0-631-19565-3, Oxford, 1996.

Compbell , J, '2011: Trouble Looms in Nigeria' *Sunday Sun* August 22, 2010:56.

Esere, M. O. (2004) 'Globalisation and the Challenges of Human Development in Nigeria: The Counsellor's Factor' *The Nigerian Journal of Guidance and Counselling, 9*(1), 71-84.

Fafunwa, B. (1974) *History of Education in Nigeria ,* George Allen & Unwin, London.

Federal Republic of Nigeria (1989), *National Policy on the Environment,* Federal Environmental Protection Agency.

Foley, G. (1991) *Global Warming,* Panos Publication, London.

Freire, P. (1970) *Pedagogy of the Oppressed,* Seabury, ISBN: 0-8164-9132-1, New York.

Freire, P. (1970 b) *Cultural Action for Freedom,* Monograph Series, No. 1 Cambridge Massachusetts: Harvest Educational

Review and Centre for the Study of Development and Social Change.

Harris, S., (2006) *The End of Faith Religion, Terror, and the Future of Reason,* Cox & Wyman, ISBN 13: 978-0-7432-6809-7, Reading

Houghton, J., (1994) *Global Warming,* Loin Publishing, Oxford.

Idowu, A. I. (2001) 'Education for National Reconstruction, Reformation and Development'. Being text of a lead paper presented at the first National Conference of the School of Education, College of Education, Oro, Kwara State, May 1- 4, 2001.

Ijere, J.A. (1989) 'Patterns of Rural Development in Borno State' in K. Swindell (Ed.), *Inequality and Development,* Macmillan Publishers, Hongkong.

Kaya, S. 'Small Kindness Movement' in Pauling, L., Laszlo, E and Yoo, J. Y. (eds.) (1986) *World Encyclopedia of Peace Volume 2.* Pergamon Press, ISBN 0-08-032685-4, Oxford.

Lewis, B., (2003) *The Crisis of Islam: Holy War and Unholy Terror.* Modern Library, New York.

Makinda, S, (1996) 'Sovereignty and International Security: Challenges for the United Nations' in *Global Governance* Vol. 2 No 2 May-August, ISSN 1075-2846.

Natufe, O. 1 (2001) 'The Problematic of Sustainable Development and Corporate Social. Responsibility: Policy Implications for the Niger Delta'. A Conference Paper delivered at the Second Annual Conference and General Meeting of Urhobo Historical Society, Denville, New Jersey, November 2-4, 2001: 2)

Ojo, M. (2010) *Of Saints and Sinners: Pentecostalism and the Paradox of Social Transformation in Modern Nigeria,* Inaugural Lecture Series 227, Obafemi Awolowo University Press, Ile-Ife, Nigeria.

Omorogbe, J. I. (2003) 'Intellectual Giants But Moral Dwarfs: An Examination of the Role of Morality in Education' in Dukor, M. (ed) (2003 edn.) *Philosophy and Politics: Discourse on Values, Politics and Power in Africa,* Malthouse Press, Lagos.

Oyeshola, D, *(1995) Essentials of Environmental Issues: The World and Nigeria in Perspective,* Daily graphics Publications, ISBN 978 - 33609 - 0 - 6, Ibadan.

Oyeshola, D (2008), *Sustainable Development: Issues and Challenges for Nigeria,* Daily Graphics, ISBN: 978-8095-45-3, Ibadan. The Palme Commission (1989), *A World at Peace: Common Security in the Twenty-First Century,* The Palme Commission on Disarmament and Security Issues, Stockholm.

Pontifical Commission Justitia et Pax (1975) *This is Right,* Catholic Truth Society, London.

Rodda, A, (ed) (1991) Women and the Environment, Zed Books, London. UN (2007), *Millennium Development Goals Report 2007,* New York.

UNDP (2006) Human Development Report. Beyond Scarcity: Power, Poverty and Global water Crisis, New York.

Simonis, U. E. (2007) 'Kyoto 11 and Houston Protocol' - On the Future of International Climate Policy Lecture at the 12th Japanese -German Symposium

Shorter. A and Ujiru. N (2001) *New Religious Movements in Africa,* Pauline's Publications Africa, ISBN 9966-21-570-0, Nairobi.

South Letter, (2003) The South Centre, Issue 39. ISBN 1023-1366.

World Commission on Environment and Development (1987) *Our Common Future,* Oxford University Press, New York, Washington, D.C.